D0205185

Earthquake-resistant Concrete Structures

JOIN US ON THE INTERNET VIA WWW, GOPHER, FTP OR EMAIL:

WWW: http://www.thomson.com
GOPHER: gopher.thomson.com
FTP: ftp.thomson.com
EMAIL: findit@kiosk.thomson.com

A service of I(T)P®

Earthquake-resistant Concrete Structures

GEORGE G. PENELIS

Department of Civil Engineering, University of Thessaloniki, Greece

and

ANDREAS J. KAPPOS

Department of Civil Engineering, Imperial College, London
and Department of Civil Engineering, University of Thessaloniki, Greece

With a foreword by Professor P.E. Pinto

E & FN SPON

An Imprint of Chapman & Hall

London · Weinheim · New York · Tokyo · Melbourne · Madras

Published by E & FN Spon, an imprint of Chapman & Hall, 2–6 Boundary Row, London SE1 8HN, UK

Chapman & Hall, 2–6 Boundary Row, London SE1 8HN, UK

Chapman & Hall GmbH, Pappelalle 3, 69469 Weinheim, Germany

Chapman & Hall USA, 115 Fifth Avenue, New York, NY 10003, USA

Chapman & Hall Japan, ITP-Japan, Kyowa Building, 3F, 2-2-1 Hirakawacho, Chiyoda-ku, Tokyo 102, Japan

Chapman & Hall Australia, 102 Dodds Street, South Melbourne, Victoria 3205, Australia

Chapman & Hall India, R. Seshadri, 32 Second Main Road, CIT East, Madras 600 035, India

TA
658.44
.P46
1997

First edition 1997

© 1997 George E. Penelis and Andreas J. Kappos

Typeset in 9.5/11 pt Times by Thomson Press Ltd, Madras, India

Printed in Great Britain by
St Edmundsbury Press Limited, Bury St Edmunds, Suffolk

ISBN 0 419 18720 0

Apart from any fair dealing for the purposes of research or private study, or criticism or review, as permitted under the UK Copyright Designs and Patents Act, 1988, this publication may not be reproduced, stored, or transmitted, in any form or by any means, without the prior permission in writing of the publishers, or in the case of reprographic reproduction only in accordance with the terms of the licences issued by the Copyright Licensing Agency in the UK, or in accordance with the terms of licences issued by the appropriate Reproduction Rights Organization outside the UK. Enquiries concerning reproduction outside the terms stated here should be sent to the publishers at the London address printed on this page.

The publisher makes no representation, express or implied, with regard to the accuracy of the information contained in this book and cannot accept any legal responsibility or liabiliity for any errors or omissions that may be made.

A catalogue record for this book is available from the British Library

Library of Congress Catalog Card Number: 96-070571

DISCARDED
WIDENER UNIVERSITY
WIDENER UNIVERSITY
WOLFGRAM
LIBRARY
CHESTER, PA.

♾ Printed on acid-free text paper, manufactured in accordance with ANSI/NISO Z39.48-1992 (Permanence of Paper).

Contents

Foreword by Professor P.E. Pinto xiii

Preface xv

List of abbreviations xix

1 Introduction 1

2 Elements of engineering seismology 3

2.1 Origin – geographical distribution of earthquakes 3
2.2 Instruments for recording seismic motions 3
2.3 The magnitude and the intensity of the earthquake 4
2.3.1 Earthquake magnitude 5
2.3.2 Earthquake intensity 8
2.4 Seismicity and seismic hazard 12
2.4.1 Seismicity 12
2.4.2 Seismic hazard 13
2.5 Concluding remarks 15
2.6 References 16

3 Elements of structural dynamics 17

3.1 Introduction 17
3.2 Dynamic analysis of elastic single-degree-of-freedom (SDOF) systems 17
3.2.1 Introduction 17
3.2.2 Equations of motion 19
3.2.3 Response spectra 21
3.3 Inelastic response of SDOF systems 27
3.3.1 Introduction 27
3.3.2 Viscous damping 28

3.3.3 Hysteretic damping 29
3.3.4 Energy dissipation and ductility 32
3.4 Dynamic analysis of multidegree-of-freedom elastic systems 38
3.4.1 Introduction 38
3.4.2 The two methods of analysis 40
3.5 Dynamic analysis of MDOF inelastic systems 44
3.5.1 Introduction 44
3.5.2 Methodology of inelastic dynamic analysis with DRAIN-2D/90 44
3.6 References 50

4 Design principles and design seismic actions **51**

4.1 Introduction 51
4.2 The conceptual framework of seismic design 52
4.2.1 Basic principles and requirements of modern seismic codes 52
4.2.2 The concept of seismic isolation 55
4.3 Configuration of the structural system 56
4.3.1 Fundamental requirements 56
4.3.2 Structural systems covered by seismic codes 57
4.3.3 Recommendations concerning structural configurations 58
4.4 Design seismic actions 64
4.4.1 General 64
4.4.2 Seismic zones 64
4.4.3 The local subsoil conditions 65
4.4.4 Elastic response spectrum 66
4.4.5 Design spectrum for linear analysis 68
4.4.6 Importance factor 70
4.4.7 General remarks on the design spectrum 71
4.4.8 Alternative representation of the seismic action 72
4.5 Combination of seismic action with other actions 72
4.6 References 73

5 Analysis of the structural system **75**

5.1 Structural regularity 75
5.1.1 Introduction 75
5.1.2 Criteria for regularity in plan 75
5.1.3 Criteria for regularity in elevation 77
5.2 Modelling of the structure 78
5.2.1 General 78
5.2.2 Masses contributing to the inertia forces 78
5.2.3 Application of the design seismic actions 78
5.3 Methods of analysis 79
5.4 Simplified modal response spectrum analysis 79
5.4.1 General 79
5.4.2 Base shear forces 80
5.4.3 Distribution of the horizontal seismic forces 80

5.4.4 Estimation of the fundamental period 82
5.4.5 Torsional effects 82
5.4.6 Proposed procedure for the analysis 84
5.5 The pseudospatial structural systems under horizontal loading 85
5.5.1 General 85
5.5.2 Plane structural systems 85
5.5.3 Frame or shear system 86
5.5.4 Wall systems or flexural systems 87
5.5.5 Dual systems 87
5.5.6 The pseudospatial structural system 89
5.6 Multimodal response spectrum analysis 98
5.6.1 General 98
5.6.2 Suggested procedure for the analysis 100
5.7 Combination of the components of gravity loads and seismic action 102
5.7.1 General 102
5.7.2 Theoretical background 103
5.7.3 Simplified procedure 104
5.7.4 Code requirements 109
5.8 Second-order effects (P–Δ effects) 111
5.9 The influence of masonry infilled frames on the seismic behaviour of structures 113
5.9.1 General 113
5.9.2 Effects on the analysis 113
5.9.3 Design seismic action effects 114
5.9.4 Irregularities due to masonry infills 115
5.9.5 Remarks on infilled frames 116
5.10 General remarks on the analysis of the structural system 116
5.11 References 118

6 Design action effects–safety verifications **120**

6.1 The design action effects–capacity design procedure 120
6.1.1 General 120
6.1.2 Design criteria influencing the design action effects 121
6.1.3 Capacity design procedures for beams 122
6.1.4 Capacity design procedure for columns 123
6.1.5 Capacity design procedure for shear walls 128
6.1.6 Capacity design for connecting beams of the footings 130
6.2 Safety verifications 131
6.2.1 General 131
6.2.2 Ultimate limit state 131
6.2.3 Serviceability limit state 133
6.2.4 Specific measure 134
6.3 Application of EC8 to the design of a simple dual system 137
6.3.1 Introduction 137
6.3.2 System geometry 137
6.3.3 Characteristics of the materials 137
6.3.4 Design gravity loads 139

6.3.5 Design seismic actions 139
6.3.6 Equivalent horizontal forces 141
6.3.7 Design load combinations 142
6.3.8 Structural analysis 142
6.4 References 147

7 Earthquake-resistant properties of the materials of reinforced concrete **149**

7.1 Introduction 149
7.2 Reference to code provisions 150
7.3 Plain (unconfined) concrete 150
7.3.1 Response to monotonic loading 151
7.3.2 Response to cyclic loading 155
7.3.3 Response to multiaxial loading 158
7.3.4 Relevant code provisions 177
7.4 Confined concrete 177
7.4.1 The notion of confinement 177
7.4.2 Parameters affecting confinement 179
7.4.3 Confinement with hoops 180
7.4.4 Confinement with spirals 190
7.4.5 Relevant code provisions 194
7.5 Steel 197
7.5.1 Main requirement for seismic performance 197
7.5.2 Response to monotonic loading 199
7.5.3 Response to cyclic loading 201
7.5.4 Relevant code provisions 204
7.6 Bond between concrete and steel 205
7.6.1 Constitutive equations of bond 205
7.6.2 Bond under monotonic loading 206
7.6.3 Bond under cyclic loading 211
7.6.4 Relevant code provisions 217
7.7 References 219

8 Earthquake-resistant design of reinforced concrete linear elements **223**

8.1 Introduction 223
8.2 Seismic behaviour of beams 224
8.2.1 Behaviour under monotonic loading 224
8.2.2 Behaviour under cyclic loading 237
8.3 Seismic design of beams 251
8.3.1 Design for flexure 251
8.3.2 Design for shear 255
8.3.3 Other design requirements 258
8.4 Seismic behaviour of columns 261
8.4.1 Uncertainties regarding the capacity design of columns 261
8.4.2 Behaviour under monotonic loading 263
8.4.3 Behaviour under cyclic loading 269

8.5 Seismic design of columns 280
8.5.1 Design for flexure and axial loading 280
8.5.2 Design for shear and local ductility 285
8.5.3 Other design requirements 290
8.6 Design example 293
8.6.1 General data and analysis procedure 293
8.6.2 Design of beams 296
8.6.3 Design of columns 304
8.7 References 313

9 Earthquake-resistant design of reinforced concrete planar elements **317**

9.1 Introduction 317
9.2 Beam–column joints 317
9.2.1 Basic design principles 318
9.2.2 Behaviour of joints under cyclic shear 320
9.2.3 Design for shear 329
9.2.4 Anchorage of reinforcement in joints 333
9.2.5 Special types of joints 337
9.2.6 Design example 341
9.3 Seismic behaviour of walls 345
9.3.1 Advantage of structural walls 345
9.3.2 Behaviour under monotonic loading 346
9.3.3 Behaviour under cyclic loading 349
9.3.4 Walls with openings 356
9.4 Seismic design of walls 361
9.4.1 Design for flexure and axial loading 361
9.4.2 Design for shear and local ductility 361
9.4.3 Other design requirements 370
9.4.4 Design example 372
9.4.5 Design of walls with openings 380
9.5 Seismic design of diaphragms 383
9.5.1 Requirements regarding configuration and design actions 383
9.5.2 Behaviour under cyclic loading 384
9.5.3 Resistance verification 384
9.6 References 385

10 Seismic performance of buildings designed to modern design codes **388**

10.1 Methods for assessing the seismic performance 388
10.1.1 Introductory remarks 388
10.1.2 Performance assessment through testing of models and inspection of actual structures 389
10.1.3 Performance assessment using inelastic dynamic analysis 390
10.2 Seismic performance of frames 392
10.2.1 Selection of input motions 392
10.2.2 Modelling assumptions and failure criteria 395
10.2.3 Performance under the design earthquake 399
10.2.4 Serviceability and survival earthquake 404

10.3 Seismic performance of dual systems 406
10.3.1 Modelling assumptions and failure criteria 407
10.3.2 Performance under the design earthquake 409
10.3.3 Serviceability and survival earthquake 415
10.4 Influence of design ductility class 418
10.4.1 Influence on cost 418
10.4.2 Influence on seismic performance 419
10.5 Influence of masonry infills 424
10.6 Concluding remarks 429
10.7 References 430

11 Seismic pathology **433**

11.1 Classification of damage in R/C structural members 433
11.1.1 Introduction 433
11.1.2 Damage to columns 434
11.1.3 Damage to R/C walls 436
11.1.4 Damage to beams 440
11.1.5 Damage to beam–column joints 445
11.1.6 Damage to slabs 446
11.1.7 Damage to infill panels 447
11.1.8 Spatial distribution of damages in buildings 452
11.1.9 Stiffness degradation 454
11.2 Factors affecting the degree of damage to buildings 455
11.2.1 Introduction 455
11.2.2 Divergence between the design spectrum and the response spectrum of the earthquake under consideration 455
11.2.3 Brittle columns 456
11.2.4 Asymmetric arrangement of stiffness elements on the floor plan 457
11.2.5 Flexible ground floor 458
11.2.6 Short columns 460
11.2.7 Shape of the floor plan 461
11.2.8 Shape of the building in elevation 461
11.2.9 Slabs supported by columns without beams (flat plate system) 461
11.2.10 Damage from previous earthquakes 462
11.2.11 Pure frame systems 463
11.2.12 Number of storeys 464
11.2.13 Type of foundations 464
11.2.14 The location of adjacent structures in the block 466
11.2.15 Slab levels of adjacent structures 466
11.3 References 468

12 Emergency post-earthquake damage inspection and evaluation **469**

12.1 Introduction 469
12.2 Inspections and damage assessment 470
12.2.1 Introductory remarks 470
12.2.2 Purpose of the inspections 470

12.2.3 Damage assessment 471
12.3 Organizational scheme for inspections 473
12.3.1 Introduction 473
12.3.2 Usability classification – inspection forms 474
12.3.3 Inspection levels 476
12.4 Action plan 476
12.4.1 Introduction 476
12.4.2 State agency responsible for the operation 476
12.4.3 Inspection personnel 476
12.4.4 Pre-earthquake organizing procedures 478
12.4.5 Post-earthquake organizing procedures 479
12.5 Final remarks 481
12.6 References 481

13 Design of repair and strengthening **483**

13.1 General 483
13.2 Definitions 484
13.3 Objectives and principles of interventions 487
13.4 Criteria for repair or strengthening 488
13.4.1 Basic principles 488
13.4.2 The UNIDO/UNDP procedure 489
13.5 Design steps of intervention 495
13.5.1 General 495
13.5.2 Strengthening 495
13.5.3 Repair 498
13.5.4 Repair of the masonry infills 499
13.6 Criteria governing structural interventions 499
13.6.1 General criteria 500
13.6.2 Technical criteria 500
13.6.3 Type of intervention 500
13.6.4 Examples of repair and strengthening techniques 501
13.7 Final remarks 501
13.8 References 504

14 Technology of shoring, repair and strengthening **506**

14.1 General 506
14.2 Emergency measures for temporary supports 507
14.2.1 General 507
14.2.2 Techniques for supporting vertical loads 508
14.2.3 Techniques for resisting lateral forces 509
14.2.4 Wedging techniques 514
14.3 Materials and intervention techniques 515
14.3.1 Conventional cast-in-place concrete 515
14.3.2 High-strength concrete using shrinkage compensating admixtures 516
14.3.3 Shotcrete (gunite) 516
14.3.4 Polymer concrete 518
14.3.5 Resins 518

14.3.6 Resin-concretes 519
14.3.7 Grouts 520
14.3.8 Gluing metal sheets on concrete 520
14.3.9 Welding of new reinforcement 521
14.3.10 Gluing Fibre-Reinforced Plastic (FRP) sheets on concrete 521
14.4 Redimensioning and safety verification of structural elements 522
14.4.1 General 522
14.4.2 Revised γ_m-factors 523
14.4.3 Load transfer mechanisms through interfaces 524
14.4.4 Simplified estimation of the resistance of structural elements 528
14.5 Repair and strengthening of structural elements 529
14.5.1 General 529
14.5.2 Columns 530
14.5.3 Beams 537
14.5.4 Beam–column joints 544
14.5.5 R/C walls 546
14.5.6 R/C slabs 551
14.5.7 Foundations 553
14.5.8 Infill masonry walls 555
14.6 Addition of new structural elements 557
14.7 Quality assurance of interventions 560
14.7.1 General 560
14.7.2 Quality control of design 560
14.7.3 Quality control of construction 561
14.8 Final remarks 561
14.9 References 562

Index **565**

Foreword

By Professor P.E. Pinto, Universita Degli Studi di Roma 'La Sapienza', Rome.

To write a book on Earthquake Engineering is an arduous exercise in selection which is, by itself, revealing about the skill of the author. It would be tempting, though unrealistic, to start the book with the physics of earthquake generation and finish it with the social aspects of mitigating the effects of earthquakes. This might make interesting reading but would not be useful either as a research or professional tool.

The authors of this book are in no doubt about the character it should have. It guides the reader through the steps required for conscious design and repair of reinforced concrete building structures, following the well established sequence of the last generation of codes, with Eurocode 8 particularly in evidence.

Unlike the codes of the past, modern codes tend to be both descriptive and justificative. However this does not eliminate the need for a large amount of input on the part of the user which is, on the contrary, essential in order to appreciate and exploit the potential of these complex documents. The central part of this book is especially attractive in this respect: it gives the most recent available information on the behaviour and modelling of materials, elements and structures under cyclic actions, then explains how this knowledge translates into code provisions (particularly EC8), and finally exemplifies the use of these provisions in realistic and detailed design applications.

This systematic application of technical knowledge is not to be found in other books. Also peculiar to this book, and of important educational value, is the chapter devoted to the assessment of code-designed structures – an idea foreign to ordinary thinking by which a structure correctly designed following a code is assumed to be 'earthquake-proof'.

The final chapters of the book deal with pathology, emergency post-earthquake inspections, design principles and technology for repair and strengthening. The decision to introduce these topics, notwithstanding the limitations of space, reflects the authors' belief that the process of redesign either pre- or post-earthquake cannot be separated from that of design and that the larger number of variables and uncertainties involved should be an incentive for applying consistent rules more widely, not an excuse for ignoring them. This reflects the world-wide awareness that the problem of existing buildings is now the most pressing challenge facing Earthquake Engineering.

It is customary to end a foreword of this type by summarizing the merits of the book, and by indicating the types of reader who will benefit most from it. Answers to both aspects should emerge clearly from this short presentation: all present and future practitioners in the field who wish to operate knowing the conceptual framework, scientific background, and the correct way of applying the design procedures contained in the most recent codes.

Preface

Earthquake engineering, as an independent field of science, may by considered as a development of the last 40–50 years. The installation of dense networks of accelerographs worldwide, the feasibility of analysing complicated structures in both the elastic and the inelastic stage of their dynamic response using the recently developed powerful computers, the experimental testing of structural members and subassemblages under inelastic load reversals including inelastic response, the development of earthquake simulators for studying structural models, the refinement and the extensive use of *in situ* measuring techniques, and finally the broadening of the knowledge regarding the behaviour of soil, either in free-field conditions or in interaction with the structures, constitute significant steps towards the development of this relatively new field of engineering.

As knowledge was accumulating, it became clear that the problem of the seismic behaviour of structures is primarily an energy-related one. In order for a structure to avoid collapse, it should be in a position to absorb and dissipate the kinetic energy imparted in it during the seismic excitation. The understanding of this simple energy balance principle was the key for the development of modern earthquake-resistant design, which followed three directions:

1. design of structures with members able to dissipate significant amounts of energy through stable cycles of inelastic deformation, while sustaining a limited degree of damage;
2. seismic isolation of structures, with a view to controlling the energy imparted in them by the earthquake;
3. use of special energy-dissipating devices, for limiting the degree of damage sustained by structures.

The first direction is the traditional one, essentially followed by all modern seismic codes. The second and the third directions constitute quite novel and promising approaches, which nevertheless have known very limited application to date. These rigorous steps, both at the scientific and the technological level, are clearly reflected in the development of the seismic codes during the last 50 years.

Notwithstanding the foregoing developments at the scientific, the technological and the regulatory level, the amount of losses due to destructive earthquakes, with regard to both human lives and the built environment, remains rather high. There are various reasons for this, such as the increased urbanization of major cities often combined with uncontrolled and/or illegal building construction, the high percentage of old structures in the existing building stock, the daring use of modern construction materials in structures with large spans and heights, and last but not least the fact that after each new destructive earthquake, new lessons are learned and new knowledge on the complicated problem of earthquake resistance is acquired. A detailed discussion of these causes of earthquake losses is beyond the scope of a preface; their consequences, however, render equally important with the construction of new earthquake-resistant building, the pre-earthquake, as well as the post-earthquake assessment of the seismic capacity of existing structures, and also the technology of repair and strengthening, aiming at either restoring the structures to their pre-earthquake condition, or at preventing significant damage in important structures.

This book attempts to introduce postgraduate students of earthquake engineering, as well as undergraduate students specializing in earthquake-resistant structures, to all modern approaches to seismic design outlined previously, with a focus on reinforced concrete structures. The book was first published in Greek, under the same title, in 1989, and has been continuously used by the authors as a text book for the students of the Civil Engineering Department at the Aristotle University of Thessaloniki. The first part of the book, covering the fundamentals of engineering seismology and structural dynamics, as well as the problem of seismic actions and analysis of concrete structures subjected to these actions, has been used in the undergraduate (core) course 'Reinforced Concrete Structures III', taught in the eighth semester (spring term of the fourth year, in the five-year course in civil engineering). The second part, referring to the earthquake-related properties of the materials of reinforced concrete and the seismic behaviour of various structural elements, and the third part referring to seismic pathology, damage assessment and repair and strengthening techniques, have been used in the undergraduate optional course 'Earthquake-resistant Design and Seismic Pathology of Concrete Structures', taught in the tenth (final) semester of the civil engineering curriculum.

The experience gained from the response of the students to the material included in the Greek edition of the book, and the introduction of the new Eurocode 8 (EC8) as a European pre-standard has led the authors to a substantial revision of the material, mainly with a view to harmonizing it with EC8, which aims at becoming the reference document for the development of seismic codes in the European Union states, as well as in the rest of Europe and in several other parts of the world. The present edition of the book, which is significantly extended with regard to the initial Greek edition, has 14 chapters.

The first six chapters provide some fundamentals of engineering seismology (intended for students and/or practising engineers not familiar with the subject), a brief but quite comprehensive outline of structural dynamics with emphasis on methods for calculating the response of structures to imput accelerograms and on energy considerations, and a detailed treatment of the design of the structural configuration and the static and dynamic analysis of concrete structures subjected to seismic actions; also included in the first part is an extensive presentation of elastic and inelastic spectra, including the design spectrum specified in EC8.

The following four chapters (7–10) contain an extensive treatment of the problem of the earthquake-related properties of reinforced concrete materials (including the topic of bond under seismic conditions), as well as of the seismic behaviour of structural members, that is beams, columns, beam–column joints, walls and diaphragms. In each section referring to a specific member a detailed presentation of the related EC8 design provisions is included, as well as a fully worked design application. This second part of the book concludes with a discussion of methods of assessing the seismic performance of structures, and a detailed presentation of several case studies involving concrete buildings with frames and structural walls, with and without masonry infill panels.

The last four chapters (11–14) treat in depth the topics of seismic pathology of concrete structures (types of earthquake-induced damage), the assessment of seismic capacity of existing structures, and the procedures and techniques for repair and/or strengthening of structures. A reference to part 1–4 of EC8, as well as to other, more detailed, documents addressing the problem of repair and strengthening is made in this last part of the book.

The material included herein will hopefully appeal both to practising engineers involved in seismic design and/or retrofitting of existing structures, as well as to graduate and undergraduate students of civil engineering and/or earthquake engineering. With regard to academic curricula, it is clear that the material included in this book cannot be accommodated in a single course, even at an advanced postgraduate level. Material included in Chapters 7–10 of the book forms currently the basis of the graduate course 'Analysis and design of structures in seismic areas' which is the backbone of the M.Sc. course on earthquake engineering and structural dynamics at Imperial College, London, and of a similar course included in the postgraduate curriculum in structural engineering, started at the University of Thessaloniki in the academic year 1995–96. The latter also includes a course on repair and strengthening methods, where material included in Chapters 11–14 of this book is being used. Selected material from these chapters is also used in the aforementioned course at Imperial College. As mentioned previously, the first part (Chapters 2–6) forms the basis of the undergraduate course on seismic design at the Civil Engineering Department of Thessaloniki. Lecturers may find material included in Chapters 4–6 appropriate for postgraduate courses on seismic actions and analysis of building structures subjected to these actions.

On the other hand, the practising engineer involved in seismic design may find quite useful the detailed information included herein on both the background and the way of applying EC8 to the design of earthquake-resistant structures; the example designs in Chapters 8 and 9 are particularly useful in this context. Moreover earthquake engineers would be particularly interested in the last part of the book (Chapters 11–14), both with respect to the practical guidance provided on methods of earthquake reconnaissance and identification of the type of seismic damage, and with respect to the guidance on selecting and applying specific methods for repair and/or strengthening of seismically damaged structures.

In closing, the authors would like to acknowledge the contribution of the following people in preparing this book:

- Mr D. Baxevanis, Mrs V. Binikou and Ms D. Kakoulidou, members of the technical staff of the Department of Civil Engineering in Thessaloniki, for the careful preparation of the figures and the typed manuscript.
- Dr A. Michael for her valuable contribution in translating parts of the Greek edition of the book.
- Dr C. Athanassiadou for the careful correction of the manuscript under a critical eye.

- Mr V. Salpistis for the drafting of the cover illustration.
- Mr J.N. Clarke, Senior Editor, E & FN Spon, for his efforts in coordinating the publication of the book.

Last but not least the authors would like to thank Professor P. E. Pinto, Chairman of the Drafting Committee of Eurocode 8, both for providing earlier drafts of the code and for being kind enough to contribute a foreword to this book.

Thessaloniki and London
July 1996

G. G. Penelis
A. J. Kappos

List of abbreviations

ACI	American Concrete Institute
ASCE	American Society of Civil Engineers
ATC	Applied Technology Council (USA)
Bull.	Bulletin
Calif.	California (USA)
Cal Tech	California Institute of Technology (Pasadena, Calif.)
CEB	Comité Euro-International du Béton
CEN	Comité Européen de Normalisation
Civ.	Civil
Conf.	Conference
DC	Ductility Class
Dept	Department
Div.	Division
Earthq.	Earthquake
EC	Eurocode
ECEE	European Conference on Earthquake Engineering
EERC	Earthquake Engineering Research Center (Berkeley, Calif.)
EERI	Earthquake Engineering Research Institute (Berkeley, Calif.)
EERL	Earthquake Engineering Research Laboratory (Pasadena, Calif.)
Engng	Engineering
EPPO	Earthquake Protection and Planning Organization (Athens, Greece)
IABSE	International Association for Bridge and Structural Engineering
ICBO	International Conference of Building Officials (Whittier, Calif.)
IISEE	International Institute of Seismology and Earthquake Engineering (Japan)
Inf.	Information
Int.	International
MC	Model Code
MC/SD	Model Code for Seismic Design
Mech.	Mechanics
MIT	Massachusetts Institute of Technology (Cambridge, Mass.)
Nat.	National

NISEE	National Information Service for Earthquake Engineering (Berkeley, Calif.)
PCA	Portland Cement Association (Shokie, Illinois, USA)
PCI	Prestressed Concrete Institute
Proceed.	Proceedings
Publ.	Publication
R/C	Reinforced concrete
Rep.	Report
SANZ	Standards Association of New Zealand
Struct.	Structural
UBC	Uniform Building Code (USA)
Univ.	University
WCEE	World Conference on Earthquake Engineering

1

Introduction

The earthquake, considered as the independent natural phenomenon of vibration of the ground, in very few cases poses a threat to humans, as for example when it causes major landslides or tidal waves (tsunamis). The earthquake becomes a dangerous phenomenon only when it is considered in relation with structures. Of course, the problem is the structure under seismic excitation and not the earthquake itself. This is because the structural system is designed basically for gravity loads and not for the horizontal inertia loads that are generated due to ground accelerations during an earthquake. Therefore, the earthquake has begun to become a problem for humans since they started building. Since the early steps of the technological development of mankind the joy of creation was associated with the fear that some superior force would destroy in a few seconds what was built with great effort over a lifetime. In other words, the earthquake was always associated with the structure and therefore it mainly concerns the structural engineer.

Although destructive earthquakes are confined to certain geographical areas, the seismic zones, the large-scale damage that they cause in densely populated areas and the number of deaths are such that they have an impact on the whole world.

Earthquakes, because of the deaths and the damage to buildings that they cause, have several economic, social, psychological and even political effects in the areas and the countries where they take place. Thus, many scientists deal with this problem, such as seismologists, engineers, psychologists, economists and so on. All these scientific disciplines are coordinated by special bodies on a national level and by special institutes of interdisciplinary character, or at the university level, by interdepartmental cooperation. The goal of all these efforts is basically the **earthquake-resistant structure**, that is its improvement from the safety–cost point of view, which are two antagonistic parameters.

Given the fact that a large part of the knowledge necessary for the design of earthquake-resistant structures is covered in other books referring to the fields of engineering geology, soil mechanics, engineering seismology, earthquake engineering, structural dynamics and reinforced concrete structures, this book, although for reasons of completeness it covers many relevant topics, is focused on some special items. Thus, the structure of this book will have the following form:

- elements of engineering seismology
- elements of structural dynamics

- earthquake-resistant reinforced concrete (R/C) structures
- R/C structural members under seismic loading
- seismic pathology–damage assessment–structural repair.

Finally, it should be noted that the structure of a technical book is strongly influenced by the prevailing tendencies of the contemporary relevant codes. In particular, the structure of this book has been influenced by the recently approved EC8 (*Earthquake Resistant Design of Structures*), and by the Model Code for Seismic Design–85 of CEB.

2

Elements of engineering seismology

2.1 ORIGIN–GEOGRAPHICAL DISTRIBUTION OF EARTHQUAKES

Earthquakes are ground vibrations that are caused mainly by the fracture of the crust of the earth or by the sudden movement along an already existing fault (tectonic earthquakes). Very rarely, earthquakes may be caused by volcanic eruptions. A widely accepted and well-established theory for the origin of tectonic earthquakes is the 'elastic rebound theory' which was developed in 1906 by Reid (1911). According to this theory, earthquakes are caused by the sudden release of elastic strain energy in the form of kinetic energy along the length of a geological fault (Figure 2.1). The accumulation of strain energy along the length of geological faults can be explained by the theory of motion of **lithospheric plates** into which the crust of the earth is divided. These plates are developed in oceanic rifts and they sink in the **continental trench system** (Figures 2.2 and 2.3) (Papazachos, 1986; Strobach and Heck, 1980).

The boundaries of the lithospheric plates coincide with the geographical zones which experience frequent earthquakes (Figure 2.4). Figure 2.5 shows some characteristic terms that are related to the earthquake phenomenon.

2.2 INSTRUMENTS FOR RECORDING SEISMIC MOTIONS

There are two basic categories of instruments that facilitate the quantitative evaluation of the earthquake phenomenon:

1. The **seismographs** which record the displacement of the ground as a function of time (Figure 2.6) (Bolt, 1978) and they operate on a continuous real-time basis. Their recordings are of interest mainly to the seismologists.
2. The **accelerographs** which record the acceleration of the ground as a function of time (Figure 2.7). They are adjusted to start operating whenever a certain ground acceleration is exceeded. They are used for the recording of strong ground motions that are of interest to structural engineers to be used for the design of structures (strong motion accelerographs).

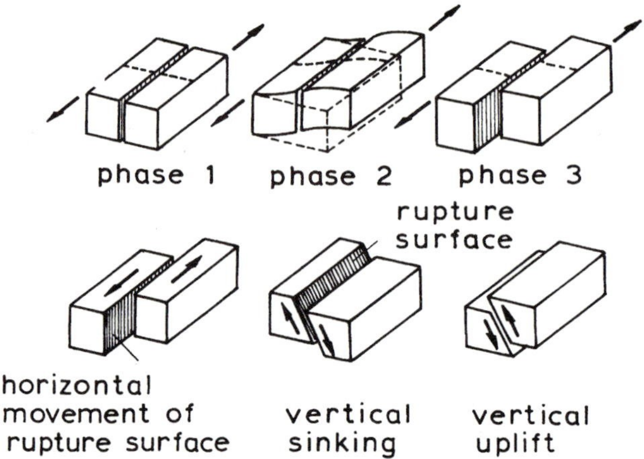

Figure 2.1 Schematic presentation of the earthquake origin.

Figure 2.2 Motion of the lithospheric plates.

2.3 THE MAGNITUDE AND THE INTENSITY OF THE EARTHQUAKE

The magnitude and the intensity of an earthquake are terms that were developed in an attempt to evaluate the earthquake phenomenon.

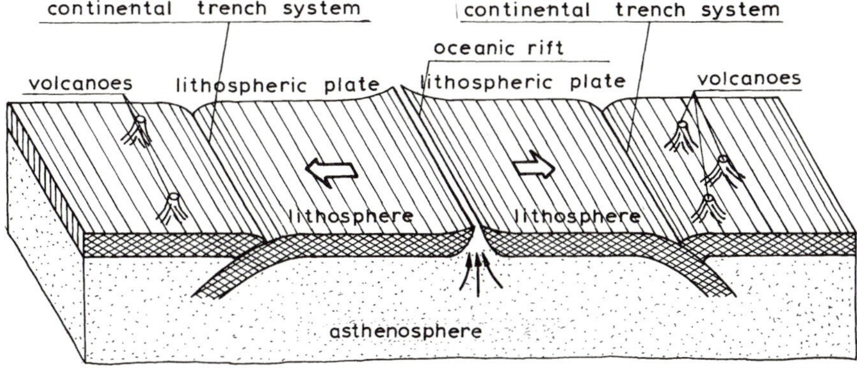

Figure 2.3 Motion system of the lithospheric plates.

2.3.1 Earthquake magnitude

The magnitude of the earthquake is a measure of this phenomenon in terms of the energy which is released, in the form of seismic waves, at its point of origin. It is measured on the **Richter scale** named after the seismologist who invented it. This scale is based on the observation that, if the logarithm of the maximum displacement amplitudes which were recorded by seismographs located at various distances from the epicentre are put on the same diagram, and this is repeated for several earthquakes with the same epicentre, the resulting curves are parallel to each other (Figure 2.8). This means that, if one of these earthquakes is taken as the basis, the coordinate difference between that earthquake and every other earthquake, measures the magnitude of the earthquake at the epicentre. Richter defined as a **zero magnitude earthquake** one which is recorded with 1 μm amplitude at a distance of 100 km. Therefore, the local magnitude M_L of an earthquake which is recorded with amplitude A at a certain distance is given by the relation

$$\boxed{M_L = \log A - \log A'} \tag{2.1}$$

where A' is the amplitude of the zero magnitude earthquake ($M_L = 0$), given in tables as a function of the distance from the epicentre.

The magnitude M_L of the earthquake is related to the energy released from the epicentre through the relation

$$\boxed{\log E = 12.24 + 1.44 M_L \text{ (erg)}} \tag{2.2}$$

which shows that a unit increase in the magnitude of the earthquake results in an increase in energy of about 28 times.

The largest earthquake that has ever been recorded was of magnitude 8.9 on the Richter scale (Colombia–Ecuador 1906, Japan 1933) and it is believed that it is the maximum value that can occur. This conclusion is based on the estimation of the maximum elastic energy that can be accumulated on the crust of the earth before fracture occurs. It is generally accepted that earthquakes with magnitude below 5 on the Richter scale are not destructive to engineered structures.

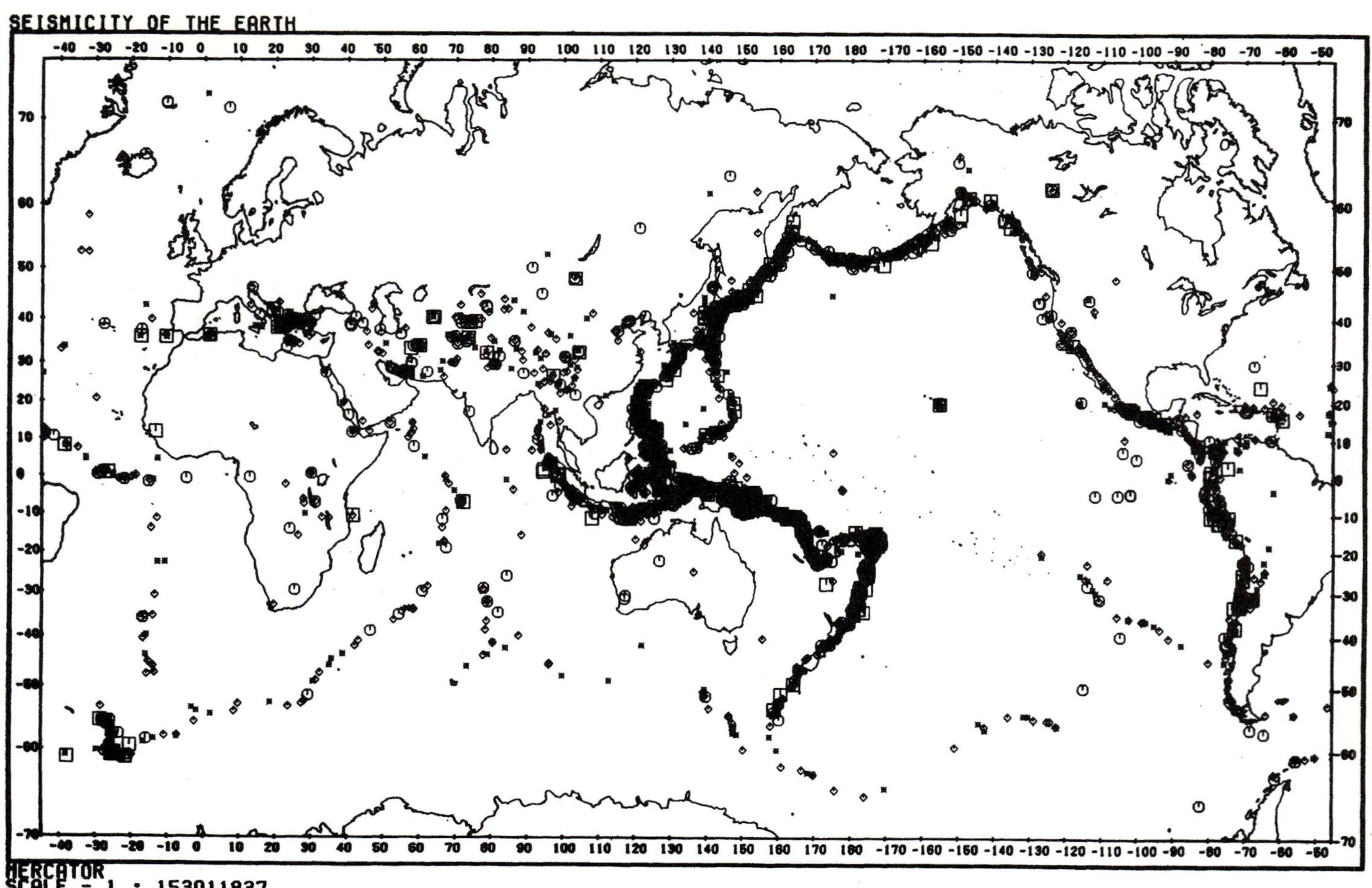

Figure 2.4 Spatial distribution of the world's strong ($M_s \geqslant 5.5$) earthquakes during the period 1966–85 (Tsapanos, Scordilis and Papazachos, 1990).

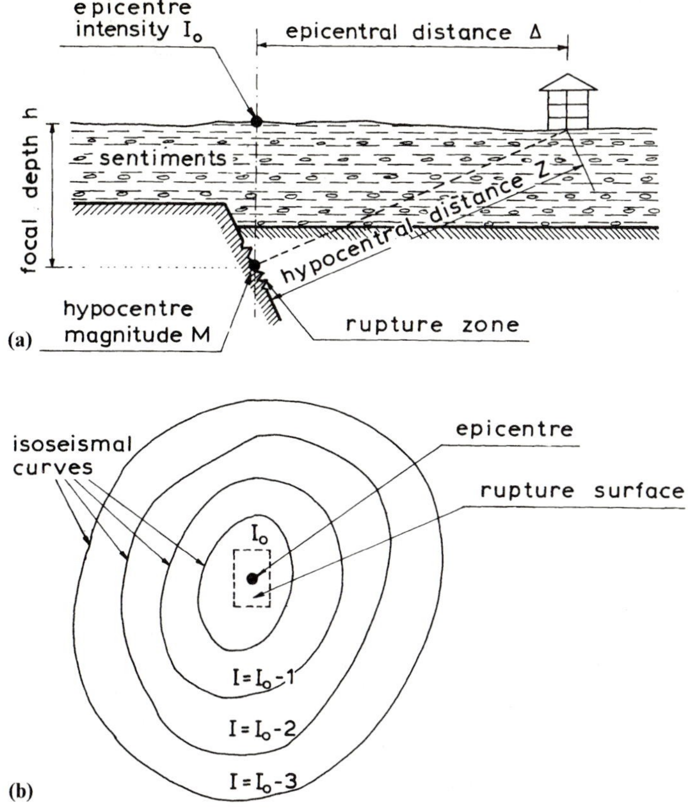

Figure 2.5 Terminology related to the natural phenomenon of the earthquake: (a) generation and propagation; (b) isoseismal curves.

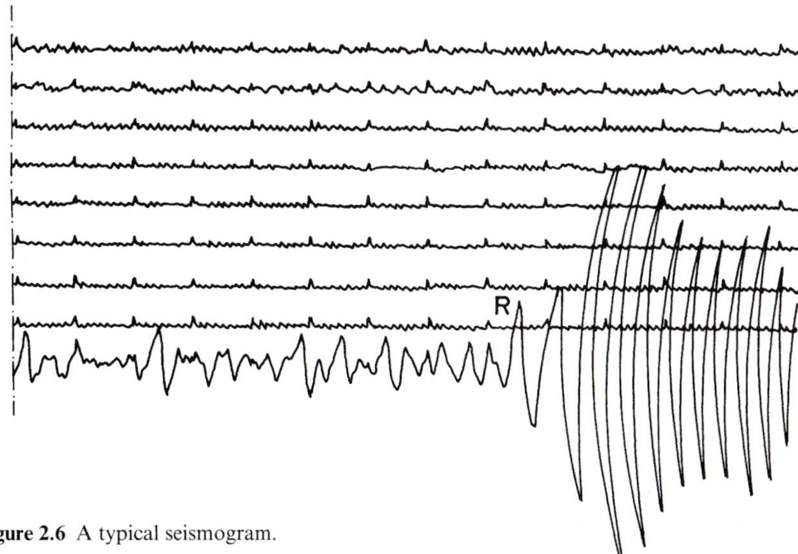

Figure 2.6 A typical seismogram.

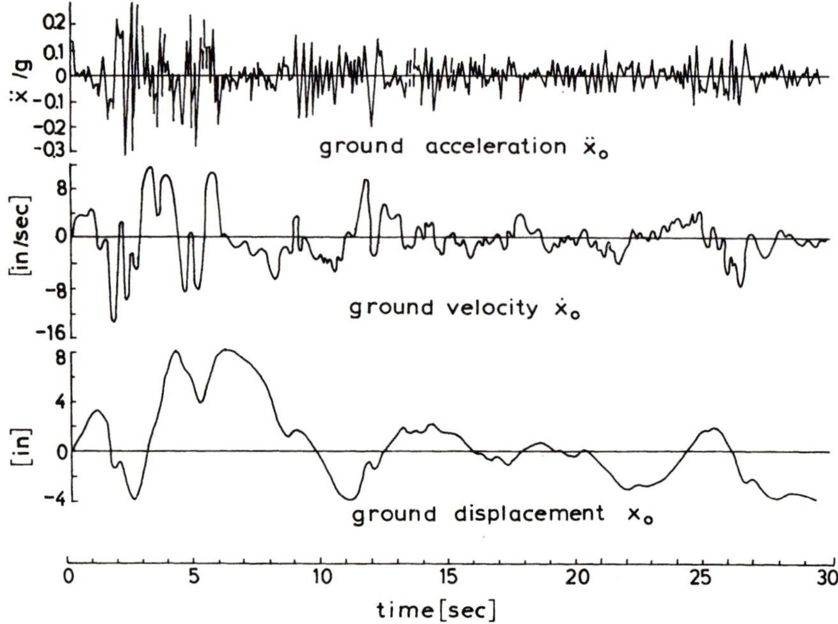

Figure 2.7 Accelerograms of the El Centro earthquake of the 18 May 1940 (N-S component) and derived diagrams of velocity and displacement.

2.3.2 Earthquake intensity

The potential destructiveness of an earthquake, although partly related to its magnitude, is also a function of other equally important factors, such as the focal depth of the earthquake, the distance from the epicentre, the soil conditions and the mechanical properties of the structures (strength, natural period, ductility and so on). The term **intensity** of the earthquake is a measure of the consequences that this earthquake has on the people and the structures of a certain area. It is obvious that it is impossible to measure the damage due to an earthquake using a single quantity system. Therefore, the damage is usually qualitatively estimated using **empirical intensity scales**. The most common macroseismic scales that are used today are the modified Mercalli (MM) scale (Table 2.1) and the Medvedev, Sponheur, Karnik (MSK) scale (Table 2.2), both of which have 12 intensity grades. Figure 2.9 shows the division of Greece into seismic zones (Papaioannou *et al.*, 1994) according to the MM scale. An earthquake has only one magnitude but different intensities from one place to another. The intensity generally attenuates as the distance from the epicentre increases. The soil conditions have a significant effect on the distribution of structural damage. This effect is estimated through so-called **microzonation** studies.

If the points of equal intensity are connected on a map, the resulting curves are called **isoseismal contours** which divide the affected area into sections of equal intensity (Figure 2.9). From the structural design point of view, the intensity of the earthquake on a macroseismic scale is not of great interest. The reason for

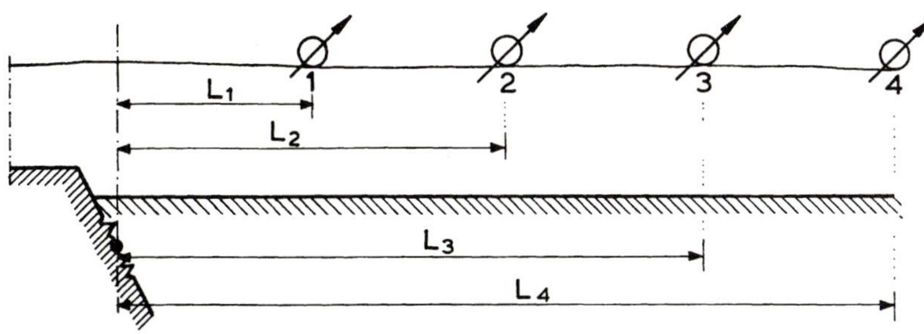

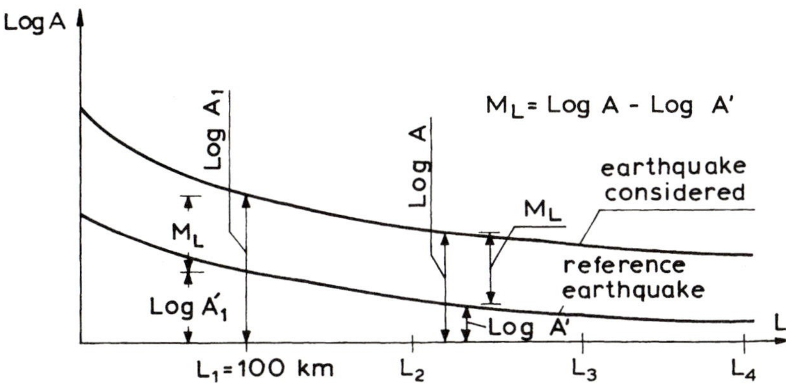

Figure 2.8 Procedure for measuring the magnitude of an earthquake M_L on the Richter scale.

this is that, on the one hand it does not provide any quantitative information about the parameters that are related to the ground motion (e.g. maximum displacement, velocity, acceleration, prevailing period, duration) and on the other because it is not an objective procedure for evaluating the exciting force (the earthquake) using the excited medium (the structure), the response of which depends on a series of variables such as strength, natural period, ductility, type of structural system and so on. It follows that the ideal way, again from the structural design point of view, to estimate the **seismic hazard** of an area is the existence of long-term records of strong seismic motions (accelerograms) and the statistical processing of their basic elements.

However, considering the fact that seismological records (from seismographs) did not exist before the present century, that records of strong earthquake motions did not exist prior to 1938 and that the number of the latter is generally limited, it is obvious that there is no other way but the one which combines the limited seismic motion records with the estimations of the intensity of previous earthquakes using scales such as the MM scale. Indeed, these macroseismic scales despite their subjective character allow:

- the use of the seismic history of a geographical area;

Table 2.1 The modified Mercalli scale (Fintel and Derecho, 1974)

		Ground acceleration a	
		$\frac{\text{cm}}{\text{sec}}$	$\frac{a}{g}$
I	Not felt except by a very few under especially favourable circumstances		
II	Felt only by a few persons at rest, especially on upper floors of buildings. Delicately suspended objects may swing	2 3	
III	Felt quite noticeably indoors, expecially on upper floors of buildings, but many people do not recognize it as an earthquake. Standing motor cars may rock slightly. Vibration like passing truck. Duration estimated	4 5 6	 0.005 g
IV	During the day felt indoors by many, outdoors by few. At night some awakened. Dishes, windows, doors disturbed: walls make creaking sound. Sensation like heavy truck striking building. Standing motor cars rocked noticeably	7 8 9 10	 0.01 g
V	Felt by nearly everyone: many awakened. Some dishes, windows, etc., broken; a few instances of cracked plaster; unstable objects overturned. Disturbances of trees, poles and other tall objects sometimes noticed. Pendulum clocks may stop	20 30	
VI	Felt by all; many frightened and run outdoors. Some heavy furniture moved: a few instances of fallen plaster or damaged chimneys. Damage slight	40 50 60	 0.05 g
VII	Everybody runs outdoors. Damage negligible in buildings of good design and construction: slight to moderate in well-built ordinary structures: considerable in poorly built or badly designed structures; some chimneys broken. Noticed by persons driving motor cars	70 80 90 100	 0.1 g
VIII	Damage slight in specially designed structures: considerable in ordinary substantial buildings, with partial collapse; great in poorly built structures. Panel walls thrown out of frame structures. Fall of chimneys, factory stacks, columns, monuments, walls. Heavy furniture overturned. Sand and mud ejected in small amounts. Changes in well water. Disturbs persons driving motor cars	200 300	
IX	Damage considerable in specially designed structures: well-designed frame structures thrown out of plumb; great in substantial buildings, with partial collapse. Buildings shifted off foundations. Ground cracked conspicuously. Underground pipes broken	400 500 600	 0.5 g
X	Some well-built, wooden structures destroyed: most masonry and frame structures destroyed with foundations; ground badly cracked. Rails bent. Landslides considerable from river banks and steep slopes. Shifted sand and mud. Water splashed over banks	700 800 900 1000	 1 g
XI	Few, if any masonry structures remain standing, Bridges destroyed. Broad fissures in ground. Underground, pipelines completely out of service. Earth slumps and landslips in soft ground. Rails bent greatly	2000 3000	
XII	Damage total. Waves seen on ground surfaces. Lines of sight and level distorted. Objects thrown upward into the air	4000 5000 6000	 5 g

Table 2.2 The MSK intensity scale

Degree	Intensity	Effect		
		on people	on structures	on the environment
1	Insignificant	Not felt		
2	Very light	Slightly felt		
3	Light	Felt mainly by people at rest		
4	Somewhat strong	Felt by people indoors	Trembling of glass windows	
5	Almost strong	Felt indoors and outdoors, awakening of sleeping people	Oscillation of suspended objects, displacement of pictures on walls	
6	Strong	Many people are frightened	Light damage to structures, fine cracks in plaster	Very few cracks on wet soil
7	Very strong	Many people run outdoors	Considerable damage to structures, cracks in plaster, walls and chimneys	Landslides of steep slopes
8	Damaging	Everybody is frightened	Damage to buildings, large cracks in masonry, collapse of parapets and pediments	Changes in well-water, Landslips of road Embankments
9	Very damaging	Panic	General damage to buildings, collapse of walls and roofs	Cracks on the ground, landslides
10	Extremely damaging	General panic	General destruction of buildings, collapse of many buildings	Changes on the surface of the ground, appearance of new water wells
11	Destructive	General panic	Serious damage to well-built structures	
12	General destruction	General panic	Total collapse of buildings and other civil engineering structures	Changes on the surface of the ground, appearance of new water wells

- the correlation of the maximum expected intensity in a certain period of time with existing records of strong seismic motions in other areas and adoption of appropriate response spectra.

Of course, it is not unusual fo this kind of extrapolation to lead to serious estimation errors, as will be discussed in subsequent chapters

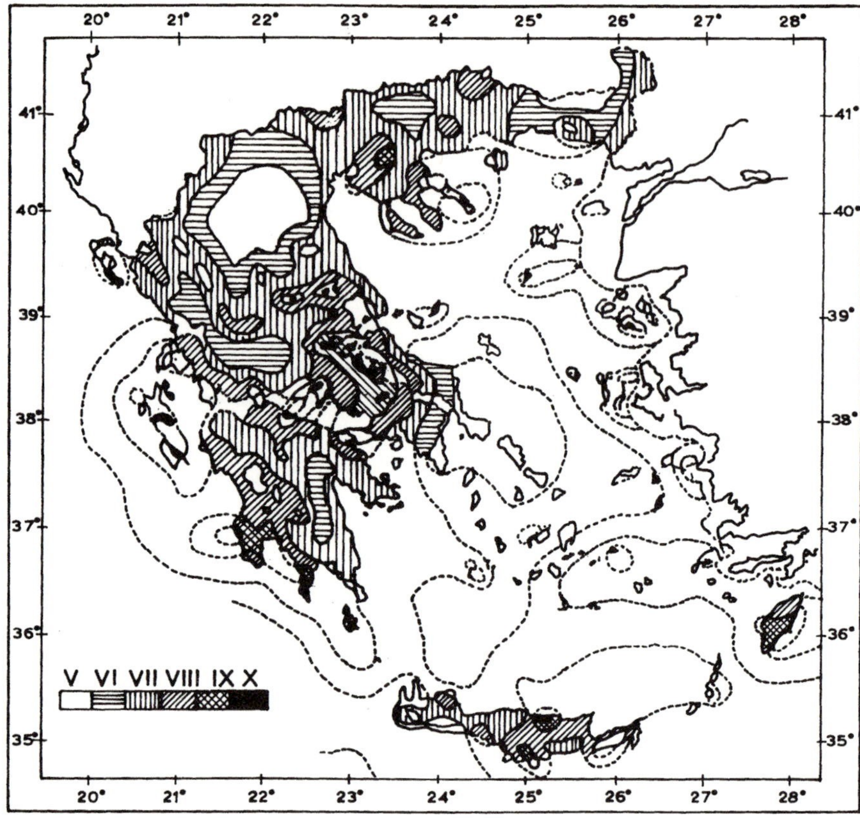

Figure 2.9 Maximum observed intensities in Greece between 1700 and 1981 on the MM scale.

2.4 SEISMICITY AND SEISMIC HAZARD

In order to design earthquake-resistant structures it is essential to know the expected ground motion due to earthquakes. The earthquake, however, is a stochastic phenomenon with a random distribution of magnitude and intensity in time and space. Therefore, even for a case in which there are long-term seismic records, statistical processing of the latter is necessary for the design earthquake to be chosen with a preselected probability of occurrence in a certain period of time (e.g. 50–60 years which is the design life of structures).

For this reason two concepts have been introduced: seismicity and seismic hazard.

2.4.1 Seismicity

Seismicity is a parameter that increases not only with the magnitude but also with the frequency of occurrence of earthquake in an area. For this reason, the defin-

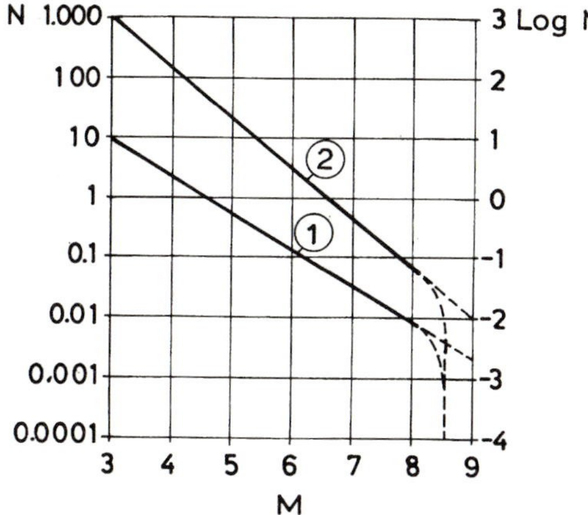

Figure 2.10 Cumulative function of earthquakes in the area of northern Greece (1) and Greece (2).

ition of the seismicity of an area is based on the statistical law of Gutenberg which gives the frequency of the earthquakes (number of earthquakes per year) as a function of their magnitude by the relation (Figure 2.10)

$$\boxed{\log N = a - bM} \tag{2.3}$$

where N is the frequency of earthquakes with magnitude M or larger, M the magnitude of the earthquake and a,b constants that are defined from the statistical processing of the seismic records.

As an example, for the area of Greece and for the period 1901–83 the values of a and b are (Papazachos, 1986)

$$a = 5.99 \qquad b = 0.94$$

Based on the a and b values, several quantities which are used as seismicity indices can be estimated. Thus, for instance, the number of earthquakes per year N_m which have magnitude M or larger and the corresponding average occurrence period T_m in years are given by the relations

$$\boxed{N_m = \frac{10^a}{10^{bM}} \qquad T_m = \frac{10^{bM}}{10^a}} \tag{2.4}$$

The results of these calculations are usually given in the form of **seismicity maps**.

2.4.2 Seismic hazard

Usually the seismic hazard in an area is expressed quantitatively either through the probability of occurrence of an earthquake with acceleration a_g or intensity I larger than a certain value in a certain period of time, or through the value of the acceleration a_g or intensity I for which the probability of exceeding that value in a certain period of time is less than a predefined limit.

It has already been mentioned that the intensity I of an earthquake or the maximum acceleration a_g generally decreases as the distance from the epicentre increases. The statistical evaluation of a large number of earthquakes has produced some empirical **attenuation laws** which relate the intensity I or the maximum acceleration a_g with the magnitude of the earthquake M and the distance Δ from the epicentre. For Europe the following attenuation law for a_g has been proposed by Ambraseys and Bommer (1991):

$$\boxed{\log a_g = -0.87 + 0.217 M_s - \log r - 0.00117\, r + 0.26 P} \tag{2.5}$$

where

$$r = \sqrt{\Delta^2 + h^2}$$

with Δ the source distance and h the focal depth. In equation (2.5) P is 0 for 50 percentile values and 1 for 84 percentiles.

For Greece, which is the most seismically active country in Europe, the following attenuation laws for the intensity I and the maximum ground acceleration a_g have been proposed (Papaioannou *et al.*, 1994; Papazachos, 1986):

$$I = 6.362 + 1.20M - 4.402 \log(\Delta + 15) \tag{2.6}$$

$$\boxed{\log a_g = 3.775 + 0.38M - 2.370 \log(\Delta + 13)} \tag{2.7}$$

According to the above, when the potentiality of existing faults influencing an area is known, it is easy to estimate the statistical distribution of a parameter of the seismic motion in the area (e.g. the maximum acceleration a_g or intensity I in MM), since the epicentral distances Δ and the statistical distribution in time of the magnitude M at every epicentre are known.

Seismic hazard tables (Table 2.3) (Papaioannou *et al.*, 1994) and maps (Figure 2.11) (Drakopoulos and Makropoulos, 1983) are derived from the statistical distributions of a_g and I. Such maps and tables constitute for the time being a major contribution of engineering seismology to structural design, as they provide in effect the maximum design acceleration for an area. However, the designer should not overlook the high degree of uncertainty not only in the maximum acceleration but also in other earthquake characteristics, which are not included in the maps, such as the prevailing period of strong motions, the duration of strong motions and so on.

Table 2.3 Values of maximum expected intensities I and accelerations a_g in 10 Greek cities for an 80-year return period

Town	I (MM)	a_g/g
Rhodes	8.0	0.38
Larissa	7.8	0.37
Patra	7.6	0.37
Mitilini	7.6	0.30
Thessaloniki	7.3	0.26
Kalamata	7.2	0.24
Iraklion	7.1	0.23
Ioannina	7.1	0.20
Athens	6.7	0.17
Kavala	6.5	0.11

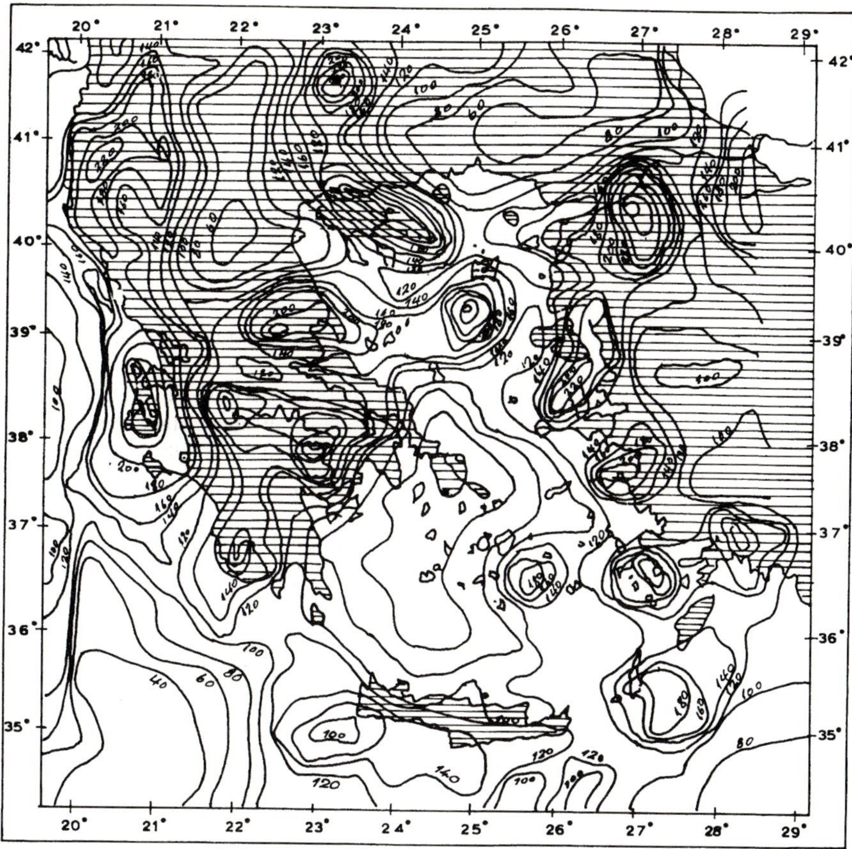

Figure 2.11 Maximum accelerations in gal (1000 gal = g) with 90% probability not to be exceeded in 25 years.

2.5 CONCLUDING REMARKS

Summarizing the material presented above, we should focus on the following points:

1. An earthquake as an independent natural phenomenon in very few cases poses a threat to humans; it becomes a hazardous phenomenon primarily when it is considered in relation with structures. Therefore earthquakes are of special interest for the structural engineer working in seismic areas.
2. The magnitude of the earthquake on the Richter scale is a measure of the phenomenon in terms of energy release at its point of origin. Therefore, the destructiveness of an earthquake, although partly related to its magnitude, is also a function of many other parameters such as the focal depth, the distance from the epicentre, the soil conditions and the mechanical properties of the structures.
3. The intensity of the earthquake is a measure of the consequences that the earthquake has on the people and the structures of a certain area. For many years,

only qualitative macroseismic intensity scales have been used for the damage estimate. Only recently has quantitative information been used, if of course it is available, based on records of strong ground motion in the reference area (e.g. maximum acceleration, dominant period, duration).

4. An earthquake is a stochastic phenomenon and consequently long-term records are needed for a reliable estimate of the seismicity and the seismic hazard of an area. Taking into account that strong motion records go back only to the early 1940s and their number is rather limited, it is inevitable for a reliable estimate of the seismicity and the seismic hazard of an area to combine both seismic records and macroseismic intensity scale estimates of the past.
5. Bearing in mind that the estimate of the seismic hazard of an area is based on information of limited reliability, partly quantitative and partly qualitative, it is quite rational to base the safety of the structures in seismic areas mainly on specially designed extra reserves of strength and energy-dissipation mechanisms at low additional cost. This last comment constitutes the basic concept for the design of earthquake-resistant structures.

2.6 REFERENCES

Ambraseys, N.N. and Bommer, J.J. (1991) The attenuation of ground accelerations in Europe. *Earthq. Engng and Struct. Dynamics*, **20**(12), 1179–1202.

Bolt, B.A. (1978) *Earthquakes*, W.H. Freeman & Co., San Francisco.

Drakopoulos, J. and Makropoulos, K. (1983) *Seismicity and Hazard Analysis Studies in the Area of Greece*, Athens, Greece.

Fintel, M. and Derecho, A.R. (1974) Earthquake-resistant Structures, in *Handbook of Concrete Engineering*, Van Nostrand Reinhold Co., New York, pp. 356–432.

Papaioannou, C.A. Kiratzi, A.A., Papazachos, B.C., and Theoduiidis, N.P. (1994) Scaling of normal faulting earthquake response spectra in Greece. *Proceed. 7th Congress of Hell. Geol. Society*, Thessaloniki, May 1994 (in press).

Papazachos, B.C. (1986) Active tectonics in the Aegean and surrounding area. *Proceed. Summer School on Seismic Hazard in the Mediterranean Region*, Strasbourg, France, 21–30 July. Kluwer Academic Publishers, pp. 301–331.

Papazachos, B.C., Papaioannou, C.A., Margaris, V.N. and Theodoulidis, N. (1993). Regionalization of seismic hazard in Greece based on seismic sources. *Natural Hazards*, **8**, 1–18.

Reid, H.F. (1911) The elastic-rebound theory of earthquakes. *Bull. of Geology*, **6**, 413.

Strobach, K.L., and Heck, H. (1980) Von Wegeners Kontinental Verschiebung zur modernen Plattentektonik. *Bild der Wissenschaft*, **11**, 99–109.

Tsapanos, T.M., Scordilis, E.M., and Papazachos, B.C. (1990) Global seismicity during the time period 1966–1985. *Proceed. of the XXII Gen. Assembly of ESC*, Barcelona, 17–22 September (eds. A. Rocca and D. Mayer-Rosa), **II**, pp. 709–714.

3

Elements of structural dynamics

3.1 INTRODUCTION

The deformation of a structure during a seismic excitation is due to the forced motion of its foundations, which results in the oscillation of the structure (Figure 3.1). It is a procedure during which an amount of kinetic energy is imparted to the structure in the form of elastic deformation. This energy, during the successive phases of oscillation of the structure, alternates continuously from kinetic to potential energy and vice versa, until it is dissipated in the form of the heat through the procedure of viscous and hysteretic damping. Thus, the main problem for the structural engineer in designing an earthquake-resistant structure is to provide a structural system able to dissipate this kinetic energy through successive deformation cycles, without exceeding certain damage limits, defined for characteristic excitation levels. It is obvious that it is very important for the structure to be able to 'store' large quantities of potential energy in the form of large deformations in the plastic range of the material. In view of the above, while for the design of the structure for static loading the main consideration is strength, in seismic design equally important factors besides strength are the flexural stiffness of the structural elements, their ability to deform which is called ductility, and also the mass of the structure.

As far as the seismic excitation is concerned, this parameter is introduced in the analysis in the form of time-dependent accelerograms of the foundations of the structure. The most important parameters of the accelerogram that affect the response of the structure are the maximum acceleration, the prevailing period and the duration of the large-amplitude oscillations of the ground.

The acceleration at any point of the ground is described by two horizontal and one vertical component. However, the vertical component, being less important in the response of the structures, is not usually taken into account.

3.2 DYNAMIC ANALYSIS OF ELASTIC SINGLE-DEGREE-OF-FREEDOM (SDOF) SYSTEMS

3.2.1 Introduction

The study of the dynamic response of structures can be easily carried out by analysing the oscillation of the structural system into **normal modes of vibration**. In general, a system has the same number of normal modes as degrees of

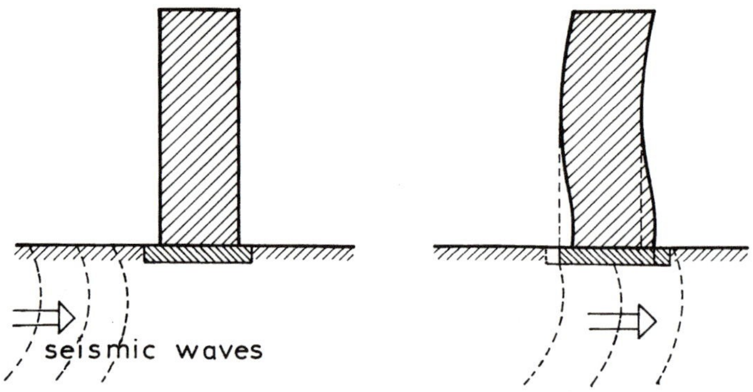

Figure 3.1 Seismic excitation of a building.

freedom. In each of these modes all the discrete masses of the system oscillate **in phase**, which means that for every oscillation, they pass through the resting point or maximum displacement at the same time (Figure 3.2). Directly related to each mode is the corresponding **period of vibration**, that is, the time which is required for a full oscillation. Thus, every normal mode can be considered as a single-degree-of-freedom (SDOF) system with its own natural period. The first or the **fundamental mode** of a system corresponds to the longest natural period.

The response of most buildings is estimated mainly by the superposition of the first few modes of vibration, given the fact that higher-order modes affect only very flexible buildings and only the response of the upper storeys. The study of the response of tall buildings, with a structural system consisting of frames, has shown that the fundamental mode contributes about 80% of the total response, and the second and third modes about 15% (Biggs, 1964). Because of the prevailing contribution of the fundamental mode to the response of the structures, the

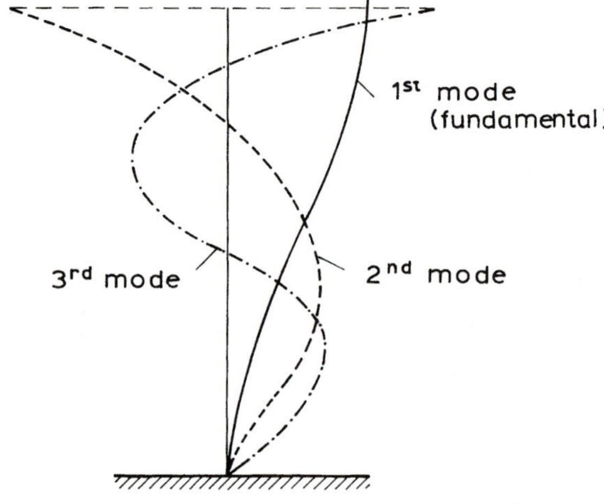

Figure 3.2 The first three normal modes of vibration of a system.

simple formulation and solution of the dynamic problem associated with it, and the extremely important role that the fundamental mode plays in the understanding of the natural phenomenon of vibration, a detailed treatment of the SDOF system will follow (Biggs, 1964; Clough and Penzien, 1975; Warburton, 1976).

3.2.2 Equations of motion

The simplest dynamic system that can be considered is that of Figure 3.3 which consists of a mass M on a spring (two columns) that remains in the linear elastic range ($V=ku$) when it oscillates under a seismic excitation $\ddot{x}_0(t)$. In this case the only 'external loading' is the base acceleration $\ddot{x}_0$; therefore the total acceleration of the system $\ddot{x}$, as well as the corresponding velocity $\dot{x}$ and displacement x, are given by the relations

$$\begin{aligned} x &= x_0 + u \\ \dot{x} &= \dot{x}_0 + \dot{u} \\ \ddot{x} &= \ddot{x}_0 + \ddot{u} \end{aligned} \tag{3.1}$$

Application of d'Alembert's principle of dynamic equilibrium results in the equation of motion for the SDOF system

$$M\ddot{x} + c(\dot{x} - \dot{x}_0) + k(x - x_0) = 0 \tag{3.2}$$

or

$$\boxed{M\ddot{u} + c\dot{u} + ku = -M\ddot{x}_0(t)} \tag{3.3}$$

In the above relation the term $c\dot{u}$ represents the viscous damping which is proportional to the relative velocity of oscillation.

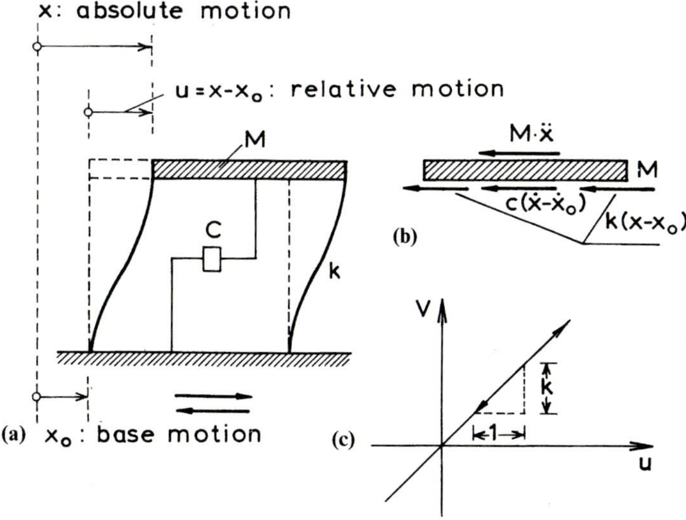

Figure 3.3 SDOF system excited by the base motion: (a) notation; (b) dynamic equilibrium condition; (c) shear force vs relative displacement diagram for the columns.

The equation of motion shows that the most significant factor related to the seismic excitation for the description of the oscillation is the **time-dependent base acceleration**, that is the accelerograph record.

If during oscillation the exciting force becomes zero ($\ddot{x}_0(t)=0$) the system continues to vibrate freely. In this case and for zero damping ($c=0$) the following relation applies:

$$u=u_0\sin\frac{2\pi}{T_0}t \tag{3.4}$$

where

$$\boxed{T_0=2\pi\sqrt{\frac{M}{k}}=\frac{2\pi}{\omega}\ \text{(s)}} \tag{3.5}$$

The **natural period** T_0 is the dynamic constant of the system, in which the characteristics of the system, that is the mass M and the spring constant k, have been incorporated, and ω is the natural circular frequency (in rad/s)

When damping is different from zero ($c\neq 0$) the following relation applies (Figure 3.4):

$$u=u_0\exp[-(c/2M)t](A\sin\gamma t+B\cos\gamma t) \tag{3.6}$$

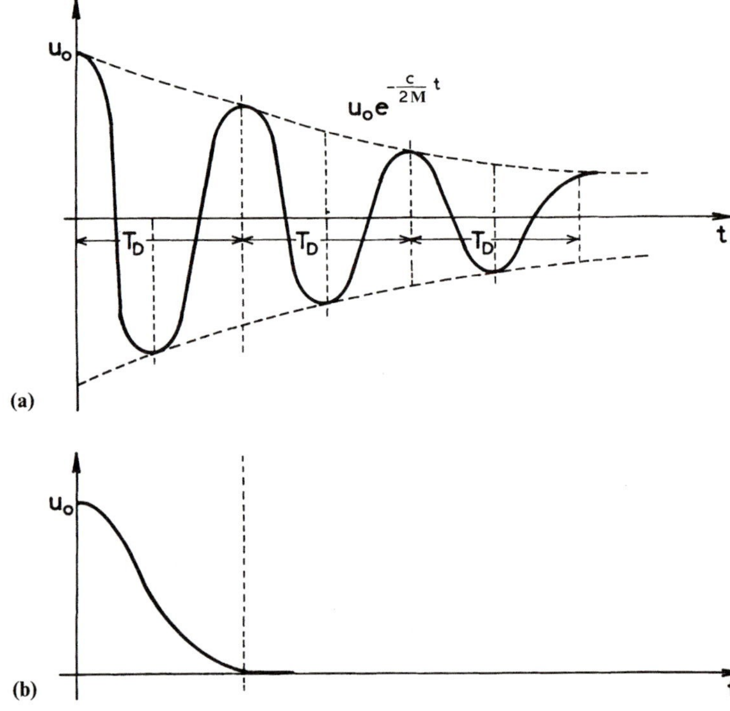

Figure 3.4 Free vibration with damping: (a) initial relative displacement u_0; (b) initial displacement and critical damping.

where

$$\gamma=\sqrt{\frac{k}{M}-\left(\frac{c}{2M}\right)^2} \tag{3.7}$$

For $\gamma=0$, that is for

$$\boxed{c=c_{cr}=2\sqrt{Mk}}$$

there is no free oscillation Figure 3.4(b). Setting $\zeta=c/c_{cr}$ (critical damping ratio) equation (3.6) becomes

$$\boxed{u=u_0\mathrm{e}^{-\zeta\omega t}(A\sin\omega_D t+B\cos\omega_D t)} \tag{3.8}$$

where

$$\omega_D=\omega\sqrt{1-\zeta^2}=\gamma \tag{3.9}$$

$$T_D=\frac{T_0}{(1-\zeta^2)^{1/2}} \tag{3.10}$$

For values of $\zeta<0.10$ the natural period of the oscillating system T_D is almost identical to that of the undamped system T_0.

In the case of an excitation ($\ddot{x}_0(t)\neq 0$) the particular integral of equation (3.3) must be added in equation (3.8).

In the case of a transient seismic excitation, the above process of summation of one general and one particular integral of the differential equation is not possible. However, the introduction of computers has made possible the application of numerical methods for the integration of equation (3.3). This integration describes the time history of the oscillation phenomenon ($u=u(t),\dot{u}=\dot{u}(t),\ddot{u}=\ddot{u}(t)$).

3.2.3 Response spectra

The time history of the oscillation phenomenon is not always needed in practice in its entirety, but it suffices to know the maximum amplitude of the relative displacement, the relative velocity and the absolute acceleration developed during a seismic excitation. This is because from these values the maximum stress and strain state of the system can be determined. For this reason the concept of the **response spectrum** has been introduced. The response spectrum of an earthquake is a diagram whose ordinates present the maximum amplitude of one of the response parameters (e.g. relative displacement, relative velocity, acceleration) as a function of the natural period of the SDOF system.

For every seismic excitation, that is for every accelerogram ($\ddot{x}_0(t)$), there is a series of response spectra whose ordinates give the maximum amplitude of the relative displacement, relative velocity or absolute acceleration of an SDOF system with a natural period defined by the corresponding abscissa (Figure 3.5). This means that the response spectrum of an earthquake reflects the behaviour of all SDOF systems with T between 0 and ∞ during that specific excitation. The first response spectra were produced experimentally by Biot in 1935 on a shaking table with SDOF oscillators with a natural period between 0.1 and 2.4 s (Figure 3.5). After the introduction of computers in 1949, Housner and Kahn produced the first response spectra analytically (Polyakov, 1974; Housner *et al.*, 1953).

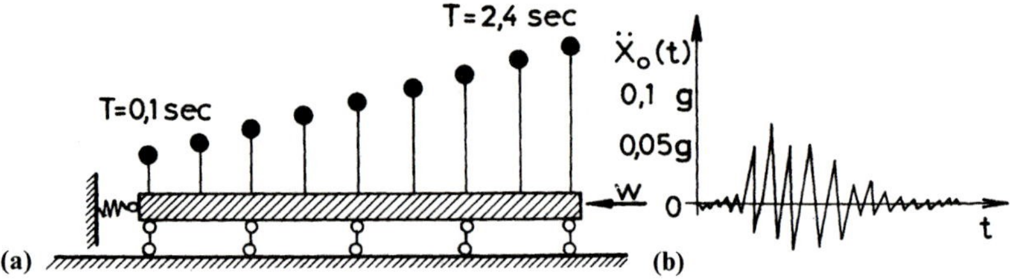
T=2,4 sec
T=0,1 sec
W
(a)
$\ddot{X}_o(t)$
0,1 g
0,05g
0
t
(b)

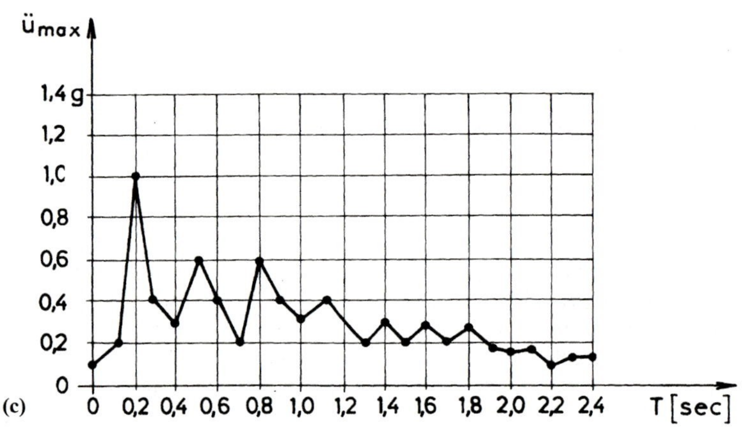
ü max
1,4 g
1,2
1,0
0,8
0,6
0,4
0,2
0
(c)
0 0,2 0,4 0,6 0,8 1,0 1,2 1,4 1,6 1,8 2,0 2,2 2,4
T [sec]

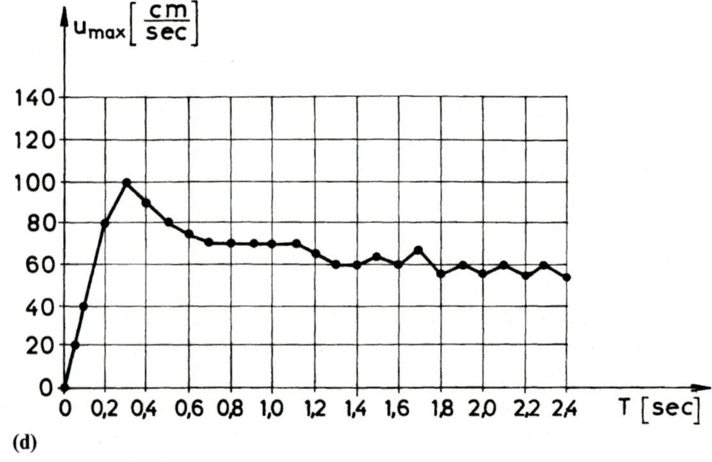
u max [cm/sec]
140
120
100
80
60
40
20
0
0 0,2 0,4 0,6 0,8 1,0 1,2 1,4 1,6 1,8 2,0 2,2 2,4
T [sec]
(d)

Figure 3.5a–d

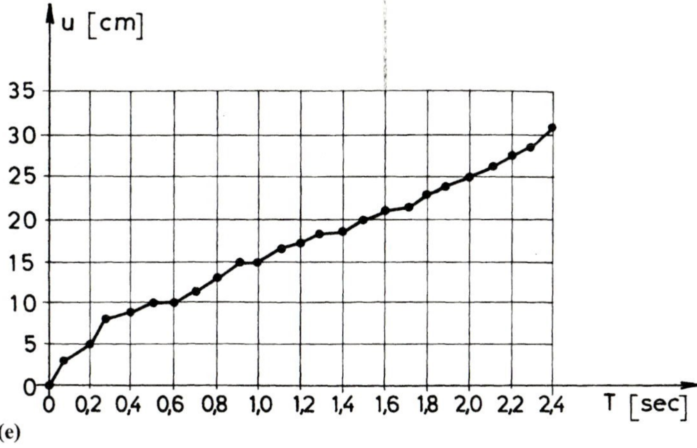

Figure 3.5a–e Response spectra of SDOF systems: (a) shaking table carrying SDOF systems with $T=0.1$ to 2.4 s; (b) accelerogram, (c) acceleration spectrum; (d) velocity spectrum; (e) displacement spectrum.

The ordinates of the spectrum become smaller as the damping ratio ζ is increased (Figure 3.6). For high values of damping the spectra become smooth. In practice, for reinforced concrete structures, ζ is taken between 0.05 and 0.10. Studying the variations of the acceleration spectrum as a function of T, one can see that for $T=0$, that is for a completely rigid structure, the maximum acceleration that is developed on the oscillating mass is equal to $\ddot{x}_0(t)_{max}$. With the increase of the natural period the absolute acceleration S_a of the system increases as well and for a value of the natural period $T=T_{prevail}$, S_a reaches its maximum value which is two to six times the $\ddot{x}_0(t)_{max}$. This maximum appears in systems whose natural period falls in the vicinity of the prevailing period $T_{prevail}$ of the accelerogram ($\ddot{x}_0(t)$) which is the prevailing period of the seismic excitation. In this case the system is in resonance with the seismic excitation. For systems with natural period T larger than $T_{prevail}$ which are generally flexible systems, the value of S_a begins to decrease, and this is because the system again goes out of phase. Earthquakes of small depth (up to 60 km) which are the most frequent ones, have a prevailing period of the order of 0.2–0.4 s, therefore SDOF systems with a natural period within these limits, experience the largest acceleration of their oscillating masses. This natural period generally corresponds to the fundamental period of two- to four-storey buildings.

For earthquakes of large depth (70–300 km) or large epicentral distance, which are more rare (1977 Bucharest earthquake, 1985 Mexico City earthquake), the prevailing period of the earthquake appears to be 1.0–2.0 s, therefore the systems with a natural period within these limits experience the largest acceleration of the oscillating masses. This natural period generally corresponds to the fundamental period of 10–20 storey buildings (Figure 3.7).

Flexible (high-rise) buildings with a large fundamental period are vulnerable to earthquakes when their foundations rest on soft soil, because this kind of soil shifts the maximum of the acceleration spectrum to the right. On the contrary, stiff (low-rise) buildings appear to be vulnerable to earthquakes when they sit on

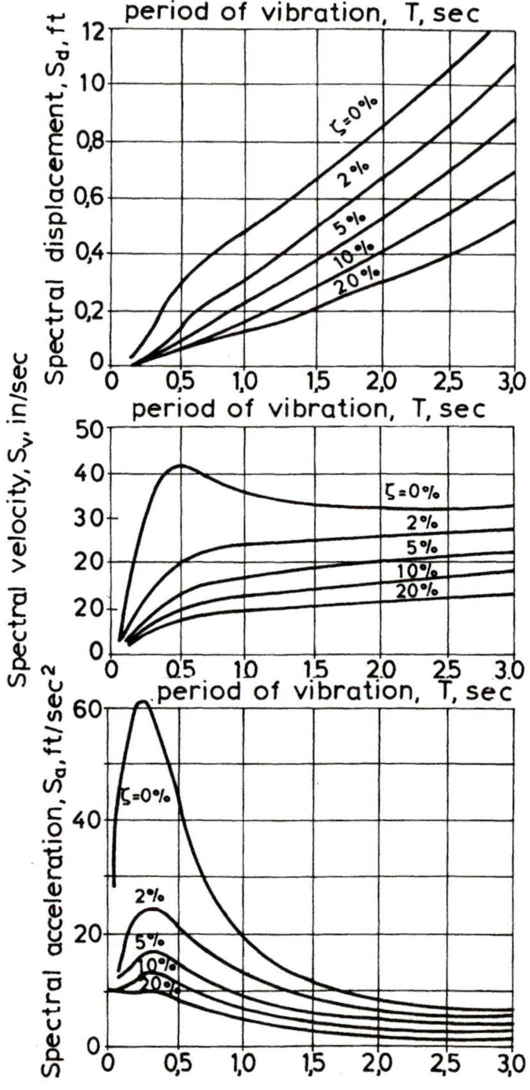

Figure 3.6 Normalized response spectrum.

firm soil, where the maximum of the acceleration spectrum is shifted to the left (Figure 3.8) (Richart, Woods and Hall, 1970).

Response spectra, in particular velocity spectra, allow quantitative evaluation of the total seismic excitation and therefore motions of different amplitude can be scaled (normalized) to the same level of intensity. This can be accomplished with the following integral, which was originally defined by Housner (Wiegel, 1970) as **spectrum intensity**

$$\mathrm{SI} = \int_{T_1}^{T_2} S_V(T,\zeta)\,\mathrm{d}T$$

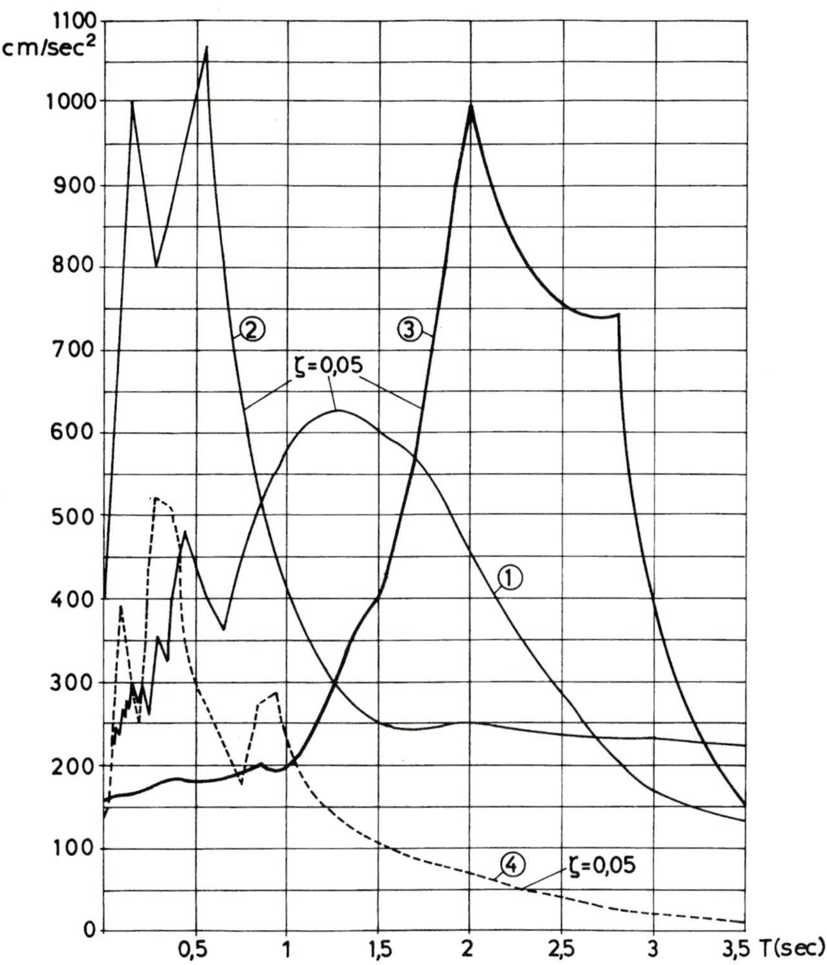

Figure 3.7 Characteristic acceleration spectra of strong earthquakes: (1) acceleration spectrum of Bucharest earthquake 4.3.77, N–S; (2) acceleration spectrum of El Centro Earthquake 18.5.1940, N–S; (3) acceleration spectrum of Mexico City earthquake 19.9.85, E–W, SCI; (4) acceleration spectrum of Thessaloniki earthquake 20.6.78, E–W city hotel.

with integration limits $T_1=0.1$ and $T_2=2.5$ s. The above integral represents the area under the velocity spectrum for a given damping coefficient ζ, between the limits T_1 and T_2, expressed in units of length. Table 3.1 gives some characteristic SI values (in mm) of several accelerograms, as well as the resulting normalization factors with regard to the El Centro spectrum. (These factors represent the ratio of the El Centro SI to the SI of the accelerogram under consideration (mean value for $\zeta=5$ and 10%). The SI values are calculated for $T_1=0.1$ and $T_2=2.6$ s.

The reliability of the normalization factors can be checked by studying the response of multi-storey structures for several accelerograms normalized to the same level of spectrum intensity. Such a response is shown in Figures 3.25 and

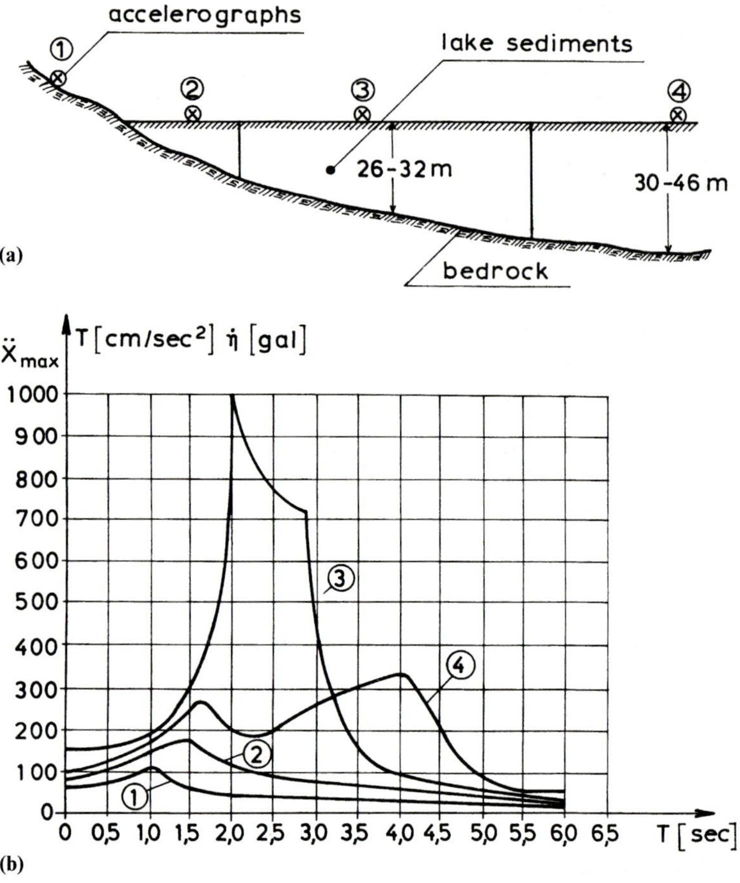

Figure 3.8 (a) A ground section of Mexico City and (b) acceleration spectra at points (1), (2), (3) and (4).

Table 3.1 Values (in mm) and normalization factors for the first 10 s of a series of accelerograms

Accelerogram		SI			Factor
		ζ=0%	ζ=5%	ζ=10%	
El Centro	S00E	2334	1479	1230	1.00
Taft	N21E	1021	651	555	2.24
Taft	S29E	1176	749	605	2.00
Cal Tech	S90W	634	414	338	3.60
Pacoima	S16E	5876	3910	3316	0.37
Thessaloniki	N30E	721	555	484	2.61
Thessaloniki	N60E	698	517	457	2.78

3.26 (Kappos, 1990) which refer to maximum response parameters of the nine-storey R/C frame system shown in Figure 3.24. It is seen that although the accelerograms are normalized, the differences in response produced by various accelerograms are significant. Recently an attempt was made to modify the limits of the spectrum intensity integral (Kappos, 1990), taking into account the period of the structures under consideration in order to reduce the scatter in the results, but the improvement was only moderate.

Acceleration spectra, besides allowing an overall picture to be gained of the response of the structures to an excitation, also make possible static consideration of the seismic excitation, which is a totally dynamic phenomenon. Such a consideration offers significant advantages to the civil engineer whose training is mainly focused on statics. Indeed, the relation between the inertia forces and the restoring forces (Figure 3.3(b)) has the following form:

$$V=k(x-x_0)=ku=-(c\dot{u}+M\ddot{x}) \tag{3.11}$$

The **base shear** V_{max} assumes its maximum value when the relative displacement also reaches its maximum value, in which case $\dot{u}=0$.

$$V_{max}=ku_{max}=M\ddot{x}_{max} \tag{3.12}$$

or

$$V_{max}=kS_d=MS_a \tag{3.13}$$

If V_{max} is expressed as a function of the weight of the structure, the following relation results:

$$\boxed{\frac{V_{max}}{W}=C=\frac{MS_a}{W}=\frac{MS_a}{Mg}=\frac{S_a}{g}} \tag{3.14}$$

The above relation expresses the following significant conclusion: in order to determine the maximum stress and strain of an SDOF system, one can statically load the concentrated mass M with a horizontal force V, which is equal to the weight W of the mass, multiplied by the seismic coefficient C that results from the response spectrum of the specific earthquake, scaled to g (Figure 3.9).

3.3 INELASTIC RESPONSE OF SDOF SYSTEMS

3.3.1 Introduction

After the introduction of the concept of the response spectrum in earthquake engineering and the development of the first elastic spectra by Housner in 1949 (Housner *et al.*, 1953), it was noticed with surprise that the maximum acceleration of the vibrating masses in structures close to resonance with the earthquake was two to six times larger than the maximum base acceleration. Thus, for $\ddot{x}_{0max}/g\simeq 0.17$ the seismic coefficient C reached the value of 0.45–1.00. However, all the existing structures at the time were designed for C values between 0.04 and 0.16, according to the codes then in force, while the damage in engineered structures from the earthquakes that occurred in the mean time was not always destructive. The difference was such that it could not be attributed to the existing safety factor or to calculation errors. Therefore, a more precise approach to the problem

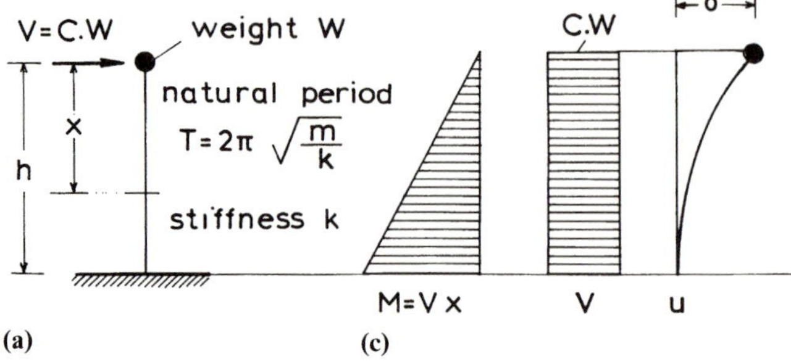

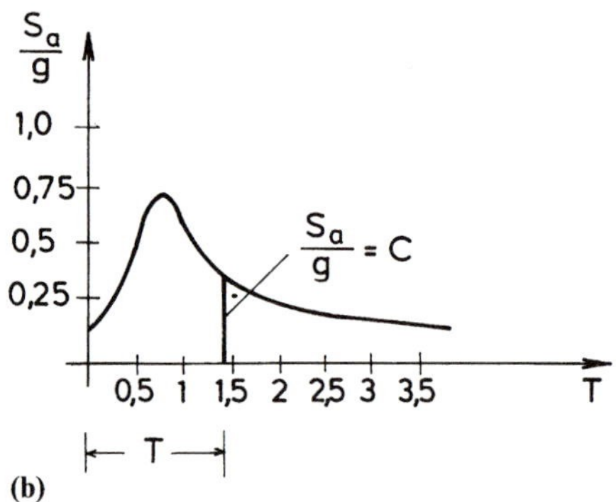

Figure 3.9 Transformation of the seismic problem to a static one: (a) an SDOF system; (b) acceleration response spectrum S_d/g; (c) maximum response of the structure.

was sought, that would take into account the inelastic behaviour of the structures which leads to the dissipation of a large percentage of the kinetic energy of the system through damping.

3.3.2 Viscous damping

Up to now, the damping phenomenon, the existence of which is observed in actual situations, has been studied using the Kelvin–Voigt model (Figure 3.10), which

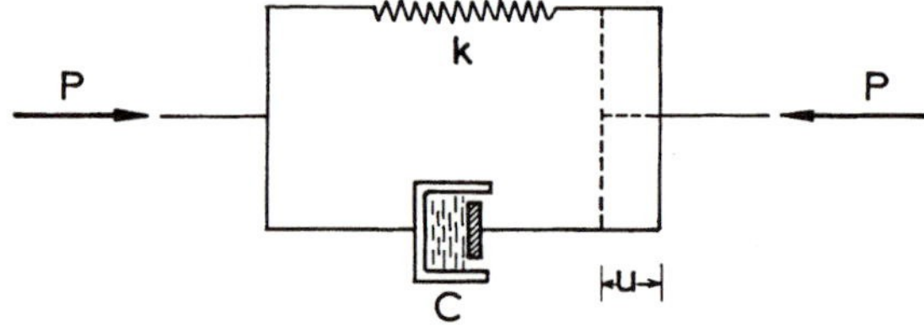

Figure 3.10 The Kelvin–Voigt viscoelastic model.

consists of a spring and an oil damper connected in parallel (Housner *et al.*, 1953). Thus, the total force P required for the displacement u of the system, is the sum of the force P_e of the Spring and P_d of the damper

$$P = P_e + P_d = ku + c\dot{u} \tag{3.15}$$

since the force that the oil damper receives is

$$P_d = c\dot{u} \tag{3.16}$$

In real structural systems, however, springs and dampers are substituted by elastic members that connect the masses to each other and to the ground. It is therefore important to evaluate whether relation (3.16) expresses with sufficient accuracy the damping phenomenon in structures. As is known, the response of a material to an external force depends on the rate of loading. The higher the rate of loading, the larger the force that is required for the same deformation (Figure 3.11).

$$P = P_e + \Delta P = P_e + P_d \tag{3.17}$$

where P_e is the elastic part of the loading and P_d the viscous one.

As Figure 3.11 shows, the viscous part of the loading is a complicated function of the deformation rate $\dot{u}$. However, if this function is expanded in a polynomial series of $\dot{u}$ and only the first is retained, P_d takes the form

$$P_d = c\dot{u}$$

which is a sufficiently accurate expression for P_d. Therefore the relation $P = P_e + P_d$ takes the form of (3.15).

3.3.3 Hysteretic damping

Besides the viscoelastic behaviour of materials there are other factors that lead to 'damping'. The most significant, especially for high values of deformation such

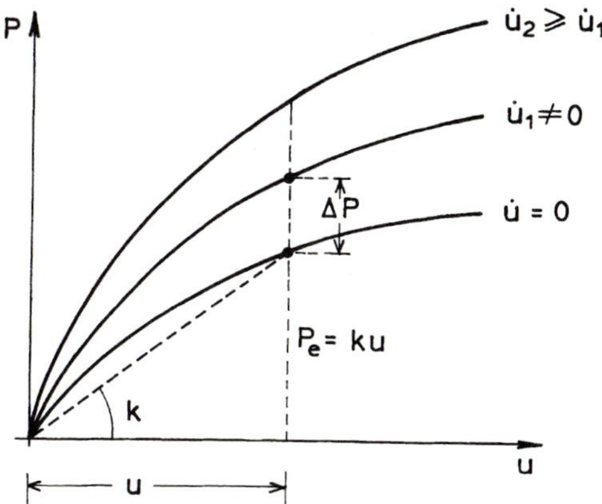

Figure 3.11 Qualitative representation of the increase of the deformation force P with the rate of loading $\dot{u}$.

as those resulting from earthquakes, is the hysteretic behaviour of materials. The stress–strain diagram of a material under cyclic loading has the form of Figure 3.12, with several variations depending on special characteristics of the material (see also Chapter 7). The area of the shaded loop of Figure 3.12 represents the energy that is dissipated in every loading cycle in the form of heat, due to the plastic behaviour of the material. It is obvious that the larger the area of the hysteresis loop, that is, the higher the deformation level of the material, the larger the dissipated energy and therefore the damping.

The question is then raised of whether or not the expression for the damping force with the form

$$P_{\mathrm{d}} = c\dot{u}$$

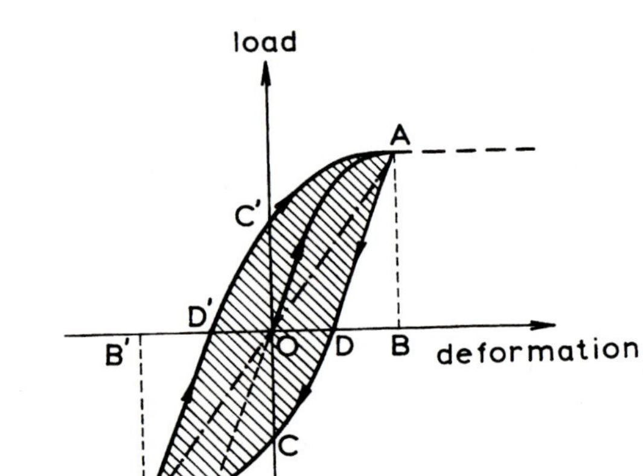

Figure 3.12 A typical hysteresis loop.

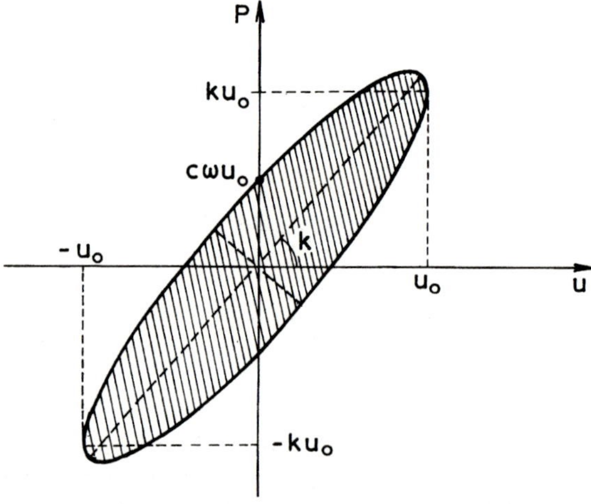

Figure 3.13 Schematic presentation of the restoring force $P = P_e + P_d$.

can also express, even approximately, the hysteretic damping. In order to investigate this, it is helpful to represent graphically the restoring force

$$P=P_e+P_d=ku+c\dot{u} \tag{3.18}$$

If the system of Figure 3.3 undergoes an oscillation of the form

$$u=u_0 \sin \omega t \tag{3.19}$$

then relation (3.18) takes the form

$$P=ku_0 \sin \omega t+c\omega u_0 \cos \omega t \tag{3.20}$$

Relations (3.19) and (3.20) define a P–u function which describes an ellipse (Figure 3.13). It is evident that the hysteretic loop of Figure 3.12 may be represented by the ellipse of Figure 3.13. Thus, nonlinear diagrams of materials and structures can be satisfactorily approximated by the differential equation

$$M\ddot{u}+c\dot{u}+ku=-M\ddot{x}_0(t) \tag{3.21}$$

which is linear, while the general form of their exact expression would be of the form

$$M\ddot{u}+V(\dot{u},u)=-M\ddot{x}_0(t) \tag{3.22}$$

where $V(\dot{u}, u)$ is a function of the restoring force which includes both viscous and hysteretic damping.

In the case where equation (3.21) is used to express the hysteretic damping of the inelastic behaviour, the equivalent hysteretic damping ratio $\zeta_e=c/c_{cr}$ results from energy criteria as follows. The dissipated energy in the case of the hysteresis loop of Figure 3.13 for a full loading cycle is equal to

$$\Delta W=\int_T^{T+\frac{2\pi}{\omega}} P(t)\,\frac{du}{dt}\,dt=\pi c\omega u_0^2 \tag{3.23}$$

On the other hand, the maximum potential energy U of the system is equal to

$$U_e=\frac{1}{2}ku_0^2$$

Therefore

$$\frac{\Delta W}{U_e}=\frac{2\pi c\omega}{k} \tag{3.24}$$

But

$$c=\zeta_e c_{cr}=2M\omega\zeta_e \tag{3.25}$$

Substituting (3.25) into (3.24), ζ_e becomes

$$\boxed{\zeta_e=\frac{1}{4\pi}\frac{\Delta W}{U_e}} \tag{3.26}$$

The availability of large computational facilities during the last 30–35 years has allowed the analysis of inelastic systems with numerical methods. However, equation (3.26) is still a basis for the qualitative understanding of the phenomenon and an indicator of energy dissipation per cycle of loading.

3.3.4 Energy dissipation and ductility

The effects of inelastic behaviour on the response of structures to strong seismic motions may be clarified by studying the SDOF system.

Consider two SDOF systems, each with the same mass M and the same spring constant k, and without damping (Figure 3.14). Suppose that both systems oscillate freely and that when they pass through their original equilibrium position they both have the same velocity $\dot{u}_{max} = v_{max}$ (Park and Paulay, 1975). Suppose also that the first one has an elastic connection of ultimate strength V_{1u}, while the second has a strength V_{2u} which is much smaller than V_{1u} (Fig. 3.15).

The first system undergoes a displacement u_{01} of its mass such that the potential energy stored in it in the form of strain energy, represented by the area of the triangle OBF, is equal to the kinetic energy of the system:

$$\frac{1}{2}Mv_{max}^2 = \frac{1}{2}ku_{01}^2 \tag{3.27}$$

therefore

$$u_{01} = \left(\frac{M}{k}\right)^{1/2} v_{max} \tag{3.28}$$

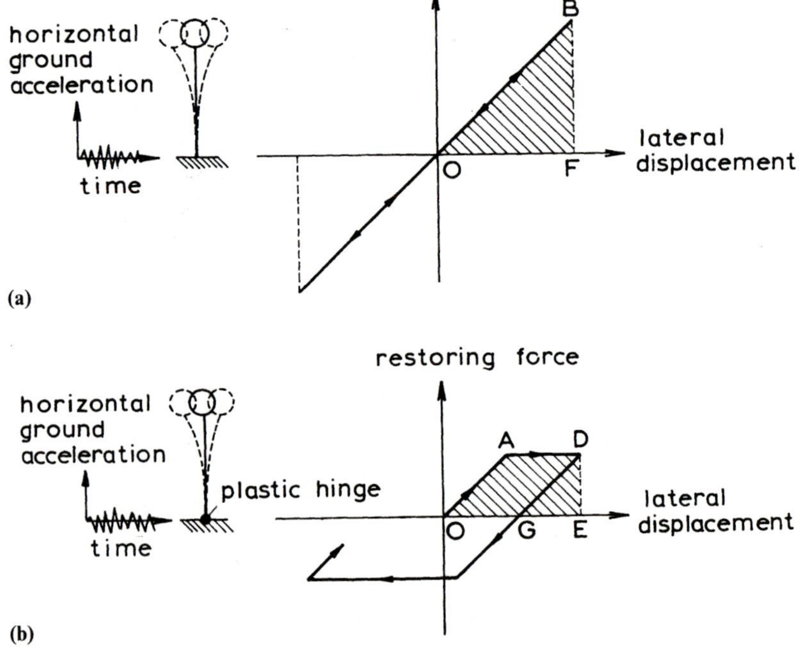

Figure 3.14 Response of SDOF systems to seismic motion: (a) elastic response; (b) elastoplastic response.

In this case, a maximum restoring force is developed in the elastic connection, equal to $V_{1max} < V_{1u}$ which coincides with the maximum inertia force $M\ddot{x}_{max}$. This restoring force, given the fact that the velocity becomes zero, starts to accelerate the system towards the opposite direction, thus causing oscillations of constant amplitude.

The second system, unable to develop a restoring force equal to the first one, is led to the creation of a plastic hinge at the base, with maximum restoring force V_{2u} and maximum displacement u_{02} such that the area of the trapezoid OADE is equal to the kinetic energy of the system. Thus

$$\frac{1}{2}Mv^2_{max} = \frac{1}{2}V_{2u}u_{y2} + V_{2u}u_{pl} \tag{3.29}$$

therefore

$$u_{pl} = \frac{1}{2V_{2u}}(Mv^2_{max} - V_{2u}u_{y2}) \tag{3.30}$$

and the total diplacement of the second system is equal to

$$u_{02} = u_{y2} + u_{pl} \tag{3.31}$$

For a displacement equal to u_{02} the system has consumed all its kinetic energy; therefore, under the influence of V_{2u} it begins to move towards its original position. At the moment when V_2 becomes zero, the potential energy which has been

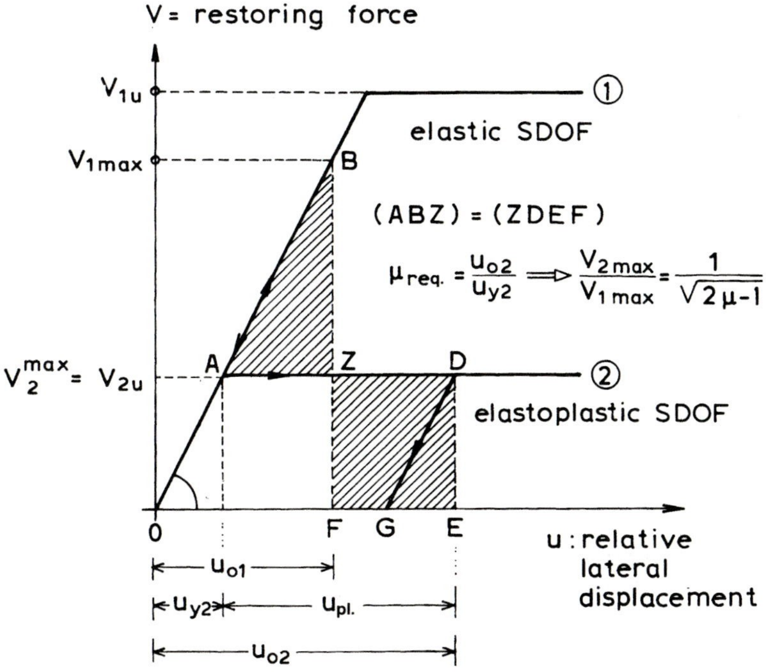

Figure 3.15 Quantitative relation between elastic and elastoplastic response of an SDOF system to an earthquake.

transformed into kinetic energy is represented by the area of the triangle EDG because the energy represented by the parallelogram OADG has been dissipated by the plastic hinge in the form of heat and other irrecoverable forms of energy.

From the above it is obvious that, while in the elastic system there is a successive interchange between kinetic and potential energy, which results in a cumulative effect of the successive excitation cycles on the relative displacements, in the elastoplastic system only part of the potential energy is transformed into kinetic energy from cycle to cycle, a fact which results in the quick damping of the phenomenon. This means that the displacement u_{02} as defined above, is an upper limit for the elastoplastic systems. Detailed analytical studies have shown that exactly because of the heavy damping the maximum displacements of the two systems described above during their excitation from a seismic motion are approximately of the same magnitude.

From the above discussion one can conclude that the action of a seismic excitation on an oscillating system can be resisted either with large restoring forces and oscillation within the elastic range or with smaller restoring forces and exploitation of the ability of the system to undergo plastic deformation, as long as the system has such abilities. The ability of the system to undergo plastic deformation is characterized as **ductility** and it is a property of paramount importance for earthquake-resistant structures, because it gives the designer the choice to design the structure for much lower forces than consideration of an elastic system would require.

The **ductility factor** is defined as the ratio of the ultimate deformation at failure to the yield deformation, δ_u/δ_y (Figure 3.16). The ultimate deformation at failure is defined for design purposes as the deformation for which the material or the structural element loses only a small, predefined percentage of its ultimate strength (e.g. 15% for concrete). The larger the available ductility factor of a structural element with constant strength, the larger the safety margins of the element against an earthquake.

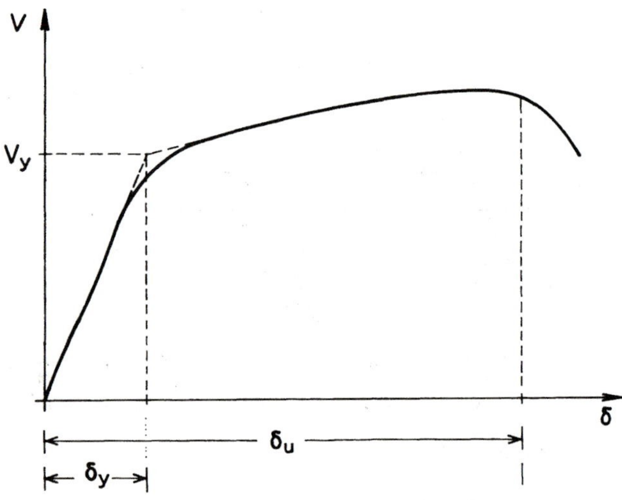

Figure 3.16 Definition of the ductility factor.

Of particular interest is the determination of the required ductility factor of a structure for a given ratio of reduction of the elastic restoring force. Consider an elastoplastic system with mass M, stiffness coefficient k and damping ζ, which is subject to a seismic excitation represented by a given acceleration spectrum. Under the assumption of fully elastic behaviour, the maximum restoring force V_{el} which would act on the system can be easily found from the above data. Now, if the system has an ultimate strength V_u which is smaller than V_{el}, the required ductility factor μ whereby the system responding with V_u would be able to resist the earthquake, results from the equation of the potential energy in the two cases (Figure 3.15). Indeed, setting the area of the triangle (ABZ) equal to the area of the rectangle (EDZF) one gets

$$\boxed{R_1 = \frac{V_u}{V_{el}} = \frac{1}{(2\mu - 1)^{1/2}}} \tag{3.32}$$

where R_1 is the reduction coefficient of the spectral value and μ the minimum required ductility factor for this reduction.

Another approach to the problem is based on the comment made earlier, that the maximum displacements that result from the analysis of different systems are of the same order of magnitude, whether the analysis is based on the assumption of elastic behaviour or of elastoplastic behaviour (Figure 3.17).

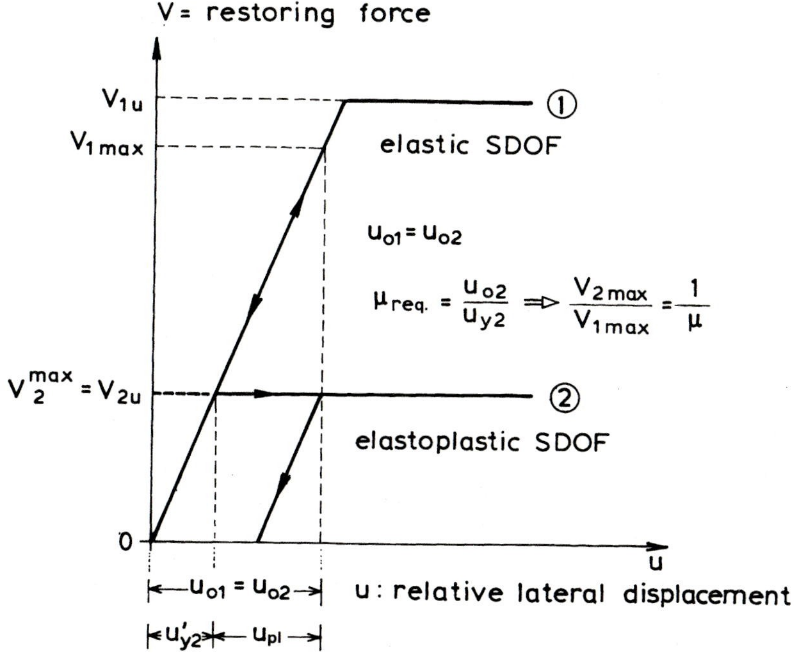

Figure 3.17 An alternative quantitative relation between elastic and elastoplastic response of an SDOF system.

From the geometry of Figure 3.17 the following relation can be obtained:

$$\boxed{R_2 = \frac{V_u}{V_{el}} = \frac{1}{\mu}} \tag{3.33}$$

Figure 3.18 shows the two curves described by relationships (3.32) and (3.33), as well as the results of an inelastic analysis of several SDOF systems for the 1940 EI Centro earthquake, N–S component (Wiegel, 1970).

At this point, it might be useful to explain in more detail what the term 'ability to behave in an elastoplastic manner' means for reinforced concrete. Consider the cantilever of Figure 3.19 which is loaded with a horizontal force V. By increasing V, the cantilever reaches the failure limit state. Failure can occur in two ways, either by yielding of reinforcement in normally reinforced sections, or by crushing of concrete in highly reinforced sections, where the strength of the compression zone is

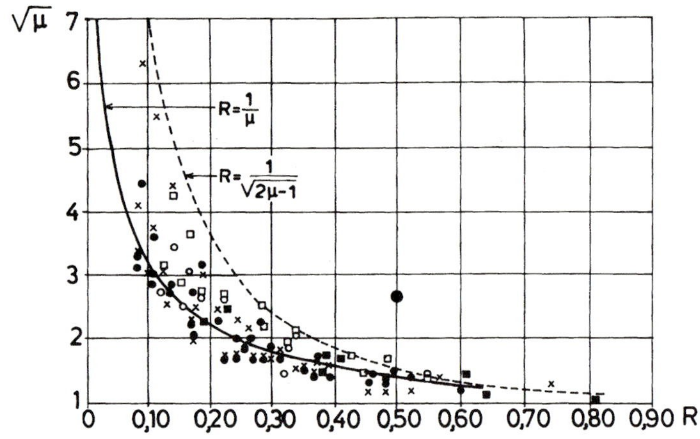

Figure 3.18 Representation of R as a function of $\sqrt{\mu}$.

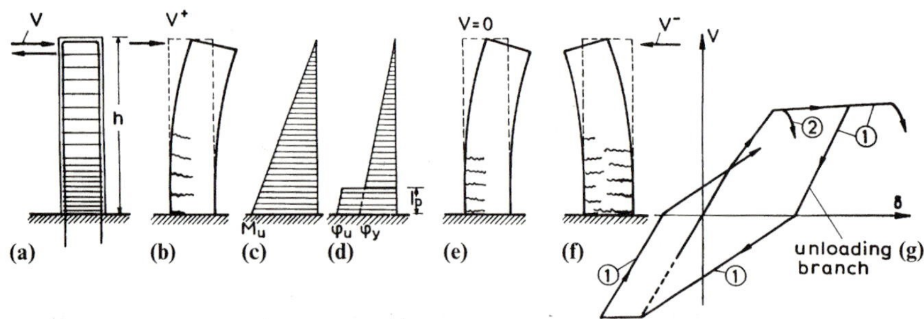

Figure 3.19 Inelastic response of a cantilever under cyclic loading: (a) arrangement of cantilever loading; (b) loading with V^+ beyond yielding; (c) ultimate failure moment diagram; (d) distribution of curvature ϕ–plastic hinge zone l_p; (e) unloading $V=0$; (f) loading with V^- beyond yielding; (g) V–δ diagram for normally reinforced (1) and over-reinforced (2) cantilever.

lower than the yield strength of the reinforcement. In the first case, for a very small increase of V the displacement δ exhibits considerable values (Figure 3.19), and this is accomplished through opening of the cracks due to yielding of the reinforcement. At some point, after large plastic deformations have developed, because of the opening of the cracks the depth of the compression zone is substantially decreased and the concrete crushes. Beyond this point there is a rapid deterioration of the structural system and a steep descending branch on the V–δ diagram (Figure 3.19(g), curve 1). In the second case (fracture of the compression zone without yielding of the reinforcement) there is a brittle failure (Figure 3.19(g), curve 2) and a steep descending branch on the V–δ diagram, without the development of plastic deformations. In the first case there is large amount of available ductility; however, in order to use it, some yielding zones in the structure must be tolerated, which of course implies accepting some degree of damage due to the appearance of wide cracks. In the second case the available ductility is very small. Therefore over-reinforced systems are not suitable for earthquake-resistant structures.

The design and detailing of R/C structural elements for high available ductility will be discussed in detail in Chapters 8 and 9.

At this point, it will be useful to approach the concept of the ductility factor and the way it is defined in some detail. It has already been explained in the previous example of the cantilever subject to horizontal loading, that in terms of the displacement at the tip of the cantilever, the ductility factor is defined by the ratio

$$\mu_\delta = \frac{\delta_u}{\delta_y}$$

If only bending deformations are considered, δ is in general the integral of the rotation θ along the axis of the cantilever, while θ is the integral of the curvature ϕ along the same axis (Figure 3.19(d)). Thus, the ductility factor in terms of horizontal displacements can be obtained by integrating the ductility factor which is expressed in terms of the rotations θ

$$\mu_\theta = \frac{\theta_u}{\theta_y}$$

at the base of the cantilever, and this can be obtained by integrating the ductility factor expressed in terms of the curvatures ϕ at the base of the cantilever

$$\mu_\phi = \frac{\phi_u}{\phi_y}$$

In Chapters 7–9 it will be explained in detail how to get from one level to the next. At this point it can only be stressed that for a given ductility factor in terms of displacements, the ductility factor in terms of rotations is larger, while the ductility factor in terms of curvatures is much larger:

$$\boxed{\mu_\delta < \mu_\theta \ll \mu_\phi} \tag{3.34}$$

Finally, it may be useful to present some inelastic response spectra for the V–δ hysteresis model of Clough (Clough and Johnston, 1966) as modified by Riddell and Newmark (1979), which has been widely used for reinforced concrete. The spectra shown are for several values of required ductility μ (Figure 3.20), so that the effect of the ability of the system to dissipate energy during its response to a seismic excitation will become clear.

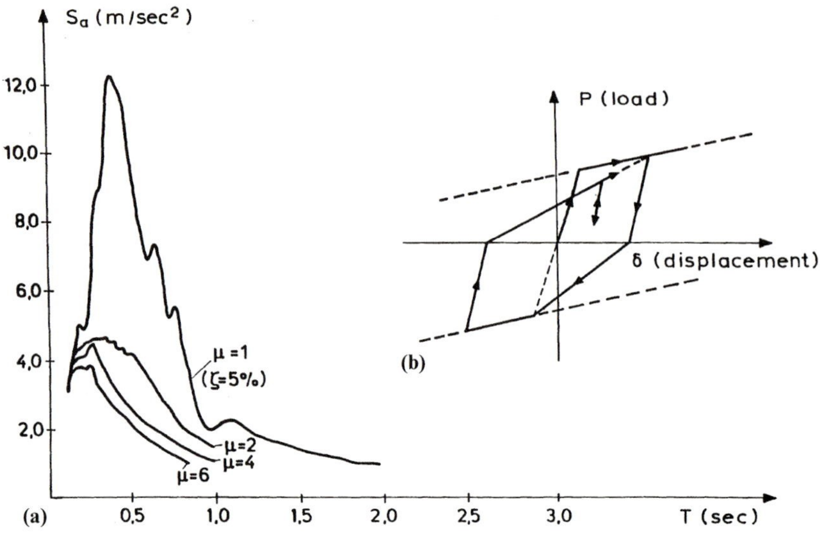

Figure 3.20 (a) Inelastic spectra of the Kalamata, Greece, earthquake, component N10°W, for various ductility factors μ; (b) Clough–Riddel–Newmark hysteresis model.

3.4 DYNAMIC ANALYSIS OF MULTIDEGREE-OF-FREEDOM ELASTIC SYSTEMS

3.4.1 Introduction

The scope of the book does not allow a detailed approach to the dynamic analysis of systems with more than one degree of freedom. Standard textbooks of structural dynamics (Biggs, 1964; Clough and Penzien, 1975; Warburton, 1976) provide a detailed study of the subject. However, some elements which refer to this problem are presented here for the sake of completeness.

The number of degrees of freedom of a lumped-mass system is determined by the minimum number of independent displacements and rotations of the lumped masses whereby their geometric position can be defined at a given moment. Thus, in a plane frame, with the mass concentrated in the beams of the floors and with large axial stiffness of the beams, which are both very realistic assumptions for typical R/C structures, the degrees of freedom are determined by the number of storeys, while the independent variables of motion are their horizontal displacements.

The **normal modes** or natural modes of a linear system are the free, undamped periodic oscillations whose linear combination represents the position of the system at every moment. For every such normal mode all the masses of the system oscillate in phase, that is, at every moment the ratio of the displacements of the discrete masses remains constant. As a result, at any time all the masses go through rest and reach maximum amplitude at the same moment (see section 3.2).

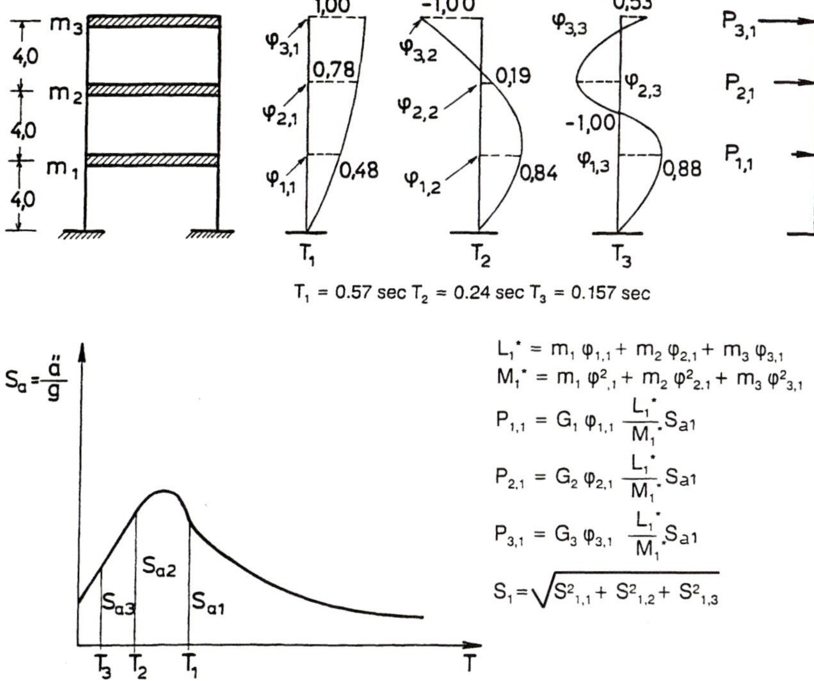

Figure 3.21 A three-storey plane frame analysed according to spectral modal analysis.

The conclusion of the above discussion is that the number of normal modes of a structure equals the number of its degrees of freedom. Every normal mode is related to a natural frequency or a period of oscillation, known as the natural period. For plane systems the normal mode with the longest natural period is by definition the first or **fundamental normal mode**, while for pseudospatial ones the fundamental period is replaced by the set of three longest periods corresponding to two translational and one rotational mode of vibration.

Figure 3.2 shows the first three normal modes of a multi-storey building. Note that the curves intersect the vertical axis at a number of points (including the one at the base) which coincides with the order of the natural mode. The amplitudes of each natural mode are normalized. Figure 3.21 shows the typical normal modes of a three-storey building for which the three eigenvectors ϕ_i are

$$\phi_1 = \begin{pmatrix} 1.00 \\ 0.78 \\ 0.48 \end{pmatrix} \qquad \phi_2 = \begin{pmatrix} -1.00 \\ 0.19 \\ 0.84 \end{pmatrix} \qquad \phi_3 = \begin{pmatrix} 0.53 \\ -1.00 \\ 0.88 \end{pmatrix}$$

The maximum relative displacement in every case was taken equal to unity. It is significant to note that the ratio of the displacements at any moment is constant for every normal mode.

3.4.2 The two methods of analysis

The term 'dynamic analysis of multidegree-of-freedom (MDOF) systems' has been related in structural dynamics with two analytical approaches slightly different from each other. According to the first one which is known as **spectral modal analysis**, the values of the maximum response parameters (e.g. displacements, bending moments) can be approximately determined as a combination of the maximum responses that correspond to every natural mode. Indeed, since every contributing normal mode behaves as an independent SDOF system with a certain characteristic natural period, the maximum response at this specific natural mode can result from the corresponding spectra for SDOF systems. There are several methods of combining modal contributions, taking into account that the maxima of the different modes do not occur at the same time. The most common method takes the square root of the sum of the squares of the maximum modal amplitudes, treating them as random quantities. In the case where the natural periods are only slightly different from each other (closely coupled modes), the square root of the sum of the squares underestimates the expected final value. Therefore, more reliable combination techniques are used, as will be discussed later. Usually, only the contribution of the first few modes is taken into account since they contribute the largest portion of the response.

Spectral modal analysis is the most widely used method for the design of structures. Normalized spectra, based on a number of seismic records and scaled to a typical reference intensity, are used as response spectra. The use of normalized spectra provides a simple means for studying the variation of the response to different seismic inputs. In the following chapters frequent reference will be made to normalized response spectra.

According to the second method known as **time-history analysis**, the evolution of the response of the model of the structure with time is determined, when this model is subjected to a base accelerogram. Either superposition of the normal modes or direct numerical integration of the equations of motion is used for the analysis. In both cases, the total response of the system is calculated at the end of a very small time step, and the analysis proceeds step by step, using the end conditions of one step as initial conditions for the next one.

(a) Spectral modal analysis

The analysis of multi-storey structures using this method can only be performed with the aid of a computer; either a plane or a pseudo-three-dimensional frame is commonly used, and the procedure is the same for both cases (Clough and Penzien, 1975; Warburton, 1976; Gupta, 1990):

1. First the normal modes and the natural periods of the system are determined (Figure 3.21), that is T_1, T_2,..., T_j.
2. From the design spectrum, the maximum accelerations (S_{a1}, S_{a2},..., S_{aj}) corresponding to the periods (T_1, T_2,..., T_j) are determined for each normal mode.
3. For each normal mode, the effective modal masses M_j^* are determined and from these the maximum inertia forces $P_{i,j}$ for each mode. That is

$$\left.\begin{aligned}
L_1^* &= m_1\phi_{1,1}+m_2\phi_{2,1}+\cdots+m_j\phi_{j,i}\\
M_1^* &= m_1\phi_{1,1}^2+m_2\phi_{2,1}^2+\cdots+m_j\phi_{j,1}^2\\
P_{1,1} &= G_1\phi_{1,1}\frac{L_1^*}{M_1^*}S_{a,1}\\
P_{2,1} &= G_2\phi_{2,1}\frac{L_1^*}{M_1^*}S_{a,1}\\
&\vdots\\
P_{j,1} &= G_j\,\phi_{j,1}\frac{L_1^*}{M_1^*}S_{a,1}
\end{aligned}\right\}\text{1st mode}$$

$$\left.\begin{aligned}
L_2^* &= m_1\phi_{1,2}+m_2\phi_{2,2}+\cdots+m_j\phi_{j,2}\\
M_2^* &= m_1\phi_{1,2}^2+m_2\phi_{2,2}^2+\cdots+m_j\phi_{j,2}^2\\
P_{1,2} &= G_1\phi_{1,2}\frac{L_2^*}{M_2^*}S_{a,2}\\
P_{2,2} &= G_2\phi_{2,2}\frac{L_2^*}{M_2^*}S_{a,2}\\
&\vdots\\
P_{j,2} &= G_j\,\phi_{j,2}\frac{L_2^*}{M_2^*}S_{a,2}
\end{aligned}\right\}\text{2nd mode}$$

$$\left.\begin{aligned}
L_j^* &= m_1\phi_{1,j}+m_2\phi_{2,j}+\cdots+m_j\phi_{j,j}\\
M_j^* &= m_1\phi_{1,j}^2+m_2\phi_{2,j}^2+\cdots+m_j\phi_{j,j}^2\\
P_{1,j} &= G_1\phi_{1,j}\frac{L_j^*}{M_j^*}S_{a,j}\\
P_{2,j} &= G_2\phi_{2,j}\frac{L_j^*}{M_j^*}S_{a,j}\\
&\vdots\\
P_{j,j} &= G_j\,\phi_{j,j}\frac{L_j^*}{M_j^*}S_{a,j}
\end{aligned}\right\}j\text{th mode}$$

where G_j is the gravity load of each storey.

4. For the maximum inertia forces of every normal mode, the maximum values of the response parameters (moments, shears, displacements and so on) are determined through a classic static analysis.

5. The above quantities for the modes under consideration are superimposed by taking the square root of the sum of their squares (SRSS) that is

$$\boxed{S_1=\sqrt{S_{1,1}^2+S_{1,2}^2+S_{1,3}^2+\cdots+S_{1,j}^2}} \tag{3.35}$$

Therefore, for the bending moments for example, the above relation takes the form

$$\boxed{M_s=\sqrt{M_{s,1}^2+M_{s,2}^2+M_{s,3}^2+\cdots+M_{s,j}^2}} \tag{3.36}$$

Thus, the superposition is based on the concept that all modes do not reach their maxima simultaneously and that the response in the vibration modes (including both translational and torsional modes) may be considered as independent from each other. Therefore, according to the probability theory (Clough, 1970), their most probable maximum values result through the SRSS. The concept of mode independence according to EC8 (CEN, 1994) is considered to be fulfilled if

$$T_j \leqslant 0.9T_i; \quad (i<j) \tag{3.37}$$

where T_i, T_j are the natural periods of any two successive modes of vibration taken into account for the determination of seismic effects.

If equation (3.37) is not satisfied, a more accurate procedure, the 'Complete quadratic combination' (CQC) must be adopted, that is (Wilson and Button, 1982)

$$\boxed{S_1=\left(\sum_{j=1}^{n} S_{1,j}^2+2\sum |S_{1,i}S_{1,j}|\right)^{1/2} \quad i\neq j} \tag{3.38}$$

It should be noted that in the case of a pseudospatial system and for a seismic action parallel to one of the two main axes, e.g. parallel to the x–x axis, in the formulation of $P_{i,j}$ the spectral coordinate $S_{a,j}$ takes the following values:

(a) the ordinate of the design spectrum ($S_{ax,j}$) corresponding to $T=T_j$, for all components of the jth mode corresponding to the x–x axis;
(b) a zero value for all other displacements ($S_{ay,j}=0$) and torsional deformations ($S_{aw,j}=0$) of the jth mode (components corresponding to the y–y direction and to torsional deformation).

Therefore, $P_{i,j}$ for a three-storey pseudo-three-dimensional frame, for seismic action parallel to the x–x direction, take the following values:

$$\left.\begin{aligned} P_{x1,j}&=G_1\,\phi_{x1,j}\frac{L_j^*}{M_j^*}S_{ax,j}\\ P_{x2,j}&=G_2\,\phi_{x2,j}\frac{L_j^*}{M_j^*}S_{ax,j}\\ P_{x3,j}&=G_3\,\phi_{x3,j}\frac{L_j^*}{M_j^*}S_{ax,j}\end{aligned}\right\}\quad x\text{–}x\text{ direction}$$

$$\left.\begin{aligned}P_{y1,j}&=G_1\,\phi_{y1,j}\frac{L_j^*}{M_j^*}S_{ay,j}=0\\P_{y2,j}&=G_2\,\phi_{y2,j}\frac{L_j^*}{M_j^*}S_{ay,j}=0\\P_{y3,j}&=G_3\,\phi_{y3,j}\frac{L_j^*}{M_j^*}S_{ay,j}=0\end{aligned}\right\}\quad y\text{–}y\text{ direction}$$

$$\left.\begin{aligned}M_{t1,j}&=G_1\,\phi_{w1,j}\frac{L_j^*}{M_j^*}S_{aw,j}=0\\M_{t2,j}&=G_2\,\phi_{w2,j}\frac{L_j^*}{M_j^*}S_{aw,j}=0\\M_{t3,j}&=G_3\,\phi_{w3,j}\frac{L_j^*}{M_j^*}S_{aw,j}=0\end{aligned}\right\}\quad \text{torsional direction}$$

(b) Time-history analysis

For time-history analysis two different procedures may be followed; modal analysis or direct numerical integration.

When the modal analysis procedure is chosen and the normal modes of the system are found, the displacement $u_i(t)$ of the ith mass of an MDOF system with N degrees of freedom can be expressed as the linear combination of the characteristic modal displacements and a function of time $q_j(t)$:

$$u_i(t)=\sum_{j=1}^{N}\phi_{ij}\,q_j(t) \tag{3.39}$$

where j is the order of the normal mode. In matrix form, this relation is expressed as follows:

$$\mathbf{u}(t)=\boldsymbol{\Phi}\,\mathbf{q}(t)$$

where $\boldsymbol{\Phi}$ is the $(N\times N)$ modal matrix, each column ϕ_j of which represents the characteristic displacements (eigenvectors) of mode j, and $\mathbf{q}(t)$ the vector of time functions, each element $q_j(t)$ of which represents the time function of the oscillation at the jth mode, for an inertia load

$$\frac{\phi_j^{\mathrm{T}}\mathbf{M}\,\mathbf{i}}{\phi_j^{\mathrm{T}}\mathbf{M}\phi_j}\ddot{x}_0(t) \tag{3.40}$$

where $\mathbf{M}$ is the $(N\times N)$ diagonal mass matrix and $\ddot{x}_0(t)$ the time history of the base acceleration.

The modal analysis briefly described above allows a significant reduction in computing time, since the determination of the normal modes and natural periods is done only once. The rest of the calculations deal with the determination of the response of the SDOF systems.

If the method of direct numerical integration is chosen, the system of differential equations for every time step is transformed into a system of algebraic equations involving the displacements. The known terms of the system are found using a certain assumption for the variation of the base acceleration during the integration interval (e.g. linear variation).

No matter which of the two procedures is chosen, the system must be analysed for a series of base accelerograms (typically four to five) scaled to a common level of spectrum intensity, so that the results of this method are sufficiently reliable.

In order to apply the method of time-history analysis for a typical multi-storey building, too much computational time is required. Thus, the use of this method can only be justified in special cases. For all other cases, if a dynamic analysis is chosen, it can be performed with the aid of spectral modal analysis.

3.5 DYNAMIC ANALYSIS OF MDOF INELASTIC SYSTEMS

3.5.1 Introduction

The dynamic analysis of MDOF inelastic systems is performed using direct step-by-step integration for small successive time steps. This method considers the response of the inelastic system within every integration time interval as linear. The value of the stiffness during the integration interval is taken equal to the slope of the local tangent to the load–deflection curve. Thus, while yielding occurs in some members and the stiffness of the structure changes, the response of the non-linear system is considered to be the response of successive linear systems, with different stiffnesses. Every change in stiffness of a member which could occur either during its yielding or during its unloading from the yielding branch, theoretically changes the stiffness of the whole system. Therefore the dynamic inelastic analysis, even for plane multi-storey systems, requires too much computational time.

The computer codes for dynamic inelastic analysis of multi-storey structures offer the possibility of obtaining significant output parameters, such as maximum deformations and corresponding internal forces at all critical sections, ductility requirements and time histories of deformations at specific points of the structure. Although there are some reservations about this method, which are mainly related to the uncertainties regarding the stiffness and the damping that have to be introduced in the model, dynamic non-linear analysis constitutes a powerful tool for the study of the response of structures. It will be useful, following the above discussion, to briefly explain the basic concepts of such a program, known as DRAIN-2D/90 (Kappos, 1991, 1993) which is an extended version of the internationally known program DRAIN-2D (Kanaan and Powell, 1973). It will also be useful to refer to some of the results of such a program and the way they are evaluated.

3.5.2 Methodology of inelastic dynamic analysis with DRAIN-2D/90

The structural systems which are analysed with this program are plane frames or dual systems (Figure 3.22). The system is discretized into the structural elements

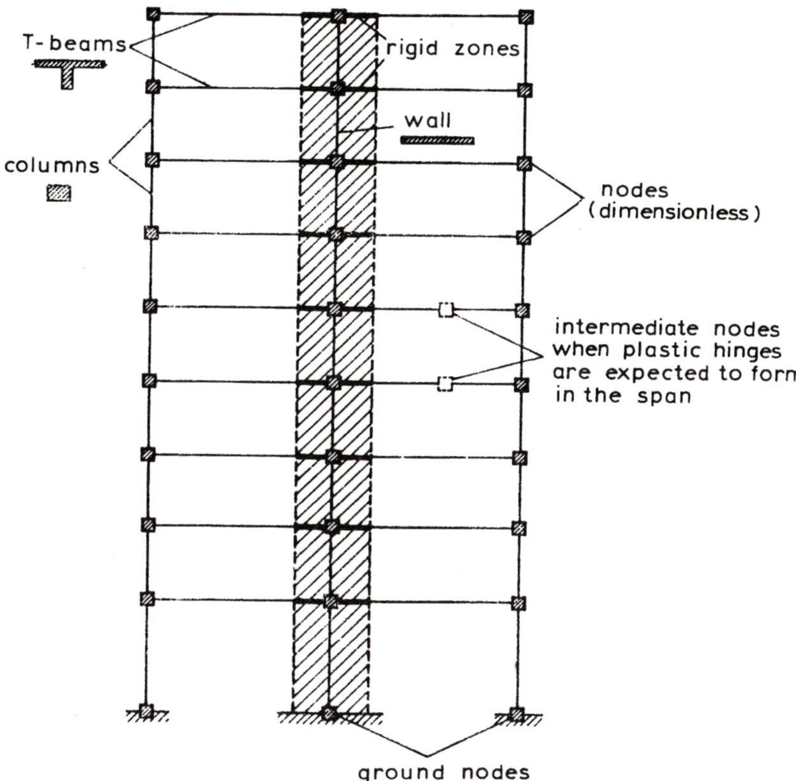

Figure 3.22 Discretization of a reinforced concrete plane structure.

forming the structure, as shown in Figure 3.22. For the analysis the direct stiffness method is used, according to which the stiffness matrix of the system is derived by appropriately adding the elements of the stiffness matrices of the individual members.

The basic equation of dynamic equilibrium of a discretized system (as the one shown in Figure 3.22) which is subject to a base seismic acceleration is given by the relationship (Clough and Penzien, 1975)

$$\mathbf{M}\Delta\ddot{\mathbf{u}}+\mathbf{C}_{\mathrm{T}}\Delta\dot{\mathbf{u}}+\mathbf{K}_{\mathrm{T}}\Delta\mathbf{u}=-\mathbf{M}\Delta\ddot{\mathbf{x}}_0 \tag{3.41}$$

where $\mathbf{M}$ is the mass matrix of the system, $\mathbf{C}_{\mathrm{T}}$ the tangent damping matrix, $\mathbf{K}_{\mathrm{T}}$ the tangent stiffness matrix of the system, which in a typical structure changes every time that one or more elements go from the elastic to the post-elastic phase or vice versa, $\Delta\mathbf{u}$, $\Delta\dot{\mathbf{u}}$, $\Delta\ddot{\mathbf{u}}$ are the vectors of incremental changes in relative displacements, velocities and accelerations of the nodes of the system, during the time step Δt of the numerical integration, and $\Delta\ddot{\mathbf{x}}_0$ is the vector of incremental changes in base seismic acceleration during the time interval Δt.

For the solution of the matrix differential equation of second order (3.41) the Newmark $\beta=1/4$ method is used (method of constant acceleration for every integration time step) in combination with the assumption that the damping matrix $\mathbf{C}_T$ is given by the relationship

$$\mathbf{C}_T = \alpha\mathbf{M} + \beta_T\mathbf{K}_T + \beta_0\mathbf{K}_0 \tag{3.42}$$

where $\mathbf{K}_0$ is the initial stiffness matrix of the system and α, β_T, β_0 are damping coefficients, proportional to the damping ratio ζ (ζ=0.02–0.05 is typically assumed for inelastic analysis).

The inelastic behaviour of R/C members is taken into consideration with the adoption of appropriate models for the moment (M)–rotation (θ) diagram at their ends. For the beams and the vertical members with approximately constant axial load, the model of Figure 3.23 is used, which is a modification of the well-known Takeda model (Takeda, Sozen and Nielsen, 1970), proposed by Otani and Sozen (1972). This model represents with adequate reliability the inelastic response of R/C members with predominantly flexural deformation to cyclic loading. For the quantitative estimation of the model parameters for every structural element, the dimensions of its cross-sections, its reinforcement and the characteristics of the construction materials are needed.

In order to analyse the response of a plane R/C structure with inelastic behaviour to a seismic excitation, the program needs the following data as input:

1. the geometry of the system (length of structural members, dimensions of cross-sections, type of connections (compare Figure 3.24));
2. the strength of the structural members (yield moments), defined as a function of the axial load, where appropriate;
3. the masses of the frame, typically assumed to be lumped at every floor level;
4. the viscous damping ratio ζ;
5. digitized time history of acceleration at the base of the structure.

Therefore, the design of a frame system should precede the inelastic dynamic analysis of the system which is subject to a seismic excitation. This means that the system needs first to be analysed for static seismic loading (or for the elastic response spectrum), and then be dimensioned and reinforced according to the internal forces that result from the combination of vertical loads and lateral (seis-

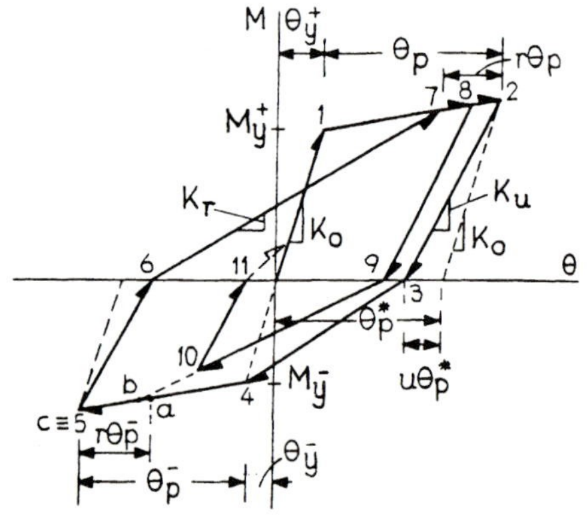

Figure 3.23 The modified Takeda model.

mic) forces. Then, the program using the above-listed input data from (1) to (5), first produces the moment–rotation diagrams for every structural member, the mass matrices, the stiffness matrices and the damping matrices, as well as the excitation vector. Once more it should be repeated that the stiffness and the damping matrices change at every time step of the numerical integration, while the mass matrix remains constant throughout the integration procedure.

In view of the above, it will be interesting to discuss the results of such an analysis. First of all, it should be noted that the inelastic analysis can give for every critical section the time history of response, such as internal forces M,N, relative displacements u, rotations θ of the nodes, relative displacements between the floors (inter-storey drifts) and so on. Given that the integration is carried out for very small time steps (1/200 to 1/50 s), and the critical sections are too many, calculated results are huge in number and their evaluation is difficult. Thus, every computer program is equipped with a series of post-processors which facilitate the evaluation of the results. DRAIN-2D/90, for example, produces an additional output, including among others the following:

- maximum floor displacements (Figure 3.25(a));
- maximum inter-storey drift ratio (deflection of columns from their vertical axis) (Figure 3.25(b));

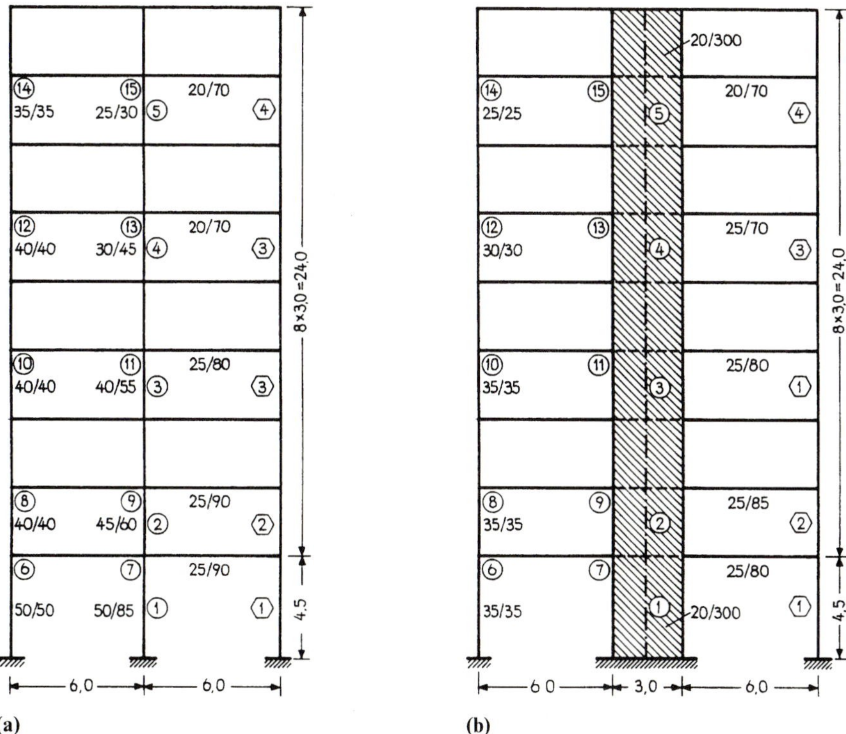

Figure 3.24 Geometric characteristics of the systems for which inelastic analysis was performed: (a) GRFR8 structure; (b) GRFW8 structure.

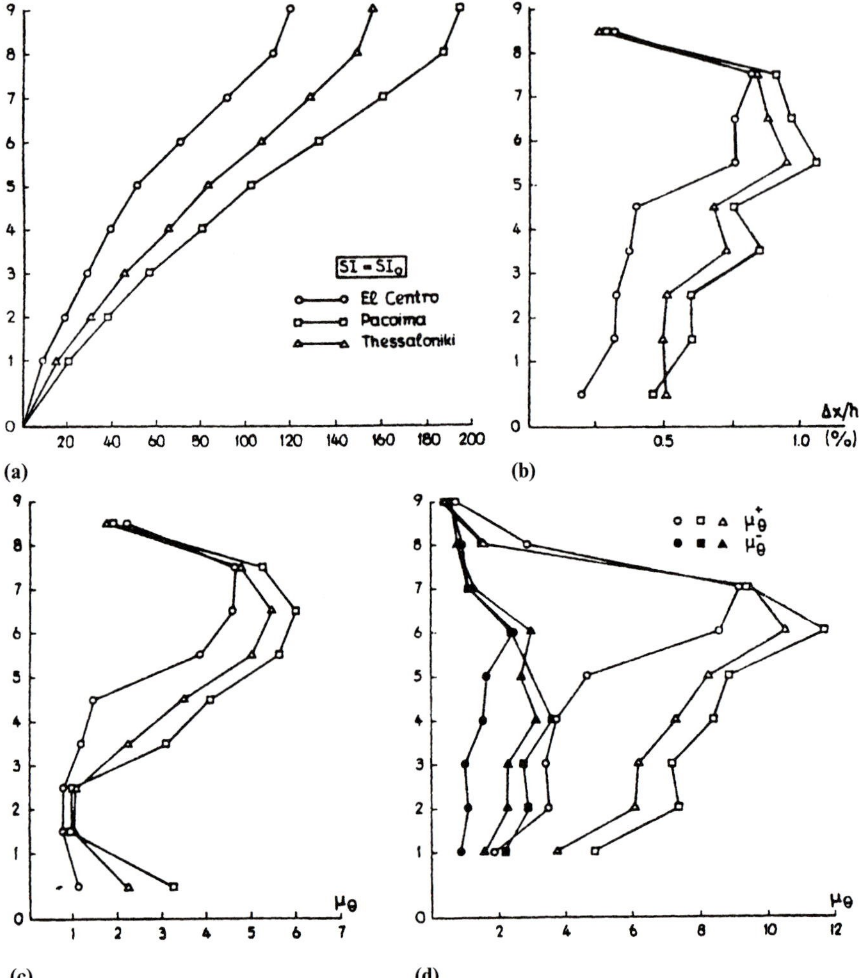

Figure 3.25 Maximum response parameters of a frame system (GRFR8) subjected to a seismic excitation with SI = 1.0 SI_0 for the El Centro × 1.0, Pacoima × 0.4, and Thessaloniki × 2.6 earthquakes: (a) floor displacements; (b) inter-storey drifts $\Delta x/h$; (c) required column ductilities; (d) required beam ductilities.

- maximum required column ductility (Figure 3.25(c));
- maximum required beam ductility (Figure 3.25(d));
- points where plastic hinges are formed during the excitation (Figure 3.26).

The maximum inter-storey drift ratios, the points where plastic hinges are formed and the correlation between required and available ductility, are critical indicators of the expected damage to the system.

It has to be mentioned here that the maximum required rotation ductilities (Figure 3.23)

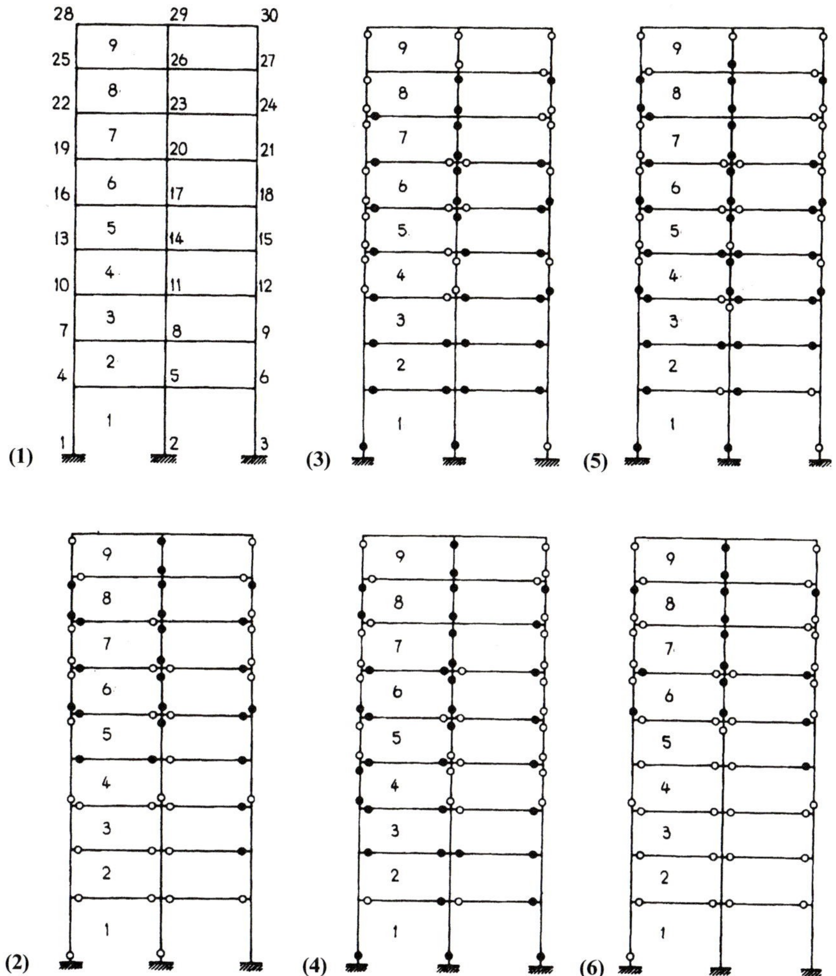

Figure 3.26 Frame system GRFR8: (1) Numbering of nodes and floors. Points where plastic hinges are formed during an excitation from the following earthquakes: (2) El Centro × 1.0, (3) Pacoima × 0.4. (4) El Centro × 1.5. (5) Thessaloniki × 2.6, (6) El Centro × 0.75. ○ = Yielding of the reinforcement at one side; ● = Yielding of the reinforcement at both sides.

$$\mu_\theta = \frac{\theta_y + \theta_p}{\theta_y} = 1 + \frac{\theta_p}{\theta_y} \tag{3.43}$$

result from the inelastic analysis of the system and they are different for the same structural system and the same structural member if the seismic excitation of the base is different. On the contrary, the available ductility for rotation or curvature of every structural member depends on its geometric properties (span, cross-section), the reinforcement, the quality of the construction materials and the level of axial loading. The procedure for determining the available ductility of structural elements will be explained in detail in Chapters 8 and 9.

3.6 REFERENCES

Biggs, J.M. (1964) *Introduction to Structural Dynamics*, McGraw-Hill, New York.

CEN Techn. Comm. 250/SC8 (1994) *Eurocode 8: Earthquake Resistant Design of Structures–Part 1: General Rules and Rules for Buildings (ENV 1998-1-1)*. CEN, Berlin.

Clough, R.W. and Johnston, S.B. (1966) Effects of stiffness degradation on earthquake ductility requirements. *Proceed. of Japan Earthq. Engng Symposium*, Tokyo, pp. 227–32.

Clough, R.W. (1970) *Earthquake Response of Structures*, Earthquake Engng, Prentice-Hall, New Jersey.

Clough, R.W. and Penzien, J. (1975) *Dynamics of Structures*, McGraw-Hill, New York.

Gupta, A.K. (1990) *Response Spectrum Method in Seismic Analysis and Design of Structures*, Blackwell Scientific Publ., Cambridge, Mass.

Housner, G.W. *et al.* (1953) Spectral analysis of strong-motion earthquakes. *Bull. of the Seismol. Society of America*, 43(2).

Kanaan, A.E. and Powell, G.H. (1973) DRAIN-2D: *A General Purpose Computer Program for Dynamic Analysis of Inelastic Plane Structures*. Rep. EERC-73/6 and EERC-73/22, Univ. of California, Berkeley.

Kappos, A.J. (1990) Sensitivity of calculated inelastic seismic response to input motion characteristics. *Proceed. of 4th U.S Nat. Conf. on Earthq. Engng*, Palm Springs, Calif., **2**, pp. 25–34.

Kappos, A.J. (1991) Analytical prediction of the collapse earthquake for R/C buildings: case studies. *Earthq. Engng and Struct. Dynamics*, **20** (2), 177–90.

Kappos, A.J. (1993) DRAIN-2D/90: *A Microcomputer Program for Dynamic Time-History Analysis of Inelastic Structures– User's Guide*, Dept of Civ. Engng, Univ. of Thessaloniki.

Otani, S. and Sozen, M.A. (1972) *Behaviour of Multistorey Reinforced Concrete Frames during Earthquakes*, Civil Engineering Studies, Structural Research Series No. 392, Univ. of Illinois, Urbana.

Park, R. and Paulay, T. (1975) *Reinforced Concrete Structures*, J. Wiley & Sons, New York.

Polyakov, S. (1974) *Design of Earthquake-resistant Structures*, Mir Publ., Moscow.

Richart, F.E., Woods, R.D. and Hall, J.R. (1970) *Vibrations of Soils and Foundations*, Prentice-Hall, New Jersey.

Riddell, R. and Newmark, N.M. (1979) Force-deformation models for nonlinear analysis (technical note). *Journal of the Struct. Div., ASCE*, **105** (12), 2773–8.

Takeda, T., Sozen, M.A. and Nielsen, N.N. (1970) Reinforced concrete response to simulated earthquakes. *Journal of the Struct. Div., ASCE*, **96** (12), 2557–73.

Warburton, G.B. (1976) *The Dynamical Behaviour of Structures*, 2nd edn, Pergamon Press, Oxford.

Wiegel, L.R. (ed.) (1970) *Earthquake Engineering*, Prentice-Hall, New Jersey.

Wilson, E.L. and Button, M.R. (1982) Three-dimensional dynamic analysis for multicomponent earthquake spectra. *Earthq. Engng and Struct. Dynamics*, **10**, 471–6.

4

Design principles and design seismic actions

4.1 INTRODUCTION

It has already been stressed that the behaviour of a structure during an earthquake depends on two basic parameters: (a) the intensity of the earthquake and (b) the quality of the structure. The quality of the structure is a parameter which exhibits a sufficient level of reliability since it depends on the configuration of the structural system, the design procedure, the detailing of structural elements and careful construction. The intensity of the earthquake, however, is a parameter with very high uncertainty, whose expected maximum value during the lifetime of the structure can be estimated based on very limited field data and on questionable evaluation of any existing historical information.

In fact, the intensity of the earthquake at a certain reference point, reflected to a degree on the maximum ground displacements, velocities, accelerations and on the relevant response spectra, is a function of several factors, such as the epicentral distance, the focal depth, the magnitude of the earthquake on the Richter scale, the geological formations between the reference point and the epicentre, the soil conditions at the area of the reference point and so on.

Thus, although the ideal solution for the estimation of the seismic hazard would have been the existence of response spectra for every geographical area based on long-term observations of seismic action, due to the lack of such material the estimation is usually based on two not particularly reliable methods:

1. estimation of the expected maximum ground acceleration with a specific probability of occurrence for a certain return period, based on geotectonic, seismological soil-dynamics data, and on a limited number of strong motion records as well, in case they exist;
2. estimation of the expected intensity measured on subjective (to a certain degree) scales or depending on the behaviour of structures to the earthquake, such as the modified Mercalli (MM) scale.

Despite its uncertainty and its subjective character, the later method allows:

- the use of the seismic history of a geographical area;
- the correlation of the maximum expected intensity expressed on the Mercalli scale or in terms of maximum acceleration with existing records of strong seis-

mic motions in the reference or other areas and adoption of their normalized response spectra (unscaled or scaled) for the area under consideration.

Of course it is not unusual for these methods to lead to serious errors of estimation, as happened in the case of Bucharest, where the adoption of the El Centro response spectrum proved to be in complete contradiction to the spectrum of the May 1977 earthquake as comparison of these two spectra shows (Figure 3.7) (Penelis, 1979).

However, considering the facts that seismological records (seismographs) did not exist a century ago, that records of strong earthquake motions in the form of accelerograms did not exist before the late 1930s, and that the number of the latter is limited in general, one can realize that there is no other way to follow but the one that combines limited seismic motion records with the procedure discussed earlier and was explained in detail in Chapter 3.

4.2 THE CONCEPTUAL FRAMEWORK OF SEISMIC DESIGN

4.2.1 Basic principles and requirements of modern seismic codes

Given the remarks made in the previous section, the most logical approach to the seismic design problem is to accept the uncertainty of the seismic phenomenon and consequently to design the structure in such a way that an adequate reserve of resistance is available to prevent failure in the case of a major earthquake, but at little or no additional cost compared to designing the structure to resist frequent earthquake motions.

Thus, the seismic design philosophy can be summarized in the following requirements.

1. *Serviceability limit state:* Structures must resist low-intensity earthquakes without any structural damage. Thus, during small and frequent earthquakes all structural components forming the structure should remain in the elastic range.
2. *Ultimate limit state:* Structures should withstand an earthquake of moderate intensity ('design earthquake' having a peak acceleration with 90% probability of not being exceeded in 50 years) with very light and repairable damage in the structural elements, as well as in the infill elements.
3. *Collapse limit state:* Structures should withstand high-intensity earthquakes with a return period much longer than their design life without collapsing.

The preceding criteria do not include any quantitative elements. However, their application implies that the maximum expected seismic loading intensity as well as its return period must be taken into consideration when designing a structure. Furthermore, it implies that the elastic limit of the structure is allowed to be exceeded during earthquakes with moderate or high intensity. This means that the structure should be able to undergo post-elastic deformations without losing a large percentage of its strength. As mentioned in Chapter 3, this structural property can be defined as ductility. Thus, one of the basic provisions of all modern structural codes refers to providing sufficient strength, and a corresponding sufficient ductility. It was made clear in Chapter 3 how these two properties are closely related. However, the above three requirements do not apply to some special types of structures for which conservative design criteria are adopted, either because of a low structural redundancy (chimneys, water-towers, core-suspended structures and so on), or because of the enormous risk that a possible failure can cause (nuclear power plants, dams and so on).

The previous thoughts constitute the guidelines of modern seismic codes. These codes prescribe for the structures under consideration a dynamic or a static analysis in the elastic range, but for seismic actions reduced to 1/2–1/5 of their elastic value, depending on the ductility level of the structure. In this way, the capacity of the structural system to resist seismic action in the inelastic range is taken into account. In particular, in the case of UBC (ICBO, 1994) where a partial safety factor for the loads equal to $\gamma_f = 1.40$ is introduced, the reduction values R_w vary between 4 and 12 (Figure 4.1).

This lower value of the seismic loading, however, is combined with the requirement for ductile behaviour of the structures which can be assured with the appropriate design of the structural system (see e.g. section 6.1.4 referring to the requirement for strong columns and weak beams) and with the appropriate detailing of structural elements (see Chapters 8 and 9).

Of course, according to modern seismic codes the local soil conditions can be taken into account in defining the design spectrum (microzonation). There is also a provision for a higher safety factor for public buildings that are of vital importance after a disastrous earthquake (e.g. hospitals) as there is for buildings of special social function (e.g. schools).

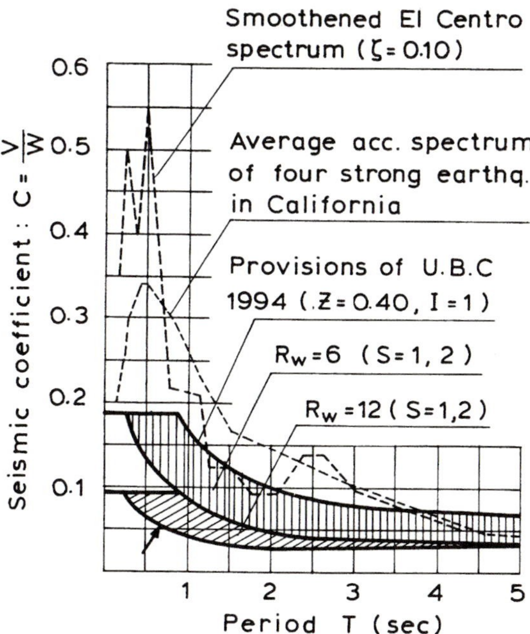

Figure 4.1 Seismic coefficient $C = V/W$ according to UBC, 1994 (ICBO, 1994).

Finally, in recent years there has been a tendency (Veletsos, 1981; ATC 3-06, 1978) for the contemporary codes to include parameters that take into account soil–structure interaction. This can be achieved by appropriately increasing the period T of the structure to take into account the flexibility of the soil, and its damping coefficient ζ, representing the soil's hysteretic damping.

Summarizing the above, one can see that the rationale underlying modern codes is the following:

1. Either a static or a dynamic analysis is prescribed; the analysis, however, is carried out in the elastic range with the seismic actions reduced to almost 1/2–1/5 of their actual values, depending on the ductility level of the structure.
2. The ultimate and the serviceability limit state design is carried out for the combination of the internal forces (M, V, N), caused by gravity and the reduced seismic loads described above.
3. The reduced seismic loading is combined with appropriate detailing of the structure and of critical regions of the structural elements in such a way that the energy of a high-intensity earthquake is dissipated through large plastic deformations without collapse.

The above-presented conceptual framework is formulated in EC8 (CEN, 1994) *Earthquake Resistant Design of Structures*, in two fundamental requirements associated with two compliance criteria and some specific measures, which are listed under the following headings.

(a) Fundamental requirements

1. *No collapse requirement.* The structure should be designed and constructed to withstand the design seismic action without partial or total collapse, thus retaining structural integrity and a residual load-bearing capacity after the seismic event.
2. *Damage limitation requirements.* The structure should be designed and constructed to withstand a seismic action which has a larger probability of occurrence than the design seismic action, without sustaining damage that could impose any limitations on the use of the structure.

(b) Compliance criteria

In order to satisfy the above fundamental requirements, the following limit states should be checked:

1. *Ultimate limit state.* The structural system should be verified to have the required strength and ductility, as specified by the appropriate sections of EC8.
2. *Serviceability limit state.* An adequate degree of reliability against intolerable damage should be ensured, by satisfying the deformation limits defined by the appropriate sections of EC8.
3. *Specific measures (collapse limit state).* Additional specific measures referring to

   (a) design
   (b) foundations and
   (c) quality system plan

should also be taken, in order to limit the uncertainties related to the behaviour of structures under the design seismic action and to promote a good response under seismic action more severe than the reference one.

In the sections that follow, the way that the preceding principles are given substance in EC8 will be presented in detail.

4.2.2 The concept of seismic isolation

The conventional approach to the seismic design of structures, which relies on the ductile behaviour of the structural system to dissipate the seismic energy, has the obvious disadvantage that the structure, during high-intensity earthquakes, suffers damage requiring costly repair. This damage can sometimes be so severe that the building may need to be demolished.

An alternative approach to this problem is the separation of the structural system from the seismic energy dissipation mechanism. In contrast to conventional design philosophy, according to which the whole structure dissipates the seismic energy through plastic deformation cycles, according to this approach the superstructure and the foundations are separated by a seismic isolation system. This can be accomplished with some special kind of pads with elastoplastic response, which are placed between the foundations and the superstructure. The yield shear of the pads is set to be slightly higher than the seismic action corresponding to the serviceability limit state. Thus, for moderate-intensity earthquakes the pads remain in the elastic range and there is no relative displacement between the foundations and the superstructure. However, during high-intensity earthquakes the pads yield, transferring to the superstructure a predefined base shear equal to the yield shear of the pads. Thus, the seismic isolation mechanism defines the upper limit of the earthquake forces that can be transferred from the foundations to the structure. The seismically isolated structure is then designed only for vertical loading and for the seismic actions that are predefined according to the yield shear of the pads (Figure 4.2) (Tarics, 1987). When subjected to a cyclic loading in the plastic range, the pads absorb and dissipate the seismic energy through elastoplastic loops (Kelly, Skinner and Beucke, 1980; Megget, 1978; Tarics, 1987). This approach to the seismic phenomenon appeared to be extremely innovative. However, the attempt to realize it leads to several technical problems. One is the production of pads capable of undergoing such cyclic elastoplastic deformations without being destroyed; the technique for replacing them in the structure when their intended life is exceeded (ageing effect) is another. Thus, realization of this idea in a practical way with pads made of reinforced neoprene, possibly with a lead core acting as a dissipative mechanism started relatively recently. It appears that in the near future seismic isolation will be an alternative to the conventional approach, at least for structures of special use, such as bridges, nuclear reactors, hospitals, schools. For the time being all the design codes are based on the conventional design philosophy and this is what will be discussed next. Only recently have the first guidelines for the design of seismically isolated buildings been published in California (Structural Engineers Association of Northern California, 1986). It should also be noted that although in EC8 guidance on base-isolated buildings is not given, the use of base isolation is not precluded, provided special studies are undertaken. Moreover, Part 5 of EC8 (Draft) contains special provisions for base-isolated bridges.

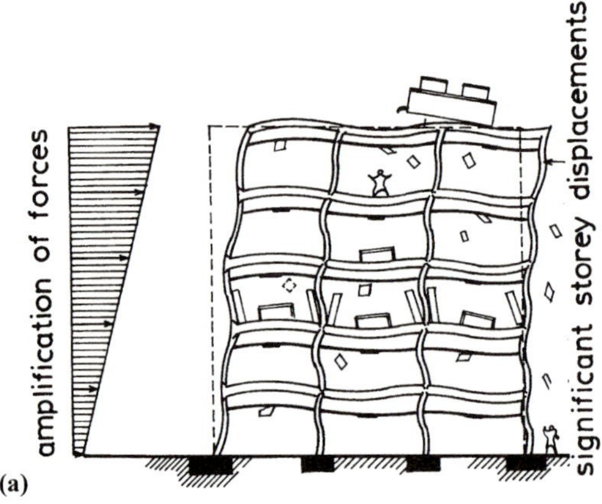

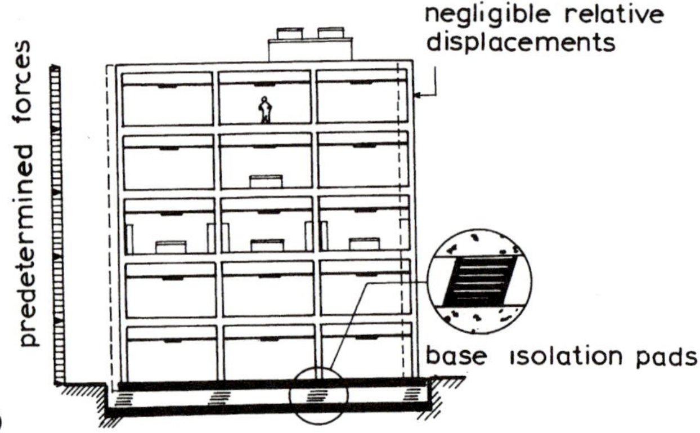

Figure 4.2 Comparative response to an earthquake of (a) a conventional earthquake-resistant structure and (b) a structure with seismic isolation.

4.3 CONFIGURATION OF THE STRUCTURAL SYSTEM

4.3.1 Fundamental requirements

One of the basic factors contributing to the proper seismic behaviour of a building is the rational conceptual design of the structural system in a way that lateral loads are transferred to the ground without excessive rotations and in a ductile manner. This cannot be achieved through strict mandatory requirements of the code. However, there are some general principles which can lead to the desirable outcome when they are followed. The guiding principles governing a conceptual design against seismic hazard are summarized as follows (EC8):

- structural simplicity
- uniformity and symmetry
- redundancy

- bidirectional resistance and stiffness
- torsional resistance and stiffness
- diaphragmatic action at storey level
- adequate foundations.

A brief discussion of these principles follows.

4.3.2 Structural systems covered by seismic codes

The structural system should preferably be composed of frames, either alone or coupled with shear walls in two directions, so that a clearly defined flow of lateral forces is achieved. The structural systems covered by EC8 should belong to one of the following structural types according to their behaviour under horizontal seismic action:

- *Frame system*: Structural system in which both the vertical and lateral loads are mainly resisted by space frames (Figure 4.3(a)).
- *Wall system* (coupled or uncoupled): Structural system in which both vertical and lateral loads are mainly resisted by vertical structural walls coupled or uncoupled, with high shear resistance (Figure 4.3(b)).
- *Dual system*: Structural system in which support for vertical loads is mainly provided by a space frame and resistance to lateral loads is provided in part by the frame system and in part by structural walls, isolated or coupled (Figure 3.22).
- *Core system*: Dual or wall system without satisfactory torsional rigidity, e.g. a structural system composed of flexible frames combined with walls concentrated near the centre of the building in plan (Figure 4.3(c)).
- *Inverted pendulum system*: Structural system where 50% of its mass is located in the upper third of the height of the structure.

Seismic resistant flat slab frames are not covered by EC8 if no additional measures are foreseen (e.g. the combination with other seismic resistant structural systems).

When shear walls are used, they should be arranged symmetrically and if possible along the perimeter of the building (see Figure 4.5). In general, the use of shear walls makes the building stiffer and reduces damage in infill elements. Furthermore, they can be used as a means to modify the stiffness of the building in order to avoid resonance with the foundation soil (for example, the combination soft soil–stiff building is desirable in seismic design).

A good criterion at the stage of conceptual design for providing a wall, a dual or a core system, with a high degree of lateral stiffness so that at least second-order effects are prevented, could be the use of the provisions of DIN 1045 (1982) as they have been adopted by the revised *Greek Code for R/C Structures* (Ministry for Environment and Public Works, *1991*), that is:

- For buildings with two or more storeys the following relation must hold:

$$h_{\text{total}}\sqrt{\frac{W_{\text{n}}}{E_{\text{cm}}J}} \leqslant 0.2+0.1n \quad \text{for } n \leqslant 3 \tag{4.1}$$

$$h_{\text{total}}\sqrt{\frac{W_{\text{n}}}{E_{\text{cm}}J}} \leqslant 0.6 \quad \text{for } n \geqslant 4 \tag{4.2}$$

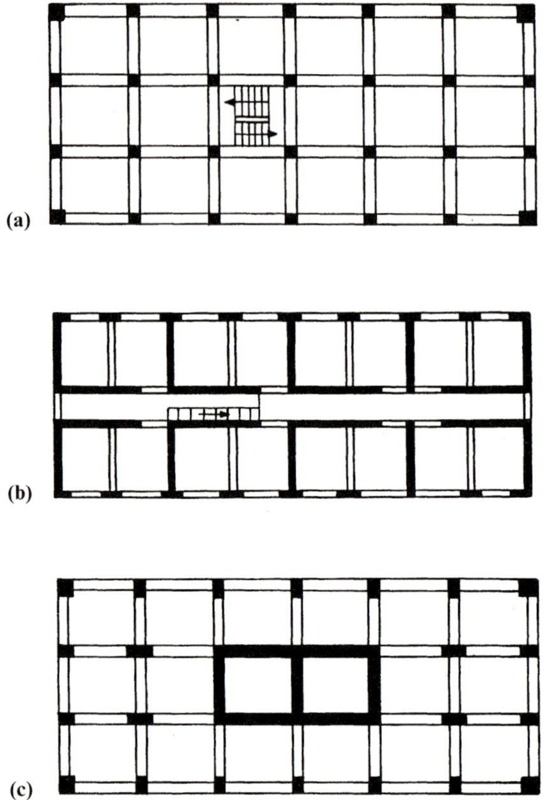

Figure 4.3 (a) A typical form of a frame system; (b) a typical configuration of R/C shear wall system; (c) a system with a core and frames.

where W_n is the total gravity load of the structure, n the number of storeys, h_{total} the total height of the structure and $E_{cm}J$ the sum of the stiffness of all R/C shear walls in the direction under consideration (assuming uncracked cross-sections).

4.3.3 Recommendations concerning structural configuration

1. Buildings regular in plan and in elevation, without re-entrant corners and discontinuities in transferring the vertical loads to the ground, display good seismic behaviour. The presence of irregularities leads to stress concentrations hazardous to the structure. Although the symmetrical arrangement of stiffness elements is not always possible, there should be a special effort in this direction so that torsion of unsymmetrical structures, which can lead to failure of the corner columns and the walls at the perimeter, will be avoided (Figures 4.4–4.7) (Baden Württemberg Innenministerium, 1985).
2. Concrete shear walls should span the whole distance between adjacent columns. In this way both strength and ductility of the structure are enhanced.

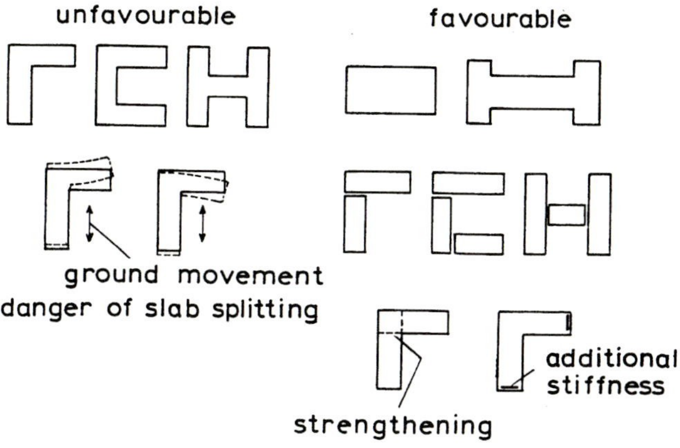

Figure 4.4 Unfavourable and favourable geometric configuration in plan.

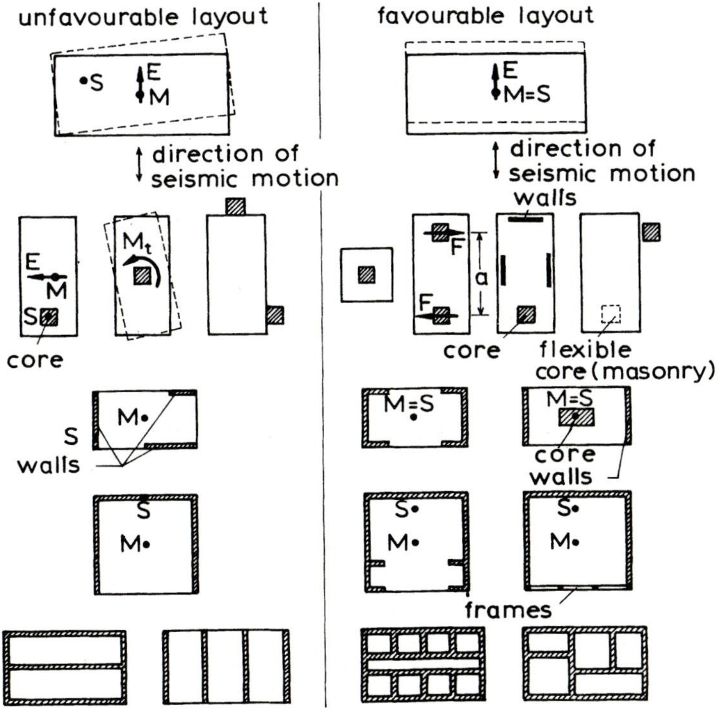

Figure 4.5 Distribution of mass and stiffness elements in plan.

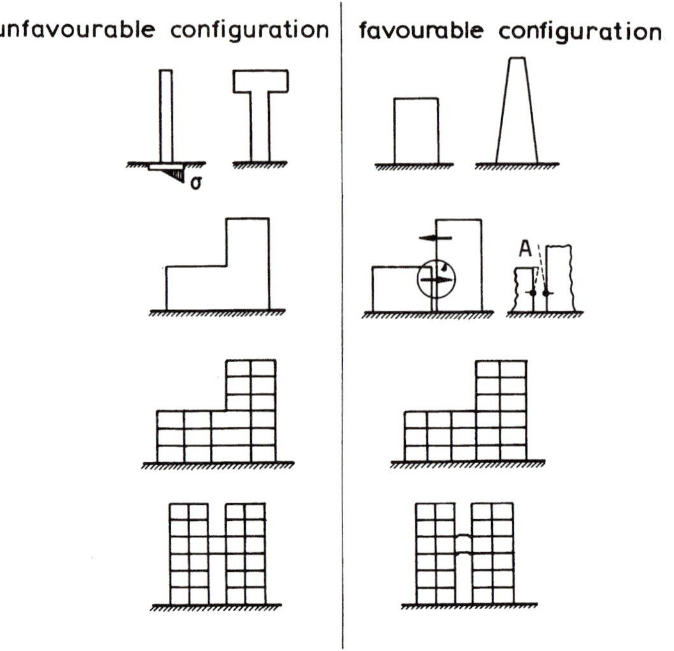

Figure 4.6 Unfavourable and favourable configuration in elevation.

3. All the structural elements forming the structure, including the foundations, should be well interconnected to build a monolithic structure (high redundancy).
4. Short columns resulting from the presence of mezzanines or stiff masonry walls below window openings should be avoided. If such arrangements cannot be avoided, their effect on the behaviour of the structure should be taken into account (Figures 4.8 and 4.9).
5. As already mentioned, flat slab systems without any beams should also be avoided. In such cases the whole seismic action has to be resisted by R/C shear walls or cores.
6. Large discontinuities in the infill system (such as open ground storeys) should be avoided (Figure 4.7).
7. Weak points in the slab endangering its diaphragmatic action should be avoided (Figure 4.4).
8. Structures have to be composed of strong columns and weak beams. In the following chapters this recommendation will be discussed in detail (capacity design procedure).
9. It is advisable for the structural system to include 'a second line of defence' formed by **ductile frames**. Thus, the dual system seems to be the most appropriate for resisting earthquake loads. ICBO (1994) and ATC (1978) required that, independent of the analysis results, 25% of the earthquake loads has to be carried by these frames.

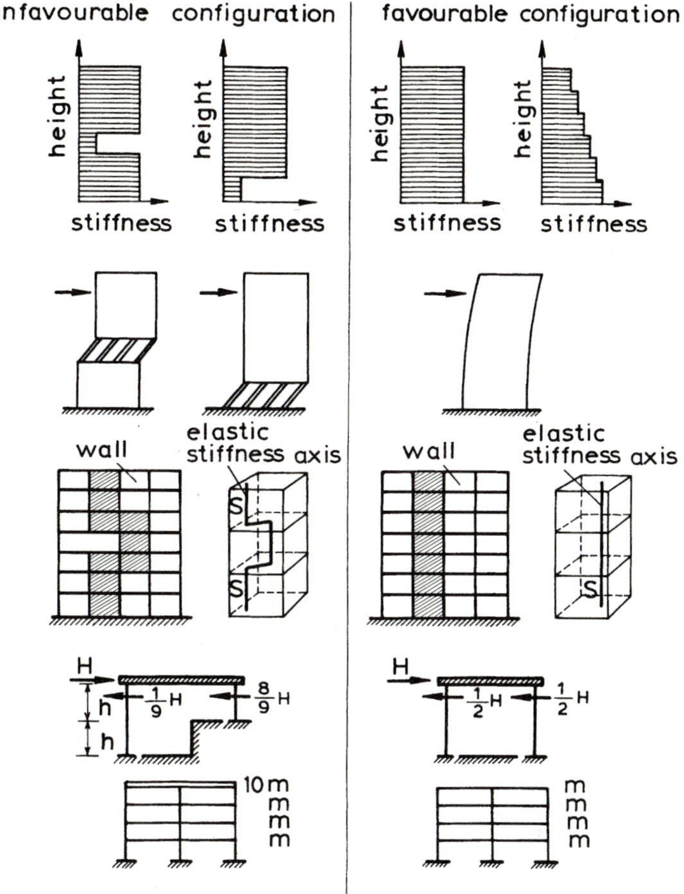

Figure 4.7 Distribution of mass and stiffness in elevation.

10. Although the subject of foundations falls outside the scope of this book, a few recommendations referring to them will be made here.

   (a) The construction site and the nature of the foundation soil should normally be free of risks of soil rupture, slope instability and permanent settlements caused by liquefaction or densification in the event of an earthquake.
   (b) The footings should be interconnected either by a mat foundation or by a grid of foundation beams, or at least with connecting beams.
   (c) All footings should rest on the same horizontal level (Figure 4.10).
   (d) Only one foundation type should in general be used for the same structure, unless the latter consists of dynamically independent units.

Referring to the second recommendation, it should not be forgotten that the seismic actions reach the foundations of a structure in the form of a wave (Figure 4.11).

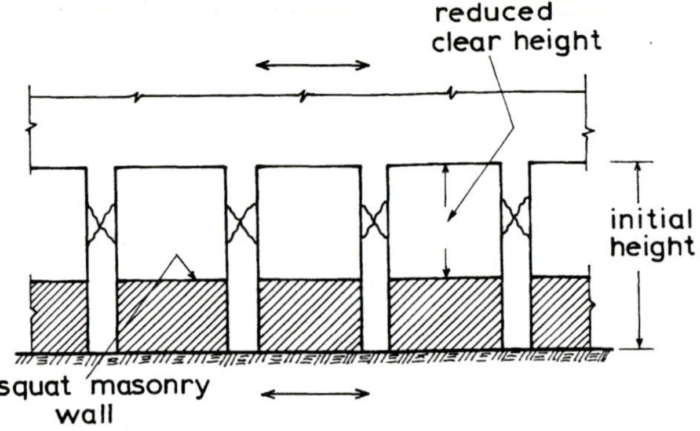

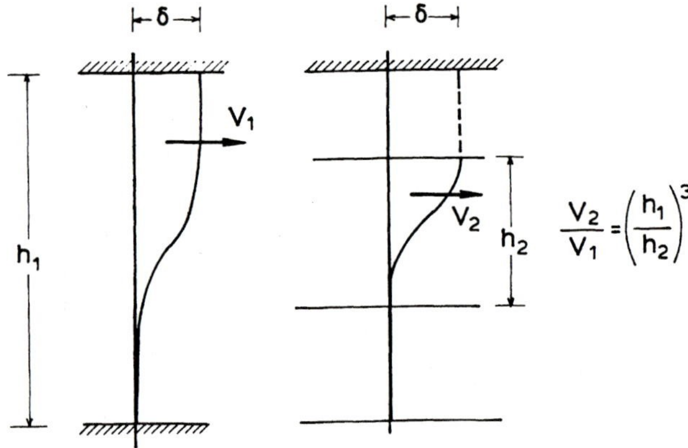

Figure 4.8 Concentration of large shear force on short columns at the perimeter of a building.

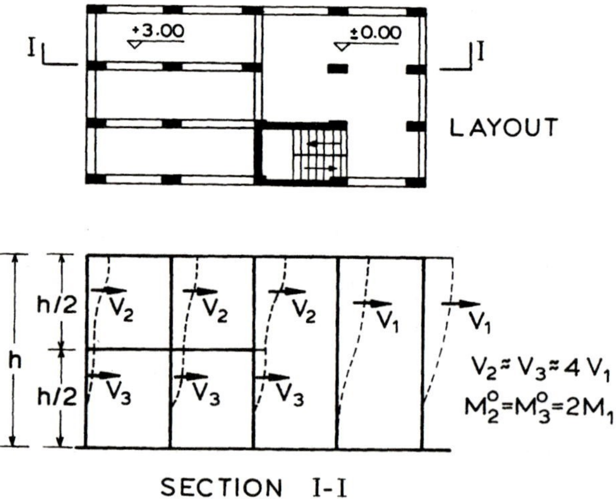

Figure 4.9 Concentration of large shear force on short mezzanine columns.

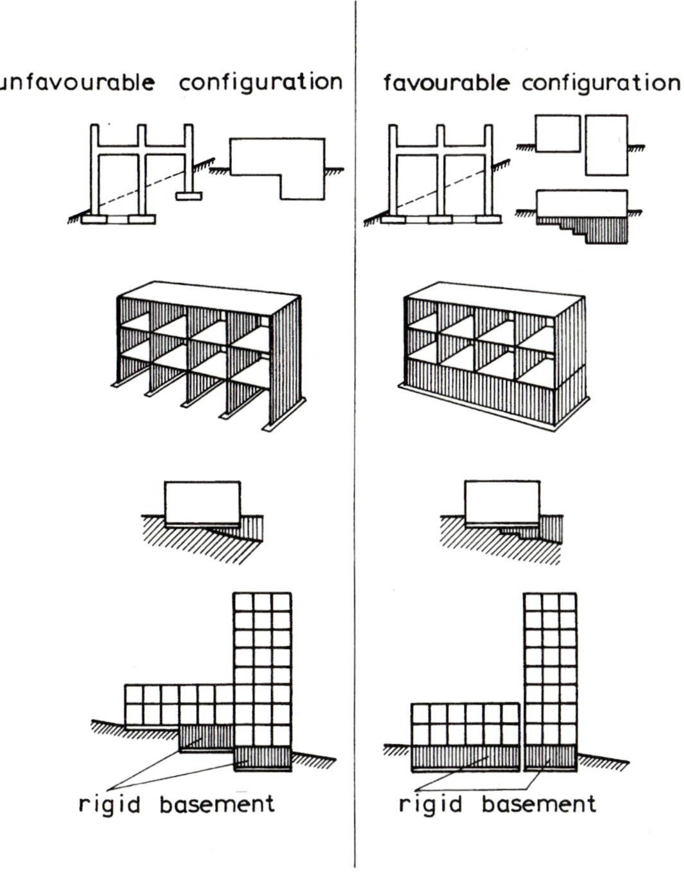

Figure 4.10 Unfavourable and favourable configuration of the foundation and the basement.

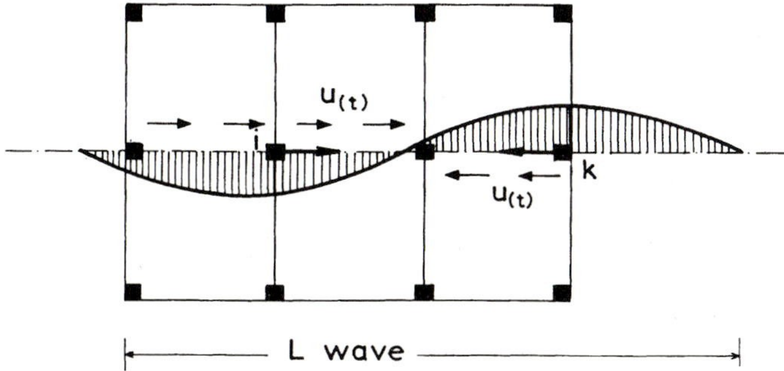

Figure 4.11 Relative displacement of the footings of columns *i* and *k* because of the phase difference of the ground motion at points *i* and *k*.

Thus, if footings are not interconnected each of them experiences a different displacement at the same instant, which contradicts the basic design principle that all footings move in phase.

4.4 DESIGN SEISMIC ACTIONS

4.4.1 General

Design seismic actions are by definition the earthquake actions which, in combination with the rest of the dead and live loads, determine the limit states for the structure.

Within the scope of EC8 the earthquake motion for the calculation of design seismic actions at a given point of the surface of the earth is generally represented by an elastic ground acceleration response spectrum, henceforth called **elastic response spectrum**.

It has already been mentioned that, among seismic actions, usually only horizontal ones are taken into account when designing a structure. These actions are described by two orthogonal components considered independent and represented by the same response spectrum (Penzien and Watabe, 1974; Rosenblueth and Contreras, 1977). However, for the design of certain structures the vertical component of the seismic action needs to be considered. These structures are:

- prestressed beams
- cantilevers
- beams supporting columns
- structural members with spans over 20 m.

Unless specific studies indicate otherwise, the vertical components of the seismic action shall be modelled by the response spectrum as defined for the horizontal seismic action, but with ordinates reduced as follows:

- for vibration periods $T \leqslant 0.15$ s, reduction factor: 0.70;
- for vibration periods $T \geqslant 0.5$ s, reduction factor: 0.50;
- for vibration periods $0.15 < T < 0.50$ s, reduction factor: through linear interpolation.

4.4.2 Seismic zones

As has already been mentioned in Chapter 3, it is necessary for the formation of the elastic response spectrum to know the effective peak ground acceleration. In this context, national territories should be divided by national authorities into seismic zones depending on the local hazard, usually described in terms of the value a_g of the effective peak ground acceleration in rock or firm soil, called **design ground acceleration**. This acceleration corresponds to a reference return period of 475 years and coincides with the peak acceleration with 90% probability of not being exceeded in 50 years. Figure 4.12 presents the zonation of Greece, included in the recently revised *National Code for Earthquake Resistant Structures* (Ministry for Environment and Public Works, *1992*) and based on the assumptions described above.

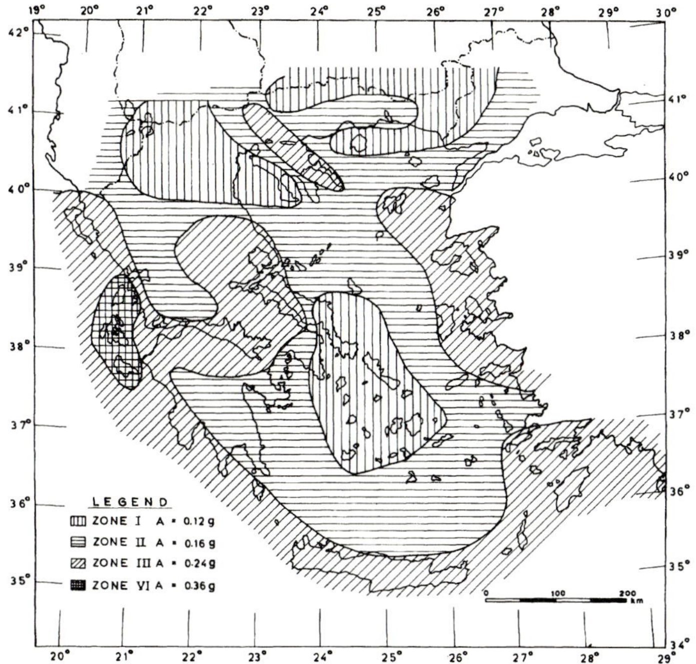

Figure 4.12 The zonation of Greece, according to the *Greek Code for Earthquake Resistant Structures* (Ministry for Environment and Public Works, 1992).

Seismic zones with a design ground acceleration a_g not greater than $0.05g$ are characterized as low seismicity zones for which reduced or simplified seismic design procedures for certain types or categories of structures may be used. In seismic zones with design ground acceleration a_g not greater than $0.02g$ the provisions of seismic codes need not be considered.

4.4.3 The local subsoil conditions

As has already been noted, the local ground conditions influence the seismic action and should therefore be taken into account. According to EC8, this is generally accounted for by considering the three subsoil classes A,B, and C described by the following different stratigraphic profiles:

1. subsoil class A:

   (a) rock or other geological formation characterized by shear wave velocity $V_S \geqslant 800$ m/s;
   (b) stiff deposits of sand, gravel or overconsolidated clay up to several tens of metres thick ($V_S \geqslant 400$ m/s at a depth of 10 m);

2. subsoil class B:
   (a) deep deposits of medium dense sand, gravel or medium stiff clays with thickness from several tens to hundreds of metres ($V_S \geq 200$ m/s at a depth of 10 m to 350 m/s at a depth of 50 m);
3. subsoil class C:
   (a) loose cohesionless soil deposits with or without some soft cohesive layers (V_S below 200 m/s in the uppermost 20 m);
   (b) deposits with predominant soft-to-medium stiff cohesive soils characterized by V_S values below 200 m/s in the uppermost 20 m.

The influence of the subsoil classes described above will be shown in section 4.4.4. It should be noted here that the treatment of local modifications of the ground motion characteristics is a complex and still controversial subject. The procedure adopted in EC8 and CEB MC/SD-85 follows to a degree that proposed by ATC 3-06 (1978).

4.4.4 Elastic response spectrum

The elastic response spectrum for the reference return period is defined in EC8 by the following expressions (see Figure 4.13):

$$\left.\begin{aligned} S_e(T) &= a_g S\left[1+\frac{T}{T_B}(n\beta_0-1)\right] \quad \text{for } 0\leqslant T\leqslant T_B \\ S_e(T) &= a_g S n\beta_0 \quad \text{for } T_B< T\leqslant T_C \\ S_e(T) &= a_g S n\beta_0\left[\frac{T_C}{T}\right]^{k_1} \quad \text{for } T_C<T\leqslant T_D \\ S_e(T) &= a_g S n\beta_0\left[\frac{T_C}{T_D}\right]^{k_2}\left[\frac{T_D}{T}\right]^{k_2} \quad \text{for } T_D<T \end{aligned}\right\} \tag{4.3}$$

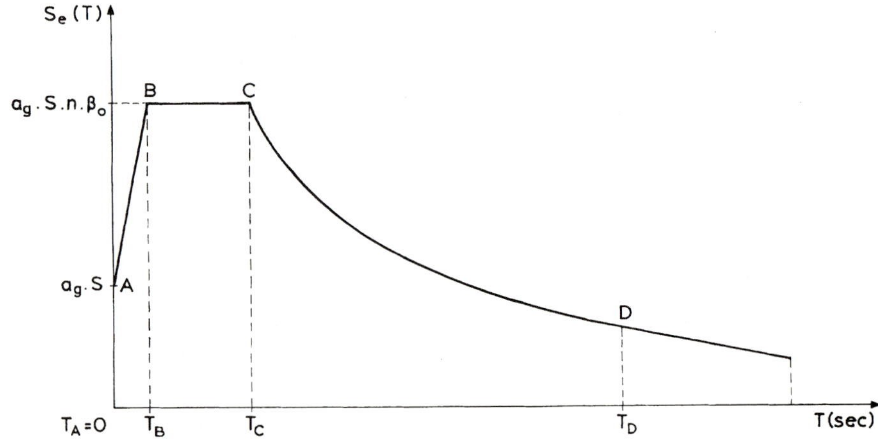

Figure 4.13 Elastic response spectrum.

where $S_e(T)$ is the ordinate of the elastic response spectrum, T the natural period of a linear single-degree-of-freedom system, a_g the design ground acceleration for the reference return period, β_0 the maximum normalized spectral value assumed constant between T_B and T_C, T_B, T_C the limits of the constant spectral acceleration branch, T_D the value defining the beginning of the constant displacement range of the spectrum, k_1, k_2 the exponents which influence the shape of the spectrum for a vibration period greater than T_C and T_D respectively, S the soil parameter with reference value 1.0 for rocky or firm soil (class A) and n the damping correction factor with reference value 1.0 for $\zeta=5\%$ viscous damping.

For the three subsoil classes A, B, C, the values of the parameters S, β_0, k_1, k_2, T_B, T_C, T_D are given in Table 4.1.

For sites with ground conditions not matching the three subsoil classes A, B or C, special studies for the definition of the seismic action may be required.

When the subsoil profile includes an alluvial surface layer with thickness varying between 5 and 20 m, underlain by much stiffer materials of class A, the spectrum shape for subsoil class B can be used together with an increased soil parameter S equal to

$$S=1.40$$

unless a special study is performed.

The value of the damping correction factor n is given by

$$n=\sqrt{7/(2+\zeta)}\geqslant 0.7 \tag{4.4}$$

where ζ is the value of the viscous damping ratio of the structure expressed in per cent. Some values of ζ are listed below:

- reinforced concrete: $\zeta=5\%$
- prestressed concrete $\zeta=3\%$
- reinforced masonry: $\zeta=6\%$.

Figure 4.14 presents the elastic response spectra for subsoil classes A,B,C, and for a damping correction factor n equal to 1.0 (5% viscous damping). It should be noted again (see section 3.3.2) that soft soils move the maxima of the response spectra to the right, effectively influencing the response of flexible structures.

The spectral amplification factor β_0 depends on the following factors: the frequency content of the motion, the ratio between the duration of the motion and the structure's fundamental period, the selected probability of being exceeded and the peak ground acceleration. An amplification factor

$$\beta_0=2.5$$

can be assumed to give a probability of not being exceeded lying between 70 and 80% in 50 years (CEB, 1985).

Table 4.1 Values of the parameters describing the elastic response spectrum

Subsoil class	S	β_0	k_1	k_2	T_B(s)	T_c(s)	T_D(s)
A	1.0	2.5	1.0	2.0	0.10	0.40	3.0
B	1.0	2.5	1.0	2.0	0.15	0.60	3.0
C	0.9	2.5	1.0	2.0	0.20	0.80	3.0

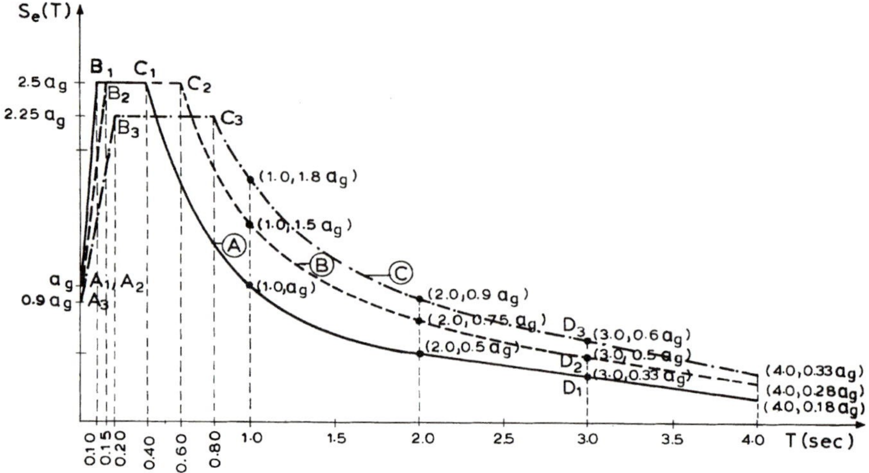

Figure 4.14 Elastic response spectra for subsoil classes A, B, C and $n=1$ ($\zeta=5\%$).

The exponent k_1 of the descending branch of the spectrum also depends on the frequency content of the motion and on the selected probability of being exceeded. For an approximately constant spectral density (band-limited white noise) as expected for a non-distant, moderate-to-large intensity motion travelling on rock, the value of k_1 consistent with $\beta_0=2.5$ would be between 1.0 and 0.9.

In the absence of special studies, the value d_g of the peak ground displacement may be estimated by means of the following expression (EC8):

$$d_g=0.05a_gST_C\,T_D \tag{4.5}$$

with the values of a_g, S, T_C, T_D as defined before.

4.4.5 Design spectrum for linear analysis

As has already been discussed in detail, the capacity of structural systems to resist seismic actions in the nonlinear range generally permits their design for forces smaller than those corresponding to a linear elastic response.

To avoid explicit nonlinear analysis in design, the energy dissipation capacity of the structure, through mainly the ductile behaviour of its elements, is taken into account by performing a linear analysis based on a reduced response spectrum, henceforth called **design spectrum**. This reduction is accomplished by introducing the **behaviour factor** q. In addition, modified exponents k_{d1} and k_{d2} may be used.

The design spectrum $S_d(T)$ for the reference return period which is normalized by the gravity acceleration g is defined (for $\zeta=5\%$) in EC8 by the following expressions:

$$\left.\begin{array}{ll} S_d(T) = \alpha S[1 + T/T_B(\beta_0/q - 1)] & \text{for } 0 \leqslant T \leqslant T_B \\ S_d(T) = \alpha S\beta_0/q & \text{for } T_B < T \leqslant T_C \\ S_d(T)\begin{cases} = \alpha S\beta_0/q[T_C/T]^{k_{d_1}} \\ \geqslant 0.20\alpha \end{cases} & \text{for } T_C < T \leqslant T_D \\ S_d(T)\begin{cases} = \alpha S\beta_0/q[T_C/T_D]^{k_{d_1}}[T_C/T_D]^{k_{d_1}} \\ \geqslant 0.20\alpha \end{cases} & \text{for } T_D < T \end{array}\right\} \quad (4.6)$$

where $S_d(T)$ is the ordinate of the design spectrum which is normalized by g, α the ratio of the design ground acceleration a_g to the acceleration of gravity g ($\alpha = a_g/g$), q the behaviour factor and k_{d_1}, k_{d_2} exponents which influence the shape of the design spectrum for a vibration period greater than T_C, T_D respectively.

Values of the parameters S, β_0, T_B, T_C, T_D are given in Table 4.1. The values of the parameters k_{d_1}, k_{d_2} are given in Table 4.2. The modification of the exponents $k_1 = 1$, $k_2 = 2$ to $k_{d_1} = 2/3$, $k_{d_2} = 5/3$ corresponds to an increase of the ordinates of the descending branch of the spectrum. The statistical treatment of the earthquake spectra has shown that the value $k_1 = 1$ is more probable than that of $k_{d_1} = 2/3$. However, the same statistical treatment has shown that there are serious uncertainties for structures with long periods T (high-rise buildings) for which there is a higher probability of having large ductility demands concentrated in a reduced number of storeys. For this reason a progressive increase of the spectral acceleration with the period T is adopted while, at the same time, a minimum value of

$$S_e(T) \geqslant 0.2\alpha$$

is introduced (CEB/MC-SD/85).

The behaviour factor q expresses to a certain degree the ability of the structure to display 'ductile' behaviour. EC8 and CEB/MC-SD/85 classify concrete structures into three ductility classes:

1. *Ductility class 'L'*: It corresponds to structures designed according to EC2(CEN, 1991) '*Design of Concrete Structures*', supplemented by rules enhancing available ductility.
2. *Ductility class 'M'*: It corresponds to structures designed, dimensioned and detailed according to specific earthquake-resistant provisions enabling the structure to enter well within the inelastic range under repeated reversed loading without suffering brittle failures.
3. *Ductility class 'H'*: It corresponds to structures for which the design, dimensioning and detailing provisions are such as to ensure, in response to the seismic excitation, the development of chosen stable mechanisms associated with large hysteretic energy dissipation.

Table 4.2 Values of k_{d_1} and k_{d_2}

Subsoil class	k_{d1}	k_{d2}
A	2/3	5/3
B	2/3	5/3
C	2/3	5/3

The behaviour factor q for concrete structures in EC8 used in design with a conventional linear model and for horizontal seismic actions is derived as follows:

$$\boxed{q = q_0 k_D k_R k_W \geq 1.5} \tag{4.7}$$

where q_0 is the basic value of the behaviour factor, dependent on the structural type, k_D the factor reflecting the ductility class, k_R the factor reflecting the structural regularity in elevation and k_W the factor reflecting the prevailing failure mode in structural systems with walls. The basic values of q_0 for the various structural types are given in Table 4.3.

The factor k_D reflecting the ductility class is taken as follows:

$$k_D = \begin{cases} 1.00 & \text{for DC 'H'} \\ 0.75 & \text{for DC 'M'} \\ 0.50 & \text{for DC 'L'} \end{cases} \tag{4.8}$$

The factor k_R reflecting the regularity in elevation (see section 4.3) is taken as follows:

$$k_R = \begin{cases} 1.00 & \text{for regular structures} \\ 0.80 & \text{for non-regular structures} \end{cases} \tag{4.9}$$

The factor k_w reflecting the prevailing failure mode in structural systems with walls is taken as follows:

$$k_w = \begin{cases} 1.00 & \text{for frame and frame equivalent dual systems} \\ 1/(2.5\text{–}0.5a_0) & \text{for wall equivalent, wall and core systems} \end{cases} \tag{4.10}$$

where a_0 is the prevailing aspect ratio h_w/l_w (h_w is the height of the wall and l_w is the length of the wall).

For the vertical component of the seismic action a behaviour factor q equal to 1.0 is in general to be adopted due to the small amount of dissipation energy in the vertical oscillations of buildings.

4.4.6 Importance factor

As has already been noted (section 4.2.1) structures in seismic regions are designed and constructed to fulfill the no collapse and the damage limitation requirements.

Table 4.3 Basic values q_0 of the behaviour factor

Structural type		q_0
Frame system		5.0
Dual system	Frame equivalent	5.0
	Wall equivalent	4.5
Wall system	Coupled walls	5.0
	Single walls	4.0
Core system		3.5
Inverted pendulum system		2.0

Target reliabilities for the above requirements are established by national authorities for different types of buildings or civil engineering works on the basis of the consequences of failure. This reliability differentiation is implemented by classifying structures into different importance categories. To each category an importance factor γ_I is assigned, which reflects a higher or a lower value of the return period of the seismic event, as appropriate for the design of the specific category of structures, in comparison to the reference value (section 4.4.5).

Therefore, the values of the design spectrum for linear analysis before they are taken into account for the determination of the seismic effects in any analytical procedure, must be multiplied by the importance factor γ_I, that is

$$\boxed{\gamma_I S_d(T)}$$

Detailed guidance on the importance categories and the corresponding importance factors according to EC8 is given below.

Buildings are classified into four importance categories depending on the size of the building, its value and importance categories depending on the size of the building, its value and importance for public safety and the probability of human losses in the case of a collapse.

The importance factor $\gamma_I = 1.0$ is associated with a design seismic event with a reference period of 475 years.

The importance factors related to the various importance categories are given in Table 4.4.

4.4.7 General remarks on the design spectrum

- The behaviour factor q is of vital importance for the determination of seismic actions, as it ranges from 1.5 to 5:

$$1.5 < q \leqslant 5 \tag{4.11}$$

  causing, therefore, a reduction of the maxima of an elastic response spectrum from 0.66 to 0.20 of their values. As was expected, the higher the ductility level, the lower the seismic action that needs to be considered for the elastic design.
- The main parameters influencing q-values are the structural type (q_0:5.0 to 2.0) and the ductility class (k_D: 1.0 to 0.5).

Table 4.4 Importance categories and importance factors for buildings according to EC8.

Importance category	Buildings	Importance factor γ_I
I	Buildings whose integrity during earthquakes is of vital importance for civil protection, e.g, hospitals, fire stations, power plants	1.4
II	Buildings whose seismic resistance is of importance in view of the consequences associated with a collapse, e.g. schools, assembly halls, cultural institutions	1.2
III	Buildings of intermediate size and normal use, e.g. apartment house, office buildings	1.0
IV	Buildings of minor importance for public safety, e.g. agricultural buildings	0.8

- For non-regular structures in elevation the seismic actions are 1.25 times greater than those for regular ones ($k_R = 0.80$).
- Ductility classes 'L','M' and 'H' are supposed to be equivalent. However, this has not yet been verified, nor has it been established which of the three is the most economical, if indeed they are equivalent. A recent study by Kappos and Papadopoulos (1995) involving 10-storey frame and dual systems has shown that structures designed for the highest ductility level (class 'H') according to CEB /MC-SD/85 are both more economical and more reliable with regard to their seismic behaviour, compared with structures designed to the other two ductility classes (see Chapter 10 for more details). Clearly, more studies involving a wider spectrum of structural configurations have to be carried out before final conclusions regarding the 'optimum' ductility class are drawn.
- The importance factor varies from 1.4 to 0.8. However, for the majority of the buildings the importance factor is equal to 1.0.
- The prescribed design spectrum multiplied by the importance factor γ_I is used both for static and dynamic analysis using the multi-modal response spectrum procedure, as will be discussed later.

4.4.8 Alternative representation of the seismic action

In EC8 procedures are also described for alternative representations of the seismic action. The suggested procedures are the following.

(a) Power spectrum representation

According to this method the seismic motion at a given point on the ground surface is represented as a random process defined by a power spectrum, i.e. the power spectral density function of the acceleration process, associated with a certain duration, consistent with the magnitude and the other relevant features of the seismic event.

(b) Time-history representation

According to this method the seismic motion is represented in terms of ground acceleration time histories (section 3.4.2) and related quantities (velocity and displacement). Two alternatives are provided:

1. generation of artificial accelerograms
2. use of recorded or simulated accelerograms.

For all cases, acceptance criteria are established, referring mainly to the consistency of the corresponding spectra to the elastic response spectrum described in detail in section 4.4.4.

4.5 COMBINATION OF SEISMIC ACTION WITH OTHER ACTIONS

The design value E_d of the action effects in the seismic design situation is determined, according to EC8, by combining the values of the relevant actions as follows:

$$\boxed{\Sigma G_{kj}\ `+'\ \gamma_I A_{E_d}\ `+'\ P_k\ `+'\ \Sigma\psi_{2i} Q_{ki}} \tag{4.12}$$

where '+' implies 'to be combined with', Σ implies 'the combined effect of', G_{kj} is the characteristic value of permanent action j, γ_I the importance factor, A_{E_d} the design value of the seismic action (e.g. design spectrum), P_k the characteristic value of prestressing action, ψ_{2i} the combination coefficient for quasi-permanent value of variable action i and Q_{ki} the characteristic value of variable action i.

The combination of actions given in expression (4.12) is used for both the ultimate limit state and the serviceability limit state. However, it should be noted that the resulting displacements are reduced before they are used for the serviceability verifications, so that a lower return period is taken into account (see also section 6.2.3).

The effects of the seismic action are evaluated by considering that all gravity loads appearing in the following combination of actions are present:

$$\boxed{\Sigma G_{kj}\ `+'\ \Sigma\psi_{Ei} Q_{ki}} \tag{4.13}$$

where ψ_{Ei} is the combination coefficient for variable action i (section 5.2.2).

Expressions (4.12) and (4.13) will be discussed in detail in sections 5.7 and 5.2 respectively.

4.6 REFERENCES

ATC 3–06 (1978) *Tentative Provisions for the Development of Seismic Regulations for Buildings*, NBS Special Publ. 510, Washington DC.

Baden Württemberg Innenministerium (1985) *Erdbebensicher Bauen*, Stuttgart.

CEB (1985) Model Code for Seismic Design of Concrete Structures. *Bulletin d'Information CEB*, **165**, Lausanne.

CEN Techn. Comm. 250/SC2 (1991) *Eurocode 2: Design of Concrete Structures – Part 1: General Rules and Rules for Buildings (ENV 1992–1–1)*, CEN, Berlin.

CEN Techn. Comm. 250/SC8 (1994) *Eurocode 8: Earthquake Resistant Design of Structures – Part 1: General Rules and Rules for Buildings (ENV 1998–1–1)*, CEN, Berlin.

DIN 1045 (1982) *Beton und Stahlbeton, Bemessung und Ausführung 1978*, Beton-Kalender, Wilh. Erns. & Sohn Verl., Berlin, Vol II, pp. 239–394.

ICBO (Int. Conf. of Building Officials) (1994) *Uniform Building Code – 1994 Edition*, Vol. 2: *Structural Engineering Design Provisions*, Whittier, Calif.

Kappos, A. and Papadopoulos, D. (1995) *Influence of design ductility level on the seismic behaviour of R/C frames. Proceed. of 10th European Conf. on Earthq. Engng*, 1994, Vienna, Austria, **2**, pp. 953–8. Balkema.

Kelly, J.M., Skinner, M.S. and Beucke, K.E. (1980) *Experimental Testing of friction-damped Aseismic Base Isolation System with Fail Safe Characteristics*, Rep. UBC/EERC-80/18, Univ. of California, Berkeley.

Megget, L.M. (1978) Analysis and design of a base-isolated reinforced concrete frame building. *Bull. of the N. Zealand Nat. Soc. for Earthq. Engng*, **11** (4), 245–54.

Ministry for Environment and Public Works (1991) *Greek Code for R/C Structures*, Athens.

Ministry for Environment and Public Works (1992) *Greek Code for Earthquake Resistant Structures*, Athens.

Penelis, G. (1979) Modern trends in the design of earthquake resistant structures – codes. *Proceed. Seminar on Repair and strengthening*, Thessaloniki, pp. 121–58 (in Greek).

Penzien, J. and Watabe, M. (1974) Simulation of 3-dimensional earthquake ground motion. *Bull. of Int. Inst. of Seism. and Earthq. Engng*, (Tsukuba, Japan), **12**.

Rosenblueth, E. and Contreras, H. (1977) Approximate design for multicomponent earthquakes. *ASCE, Journal of Engng Mech.*, **103**, EM5.

Structural Engineers Association of Northern California (1986) *Tentative Seismic Isolation Design Requirements*, San Francisco, Calif.

Tarics, A. (1987) Base isolation: a new strategy for earthquake protection of buildings. *The Journal of Arch. and Plan. Research.*

Veletsos, A.S. (1981) Seismic design provisions for soil–structure interaction. *State-of-the-Art in Earthquake Engineering*, Turkish National Committee on Earthquake Engineering, Ankara.

5

Analysis of the structural system

5.1 STRUCTURAL REGULARITY

5.1.1 Introduction

For the purpose of seismic design, building structures in all modern codes are separated into two categories:

- regular buildings
- non-regular buildings.

This distinction has implications for the structural model, the method of analysis and sometimes the value of the behaviour factor q. More specifically, according to EC8 (CEN, 1995):

- The structural model can be either a simplified plane or a spatial one.
- The method of analysis can be either a simplified modal or a multimodal analysis.
- The value of the behaviour factor q can be decreased (section 4.4.5).

Table 5.1 describes the implications of structural regularity on the design according to EC8.

5.1.2 Criteria for regularity in plan

Buildings regular in plan must fulfil the following requirements:

- The building structure, with respect to lateral stiffness and mass distribution, is approximately symmetrical in plan in two orthogonal directions.
- The plan configuration is compact with re-entrant corners not exceeding 25% of the overall external plan dimensions (Figure 5.1).
- The in-plane stiffness of the floors is sufficiently large that a rigid floor diaphragm behaviour may be assumed.
- Torsional stiffness of the system compared to the translational one is sufficiently large that at any storey the maximum displacement in the direction of the seismic forces for a 5% accidental eccentricity (of the seismic forces), does not exceed the average storey displacement by more than 20%.

Table 5.1 Consequences of structural regularity on seismic design

Regularity		Allowed simplification		Behaviour factor
Plan	Elevation	Model	Analysis	
Yes	Yes	Plane	Simplified	Reference
Yes	No	Plane	Multimodal	Decreased
No	Yes	Spatial*	Multimodal*	Reference
No	No	Spatial	Multimodal	Decreased

See section 5.4

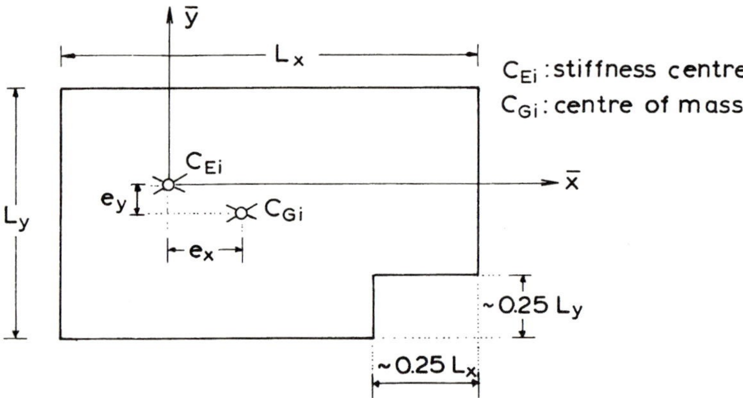

Figure 5.1 A building 'regular' in plan according to the relevant requirements.

It should be noted that the first requirement, as it is stated, is of qualitative character, while compliance with the last one can only be checked through the results of the analysis. However, this information is a prerequisite for the choice of structural system and method of analysis. A quantitative approach to the first requirement is given by CEB/MC-SD/85 (CEB, 1985) where it is stated that at any storey the distance between the centre of mass and the stiffness centre (section 5.5.3) should not exceed 15% of the 'torsional radius', defined as the square root of the ratio of the storey torsional to translational stiffness (Figure 5.1), that is

$$e_x, e_y \leqslant 0.15\left[\frac{\Sigma(K_{ix}\bar{y}_i^2 + K_{iy}\bar{x}_i^2 + T_i)}{\Sigma K_{ix} \text{ or } K_{iy}}\right]^{1/2} \tag{5.1}$$

where K_{ix}, K_{iy} are the translational stiffness of the ith vertical element along the x–x and y–y directions respectively, $\bar{x}_i$ and $\bar{y}_i$ the coordinates of the ith vertical element with respect to the stiffness centre, and T_i the torsional stiffness of the ith vertical element.

For the purpose of checking compliance with this clause, floor torsional and translational stiffness can be replaced by the sectional inertias of all vertical structural elements, and the stiffness centre by the centre of gravity of the sectional inertias. The above quantitative criterion ensures that for a quadrilateral in plan building with uniform distribution of stiffness in both main directions, the maximum torsional displacement in the direction of the seismic forces does not exceed the translational displacement of the stiffness centre by more than 18–20%. In this case, the corresponding eccentricity e_x or e_y is smaller than

$$e_x, e_y \leqslant 0.06L \tag{5.2}$$

where L is the dimension of the corresponding side of the quadrilateral floor.

A quantitative approach to the last requirement, useful for a preliminary design, is given as follows. Consider a single-storey building with rigid girders (section 5.5.6, equation (5.47)). The translational displacement for seismic forces H_x parallel to the x direction is given by the expression

$$\delta_x^0 = \frac{H_x}{\Sigma K_{ix}}$$

The torsional displacement parallel to the x direction at the perimeter of the building, for an accidental eccentricity equal to 5%, is given by the expression

$$\theta^T = \frac{M_T}{\Sigma (K_{ix} y_i^2 + K_{iy} x_i^2 + T_i)} = \frac{0.05 L_y H_x}{\Sigma (K_{ix} y_i^2 + K_{iy} x_i^2 + T_i)}$$

$$\delta^T_{x\,\mathrm{extr}} = \frac{L_y}{2}\theta^T = \frac{0.05 L_y^2 H_x}{2\Sigma (K_{ix} y_i^2 + K_{iy} x_i^2 + T_i)}$$

Taking into account that

$$\frac{\delta_{x\,\mathrm{extr}}}{\delta_x^0} \leqslant 0.20$$

The following criterion results

$$\boxed{L_x, L_y \leqslant \sqrt{8}\left[\frac{\Sigma (K_{ix} y_{iG}^2 + K_{iy} x_{iG}^2 + T_i)}{\Sigma K_{ix} \text{ or } K_{iy}}\right]^{1/2}} \tag{5.3}$$

where L_x, L_y are the floor dimensions.

For a preliminary design, floor torsional and translational stiffnesses can be replaced by the sectional inertias of all vertical structural elements as was explained in the previous paragraph (equations (5.46) and (5.47)).

5.1.3 Criteria for regularity in elevation

Buildings regular in elevation must fulfil the following requirements:

- All lateral load-resisting systems, such as cores, structural walls or frames, run from their foundations to the top of the building.

- Both the lateral stiffness and the mass remain constant or reduce gradually from the base to the top.
- When setbacks are present, special additional provisions apply.

5.2 MODELLING OF THE STRUCTURE

5.2.1 General

Analysis of the structural system according to EC8 is performed on a linear elastic model assuming uncracked sections in general. When the floor diaphragms are sufficiently rigid in their plane, the masses and the moments of inertia of each floor may be lumped at the centre of gravity, thus reducing the dynamic degrees of freedom to three per floor (two horizontal displacements and a rotation about the vertical axis).

The structural model must be reliable and able to account for the stiffness of the infill walls whenever they can affect the seismic response. This subject will be treated in detail later (section 5.4).

5.2.2 Masses contributing to the inertia forces

As already mentioned in section 4.5, the masses contributing to the inertia forces are calculated from all the gravity loads appearing in the following combination of actions:

$$\boxed{\Sigma G_{kj} \text{ `+' } \Sigma \psi_{Ei} Q_{ki}} \tag{5.4}$$

where G_{kj} is the characteristic value of permanent action j (dead loads), Q_{ki} the characteristic value of variable action i (live loads) and ψ_{Ei} the combination coefficient for variable action i.

The combination coefficients ψ_{Ei} take into account the probability of loads ψ_{2i} Q_{ki} not being present over the entire structure during the occurrence of the earthquake. Values of ψ_{2i} are given in Part 1 of EC1 and they vary from

$$\psi_{2i} = 0.2 \text{ to } 0.60$$

Values of ψ_{Ei} are given in EC8 Part 1–2 in the form

$$\psi_{Ei} = \varphi \psi_{2i}$$

The values of φ vary from

$$\varphi = 1.0 \text{ to } 0.5$$

and may be obtained from Table 3.2 of EC8, Part 1–2.

5.2.3 Application of the design seismic actions

The seismic actions are applied along the two main directions of the structure at the centre of mass of each floor. In order to cover uncertainties about the location of masses, the calculated centre of mass of each floor i is considered displaced from its nominal location in each direction by an accidental eccentricity:

$$\boxed{e_{1i} = \pm 0.05\ L_i} \quad (5.5)$$

where L_i is the floor dimension perpendicular to the direction of the seismic action. It is pointed out that e_{1i} is additional to the effect of actual eccentricities.

The accidental torsional effects generally render obligatory the analysis of the structural system for at least four seismic load cases, that is the four alternative positions of the centre of mass.

5.3 METHODS OF ANALYSIS

The reference method according to EC8 for determining the seismic effects is modal response spectrum analysis using a linear-elastic model of the structure and the design spectrum given in section 4.4.5, modified by the importance factor (section 4.4.6).

Depending on the structural characteristics of the building, one of the following two types of analysis is used:

1. 'Simplified modal response spectrum analysis' where a static simulation of the seismic action is adopted, and
2. 'multi-modal response spectrum analysis' already discussed in Chapter 3.

The cases where each method is applied are given in Table 5.1.

As alternatives to these basic methods, other methods of structural analysis such as

- power spectrum analysis
- nonlinear time-history analysis (section 3.4.2 (b))
- frequency domain analysis

are allowed under conditions specified in EC8.

5.4 SIMPLIFIED MODAL RESPONSE SPECTRUM ANALYSIS

5.4.1 General

This method takes into account for both main directions of the building, only the fundamental mode of vibration. Based on the above modes of vibration, the respective fundamental periods T_{1x}, T_{1y}, and the relevant design spectrum, modified by the importance factor, the total inertia forces in the two main directions and their contribution along the height of the structure are defined (section 3.2.3). For these loads a static analysis of the structural system is carried out. In this context this method might be characterized as an **equivalent static analysis**. From the above presentation it is concluded that this type of analysis can only be applied to buildings that can be analysed by two plane models, and whose response is not expected to have any essential contribution from higher modes of vibration.

These requirements seem to be satisfied by buildings which:

1. (a) meet the criteria for regularity in plan and in elevation given in section 5.1 or
   (b) meet only the criteria for regularity in elevation in combination with the condition that the centres of lateral stiffness and mass for every floor are each approximately located on a vertical line;
2. have fundamental periods of vibration T_1 in two main directions less than the following values:

$$T_1 \leqslant \begin{cases} 4T_C \\ 2.0\ \text{s} \end{cases} \tag{5.6}$$

where T_C is given in Table 4.1.

According to EC8 these types of structures may be replaced by two plane models, one for each main direction, and analysed for the respective inertia forces H_{ix}, H_{iy} independently. Torsional effects in this case may be taken into account in a simplified way as will be discussed later in this section.

The determination of the total inertia forces (base shear forces) and their distribution along the height of the structure is given below.

5.4.2 Base shear forces

The seismic base shear force V_B for each main direction is determined as follows:

$$\boxed{V_B = \gamma_I S_d(T_1) W} \tag{5.7}$$

where $S_d(T_1)$ is the ordinate of the design spectrum at period T_1, T_1 the fundamental period of vibration of the building for translational motion in the direction under consideration, W the total weight of the building computed in accordance with section 5.2.2 and γ_I the importance factor of the building.

5.4.3 Distribution of the horizontal seismic forces

The base shear is distributed among the storeys in the same proportion as the inertia forces that correspond to the fundamental period of the structural system, which is homologous to the characteristic shape of the fundamental mode. Given the fact that the first mode of a multi-storey, multi-column system, with a limited number of storeys and sufficient lateral stiffness, appears to be linear (Figure 5.2) (Biggs, 1964; Polyakov, 1974), the following relationships apply:

$$\varphi_{i1} = \frac{h_i}{h_n} \tag{5.8}$$

$$H_i = \gamma_I \frac{L_1^*}{M_1^*} W_i S_d \varphi_{i1} \tag{5.9}$$

(section 3.4.2 (a))

$$\Sigma H_i = V_B = \gamma_I \frac{L_1^*}{M_1^*} S_d \Sigma W_i \varphi_{i1} \tag{5.10}$$

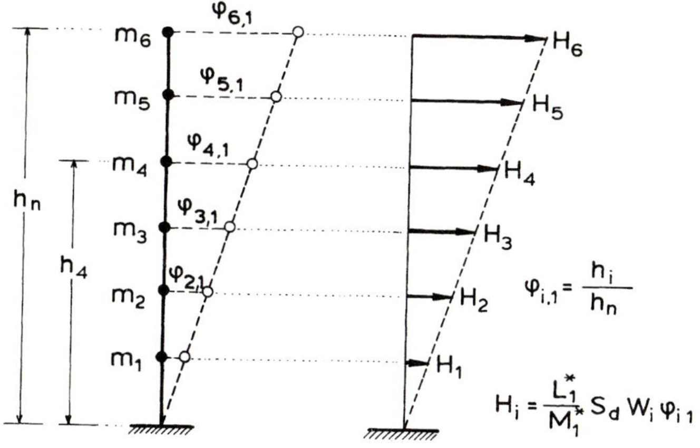

Figure 5.2 The fundamental mode of a multi-storey system and the inertia forces that correspond to it.

and

$$\gamma_1 \frac{L_1^*}{M_1^*} S_d = \frac{V_B}{\Sigma W_i \varphi_{i1}} \tag{5.11}$$

Substitution of equations (5.8) and (5.11) in equation (4.9) results in the following:

$$H_i = \frac{W_i \varphi_{i1}}{\Sigma W_i \varphi_{i1}} V_B = \frac{W_i h_i}{\Sigma W_i h_i} V_B \tag{5.12}$$

Recalling equation (5.7) which states that

$$V_B = \gamma_1 S_d \Sigma W_i$$

equation (5.12) takes the form

$$H_i = \gamma_1 S_d \frac{\Sigma W_i}{\Sigma W_i h_i} h_i W_i \tag{5.13}$$

Based on the above relationship EC8 defines the seismic actions for every storey as follows:

$$\boxed{H_i = \gamma_1 S_d \lambda_i W_i} \tag{5.14}$$

where

$$\boxed{\lambda_i = h_i \frac{\Sigma W_i}{\Sigma W_i h_i}} \tag{5.15}$$

with h_i the height of storey i from the ground. Equation (5.14) for constant values of W_i and storey heights, according to the above discussion, yields a triangular distribution of seismic loading.

5.4.4 Estimation of the fundamental period

The fundamental period which is needed for the determination of S_d may be derived either from a dynamic analysis or, for preliminary design, by using one of the following relations (EC8, CEB/MC-SD/85):

- For moment-resistant space concrete frames:

$$\boxed{T_1 = 0.075h^{3/4}\ \text{(s)}} \tag{5.16}$$

where h is the height of the building in metres.

- For structures with concrete shear walls:

$$\boxed{T_1 = 0.05h^{3/4}\ \text{(s)}} \tag{5.17}$$

or

$$\boxed{T_1 = \frac{0.09h}{L^{1/2}}\ \text{(s)}} \tag{5.18}$$

where h is the height of the building (in metres) and L the dimension of the building parallel to the applied forces (in metres).

- For all types of structures, by the expression:

$$\boxed{T_1 = 2\sqrt{d}\ \text{(s)}} \tag{5.19}$$

where d is the lateral displacement (in metres) of the top of the building due to the total gravity loads applied horizontally.

The dynamic analysis for the determination of the fundamental period T_1 may be substituted by the Rayleigh method properly conformed. In this case:

$$\boxed{T_1 = 2\pi\left(\frac{1}{g}\,\frac{\Sigma W_i\delta_1^2}{\Sigma H_i\delta_1}\right)^{1/2}\ \text{(s)}} \tag{5.20}$$

where H_i $(i=1,2\ldots, N)$ is a group of forces at the level of the floors, with a triangular distribution, δ_i $(i=1,2,\ldots,N)$ the corresponding displacements of the floors and W_i $(i=1,2,\ldots,N)$ the vertical loads at each storey i. For $N=1$, from equation (5.20) equation (5.19) can be derived.

5.4.5 Torsional effects

In the case of systems regular in plan and in elevation as these properties have been defined in section 5.1, the torsional effects taken into account in the 'simplified modal analysis' are only those related to an accidental eccentricity which is equal to

$$e_{1j} = \pm 0.05L_j$$

(section 5.2.3). In this case EC8 allows the torsional effects to be taken into account by amplifying the action effects on the individual load-resisting plane elements parallel to the seismic action using an amplification factor δ equal to (Figure 5.3):

$$\boxed{\delta = 1 + 0.6\frac{x_j}{L_j} \leqslant 1.30} \tag{5.21}$$

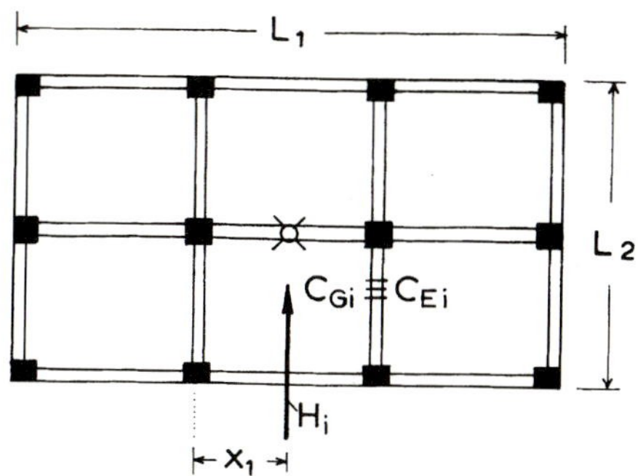

Figure 5.3 Evaluation of the torsional effects on a symmetric system with the aid of an amplification factor.

Therefore, when this simplification is applied the torsional effects are overestimated by more than 10% (section 5.1.2).

In the case of systems meeting only the criteria of regularity in elevation, and also some special criteria on regularity in plan (that is, the centres of lateral stiffness and mass of the floors are located approximately on a vertical line or the building height does not exceed 10.0 m etc.), the torsional effects taken into account in the 'simplified modal analysis' are those resulting from the sum of the accidental eccentricity e_1 and an additional eccentricity e_2 taking into account the dynamic effect of simultaneous translational and torsional vibrations (Figure 5.4) (Rutenberg, 1992). Thus, the torsional effects may be determined as the envelope of the effects resulting from an analysis of two static loadings consisting of torsional moments M_i due to the two eccentricities:

$$M_i = H_i e_{\max} = H_i(e_0 + e_1 + e_2) \tag{5.22}$$

$$M_i = H_i e_{\min} = H_i(e_0 - e_1) \tag{5.23}$$

where e_0 is the actual eccentricity between the stiffness centre C_{Ei} and the nominal mass centre C_{Gi} (Figure 5.4).

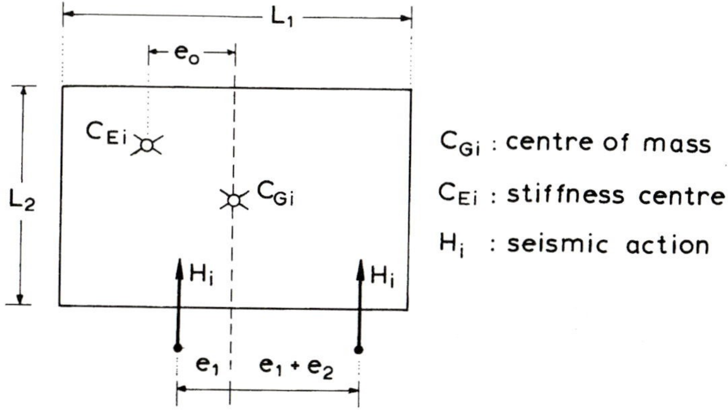

Figure 5.4 Loading centre according to EC8.

The eccentricity e_2 according to MC-SD/85 is taken as

$$e_2 = 0.50e_0 \tag{5.24}$$

while according to the OBC (ICBO 1994):

$$e_2 = 0$$

According to EC8, e_2 can be approximated as the lower of the following values:

$$e_2 = 0.10(L_1 + L_2)\left(10\frac{e_0}{L_1}\right)^{1/2} \leqslant 0.10(L_1 + L_2) \tag{5.25}$$

and

$$e_2 = \frac{1}{2e_0}\left[l_s^2 - e_0^2 - r^2 + \sqrt{(l_s^2 + e_0^2 - r^2)^2 + 4e_0^2 r^2}\right] \tag{5.26}$$

where $l_s^2 = (L_1^2 + L_2^2)/12$ (square of the 'radius of gyration') and r^2 is the ratio of the storey torsional to translational stiffness (square of 'torsional radius') as defined in section 5.1.2. In the case that the ratio of the storey torsional to lateral stiffness r^2 exceeds a certain value, that is if

$$r^2 \geqslant 5(l_s^2 + e_0^2) \tag{5.27}$$

the additional eccentricity according to EC8 may be neglected.

5.4.6 Proposed procedure for the analysis

Bearing in mind that nowadays strong computational tools are available at low cost, it is proposed that, in the case that 'the simplified modal response spectrum analysis' is allowed, a spatial structural system with rigid in-plane floor diaphragms is used (pseudospatial structural system).

This system will be analysed for the following loading cases:

$$\begin{array}{lll}
\bullet\; W = G \text{ '+' } \psi_{2i} Q_i & \text{gravity loading} & \\
\bullet\; H_{ix}: \text{acting horizontally at the centre of mass} & \left.\begin{matrix}\\ \\ \\ \end{matrix}\right\} x\text{–}x \text{ loading} & \\
\bullet\; \overline{M}_i^{\max} = (e_{1y} + e_{2y})H_{ix} & & \\
\bullet\; \overline{M}_i^{\min} = (-e_{1y})H_{ix} & & (5.28) \\
\bullet\; H_{iy}: \text{acting horizontally at the centre of mass} & \left.\begin{matrix}\\ \\ \\ \end{matrix}\right\} y\text{–}y \text{ loading} & \\
\bullet\; \overline{M}_i^{\max} = (e_{1x} + e_{2x})H_{iy} & & \\
\bullet\; \overline{M}_i^{\min}(-e_{1x})H_{iy} & &
\end{array}$$

The loading effects (E) from the above analysis, for each loading direction, will be derived as follows (Figure 5.5):

$$\begin{array}{ll}
E_W \Rightarrow G \text{ '+' } \psi_{2i} Q_i & \text{gravity loading} \\
\left.\begin{array}{l} E_u^x \Rightarrow H_{ix} \text{ '+' } (e_{1y} + e_{2y})H_{ix} \\ E_0^x \Rightarrow H_{ix} \text{ '+' } (-e_{1y})H_{ix} \end{array}\right\} & x\text{–}x \text{ direction} \\
\left.\begin{array}{l} E_r^y \Rightarrow H_{iy} \text{ '+' } (e_{1x} + e_{2x})H_{iy} \\ E_l^y \Rightarrow H_{iy} \text{ '+' } (-e_{1x})H_{iy} \end{array}\right\} & y\text{–}y \text{ direction}
\end{array} \tag{5.29}$$

where '+' means 'combined with'.

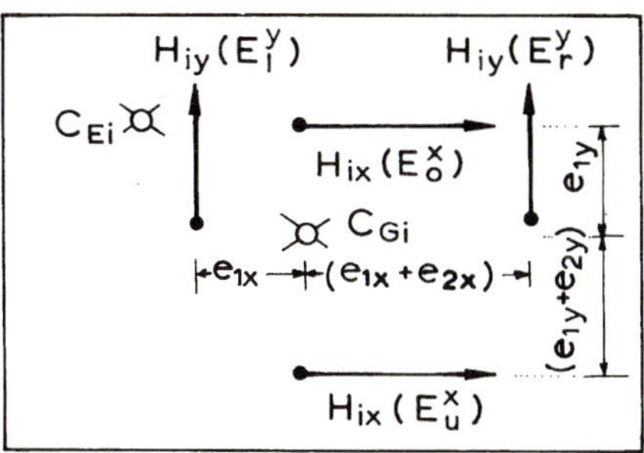

Figure 5.5 The loading cases and load effects for a building regular in elevation but not regular in plan.

The above procedure seems to be the most convenient because the same structural system is used for vertical and seismic loads. Therefore, only one set of data is needed as input for the computer and the whole procedure runs automatically. The procedure is somewhat simpler in the case of $e_2 = 0$.

All other options given by the various codes and particularly by EC8, for example the use of two plane models – one for each direction – combined with the simplified torsional analysis, are recommended only for cases where no efficient computational means are available.

5.5 THE PSEUDOSPATIAL STRUCTURAL SYSTEMS UNDER HORIZONTAL LOADING

5.5.1 General

From the preceding discussion it can easily be concluded that simplified modal response spectrum analysis is an equivalent static analysis for horizontal loading, parallel to the two main directions of the structure. Therefore, thorough discussion on the behaviour of the structural system under horizontal loading is considered to be of major importance. The significance of this approach appears to be more profound if we take into account the fact that even multimodal response spectrum analysis results in a series of inertial horizontal loadings, one for each mode (section 3.4.2), the final resultant of which is in general similar to the result of the simplified method. In this respect, knowledge of the structural behaviour of the pseudospatial systems under horizontal loading turns out to be a useful tool for:

- the conceptual design of the structure;
- the qualitative evaluations of the computational output of the analysis for seismic loading, no matter which method has been used.

5.5.2 Plane structural systems

As already stated elsewhere (section 4.3), the structural systems mainly used as earthquake-resistant structures are:

- frame systems

- wall systems
- dual systems
- core systems.

Core systems for translational displacement present the same behaviour as dual systems. Therefore, before the pseudospatial systems are discussed, it would be interesting to study the above systems in plane configuration, since the spatial systems are composed of plane ones.

5.5.3 Frame or shear systems

Frames with rigid girders subjected to lateral forces (Figure 5.6) exhibit zero moments at the mid-height of the columns, shear distribution proportional to the moments of inertia of the columns and relative displacements (or inter-storey drifts) proportional to the shear forces:

$$u_i = \left[\sum_j V_{ij}\right] \frac{h^3}{12E\sum_j J_{ij}} = \left[\sum_{i=1}^{i=n} H_j\right] \frac{h^3}{12E\sum_j J_{ij}} \tag{5.30}$$

$$V_{ij} = \left(\sum_{i=i}^{i=n} H_i\right) \frac{J_{ij}}{\sum_j J_{ij}} \tag{5.31}$$

From equation (5.30) it is concluded that the relative displacements are proportional to the shear forces; this is the reason why these systems are also called 'shear systems'. The deformation of these systems is such that they present a concave form on the side of the loading. In real frames, the girders due to their *T*-section exhibit in general much larger stiffness compared to that of the columns. Therefore their behaviour is very similar to the behaviour of shear systems.

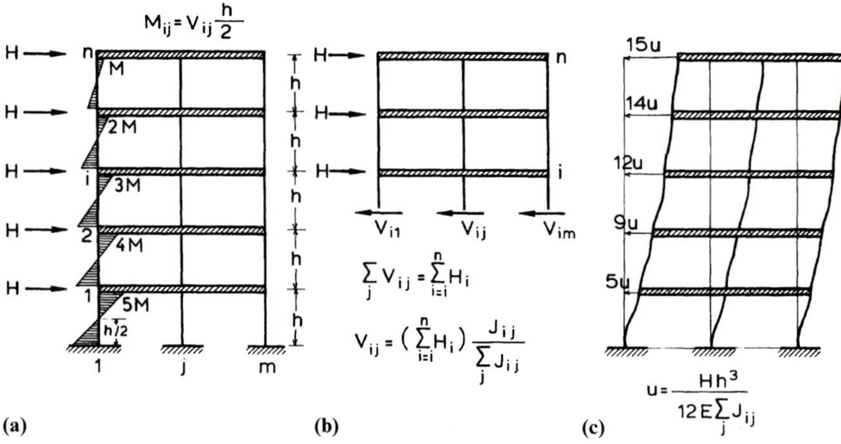

Figure 5.6 Action effects of shear frames under lateral loading: (a) *M*-diagram; (b) equilibrium between shear and horizontal loads; (c) storey displacements.

5.5.4 Wall systems or flexural systems

Isolated walls or ones coupled with beams of low-flexural stiffness behave under the action of lateral forces as cantilevers (Figure 5.7). The shear distribution is proportional to the moments of inertia of the cross-sections of the walls. The relative displacements of the floors result from bending deformation of the walls and therefore they present a convex form on the side of the loading.

5.5.5 Dual systems

The coupling of the two systems analysed above into a dual system under lateral loading, because of the completely different deformation shape of the individual components, results in such interaction forces that alter the moment and shear diagrams of both the frame and the wall (Figure 5.8). The characteristic of this combination is that in the lower floors the wall retains the frame while in the upper floors the frame inhibits the large displacements of the wall. As a result, the frame exhibits a small variation in storey shear V between the first and the last floors.

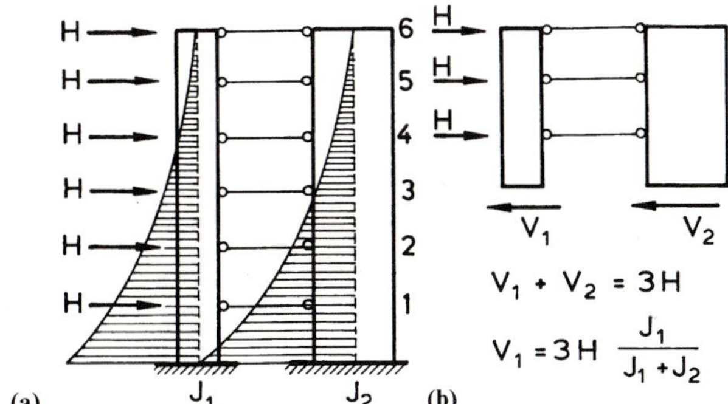

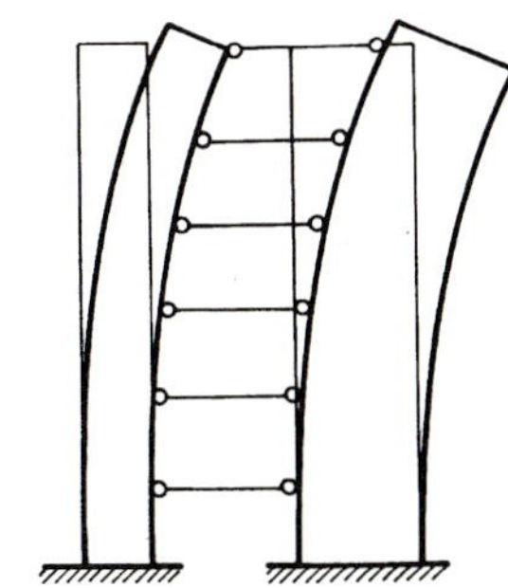

Figure 5.7 Action effects of shear wall system under lateral loading: (a) M-diagram; (b) equilibrium between shear and horizontal loads; (c) storey displacements.

Therefore, the moment diagram of the columns is antisymmetric with small variation from storey to storey. This observation allows the simulation of the dual system with the frame and the wall coupled only at the top of the building (Figure 5.9) (Macleod, 1970). The basic conclusion of the analysis of the dual system is

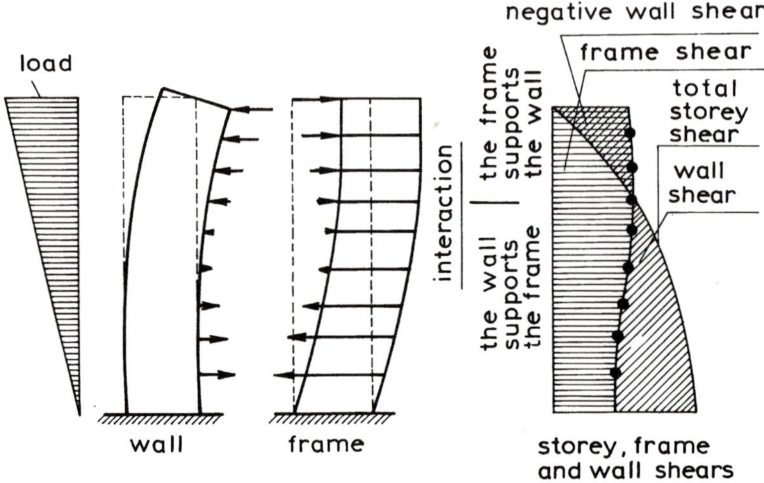

Figure 5.8 Interaction between frame and shear wall in a dual system under lateral loading.

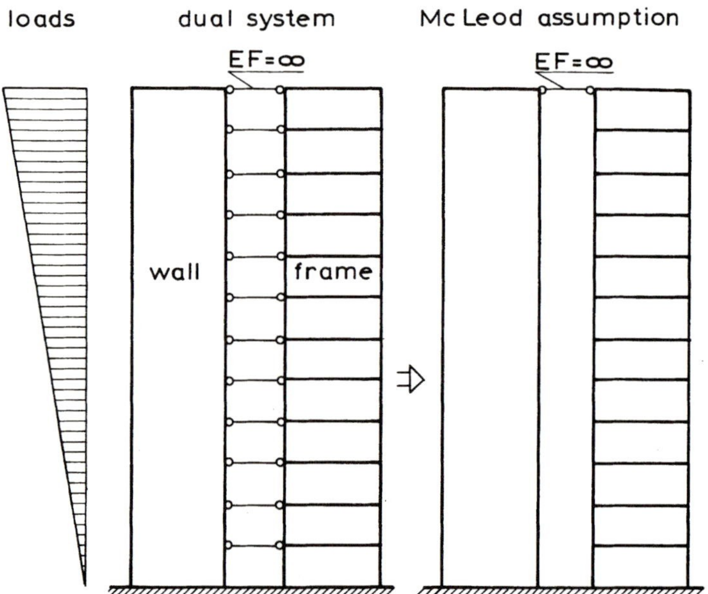

Figure 5.9 Simulation of the coupling between frame and shear wall in a dual system with only one bar at the top.

that the function of the wall resembles a beam which is fixed at the bottom and has an elastic support at the top, therefore the fixed-end moment is sufficiently large, but less than the fixed-end moment of a cantilever. These diagrams can be considered realistic only if the wall is completely fixed to the ground. Otherwise, the fixed-end moment of the shear wall is further reduced without any significant reduction of the shear forces while the moments of the beams coupling the frame and the wall are increased. The diagrams of Figure 5.10 correspond to 100, 50 and 25% rotational restraint of the fixed end. The fact that shear wall failures during strong earthquakes are almost always due to shear (exhibit X-shaped cracks) and seldom to flexure, should be attributed to this phenomenon, that is to the elastic rotation of the foundations.

5.5.6 The pseudospatial structural system

The analysis and design of a spatial system under horizontal forces are in general, a very complicated static problem because of the high degree of redundancy involved. For example, an 8-storey building with 28 joints on every floor (Figure

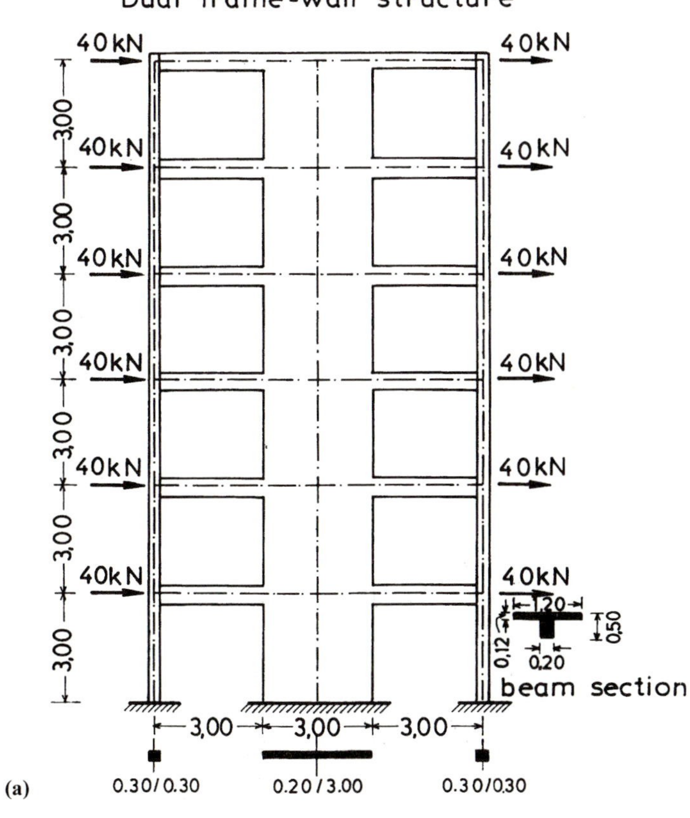

Figure 5.10a

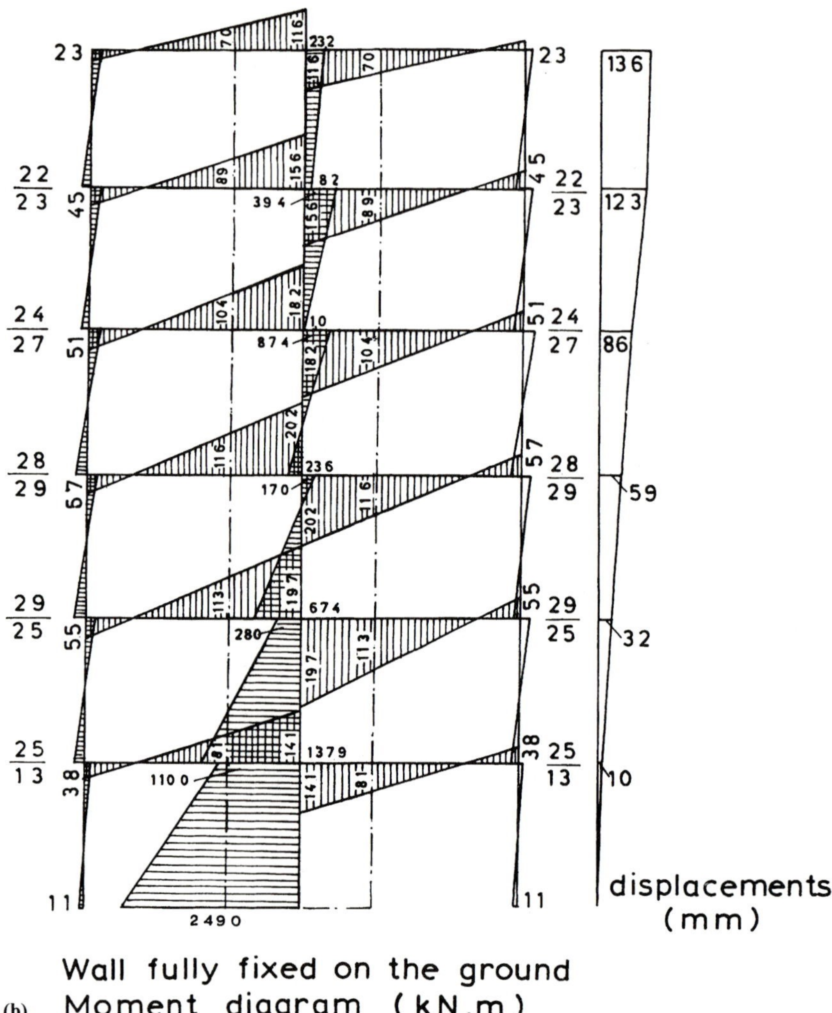

Figure 5.10b

4.3(a)) exhibits 6 × 8 × 28 = 1344 unknown rotations and displacements, with a corresponding effect not only in computer time for the analysis of the system but also in the time required for input data preparation and result evaluation.

In R/C structures, the in-plane stiffness of the floors is usually large enough in comparison with the lateral stiffness of the vertical structural elements, that a rigid floor diaphragm behaviour may be assumed. Therefore, the displacement and rotation of any joint on the plane of the floor can be expressed as a function of two displacements, u_x and u_y and a rotation, ϕ_z (displacement of a rigid disc). Thus, the independent deformations of every joint are limited to two rotations, ϕ_x and ϕ_y and a vertical displacement u_z. This system is known as a **pseudospatial struc-**

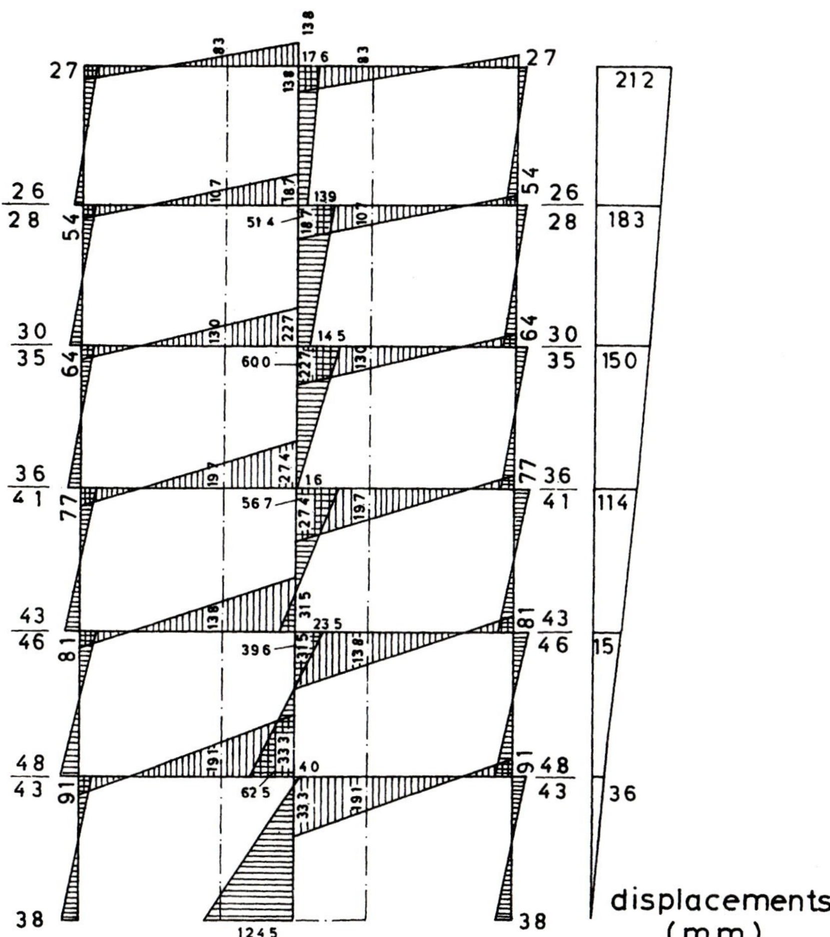

Wall flexibly supported on the ground (50% fixity)
(c) Moment diagram (kN.m)

Figure 5.10c

tural system. In this system, if the elastic deformation work of axial forces in the columns is assumed to be zero ($u_z = 0$), the unknowns are drastically limited. In the previous example of the eight-storey building the unknown deformations are limited to:

8 storeys × 28 joints × 2 rotations	=	448 unknowns
8 storeys × 3 (2 displacements + 1 rotation)	=	24 unknowns
Total	=	472 unknowns

Finally, in the pseudospatial system, if the torsional stiffness of the beams is assumed to be zero, which is a very reasonable assumption, given the fact that the

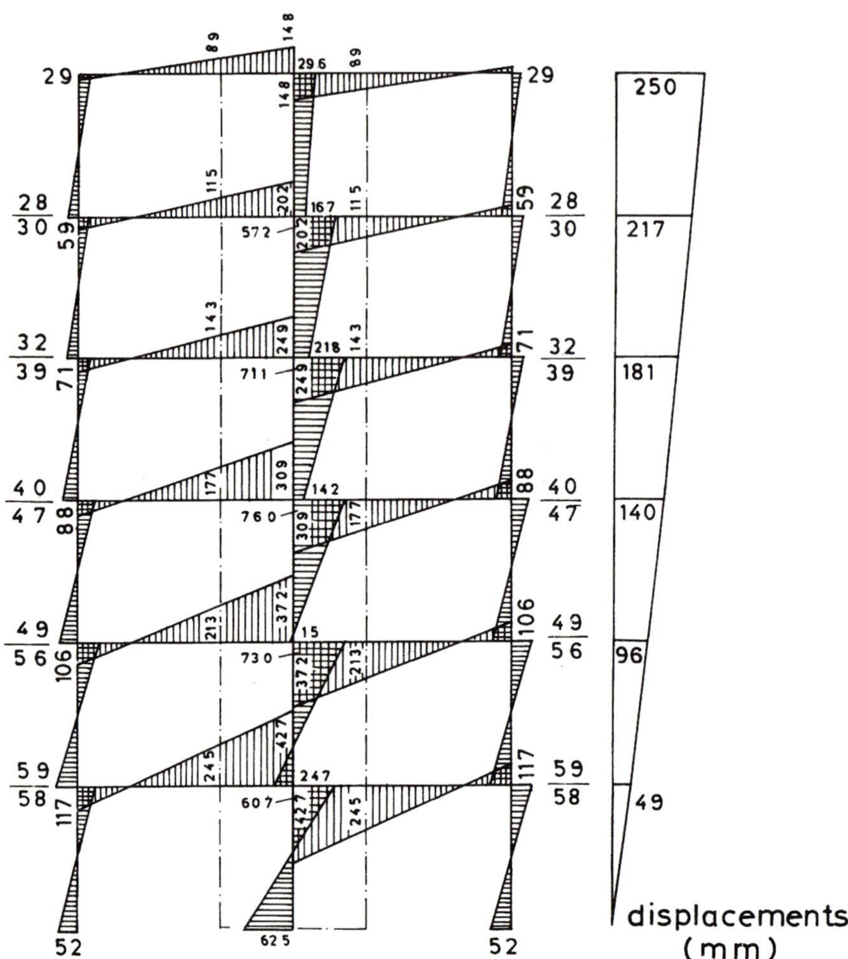

Figure 5.10a–d Action effects of a dual system under horizontal loading.

torsional stiffness of a cracked structural element is very low compared to its flexural stiffness, then the rotation ϕ_x, for example, of any joint in a plane x–x frame of the spatial system, is independent of the rotation of the respective joint of the adjacent x–x plane frame (Figure 5.11). In this respect, the structure may be simplified and broken into a number of separate plane frames in each of the two main directions, whose stiffnesses are coupled to form a system of $3i$ equations equal to the number of independent displacements and rotations of the i floors of the structure. The pseudospatial system described above will now be examined.

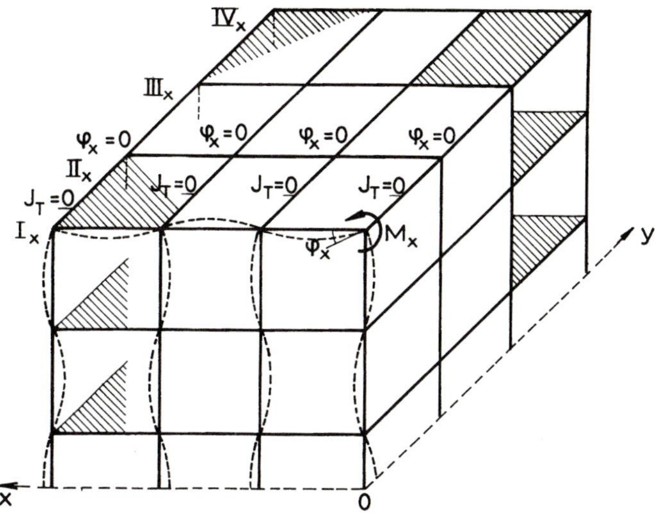

Figure 5.11 Pseudospatial system with parallel plane frames with no interaction between them because of the zero torsional stiffness of the connecting beams.

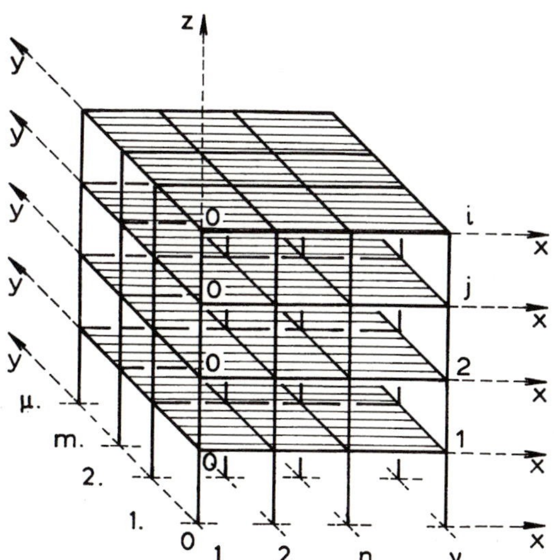

Figure 5.12 The geometry of a pseudospatial frame.

Consider the system of Figure 5.12 (Roussopoulos, 1956; Penelis, 1971). Under the action of lateral forces each floor sustains a relative displacement with respect to the floor below, which can be described by three independent variables, the horizontal relative displacements, u_{0j} and v_{0j}, of the origin of the coordinate system and the rotation ω_j of the floor. Thus, the relative displacement of the frame

m along the x-axis on the floor j is determined by the relationship

$$u_{jm}=u_{j0}-\omega_j y_m \tag{5.32}$$

while the relative displacement of the frame n along the y-axis on the same floor by the relationship

$$v_{jn}=v_{j0}+\omega_j x_n \tag{5.33}$$

The above relationships determine the displacements of the joint n, m on the floor j. In matrix form they can be written as follows:

$$\begin{aligned}\mathbf{u}_m&=\mathbf{u}_0-\omega y_m\\ \boldsymbol{v}_n&=\boldsymbol{v}_0+\omega x_n\end{aligned} \tag{5.34}$$

Next, the lateral stiffness of the plane frames will be defined. Consider the frame of Figure 5.13 which is loaded with horizontal forces H_j. **Storey shear** V_j is called the sum of the shears of the columns of the storey j, that is

$$V_j=\sum_{j=j}^{j=i} H_j \tag{5.35}$$

If $u_1, u_2 \ldots u_j \ldots u_i$ are the relative displacements of the floors due to the action of H_j then the shear of the storey j is related to the u_j through the relationship

$$V_j=k_{j1}u_1+k_{j2}u_2+\cdots+k_{jj}u_j+\cdots+k_{ji}u_i \tag{5.36}$$

or in matrix form

$$\mathbf{V}=\mathbf{K}\cdot\mathbf{u} \tag{5.37}$$

The above relationship, for $u_j=1$ and $u_1=u_2\ldots=u_{j-1}=u_{j+1}=u_i=0$ (Figure 5.14) results in

$$V_1=K_{1j} \quad V_2=K_{2j}, \ldots \; V_i=K_{ij}$$

which means that the elements of the matrix **K** can be considered as the storey shears for a unit relative displacement of the storey. In the case of rigid girders and $s\neq j$, K_{sj} are zero, and the matrix **K** becomes diagonal.

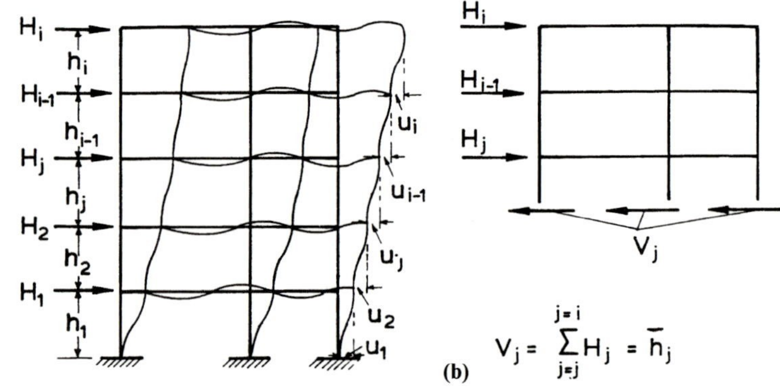

Figure 5.13 Deformation shape of a plane frame under lateral loading: (a) notation; (b) equilibrium condition of horizontal forces.

From the equilibrium conditions of the shear forces of every storey towards the lateral forces that act on the floor under consideration, $3i$ equations with $3i$ unknowns result, from which the relative displacements u, v and ω of the floors can be determined.

Indeed, for every storey three equilibrium conditions are stated, that is

$$\left.\begin{aligned} \Sigma V_{mj} &= \Sigma H_{xj} = \bar{h}_{xj} \\ \Sigma V_{nj} &= \Sigma H_{yj} = \bar{h}_{yj} \\ \Sigma V_{nj} x_n - \Sigma V_{mj} y_m &= \bar{h}_{yj} x_G - \bar{h}_{xj} y_G \end{aligned}\right\} \tag{5.38}$$

where x_G, y_G are the coordinates of the centre of mass, or in matrix form

$$\left.\begin{aligned} \Sigma \mathbf{V}_m &= \bar{\mathbf{h}}_x \\ \Sigma \mathbf{V}_n &= \bar{\mathbf{h}}_y \\ \Sigma \mathbf{V}_n x_n - \Sigma \mathbf{V}_m y_m &= \bar{\mathbf{h}}_y x_G - \bar{\mathbf{h}}_x y_G \end{aligned}\right\} \tag{5.39}$$

Substituting equations (5.34) and (5.37) into equation (5.39) we have

$$\left.\begin{aligned} (\Sigma \mathbf{K}_m)\mathbf{u}_0 + 0 \quad &- (\Sigma \mathbf{K}_m y_m)\,\boldsymbol{\omega} &&= \bar{\mathbf{h}}_x \\ 0 + (\Sigma \mathbf{K}_n)\mathbf{v}_0 \quad &+ (\Sigma \mathbf{K}_n x_n)\,\boldsymbol{\omega} &&= \bar{\mathbf{h}}_y \\ -(\Sigma \mathbf{K}_m y_m)\mathbf{u}_0 + (\Sigma \mathbf{K}_n y_n)\mathbf{v}_0 \quad &+ (\Sigma \mathbf{K}_n x_n^2 + \Sigma \mathbf{K}_m y_m^2)\,\boldsymbol{\omega} &&= \bar{\mathbf{h}}_y x_G - \bar{\mathbf{h}}_x y_G \end{aligned}\right\} \tag{5.40}$$

These equations allow the determination of the relative displacements and rotations of the floors in their plane, and consequently the effects of the horizontal forces on the system. From the above presentation it is obvious that for an efficient treatment of a pseudospatial system under horizontal loading computational aid is needed (AUT Reinforced Concrete Lab, 1986). However, methods have been developed (Valanidis and Penelis, 1987) that allow a quick but approximate approach to the problem without the use of strong computational means, for a preliminary design.

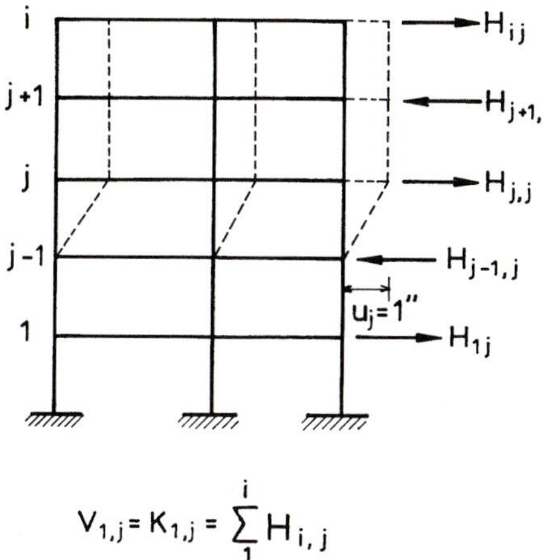

Figure 5.14 Structural meaning of the elements of matrix **K**.

If the origin of the coordinate system at every floor is replaced by a new point, called the stiffness centre (section 5.1.2) such that the following relationships are fulfilled:

$$\sum_m \mathbf{K}_m y_m = 0 \quad \sum_n \mathbf{K}_n x_n = 0 \tag{5.41}$$

then equations (5.40) take the following form:

$$\left.\begin{aligned} (\Sigma\mathbf{K}_m)\mathbf{u}_0 &= \overline{\mathbf{h}}_x \\ (\Sigma\mathbf{K}_n)\mathbf{v}_0 &= \overline{\mathbf{h}}_y \\ (\Sigma\mathbf{K}_n\overline{x}_n^2 + \Sigma\mathbf{K}_m\overline{y}_m^2)\boldsymbol{\omega} &= \overline{\mathbf{h}}_y\overline{x}_G - \overline{\mathbf{h}}_y\overline{y}_G \end{aligned}\right\} \tag{5.42}$$

All the coordinates $\overline{x}$ and $\overline{y}$ refer to the new systems that originate from the stiffness centres of the floors. The coordinates x_E and y_E of the stiffness centre for every floor are derived from the relationships

$$x_{Ej} = \frac{\Sigma\mathbf{K}_n^j x_n}{\Sigma\mathbf{K}_n^j} \quad y_{Ej} = \frac{\Sigma\mathbf{K}_m^i y_m}{\Sigma\mathbf{K}_m^j} \tag{5.43}$$

In the case of a symmetric system (for example along the x-axis) both in geometry and loading, the stiffness centres are on the symmetry plane along which the loading $\overline{\mathbf{h}}_x$ also occurs. Consequently, equations (5.42) become

$$\left.\begin{aligned} (\Sigma\mathbf{K})\mathbf{u}_0 &= \overline{\mathbf{h}}_x \\ (\Sigma\mathbf{K})\mathbf{v}_0 &= 0 \\ \boldsymbol{\omega}\,[\Sigma\mathbf{K}_n\overline{x}_n^2 + \Sigma\mathbf{K}_m\overline{y}_m^2] &= 0 \end{aligned}\right\} \tag{5.44}$$

That is, $\mathbf{v}$ and $\boldsymbol{\omega}$ are equal to zero and the system is subjected to a translational displacement $\mathbf{u}_0$ only, so the problem is simplified into a plane one (Figure 5.15).

In the case that the system is symmetric in both its main directions equations (5.42) yield

$$\left.\begin{aligned} (\Sigma\mathbf{K}_m)\mathbf{u}_0 &= \overline{\mathbf{h}}_x \\ (\Sigma\mathbf{K}_n)\mathbf{v}_0 &= \overline{\mathbf{h}}_y \\ \omega[\Sigma\mathbf{K}_n\overline{x}_n^2 + \Sigma\mathbf{K}_m\overline{y}_m^2] &= 0 \end{aligned}\right\} \tag{5.45}$$

That is, the system is subjected only to translational displacements along the x–x and y–y directions, and the deformations $\mathbf{u}_0$, $\mathbf{v}_0$ are completely uncoupled. Therefore the system may be replaced by two independent plane systems (section 5.1.2).

For a qualitative understanding of the behaviour of the pseudospatial system under horizontal loading, the case of a single-storey structure with rigid girders will be treated below (Figure 5.16). In this case the coordinates of the stiffness centre can be derived from the relationships

$$x_E = \frac{\Sigma J_n x_n}{\Sigma J_n} \quad y_E = \frac{\Sigma J_m y_m}{\Sigma J_m} \tag{5.46}$$

while equations (5.42) take the form

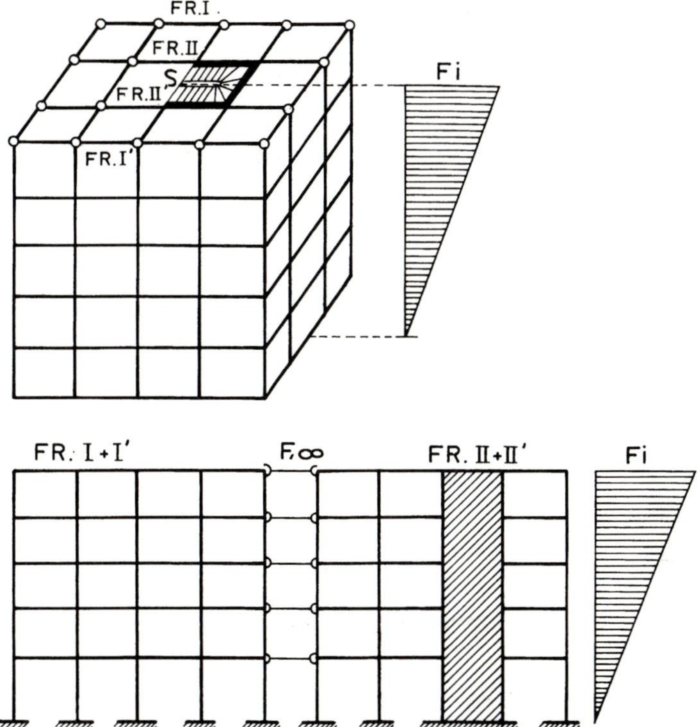

Figure 5.15 Analysis of a symmetrical system.

$$\left.\begin{aligned}(\Sigma J_m)u_0 &= H_x\\(\Sigma J_n)v_0 &= H_y\\(\Sigma J_n\bar{x}_n^2+\Sigma J_m\bar{y}_m^2)\omega &= H_y\bar{x}_G-H_x\bar{y}_G\end{aligned}\right\}\qquad(5.47)$$

where J_m, J_n are the moments of inertia of the sections of the lateral load-resisting vertical elements along the x and y-axes respectively.

Equations (5.47) show that if the centre of mass G coincides with the stiffness centre E (i.e. $\bar{x}_G=0$, $\bar{y}_G=0$), then the system under H_x and H_y loading experiences only a translation along the x or y-axis, depending on the direction of the loading. If the centre of mass and the centre of stiffness have different positions, then the rotation of the system is proportional to the distance between the two centres and inversely proportional to the storey torsional stiffness, which is given by the formula (section 5.1.2)

$$J_p=\sum_n J_n\bar{x}_n^2+\sum_m J_m\bar{y}_m^2 \qquad (5.48)$$

It is obvious that the value of J_p depends mainly on the existence of elements with high flexural stiffness (i.e. shear walls) along the perimeter of the structure (Figure 5.16).

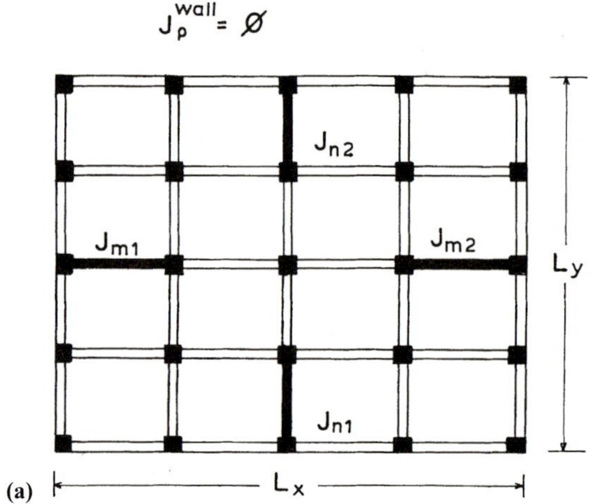

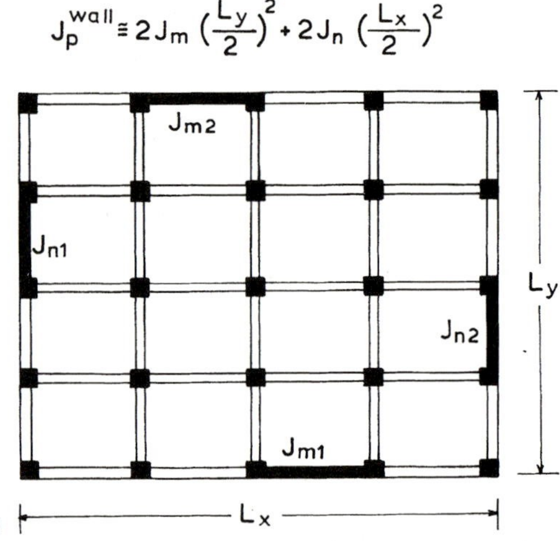

Figure 5.16 Effect of the arrangement of the shear walls on the torsional stiffness of the system: (a) system with low torsional stiffness; (b) system with high torsional stiffness.

5.6 MULTIMODAL RESPONSE SPECTRUM ANALYSIS

5.6.1 General

As already mentioned, this type of analysis, according to EC8 is required for structures which do not satisfy the conditions for applying the simplified modal response spectrum analysis (section 5.1.1).

For buildings complying with the criteria of regularity in plan but not in elevation, the analysis can be performed using two plane models, one for each main direction. Otherwise, the system must be analysed using a spatial model. Whenever a spatial model is used, the design seismic action will be applied along its two main directions determined by the resisting elements of the system. Otherwise the design seismic action will be applied along all relevant horizontal directions and their orthogonal horizontal axes.

In a multimodal analysis the responses of all modes of vibration contributing significantly to the global response are taken into account (Clough and Penzien, 1975). This may be satisfied by either of the following:

- by demonstrating that the sum of the effective modal masses for the modes considered, amounts to at least 90% of the total mass of the structure, that is

$$\boxed{\sum_{i=1}^{k} \frac{L_i^{*2}}{M_i^*} \geqslant 0.9 \sum_{i=1}^{n} M_i} \tag{5.49}$$

  where k is the number of modes considered and n the number of masses;
- by demonstrating that all modes with effective modal masses greater than 5% of the total mass are considered, that is

$$\boxed{\frac{L_j^{*2}}{M_j^*} \leqslant 0.05 \sum_{i=1}^{n} M_i} \tag{5.50}$$

  where j is the index of the modes not considered.

In the case of a spatial model the above conditions must be verified for each main direction.

In buildings with a significant contribution from torsional modes, if the above conditions cannot be satisfied the minimum number of modes k to be considered in a spatial analysis should satisfy the following conditions.

$$\boxed{k \geqslant 3\sqrt{n}} \tag{5.51}$$

and

$$\boxed{T_k \leqslant 0.20 \text{ s}} \tag{5.52}$$

where k is the number of modes considered, n the number of storeys and T_k the period of vibration of mode k. This means that if the period T_k of mode k continues to be greater than 0.2 s, all additional modes with period T_k greater than 0.2 s should be taken into account.

The multimodal response spectrum analysis of either a plane or a spatial model has already been presented in detail (sections 3.4 and 3.5). There, special reference was also made to alternative methods of analysis, e.g. time-history analysis for linear and nonlinear behaviour of the structure.

Here it should only be added that whenever a spatial model is used,

- the floor masses will be considered as lumped masses concentrated at the centre of gravity of each floor; and
- the accidental torsional effects may be determined as the envelope of the effects resulting from an analysis for static loading consisting of torsional moments

M_{1i} about the vertical axis of each storey i

$$\boxed{M_{1i}=e_{1i}H_i} \tag{5.53}$$

where M_{1i} is the torsional moment of storey i about its vertical axis, e_{1i} the accidental eccentricity of storey mass i accounting for the two main directions and H_i the horizontal force acting at storey i as derived from the application of simplified modal analysis for the two main directions.

The effect of the loading described above is considered with alternating signs, the same for all storeys.

Whenever two separate plane models are used for the analysis, the torsional effects may be considered by applying the procedure presented in section 5.4.5.

5.6.2 Suggested procedure for the analysis

Bearing in mind, as already mentioned, that strong computational tools are available at low cost, it is suggested that, in the case that multimodal response spectrum analysis is applied, a spatial system with diaphragms at floor levels is used. This system will be analysed for the following actions:

- $W=G$ '+' $\psi_{2i}Q_i$ — gravity load
- design spectrum (S_{dx}) modified by the importance factor for horizontal excitation parallel to the x–x axis and masses at the centre of gravity of the floors
- $M_i=\pm e_{1y}H_{ix}$

  } x–x seismic action
- design spectrum (S_{dy}) modified by the importance factor for horizontal excitation parallel to the y–y axis and masses at the centre of gravity of the floors
- $M_i=\pm e_{1x}H_{iy}$

  } y–y seismic action (5.54)

The loading effects (E) from the above analysis, for each loading direction, will be derived as follows (Figure 5.17):

$$E_W \Rightarrow G \text{ '+' } \psi_{2i}Q_i \qquad \text{gravity load}$$

$$\left.\begin{aligned} E_u^x &\Rightarrow S_{dx} \text{ '+' } e_{1y}H_{ix} \\ E_o^x &\Rightarrow S_{dx} \text{ '+' } (-e_{1y})H_{ix} \end{aligned}\right\} \quad x\text{–}x \text{ seismic action}$$

$$\left.\begin{aligned} E_r^y &\Rightarrow S_{dy} \text{ '+' } e_{1x}H_{iy} \\ E_l^y &\Rightarrow S_{dy} \text{ '+' } (-e_{1x})H_{iy} \end{aligned}\right\} \quad y\text{–}y \text{ seismic action} \tag{5.55}$$

where '+' means 'combined with'.

For the reasons mentioned in section 5.4.6 the above procedure seems to be the most convenient. All other options, e.g. the use of two plane models combined with simplified torsional analysis, are proper only in the case that no efficient computational tools are available.

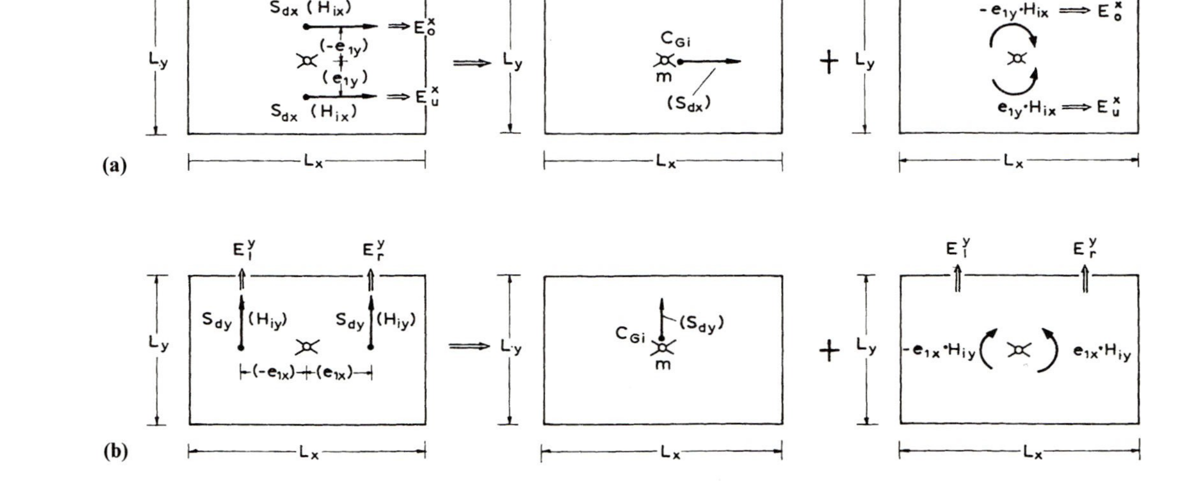

Figure 5.17 Seismic action effects E resulting from a multimodal response spectrum analysis; (a) seismic action in the x–x direction; (b) seismic action in the y–y direction.

5.7 COMBINATION OF THE COMPONENTS OF GRAVITY LOADS AND SEISMIC ACTION

5.7.1 General

In general, no matter which one of the two procedures presented above has been used, the horizontal components of the seismic action should be considered, according to EC8, as acting simultaneously in the two main directions. These two components may also be considered as having equal and uncorrelated intensities (Rosenblueth and Contreras, 1977).

The combination of these two horizontal components for the determination of maximum seismic effects and further on, their combination with the permanent gravity loads, may be carried out as follows:

1. At first the structural response to each horizontal component shall be computed by means of the combination rules for modal responses given in section 3.4.2(a) or by means of simplified modal response spectrum analysis (sections 5.4.2 and 5.4.6).
2. Then the maximum value of each action effect on the structure due to the two horizontal components of the seismic action may be estimated by the square root of the sum of the squared responses to each component of the seismic action, that is

$$\boxed{E_{max} = \pm\sqrt{E_x^2+E_y^2}} \tag{5.56}$$

   where E_{max} is the maximum action effects (M_x, M_y, M_z, V_x, V_y, N) due to the simultaneous action of the earthquake in both main directions, E_x the maximum action effects due to the application of the seismic action along the horizontal axis x–x of the structure and E_y the maximum action effects due to the application of the seismic action along the horizontal axis y–y of the structure.
3. The seismic action effects must be superimposed on the gravity load effects, that is on

$$E_W = E(G\text{ `+' }\psi_{2i}Q_i)$$

Therefore, the final action effects due to gravity loads and earthquake will have the form (section 4.5)

$$\boxed{E_S = E(G\text{ `+' }\psi_{2i}Q_i)\text{ `+' }E(\gamma_I S_{dx}\text{ `+' }\gamma_I S_{dy})} \tag{5.57}$$

where '+' implies 'to be combined with', G are the dead loads, Q_i the characteristic value of variable action i, $S_{dx,y}$ the design value of the seismic action parallel to x–x and y–y respectively, γ_I the importance factor and ψ_{2i} the combination coefficient for quasi-permanent value of variable action i.

It should be noted that the extreme values of seismic effects ($M_{x,ex}$, $M_{y,ex}$, $M_{z,ex}$, $V_{x,ex}$, $V_{y,ex}$, N_{ex}) determined above do not act simultaneously. Therefore, in the case that more than one load effects are needed for the safety verification at ultimate limit state (i.e. M_x, M_y, N for the cross-section of a column) the combination of the extreme values of all relevant load effects would be, at first glance, conservative.

In the following a theoretical approach to the problem will be presented so that the reader may have a global view of the approximations involved in various procedures.

5.7.2 Theoretical background

Let n be the number of the load effects defining the response state of an R/C structural element (i.e. $n=3$ for a column under M_x. M_y, N). Its response to gravity and earthquake loading acting parallel to x and y-axes simultaneously is defined at an n-dimensional **response space** of the interacting load effects (i.e. M_x, M_y, N) by an ellipsoid (Rosenblueth and Contreras, 1977; Gupta, 1990) with its centre at $\mathbf{r}_0$ (M_{x0}, M_{y0}, N_0) (Figure 5.18), described at a local reference coordinate system by the equation

$$\bar{\mathbf{x}}^T\mathbf{G}^{-1}\bar{\mathbf{x}}=1$$

where M_{x0}, M_{y0}, N_0 are the responses to gravity loads,

$$\bar{\mathbf{x}}^T=\{\bar{X}_c, \bar{Y}_c, \bar{Z}_c\}=\{\bar{M}_{xc}, \bar{M}_{yc}, \bar{N}_c\}$$

the vector of the most probable simultaneously acting relevant load effects at the

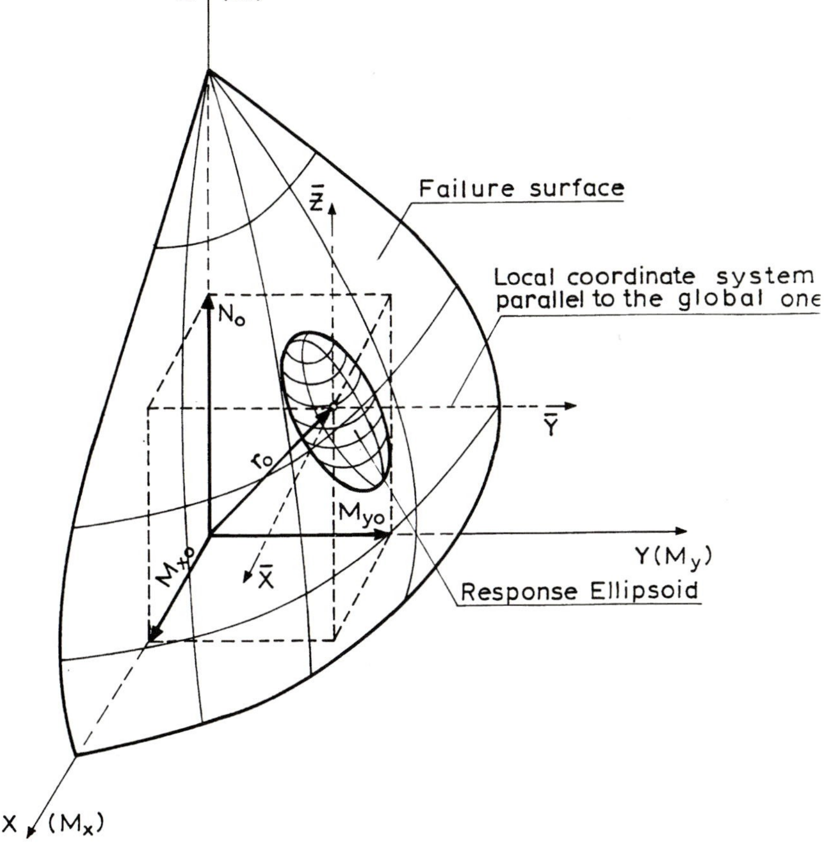

Figure 5.18 Response ellipsoid and failure curve: (a) various simplifications; (b) response ellipse in the safe domain.

local reference system, and

$$\mathbf{G}=\begin{bmatrix} X^2_{ex} & \rho_{xy} & \rho_{xz} \\ \rho_{yx} & Y^2_{ex} & \rho_{yz} \\ \rho_{zx} & \rho_{xy} & Z^2_{ex} \end{bmatrix}=\begin{bmatrix} M^2_{X_{ex}} & \rho_{xy} & \rho_{xz} \\ \rho_{yx} & M^2_{Y_{ex}} & \rho_{yz} \\ \rho_{zx} & \rho_{zy} & N^2_{ex} \end{bmatrix} \qquad (5.58)$$

In the above matrix **G**, quantities X_{ex}, Y_{ex}, Z_{ex}, are derived from equation (5.56), while quantities $\rho_{xy}=\rho_{yx}$, $\rho_{xz}=\rho_{zx}$, $\rho_{yz}=\rho_{zy}$ are derived from the following expressions:

$$\left.\begin{aligned} \rho_{xy}&=\sum_i(X_{i,x}\,Y_{i,x}+X_{i,y}\,Y_{i,y})=\sum_i(M_{xi,x}\,M_{yi,x}+M_{xi,y}\,M_{yi,y}) \\ \rho_{xz}&=\sum_i(X_{i,x}\,Z_{i,x}+X_{i,y}\,Z_{i,y})=\sum_i(M_{xi,x}\,N_{i,x}+M_{xi,y}\,N_{i,y}) \\ \rho_{yz}&=\sum_i(Y_{i,x}\,Z_{i,x}+Y_{i,y}\,Z_{i,y})=\sum_i(M_{yi,x}\,N_{i,x}+M_{yi,y}\,N_{i,y}) \end{aligned}\right\} \qquad (5.59)$$

where $X_{i,x}$, $Y_{i,x}$, $Z_{i,x}$ are the response spectrum values of the interacting response in the *i*th mode of vibration due to the *x–x* earthquake component ($i=1, 2\ldots k$) and $X_{i,y}$, $Y_{i,y}$, $Z_{i,y}$ the response spectrum values of the interacting response in the *i*th mode of vibration due to the *y–y* earthquake component ($i=1, 2\ldots k$).

In general the state of ellipsoid (or response ellipsoid) which presents the simultaneous variation of the values of the three responses (M_x, M_y, N) to gravity and earthquake loading (equation (5.57)) has inclined axes, while the failure surface for M_x, M_y, N of the R/C structural element is not susceptible to simple description. The task of investigating whether the ellipsoid lies entirely within the safe domain, and that of selecting a failure surface which will lie just outside the ellipsoid are excessively complicated for routine design. Consequently, based on the above theoretical background, a series of simplified procedures have been developed.

5.7.3 Simplified procedures

Some of these simplified procedures will be presented here, mainly for a two-dimensional response space (i.e. M and N on the cross-section of an R/C wall), so that a plane schematic presentation of the various approaches can be feasible (Figure 5.19(a)). In this case the response, or interaction, ellipsoid is reduced to an interaction ellipse defined at the local coordinate system by the following equation (Gupta and Singh, 1977; Panetsos and Anastassiadis, 1994):

$$\left.\begin{aligned} &\frac{X_c^2}{X_{ex}^2}+\frac{Y_c^2}{Y_{ex}^2}-2\rho_{xy}\frac{X_cY_c}{X_{ex}^2Y_{ex}^2}=1-\left(\frac{\rho_{xy}}{X_{ex}Y_{ex}}\right)^2 \\ &\text{or} \\ &\frac{M_c^2}{M_{ex}^2}+\frac{N_c^2}{N_{ex}^2}-2\rho_{xy}\frac{M_cN_c}{M_{ex}^2N_{ex}^2}=1-\left(\frac{\rho_{xy}}{M_{ex}N_{ex}}\right)^2 \end{aligned}\right\} \qquad (5.60)$$

where

$$\rho_{xy}=\sum_i(X_{i,x}Y_{i,x}+X_{i,y}Y_{i,y})$$

The centre of this ellipse must be placed in the global coordinate system at the point $\mathbf{r}_0(X_0=M_0, Y_0=N_0)$ representing the gravity load effect vector (Figure 5.19(b)).

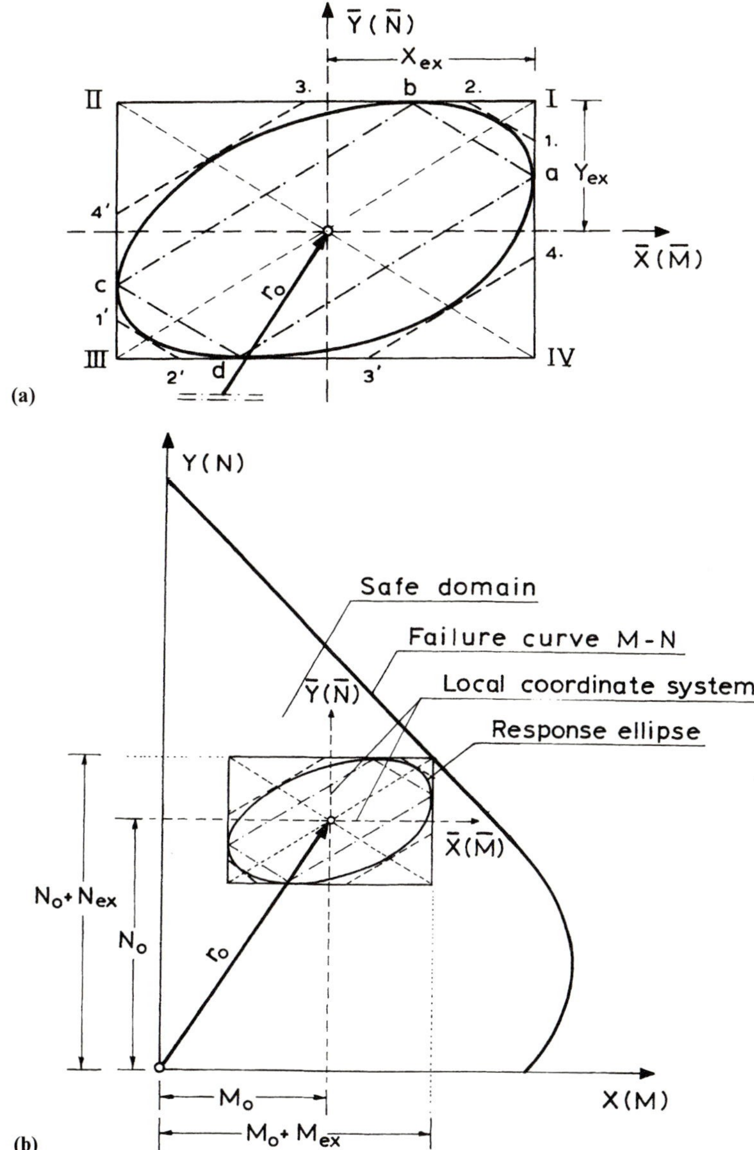

Figure 5.19 Response ellipse and failure curve: (a) various simplifications; (b) response ellipse in the safe domain.

(a) Combination of the extreme values of the interacting load effects

In the case that for safety verfication the maximum values resulting from equation (5.56) are combined, the ellipse of equation (5.60) is replaced by the rectangle I, II, III, IV (Figure 5.19). It is obvious that this approach is on the safe side.

In most conventional design procedures, it is implicitly assumed that the maxima do occur simultaneously. This assumption introduces an error on the safe side which can be significant. From various case studies conducted so far (Panetsos and Anastassiadis, 1994; Zararis, Salonikios and Botis, 1994; Leblond, 1980) it is concluded that for R/C columns and shear walls this error ranges from 15 to 35%, measured as a percentage of the 'exact' reinforcement of the R/C element.

The number of combinations in this case is four for a two-load effect component interaction, and generally

$$\lambda = 2^n$$

where n is the number of load effect components involved.

(b) Combination of each extreme load effect with the corresponding values of the interacting ones

In this case the ellipse (Figure 5.19(a)) is replaced by the parallelogram a, b, c, d. It is obvious that this approach is on the unsafe side. The relevant combinations for the load effect components (i.e. M and N) are listed in Table 5.2 (Gupta and Chu, 1977).

In the case of a three-component interaction problem (i.e. M_x, M_y, N) the simultaneously acting components are given in Table 5.3.

According to the conclusions of various case studies conducted so far, the error for R/C columns and walls ranges from −5 to −10%, measured as a percentage of the exact reinforcement of the R/C element.

(c) The Gupta–Singh procedure

According to this procedure (Gupta and Singh, 1977), the ellipse is approximated by the circumscribed octagon 1, 2, 3, 4, 1′, 2′, 3′, 4′ (Figure 5.19(a)). It is obvious that this approach is on the safe side, while the error is not so significant as in case (a). The coordinates of the above-designated points are as follows:

Point 1: $X_{1c} = X_{ex}$ $\quad Y_{1c} = Y_{ex}\,(\mu_{xy} - 1)$
Point 2: $X_{2c} = X_{ex}\,(\mu_{xy} - 1)$ $\quad Y_{2c} = Y_{ex}$
Point 3: $X_{3c} = X_{ex}(1 - r_{xy})$ $\quad Y_{3c} = Y_{ex}$
Point 4: $X_{4c} = X_{ex}$ $\quad Y_{4c} = Y_{ex}(1 - r_{xy})$

Table 5.2 Simultaneously acting X_c, Y_c values

	Points a,c	Points b,d
X_c	$\pm X_{ex}$ (Eq. 5.56)	$\pm \rho_{xy}/X_{ex}$
Y_c	$\pm \rho_{xy}/X_{ex}$	$\pm Y_{ex}$ (Eq. 5.56)

Table 5.3 Simultaneously acting X_c, Y_c, Z_c values

	2 points	2 points	2 points
X_c	$\pm X_{ex}$	$\pm \rho_{xy}/Y_{ex}$	$\pm \rho_{xz}/Z_{ex}$
Y_c	$\pm \rho_{xy}/X_{ex}$	$\pm Y_{ex}$	$\pm \rho_{yz}/Z_{ex}$
Z_c	$\pm \rho_{xy}/X_{ex}$	$\pm \rho_{zy}/Y_{ex}$	$\pm Z_{ex}$

where

$$\mu_{xy}=\left[2\left(1+\frac{\rho_{xy}}{X_{ex}Y_{ex}}\right)\right]^{1/2} \quad r_{xy}=\left[2\left(1-\frac{\rho_{xy}}{X_{ex}Y_{ex}}\right)\right]^{1/2}$$

The other four vertices are symmetric to the previous ones with respect to the centre of the octagon.

In the case of a three-component interaction problem (i.e. M_x, M_y, N) the ellipse is approximated by a polyhedron with 24 vertices. Their coordinates are as follows:

Point 1:	$X_{1c}=X_{ex}$	$Y_{1c}=Y_{ex}(\mu_{yz}-1)$	$Z_{1c}=Z_{ex}(\mu_{xz}-1)$
Point 2:	$X_{2c}=X_{ex}(\mu_{yz}-1)$	$Y_{2c}=Y_{ex}$	$Z_{2c}=Z_{ex}(\mu_{xy}-1)$
Point 3:	$X_{3c}=X_{ex}(\mu_{zx}-1)$	$Y_{3c}=Y_{ex}(\mu_{xy}-1)$	$Z_{3c}=Z_{ex}$
Point 4:	$X_{4c}=X_{ex}(1-r_{yz})$	$Y_{4c}=Y_{ex}$	$Z_{4c}=Z_{ex}(1-r_{xy})$
Point 5:	$X_{5c}=X_{ex}(1-r_{xz})$	$Y_{5c}=Y_{ex}(1-r_{xy})$	$Z_{5c}=Z_{ex}$
Point 6:	$X_{6c}=X_{ex}$	$Y_{6c}=Y_{ex}(1-r_{yz})$	$Z_{6c}=Z_{ex}(1-r_{xz})$
Point 7:	$X_{7c}=X_{ex}$	$Y_{7c}=Y_{ex}(\mu_{yz}-1)$	$Z_{7c}=Z_{ex}(1-r_{xz})$
Point 8:	$X_{8c}=X_{ex}(1-r_{yz})$	$Y_{8c}=Y_{ex}$	$Z_{8c}=Z_{ex}(\mu_{xy}-1)$
Point 9:	$X_{9c}=X_{ex}(\mu_{xz}-1)$	$Y_{9c}=Y_{ex}(1-r_{xy})$	$Z_{9c}=Z_{ex}$
Point 10:	$X_{10c}=X_{ex}$	$Y_{10c}=Y_{ex}(1-r_{yz})$	$Z_{10c}=Z_{ex}(\mu_{xz}-1)$
Point 11:	$X_{11c}=X_{ex}(\mu_{yz}-1)$	$Y_{11c}=Y_{ex}$	$Z_{11c}=Z_{ex}(1-r_{xy})$
Point 12:	$X_{12c}=X_{ex}(1-r_{xz})$	$Y_{12c}=Y_{ex}(\mu_{xy}-1)$	$Z_{12c}=Z_{ex}$

where

$$\mu_{xy}=\left[2\left(1+\frac{\rho_{xy}}{X_{ex}Y_{ex}}\right)\right]^{1/2} \quad r_{xy}=\left[2\left(1-\frac{\rho_{xy}}{X_{ex}Y_{ex}}\right)\right]^{1/2}$$

$$\mu_{yz}=\left[2\left(1+\frac{\rho_{yz}}{X_{ex}Y_{ex}}\right)\right]^{1/2} \quad r_{yz}=\left[2\left(1-\frac{\rho_{yz}}{X_{ex}Y_{ex}}\right)\right]^{1/2}$$

$$\mu_{xz}=\left[2\left(1+\frac{\rho_{xz}}{X_{ex}Y_{ex}}\right)\right]^{1/2} \quad r_{xz}=\left[2\left(1-\frac{\rho_{xz}}{X_{ex}Y_{ex}}\right)\right]^{1/2}$$

The other 12 vertices are symmetric to the previous ones with respect to the centre of the polyhedron.

Various case studies (Panetsos and Anastassiadis, 1994; Leblond, 1980) have shown that for R/C columns and walls the error ranges from 2 to 6%, measured as a percentage of the exact reinforcement of the R/C element.

(d) The Rosenblueth and Contreras procedure

Rosenblueth and Contreras (1977) have replaced the ellipsoid by a vector (Figure 5.20);

$$\mathbf{r}_c=\mathbf{r}_0+a_x\mathbf{r}_x+a_y\mathbf{r}_y \tag{5.61}$$

where $\mathbf{r}_c$ is the most probable extreme response vector of the R/C element to gravity and earthquake loading, $\mathbf{r}_0$ the response vector of the R/C element to gravity loading only (i.e. M_{x0}, M_{y0}, N_0), $\mathbf{r}_x$ the most probable extreme response vector of the R/C element to earthquake loading parallel to the x–x–axis (i.e. $M_{x,ex,x}$, $M_{y,ex,x}$, $N_{ex,x}$), $\mathbf{r}_y$ the most probable extreme response vector of the R/C element to earth-

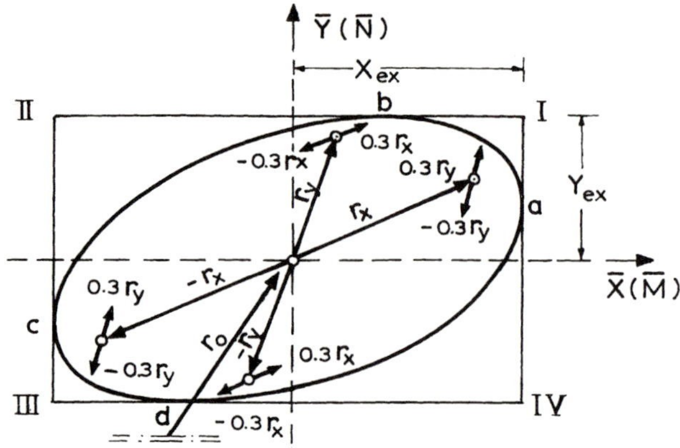

Figure 5.20 The Rosenblueth and Contreras procedure.

quake loading parallel to the y–y-axis (i.e. $M_{x,ex,y}$, $M_{y,ex,y}$, $N_{ex,y}$) and a_x, a_y are constant coefficients to be determined, so that the probable error on the safe side will be equal to that on the unsafe side.

Through this approach it has been concluded that for

$$a_x = \begin{cases} 1.00 \\ 0.336 \end{cases} \quad \text{and} \quad a_y = \begin{cases} 0.336 \\ 1.00 \end{cases}$$

respectively, the maximum error is $\pm 5.5\%$. This simplified procedure had also been proposed earlier in 1975 (with $a_{x,y} = 1.00$ and 0.33) by A.S. Veletsos, and it has served as a basis for code requirements in the USA (UBC, 1988) and recently in the European Union (EC8).

In both codes the values of a_x, a_y introduced are the following:

$$a_x = \begin{cases} \pm 1.00 \\ \pm 0.30 \end{cases} \quad \text{and} \quad a_y = \begin{cases} \pm 0.30 \\ \pm 1.00 \end{cases}$$

respectively. For these values the maximum error is 4.4% on the safe side and 8.1% on the unsafe side. The number of combinations is derived by the formula

$$\lambda = 2 \times 2^c \tag{5.62}$$

where c is the number of not correlated earthquake actions. For $c = 2$ (seismic action parallel to x and y-axes) $\lambda = 8$ combinations, no matter how many interaction response components are needed for the design.

In the case of a two-dimensional response space (i.e. M and N on an R/C wall) the ellipsoid is reduced to an interaction ellipse and equation (5.61) takes the following algebraic form:

$$\left.\begin{aligned} M_{1c} &= M_0 + M_{ex,x} + 0.30M_{ex,y}, & N_{1c} &= N_0 + N_{ex,x} + 0.30N_{ex,y} \\ M_{2c} &= M_0 + M_{ex,x} - 0.30M_{ex,y}, & N_{2c} &= N_0 + N_{ex,x} - 0.30N_{ex,y} \\ M_{3c} &= M_0 - M_{ex,x} + 0.30M_{ex,y}, & N_{3c} &= N_0 - N_{ex,x} + 0.30N_{ex,y} \\ M_{4c} &= M_0 - M_{ex,x} - 0.30M_{ex,y}, & N_{4c} &= N_0 - N_{ex,x} - 0.30N_{ex,y} \end{aligned}\right\} \text{4 combinatons}$$

$$\left.\begin{array}{ll} M_{5c}=M_0+0.30M_{ex,x}+M_{ex,y}, & N_{5c}=N_0+0.30N_{ex,x}+N_{ex,y} \\ M_{6c}=M_0+0.30M_{ex,x}-M_{ex,y}, & N_{6c}=N_0+0.30N_{ex,x}-N_{ex,y} \\ M_{7c}=M_0-0.30M_{ex,x}+M_{ex,y}, & N_{7c}=N_0-0.30N_{ex,x}+N_{ex,y} \\ M_{8c}=M_0-0.30M_{ex,x}-M_{ex,y}, & N_{8c}=N_0-0.30N_{ex,x}-N_{ex,y} \end{array}\right\} \text{4 combinatons}$$

(e) The extreme stress procedure

Anastassiadis (1993) has adopted the procedure mainly used for steel structures (Wilson and Button, 1982) for the design of R/C structures. According to this procedure, it is assumed that the cross-section of the R/C element is homogeneous and uncracked. Therefore the extreme values of the stresses at the vertices of the R/C cross-section may be computed as if it were a steel cross-section. The computed extreme stresses themselves are of no significance but they are used only as a vehicle for the determination of the components that should be combined.

5.7.4 Code requirements

1. In most conventional design procedures (see EC8), it is implicitly assumed that the maximum responses do occur simultaneously. This assumption introduces an error on the safe side ranging from 15 to 35%. In this case, bearing in mind that equation (5.56) produces the extreme values of the load effects with plus or minus signs, it is concluded that the number of combinations must be equal to

$$\lambda=2^n$$

where n is the number of load effects that must be taken into account.For example, in the case of a column, where three load effects must be considered for the design, that is M_x, M_y and N, the number of combinations is eight.

In addition, it should be noted that due to the accidental eccentricities introduced, four different centres of masses must be considered. Therefore, it is concluded that the total number of combinations for the design of the cross-section of a column is 32 (4×2^n). In fact, substituting equation (5.55) in equation (5.56) and superimposing the gravity load effects, the following combinations result:

$$E=E_w\pm\sqrt{(E_u^x)^2+(E_r^y)^2}=\begin{cases}E_{1,I}\\E_{1,II}\end{cases}$$

$$E=E_w\pm\sqrt{(E_u^x)^2+(E_l^y)^2}=\begin{cases}E_{2,I}\\E_{2,II}\end{cases}$$

$$E=E_w\pm\sqrt{(E_o^x)^2+(E_r^y)^2}=\begin{cases}E_{3,I}\\E_{3,II}\end{cases}$$

$$E=E_w\pm\sqrt{(E_o^x)^2+(E_l^y)^2}=\begin{cases}E_{4,I}\\E_{4,II}\end{cases}$$

that is eight different combinations for each load effect (M_x, M_y, M_z, V_x, V_y, N). In the case of a column where three load effects must be considered, that is M_x, M_y and N, the number of combinations is 32:

$$\left.\begin{matrix} M_x = M_{1,Ix} & M_y = M_{1,Iy} & N = N_{1,I} \\ M_x = M_{1,IIx} & M_y = M_{1,IIy} & N = N_{1,II} \end{matrix}\right\} \quad \text{8 combinations}$$

$$\left.\begin{matrix} M_x = M_{2,Ix} & M_y = M_{2,Iy} & N = N_{2,I} \\ M_x = M_{2,IIx} & M_y = M_{2,IIy} & N = N_{2,II} \end{matrix}\right\} \quad \text{8 combinations}$$

$$\left.\begin{matrix} M_x = M_{3,Ix} & M_y = M_{3,Iy} & N = N_{3,I} \\ M_x = M_{3,IIx} & M_y = M_{3,IIy} & N = N_{3,II} \end{matrix}\right\} \quad \text{8 combinations}$$

$$\left.\begin{matrix} M_x = M_{4,Ix} & M_y = M_{4,Iy} & N = N_{4,I} \\ M_x = M_{4,IIx} & M_y = M_{4,IIy} & N = N_{4,II} \end{matrix}\right\} \quad \text{8 combinations}$$

Total 32 combinations

2. As an alternative to the above procedure, it is allowed according to EC8 and UBC 88 to compute the action effects due to both components using the following formulae:

$$\left.\begin{matrix} E = E_x\text{‘}+\text{’}0.30E_y \\ E = 0.30E_x\text{‘}+\text{’}E_y \end{matrix}\right\} \tag{5.63}$$

This assumption introduces a maximum error of 4.4% on the safe side and 8.1% on the unsafe side. The load effect combinations according to equation (5.62) that must be considered are

$$\lambda = 2 \times 2^c = 2 \times 2^2 = 8$$

that is eight combinations for the case of two simultaneous horizontal seismic actions parallel to the x and y-axes.

Again because of the obligation to introduce the accidental eccentricities the number of the above combinations is multiplied by four, therefore the total number of combinations that must be taken into account becomes

$$\lambda = 4 \times 2 \times 2^c = 4 \times 2 \times 2^2 = 32$$

However, parametric studies (Stylianidis *et al.*, 1994) have shown that the number of combinations can be reduced to six for an R/C column and to four for an R/C wall, if just the maximum and minimum values of each load effect and the corresponding values of the interacting ones were considered.

Statistical evaluation of the results of a study conducted for a sample of 150 cases revealed the following:

- For a 95% fractile, the confidence limit was at the level of 0.97 of the ‘exact’ evaluation.
- For a 100% fractile, the confidence limit was at the level of 0.87 of the ‘exact’ evaluation.

The above simplification was introduced in the new *Greek Code for Earthquake-Resistant Structures* (Ministry for Environment and Public Works, 1992).

3. From the above analysis it is concluded that both methods introduced in EC8 are operational. The application of the other three procedures also presented in section 5.7.2 is not usual in routine design.

4. For buildings satisfying the regularity criteria in plan, and in which walls are the only horizontal load-resisting components, the seismic action may be assumed to act separately along the two main orthogonal horizontal axes of the structure.
5. It has already been mentioned (section 4.4.1) that the vertical component of the seismic action has to be considered only for certain structures (Luft, 1989). The effects of the vertical component according to EC8 need only be taken into account for the elements under consideration and their directly associated supporting elements or substructures.

In the case that the horizontal components of the seismic action are also relevant for these elements, EC8 introduces the following combinations:

$$\left.\begin{aligned} E &= 0.30E_x \text{ `+' } 0.30E_y \text{ `+' } E_z \\ E &= E_x \text{ `+' } 0.30E_y \text{ `+' } 0.30E_z \\ E &= 0.30E_x \text{ `+' } E_y \text{ `+' } 0.30E_z \end{aligned}\right\} \tag{5.64}$$

where E_z is the action effect due to the application of the vertical component of the design seismic action.

The number of combinations needed for the design of R/C sections is derived from equation (5.62) and is equal to

$$\lambda = 4 \times 2 \times 2^c = 4 \times 2 \times 2^3 = 64$$

This number includes the displacements of the centre of mass due to accidental eccentricities.

5.8 SECOND-ORDER EFFECTS (P–Δ EFFECTS)

Most structural systems under the action of seismic forces, because of their inelastic response, sustain large horizontal displacements resulting in the creation of large secondary effects (Paulay, Bachmann and Moser, 1990; Luft, 1989; Wilson and Habibullah, 1987). Consider the frame of Figure 5.21. When this frame, for some external reason (an earthquake in this case), is displaced by Δ, each of the two $W/2$ column loads can be analysed into an axial force on the column with a

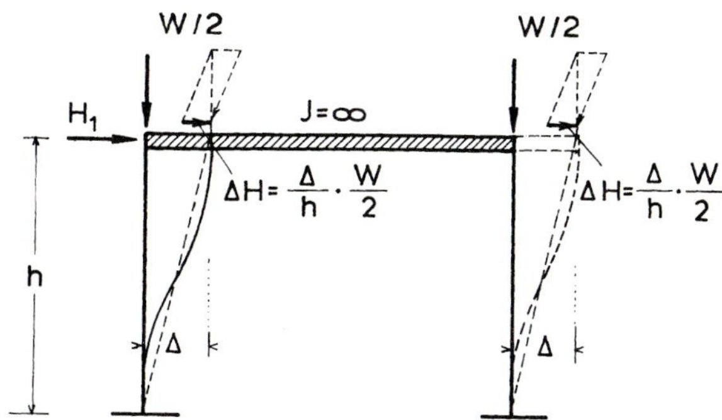

Figure 5.21 Second-order effects on one-storey, two column frame.

value $W/2$ and a horizontal one

$$\Delta H_{1,2} = \frac{\Delta}{h}\,\frac{W}{2}$$

Thus the floor is loaded with an additional (second-order) horizontal force equal to

$$\Delta H = \frac{\Delta}{h}\,W \tag{5.65}$$

In the case of a seismic action the displacement Δ, according to what has been explained in the chapter about ductility, is equal to Δ_{el} which results from the seismic loading of the code, multiplied by a behaviour factor, q, of the structure:

$$\Delta = \Delta_{el} q$$

Therefore, the additional shear force of the storey, because of the second order effect, is equal to

$$\boxed{\Delta V = \frac{\Delta_{el}\,q}{h}\,W} \tag{5.66}$$

EC8 specifies that:

1. For

$$\theta = \frac{\Delta V}{V} = \frac{\Delta_{el}\,q}{h}\,\frac{W}{V} \leqslant 0.10 \tag{5.67}$$

   a second-order analysis is not required.
2. For

$$0.10 \leqslant \theta \leqslant 0.20 \tag{5.68}$$

   the P–Δ effect must be taken into account. In this case an acceptable approximation could be to increase the relevant seismic action effects by a factor equal to $1/(1-\theta)$
3. For

$$0.20 \leqslant \theta \tag{5.69}$$

   the lateral stiffness of the system must be increased.

In the above relations, V is the shear force of the storey due to the seismic actions, Δ_{el} the relative lateral displacement of the top in relation to the bottom of the storey, also known as **inter-storey drift**, q the behaviour factor of the structure, h the storey height and W the total gravity load above the storey under consideration.

The provisions of EC8 as they are stated assume that a plane model in the two main directions is used. For application in three-dimensional systems appropriate modification is needed (Penelis, 1971). A modification on the safe side could be the application of the previously presented provisions on each plane constituent frame of the spatial system, for W, V and Δ_{el} resulted from the analysis of the spatial system for the frame under consideration.

It it again recommended that a high degree of lateral stiffness be provided for the structural system, so that at least second-order effects are prevented (secion 4.3).

5.9 THE INFLUENCE OF MASONRY INFILLED FRAMES ON THE SEISMIC BEHAVIOUR OF STRUCTURES

5.9.1 General

The masonry infills dealt with here are those constructed after hardening of the concrete frames, in contact with them, but without special connection to them, and are considered in the first instance as non-structural elements. This type of masonry infill is very common in southern European countries where seismicity is very high. Given the fact that infills of this type have a considerable strength and stiffness, they have a marked effect on the seismic response of the structural system. In general, the presence of masonry infills affects the seismic behaviour of buildings in the following ways (Dowrick, 1987; Tassios, 1984):

- The stiffness of the building is increased, the fundamental period is decreased and therefore the base shear due to seismic action is increased.
- The distribution of the lateral stiffness of the structure in plan and elevation is modified.
- Part of the seismic action is carried by the infills, thus relieving the structural system.
- The ability of the building to dissipate energy is substantially increased.

The more flexible the structural system, the greater the above effects of the infills. The masonry infills have less strength and deformability than the structural system, and therefore they fail first, presenting separations from the frame and x-shaped cracks (Figure 6.9) Thus, infills absorb and dissipate large quantities of seismic energy, acting as the first line of seismic defence of the building. This failure, though, leads to a decrease in strength and stiffness of the masonry-infilled frame at an early deformation stage and to a parallel load transfer from the infill to the structural system in the form of impulse loading (Figure 6.10). The whole response mechanism is very complicated and is still under research (Michailidis, Stylianidis and Kappos, 1995).

5.9.2 Effects on the analysis

The effects of the infills on the analysis must be considered together with the high degree of uncertainty related to their behaviour, namely;

- the variability of their mechanical properties, and therefore the low reliability in their strength and stiffness;
- their wedging condition, that is how tightly they are connected to the surrounding frame;
- the potential modification of their integrity during the use of the building;
- the non-uniform degree of their damage during the earthquake.

Thus, the safety of the structure cannot rely, not even partly, upon the infills and only their probable negative influence is taken into account. Because of their high lateral stiffness a large percentage of the seismic effects would have been transferred to them in the case they were taken into account as structural elements, but for the reasons stated above this should be avoided.

According to EC8 the seismic analysis, in general, is carried out on the bare structural system and only additional measures are taken for the influence of the infills. These measures are given in the following, and they refer only to frame systems and frame equivalent dual systems for DC 'M' and 'H'. For all wall systems and wall equivalent dual systems, the interaction between the concrete structural system and the infills may be neglected. It should be noted here that in addition to these measures some specified structural elements and the infills themselves are protected through additional design measures described in Chapter 8.

5.9.3 Design seismic action effects

The design seismic action effects are modified because of the reduced natural period produced by the addition of infills. This requirement is deemed to be satisfied by applying the following rules:

1. A new ordinate $S_d\,(T_1)$ of the design spectrum is calculated using an average value T_1 of the mode period of the structure, which is derived as follows:

$$T_1 = \frac{T_{1b} + T_{1i}}{2} \tag{5.70}$$

where T_{1b} is the first mode period of the bare structure according to section 5.4.4 and T_{1i} the first mode of the structure taking into account the infills as structural seismic resistant elements.

2. All seismic action effects determined with the model of the bare structure are multiplied by the ratio of

$$\boxed{\lambda_1 = \frac{S_d(T_1)}{S_d(T_{1b})}} \tag{5.71}$$

3. The first mode period T_{1i} may be estimated by one of the following expressions:

$$T_{1i} = \frac{T_{1b}}{[1+T_{1b}^2 A_w Gg/(16h\ W)]^{1/2}} \tag{5.72}$$

where A_w is the average horizontal cross-sectional area of infill walls per storey, G the shear modulus of infill walls, g the acceleration of gravity, h the height of the building and W the weight of the structure taken into account for the seismic analysis, and

$$T_{1i} = \min \begin{cases} 0.065n \\ 0.080(h/\sqrt{B})\,[h/(h+B)] \\ 0.075h^{3/4} \end{cases} \tag{5.73}$$

where n is the number of storeys, h the height of the building in metres and B the width of the building in metres in the direction considered.

In general, the ratio $S_d\,(T_1)/S_d\,(T_{1b})$ is:

- for low-rise and rigid structures equal to one;

- for high-rise and flexible structures greater than one, which means that the seismic effects on the structure are increased by this factor.

5.9.4 Irregularities due to masonry infills

(a) Irregularities in plan

When the distribution of infills is sufficiently uniform, the existing irregularities may be taken into account by increasing the accidental eccentricity by a factor

$$\boxed{\lambda = 2.0}$$

In the case of severe irregularities in plan due to the excessively unsymmetrical arrangement of the infills (e.g. mainly along two adjacent sides of the building only) several spatial models are used for the analysis of the structure taking into account the possible limits of stiffness distribution related to the uncertain conditions of the infills. In this case the masonry infill could be taken into account in the estimation of the stiffness of the frame through the compression diagonal model (Figure 5.22), that is (Michailidis, Stylianidis and Kappos, 1995)

$$K_W = \frac{H}{\delta} = \frac{E_W w t \cos^2\theta}{d} \tag{5.74}$$

where H is the horizontal action, δ the horizontal relative displacement corresponding to H, E_W the modulus of elasticity of the masonry, $w \approx 0.20d$, t is the masonry thickness, d the length of the diagonal and θ the angle of the diagonal with the horizontal.

(b) Irregularities in elevation

In the case of considerable irregularities in elevation (e.g. drastic reduction of infills in one or more storeys) a local increase of seismic action effects on the respec-

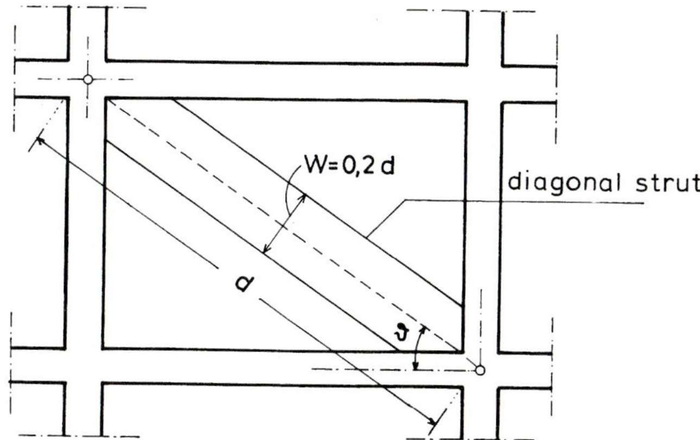

Figure 5.22 Compression diagonal model for the estimation of the infill stiffness.

tive storeys is imposed. Therefore, the calculated action effects on the bare structure are increased by a multiplication factor λ_3 as follows:

$$\boxed{\lambda_3 = 1 + \frac{\Delta V_{RW}}{\Sigma V_{S_d}}} \qquad (5.75)$$

where ΔV_{RW} is the total reduction of the resistance of masonry walls in the storey considered, compared to the most infilled storey closest to it and ΣV_{S_d} the sum of the seismic shear forces acting on all structural vertical elements of the storey considered. In the case $\lambda_3 < 1.10$ there is no need for such action effects modification.

5.9.5 Remarks on infilled frames

The approach of EC8 to the masonry-infilled frames and masonry-infilled frame equivalent dual systems of DC 'M' or 'H' seems to be rather prohibitive. In fact:

- the increase of all seismic action effects on the structure by a factor influenced by the decrease of the fundamental period of the bare structure due to the masonry infills,
- the increase of the accidental eccentricity by two, even for buildings with sufficiently uniform distribution of the infills, and finally
- the increase of the calculated action effects on the bare structure by a factor varying from $\lambda = 1.10$ to $\lambda = 1.70$ for irregularities in elevation

drive the designer to avoid the use of masonry infills or to provide separation joints between masonry and the R/C structural system. However, extensive research (Valiasis and Stylianidis, 1989; Valiasis, Stylianidis and Penelis, 1993; Michailidis, Stylianidis and Kappos, 1995) has shown that masonry infills constitute a highly effective dissipation mechanism. In addition, extensive statistical damage evaluation in areas affected by strong earthquakes (Penelis *et al.*, 1989) has in general shown a favourable influence of the masonry infills on the behaviour of the buildings during earthquakes. In this respect, infill masonry walls in a way constitute a first line of defence for the building against earthquakes (a type of damper).

Therefore the subject, even for the frame or frame equivalent dual systems, should in the future be reconsidered in the focus of additional prenormative research, so that methods can be found which will allow on the one hand local improvement of the structural elements suffering from the presence of masonry infills, while favouring and promoting the extensive use of masonry infills in building construction on the other.

5.10 GENERAL REMARKS ON THE ANALYSIS OF THE STRUCTURAL SYSTEM

1. During recent years there has been a tendency towards dynamic analysis instead of the equivalent static analysis, perhaps because of the availability of computational tools which have made such an approach feasible.

2. Therefore, EC8 sets very tight limits on the cases for which the equivalent static analysis is allowed, tighter than in any other national code.
3. These requirements, in combination with the imposed accidental eccentricities and the consideration of the seismic excitation as acting simultaneously in two directions, render the dimensioning of a structural element the result of multiple loading combinations (32 for the dimensioning of each section of a column subjected to biaxial bending). Therefore, the whole procedure can effectively be realized only with the aid of efficient computational tools.
4. The above, combined with the traditional educational background of the structural engineer, consisting mainly of statics, lead to the loss of the ability of the engineer to check the reliability of the results of the analysis.
5. Thus, there is a great need for developement of methods of analysis which would provide intermediate key results for checking the output of dynamic analysis or methods which would provide upper and lower limits on the expected results. In fact, we must not forget that civil engineering works are mainly works of large scale, each with its own characteristics. Therefore, methods used for the analysis and design of industrial products (i.e. cars, aircrafts, etc.) where the cost of multiple cross-checkings and laboratory tests are distributed among thousands of products, cannot be applied to civil works. This necessitates in structural engineering the development of methods for easy and low cost control of the resulting analytical outputs.
6. A lot of research must be carried out on the degree of improvement of the final confidence limits of the resulting action effects through the application of the provisions of modern codes, particularly if an integrated approach is used, which takes into account not only the reliability of action effects but also the reliability of the seismic actions introduced in the analysis.
7. The modern tendency towards complicated analysis procedures should not create in the engineer the impression that this is the key to earthquake resistant structures. It should not be forgotten that analysis is only one component among a large number of factors which affect the behaviour of the structure, such as the materials, the supervision of the works, the execution and so on, and therefore only equally weighted attention to all these factors will guarantee the quality of the structure. It has to be understood that modern analysis approaches merely take advantage of the computing potential provided by computers for a more precise approach to the problem from only one point of view.
8. Independently of the general remarks above, it is noted that, given the possibilities provided by modern computational tools, it seems that the most realistic approach for the analysis of concrete structures is that of the pseudospatial frame to static or dynamic seismic action accompanied by an automatic superposition of load effects. The code permission for use of plane models can be used efficiently only in special cases.
9. The approach of EC8 towards masonry infills seems to be completely negative, forcing engineers to avoid the use of this type of infill in construction, despite the favourable, so far, results of laboratory and field research on the subject. It is the authors' opinion that a reconsideration of the whole issue will be inevitable in the near future.
10. During the last few years there has been a systematic attempt to develop inelastic methods of analysis of plane or three-dimensional systems which are

subject to a seismic excitation (Kanaan and Powell, 1975; Kappos and Penelis, 1986, 1989; Michailidis, Stylianidis and Kappos, 1995). The results of these attempts are being used, for the time being, for parametric analysis of frame or dual systems, aiming at the evaluation of the reliability of the equivalent static or the multimodal response spectrum analysis. They are also used for the estimation of several other values of special interest in analysis and design, as for example the influence of masonry infills (Michailidis, Stylianidis and Kappos, 1995), the behaviour factor q' (Kappos, 1991), the 'damage coefficient' (Park *et al.*, 1985; Kappos, Stylianidis and Penelis, 1991) and so on.

5.11 REFERENCES

Anastassiadis, K. (1993) Directions sismiques défavorables et combinaisons défavorables des efforts. *Annales de l' ITBTP*, No. 512.

AUT, Reinforced Concrete Lab (1986) *ANTIS-1-2-3. A Computer Code for Aseismic Analysis and Design*, Aristotle Univ. of Thessaloniki, Greece.

Biggs, J.M. (1964) *Introduction to Structural Dynamics*, McGraw-Hill, New York.

CEB (1985) Model code for seismic design of concrete structures. *Bulletin d' Information CEB*, **165**, Lausanne.

CEN Techn. Comm. 250/SC8 (1995) *Eurocode 8: Earthquake Resistant Design of Structures–Part 1: General Rules and Rules for Buildings (ENV 1998-1-1)*, CEN, Berlin.

Clough, R.W. and Penzien, J. (1975) *Dynamics of Structures*, McGraw-Hill, New York.

Dowrick, D.J. (1987) *Earthquake Resistant Design for Engineers and Architects*, John Wiley & Sons, New York.

Gupta, A.K. (1990) *Response Spectrum Method, in Seismic Analysis and Design of Structures*, Blackwell Scientific Publ., Cambridge, Mass.

Gupta, A.K. and Chu, S.L. (1977) Probable simultaneous response by the response spectrum method of analysis. *Nuclear Engng and Design*, **44**, 93–7.

Gupta, A.K. and Singh, M.P. (1977) Design of column sections subjected to three components of earthquakes, *Nuclear Engng and Design*, **41**, 129–33.

ICBO (Int. Conf. of Building Officials) (1994) *Uniform Building Code–1994 Edition*, Vol **2**: *Structural Engineering Design Provisions*, Whittier, Calif.

Kanaan, A.E. and Powell, G.H. (1973, 1975) *DRAIN-2D: A General Purpose Computer Program for Dynamic Analysis of Inelastic Plane Structures,* Rep. EERC-73/6, EERC-73/22, Univ. of California, Berkeley.

Kappos, A.J. (1991) Analytical prediction of the collapse earthquake for R/C buildings: suggested methodology. *Earthq. Engng and Struct. Dynamics*, **20**(2),167–76.

Kappos, A.J. and Penelis, G.G. (1986) Discussion of influence of concrete and steel properties on calculated inelastic seismic response of reinforced concrete frames. *ACI Journal*, **83**(1), 167–69.

Kappos, A.J. and Penelis, G.G. (1989) Evaluation of the inelastic seismic behaviour of R/C buildings designed by CEB model code. *Proceed. 9th World Conf. on Earthq. Engng,* Tokyo-Kyoto, Japan, 1988, Maruzen, **V**, pp.1137–42.

Kappos, A.J., Stylianidis, K.C. and Penelis, G.G. (1991) Analytical Prediction of the Response of Structures to Future Earthquakes. *European Earthq. Engng*, **5**(1), 10–21.

Leblond, L.(1980) Calcul sismique par la méthode module utilisation de réponses pour le dimensionnement. *Annales de l' ITBTP*, No. 380.

Luft, R. (1989) Comparisons among earthquake codes, *Earthq. Spectra*, **5**(4), 767–89.

Macleod, I.A. (1970) *Shear Wall–Frame Interaction*, Portland Cement Association, Skokie, Ill.

Michailidis, C.N., Stylianidis, K.C. and Kappos, A.J. (1995) Analytical modelling of masonry infilled R/C frames subjected to seismic loading. *Proceed. 10th European Conf. on Earthq. Engng*, Vienna, Austria, 1994, Balkama, **3**, pp.1519–24.

Ministry for Environment and Public Works (1992) *Greek Code for Earthquake Resistant Structures*, Athens.

Panetsos, P. and Anastassiadis, K. (1994) Design of R/C elements under seismic action. *Proceed. of 11th Greek Conf. on Concrete*, Technical Chamber of Greece, Corfu, **2**, pp. 267–81 (in Greek).

Park, Y.-J. *et al.* (1985) Mechanistic seismic damage model for reinforced concrete. *ASCE Journal of Struct. Engng*, **111**(4), 722–39.

Paulay, T., Bachmann, H. and Moser, K. (1990) *Erdbebenbemessung von Stahlbetonhochbauten*, Birkhäuser, Basle.

Penelis, G.G. (1971) Die Knickung Raümlicher Mehrstockiger Rahmenträger. *Der Bauingenieur*, **11**, 393–9.

Penelis, G.G., Sarigiannis, D., Stavrakakis, E. and Stylianidis, K.C. (1989) A statistical evaluation of damage to buildings in the Thessaloniki, Greece earthquake of June 20, 1978. *Proceed of 9th World Conf. on Earthq. Engng*, Tokyo-Kyoto, Japan, 1989, Maruzen, **7**, pp. 187–92.

Polyakov, S. (1974) *Design of Earthquake Resistant Structures*, Mir Publ., Moscow.

Rosenblueth, E. and Contreras, H. (1977) Approximate design for multicomponent earthquakes. *Journal of Engng Mech., ASCE*, **103**, EM5.

Roussopoulos, A. (1956) *Earthquake Resistant Structures*. NTU, Athens (in Greek).

Rutenberg, A. (1992) Nonlinear response of asymmetric building structures and seismic codes: a state of the art review. *European Earth. Engng*, **6**(2), 3–19,

Stylianidis, K., Athanassiadou, C., Athanassiadou, L. and Prountzos, K. (1994) Parametric study and evaluation of some provisions of the Greek code for earthquake resistant structures. *Proceed. 11th Greek Conf. on Concrete*, Technical Chamber of Greece, Corfu, **3**, pp. 235–50 (in Greek).

Tassios, T.P. (1984) Masonry infill and R/C walls under cyclic actions. *CIB Symposium on Wall Structures*, Invited State-of-the-Art Report, Warsaw.

Valanidis, C. and Penelis, G.G. (1987) Approximate analysis of pseudospatial systems under lateral loading. *8th Greek Conference on Concrete*, Technical Chamber of Greece, Kavala-Xanthi, **2**, pp. 174–92 (in Greek).

Valiasis, T. and Stylianidis K.C. (1989) Masonry infilled R/C frames under horizontal loading. Experimental results. *European Earthq. Engng*, **3**(3), 10–20.

Valiasis, T., Stylianidis, K.C. and Penelis, G.G. (1993) Hysteresis model for weak brick masonry infills in R/C frames under load reversals. *European Earthq. Engng*, **7**(1), 3–9.

Wilson, E. and Button, M. (1982) Three-dimensional dynamic analysis for multicomponent earthquake spectra. *Journal of Earthq. Engng and Struct. Dynamics*, **10**, 471–6.

Wilson, E. and Habibullah, A. (1987) Static and dynamic analysis of multistorey buildings including P–δ effects. *Earthq. Spectra*, **3**(2), 289–98.

Zararis, P., Salonikios, T. and Botis, K. (1994) Dynamic analysis and comparative design of R/C structures. *Proceed. 11th Greek Conf. on Concrete*, Technical Chamber of Greece, Corfu, **1**, pp. 411–21 (in Greek).

6

Design action effects – safety verifications

6.1 THE DESIGN ACTION EFFECTS – CAPACITY DESIGN PROCEDURE

6.1.1 General

Earthquakes belong to the category of accidental actions, therefore:

- They are not combined with other accidental actions.
- The earthquake loading is combined with gravity loads (equation (5.57).

According to the procedure described in detail in Chapters 4 and 5, the action effects are found in a deterministic way, as if the loads were statistically reliable and the response of the structure was in the elastic range. These two weak points in the calculation of the action effects render necessary a more reliable approach to the problem, which would ensure the existence of adequate strength and ductility in the structure.

Indeed, since it is impossible to predict with accuracy the characteristics of the ground motion due to a large earthquake, it is impossible to estimate with accuracy the response of the R/C structure to this earthquake. However, it is possible to provide the structure with the features that will ensure the most desirable behaviour. In terms of ductility, energy dissipation, damage or failure, this means that the sequence in the breakdown of the chain of resistance of the structure will follow a desirable hierarchy (Park and Paulay, 1975). In order to ensure a certain sequence in the failure mechanism of the resistance chain, the resistance of every link should be known. This knowledge should not be based on assumptions of disputable reliability, but on the calculated strength of the structural elements which will be subjected to very large deformations (due to formation of plastic hinges) during a catastrophic earthquake.

Although the nature of the design actions is probabilistic, the ability to have a deterministic allocation of strength and ductility in the structural elements provides an effective tool for ensuring a successful response and prevention of collapse during a catastrophic earthquake. Such a response can be achieved if the successive regions of energy dissipation are rationally chosen and secured through a proper design procedure, so that the predecided energy dissipation mechanism would hold throughout the seismic action. This design concept can be included in a procedure which is called the **capacity design procedure**.

According to the capacity design procedure, the structural elements which are designated to dissipate the seismic energy are reinforced accordingly, while other members with adequate reserve strength are provided so that it is ensured that the chosen dissipating mechanism is preserved during the seismic cyclic deformation of the structure, without serious reduction of strength in the critical regions. This means that the action effects which have resulted from the analysis serve only as a guide and they are properly modified in order to accommodate the capacity design of the structure. Of course this modification is made in a way that the cost increase is kept within acceptable limits. It is evident that this modification should also be a function of the selected design ductility class as will be explained below.

The aforementioned concepts have been embodied in all modern codes for earthquake-resistant design of structures and, of course, in EC8 (CEN, 1995).

6.1.2 Design criteria influencing the design action effects

From the design criteria for R/C structures included in EC8 only

- the local resistance criteria and
- the capacity design criteria

influence the determination of the design action effects. All the others refer to dimensioning and detailing of the R/C structural elements, therefore they will be dealt with in Chapters 8 and 9. The design criteria influencing the design action effects are, in detail, the following:

1. All critical regions of the structure must exhibit resistance adequately higher than the action effects produced in these regions under the seismic design situation.
2. Brittle or other undesirable failure modes, i.e.

   (a) shear failure of the structural elements
   (b) failure of beam–column joints
   (c) yielding of foundations, or
   (d) yielding of any other element intended to remain elastic

   must be excluded. This can be ensured if the design action effects of purposely selected regions are derived from equilibrium conditions when flexural plastic hinges with their possible overstrengths have occurred in adjacent areas.
3. Extensive distribution of plastic hinges, avoiding their concentration in any single storey ('soft storey' mechanism) is ensured if the formation of plastic hinges at both ends of at least some columns on the same storey is prevented. This can be achieved if – with sufficient reliability – it is ensured that plastic hinges develop only in beams and not in columns, except for the unavoidable formation of plastic hinges at the base of the building.

The implementation of these criteria for the determination of the design action effects of the various structural elements of a structure is given below.

6.1.3 Capacity design procedure for beams

According to EC8, CEB/MC-SD/85 (CEB, 1985) and SEAOC (1990) the design values of the bending moments of beams for all ductility classes are obtained from the analysis of the structure for the seismic loading combinations, as described in detail in Chapter 5 without any modification, except for a possible redistribution.

However, according to all relevant codes beams need an additional reinforcement at their support, a compression reinforcement equal to 50% of the corresponding tension reinforcement, in order to ensure an adequate ductility level (Chapter 8). Based on the capacity design concept these reinforcement bars are appropriately anchored in concrete, so that they can operate as tension reinforcement in case of moment reversal. Therefore, the moment resistance envelope of the beams is considerably improved at low cost (the cost of anchorages of the compression reinforcement) no matter what are the values of the design action effects which have been derived from the analysis (Figure 6.1(a)). This means that the beam, as it is designed, can carry much larger moment fluctuations generated

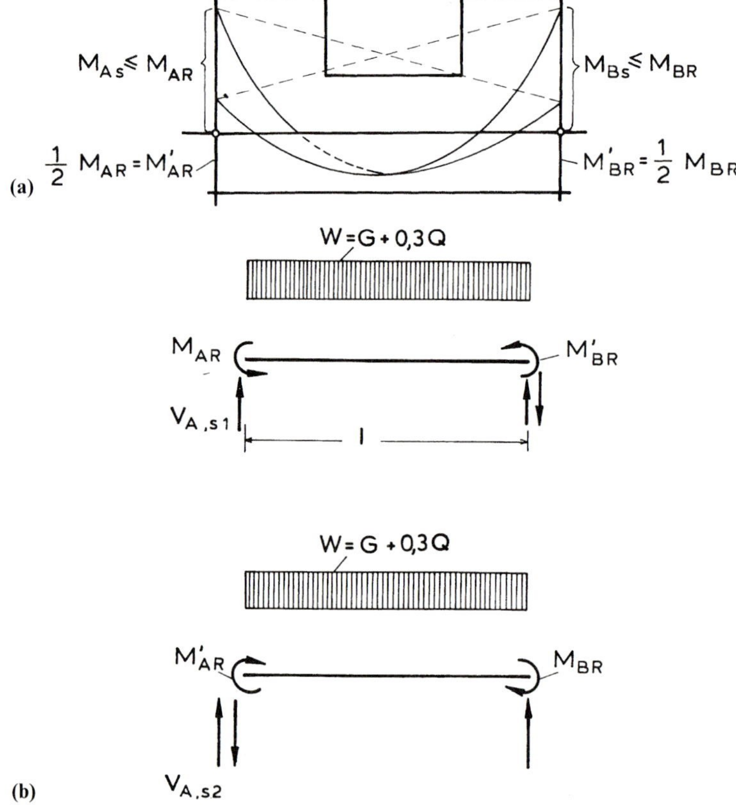

Figure 6.1 Capacity design values of shear forces acting on beams: (a) resisting moment envelope; (b) equilibrium conditions for the determination of shear forces.

by an earthquake than the design action moments. However, in order to ensure this behaviour, the structural element has to be secured against premature shear failure, because it is well known (see also Chapter 8) that shear failure does not present a ductile mode. Therefore, the design shear, at least for DC 'H' should not be that resulting from the analysis, but the shear corresponding to the equilibrium of the beam under the appropriate gravity load and a rational adverse combination of the actual bending resistances of the cross-sections (Figure 6.1(b)).

$$\left.\begin{aligned} V_{A,S1} &= \frac{wl}{2} + \gamma_{Rd}\frac{M_{AR}+M'_{BR}}{l} \\ V_{A,S2} &= \frac{wl}{2} - \gamma_{Rd}\frac{M_{BR}+M'_{AR}}{l} \\ V_{B,S1} &= -\frac{wl}{2} + \gamma_{Rd}\frac{M_{AR}+M'_{BR}}{l} \\ V_{B,S2} &= -\frac{wl}{2} - \gamma_{Rd}\frac{M_{BR}+M'_{AR}}{l} \end{aligned}\right\} \qquad (6.1)$$

where M_{AR}, M'_{AR}, M_{BR}, M'_{BR} are the actual resisting moments at the hinges accounting for the actual area of the reinforcing steel (all positive) and γ_{Rd} the amplification factor taking into account the reduced probability that all end cross-sections exhibit simultaneously the same overstrength. This γ_{Rd}-factor also counterbalances the partial safety factor γ_s of steel chosen for the fundamental load combination (section 6.2.2) and covers the hardening effects as well. In the absence of more reliable data, γ_{Rd} may be taken as

$$\gamma_{Rd} = 1.25$$

The sign of the ratio

$$\boxed{\zeta = \frac{V_{A,S2}}{V_{A,S1}} \quad \text{or} \quad \frac{V_{B,S1}}{V_{B,S2}}} \qquad (6.2)$$

has a considerable effect on the shear design of the beam, as will be explained in Chapter 8.

This capacity design procedure, according to EC8, applies only to DC 'H'. For DC 'M' and 'L' the design values of the acting shear forces are obtained from the analysis of the structure for the seismic load combination, as described in detail in Chapter 5.

6.1.4 Capacity design procedure for columns

(a) Bending

It has already been stressed that the formation of plastic hinges in the columns during an earthquake should be avoided, in order to make sure that the seismic energy is dissipated by the beams only (Park, 1986). The reasons for this requirement are the following:

1. Due to axial compression, columns have less available ductility than beams. On the other hand, for the same displacement of the frame (Figure 6.2), that

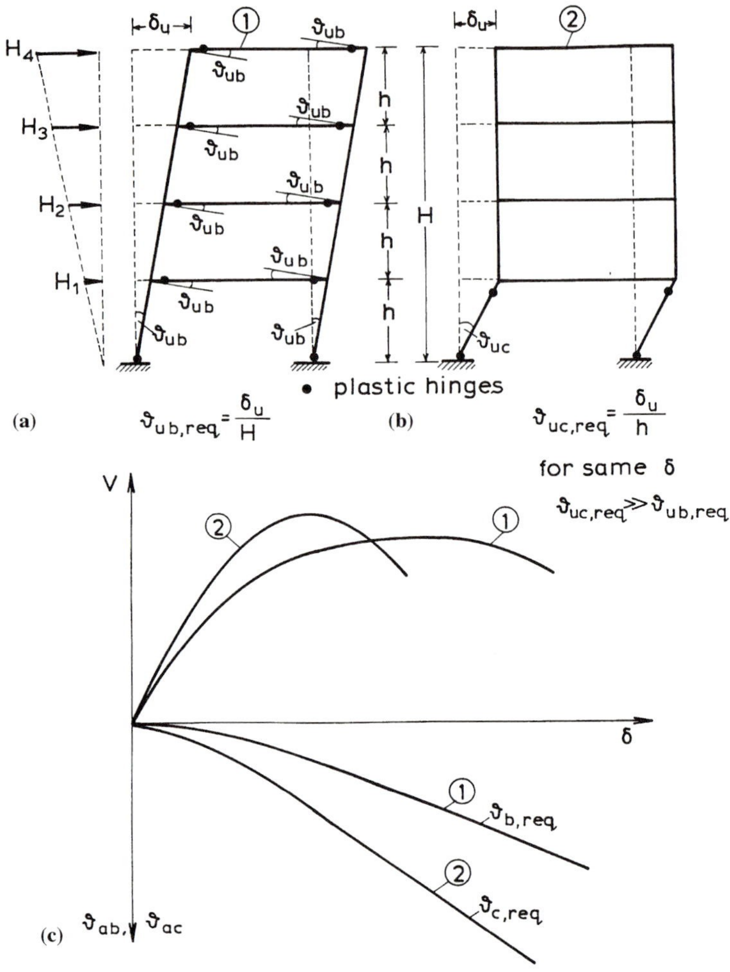

Figure 6.2 Failure mechanisms of a frame: (a) beam mechanism; (b) storey mechanism; (c) V–δ–θ diagram.

is, for the same frame ductility expressed in terms of displacements, much larger plastic column rotations are required than beam rotations. Therefore, for the same frame ductility, a larger column ductility expressed in rotations is required for the creation of a column failure mechanism than the beam ductility needed for the creation of a beam failure mechanism. Thus, while (Park and Paulay, 1975)

$$\theta_{\mathrm{uc}}^{\mathrm{avail}} \ll \theta_{\mathrm{ub}}^{\mathrm{avail}}$$

for the same δ_{u} (i.e. the same μ_{req})

$$\theta_{\mathrm{uc}}^{\mathrm{required}} = \frac{\delta_{\mathrm{u}}}{h} \gg \theta_{\mathrm{ub}}^{\mathrm{required}} = \frac{\delta_{\mathrm{u}}}{H}$$

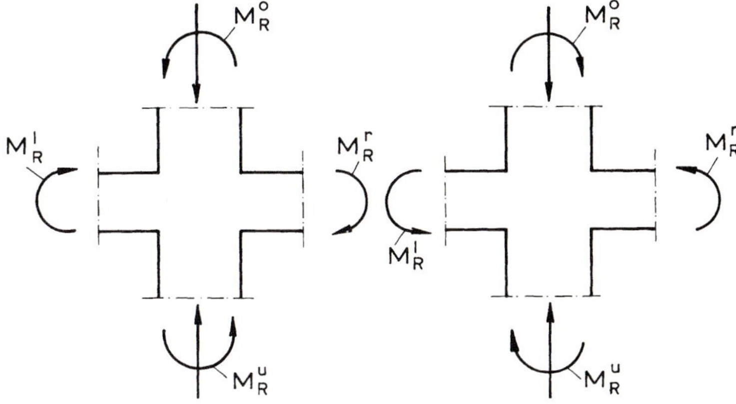

Figure 6.3 Strong columns–weak beams.

2. While beam failure exhibits extended cracking only in the tension zones due to the yielding of the reinforcement, column failure mode successively presents spalling of the concrete, breaking of the ties, crushing of the concrete core and buckling of the longitudinal reinforcement bars. This process leads to the creation of a collapse mechanism due to the inability of the columns to carry the axial gravity loads after their failure. Therefore, avoiding column failure is much more crucial for the overall safety of the structure than avoiding beam failure.
3. The formation of plastic hinges in the columns may lead to significant interstorey drifts, so that the relevant second-order effects may cause the collapse of the structure.

In order to decrease the probability of plastic hinge formation in the columns, frames must be designed to have 'strong columns and weak beams' (Park, 1986; Paulay, Bachmann and Moser, 1990; Priestley and Calvi, 1991). This concept is realized in the requirements of EC8 and other relevant codes stating that the sum of the resisting moments of the columns, taking into account the action of N, should be greater than the sum of the resisting moments of all adjacent beams for each (positive or negative) direction of the seismic action (Figure 6.3), that is

$$\left.\begin{aligned}|M^{o}_{R1}|+|M^{u}_{R1}|&\geq \gamma_{Rd}(|M^{l}_{R1}|+|M^{r}_{R1}|)\\ |M^{o}_{R2}|+|M^{u}_{R2}|&\geq \gamma_{Rd}(|M^{l}_{R2}|+|M^{r}_{R2}|)\end{aligned}\right\} \tag{6.3}$$

where γ_{Rd} is a factor which takes into account the variability of the yield stress f_y and the probability of strain-hardening effects in the reinforcement (overstrength factor).

Therefore, the capacity design is satisfied if the columns are designed for the following moments:

$$\left.\begin{aligned}M_{S1,CD}&=\alpha_{CD,1}M_{S1}\\ M_{S2,CD}&=\alpha_{CD,2}M_{S2}\end{aligned}\right\} \tag{6.4}$$

where

$$\left.\begin{aligned}\alpha_{\mathrm{C.D.1}} &= \gamma_{\mathrm{Rd}} \frac{|M^{\mathrm{l}}_{\mathrm{R1}}| + |M^{\mathrm{r}}_{\mathrm{R1}}|}{|M^{\mathrm{o}}_{\mathrm{S1}}| + |M^{u}_{\mathrm{S1}}|} \\ \alpha_{\mathrm{C.D.2}} &= \gamma_{\mathrm{Rd}} \frac{|M^{\mathrm{l}}_{\mathrm{R2}}| + |M^{\mathrm{r}}_{\mathrm{R2}}|}{|M^{\mathrm{o}}_{\mathrm{S2}}| + |M^{u}_{\mathrm{S2}}|}\end{aligned}\right\} \tag{6.5}$$

EC8 allows a relaxation of the above capacity design criterion whenever the probability of full reversal of beam end-moments is relatively low (see application in section 8.6.3). The following cases are also exempted from the requirements of the above procedure:

- in single or two-storey buildings and in the top storey of multi-storey buildings;
- in one-quarter of the columns of each storey in plane frames with four or more columns.

The design bending moments for DC 'H' are determined according to the above described capacity design criterion with

$$\boxed{\gamma_{\mathrm{Rd}} = 1.35}$$

For DC 'M' the design bending moments are determined according to the same procedure, with

$$\boxed{\gamma_{\mathrm{Rd}} = 1.20}$$

Finally, for DC 'L' the design bending moments are determined from the analysis of the structure for the seismic load combination without any application of the capacity design criterion.

The magnification factor α_{CD} (equation (6.5)) takes rather high values. In the example of Figure 6.4 where a plane frame has been analysed for gravity loads '+' seismic action, the values of α_{CD}, for DC 'M' range from 1.35 to 1.56.

(b) Shear

Shear forces according to the capacity design criterion, and following the rationale developed for the beams (section 6.1.3), are determined by considering the equilibrium of the column under the actual resisting moments at its ends, as follows (Figure 6.5):

$$V_{\mathrm{Sd.CD}} = \gamma_n \frac{M_{\mathrm{DRd}} + M_{\mathrm{CRd}}}{l_{\mathrm{C}}} \tag{6.6}$$

where γ_{n} accounts for the lower probability of all failure modes accepted for the columns, even if their ends exhibit flexural plastification. Practically, γ_{n} may take the values of γ_{Rd} used in each case.

The design shear forces for DC 'H' are determined according to the capacity design criterion developed above with

$$\boxed{\gamma_{\mathrm{Rd}} = 1.35}$$

For DC 'M' shear forces are determined according to the same procedure

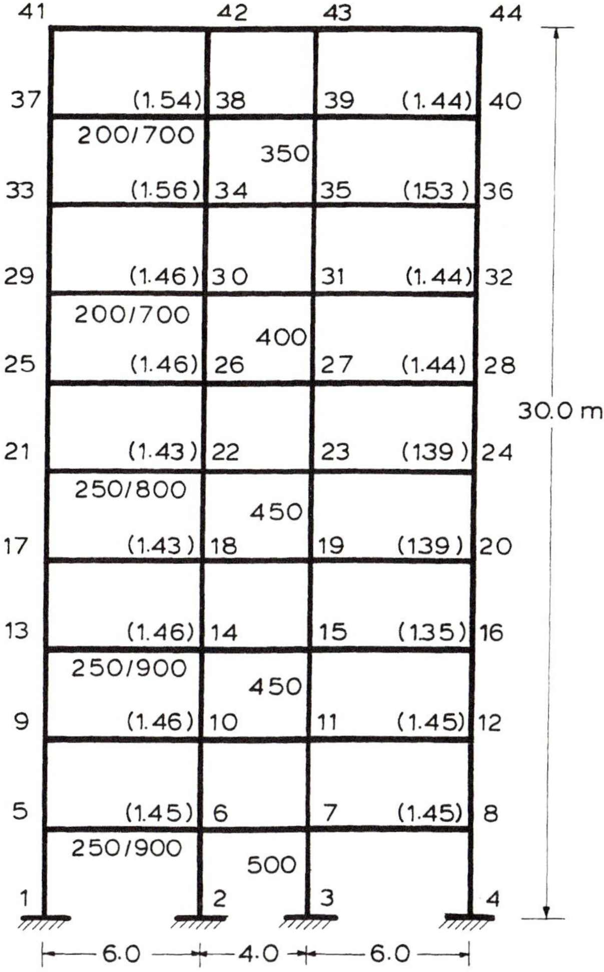

Figure 6.4 Values of the magnification factors a_{CD} for a 10-storey 4-column R/C frame.

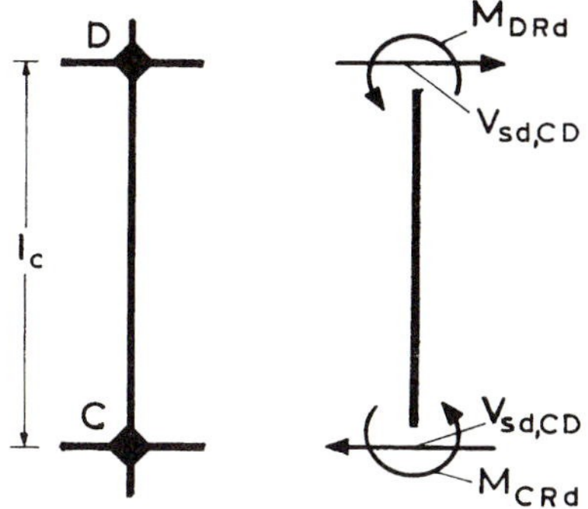

Figure 6.5 Capacity design value of shear forces acting on columns.

with

$$\boxed{\gamma_{Rd} = 1.20}$$

Finally, for DC 'L' the design action shear forces are determined by the analysis of the structure for the seismic load combination without any application of the capacity design criterion.

6.1.5 Capacity design procedure for shear walls

(a) Bending

The moment diagrams for slender shear walls ($h_w/l_w > 2$) in dual systems under static seismic action have the form of Figure 6.6. However, the dynamic response analysis results in moment diagrams with approximately linear variation. Thus, the design moment diagram introduced by EC8 has the form of a trapezoid covering the saw-like M-diagram (Paulay, 1986). The vertical displacement h_{cr} of the envelope aims at ensuring that inelastic deformations, i.e. curvature ductility demands during an earthquake, will be restricted to the base of the wall (capacity design). The values of h_{cr} above the base of the wall may be estimated as follows (Figure 6.7):

$$h_{cr} = \max(l_w, h_w/6) \tag{6.7}$$

but

$$h_{cr} \leq \begin{cases} 2l_w \\ \begin{cases} h_s \text{ for } n \leq 6 \text{ storeys} \\ 2h_s \text{ for } n \geq 7 \text{ storeys} \end{cases} \end{cases} \tag{6.8}$$

The above design moment envelope applies to all ductility categories.

For squat walls ($h_w/l_w \leq 2$) there is no need for a design envelope of bending moments.

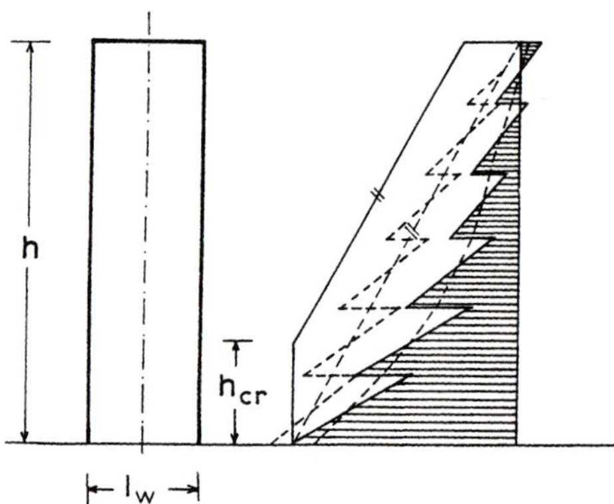

Figure 6.6 Shear wall moment diagram for capacity design procedure.

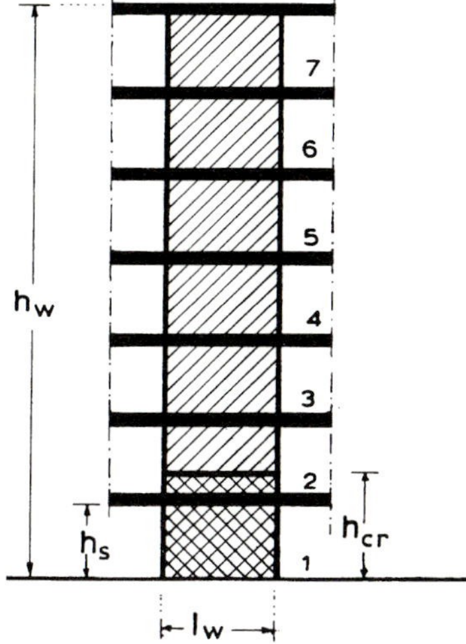

Figure 6.7 Critical region at the base of the wall.

(b) Shear

In dual systems containing slender walls – in order to account for the uncertainties in the contribution of higher modes – a modified design envelope of the shear forces is adopted (Figure 6.8). In addition, inelastic analyses performed so far (Eibl and Keintzel, 1988) have shown that the resulting shears are much higher than the shears derived from an elastic response analysis. For this reason the design envelope of the shear forces along the height of the wall is derived as fol-

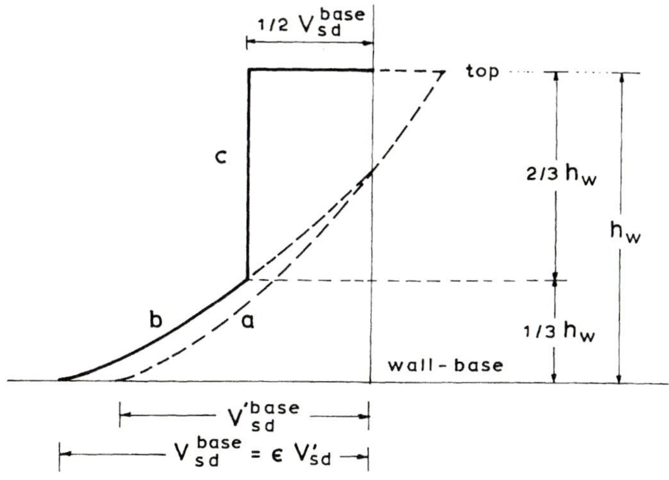

Figure 6.8 Design envelope for shear forces in slender walls of dual systems.

lows (application of the capacity design criterion):

$$V_{Sd} = \varepsilon V'_{Sd} \tag{6.9}$$

for $z < 1/3h_w$ while for $1/3h_w < z < h_w$ the variation is linear according to Figure 6.8. In the above equation V'_{Sd} is the shear force along the height of the wall, obtained from the analysis and ε the magnification factor, depending on the ductility class as follows:

- For DC 'H' and 'M' the magnification factor ε may be estimated as follows

$$\varepsilon = q\left[\left(\frac{\gamma_{Rd}}{q}\frac{M_{Rd}}{M_{Sd}}\right)^2 + 0.1\left(\frac{(S_e(T_c)}{S_e(T_1)}\right)^2\right]^{1/2} \leqslant q \tag{6.10}$$

  where M_{Sd} is the design bending moment at the base of the wall, M_{Rd} the design flexural resistance at the base of the wall, γ_{Rd} a factor equal to 1.25 for DC 'H' and 1.15 for DC 'M', T_1 the fundamental period of vibration of the building, T_c the upper limit period of the constant spectral acceleration branch (section 4.4.4) and S_e (T) the ordinate of the elastic response spectrum.
- For DC 'L' the magnification factor ε may be taken equal to 1.3.

In the case of squat walls the design shears V_{Sd} should be taken as follows:

- For DC 'H' and 'M'

$$V_{Sd} = \gamma_{Rd}\left(\frac{M_{Rd}}{M_{Sd}}\right)V'_{Sd} \leqslant qV'_{Sd} \tag{6.11}$$

- For DC 'L' the shear force V'_{Sd} may be increased by a magnification factor equal to 1.3.

6.1.6 Capacity design for the connecting beams of the footings

It has already been discussed in detail that seismic design, dynamic or equivalent static, is based on the assumption that the motion of all footings is in phase. As a result, there are no relative displacements between the footings and consequently there are no action effects (M, N, V) on the connecting beams, therefore their existence is not justified from the seismic design point of view. However, according to what has been discussed in section 4.3 their existence is necessary to guarantee to a certain degree that the assumptions of the design will be satisfied. MC-SD/85 does not provide any suggestions for the design procedure of these beams. However, a series of codes such as EC8, SEAOC (1975, 1990), the Greek Seismic code (Ministry for Environment and Public Works, 1992) and others include special provisions for these connecting beams.

The simplest one is the SEAOC code specifying for each connecting beam a tension or compression normal force equal to 10% of the largest axial load of the columns that the beam connects. EC8 specifies that every connecting beam should be designed for a tension or compression normal force equal to the largest shear that the connected columns can carry (capacity design).

6.2 SAFETY VERIFICATIONS

6.2.1 General

As already mentioned in section 4.2, in order to satisfy the fundamental requirements of 'no collapse' and 'damage limitation', three compliance criteria should be considered, that is:

- ultimate limit state
- serviceability limit state
- specific measures.

For buildings of importance categories II to IV (section 4.4.6) the requirements of earthquake-resistant design may be considered satisfied when the total base shear due to seismic action combination, calculated with a behaviour factor $q = 1.0$, is less than that due to the other relevant combinations. This means that in regions of low seismic hazard, where wind or other horizontal loading subjects the structure to base shears higher than that caused by the seismic actions (for $q = 1.0$), the design of the structure may be carried out without taking into account the seismic actions, except for some special measures specified in EC8. In seismic regions where the preceding conditions are not fulfilled, the following verifications must be considered.

6.2.2 Ultimate limit state

Safety against collapse is considered to be ensured if the following conditions are met:

- resistance condition
- second-order effects
- ductility condition
- equilibrium condition
- resistance of horizontal diaphragms
- resistance of foundations
- seismic joint conditions.

The resistance condition is satisfied if for every structural element the following relations are valid:

$$\boxed{\begin{aligned} E_{d1} &\leqslant R_d \\ E_{d2} &\leqslant R_d \end{aligned}} \qquad (6.12)$$

where E_{d1} are the design action effects for regular load combinations without seismic action (EC2–CEN, 1991) The simplest form of $E_{d1} = E(1.35G\text{ '+' }1.50Q)$. E_{d2} are the design action effects on the structural element for gravity loads and seismic actions that have resulted by also taking into account the capacity design conditions and R_d is the corresponding design resistance of the same element.

For the determination of R_d the partial safety factors for concrete γ_c and for steel γ_s may be taken from Eurocode 2 for the fundamental load combination, that is

$$\gamma_s = 1.15$$
$$\gamma_c = 1.50$$

The choice of the above values is based on the assumption that, due to the local ductility provisions, the ratio between the residual strength after degradation and the initial strength is roughly equal to the ratio between the γ_n-values for accidental and fundamental load combinations.

The consideration of second-order effects (P–δ effects) has been thoroughly discussed in section 5.8.

The ductility condition is satisfied through:

- specific material-related requirements;
- maximum–minimum requirements for the reinforcement percentage of the R/C elements;
- appropriate reinforcement detailing;
- application of capacity design procedure.

The above requirements are closely related to the selected ductility class of the structure. These requirements will be presented and discussed in detail in Chapters 8 and 9 for every type of structural component, in connection with all necessary experimental and analytical evidence.

Equilibrium condition refers to the stability of the building under the set of actions given by the combination rules described in section 4.5. Effects such as overturning and sliding are also included. It is obvious that these types of verification refer mainly to slender structures (overturning problems), or to structures or structural elements susceptible to sliding.

The resistance of horizontal diaphragms refers to their ability to transmit with sufficient overstrength the design seismic action effects to the various lateral load-resisting systems to which they are connected. As already discussed in section 4.3, R/C slab diaphragms without re-entrant corners or discontinuities in plan, present a high level of diaphragmatic resistance. In the case, however, that a resistance verification of the diaphragm has been decided, the forces obtained from the analysis as acting on the diaphragm must be multiplied by an overstrength factor

$$\gamma_f = 1.30$$

for the relevant resistance verifications.

Resistance of foundations is verified for action effects derived on the basis of capacity design considerations which take into account the development of possible overstrength, but they need not exceed the action effects which correspond to the response of the structure for $q = 1.0$ (elastic range).

Seismic joint condition is satisfied through proper joints between adjacent buildings necessary to protect them from earthquake-induced collisions (section 6.2.3).

At this point, it should be mentioned that for the time being, there is no code which provides a direct computational procedure for securing a structure against collapse during a high-intensity earthquake. However, it is considered that a series of provisions for the design of a structure that have already been analysed, such as appropriate configuration in plan and in elevation, correct analysis based on a reliable design spectrum, conformation of the seismic design effects to the capacity design considerations, and design verifications according to the code provisions, constitute a reliable set of conditions that secure the structure against collapse. During the last 10–15 years an attempt has been made to control the collapse limit state quantitatively, using inelastic seismic analysis combined with the introduction of 'damage indices'; however, these attempts are still at a research stage (Park *et al.*, 1985; Kappos, Stylianidis and Penelis, 1991).

6.2.3 Serviceability limit state

The task of the serviceability limit state verification is to ensure the protection of non-structural elements (masonry, glass panels, tiles and so on) from premature failure under seismic actions with a higher probability of occurrence than the design seismic action. The main factor that affects the behaviour of these elements is the ratio of the inter-storey drift d_r to the interstorey height h, which is induced in them by the attached R/C components of the structure (Figure 6.9) (Uang and Bertero, 1991). The reason for this is the fact that masonry and other brittle materials fail at much lower inter-storey drift d_r/h than the surrounding R/C frames (Figure 6.10).

From the above it is concluded that, for the verification of the serviceability limit state, the displacement due to the design seismic actions must be calculated and then reduced so that a lower return period is taken into account.

The displacements due to the design seismic action could be the result of the elastic deformation of the structural system magnified by the behaviour factor q, since the ordinates of the design response spectrum have been divided by q,

$$d_s = q\, d_e \tag{6.13}$$

where d_s is the displacement of a point of the structural system induced by the design seismic action, q the behaviour factor specified in section 4.4.5 and d_e the

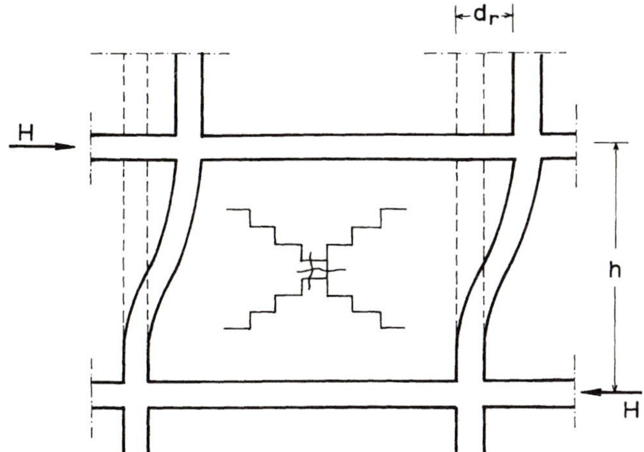

Figure 6.9 Masonry failure with X-shaped cracks due to the R/C frame inter-storey drift.

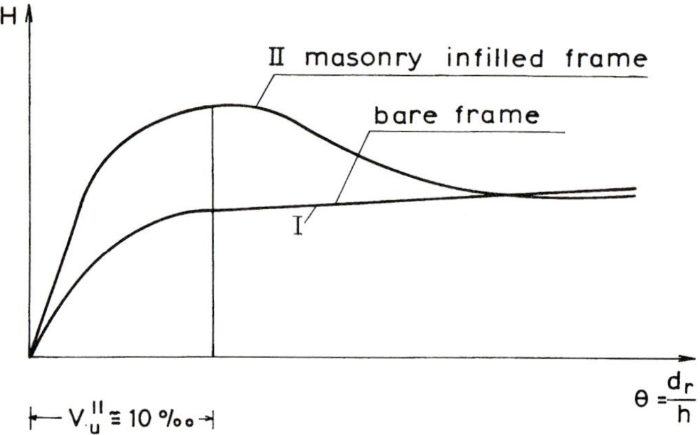

Figure 6.10 H–θ diagram of a masonry-infilled frame.

displacement of the same point as determined by the linear analysis based on the design response spectrum (elastic response spectrum divided by q and multiplied by the importance factor γ_I).

In order to ensure the serviceability limit state, according to EC8 the interstorey drift must fulfil the following limits:

- for buildings having non-structural elements of brittle materials (e.g. tiles, masonry) attached to the structure

$$\frac{d_s}{v} \leqslant 0.004h \tag{6.14}$$

- for buildings having non-structural elements fixed in such a way that they do not interfer with structural deformations (Figure 6.11)

$$\frac{d_s}{v} \leqslant 0.006h \tag{6.15}$$

where d_s is the design inter-storey drift, h the storey height and v the reduction factor taking into account the lower return period of the seismic event associated with the serviceability limit state.

The reduction factor v also depends on the importance category of the building. Values of v are given in Table 6.1.

It should be noted that the limits on the inter-storey drift adopted by EC8 are more strict than those of CEB/MC-SD/85 and of many other national codes (Greek Code for Earthquake Resistant Structures – Ministry for Environment and Public Works, *1992*; Luft, *1989*).

6.2.4 Specific measures

It has already been mentioned in section 4.2 that in order to limit the uncertainties related to the behaviour of structures under the design seismic action and to promote good behaviour under seismic actions more severe than the reference

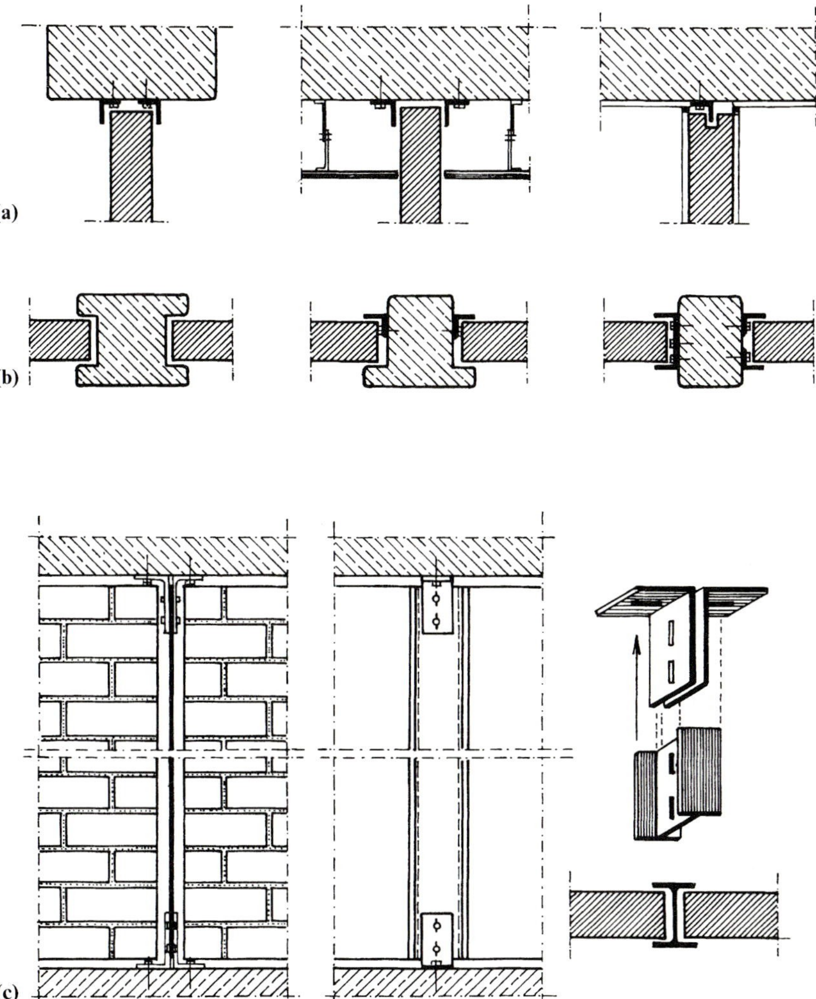

Figure 6.11 Detailing of separation joints between masonry and R/C frame: (a) upper boundary; (b) lateral boundary; (c) reinforcement for out-of-plane overturning.

Table 6.1 Values of reduction factor ν

Importance category	I	II	III	IV
Reduction factor	2.5	2.5	2.0	2.0

one, a number of specific measures are also taken. Some of these measures have already been presented in detail, while some of them will be presented in Chapters 7–9. However, it would be useful to summarize these measures here, so that a global view of the three compliance criteria mentioned above can be obtained. The specific measures refer to

- design
- foundations
- quality system plan
- resistance uncertainties
- ductility uncertainties.

(a) Design

- structures should have a simple and regular form in plan and elevation (section 4.3).
- In order to ensure an overall ductile behaviour, brittle failure or premature formation of unstable mechanisms must be avoided. For this reason, the capacity design procedure should be adopted (section 6.1).
- Since the seismic performance of the structure depends on the behaviour of its critical regions, the detailing of these regions must be such as to maintain under cyclic conditions the ability to transmit the necessary forces and to dissipate energy. For this reason, the detailing of connections between structural elements and of regions where non-linear behaviour is foreseeable deserves special care in design (Chapters 8 and 9).
- In order to limit the consequences of damage, national authorities may specify restrictions on the height or other characteristics of the structure, depending on the local seismicity, the importance of the structure and so on.
- The analysis must be based on an appropriate structural model (Chapter 5).
- Buildings must be protected from earthquake-induced collisions with adjacent structures, either through joints between adjacent buildings equal to the sum of their maximum horizontal displacements d_s according to equation (6.13) or through appropriate collision walls to act as 'bumpers' (EC8 Part 1–2, section 4.2.7).
- No changes in the structural system are allowed during the construction phase or during the lifetime of the structure, unless proper justification and verification are provided.

(b) Foundations

- The stiffness of the foundation must be adequate in terms of transmitting to the ground as uniformly as possible the actions received from the superstructure (section 4.3).
- Only one foundation type must in general be used for the same structure (section 4.3).

(c) Quality system plan

- The choice of materials and construction techniques must be in compliance with the design assumptions. The design documents must indicate structural details, sizes and quality provisions.
- Elements of special structural importance requiring special checking during construction should be identified on the design drawings. In such cases, the checking methods to be used should also be specified.

(d) Resistance uncertainties

Important resistance uncertainties could be produced by geometric errors. To avoid these uncertainties, rules referring to the following items should be applied:

- Certain minimum dimensions of structural elements (Chapters 8 and 9).
- Appropriate limitations of column drifts must be provided (Chapter 8).
- Special detailing rules should be applied in reinforcing R/C elements, so that unpredictable moment reversals and uncertainties related to the position of the inflection point are taken into account (Chapters 8 and 9).

(e) Ductility uncertainties

In order to minimize ductility uncertainties the following rules must be applied:

- An appropriate minimum local ductility is needed in every seismic resistant part of the structure (Chapters 7–9); thus, by enhancing the redistribution capacity of the structure some of the model uncertainties are alleviated.
- Minimum–maximum reinforcement percentages in all critical regions are specified to take into account ductility requirements and to avoid brittle failure upon cracking (Chapters 8 and 9).
- The normalized design axial force values are kept at a low level to avoid decrease of local ductility at the top and bottom of the columns (Chapter 8).

6.3 APPLICATION OF EC8 TO THE DESIGN OF A SIMPLE DUAL SYSTEM

6.3.1 Introduction

The example presented in this section has been prepared to serve the purpose of clarification of EC8 and of various other aspects discussed so far. It consists of the design of an isolated, 10-storey, two-span dual system, assumed to belong to the category of 'regular' buildings.

To simplify both the analysis and the presentation of the results, only one ductility class has been considered, namely DC 'M'.

6.3.2 System geometry

The geometrical dimensions of the system are illustrated in Figure 6.12. The dual system is supposed to be the central one in a series of dual systems, equally spaced, at a distance of 3.0 m.

6.3.3 Characteristics of the materials

(a) Concrete

- concrete class: C20/25 ($f_{ck} = 20$ MPa);
- specific weight: 25 kN/m^2 [EC2, section 3.1.2.4];

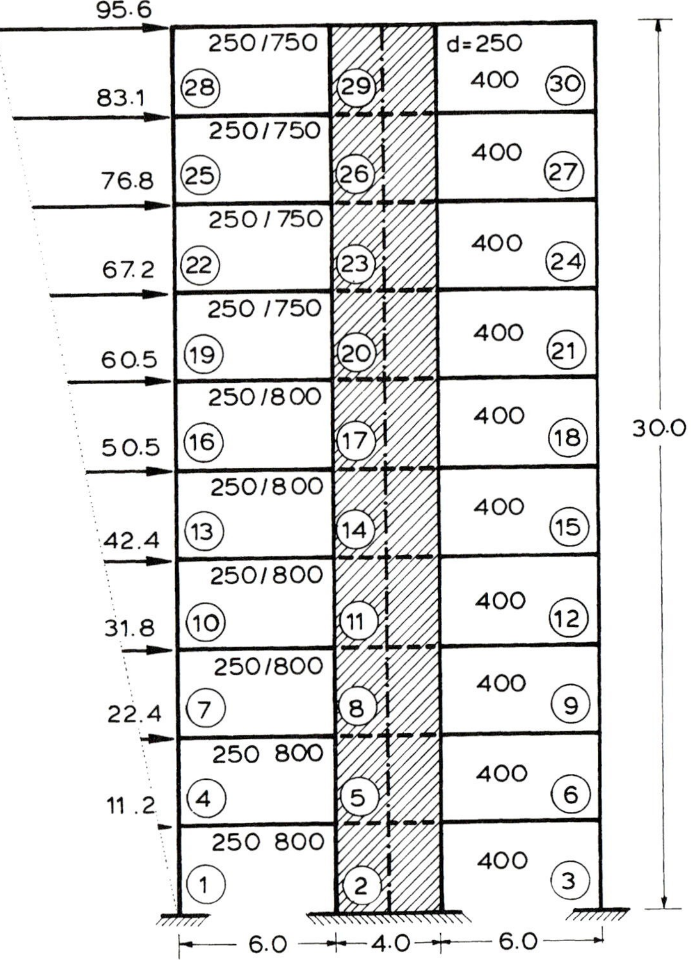

Figure 6.12 Geometry of the dual system and seismic loads at the storeys.

- Young's modulus: $E_{cm}=29.0$ GPa [EC2, section 3.1.2.5];
- Poisson's ratio: $v=0.2$;
- design compressive strength: $f_{ck}/\gamma_c=20/1.5=13.33$ MPa [EC2, section 2.3.3.2];
- design shear strength: $\tau_{Rd}=0.26$ MPa [EC2, section 4.3.2.3].

(b) Steel

- steel class:S400
- design tensile strength: $f_{yd}=f_{yk}/\gamma_s=400/1.15=347.8$ MPa [EC2, section 2.3.3.2].

6.3.4 Design gravity loads

(a) Dead load

Unit weight of the floor, consisting of the concrete slab, the beams and architectural finishes. It varies from $g_1 = 16.2$ kN/m (first floor) to $g_{10} = 14.7$ kN/m (tenth floor).

(b) Live load

Assumed as uniformly distributed and the same for all floors (2 kN/m^2), leading to a distributed beam load of $q = 6.0$ kN/m.

6.3.5 Design seismic actions

(a) Design spectrum for linear analysis

The design spectrum $S_d(T)$ is defined by the following expressions (section 4.4.5, equation (4.6)).

$$S_d(T) = \alpha S\left[1 + \frac{T}{T_B}\left(\frac{\beta_0}{q} - 1\right)\right] \quad \text{for } 0 \leqslant T \leqslant T_B$$

$$S_d(T) = \alpha S\frac{\beta_0}{q} \quad \text{for } T_B < T \leqslant T_C$$

$$S_d(T)\begin{cases} = \alpha S\dfrac{\beta_0}{q}\left[\dfrac{T_C}{T}\right]^{k_{d_1}} \\ \geqslant 0.20\,\alpha \end{cases} \quad \text{for } T_C < T \leqslant T_D$$

$$S_d(T)\begin{cases} = \alpha S\dfrac{\beta_0}{q}\left[\dfrac{T_C}{T_D}\right]^{k_{d_1}}\left[\dfrac{T_D}{T}\right]^{k_{d_2}} \\ \geqslant 0.20\alpha \end{cases} \quad \text{for } T_D < T$$

The following values have been assumed for the various parameters:

- importance category III: $\gamma_I = 1.0$
- effective peak ground acceleration normalized by the acceleration of gravity: $\alpha = 0.25$
- subsoil class: A
- amplification factor: $\beta_0 = 2.5$ (Table 4.1)
- corner periods: $\begin{cases} T_B = 0.1 \text{ s} \\ T_C = 0.4 \text{ s} \quad \text{(Table 4.1)} \\ T_D = 3.0 \text{ s} \end{cases}$
- exponents: $\begin{cases} k_{d1} = 2/3 \quad \text{(Table 4.2)} \\ k_{d2} = 5/3 \end{cases}$

- soil parameter: $S = 1.0$ (Table 4.1)
- behaviour factor: $q = q_0 k_D\, k_R\, k_W \geqslant 1.5$
- basic value: $q_0 = 4.5$ (Table 4.3)
- factor reflecting the ductility class: $k_D = 0.75$ (equation (4.8))
- factor reflecting the structural regularity in elevation: $k_R = 1.0$ (equation (4.9))
- factor reflecting the prevailing failure mode: $k_W = 1/(2.5 - 0.5h/l_w) \leqslant 1 = 1.0$

Thus, the expression for q becomes

$$q = 4.5 \times 0.75 \times 1 \times 1 = 3.375 > 1.5$$

while the expressions for $S_d(T)$ take the following form:

$$S_d(T) = 0.25 \times 1 \times \left[1 + \frac{T}{0.10}\left(\frac{2.5}{3.375} - 1\right)\right] \quad \text{for } 0 \leqslant T \leqslant 0.1 \text{ s}$$

$$S_d(T) = 0.25 \times 1 \times \frac{2.5}{3.375} = 0.185 \quad \text{for } 0.10 \text{ s} < T \leqslant 0.4 \text{ s}$$

$$S_d(T) \begin{cases} = 0.185\left(\dfrac{0.4}{T}\right)^{2/3} \\ \geqslant 0.20 \times 0.25 = 0.05 \end{cases} \quad \text{for } 0.4 \text{ s} < T \leqslant 3.0 \text{ s}$$

$$S_d(T) \begin{cases} = 0.185\left[\dfrac{0.4}{3}\right]^{2/3}\left[\dfrac{3}{T}\right]^{5/3} \\ \geqslant 0.05 \end{cases} \quad \text{for } 3.0 \text{ s} < T$$

(b) Loads contributing to the inertia forces

According to section 5.2.2 the loads to be considered for the evaluation of inertial effects are (equation (5.4)):

$$\sum G_{kj} \text{ `+' } \sum \psi_{Ei} Q_{ki}$$

where

$$\psi_{Ei} = \varphi \psi_{2,i}$$

In the above equation we have

$$\left.\begin{array}{l} \varphi = 1.0 \text{ for the last floor} \\ \varphi = 0.5 \text{ for every other floor} \end{array}\right\} \quad \text{[EC8/Part 1.2/Table 3.2]}$$

$$\psi_{2,i} = 0.30$$

The total load value per floor is derived by taking into account the above loads on the beams plus the dead load of columns and walls. These total load values are given in Table 6.2.

It should be noted that the gravitational load on the beams taken into account for the calculation of gravity load effects interacting with seismic actions is

$$\sum G_{ki} \text{ `+' } \sum \psi_{2,i} Q_{ki}$$

that is

$$16.2+0.3\times 6.0=18.0\ \text{kN/m}$$

6.3.6 Equivalent horizontal forces

Since the building has been assumed to be regular in plan and in elevation, the 'simplified response spectrum analysis' or otherwise the 'equivalent static analysis' will be used for this example. For the static simulation of the seismic actions, the first natural period T_1 of the structure is needed (equation (5.7)).This will be derived by using equations (5.17) and (5.18).

$$T_1^{(a)}=\frac{0.09h}{L^{1/2}}=\frac{0.09\times 30}{16^{1/2}}=\frac{0.09\times 30}{4}=0.67\ \text{s}$$

$$T_1^{(b)}=0.05\times h^{3/4}=0.05\times 30^{3/4}=0.05\times 12.81=0.64\ \text{s}$$

$$T_{1,\text{est}}=\frac{0.67+0.64}{2}=0.655\ \text{s}$$

Introducing $T_{1,\text{est}}$ in the proper $S_d(T)$ expression of the previous section ($0.4\leqslant T\leqslant 3.0$ s) we get

$$S_d(T)=0.185\left(\frac{0.4}{0.65}\right)^{2/3}=0.134$$

Taking into account the distribution factor λ_i (equation (5.15)) (Table 6.2)

$$\lambda_i=\frac{h_i\Sigma W_i}{\Sigma W_i h_i}=\frac{4.0112}{64.005}h_i=0.062h_i$$

we define the seismic actions for every storey (equation (5.14))

$$H_i=\gamma_1 S_d(T)W_i\lambda_i=1\times 0.134\times 0.062h_iW_i=8.4\times 10^{-3}h_iW_i$$

The results are presented in Table 6.2.

The base shear V_B is found either as the sum of H_i or by equation (5.7):

$$V_B=\gamma_1 S_d(T_1)W=0.134\times 4011.2=539.4\ \text{kN}$$

In Table 6.2 the resulting displacements δ_i due to the horizontal forces H_i are also included as they have been produced from the analysis of the system. The interstorey drifts d_{e_i} and the quantities $H_i\delta_i$ and $W_i\delta^2$ are also presented in the same table for further computations. The fundamental period of the system may now be re-estimated more accurately using the Rayleigh method (equation (5.20).

$$T_1=2\pi\left(\frac{1}{g}\ \frac{\Sigma W_i\delta_i^2}{\Sigma H_i\delta_i}\right)^{1/2}=2\pi\left(\frac{0.8389}{9.81\times 8.87}\right)^{1/2}\cong 0.62\ \text{s}$$

Therefore the preliminary estimation of $T_1=0.65$ s was adequately accurate and the whole procedure should not be repeated.

Table 6.2 Loads and displacements of the storeys

Level	W_i (kN)	h_i (m)	W_ih_i	H_i (kN)	δ_i (mm)	d_{e_i} (mm)	$H_i\delta_i$ (kN m)	$W_i\delta_i^2$
0		0.0						
1	442.1	3.0	1326.3	11.2	0.8	0.8	0.00896	0.00028
2	442.1	6.0	2652.6	22.4	2.7	1.9	0.06048	0.00322
3	417.6	9.0	3758.4	31.8	5.2	2.5	0.16536	0.01129
4	417.6	12.0	5011.2	42.4	8.1	2.9	0.34340	0.02739
5	397.6	15.0	5964.0	50.5	11.1	3.0	0.56055	0.04899
6	397.6	18.0	7156.8	60.5	14.2	3.1	0.85910	0.08017
7	378.1	21.0	7940.1	67.2	17.1	2.9	1.14910	0.11056
8	378.1	24.0	9074.4	76.8	19.8	2.7	1.52060	0.14820
9	363.6	27.0	9817.2	83.1	22.3	2.5	1.85313	0.18080
10	376.8	30.0	11304.0	95.6	24.6	2.3	2.35176	0.22802
Total	4011.2		64005.0	541.5			8.87240	0.83890

6.3.7 Design load combinations

Two basic load combinations have to be considered:

- one with the live load as the main variable action (section 6.2.2)

$$E_{d1} = E(1.35G + 1.5Q)$$

- one with the seismic action as the main action (section 5.4.6, equation (5.29))

$$E_{d2} = E(G + \Sigma\psi_{2,i}Q_i + H_i)$$

The gravity load $(G+\Sigma\psi_{2,i}Q_i)$ has been evaluated in section 6.3.5(b), and the seismic loads (H_i) in section 6.3.6.

6.3.8 Structural analysis

The dual system in Figure 6.12 has been analysed for the two loading cases mentioned above, using a structural analysis program implemented on a PC.

For the stiffness of the various structural elements the following assumptions have been made, taking into account the effect of axial loading on the degree of cracking:

- columns: $EJ_{ef} = 0.80EJ_g$
- shear-walls: $EJ_{ef} = 0.60EJ_g$
- beams: $EJ_{ef} = 0.40EJ_g$

where EJ_{ef} is the stiffness introduced in the analysis and EJ_g the stiffness of the uncracked sections.

The system has been considered free to deform without appreciable interactions with non-structural elements. The load effects obtained from the analysis are presented in Tables 6.3–6.6 and in Figures 6.13–6.15.

Table 6.3 Load effects for $E_d = (1.35G + 1.5Q)$

Storey	Joint	Columns			Beams		Wall
		M(kN m)	V (kN)	N (kN)	M (kN m)	V (kN)	N (kN)
1	i*[a]	9.29	9.29	−1059.86	−44.21	112.27	−3568.07
	j*[a]	18.58			−130.60		
2	i	25.63	16.34	−947.39	−43.85	113.23	−3214.53
	j	23.40			−136.00		
3	i	20.45	13.39	−835.87	−37.97	115.46	−2859.06
	j	19.73			−143.50		
4	i	18.24	11.90	−726.58	−33.57	117.22	2499.13
	j	17.47			−149.63		
5	i	16.10	10.53	−619.05	−29.60	118.77	−2135.70
	j	15.50			−154.96		
6	i	14.10	9.13	−513.07	−27.66	119.72	−1769.16
	j	13.30			−158.70		
7	i	14.36	10.19	−408.04	−33.80	109.55	−1400.72
	j	16.22			−140.31		
8	i	17.58	11.56	−304.99	−33.05	109.93	−1052.62
	j	17.09			−141.84		
9	i	15.96	10.43	−202.32	−33.54	109.93	−703.76
	j	15.33			−142.53		
10	i	18.21	13.32	−99.65	−21.73	112.95	−354.89
	j	21.73			−148.61		

[a] For columns, j is the upper joint of the storey and i is the lower joint of the storey. For beams, i is the joint on the column and j the joint on the wall.

Table 6.4 Load effects on beams for $E_d = E(G + 0.30Q + H_i)$

Storey	Joint		M_d^+ (kN m)	M_d^- (kN m)	$V_{d,\max}$ (kN)
1	i*[a]	4,6	43.40	−94.99	93.65
	j*[b]	5	15.62	−171.13	
2	i	7,9	85.79	−136.31	110.89
	j	8	69.41	−232.18	
3	i	10,12	113.07	−156.09	121.62
	j	11	96.83	−269.23	
4	i	13,15	126.41	−163.76	126.87
	j	14	107.12	−287.42	
5	i	16,18	130.99	−163.25	128.67
	j	17	106.49	−293.63	
6	i	19,21	124.10	−153.68	126.50
	j	20	95.48	−287.52	
7	i	22,24	90.15	−126.33	104.66
	j	23	50.15	−217.39	
8	i	25,27	75.64	−110.76	99.70
	j	26	32.89	−202.13	
9	i	28,30	69.22	−104.61	96.85
	j	29	21.42	−191.49	
10	i	31,33	42.72	−65.58	89.63
	j	32	−2.11	−174.69	

[a]* i: joint on the column.
[b]* j: joint on the wall.

Table 6.5 Load effects on columns for $E_d = E(G+0.3Q+H_i)$

Storey	Joint	M_{min}	N_{min}	M_{max}	N_{max}	V_{max}
1	i^{*a}	18.22	−312.95	29.09	−1162.44	20.06
	j^{*b}	9.34		31.08		
2	i	34.05	−261.81	63.91	−1057.11	41.40
	j	33.18		60.28		
3	i	52.61	−227.90	76.03	−935.92	49.97
	j	51.46		73.94		
4	i	61.61	−204.72	82.15	−806.87	54.22
	j	60.96		80.52		
5	i	65.45	−186.79	83.24	−674.82	55.12
	j	65.12		82.14		
6	i	65.86	−170.66	81.11	−542.97	54.03
	j	66.70		80.96		
7	i	57.40	−152.36	72.72	−414.54	46.92
	j	50.68		68.05		
8	i	39.47	−121.22	58.28	−302.73	39.34
	j	41.53		59.74		
9	i	34.12	−85.11	51.03	−196.38	33.68
	j	33.80		50.01		
10	i	35.42	−46.17	54.60	−92.98	40.06
	j	42.72		65.58		

[a]$*i$: lower joint of the storey.
[b]$*j$: upper joint of the storey.

Table 6.6 Load effects on the shear wall for $E_d = E(G+0.3Q+H_i)$

z	Storey	M_{cal}	M_{Sd}	$\lvert N_d \rvert$
0.0	1	4083.06	4083.06	2638.60
3.0	2	2841.42	4083.06	2376.48
6.0	3	1918.09	3955.83	2112.98
9.0	4	1226.22	3574.14	1846.61
12.0	5	705.95	3192.46	1577.99
18.0	6	522.26	2429.09	1307.38
21.0	7	638.61	2047.40	1035.50
24.0	8	777.22	1665.72	777.86
27.0	9	519.57	1284.03	519.70
30.0	10	266.21	902.35	261.46

For the determination of the design action effects that also takes into account the capacity design procedure, it is first necessary to calculate the flexural reinforcement of the beams. This type of procedure will therefore be presented in Chapters 8 and 9.

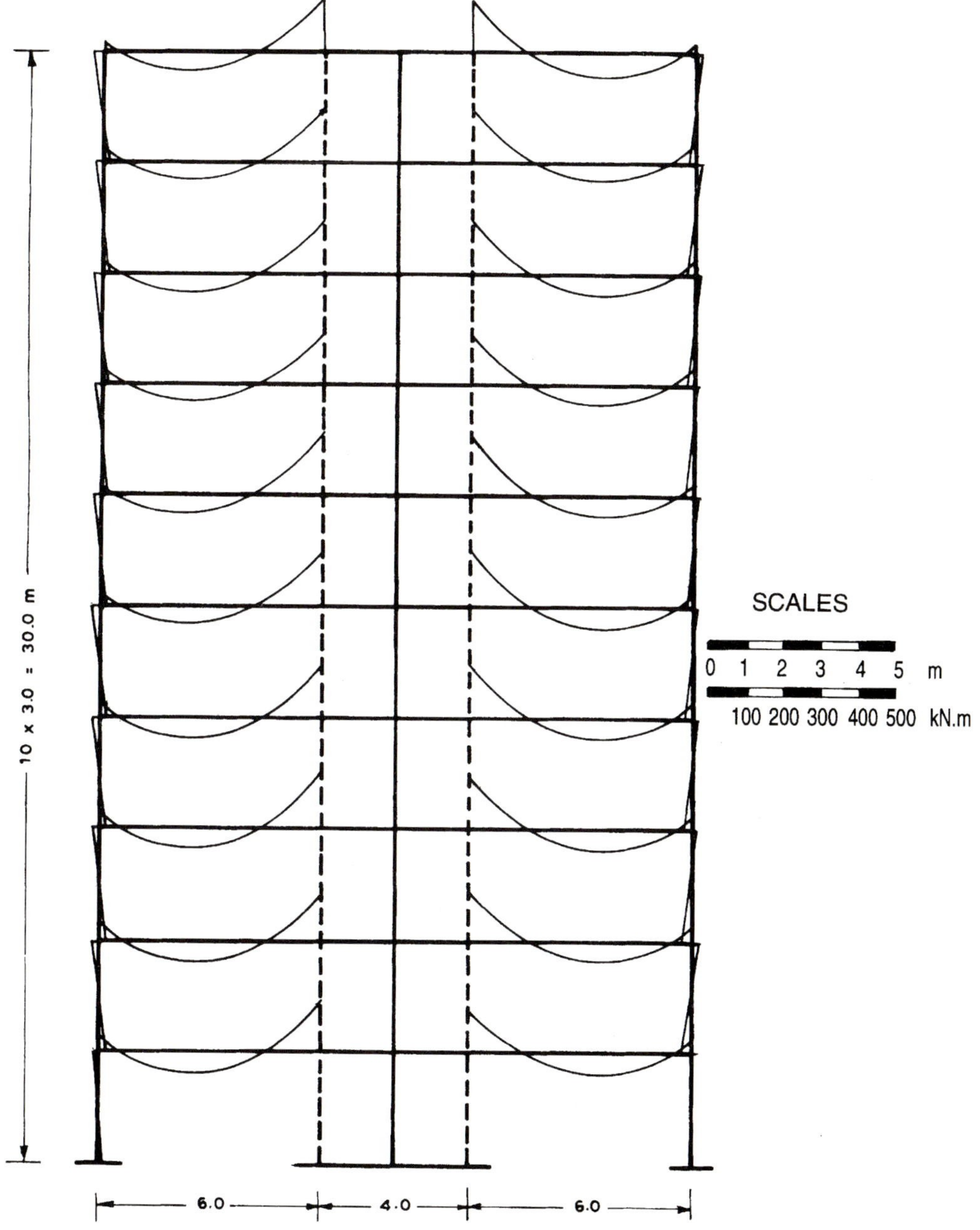

Figure 6.13 *M*-diagrams of the dual system for $E_{d_1} = E\,(1.35G_i$ '+' $1.50Q_i)$.

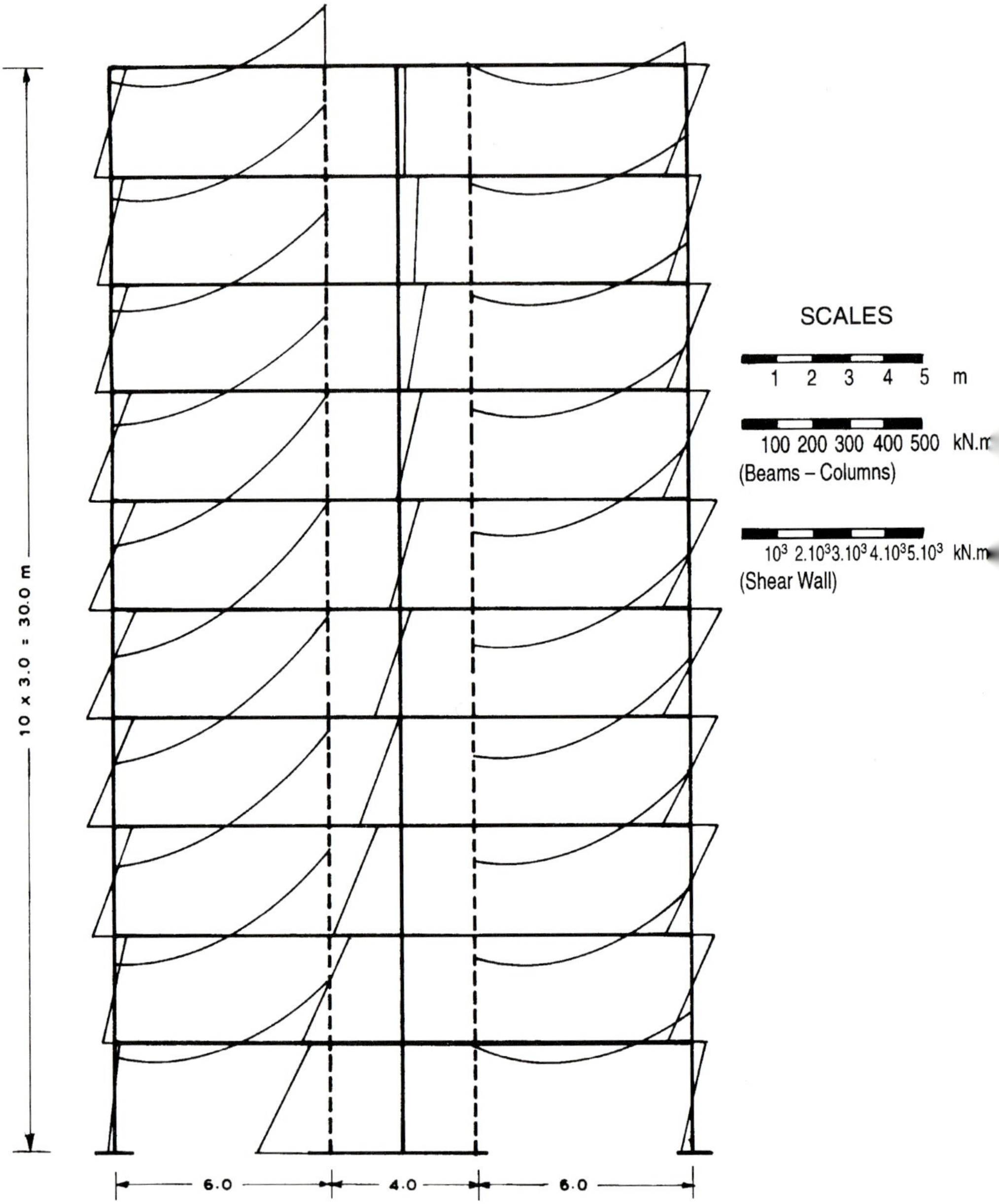

10 x 3.0 = 30.0 m

6.0

4.0

6.0

SCALES

1 2 3 4 5 m

100 200 300 400 500 kN.m

(Beams – Columns)

10^3 2.10^3 3.10^3 4.10^3 5.10^3 kN.m

(Shear Wall)

Figure 6.14 *M*-diagrams of the dual system for $E_{d_2} = E\,(G_i$ '+' $0.3Q_i$ '+' $H_{ix})$.

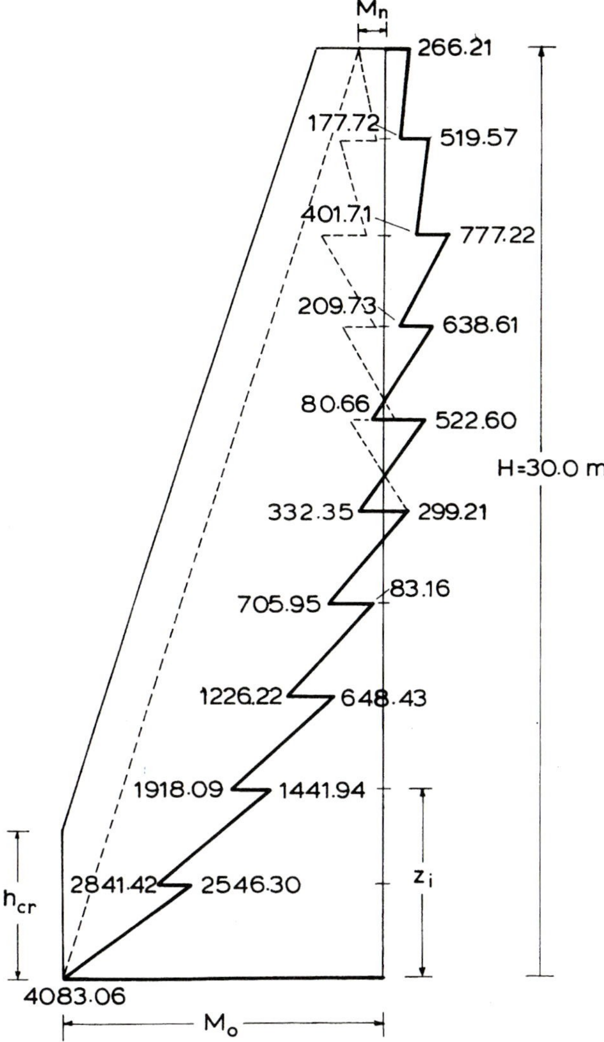

Figure 6.15 *M*-diagrams of the shear wall for $E_d = E\,(G+0.3Q+H_i)$.

6.4 REFERENCES

CEB (1985) Model Code for Seismic Design of Concrete Structures. *Bulletin d' Information CEB*, **165**, Lausanne.

CEN Techn. Comm. 250/SC2 (1991) *Eurocode 2: Design of Concrete Structures–Part 1: General Rules and Rules for Buildings (ENV 1992-1-1)*. CEN, Berlin.

CEN Techn. Comm. 250/SC8 (1994) *Eurocode 8: Earthquake Resistant Design of Structures–Part 1: General Rules and Rules for Buildings (ENV 1998-1-12)*. CEN, Berlin.

Eibl, J. and Keintzel, E. (1989) Seismic shear forces in RC cantilever shear walls. *Proceed. 9th World Conf. on Earthq. Engng*, Tokyo-Kyoto, Japan, Aug. 1988, Maruzen (Tokyo), VI, pp. 5–10.

Kappos, A., Stylianidis, K. and Penelis, G. (1991) Analytical prediction of the response of structures to future earthquakes. *European Earthq. Engng.*, **1**, 10–21.

Luft, R. (1989) Comparison among earthquake codes. *Earthq. Spectra*, **5**(4), 767–89.

Ministry for Environment and Public Works (1992) *Greek Code for Earthquake Resistant Structures*, Athens.

Park, R. (1986) Ductile design approach for reinforced concrete frames. *Earthq. Spectra*, Earthq. Engng Research Institute, **2**(3), 565–619.

Park, R. and Paulay, T. (1975) *Reinforced Concrete Stuctures*, J. Wiley & Sons, New York.

Park, Y.-J. *et al.* (1985) Mechanistic seismic damage model for reinforced concrete. *ASCE, Journ. of Struct. Engng.*, **111**(4), 722–39.

Paulay, T. (1986) The design of ductile reinforced concrete structural walls for earthquake resistance. *Earthq. Spectra*, Earthq. Engng Research Inst., **2**(4), 783–823.

Paulay, T., Bachmann, H. and Moser, K. (1990) *Erdbebenbemessung von Stahlbeton Hochbauten*, Birkhäuser.

Priestley, M.J.N. and Calvi, E.M. (1991) Towards a capacity-design assessment procedure of reinforced concrete frames. *Earthq. Spectra*, **7**(3), 413–38.

SEAOC (1975, 1990) *Recommended Lateral Force Requirements and Commentary*, San Francisco.

Uang, C.-M. and Bertero, V. (1991) UBC seismic serviceability regulations: critical review. *ASCE Journ. of Struct. Engng.*, **117**(7), 2055–68.

7

Earthquake-resistant properties of the materials of reinforced concrete

7.1 INTRODUCTION

As pointed out in Chapter 6, **ductility** constitutes one of the fundamental requirements regarding the mechanical behaviour of structural elements (members) in an earthquake-resistant structure.

The ductility of an element should be conceived at the same time as: (a) ability to sustain large inelastic deformations without substantial reduction in strength, and (b) ability to absorb and dissipate seismic energy through relatively stable hysteresis loops.

Although the concept of ductility is quite clear, translating it into specific performance criteria for structural elements is still a controversial matter. The previous version of the New Zealand Loadings Code (Standards Association of New Zealand, SANZ, 1984) was the first one to introduce a conventional criterion to quantify ductility requirements. According to this criterion a structural element should be able to develop four complete hysteresis loops at a displacement ductility factor (section 3.3.4) $\mu_\delta = 4$, while its strength is not reduced by more than 30%. On the other hand, in the more recent EC8 (CEN, 1995) member ductility requirements are expressed in terms of conventional curvature ductility factors, defined with reference to the 85% strength level on the descending branch of the moment–curvature diagram (specific values for R/C members are given in Chapter 8). Contrary to the New Zealand criterion (which appears to be appropriate for high ductility structures), EC8 specifies different ductility requirements for each ductility class (section 4.4.3).

For the ductility of an R/C structural element to be ensured, it is first necessary to obtain a ductile behaviour of its constituent materials (concrete and steel), and also an adequate composite action of the two materials under seismic conditions. Described in the present chapter are the mechanical properties of concrete and steel under **cyclic loading**, where the sign of the acting force is changing (commonly referred to as **load reversal**). Special emphasis is placed on stress–strain diagrams of the two materials under this type of loading. Also examined is the problem of **bond** between steel and concrete under cyclic loading conditions, which is a critical one regarding the seismic response of an R/C element.

As a final introductory remark, it should be emphasized that there is a need in earthquake-resistant structures to use materials which are relatively easy to apply and, more important, which are easily subjected to quality control. The need to control the quality of construction materials which is present in all structural projects, is particularly imperative in the case of earthquake-resistant structures. As specified in EC8 (section 2.2.3 of Part 1.1), national authorities may, in the case of important structures, prescribe formal quality plans, covering both design and construction, supplementary to those prescribed in other relevant Eurocodes.

7.2 REFERENCE TO CODE PROVISIONS

Throughout this book reference to specific code provisions is based mainly on codes of direct interest to the practising engineer, in particular the new *Eurocode 8 for Seismic Design* (CEN, 1995) and the *Eurocode 2 for Concrete Structures* (CEN, 1991), on which the seismic design of concrete structures will be based, in Europe, as well as in many other countries which adopted the Eurocodes.

Some additional reference will be made to the CEB Model Code for Seismic Design (CEB, 1985) which, in many respects, has paved the way for EC8 and was used as a basis for numerous interesting studies, some of which are presented in the following chapters. With regard to constitutive relations for the materials of reinforced concrete and to related phenomena such as bond, reference will be made to the 1990 CEB Model Code (CEB, 1993) which, to the best of the authors' knowledge, is the most up-to-date and comprehensive document on concrete design. Finally some complementary reference will be made to the New Zealand Concrete Code (SANZ, 1982), wherein the capacity design procedures were first introduced and the most strict provisions regarding detailing for ductility were adopted.

Throughout the following chapters an effort will be made to avoid simultaneous reference to similar provisions in different codes, which although of some interest from the comparison point of view, could easily lead to confusion and also unnecessarily increase the volume of the book. Seen from this perspective, references to codes other than the Eurocodes will only be made on a complementarity basis, with a view to filling in 'blanks' or, more often, to facilitate understanding of the background of certain Eurocode provisions.

7.3 PLAIN (UNCONFINED) CONCRETE

The main factor influencing the seismic behaviour of concrete is lateral confinement. The term **confinement** refers to the influence that lateral reinforcement (in the form of hoops or spirals) exercises on concrete, which leads to a modification of the compression stress state from uniaxial to multiaxial. As will be shown subsequently, the presence of confinement has a favourable effect on the strength, as well as on the ductility of concrete. For this reason it was deemed appropriate to examine separately the earthquake-resistant properties of confined and of unconfined (or plain) concrete.

The earthquake-resistant properties of a material can be evaluated using their stress–strain diagrams, where both strength and deformation characteristics are reflected. The determination of stress–strain diagrams (σ_c–ε_c) for concrete under seismic loading is commonly obtained using experimental set-ups, whereby the material is subjected to repeated loading and unloading without change of sign, or to cyclic loading (involving load reversals); as a rule both tests are of the static type, that is they are carried out at very low loading rates. Correlating the results from such tests with the actual behaviour under seismic loading, which is characterized by deformations induced at a very high loading rate (typical rates $\dot{\varepsilon}$ range from 0.01 to 0.02 s^{-1}, i.e. induced strains of 1–2% s^{-1}), is always a safe procedure as far as strength of the material is concerned. On the contrary, using the corresponding results regarding ultimate deformation does not lie, as a rule, on the safe side.

7.3.1 Response to monotonic loading

In Figure 7.1 are shown the diagrams of stress (σ_c) versus strain (ε_c) for monotonic compression, resulting from tests on cylinders of various concrete grades (CEB, 1993), carried out with strain control after the development of maximum strength. It is clearly seen that as strength increases, the ultimate strain of concrete decreases, in other words low-grade concrete is more ductile than high-grade concrete. The curves shown in Figure 7.2 resulted from the application of three analytical models proposed for monotonic compression of plain (unconfined) concrete.

From Figures 7.1 and 7.2 it is seen that the σ_c–ε_c curve consists of three parts:

1. The initial, almost linear part (on which the permissible stresses theory was based), which corresponds to an elastic behaviour;

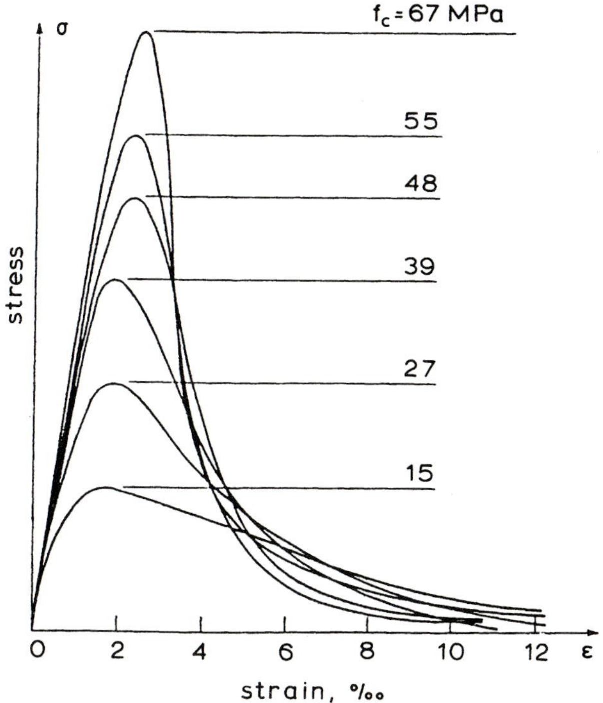

Figure 7.1 Stress–strain curves for cylinders of concrete subjected to uniaxial compression.

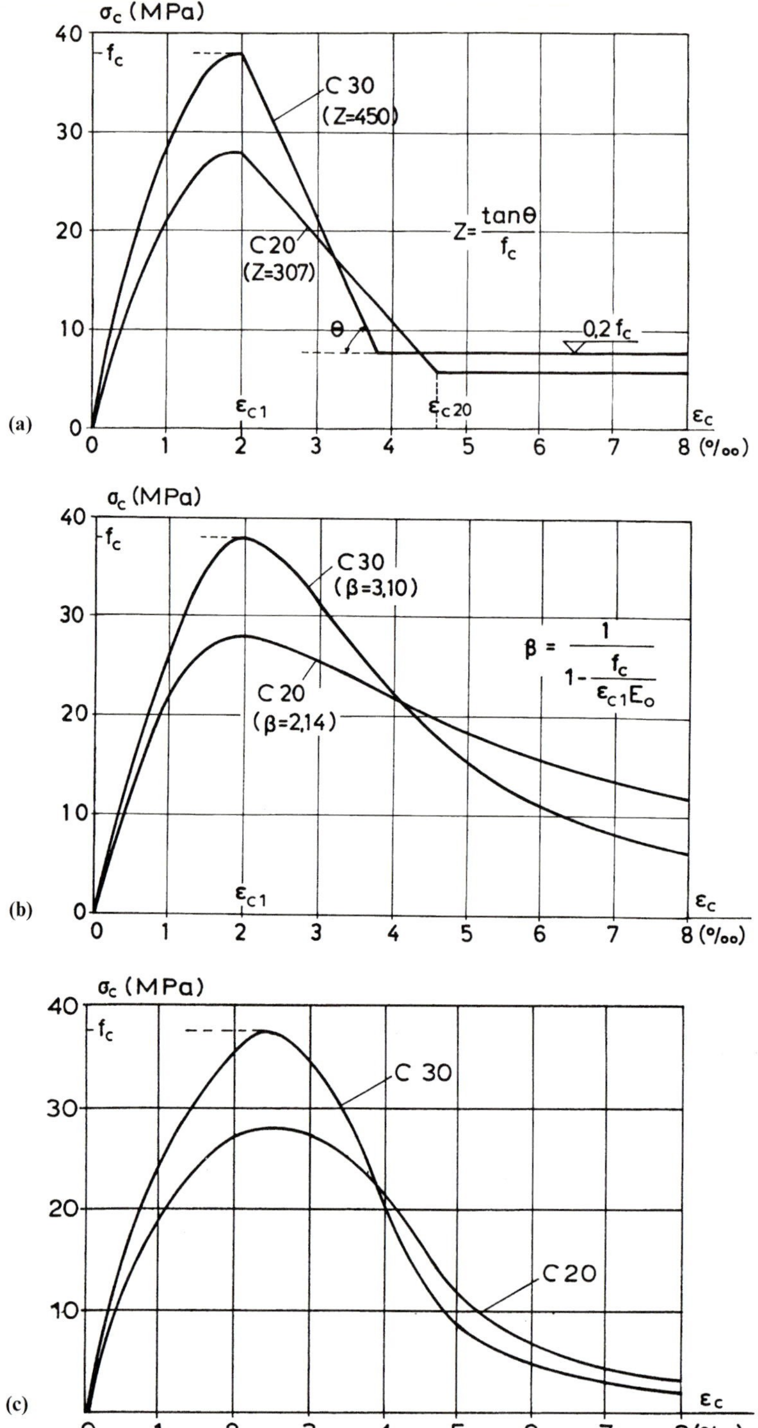

Figure 7.2 Analytical models for estimating the stress–strain curve for concrete subjected to uniaxial compression: (a) Park–Kent (see Park and Paulay, 1975); (b) Carreira–Chu (1985); (c) CEB 1990 Model Code (CEB, 1993).

2. A second part defined by strains corresponding to stresses equal to about 70% up to 100% of the cylinder (or prism) strength, which is characterized by a nonlinear behaviour of the material, indicated by a gradual reduction of its tangent modulus. Strictly speaking, some nonlinearity appears at strains as low as $0.3f_c$, beyond which the **microcracks** already present in the unloaded concrete (due to shrinkage, temperature effects and other causes) start to propagate. At stresses between 0.5 and $0.7f_c$ adjacent **bond cracks** at the interface of mortar and aggregates, caused by the different stiffness of the two materials, start to bridge in the form of **mortar cracks**, due to stress concentrations at the tips of bond cracks (Aoyama and Noguchi, 1979; Chen, 1982).
3. A third part, along which strain increases while stress decreases (descending branch). This phenomenon, called **strain softening** is attributed to the unstable propagation of the aforementioned internal cracks which tend to become macroscopic.

In calculations for design purposes it is customary to use only a portion of the descending branch, determined by the maximum 'usable' strain or the conventional 'failure' strain (ε_{cu}), beyond which damage to the material is no longer acceptable. For instance EC2 (CEN, 1991) defines a value of 0.35% as the maximum usable strain for concrete, in combination with the assumption that no stress reduction takes place up to this level of deformation. It is seen from Figures 7.1 and 7.2 that neither assumption is strictly valid, nevertheless the area under the $\sigma_c - \varepsilon_c$ curve assumed by the code is clearly smaller than that under the actual curve.

For the ascending branch of the stress–strain curve the best known analytical expression (which has been adopted, with slight modification, by most codes, including Eurocode 2 and the CEB Model Code (CEB, 1993)) is Hognestad's parabola (see also Park and Paulay, 1975).

$$\sigma_c = f_c \left[\frac{2\varepsilon_c}{\varepsilon_{c1}} - \left(\frac{\varepsilon_c}{\varepsilon_{c1}} \right)^2 \right] \tag{7.1}$$

where $\varepsilon_{c1} = 2f_c/E_{c0}$ is the strain corresponding to a stress $\sigma_c = f_c$, where f_c is the cylinder strength of concrete, and E_{c0} is the initial slope of the curve (initial tangent modulus of elasticity).

For plain concrete the value of strain ε_{c1} (corresponding to the strength f_c) is of the order of 0.2%. A constant value $\varepsilon_{c1} = 0.22\%$ has been adopted by the new CEB Code (CEB, 1993), although experiments have shown increased values for high strength grades (Mueller and Hilsdorf, 1993).

For the descending branch one of the simplest, nevertheless widely used in the literature, expressions is that of Kent and Park (see Park and Paulay, 1975), who proposed a straight line defined by the point (f_c, ε_{c1}) and the point corresponding to a strength reduction of 50%. By fitting to experimental results (available up to 1970) the strain corresponding to the aforementioned point ($\sigma_c = 0.5f_c$) was found to be

$$\varepsilon_{c50} = \frac{3 + 0.29f_c}{145f_c - 1000} \tag{7.2}$$

where the compressive strength f_c is given in MPa (meganewtons per m^2). Therefore the equation for the descending branch ($\varepsilon_c > \varepsilon_{c1}$) becomes (see also

Figure 7.2(a))

$$\sigma_c = f_c \, [1 - Z\,(\varepsilon_c - \varepsilon_{c1})] \tag{7.3}$$

with

$$Z = \frac{0.5}{\varepsilon_{c50} - \varepsilon_{c1}} \tag{7.4}$$

Finally Kent and Park suggested that for large strains a residual strength (attributed to friction along failure surfaces) equal to $0.2f_c$ be taken into account (horizontal branch in Figure 7.2(a)).

A somewhat more refined model (in the sense that it fits the experimental results better than the previous one) has been adopted by the new CEB Model Code (CEB, 1993). For the ascending, as well as for the descending, branch up to a value of $0.5f_c$, the following equation is suggested:

$$\sigma_c = \frac{\dfrac{E_{c0}}{E_{c1}}\dfrac{\varepsilon_c}{\varepsilon_{c1}} - \left(\dfrac{\varepsilon_c}{\varepsilon_{c1}}\right)^2}{1 + \left(\dfrac{E_{c0}}{E_{c1}} - 2\right)\dfrac{\varepsilon_c}{\varepsilon_{c1}}} f_c \tag{7.5}$$

where $E_{c1} = f_c/\varepsilon_{c1} = f_c/0.0022$ and the rest of the symbols have the same meaning as in equation (7.1). For the tangent modulus at the origin E_{c0} the following formula is suggested by CEB:

$$E_{c0} = 2.15 \times 10^4 \, (f_c/10)^{1/3} \tag{7.6}$$

where E_{c0} is in MPa. Note that for $E_{c1} = E_{c0}/2$, it is $\varepsilon_{c1} = 2f_c/E_{c0}$ and equation (7.5) becomes identical to Hognestad's parabola (equation (7.1)). In fact for relatively low concrete grades E_{c1} is quite close to $E_{c0}/2$, but for high concrete grades E_{c1} assumes much higher values. It is also pointed out that for stresses lower than $0.5f_c$ along the descending branch equation (7.5) is no longer applicable (an additional, rather complicated expression is suggested by CEB for this range). Furthermore, where a purely elastic analysis is to be carried out, the value of E_{c0} from equation (7.6) should be reduced by 15%, to account for the initial plastic strain.

Numerous similar models have been suggested in the literature. Most of them use as parameters the cylinder strength f_c and the corresponding strain ε_{c1}, and also the initial tangent modulus E_{c0}. For estimating the latter, various procedures have been proposed (such as equation (7.6)). Carreira and Chu (1985) have suggested a procedure based on concrete density, which is relatively easy to determine, but the resulting equation, which is of the exponential type, requires a numerical solution. Comparative studies using all the aforementioned models, which were incorporated in a computer program for the analysis of R/C sections subjected to monotonic loading until failure (Kappos, 1993) have shown that differences between them are practically negligible up to the attainment of maximum strength, but relatively significant thereafter. Among the previously mentioned models, that by Kent and Park appears to be the most conservative regarding the descending branch of the σ_c–ε_c curve (see also Figure 7.2).

The aforementioned analytical models, as well as the tests on which they were based, refer to static loading conditions, involving strain rates of the order of 10^{-5}

s^{-1}. However, as already pointed out, concrete strength is higher in the case of strain rates higher than the static one. This increase in strength can be estimated as a function of the strain rate ε_c, for instance using the expression adopted by the CEB Code (CEB, 1993)

$$f_{c.dyn} = f_{c.stat}\,(\dot{\varepsilon}_c/\dot{\varepsilon}_{c0})^{1.026a_s} \tag{7.7}$$

where the coefficient a_s can be determined from the following equation:

$$a_s = \frac{1}{5+0.9f_c} \tag{7.8}$$

In equation (7.7) the index 'dyn' refers to the strength under dynamic loading ($\dot{\varepsilon}_c > \dot{\varepsilon}_{c0}$) and the index 'stat' to the strength under static loading ($\varepsilon_{c0} = 3\times10^{-5}$ s^{-1}). Similar expressions have been suggested by various investigators (see for instance Soroushian and Sim, 1986). Equations for the strain corresponding to dynamic strength have also been suggested; the one adopted by CEB (1993) is

$$\varepsilon_{cl.dyn} = \varepsilon_{cl.stat}\,(\dot{\varepsilon}_c/\dot{\varepsilon}_{c0})^{0.2} \tag{7.9}$$

Using the aforementioned expressions it is estimated that for a strain rate of $\dot{\varepsilon}_c = 0.01$, concrete strength increases by about 20% (compared with the static one), for normal concrete grades, but by less than 10% for high concrete grades ($f_c > 60$); the strain at maximum strength increases by 12%, irrespective of the concrete grade. The corresponding values for $\dot{\varepsilon}_c = 0.02$ are about 25% for normal grades, 10% for high grades, and 14% for ε_{cl}. It should be remembered that typical values of strain rates induced in structural members by earthquake motions with normal frequency content range between 0.01 and 0.02 s^{-1}.

Soroushian and Sim (1986) have modified the Park–Kent model to account for strain rate effects. Expressions analogous to (7.7) and (7.9) have been used for $f_{c.dyn}$ and $\varepsilon_{cl.dyn}$, which yield results very close to the CEB equations as far as strength of normal grade concrete is concerned, but substantially different in the case of high grade concrete strength and of strain at maximum strength. For the slope (Z) of the descending branch it was assumed that it increases by the same amount as concrete strength (compare equation (7.7); this assumption is basically in agreement with the (rather limited) corresponding test results. It is pointed out that due to scarcity of data, no expressions for the strain-softening region in the case of high strain rates are included in the CEB Code (CEB, 1993).

7.3.2 Response to cyclic loading

In Figure 7.3 are given σ_c–ε_c diagrams for repeated compression involving loading and complete unloading, referring to cylinders of plain concrete (Karsan and Jirsa, 1969). It is seen in the figure that the slope of the unloading and reloading branches decreases as inelastic deformation increases, which is an indication of the softening of the material due to alternating load cycles. It is also seen that the envelope curve for cycling loading, that is the curve below which lie all σ_i–ε_i points corresponding to successive loading and unloading cycles, practically coincides with the curve resulting from the **monotonic** (i.e. without unloading) application of loading, up to the point of failure, with some discrepancy in the range of large inelastic deformations.

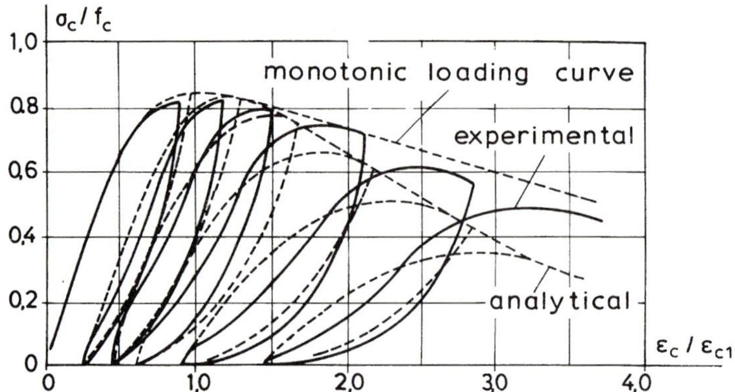

Figure 7.3 Stress–strain diagrams for concrete cylinders subjected to repeated uniaxial compression with full unloading (Karsan and Jirsa, 1969).

Results from experimental studies, such as those shown in Figure 7.3, show that repeated loadings and unloadings do not influence the behaviour of concrete so long as the value of stress σ_c does not exceed about 50% of the dynamic strength in compression, while a substantial decrease in strength as well as in stiffness is observed whenever stress exceeds about 85% of $f_{c,dyn}$ (Aoyama and Noguchi, 1979; Taylor, 1978). Strength and stiffness deterioration becomes more pronounced as the number of loading cycles increases.

Among the earlier analytical models proposed for estimating the response of concrete to cyclic loading, reference will be made here to that of Blakeley and Park (1973), which combines simplicity with a reasonably accurate description of the basic characteristics of the actual behaviour. The basic (skeleton) curve of the model, which forms the envelope of all cycles, is identical to that suggested by Kent and Park (equations (7.1)–(7.4)). As shown in Figure 7.4, for strains $\varepsilon_c \leqslant \varepsilon_{cl}$, unloading as well as reloading, take place along a line (ED in the figure) parallel to the tangent at the origin of the curve (slope equal to E_{c0}). This rule implies that prior to the development of maximum stress reloading takes place without energy dissipation and without stiffness deterioration. In the region of tensile stresses loading and unloading also take place along straight lines with a slope E_{c0}, until the tensile strength is attained. According to the CEB Code (CEB, 1993) this value is $f_{ct} = 1.4\,(f_c/10)^{2/3}$ (MPa). Beyond this point it is assumed that concrete no longer carries any tensile stresses. It is pointed out that for the analysis of flexural members appropriately increased values should be used for the flexural tensile strength (CEB, 1993).

For strains $\varepsilon_c > \varepsilon_{cl}$ the model takes into account stiffness degradation by introducing the reduction factor

$$F_c = 0.8 - \frac{0.7\,(\varepsilon_{cm} - \varepsilon_{cl})}{\varepsilon_{c20} - \varepsilon_{cl}} \geqslant 0.1 \tag{7.10}$$

where ε_{cm} is the maximum attained strain at the instant that unloading takes place, and ε_{c20} is the strain corresponding to a stress reduction of 80% (compare Figure 7.2(a)). During unloading it is assumed that 50% of the stress is lost without any

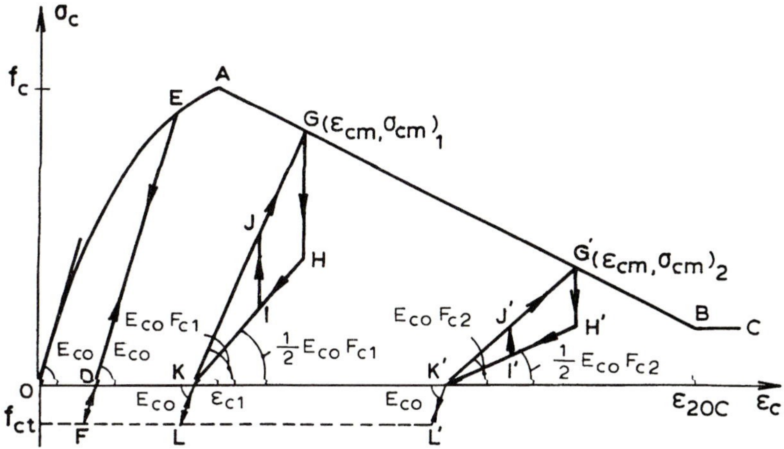

Figure 7.4 Idealized stress–strain curve for concrete subjected to cyclic loading, suggested by Blakeley and Park (1973).

reduction in strain and the subsequent slope of the unloading branch is equal to $0.5F_cE_{c0}$ until stress reaches zero (portions GHK and G′H′K′ in Figure 7.4). If cracking has not occurred, tensile stresses can develop (branches KL and K′L′ in the figure), otherwise strain decreases under zero stress. Reloading takes place with a slope F_cE_{c0} until the envelope is reached (branches KG and K′G′). If reloading takes place before stress reaches zero, somewhere between H and K, the line to be followed is parallel to the initial unloading, that is perpendicular to the strain axis (portion GHIJG in Figure 7.4).

It is clear from the aforementioned hysteresis rules that the Blakeley–Park model takes into account energy dissipation during the loading cycles, as well as stiffness degradation (through factor F_c), although in an approximate way. More refined, but also more complex, models have subsequently been suggested, such as that of Karsan and Jirsa (1969), shown in Figure 7.3 together with corresponding experimental data, where it is seen that the unloading point and the point at which the reloading curve again reaches the envelope do not coincide (their distance increases with the maximum strain), a feature which is not captured by the Blakeley–Park model. More recently proposed models include those by Mander, Priestley and Park (1988) and by Otter and Naaman (1989). Closer fits to the experimental results may be obtained using those models, nevertheless the level of sophistication warranted depends on the goal of the analysis; usually the most important aspect in modelling concrete under compression for practical purposes, is the accurate description of the envelope curve rather than the detailed shape of the reloading and unloading curves.

The response of plain concrete to cyclic loading has a practical significance mainly in the case of repeated compression, studied previously. Indeed, repeated loading in tension has only to be taken into account whenever stress does not exceed the tensile strength (f_{ct}) of concrete. As soon as f_{ct} is exceeded, cracking occurs and subsequent energy dissipation through hysteresis loops in tension is almost negligible, and as a rule is ignored in design-oriented analysis of concrete

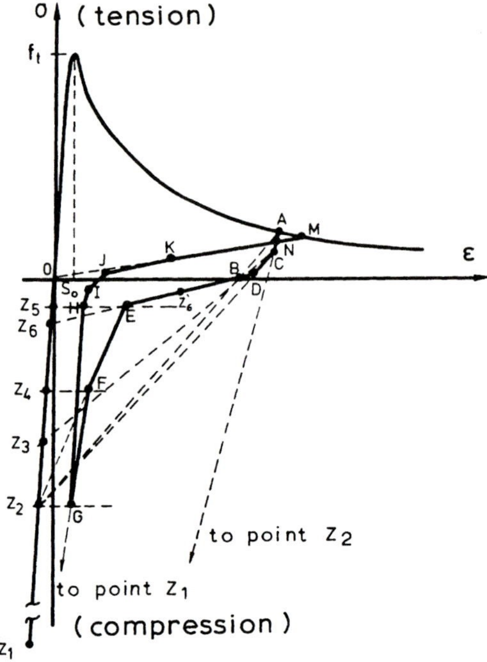

Figure 7.5 Idealized stress–strain curve for concrete under cyclic tension, suggested by Yankelevsky and Reinhardt (1989).

stuctures subjected to seismic actions. However, if a more refined analysis is sought, appropriate models may be used, describing the behaviour of plain concrete under cyclic tension. One of the possible choices is the powerful, yet rather complicated, phenomenological model by Yankelevsky and Reinhardt (1989), the main characteristics of which can be seen in Figure 7.5.

7.3.3 Response to multiaxial loading

Although the study of the response of plain concrete to uniaxial monotonic and cyclic loading, based on data derived from uniaxial tests, is valuable in evaluating the earthquake-resistant properties of the material, it is insufficient with regard to understanding the behaviour of concrete bounded by transverse reinforcement, such as hoops or spirals, which is of paramount importance for elements subjected to seismic loading, as will be explained in detail in section 7.4. The behaviour of concrete confined by transverse reinforcement can be clearly understood if its response to multiaxial loading, in particular the combination of a principal compressive stress and a lateral confining pressure ('hydrostatic' pressure), is known. Furthermore, there are many types of structural elements, such as panels, squat shear walls, low slenderness beams, thin shells, pressure vessels, dams, offshore platforms, and so on, whose stress state under normal, as well as seismic loading is clearly not uniaxial; of particular practical importance in these structures is the existence of tensile stresses, which in most cases are the cause of failure, typically in a splitting mode.

The foregoing remarks clearly point to the need to study the response of concrete under multiaxial states of stress, and this will be done in the following. Clearly a complete treatment of the subject is beyond the scope of this book, which focuses on the seismic behaviour of R/C structural elements, which is usually governed by the presence of reinforcement (longitudinal, as well as transverse), rather than by cracking of concrete. Therefore, the interested reader is referred to the existing literature in the field of multiaxial loading of concrete and the corresponding modelling techniques, developed in the framework of the finite element method (Chen,1982; ASCE, 1982; CEB, 1983, 1991; Meyer and Okamura, 1986).

(a) Biaxial loading

By far the most commonly considered case of multiaxial loading is the biaxial one, in particular that of **plane stress**. This includes, among others, the cases of walls (in particular the squat ones, see also section 9.3.3(b), panels (typically found in precast construction) and deep beams. Depending on the distance between adjacent expansion and/or seismic joints, other types of concrete structures, such as gravity dams, may also fall in this category, rather than in that of **plane strain** (typical of cylinder-like structures).

Before discussing strength and deformation characteristics of plain concrete under biaxial loading conditions, it is worth recalling the fundamental matrix equation of plane stress ($\sigma_z = 0, \tau_{yz}, \tau_{zx} = 0$, with reference to a conventional x-y-z system), for an isotropic element

$$\begin{bmatrix} \sigma_x \\ \sigma_y \\ \tau_{xy} \end{bmatrix} = E/(1-v^2)\begin{bmatrix} 1 & v & 0 \\ v & 1 & 0 \\ 0 & 0 & (1-v)/2 \end{bmatrix}\begin{bmatrix} \varepsilon_x \\ \varepsilon_y \\ \gamma_{xy} \end{bmatrix} \tag{7.11}$$

In equation (7.11) E is the well-known Young's modulus (constant if isotropy is assumed) and v the Poisson's ratio. In terms of principal stresses the equation is written as

$$\begin{bmatrix} \sigma_1 \\ \sigma_2 \end{bmatrix} = E/(1-v^2)\begin{bmatrix} 1 & v \\ v & 1 \end{bmatrix}\begin{bmatrix} \varepsilon_1 \\ \varepsilon_2 \end{bmatrix} \tag{7.12}$$

where ε_1 and ε_2 are the principal strains in the element.

Strength and failure modes It has long been recognized (Kupfer, Hilsdorf and Ruesch, 1969) that the strength of concrete under biaxial stresses depends on the ratio of principal stresses $\alpha = \sigma_2/\sigma_1$; Figure 7.6 shows the strength envelope obtained by Kupfer, Hilsdorf and Ruesch (1969) from biaxial testing of 200 × 200 × 50 mm plates of plain concrete subjected to all possible stress combinations, from biaxial compression to biaxial tension. Some important remarks can be made with reference to this envelope:

1. Under **biaxial compression** the strength of concrete increases up to about 25% for $\alpha \approx 0.5$, for higher values of α strength decreases, and for $\alpha = 1.0$ the strength increase with respect to uniaxial compression is approximately 16%.

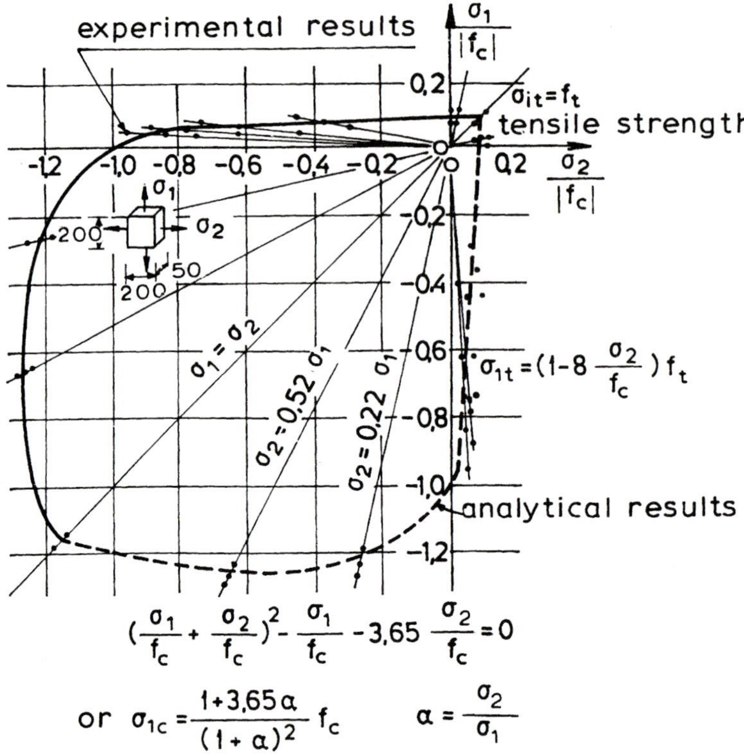

Figure 7.6 Biaxial strength envelope for concrete (adopted from Kupfer, Hilsdorf and Ruesch, 1969).

2. Under **biaxial tension** the tensile strength of concrete is not substantially influenced by the stress ratio α, and is approximately equal to its uniaxial tensile strength; it is noted that other investigations (Tasuji, Slate and Nilson, 1978) have reported strength increases of the order of 10–20% for $\alpha = 0.5$.
3. Under **biaxial tension–compression** the compressive strength decreases with the applied tensile strength; Tasuji, Slate and Nilson (1978) have found an almost linear decrease, while in the envelope of Figure 7.6 a more gradual decrease is observed for low tensile stresses.

Monotonic stress–strain relationship Typical stress–strain curves for concrete subjected to various combinations of biaxial loading, as obtained by Kupfer, Hilsdorf and Ruesch (1969), are shown in Figure 7.7–7.9; note that curves for the strain ε_3 which corresponds to extension of the specimen in the direction orthogonal to its plane, mainly due to Poisson effects, are also given in the figures.

Whenever one of the principal stresses is compressive, the shape of the σ–ε curve is similar to that corresponding to uniaxial compression ($\sigma_2 = 0$), while for the biaxial tension state (Figure 7.9) an almost linear curve up to failure was recorded (a similar shape can be seen in Figure 7.8 for $\alpha = -0.204$, that is for high ten-

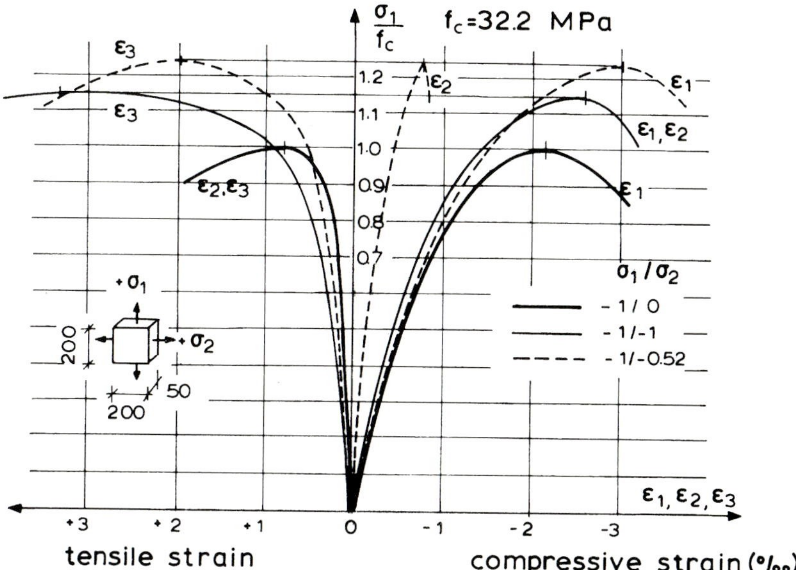

Figure 7.7 Stress–strain relationships for concrete subjected to biaxial compression (adopted from Kupfer, Hilsdorf and Ruesch, 1969).

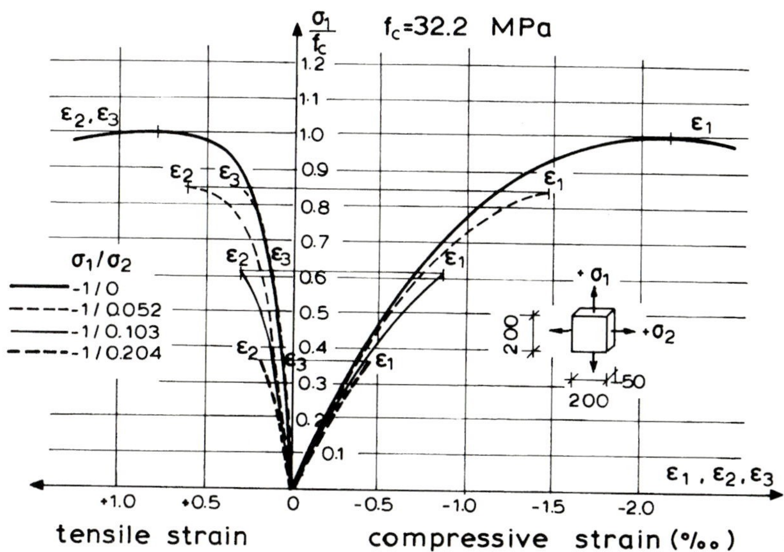

Figure 7.8 Stress–strain relationships for concrete subjected to combined tension and compression (adopted from Kupfer, Hilsdorf and Ruesch, 1969).

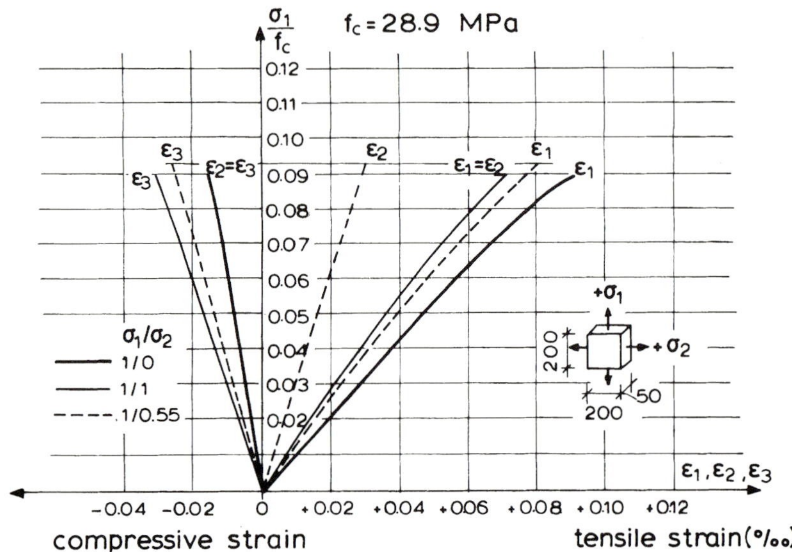

Figure 7.9 Stress–strain relationships for concrete subjected to biaxial tension (adopted from Kupfer, Hilsdorf and Ruesch, 1969).

sile stress in one direction). The descending branch of the biaxial σ–ε curves is either shorter than that recorded for uniaxial compression (Figure 7.1), or non-existent when a predominantly tensile state of stress is applied. This is partly due to the testing procedure, as can be inferred from comparing the uniaxial compression curves of Figures 7.7 and 7.8 with the corresponding curves of Figure 7.1.

Given this, it is probably more appropriate to compare the ductility of biaxially loaded specimens in terms of the strain value at maximum stress rather than in terms of ε_{c50} (equation(7.2)), which is not usually available in biaxial tests. The following remarks can be made in this respect:

1. In biaxial compression the strains ε_1 in the direction of the larger principal stress σ_1 (in absolute terms) increase in a way analogous to strength, that is the strain at maximum stress increases by about 35% with respect to the uniaxial case when $\alpha \approx 0.5$ (from 0.22 to 0.30% in Figure 7.7), but only by about 18% when $\alpha = 1.0$.
2. For combined tension and compression, strains ε_1 decrease with increasing tensile stresses (in the σ_2 direction). As shown in Figure 7.8, a reduction from about 0.22% to only 0.05% was recorded by Kupfer, Hilsdorf and Ruesch (1969) when α was changed from zero to -0.204; very similar values were also recorded in the tests by Tasuji, Slate and Nilson (1978).
3. In biaxial tension there is also a decrease in the strain at maximum stress as the transverse tensile stress increases, but less marked than in the previous case. Average values of the maximum tensile strain ranging from 0.008% (Kupfer, Hilsdorf and Ruesch, 1969) to 0.0015% (Tasuji, Slate and Nilson 1978) have been reported; the latter value has also been adopted by the CEB (1993) Model Code, where analytical expressions for the strain at maximum stress for all combinations of biaxial stress may also be found.

With regard to failure modes, it is recognized that for all biaxial stress combinations failure occurs by **tensile splitting (cleavage)**, with the fractured surface orthogonal to the direction of maximum tensile stress or strain (Tasuji, Slate and Nilson, 1978; Chen, 1982). The magnitude of the tensile strain at failure increases with the degree of compression, which is an indication that concrete can sustain higher indirect tensile strains than direct tensile strains. Tasuji, Slate and Nilson (1978) have suggested linear expressions of the form

$$\varepsilon_u = C_1 + C_2\sigma_m \tag{7.13}$$

for the principal tensile strain ($\times 10^{-3}$) at failure, where $\sigma_m = (\sigma_1 + \sigma_2 + \sigma_3)/3$ and the empirical coefficients C_1, C_2 depend on the biaxial stress state; for combined tension and compression $C_1 = 0.11 \times 10^{-3}$ and $C_2 = -0.024$, if σ_m is expressed in MPa.

Modelling The assumption of isotropy on which the standard equation (7.11) is based, applies only to uncracked concrete and for compressive stress values less than about 80% of the maximum strength, since loading at higher levels induces a strongly inelastic response as indicated by Figures 7.7 and 7.8. For a full-range modelling of concrete subjected to biaxial stress states, various approaches are possible (Chen, 1982; Noguchi, 1986). A classification of the various models used for the description of the behaviour of plain concrete under multiaxial stress states is presented in the next section (referring to triaxial loading, wherein biaxial loading may be treated as a special case). A rather simple model which produces reasonable results, at least for the case of R/C panels, has been proposed by Darwin and Pecknold (1976) and recently been adopted (with minor modifications) by the CEB (1993) Model Code. The Model has the capability of accounting for cyclic loading, thus it will now be presented in some detail.

Stress–strain curves for monotonic biaxial loading (Figures 7.7–7.9) suggest a stress-induced orthotropic behaviour; depending on the ratio $\alpha = \sigma_2/\sigma_1$ and the current values of principal stresses, both the Young's moduli and the Poisson's ratios may be different in each direction, that is $E_1 \neq E_2$ and $v_1 \neq v_2$, as long as microcracking of concrete has started. In this case it can easily be shown that the matrix equation (7.12) can be rewritten in incremental form as

$$\begin{bmatrix} d\sigma_1 \\ d\sigma_2 \end{bmatrix} = 1/(1 - v_1 v_2) \begin{bmatrix} E_1 & v_2 E_1 \\ v_1 E_2 & E_2 \end{bmatrix} \begin{bmatrix} d\varepsilon_1 \\ d\varepsilon_2 \end{bmatrix} \tag{7.14}$$

From energy considerations it may be shown that $v_1 E_2 = v_2 E_1$. To simplify the calculations and ensure that neither principal stress direction is favoured, Darwin and Pecknold (1976) have introduced an equivalent Poisson's ratio $v^2 = v_1 v_2$, which in combination with the aforementioned symmetry condition leads to the following alternative form of equation (7.14):

$$\begin{bmatrix} d\sigma_1 \\ d\sigma_2 \end{bmatrix} = 1/(1 - v^2) \begin{bmatrix} E_1 & v\sqrt{E_1 E_2} \\ v\sqrt{E_1 E_2} & E_2 \end{bmatrix} \begin{bmatrix} d\varepsilon_1 \\ d\varepsilon_2 \end{bmatrix} \tag{7.15}$$

For non-proportional loading the principal stress axes are rotating and a generalized form of equation (7.15) is required to describe the behaviour of a concrete (finite) element. Darwin and Pecknold (1976) suggest the following generalization of equation (7.11) to include orthotropic behaviour:

$$\begin{bmatrix} d\sigma_x \\ d\sigma_y \\ d\tau_{xy} \end{bmatrix} = 1/(1-\nu^2) \begin{bmatrix} E_x & \nu\sqrt{E_xE_y} & 0 \\ \nu\sqrt{E_xE_y} & E_y & 0 \\ 0 & 0 & (1-\nu^2)G \end{bmatrix} \begin{bmatrix} d\varepsilon_x \\ d\varepsilon_y \\ d\gamma_{xy} \end{bmatrix} \tag{7.16(a)}$$

where

$$G = \frac{E_x + E_y - 2\nu\sqrt{E_xE_y}}{4(1-\nu^2)} \tag{7.16(b)}$$

is the shear modulus and ν is the previously defined equivalent Poisson's ratio for the orthotropic material. A complete derivation of equation (7.16), with an extension to the axisymmetric case, may be found in Chen (1982). It is pointed out that the moduli E_1, E_2 and $\nu_1^{1/2}\nu^2$ are stress dependent, and the shear modulus defined in equation (7.16) is independent of direction; the latter is just an approximation to the real behaviour of cracked concrete.

The strain–stress relationship of concrete in biaxial loading is affected by the Poisson's ratio, as well as by cracking. For elastic concrete (prior to microcracking), the slope of the biaxial σ_1–ε_1 curve is equal to

$$E' = \frac{E}{1-\alpha\nu} \tag{7.17}$$

as can be seen from equation (7.12), that is E' varies with the stress ratio α, and for non-proportional loading (for instance, seismic loading) it is not constant. After the onset of microcracking, the tangent modulus $E'_{ti} = d\sigma_i/d\varepsilon_i$ ($i = 1.2$) is also dependent on the stress level (even negative values of E'_{ti} may arise if the descending branch of the σ_i–ε_i curve is taken into account). Various researchers have pointed to the necessity of separating the Poisson effect on E'_i from that of material nonlinearity (cracking); the most efficient proposal appears to be that of Darwin and Pecknold (1976) who defined the **equivalent uniaxial strain** $\varepsilon_{i,un}$ as the strain on the uniaxial loading curve (σ_1–ε_1 or σ_2–ε_2) corresponding to the current stress (on the actual biaxial loading curve), as shown in Figure 7.10. In other words $\varepsilon_{i,un}$ is the strain developing in the direction i, when the stress in the transverse direction is equal to zero. For linear elastic behaviour $\varepsilon_{i,un} = \sigma_i/E_{ti}$, where E_{ti} is the tangent modulus estimated from the equivalent uniaxial (σ_i–$\varepsilon_{i,un}$) curve. Equation (7.14) can now be written as

$$\begin{bmatrix} d\sigma_1 \\ d\sigma_2 \end{bmatrix} = \begin{bmatrix} E_1 & 0 \\ 0 & E_2 \end{bmatrix} \begin{bmatrix} d\varepsilon_{1,un} \\ d\varepsilon_{2,un} \end{bmatrix} \tag{7.18}$$

where it is understood that in the general (inelastic) case E_1 and E_2 are tangent values (E_{t1},E_{t2}). A family of equivalent uniaxial curves σ_i–$\varepsilon_{i,un}$ (each one corresponding to one value of the stress ratio α) are required for defining the material properties in each principal stress direction, in the general case of non-proportional loading, $\alpha = \alpha(t)$. Any appropriate equation can be used for describing these curves, for instance equation (7.5) suggested in the CEB (1993) Code. The maximum stress in each of these curves is estimated by entering biaxial strength envelopes (such as that of Figure 7.6) with the appropriate stress ratio α. The strains corresponding to the biaxial strengths may also be determined on the basis of empirical equations, such as those of the CEB (1993) Model Code, or those suggested by Darwin and Pecknold (1976). Finally, the Poisson's ratio

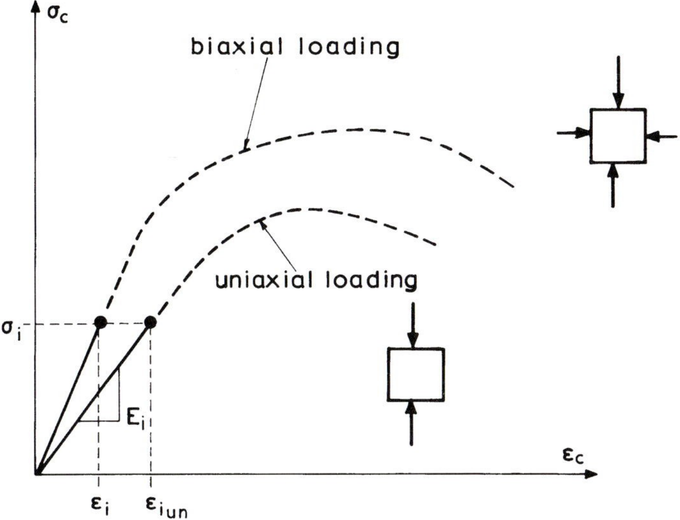

Figure 7.10 Definition of equivalent uniaxial strain.

ν may be assumed constant and equal to 0.2 for biaxial compression and biaxial tension, while for compression (σ_2)–tension (σ_1) the following expression is suggested by the CEB (1993) Code:

$$\nu(\sigma_1,\sigma_2)=0.2+0.6\left(\frac{-\sigma_2}{f_{\text{ctm}}}\right)^4+0.4\left(\frac{\sigma_1}{f_{\text{ctm}}}\right)^4 \not> 0.99 \qquad (7.19)$$

The upper limit to ν is set in order to avoid numerical difficulties, since for $\nu=1$ the term $1/(1-\nu^2)$ in equations (7.15) and (7.16) assumes an infinite value.

Cyclic loading Biaxial cyclic loading tests are much more scarce than corresponding uniaxial tests (section 7.3.2), apparently due to the difficulty in developing appropriate experimental set-ups in the former case. An early investigation by Okajima, involving hollow cylinders subjected to torsion and axial force, is summarized by Aoyama and Noguchi (1979). Figure 7.11 shows **biaxial cyclic compression** curves obtained during a more recent study (Buyukozturk and Tseng, 1984), involving 127 × 127 × 25 mm flat concrete plates subjected to a constant horizontal strain (ε_h) and an alternating vertical strain (ε_v); note that due to the Poisson effect the horizontal stress (σ_h) is not constant and the values (σ_h) shown in the figure refer to initial horizontal stress. These curves are quite similar to those obtained for uniaxial loading (Figure 7.3), in particular there is a gradual decrease of the slopes of the unloading and reloading curves, which is an indication of the progressive degradation of the material. This degradation may be attributed both to microcracking in the unconfined (out of plane) direction, and to the inelastic behaviour of mortar.

The envelope of the biaxial cyclic stress–strain curves was found to be very close to that of the uniaxial curve initially, but for higher strains the envelope lay above the uniaxial curve. A σ_v–σ_h envelope for biaxial compression is shown in Figure

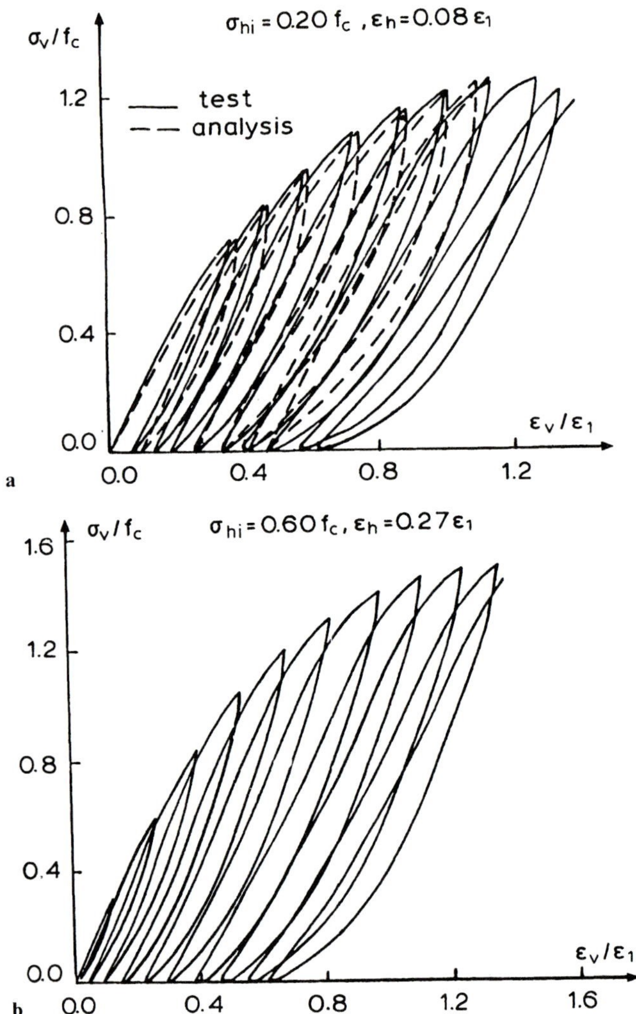

Figure 7.11 Stress–strain curves for concrete subjected to biaxial cyclic loading (adopted from Buyukozturk and Tseng, 1984).

7.12, where the curve suggested by Kupfer, Hilsdorf and Ruesch (Figure 7.6) is plotted, together with results by Buyukozturk and Tseng (1984) from monotonic and cyclic tests. It appears that for the load histories studied, the biaxial strength envelope is practically the same for both monotonic and cyclic loading. Moreover non-proportional loading (variable σ_h/σ_v) appears to result in strengths greater than those under proportional loading; as shown in Figure 7.12 strength increases up to approximately 40% with regard to the unconfined strength (f_c) were recorded for cyclic non-proportional loading.

Based on the foregoing discussion, it appears that models for concrete subjected to uniaxial cyclic loading might also be used in the case of biaxial loading, pro-

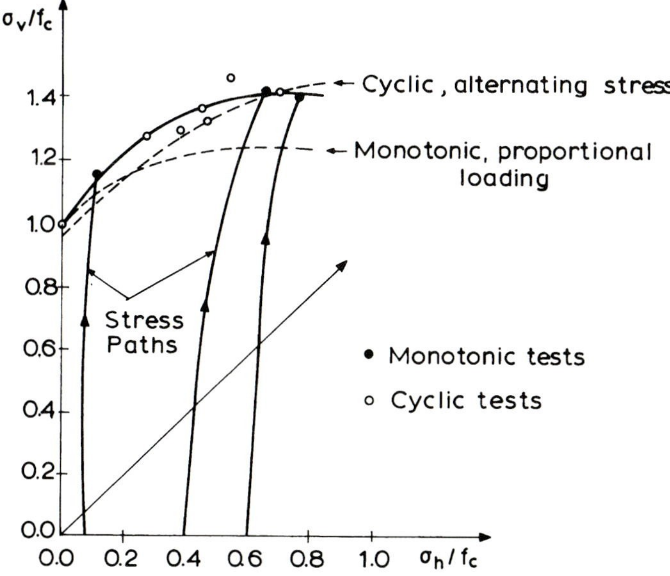

Figure 7.12 Biaxial strength envelope for monotonic and cyclic loading (Buyukozturk and Tseng, 1984).

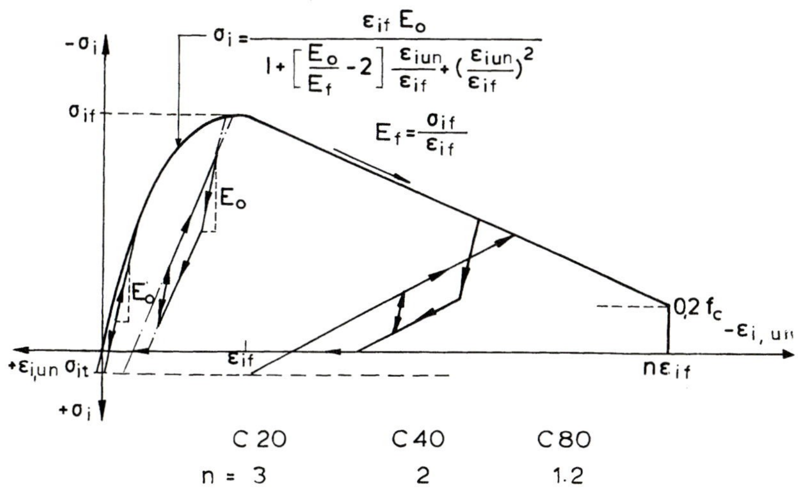

Figure 7.13 Equivalent uniaxial curve for cyclic loading (CEB, 1993).

vided an appropriate envelope curve is defined. The previously described model by Darwin and Pecknold (1976) takes into account cyclic loading, based on the (stress)–(equivalent uniaxial strain) curve shown in Figure 7.13; the hysteresis rules of the model are a simplified (piecewise linear) version of the Karsan–Jirsa

(1969) model, shown in Figure 7.3. For an incremental analysis the equivalent uniaxial strain for cyclic loading is calculated from the equation

$$\varepsilon_{i,\mathrm{un}} = \Sigma\,(\Delta\sigma_i / E_i) \tag{7.20}$$

where $\Delta\sigma_i$ is the incremental change in stress, E_i the tangent modulus estimated from the curve of Figure 7.13, and summation is carried out for all load increments. The skeleton curve shown in Figure 7.13 is not the one originally proposed by Darwin and Pecknold (1976), but its extension included in the CEB (1993) Model Code, which accounts for the effect of concrete grade on the slope of the descending branch.

(b) Triaxial loading

Although virtually all structures are typically subjected to a triaxial state of stress, it is only for certain special structures such as containment vessels, prestressed concrete reactor vessels, offshore platforms, submerged structures, and concrete gravity dams, that triaxial loading is actually accounted for in analysis. As these structures fall outside the scope of this book, only a brief reference will be made here to the demanding and complex subject of concrete under triaxial loading, the emphasis being on triaxial compression, which is the basis for understanding the behaviour of confined concrete (dealt with in the next section).

Experimental techniques for studying the behaviour of plain concrete under triaxial states of stress are quite sophisticated and expensive, and a considerable amount of scatter is usually found. This was clearly demonstrated in the course of a major joint research programme undertaken by seven groups from Europe and the United States, wherein different techniques for applying triaxial loading to concrete cubes or (less often) cylinders were used and critically evaluated (Gerstle *et al.*, 1980). The main conclusion from this cooperative study, regarding experimental procedures, was that the scatter diminishes considerably when no constraints are present on the specimen boundaries. In order to obtain unconstrained specimen boundaries the platens through which loading is applied on the specimen should be of the 'brush' or the 'flexible platen' type; these platens present virtually no resistance in the transverse direction, but are stiff enough not to buckle when applying longitudinal compression. A further conclusion reached during that study was that because of the brittle failure associated with unconstrained specimen boundaries, it was difficult to obtain strain measurements near or past maximum strength, therefore the post-peak behaviour of concrete (which is of great interest with regard to seismic response) was not studied. It is worth pointing out that so far no data on the post-peak softening behaviour of plain concrete under triaxial loading are available (CEB, 1991).

Strength and failure modes The strength of concrete under triaxial loading is a function of the three principal stresses σ_1, σ_2, σ_3. Depending on the amount of tension present, the failure mode can be quite different, as shown in Figure 7.14 (CEB, 1991). For predominantly tensile stresses failure occurs along a well-defined direction and is characterized by a single (localized) crack; in this case concrete behaves as a brittle softening material. For predominantly compressive stresses a more ductile behaviour is exhibited, characterized by more cracks distributed along a broader failure zone. The transition point (TP in Figure 7.14) separates the brit-

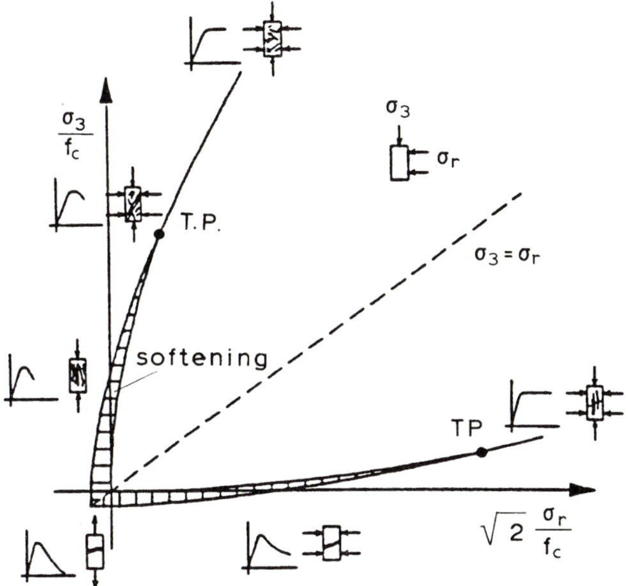

Figure 7.14 Failure modes of concrete under triaxial loading (uniform confining pressure $\sigma_1 = \sigma_2 = \sigma_r$).

tle from the ductile region, while the hatched area represents post-peak softening behaviour, which nevertheless is quite different depending on whether the principal stress σ_3 is compressive or tensile. Under hydrostatic compression ($\sigma_1 = \sigma_2 = \sigma_3$) extremely high values of concrete strength can be achieved; the compaction of the cement paste becomes increasingly pronounced, but this alone cannot lead to complete failure (disruption) of the material. In general, as the confining stress (σ_r) increases, the failure mode changes from cleavage (splitting) to crushing of cement paste; the latter is an uncommon behaviour in most practical situations.

In the case that all three stresses vary independently ($\sigma_1 \neq \sigma_2 \neq \sigma_3$) three invariants are required to describe the failure surface of plain concrete (Chen, 1982) and the resulting criteria become rather complicated. The failure criterion suggested by Ottosen (1977), which involves the first invariant of the stress tensor (I_1), and the second and third invariant of the stress deviator tensor (J_2,J_3), was adopted in the CEB (1993) Model Code, mainly because it agrees well with test data. A graphic representation of this criterion (adopted from Mueller and Hilsdorf, 1993) is shown in Figure 7.15, while the detailed equations may be found in the CEB (1993) Code, as well as in the aforementioned paper by Ottosen (1977). It is worth pointing out that this criterion does not allow the derivation of explicit expressions for the principal stresses at failure for a given stress state; this is the main reason why for the common case of biaxial loading the CEB Code proposes the use of the simpler criterion of Kupfer, Hilsdorf and Ruesch (1969), already presented in the previous section (Figure 7.6). The Kupfer criterion gives more conservative values than the Ottosen criterion, but the agreement between both predictions is acceptable (Mueller and Hilsdorf, 1993).

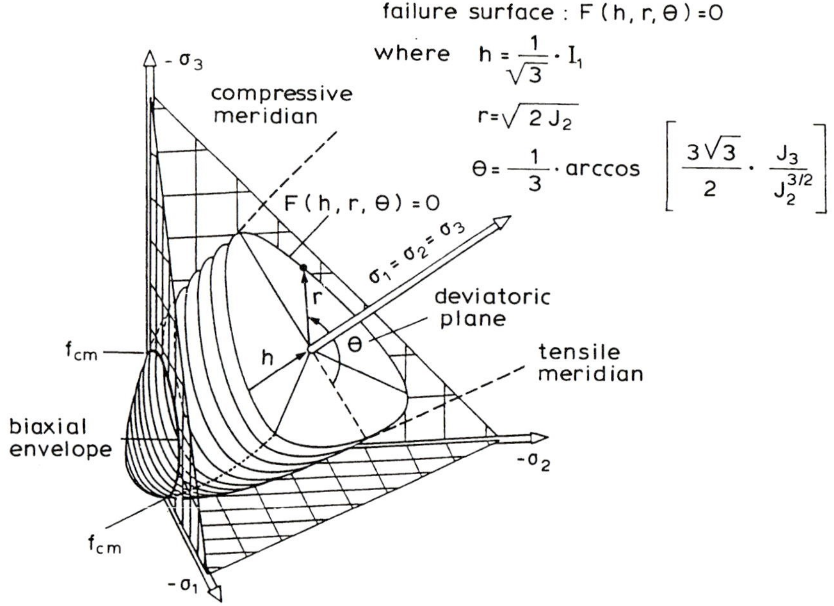

Figure 7.15 Ottosen (1977)–CEB (1993) failure criterion, presented in the three-dimensional principal stress state.

Monotonic stress–strain relationship A variety of load paths involving hydrostatic stresses (σ_0), as well as deviatoric stresses (τ_0) can be followed in a triaxial loading test, and stress–strain relationships may be drawn either for the principal stresses or for σ_0 and τ_0, which are commonly expressed in their 'octahedral' form, that is

$$\sigma_0 = \frac{1}{3}(\sigma_1 + \sigma_2 + \sigma_3) \tag{7.21}$$

$$\tau_0 = \frac{1}{3}[(\sigma_1 - \sigma_2)^2 + (\sigma_2 - \sigma_3)^2 + (\sigma_3 - \sigma_1)^2]^{1/2} \tag{7.22}$$

By far the most commonly used load path is a hydrostatic compression up to a certain stress $\sigma_1 = \sigma_2 = \sigma_3 = \sigma_0$, followed by an increase of only one principal stress (say, of σ_3) while the lateral confining pressure $\sigma_1 = \sigma_2$ is held constant (triaxial compression test). Typical stress–strain diagrams from such tests (Hobbs, Newman and Pomeroy, 1977) are shown in Figure 7.16, where it is seen that both the strength and the deformation corresponding to peak stress (which is an indication of the ductility of the material) increase significantly with the amount of the lateral stress ($\sigma_1 = \sigma_2$). For a relatively high value of lateral stress $\sigma_1 = 0.7f_c$ the strength in the major principal direction (σ_3) is more than four times the uniaxial strength (f_c), while the corresponding strain at peak stress ($\varepsilon \approx 2.5\%$) is about 10 times larger than that measured in uniaxial compression tests. It will be seen in the following section (7.4) that these characteristics are particularly favourable with regard to the earthquake response of R/C structural members.

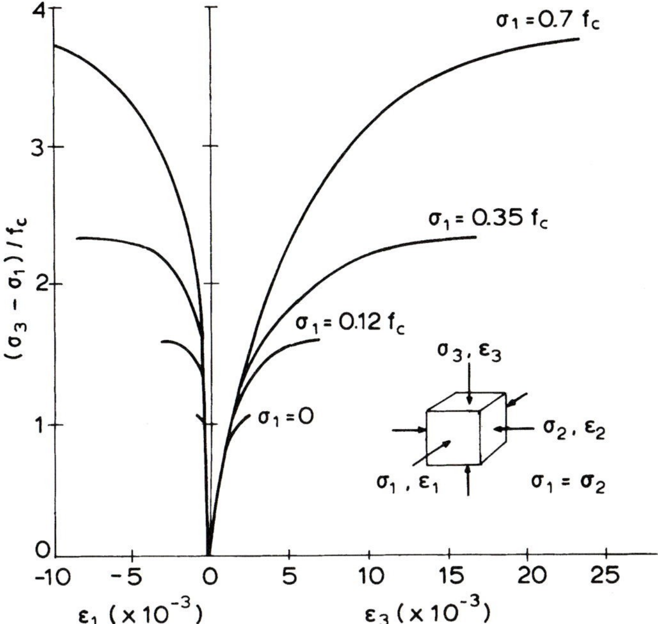

Figure 7.16 Typical stress–strain curves for plain concrete subjected to triaxial compression (Hobbs, Newman and Pomeroy, 1977).

Notwithstanding the fact that very high strengths can be obtained under hydrostatic compression, the σ_0–ε_0 relationship (ε_0 is the octahedral normal or volumetric strain) is clearly nonlinear, with initially softening behaviour (compare Figure 7.16), a possible reversal in curvature for higher σ_0 (Chen, 1982), and a residual slope equal to 15–20% of the initial value (CEB, 1991).

Another type of triaxial test, which is necessary for determining the failure surfaces on the deviatoric planes (Figure 7.15), involves hydrostatic stressing up to a level of σ_0, followed by simultaneous increase of one principal stress and decrease of the other two, so that σ_0 remains constant and only τ_0 varies (Figure 7.17(a)). The shape of the τ_0–γ_0 curves (Figure 7.17(b)) is similar to that of the σ_0–ε_0 curves (compare Figure 7.16), characterized by a gradually softening behaviour. Both the maximum shear stress and the corresponding strain increase when the confining pressure (σ_0) increases. It is worth pointing out that the volumetric strain ε_0 does not remain constant during the aforementioned test (σ_0 = constant), instead ε_0 increases for increasing τ_0, the τ_0–ε_0 coupling being stronger for higher confining pressures σ_0. Moreover, beyond a level of τ_0 which is a function of σ_0, the incremental deformation $d\varepsilon_0$ changes from compaction to dilatancy; depending on the value of σ_0 the volume change prior to failure may be either negative (compaction) or positive (CEB, 1991).

Modelling Different approaches of varying complexity are possible for modelling the behaviour of plain concrete subjected to triaxial loading. The models that

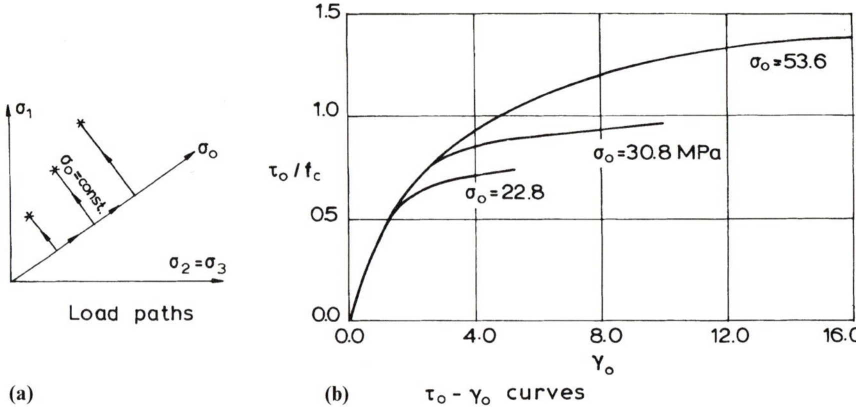

Figure 7.17 Typical deviatoric stress–strain curves for plain concrete subjected to triaxial shear loading (Kotsovos and Newman, 1979; adopted from CEB, 1991).

have been used so far can be classified into the following categories (Chen, 1982; CEB, 1991):

1. *Linear–elastic fracture models* are the simplest ones, since concrete is treated as an elastic material until it reaches its ultimate strength (according to the adopted 'failure' criterion), subsequent to which it fails in a brittle manner. These simple models can be quite accurate in cases of proportional loading whenever the tensile failure of concrete dominates the inelastic response of the structure, but they are not capable of identifying inelastic deformations and load reversals and therefore are inappropriate for seismic analysis.
2. *Nonlinear–elastic models*, especially those of the incremental or hypoelastic type, using variable tangent moduli for describing material stiffness, can take into account inelastic deformations and cyclic loading, hence they are suitable for describing the behaviour of concrete under seismic loading. These models can further be classified into those based on the equivalent uniaxial strain concept (section 7.3.3 (b)), and those based on stress invariants such as the octahedral stresses defined by equations (7.21) and (7.22) (Gerstle *et al.*, 1980).
3. *Plasticity-based models*, which in their simpler form use the perfect plasticity concepts of plastic yield surfaces (in the stress space) and the assumption that the plastic strain increment vector is normal to the yield surface (associated flow rule). More advanced (and more complicated) models are based on strain or work-hardening plasticity, wherein inelastic behaviour initiates whenever the initial yield surface is exceeded, and further loading, involving inelastic deformations, is controlled by subsequent loading surfaces and corresponding flow rules (Han and Chen, 1985). In general these models are not capable of describing inelastic unloading and reloading, strength and stiffness degradation and post-peak softening behaviour, therefore they are not appropriate for seismic analysis.

   The recent introduction of internal state variables, the **plastic damage** variables which depend on the plastic deformations only, has extended the application of plasticity models to cases that were previously beyond reach, such as the post-peak softening behaviour (descending branches of the σ_c–ε_c curve), the inelastic unloadings and reloadings, and the coupling between elastic and plastic deformations. Furthermore, the concept of the **bounding surface**, which encloses all the possible stress points and shrinks in size as damage accumulates, is often (but not necessarily) used in conjunction with the plastic damage variables, and proved to be quite effective for describing strain hardening and softening, as well as unloading and reloading (Yang, Dafalias and Herrmann, 1985; Chen and Buyukozturk, 1985).

4. *Models based on the endochronic theory of inelasticity*, which was an attempt to develop a continuous model for the inelastic behaviour, which did not require the existence of a yield condition. The theory is based on the concept of intrinsic (or endochronic) time, defined in terms of strain or stress and used to measure the degree of damage to the internal structure of the materials, and was originally developed for metals. Bazant and Bhat (1977) have subsequently extended the theory to rock, sand, plain concrete and reinforced concrete under various loading conditions. The original theory was able to describe, among others, strain hardening, unloading, and the pinching of hysteresis loops under cyclic loading. The main extensions introduced by Bazant and Bhat were the description of the coupling between inelastic shear deformation and hydrostatic stress, the post-peak strain-softening behaviour, and the inelastic dilatancy of concrete. The numerous numerical coefficients required for the development of a corresponding constitutive low were estimated by curve fitting of available experimental data. Notwithstanding the criticism regarding the stability of the model during small-amplitude stress and strain cycling (Chen, 1982) and the use of a constant shear retention factor at the cracks, it appears that the main obstacle to the application of the model, which has not undergone further development in the last 15 years, is the large number of parameters required (CEB, 1991).
5. *Fracturing models* and *continuum damage models* are both based on the concept of gradual growth of microcracks within concrete under increasing stress. The first class of models bears a strong resemblance to plasticity models, as it associates an elastic component of stress to a 'fracturing' stress decrement, governed by a potential function defined in the strain space (Dougill, 1976). The second class of models is based on the use of a set of state variables quantifying the internal damage resulting from a certain loading history. It is pointed out that the models of this category describe progressive damage of concrete occurring at the microscopic level, through variables defined at the level of the macroscopic stress–strain relationship (Krajcinovic and Fonseka, 1981).
6. *Micromechanics models* represent an attempt to develop the macroscopic σ_c–ε_c relationship starting from the mechanics of the microstructure. The only model of this type that appears to have reached a stage of practical implementation is the **microplane** model, suggested by Bazant and Prat (1988). It is a highly simplified representation of the microstructure of concrete, in which the contiguous grains of the material exchange forces on (micro) planes passing through their contact points. The stress–strain relationships for each force component acting on these planes determine the behaviour of the model at the macroscopic level.

A detailed presentation, or even a comparative assessment of the aforementioned models, is clearly beyond the scope of this book; detailed reviews may be found in Chen (1982), which covers the work up to about 1981 in a detailed and systematic way, and in the CEB Task Group 22 Report (CEB, 1991), where recent (up to about 1989) advances in modelling of concrete under multiaxial loading are outlined with emphasis on alternating (cyclic) actions. In the following a relatively simple model (Elwi and Murray, 1979), falling in the category of hypoelastic models mentioned previously, will be briefly presented. This model, which is an extension of that by Darwin and Pecknold (1976) for biaxial loading (see previous section), offers the advantage that it can treat in a straightforward way the case of cyclic loading.

Assuming that inelastic concrete behaves as an orthotropic material, the following constitutive relationship can be derived for the triaxial case, in a way similar to that previously described for the biaxial case (equation(7.16)):

$$\begin{bmatrix} d\sigma_x \\ d\sigma_y \\ d\sigma_z \\ d\tau_{xy} \\ d\tau_{yz} \\ d\tau_{zx} \end{bmatrix} = 1/\phi \begin{bmatrix} (1-\nu^2)E_x & \nu(1+\nu)\sqrt{E_xE_y} & \nu(1+\nu)\sqrt{E_xE_z} & 0 & 0 & 0 \\ & (1-\nu^2)E_y & \nu(1+\nu)\sqrt{E_yE_z} & 0 & 0 & 0 \\ & & (1-\nu^2)E_z & 0 & 0 & 0 \\ & & & \phi G_{xy} & 0 & 0 \\ & \text{(symmetrical)} & & & \phi G_{yz} & 0 \\ & & & & 0 & \phi G_{zx} \end{bmatrix} \begin{bmatrix} d\varepsilon_x \\ d\varepsilon_y \\ d\varepsilon_z \\ d\gamma_{xy} \\ d\gamma_{yz} \\ d\gamma_{zx} \end{bmatrix} \tag{7.23}$$

where $\phi = 1 - 3\nu^2 - 2\nu^3$.

In equation (7.23) it is assumed that a constant Poisson's ratio $\nu_{xy} = \nu_{yz} = \nu_{zx} = \nu$ can be used for all directions; a more general form of the orthotropic stiffness matrix involving $\nu_{xy} \neq \nu_{yz} \neq \nu_{zx}$ may be found in Elwi and Murray (1979) and CEB (1991). The shear modulus in equation (7.23) may be derived from the assumption that it is an invariant under a transformation of axes, resulting in the following expression (analogous to equation (7.16(a)):

$$G_{xy} = \frac{1}{4\phi}\left[E_x + E_y - 2\nu\sqrt{E_xE_y} - \nu(\sqrt{E_x} + \sqrt{E_y})^2\right] \tag{7.24}$$

Similar expressions may be derived for G_{yz} and G_{zx}.

Using the concept of equivalent uniaxial strain (Darwin and Pecknold, 1976), already described in the previous section for biaxial loading, it is possible to write equation (7.23) in the form

$$\begin{bmatrix} d\sigma_1 \\ d\sigma_2 \\ d\sigma_3 \end{bmatrix} = \begin{bmatrix} E_1 & 0 & 0 \\ 0 & E_2 & 0 \\ 0 & 0 & E_3 \end{bmatrix} \begin{bmatrix} d\varepsilon_{1,\mathrm{un}} \\ d\varepsilon_{2,\mathrm{un}} \\ d\varepsilon_{3,\mathrm{un}} \end{bmatrix} \tag{7.25}$$

where E_1, E_2 and E_3 are the tangent moduli in each principal stress direction, defined by a family of equivalent uniaxial curves σ_i–$\varepsilon_{i,\mathrm{un}}$ ($i = 1,2,3$), each one corresponding to a specific combination of σ_1, σ_2, σ_3. Equations such as the one suggested by Elwi and Murray (1979) or the one (7.5) included in the CEB (1993) Model Code may be used for describing the σ_i–$\varepsilon_{i,\mathrm{un}}$ curves. The peak stress σ_{if} in each curve has to be estimated from an appropriate ultimate strength surface, such as that shown in Figure 7.15 (Ottosen, 1977; CEB, 1993). In addition to σ_{if}, the corresponding strains $\varepsilon_{if,\mathrm{un}}$ have to be estimated, in order to define the σ_i–$\varepsilon_{i,\mathrm{un}}$ curve. These hypothetical quantities (as explained in the previous section $\varepsilon_{i,\mathrm{un}}$ is a fictitious parameter) can be estimated by assuming that there is a surface in the equivalent uniaxial strain space, which has the same form as the ultimate strength surface (Elwi and Murray, 1979). Alternatively, the expressions included in the CEB (1993) Model Code, which are of the type

$$\varepsilon_1 = \frac{1}{E_{csa}}[\sigma_1 - v_{csa}(\sigma_2+\sigma_3)] \tag{7.26}$$

may be used (expressions for ε_2, ε_3 are obtained from equation (7.26) by permutation of the indices 1, 2, 3). In equation (7.26) E_{csa} is the actual secant modulus of elasticity, which is assumed to depend on the major principal stress ratio σ_3/σ_{3f} and on the ultimate value E_{cf} which is defined on the basis of the second invariant (J_2) of the deviatoric stress tensor; detailed equations may be found in CEB (1993). The actual Poisson's ratio v_{csa}, which also depends on the stress level, may be taken equal to its initial value $v_0 = 0.1$–0.2 for $\sigma_3/\sigma_{3f} \leqslant 0.8$, while for higher stress levels it can be estimated from equation

$$v_{csa} = 0.36 - (0.36 - v_0)\sqrt{1-(5\sigma_3/\sigma_{3f} - 4\ )^2} \tag{7.27}$$

Reasonable agreement with test data for both biaxial and triaxial loading was found by Elwi and Murray (1979) when they used the aforementioned orthotropic triaxial model.

Cyclic loading While the definitions of unloading and reloading are clear in uniaxial loading (section 7.3.2), this is not always the case in multiaxial non-proportional loading, where a loading–unloading criterion is required. This criterion may be different, depending on whether it refers to a principal stress–strain relationship σ_i–ε_i (i = 1, 2, 3), or to an octahedral relationship σ_0–ε_0 or τ_0–γ_0 (see previous section). In the latter approach, which appears to be the most commonly adopted one, both unloading and reloading follow an essentially elastic path, in other words the slopes of the σ_0–ε_0 or τ_0–γ_0 curves during unloading and reloading are very close to the initial ones (CEB, 1991). It is pointed out, however, that the validity of the previous remark has only been verified for a multiaxial loading prior to attaining ultimate conditions. Considering the behaviour of plain concrete in uniaxial cyclic loading (Figure 7.3), it can be anticipated that some stiffness degradation during repeated loading and unloading cycles in the post-peak range will also be present in triaxial loading; it is understood that the degree of degradation depends on the amount of hydrostatic compression.

It has been noted previously that there is a coupling between τ_0 and ε_0, as well as between σ_0 and γ_0. Available experimental evidence (Kotsovos and Newman, 1979; Stankowski and Gerstle, 1985) indicates that in these relationships (τ_0–ε_0 and σ_0–γ_0) no deformation is recovered upon unloading, and no further deformation occurs before τ_0 (or σ_0) exceeds its value at the beginning of unloading.

In Figure 7.18 are shown σ_i–ε_i ($i = x, y, z$) curves for more complex triaxial loading histories involving partial unloading and reloading (Stankowski and Gerstle, 1985); note that x, y, z typically coincide with the principal stress directions ($i = 1$, 2, 3) but the x, y, z notation is preferred to avoid confusion with the stress path numbers (1, 2, 3, 4, 5, see Figure 7.18(a)). An initial hydrostatic stress $\sigma_0 = 55.2$ MPa was first applied and induced inelastic deformation in the specimen; different paths involving the deviatoric stresses $s_i = \sigma_i - \sigma_0$ were followed thereafter. Path 1

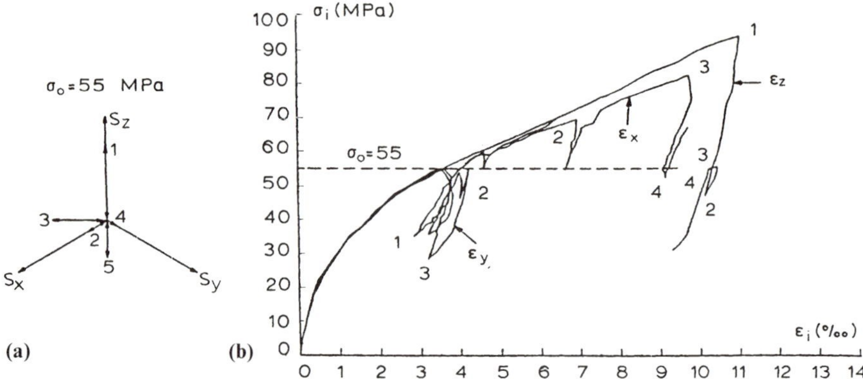

Figure 7.18 Triaxial repeated loading of plain concrete: (a) stress paths in terms of the deviatoric stresses s_x, s_y, s_z; (b) stress–strain relationships σ_i–ε_i, $i=x$, y, z (adopted from Stankowski and Gerstle, 1985).

consisted in increasing σ_z, while σ_x and σ_y were decreased in order to keep σ_0 constant. As shown in Figure 7.18(b), the σ_z–ε_z curve continued to proceed inelastically, while the slopes of the unloading curves σ_x–ε_x were close to their initial elastic values. Also close to the elastic value was the slope of the σ_z–ε_z curve when unloading occurred. In path 2 σ_y and σ_z decreased and σ_x increased (with $\sigma_0=$constant), but the increase of σ_x was kept below the value previously attained by σ_z. Again σ_x–ε_x proceeded inelastically, while unloading in the other two directions took place at slopes close to the elastic ones (Figure 7.18(b)). Paths 3, 4 and 5 confirmed that loading (and corresponding inelastic deformation) occurs whenever at least one principal stress exceeds its previously attained peak value. Chen and Buyukozturk (1985) suggested the criterion $dI_1 \geqslant 0$ for hydrostatic loading (I_1 is the first invariant of the stress tensor, $I_1=\sigma_1+\sigma_2+\sigma_3$), and $dD_i \geqslant 0$ for deviatoric loading, where D_i is the normalized distance from the current stress point to the bounding surface (see previous section on classification of models for triaxial loading), along the direction of the deviatoric stress s_i.

Currently available models for triaxial cyclic loading of concrete incorporate quite sophisticated criteria for loading and unloading (such as the aforementioned one), as well as rules for describing the post-peak behaviour, the stiffness degradation during repeated cycling, and so forth. The previously described model based on the equivalent uniaxial strain concept may incorporate any type of σ_i–ε_i relationship for cyclic loading (compare section 7.3.2). Stankowski and Gerstle (1985) have used a different approach involving the increments $d\sigma_0$ and $d\tau_0$ of the octahedral stresses and accounting for the experimentally verified fact that the directions of the deviatoric stress and strain increments ($\Delta\tau_0$ and $\Delta\gamma_0$) do not coincide, unless the stress state approaches the failure surface.

As pointed out in the 1991 report of the CEB Task Group 22, the calibration of the existing triaxial models has been done on the basis of (mostly) uniaxial and biaxial cyclic tests, hence they can be essentially considered as reasonable extrapolations from known data.

7.3.4 Relevant code provisions

Very little is said in existing seismic codes regarding requirements for concrete properties necessary to achieve the desired seismic behaviour. EC8 (CEN, 1995) specifies that concrete classes (grades) lower than C16 (characteristic strength $f_{ck} = 16$ MPa) for low ductility (ductility class (DC) 'L'), or C20 for medium and high ductility structures (DC 'M' and 'H') are not allowed. As mentioned in section 7.3.1, higher strength concrete is less ductile, therefore the previous requirements can only be related to bond stress requirements (section 7.5) and possibly to minimum requirements for member strength. However, it has to be pointed out that EC8 does not apply to lightweight concrete, whose seismic behaviour is generally inferior to that of normal weight concrete. Indeed the CEB Seismic Code (CEB, 1985) does not allow the use of lightweight concrete grades higher than LC30 ($f_{ck} = 30$ MPa), unless special proof of their adequate ductility is provided.

An increasing trend towards the use of very high strength concrete ($f_{ck} \geqslant 80$ MPa), exists mainly in high-rise R/C buildings, for which seismic design is a major consideration. It is well known (CEB, 1989) that the ductility of these concretes is lower than that of normal grade concrete, however the question is still open as to whether minimum ductility requirements for seismic design could be met by high (or 'ultra-high') strength concrete (Aoyama *et al.*, 1992). Undoubtedly the main consideration with regard to this question is not the behaviour of the material alone, but rather its combined behaviour with tranverse reinforcement. The issue of confinement provided by transverse reinforcement is dealt with in the following sections. Here it will be noted that while most national codes, as well as EC2 (CEN, 1991) are applicable for concrete grades up to 50 MPa, the new CEB Model Code (CEB, 1993) has extended its range of applicability to concrete strengths up to 80 MPa. This clearly reflects the importance of high strength concrete in structures such as high-rise buildings, bridges and offshore structures (CEB, 1989).

7.4 CONFINED CONCRETE

7.4.1 The notion of confinement

It has long been recognized that strength, as well as deformability (ductility) of concrete, substantially increase whenever its state of stress is triaxial compression (section 7.3.3(b)). In practice a loading condition equivalent to hydrostatic compression results when transverse reinforcement in the form of closed ties (hoops) or spirals, prevent lateral 'swelling' of an element subjected to axial compression. The concrete which is affected by this favourable action of the transverse reinforcement is called **confined** concrete. It has to be noted here that some degree of confinement is contributed from longitudinal reinforcement bars, in particular those of large diameter and/or with close spacing. Furthermore a role similar to confinement can be played in certain cases by axial loading (development of a triaxial stress state).

As already mentioned in section 7.3.1, the inelastic behaviour of concrete is initiated by the formation of internal bond cracks at the interface between aggregates and mortar, a phenomenon which influences the descending branch of the σ_c–ε_c diagram. The behaviour of the material is affected by confinement from the instant that internal cracking causes an increase of volume in the element ('passive' confinement, as opposed to active confinement by hydrostatic pressure). It follows that transverse reinforcement does not affect the first part of the σ_c–ε_c curve, but its contribution becomes increasingly significant as maximum strength is approached, and it dominates the response in the region of the descending branch.

(a) Advantages of confinement

Confinement offers two main advantages regarding the seismic behaviour of concrete structural elements:

1. It increases strength of concrete, which compensates for possible losses caused by **spalling**, i.e. failure of the cover concrete in an element, which occurs whenever compressive strains in the cover exceed about 0.4%
2. It reduces the slope of the descending branch of the σ_c–ε_c curve, therefore it increases the maximum usable strain ε_{cu} to values much higher than the 0.35% accepted by codes (CEB, 1985; CEN, 1991) for flexural design; in other words the ductility of concrete is increased by confinement. This is the most important effect of the transverse reinforcement and it constitutes the key to satisfying the requirements of modern seismic codes regarding local ductility.

(b) Types of confinement

The numerous experimental studies on the role of confinement (see reviews in Park and Paulay, 1975; Aoyama and Noguchi, 1979; Sakai and Sheikh, 1989) have confirmed that confinement by circular spirals is, in general, more effective than that provided by square or rectangular hoops. As shown in the qualitative diagram of Figure 7.19, confinement by circular spirals can lead to a behaviour close to that caused by a moderate hydrostatic pressure (see also Figure 7.16). This effect is due to the fact that circular spirals, by virtue of their shape, are subjected to hoop tension, creating an uninterrupted confinement pressure (σ_l) along the whole circumference, as shown in Figure 7.20(a). On the other hand square or rectangular hoops can produce substantial amounts of pressure only at their corners, given that lateral expansion of concrete enclosed by the hoops causes an

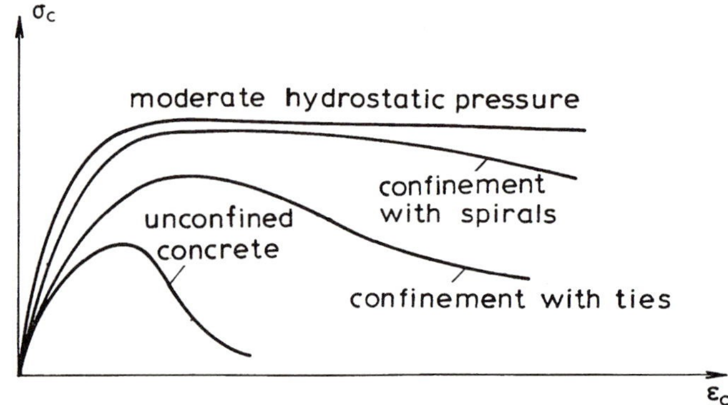

Figure 7.19 Stress–strain diagrams for concrete subjected to various types of confinement.

outward deflection of the hoop legs, leaving parts of the section (arrow-shaped, as shown in Figure 7.20(b)) without confinement. These parts are even larger at the sections between the hoops, as can also be seen in Figure 7.20(b).

7.4.2 Parameters affecting confinement

The main parameters involved in the problem of confinement are the following:

1. *The ratio of transverse reinforcement.* Typically this is expressed as the volumetric ratio ρ_w, defined as the ratio of the volume of hoops to the volume of the confined core of the member. The **core** is the part of the section enclosed by the centroidal axis[1] of the hoop (of the perimeter hoop if multiple hoop patterns are used) (Sheikh and Uzumeri, 1980; CEB, 1985; CEN, 1995). Based on the remarks made in the previous section, it is clear that with increasing ρ_w both the strength and the ductility of confined concrete increase. The quantification of this feature will be given in the next section.
2. *The yield strength of the transverse reinforcement* (f_{yw}). It is understood that the higher the strength of stirrups, the higher the confining pressure they can exert. It is pointed out that in confinement calculations (especially for code purposes) the increase of stress in the transverse reinforcement above f_{yw} (due to strain hardening) is typically ignored.

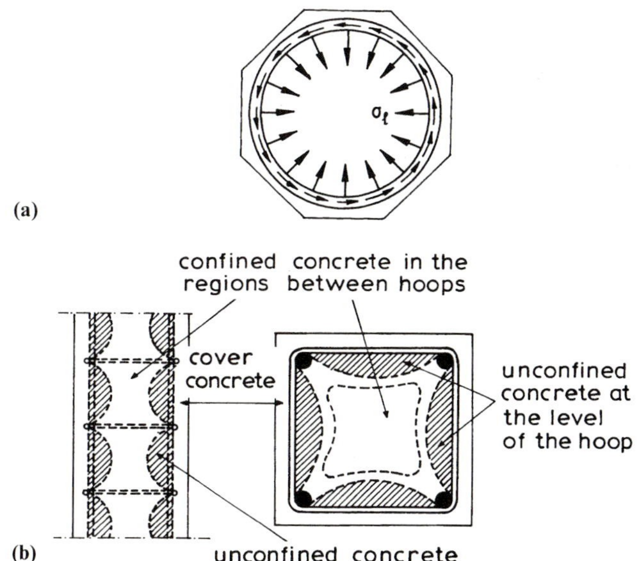

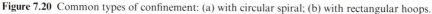

Figure 7.20 Common types of confinement: (a) with circular spiral; (b) with rectangular hoops.

[1] Some investigators (Park and Paulay, 1975; Vallenas and Bertero, 1977) define the confined core with reference to the outside diameter of the hoop. The resulting differences are insignificant, unless the member section is very small.

3. *The compressive strength of concrete* (f_c). As already mentioned in section 7.3.1, higher strength concrete is less ductile than lower strength concrete. Moreover, for the same amount of axial loading the lateral expansion (due to the Poisson effect) of a concrete member is larger in the case of low strength, therefore it is anticipated that (passive) confinement will be more efficient in this case, since the hoops will be stressed more than in a high strength concrete member.
4. *The spacing of hoops* (s). For a given volumetric ratio of hoops (ρ_w), the efficiency of confinement increases as the spacing becomes closer, since the regions of the member which remain without confinement become smaller (Figure 7.20(b)). It is worth pointing out here that closer spacing of the stirrups favourably affects the ductility of a member subjected to compression, since it prevents premature buckling of the longitudinal bars after the spalling of cover concrete.
5. *The hoop pattern*. When multiple hoop patterns (Figure 7.22) are used in a member, the regions of effectively unconfined concrete become smaller (compare Figure 7.20(b)) and strength and ductility increase.
6. *The longitudinal reinforcement*. As already mentioned at the beginning of this section, longitudinal bars (especially if closely spaced) also contribute, to a certain extent, in preventing the lateral expansion of the core, hence they increase confinement effects. The larger the diameter of the bars (d_{bl}) and their ratio ρ_l), the larger their contribution to confinement.

The following two factors, although not directly influencing confinement, are nevertheless important parameters that should be accounted for in modelling the stress–strain characteristics of confined concrete.

1. *The rate of loading*. In the case of seismic actions, it is more correct to refer to the strain rate $\dot{\varepsilon}$, rather than to the loading rate (section 7.3.1). An increase in $\dot{\varepsilon}$. with respect to static conditions leads to a moderate increase in the strength of concrete (equation(7.7)), a reduction in the corresponding strain (equation (7.9)) and a steeper slope of the descending branch of the stress–strain diagram. The foregoing imply that the dynamic strain rates associated with earthquakes have positive, as well as negative, effects on the response of confined concrete.
2. *The strain gradient*. An eccentricity in the axial load, which is typical in R/C columns, that is the presence of a gradient in the strain profile, does not have a significant influence on the strength of confined concrete, nevertheless it improves its overall ductility, since a part of the section is under a more favourable state of deformation (lower strains) than the extreme compression fibre; parts of the section which are in tension are not affected by confinement.

An efficient analytical model for confined concrete should account for all the aforementioned parameters, in the simplest possible way (section 7.4.3(b)).

7.4.3 Confinement with hoops

(a) Monotonic loading

In Figure 7.21 are shown typical (mean) stress–strain diagrams for the confined core of column specimens with various amounts of hoop reinforcement, test-

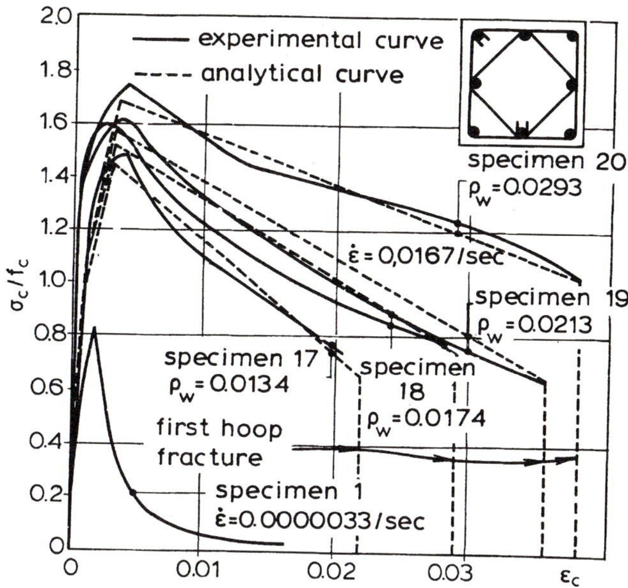

Figure 7.21 Stress–strain diagrams for concrete confined by different amounts of hoops (Scott, Park and Priestley, 1982).

ed by Scott, Park and Priestley (1982) under concentric and eccentric compression, and strain rates ranging from 0.33×10^{-5} s^{-1} (static loading) to 0.0167 s^{-1} (seismic loading). It is first pointed out that all curves for confined concrete are significantly different from the curve for unconfined concrete, the latter corresponding to a specimen identical to the others but without reinforcement. It is pointed out that the strain rate for the confined concrete specimens corresponded to seismic loading conditions, while the plain concrete specimen was tested under static loading, as indicated in Figure 7.21. Therefore, to obtain a direct comparison, the ordinates of the stress–strain diagrams for confined concrete should be reduced by about 24%. Even after this adjustment, the differences between confined and unconfined concrete remain quite remarkable, as both strength and ductility are substantially larger in the case of the confined specimens. The plain concrete specimen reached a strength which did not exceed 86% of the cylinder strength f_c, while the specimens with hoops showed an increase in strength which ranged (after adjustment for strain rate) from 19 to 41%, with respect to the cylinder strength; it is worth pointing out that a column with the typical double hoop pattern shown in Figure 7.21 may reach a strength (in its confined core) up to 80% higher than f_c, when subjected to seismic loading (high strain rate).

The differences between confined and unconfined concrete are even more marked with regard to ductility, measured in terms of ultimate concrete strain. The σ_c–ε_c curves for the specimens of Figure 7.21 are terminated at the point corresponding to the first hoop fracture detected during the test. This point may be used for defining the limiting (or 'ultimate') strain of concrete (ε_{cu}); alternative definitions of ε_{cu} based on specified drops in strength along the descending branch of the σ_c–ε_c curve and on buckling of longitudinal bars are discussed in section 8.4.2(a). As can be seen in Figure 7.21, the recorded values of ε_{cu} ranged from about 2.5 to 4.0%, which means that they are up to an order magnitude higher than the values (0.35–0.40%) for unconfined concrete. The foregoing clearly show the paramount importance of confinement with regard to the earthquake-resistant properties of R/C members.

The σ_c–ε_c diagrams of Figure 7.21 are also useful for understanding the influence of some basic parameters of confinement. Firstly, as expected, both strength and ductility increase as the volumetric ratio (ρ_w) of hoops increases. The influence of hoop spacing can be assessed if the behaviour of specimens 18 and 19 is compared. For the former $\rho_w = 1.74\%$ and $s = 72$ mm, while for the latter the volumetric ratio was higher ($\rho_w = 2.13\%$), but the hoop spacing was also larger. $s = 88$ mm. As shown in Figure 7.21, the peak stress was approximately the same for both specimens, while the differences in the descending branch were not significant. It appears, therefore, that the effect of reducing the hoop ratio in specimen 18 is outweighed to a large extent by the closer spacing used.

The effect of hoop pattern can be seen in the diagrams of Figure 7.22 derived from the tests of Sheikh and Uzumeri (1980) on 305 mm square column specimens with different arrangements of multiple hoops. It is pointed out that according to modern codes (SANZ, 1982; CEB, 1985; CEN, 1995) these patterns are compulsory for potential plastic hinge regions of R/C columns, as will be discussed in section 8.5.2(c). As expected, the relatively most inferior performance was recorded for specimen 4A4-8 which had a single ('diamond' shaped) interior hoop, while the best performance with regard to ductility was achieved by specimen 4C4-12 which had three overlapping interior hoops. It has to be emphasized that the performance of all columns shown in Figure 7.22 can be considered as satisfactory, and differences between specimens with two or more intermediate bars (at each side) supported by a hoop angle were relatively insignificant. It is

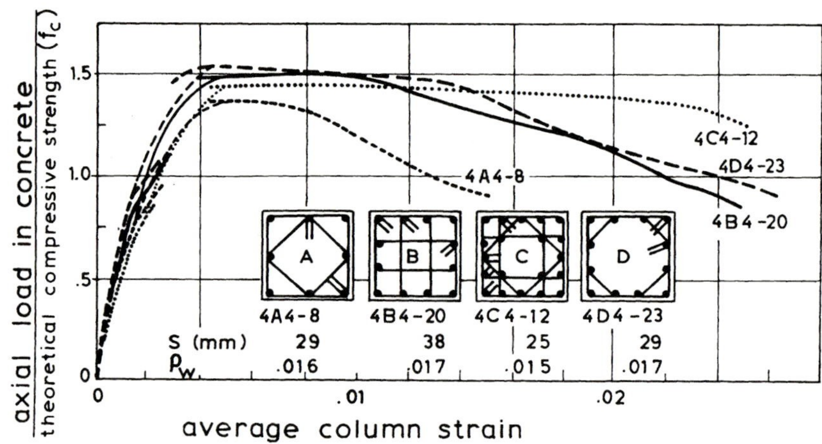

Figure 7.22 Axial load-axial deformation diagrams for columns with different multiple hoop patterns (Sheikh and Uzumeri, 1980).

also worth pointing out that in the test programme from which Figure 7.22 was derived, no single hoop patterns were studied. Columns with a single hoop, which are quite common in existing structures, were found to have a clearly inferior performance (Moehle and Cavanagh, 1985), compared with identical columns with similar ρ_w ratios, but with multiple hoop patterns. Finally, with regard to the influence of longitudinal bars on confinement, it was found (Sheikh and Uzumeri, 1980) that, at least in the case that adequate hoop reinforcement was present, the influence of these bars on the performance of the column was almost negligible.

(b) Analytical modelling

Early models for confined concrete are summarized in Park and Paulay (1975), while early, as well as more recent, models are reviewed by Sakai and Sheikh (1989), and a comprehensive list of references is given. With regard to monotonic σ_c–ε_c relationships, it appears that the most commonly used models are those suggested by Park, Priestley and Gill (1982) and Sheikh and Uzumeri (1982), possibly because they were based on adequate in number, as well as reliable, experimental data. As shown in Figure 7.23, both models are using a parabolic form of the ascending branch (equation (7.1)), the difference with unconfined concrete being that both the peak stress (f_{cc}, the second index standing for confinement)

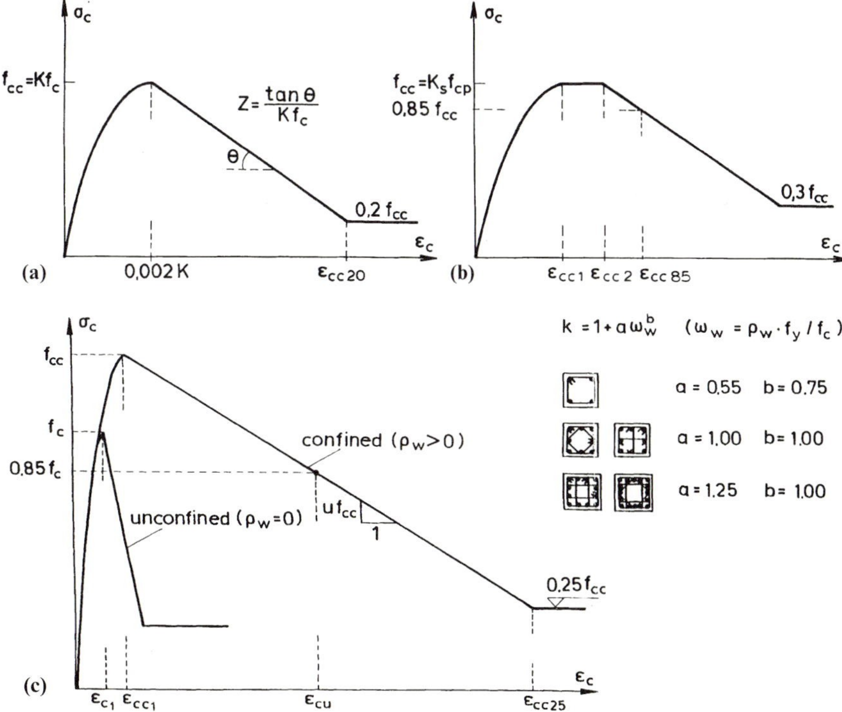

Figure 7.23 Stress–strain models for confined concrete subjected to uniaxial compression: (a) Park, Priestley and Gill (1982) model; (b) Sheikh and Uzumeri (1982); (c) Kappos (1991).

and the corresponding strain (ε_{cc1}) are increased by introducing the **confinement index** K. According to Park, Priestley and Gill (1982) the increased strength of confined concrete is $f_{cc}=Kf_c$, where

$$K=1+\rho_w f_{yw}/f_c \tag{7.28}$$

Equation (7.28) implies that the strength increase due to the presence of transverse reinforcement is proportional to the mechanical ratio (ω_w) of this reinforcement (second term in the right-hand side of equation (7.28)). It will be shown in the next section (7.4.4) that the previous assumption is equivalent to assuming that the efficiency of hoops as confinement reinforcement, with regard to strength of concrete, is approximately equal to 50% of the efficiency of circular spirals.

Sheikh and Uzumeri (1982) attempted to estimate analytically the portion of the core which is under confinement (Figure 7.20) and express the strength increase in terms of the area of the effectively confined section. The resulting relationship for the case of square cross-sections is

$$K=1+\frac{b_c^2}{140P_{0cc}}\left[\left(1-\frac{nb_i^2}{5.5b_c^2}\right)\left(1-\frac{s}{2b_c}\right)^2\right]\sqrt{\rho_w f_{yw}} \tag{7.29}$$

while for the case of rectangular cross-sections a somewhat more complicated expression applies. In equation (7.29) b_c is the size of the (square) confined core, measured to the centroid of the peripheral hoop, b_i is the length of the n equal parts of the core perimeter, defined by longitudinal bars which are supported by a hoop angle, s is the hoop spacing, f_{yw} is the yield strength of the hoops (in MPa) and

$$P_{0cc}=0.85f_c\,(b^2-A_s) \tag{7.30}$$

where A_s is the total area of longitudinal bars (P_{0cc} in kN). The strength of confined concrete is $f_{cc}=K_s f_{cp}$, where $f_{cp}\approx 0.85f_c$ is the compressive strength of a column without transverse reinforcement (Figure 7.19).

It is pointed out that while equation (7.28) oversimplifies the problem by ignoring some important parameters such as the spacing and the pattern of the hoops, equation (7.29) is rather complicated and not suitable for hand calculations. Moreover, both the foregoing models, although predicting reasonably well the behaviour of columns with multiple hoop patterns, lead to very poor results in cases of single tie columns, apparently because they were not calibrated for this case. In a recent study by Kappos (1991), an attempt was made to develop an expression for the confinement index which is simple and applicable for all hoop patterns. Using experimental data from a total of 63 column specimens tested within five different programmes (including the previously mentioned ones by Scott, Park and Priestley, Sheikh and Uzumeri, and Moehle and Cavanagh), the following general equation was suggested:

$$K=1+\alpha(\omega_w)^b \tag{7.31}$$

The following values of the empirical coefficients α and b are suggested (see also Figure 7.24(c))

$\alpha=0.55$ $b=0.75$ for single hoop patterns
$\alpha=1.00$ $b=1.00$ for double hoop patterns (diamond shaped internal hoop or two orthogonal cross-ties)
$\alpha=1.25$ $b=1.00$ for multiple hoop patterns (three or more)

Equation (7.31) provided much better results than the models of Figure 7.23 (a), (b), in the case of single tie columns, and comparable results in the case of double and multiple hoops, for the 63 specimens considered.

With regard to the strain (ε_{cc1}) corresponding to peak stress (Figure 7.23), Park, Priestley and Gill (1982) suggest a value $\varepsilon_{cc1} = K\varepsilon_{c1}$, where $\varepsilon_{c1} = 0.2\%$ is the corresponding value for unconfined concrete, while Kappos (1991) found that values $\varepsilon_{cc1} = K^2\varepsilon_{c1}$ match better the experimentally recorded values. Sheikh and Uzumeri (1982) consider that the stress remains constant between the values ε_{cc1} and ε_{cc2} (Figure 7.23(b)), where

$$\varepsilon_{cc1} = 80K_s f_c \times 10^{-6} \tag{7.32}$$

with f_c in MPa, and

$$\varepsilon_{cc2} = 0.0022\left\{1 + \frac{248}{b_i}\left[1 - 5\left(\frac{s}{b_c}\right)^2\right]\frac{\rho_w f_{yw}}{f_c^{1/2}}\right\} \tag{7.33}$$

with b_i in mm and f_c, f_{yw} in MPa.

With regard to the slope of the descending branch (assumed linear in all the aforementioned models), Park, Priestley and Gill (1982) propose the following generalization of equations (7.2) and (7.4):

$$Z = \frac{0.5}{\varepsilon_{cc50} - \varepsilon_{cc1}} \tag{7.34}$$

where

$$\varepsilon_{cc50} = \frac{3 + 0.29f_c}{145f_c - 1000} + 0.75\rho_w\left(\frac{b_c}{s}\right)^{1/2} \tag{7.35}$$

The same equations, but with f_c replaced by f_c/K, have been adopted in the Kappos (1991) model. A comparison of equations (7.4) and (7.35) shows that the contribution of confinement in reducing the steepness of the descending branch is represented by the second term of (7.35), that is both the volumetric ratio and the spacing of the hoops are taken into account. A similar expression has been used by Sheikh and Uzumeri (1982), where the slope of the descending branch, which starts at a deformation $\varepsilon_{cc2} > \varepsilon_{cc1}$ (Figure 7.23(b)), is given by

$$Z = \frac{0.5}{0.75\rho_w(b_c/s)^{1/2}} \tag{7.36}$$

It will be recalled that the volumetric ratio of hoops (ρ_w) is defined with respect to the centroid of the perimeter hoop in the Sheikh–Uzumeri (1982) and Kappos (1991) model, while Park, Priestley and Gill (1982) define ρ_w with respect to the outside perimeter of the hoop. On the basis of equation (7.36) it follows that the point along the falling branch of the σ_c–ε_c diagram which corresponds to a drop in strength equal to 15% of the peak value (this was the point where the corresponding tests were terminated) has the following abscissa (Figure 7.23(b)):

$$\varepsilon_{cc85} = 0.225\rho_w\sqrt{b_c/s} + \varepsilon_{cc2} \tag{7.37}$$

As shown in Figure 7.23, all the models under consideration suggest a residual strength at high inelastic strains which varies from 20 to 30% of f_{cc}. Such strains can only be achieved in carefully executed displacement-controlled tests; for ductility calculations it is common to define the 'ultimate' strain ε_{cu} at earlier stages, typically on the basis of a 15–50% drop in stress along the descending branch (see also section 8.4.2(a)).

EXAMPLE

The square column section shown in Figure 7.24 is reinforced with eight bars of 22 mm diameter (longitudinal reinforcement ratio equal to 1.5%), the materials used are C16 concrete and S400 steel, and the cover to the bars is equal to 20 mm. Assuming mean values of the material strengths $f_{cm} = 24$ MPa and $f_y = f_{yw} = 420$ MPa, the stress–strain diagrams for the confined core are calculated as follows,

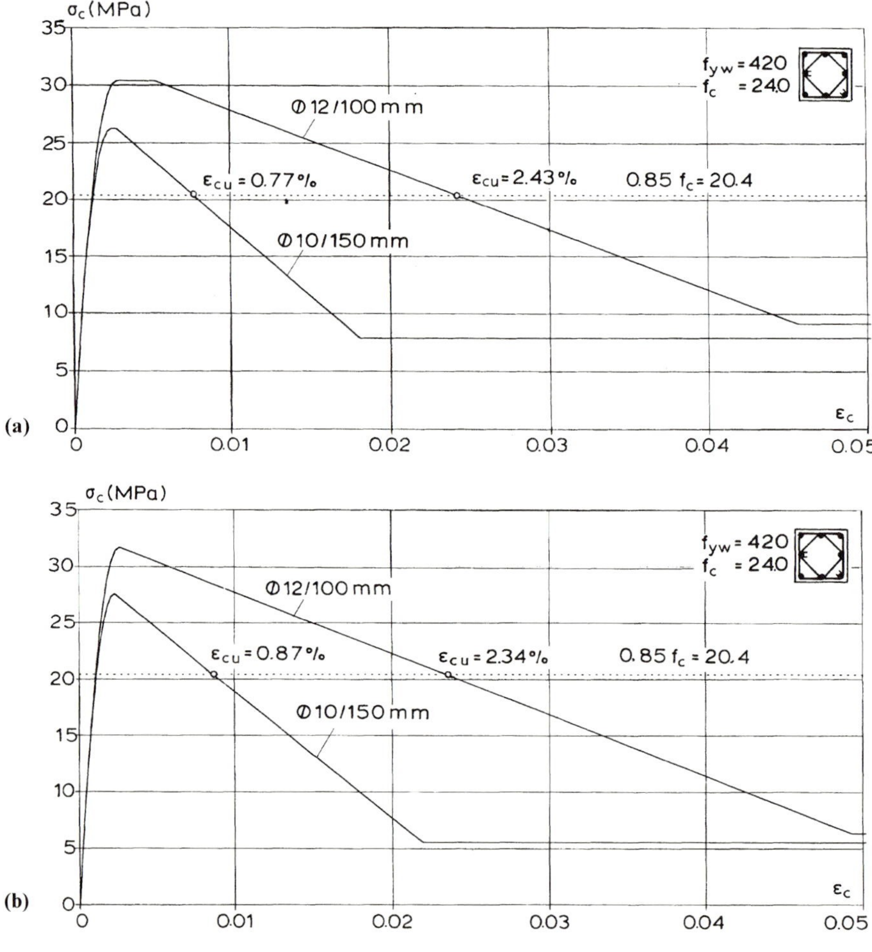

Figure 7.24a,b

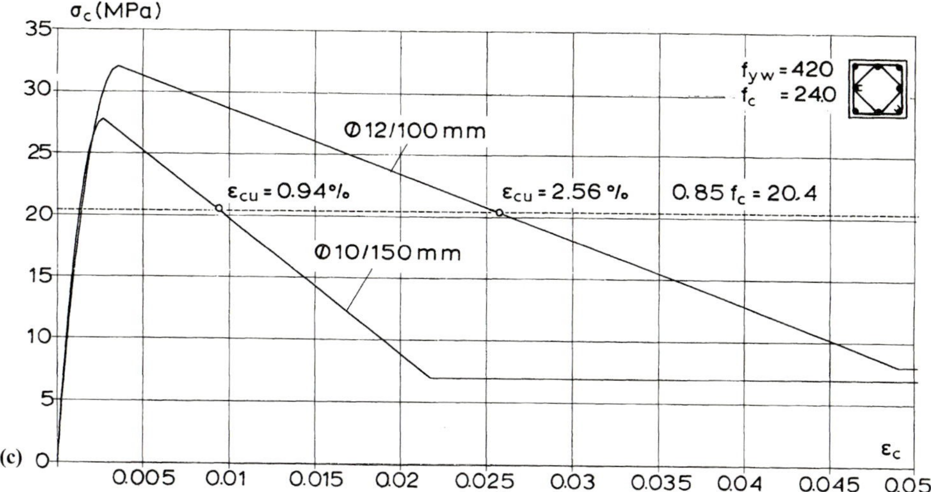

Figure 7.24 Stress–strain diagrams for the confined core of a 450 mm square column, according to: (a) Sheikh und Uzumeri (1982); (b) Park, Priestley and Gill (1982); (c) Kappos (1991).

first for the case of 10 mm double hoops spaced at 150 mm, which is close to the minimum specified by modern codes.

1. According to the Sheikh–Uzumeri (1982) model, the dimension of the core is $b_c = 450 - 2 \times 20 - 10 = 400$ mm, measured to the centroid of the perimeter square hoop. The corresponding volumetric ratio of hoops is

$$\rho_w = \frac{(4 \times 400 + 4 \times 200\sqrt{2}) \times 78.5}{400 \times 400 \times 150} = 0.0089$$

The confinement index K_s can now be calculated from equation (7.29)

$$K_s = 1 + \frac{400^2}{140 \times 3202}\left[\left(1 - \frac{8 \times 200^2)}{5.5 \times 400^2}\right)\left(1 - \frac{150}{2 \times 400}\right)\right]^2 \sqrt{0.0089 \times 420} = 1.288$$

wherein P_{0cc} has been calculated from equation (7.30)

$$P_{0cc} = 0.85 \times 24\,000 \times (0.40^2 - 0.015 \times 0.45^2) = 3202 \text{ kN}$$

Therefore, the strength of confined concrete is

$$f_{cc} = 1.288 \times (0.85 \times 24.0) = 26.3 \text{ MPa}$$

which is 9% higher than the (mean) unconfined strength (63% higher than the code-specified value $f_{ck} = 16$ MPa).

The strains ε_{cc1} and ε_{cc2} can be calculated from equations (7.32) and (7.33), respectively

$$\varepsilon_{cc1} = 80 \times 1.288 \times 24.0 \times 10^{-6} = 0.002\,47 \; (0.25\%)$$

$$\varepsilon_{cc2} = 0.0022\left\{1 + \frac{248}{200}\left[1 - 5\left(\frac{150}{400}\right)^2\right]\frac{0.0089 \times 420}{(24.0)^{1/2}}\right\} = 0.00282$$

while ε_{cc85} is calculated from equation (7.37)

$$\varepsilon_{cc85} = 0.225 \times 0.0089 \times \sqrt{400/150} + 0.00282 = 0.00609 \ (0.61\%)$$

The resulting σ_c–ε_c diagram is shown in Figure 7.24(a).

2. According to the Park, Priestley and Gill (1982) model, the volumetric ratio of hoops defined with respect to the outside of the perimeter hoop is

$$\rho_w = 0.0089 \times \frac{400^2}{410^2} = 0.0085$$

The confinement index is calculated from equation (7.28)

$$K = 1 + 0.0085 \times \frac{420}{24} = 1.149$$

Hence, the strength of the confined core is

$$f_{cc} = 1.149 \times 24.0 = 27.6 \text{ MPa}$$

that is 15% higher than the mean cylinder strength (72% higher than the characteristic strength).
The strain corresponding to peak stress is

$$\varepsilon_{cc1} = 1.149 \times 0.002 = 0.00230 \ (0.23\%)$$

Finally, the slope of the descending branch is calculated from equations (7.34) and (7.35)

$$Z = 0.5\left(\frac{3 + 0.29 \times 24.0}{145 \times 24.0 - 1000} + 0.75 \times 0.0085 \times \sqrt{410/150} - 0.0023\right)^{-1} = 40.8$$

The resulting $\sigma_c - \varepsilon_c$ diagram is shown in Figure 7.24 (b).

3. According to the Kappos (1991) model, for $\rho_w = 0.0089$ (same as in the Sheikh–Uzumeri model), the confinement index may be calculated from equation (7.31):

$$K = 1 + 0.0089 \times 420/24 = 1.156$$

Therefore $f_{cc} = 1.156 \times 24.0 = 27.7$ MPa and $\varepsilon_{ccl} = 1.156^2 \times 0.002 = 0.00267$ (0.27%)
The slope of the descending branch is

$$Z = 0.5\left(\frac{3 + 0.29 \times 24.0/1.156}{145 \times 24.0/1.156 - 1000} + 0.75 \times 0.0089 \times \sqrt{400/150} - 0.00267\right)^{-1} = 39.3$$

The corresponding σ_c–ε_c diagram is shown in Figure 7.24(c).

Also shown in Figure 7.24 are the 'ultimate' strains ε_{cu}, defined on the basis of the $0.85f_c$ stress level (f_c is the strength of unconfined concrete) along the descending branch, which is the conventional strength commonly used for design purposes (CEN, 1991). This value can be found from the following relationship, derived from geometry considerations:

$$\varepsilon_{cu} = \varepsilon_{cc1} + \frac{K - 0.85}{ZK}$$

Thus, for the Kappos (1991) model

$$\varepsilon_{cu} = 0.00267 + \frac{1.156 - 0.85}{39.3 \times 1.156} = 0.0094\ (0.94\%)$$

which is equal to 2.8 times the value (0.35%) commonly used for design calculations. For the Sheikh–Uzumeri model the corresponding value is $\varepsilon_{cu} = 0.77\%$, and for the Park, Priestley and Gill model $\varepsilon_{cu} = 0.87\%$. It is seen that the ratio of maximum to minimum predicted value of ε_{cu} is 22%, and even the minimum ε_{cu} is equal to 2.2 times the conventional value of 0.35%.

The EC8 (CEN, 1995) requirements for confinement in columns with high axial loading are very severe (section 8.5.2(c)), thus it is possible that quite heavier hoop reinforcement might be required for the column under consideration, for instance 12 mm hoops at 100 mm spacing. In this case $\rho_w = 0.0192$ according to the definition used by Sheikh–Uzumeri (1982) and Kappos (1991), while $\rho_w = 0.0183$ according to the Park, Priestley and Gill (1982) definition. The main results for each model are summarized as follows, and the corresponding stress–strain diagrams are shown in Figure 7.24.

1. Sheikh-Uzumeri (1982) model:

$$K_s = 1.491$$
$$f_{cc} = 30.4 \text{ MPa}$$
$$\varepsilon_{cu} = 2.43\%$$

2. Park, Priestley and Gill (1982) model:

$$K = 1.320$$
$$f_{cc} = 31.7 \text{ MPa}$$
$$\varepsilon_{cu} = 2.34\%$$

3. Kappos (1991) model:

$$K = 1.340$$
$$f_{cc} = 32.0 \text{ MPa}$$
$$\varepsilon_{cu} = 2.56\%$$

It is noted that in the case of heavy confinement all models give similar predictions for both strength and ductility (differences do not exceed 9%); this should be attributed primarily to the fact that all the foregoing models were calibrated on the basis of test data mainly involving columns with high ratios of hoop reinforcement. The main conclusion from the example presented is that in cases of heavy confinement (such as that commonly required by EC8), strength increases of the order of 30% and ductility increases of 600 or 700% might reasonably be expected, as also confirmed by the diagrams of Figures 7.21 and 7.22.

Other methods based on similar (Vallenas and Bertero, 1977) or different (Mander, Priestley and Park, 1988) approaches have been suggested in the literature for confined concrete subjected to uniaxial concentric compression. An energy balance approach was used by Mander, Priestley and Park (1988) to estimate the compressive strain in the concrete (ε_{cu}) corresponding to the first fracture of the transverse reinforcement, by equating the strain energy capacity of the transverse reinforcement to the strain energy stored in the concrete of the confined core (equation (8.75)). Other investigators (Soroushian and Sim, 1986) have introduced in the Park, Priestley and Gill (1982) model the effect of strain rate ($\dot{\varepsilon}$). The confined concrete strength f_{cc} and the corresponding strain ε_{cc1} are calculated as functions of $\dot{\varepsilon}$ (compare equations (7.7) and (7.9)). The same function used for estimating the strength increase is also used for calculating the increased (steeper) slope of the descending branch.

With regard to the eccentricity of axial loading (strain gradient) Sheikh and Yeh (1986) have suggested a shift of the descending branch of the σ_c–ε_c curve, based on an increased value of the strain ε_{cc2} (Figure 7.23(b), calculated from the equation

$$\varepsilon_{cc2}=0.0022\left(1+\left\{\frac{248}{b_i}\left[1-5\left(\frac{s}{b_c}\right)^2\right]+3\left(\frac{b_c}{x}\right)^{1/2}\right\}\frac{\rho_w f_{yw}}{f_c^{1/2}}\right) \tag{7.38}$$

where x is the neutral axis depth and the rest of the symbols are the same as those used in equation (7.33). Sheikh and Yeh (1986) did not propose any strength increase due to eccentric loading, since they considered that the available experimental data were inconclusive. However the tests by Scott, Park and Priestley (1982) have indicated that the presence of a strain gradient increases the ductility, as well as the load-bearing capacity of confined columns, with respect to concentric loading.

(c) Cyclic loading

Cyclic tests on confined concrete specimens are less common than monotonic tests and possibly because of the limited data available it has long been assumed that the hysteretic behaviour of confined concrete is the same as that of unconfined concrete (Figures 7.3 and 7.4), except for the envelope curve which is modified as described in the previous paragraphs. Tests by Mander, Priestley and Park (1988) on columns of various shapes (including square and elongated rectangular ones) with various amounts of transverse reinforcement, have shed some more light on the behaviour of confined concrete in cyclic compression. These tests confirmed that the monotonic curve is indeed the envelope of the cyclic loading curves, and also that the shape of the unloading and reloading curves is similar to that observed for unconfined concrete (Figure 7.3).

Mander Priestley and Park (1988) have suggested the model shown in Figure 7.25 for confined concrete subjected to arbitrary cyclic loading histories, including loading in tension; for the latter case more refined models, such as that shown in Figure 7.5, may be used (tension loading cycles are not affected by confinement). The envelope curve is the one proposed by Mander, Priestley and Park (1988) for monotonic loading, extending up to ε_{cu} defined on the basis of the aforementioned energy considerations. The similarity between the hysteresis loops of Figure 7.25 and those of Figure 7.3 (referring to unconfined concrete) is pointed out.

7.4.4 Confinement with spirals

As already mentioned, the confinement provided by circular spirals is generally more efficient than that due to rectangular ties; for close spacing of the spirals the

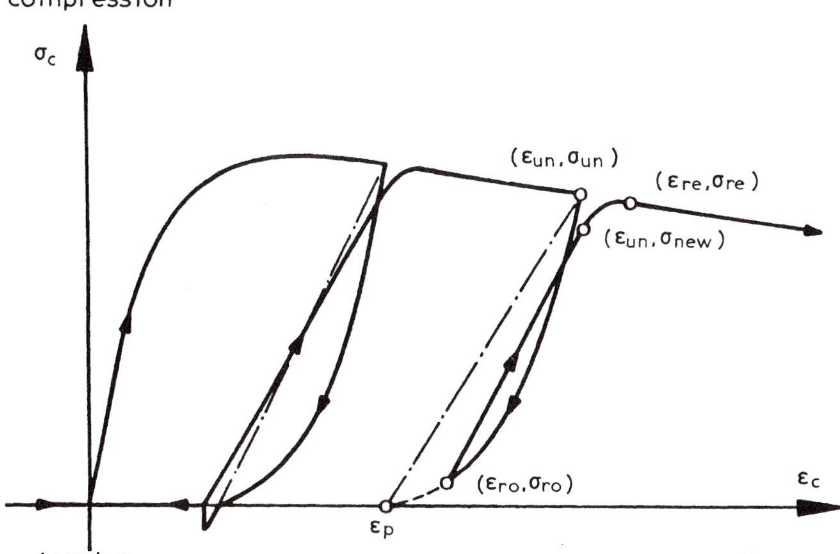

Figure 7.25 Stress–strain model for confined concrete subjected to cyclic loading (Mander, Priestley and Park, 1988).

behaviour of the confined core is similar to that of concrete under moderate amounts of hydrostatic pressure (Figure 7.19).

The lateral pressure exerted on the concrete in the interior of the spirals (Figure 7.20(a) reaches its peak value when the hoop tension which develops in the spiral attains its maximum value. If it is assumed that the spiral steel has not entered the strain-hardening range, the maximum tensile stress would be equal to the yield strength of the spiral f_{yw}. If the pitch (spacing) of the spiral is s, it is seen from the free body diagram of Figure 7.26 that equilibrium of horizontal forces results in the following relationships:

$$2f_{yw}A_{sw} = d_w s\sigma_l$$

where d_w is the diameter of the spiral and A_{sw} its area. If this relationship is solved for the pressure σ_l the following equation results:

$$\boxed{\sigma_l = \frac{2f_{yw}A_{sw}}{d_w s}} \tag{7.39}$$

As discussed in section 7.3.3(b), the strength of plain concrete subjected to triaxial compression is significantly enhanced (Figure 7.16); this increase in strength may be expressed in the form

$$f_{cc} = f_c + \lambda\sigma_l \tag{7.40}$$

with values of λ varying from 4 to 7, based on the available test data (Park and Paulay, 1975; Mander, Priestley and Park, 1988). Now, if the lateral (confining)

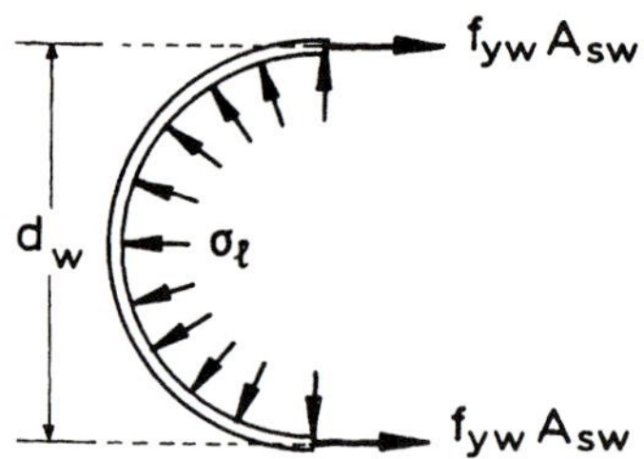

Figure 7.26 Confinement stressess in a circular spiral.

pressure σ_l in equation (7.40) is taken equal to the value given by (7.39), it follows that the increased compressive strength due to lateral confinement by circular spirals is

$$f_{cc}=f_c\left(1+\lambda\ \frac{2f_{yw}\ A_{sw}}{d_w s f_c}\right) \tag{7.41}$$

It is clear from equation (7.41) that compressive strength increases with increasing area and/or with decreasing spacing of the spiral reinforcement. It has to be pointed out that the inherent assumption in the foregoing equation is that σ_l is constant, which is actually the case only when active confinement is present, for instance by hydrostatic fluid pressure. In the case of passive confinement (provided by hoops or spirals) the pressure σ_l is obviously a function of the lateral expansion of the confined core which creates hoop tension in the transverse reinforcement.

If the value of the mechanical ratio of spiral reinforcement ($\omega_w=\rho_w f_{yw}/f_c$) which is given by the equation

$$\omega_w=\frac{4A_{sw}}{d_w s}\ \frac{f_{yw}}{f_c} \tag{7.42}$$

is substituted in (7.41), the following expression results:

$$f_{cc}=f_c[1+(\lambda/2)\omega_w] \tag{7.43}$$

It is seen that for $\lambda\approx 4$ equation (7.43) implies that confinement with spirals leads to twice the strength increase caused by confinement with rectangular hoops as expressed by equation (7.28). This is attributed primarily to the fact that the effectively confined area is substantially larger if closely spaced continuous spirals are used (Figure 7.20).

Ductility of concrete is also increased in the presence of spiral reinforcement. In Figure 7.27 are shown typical stress–strain diagrams for concrete cylinders confined with circular spirals (Shah, Fafitis and Arnold, 1983). A direct comparison with the curves of Figure 7.3 clearly shows the effect of the spirals on the ductility of concrete. More marked improvements with respect to plain concrete were recorded in the tests by Mander, Priestley and Park (1988) which involved actual circular columns with both longitudinal reinforcement and spirals; the ductility of these columns measured in terms of the concrete strain ε_{cu} at first hoop fracture ranged from 3.5 to 6.0%, which is one order of magnitude or more the corresponding value for unconfined concrete.

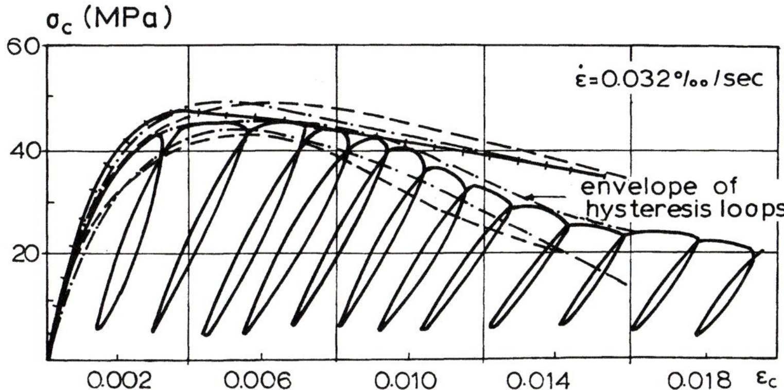

Figure 7.27 Stress–strain diagram for concrete confined with circular spirals (Shah, Fafitis and Arnold, 1983).

Various analytical models have been suggested in the literature for concrete confined by spirals. On the basis of the experiments shown in Figure 7.27, Shah, Fafitis and Arnold (1983) have suggested the following expression for the ascending branch of the σ_c–ε_c curve:

$$\sigma_c = f_c[1 - \varepsilon_c/\varepsilon_{ccl})^{\beta}] \quad (7.44)$$

and for the descending branch

$$\sigma_c = f_{cc} \exp\left[-k(\varepsilon_c - \varepsilon_{ccl})^{1.15}\right] \quad (7.45)$$

where the various symbols have the meaning previously explained and the parameters β and k are given by the relationships

$$\beta = E_c \varepsilon_{ccl}/f_{cc} \quad (7.46)$$

$$k = 24.7\, f_c \exp[-1.45\sigma_l] \quad (7.47)$$

where E_c is the initial value of the modulus of elasticity (corresponding to a stress $\sigma_c \approx 0.4f_c$), and the confining pressure σ_l is given by equation (7.39). The stresses f_c and σ_l are expressed in MPa. The value suggested by Shah, Fafitis and Arnold (1983) for the strength of confined concrete is

$$f_{cc} = f_c + (1.15 + 21/f_c)\sigma_l \quad (7.48)$$

The term in parentheses corresponds to the λ coefficient in equations (7.40) and (7.41) and it reflects the fact that confinement is less effective for higher strength concrete (see also section 7.4.2). However equation (7.48) is substantially more conservative than (7.40) if $\lambda \geqslant 4$ is assumed. The analytical curve resulting from equations (7.44)–(7.48) is compared in Figure 7.27 with the envelopes of the experimental σ_c–ε_c curves for monotonic and repeated loading, which lie quite close to each other. The analytical curve lies closer to the upper limit of the envelopes, and it is worth pointing out that in certain regions the envelope of the repeated loading lies above the corresponding curve for monotonic loading. The previously mentioned model by Mander, Priestley and Park (1988), which also assumes that the (constant) value of σ_l is calculated from (7.39) if circular spirals or hoops are used, was found to give good predictions of the experimentally derived curves for circular columns with various amounts of spiral reinforcement (Mander, Priestley and Park, 1988).

A conceptually attractive approach was adopted by Ahmad and Shah (1982) who, instead of assuming a constant σ_l (equation (7.39)), have expressed it as a function of the stress in the spiral (σ_w) using the expression

$$\sigma_l = \frac{\rho_w \sigma_w}{2}\left[1 - \left(\frac{s}{1.25 d_w}\right)^{1/2}\right] \tag{7.49}$$

where the stress σ_w is calculated from the stress–strain diagram of the spiral steel, as the value corresponding to the current lateral strain ε_l, which is determined from the current axial strain ε_1, using the octahedral stress theory briefly discussed in section 7.3.3(b). A value of the confining pressure has first to be assumed for each value of the axial strain, hence an iterative procedure is required for calculating the complete stress–strain curve. A more efficient procedure, which does not require iterations for each ε_1, has recently been suggested by Madas and Elnashai (1992), and was found to give good predictions of the experimental results of Scott, Park and Priestley (1982) and Ahmad and Shah (1982).

7.4.5 Relevant code provisions

The importance of confinement with regard to increasing the strength and, in particular, the ductility of R/C members has long been recognized by seismic codes, and detailed provisions for confinement reinforcement have been given for the critical regions of beams (section 8.3.2(b)), columns (section 8.5.2(c)) and walls (section 9.4.2(e)). However with the exception of the CEB-FIP Model Code 1990 (CEB, 1993), no specific models for confined concrete are explicitly adopted in codes of practice (which, nevertheless, do include refined or simplified σ_c–ε_c models for unconfined concrete, for use in flexural design).

According to the CEB (1993) Model Code, the effective lateral stress due to confinement is given by the equation

$$\boxed{\sigma_l / f_c = \frac{1}{2}\,\alpha_n \alpha_s \omega_w} \tag{7.50}$$

where ω_w is the previously defined mechanical volumetric ratio of hoops or spirals, and α_n and α_s are coefficients accounting for the reduction in confinement at section level and midway between hoops, respectively (Figure 7.20). The confinement effectiveness coefficient α_n, which is the ratio of the area of the effectively confined section to the area of the core, is given by the expression

$$\alpha_n = 1 - \frac{n(b_i^2/6)}{b_c^2} \tag{7.51(a)}$$

where the meaning of the various symbols is the same as in equation (7.29) suggested by Sheikh and Uzumeri (1982). Note that for square cross-sections $nb_i = 4b_c$ and the previous equation reduces to

$$\alpha_n = 1 - (8/3)\,(1/n) \tag{7.51(b)}$$

The effectiveness of confinement between adjacent stirrups is expressed through the coefficient

$$\alpha_s = \left(1 - \frac{s}{2b_c}\right)^2 \tag{7.52(a)}$$

which is applicable for $s \not> b_c/2$. As the comparison of equation (7.29) with (7.51) and (7.52) shows, the CEB model is based on the 'effectively confined area' approach suggested by Sheikh and Uzumeri (1982), which was discussed in section 7.4.3(b).

For circular columns with spiral reinforcement it is $\alpha_n = 1$, that is the entire core is assumed to be fully confined, while

$$\alpha_s = 1 - \frac{1}{2}\frac{s}{b_c} \tag{7.52(b)}$$

to reflect the increased effectiveness of continuous spirals in elevation (for circular hoops equation (7.52(a) applies).

The strength of confined concrete can be calculated from the following equation:

$$\boxed{f_{cc} = f_c\,(1.00 + 2.50\alpha_n\alpha_s\omega_w)} \tag{7.53(a)}$$

for $\sigma_l/f_c \leqslant 0.05$ ($\alpha_n\alpha_s\omega_w \leqslant 0.1$) and from

$$\boxed{f_{cc} = f_c\,(1.125 + 1.25\alpha_n\alpha_s\omega_w)} \tag{7.53(b)}$$

for $\sigma_l/f_c > 0.05$, that is the nonlinear relationship between σ_l and f_{cc}/f_c is approximated by a bilinear curve.

The strain corresponding to peak stress is given by the following equation:

$$\varepsilon_{cc1} = \varepsilon_{c1}\,(f_{cc}/f_c)^2 \tag{7.54}$$

The descending branch of the σ_c–ε_c curve for confined concrete is defined in the CEB (1993) Code on the basis of the $0.85f_c$ stress level to which the following strain corresponds:

$$\varepsilon_{cc.85} = \varepsilon_{c.85} + 0.1\alpha_n\alpha_s\omega_w \tag{7.55}$$

where $\varepsilon_{c.85} \approx 0.35\%$ (unconfined concrete).

Applying the CEB model to the 450 mm column studied in section 7.4.3(b), yields the following results for the case of 10 mm diameter hoops at 150 mm spacing ($\rho_w = 0.0089$):

$$\left.\begin{aligned} \alpha_n &= 1-(8/3)\,(1/8)=0.667 \\ \alpha_s &= (1-0.5\times 150/400)^2 = 0.817 \end{aligned}\right\} \alpha_n\alpha_s = 0.545$$

$$\alpha_n\alpha_s\omega_w = 0.545\times 0.0089\times 420/24 = 0.085 < 0.1$$

Hence, from equation (7.53(a))

$$f_{cc} = 24.0\,(1.0+2.5\times 0.085) = 29.1\text{ MPa}\quad (f_{cc}/f_c = 1.21)$$

and from equations (7.54) and (7.55)

$$\begin{aligned} \varepsilon_{cc1} &= (1.21)^2\times 0.002 = 0.29\% \\ \varepsilon_{cc,85} &= 0.35+0.1\times 0.545\times 0.156\times 10^2 = 1.2\%\quad (=\varepsilon_{cu}) \end{aligned}$$

Comparisons of the f_{cc} and ε_{cu} values calculated on the basis of the CEB model with the corresponding values estimated from the other confinement models in the example of section 7.4.3(b), indicate that the CEB model generally gives a more 'optimistic' picture of the confined core characteristics of the column under consideration, especially with regard to the ε_{cu} value which is 29% higher than the largest value calculated previously (Kappos model, $\varepsilon_{cu} = 0.94\%$).

A simplified parabola–rectangle diagram is also included in the CEB (1993) Model Code, analogous to the standard diagram used for flexural design with unconfined concrete. As shown in Figure 7.28, the design strength for confined concrete is $0.85f_{ccd}$ (the coefficient 0.85 accounting for long-term loading), where $f_{ccd} = f_{cck}/\gamma_c$ ($\gamma_c \approx 1.5$), f_{cck} being calculated from equations (7.53) using characteristic values for concrete (f_{ck}) and $\omega_{wd} = \rho_w f_{yd}/f_{cd}$; the maximum design strain ε_{cu} is calculated from equation (7.55) using the previously defined ω_{wd} value. It is pointed out that the calculation of the mechanical ratio of transverse reinforcement (ω_w) using design values for materials leads to significantly higher values compared to those resulting when mean values of material strengths are used. In the foregoing example of the 450 mm square column

$$\omega_{wd} = 0.0089\times(400/1.15)\,/\,(16.0/1.50) = 0.289$$

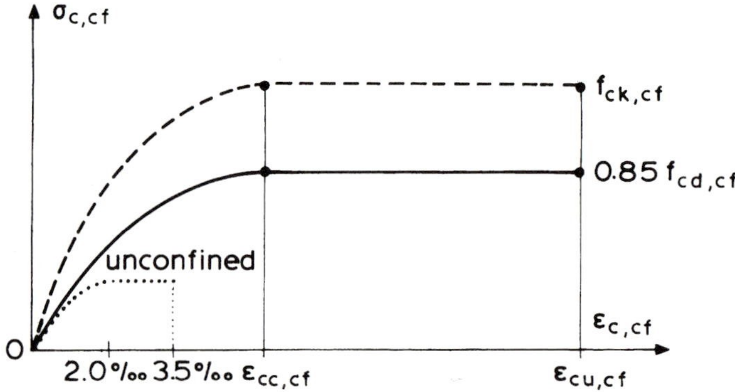

Figure 7.28 Design σ_c–ε_c diagram for confined concrete, according to the CEB (1993) Model Code.

which is 85% higher than the value ($\omega_w = 0.156$) calculated on the basis of mean strengths; this implies that safety factors here have exactly the opposite effect to that in normal design. Introducing the value $\omega_{wd} = 0.289$ in equation (7.55) results in $\varepsilon_{cu} = 1.93\%$, which is clearly an unrealistically high value of ductility, particularly for design calculations! It is believed that a revised definition of the parameters used in the CEB design model for confined concrete is required.

A constitutive law for confined concrete is not explicitly included in the final text of EC8, while the 1989 version did include such a model, essentially the same one as in the CEB Code (see Tassios, 1989). The relationships used for calculating confining reinforcement in critical areas of columns (section 8.5.2(c)) are in fact using the coefficients α_n, α_s which reflect the effectiveness of confinement. Hence, it may be stated that the model included in the CEB (1993) Model Code is also implicitly adopted by EC8.

7.5 STEEL

7.5.1 Main requirement for seismic performance

The main requirement for steel bars used as reinforcement in earthquake-resistant structural members, may be summarized as follows.

1. The ultimate strain (ε_{su}) of steel, that is the value corresponding to fracture of a bar in tension, has to be large, so that sufficient ductility of the R/C structural member is ensured. This requirement is generally satisfied since, as shown in Figure 7.29, the ultimate deformation of reinforcing steel is 12% or more (Park and Paulay, 1975; McDermott, 1978), the general tendency being that ε_{su} decreases with increasing yield strength. In the case of prestressing steel the ductility may not always be sufficient for adequate seismic performance; minimum code requirements regarding ε_{su} are given in section 7.5.4.
2. The actual yield stress (f_y) of steel should not significantly exceed its specified value, since an increase in the resistance of a structural member results in the development of a shear force higher than the one estimated during the design. As will be discussed in detail in Chapter 8, high values of shear have an unfavourable effect on the seismic behaviour of a R/C member, since they significantly reduce its ductility, Moreover, an increase in the specified strength of the reinforcement in a beam results in higher moments developing at the adjacent columns and a consequent risk of plastic hinges developing in these elements as well, which is an undesired behaviour in earthquake-resistant buildings (see also section 6.1.4).
3. Strain hardening in steel has a generally favourable effect with regard to the behaviour of plastic hinge zones, since it allows the development of bending moments higher than those corresponding to first yielding at sections beyond the critical one ($M = M_{max}$), which results in spreading of plastification in larger parts of the member, with a consequent favourable effect on its seismic behaviour (Chapter 8). However, strain hardening should not start prematurely (that is immediately after yielding), because in such a case there is a risk of affecting the strength hierarchy among structural elements which is established by the application of capacity design procedures (section 6.1). This means

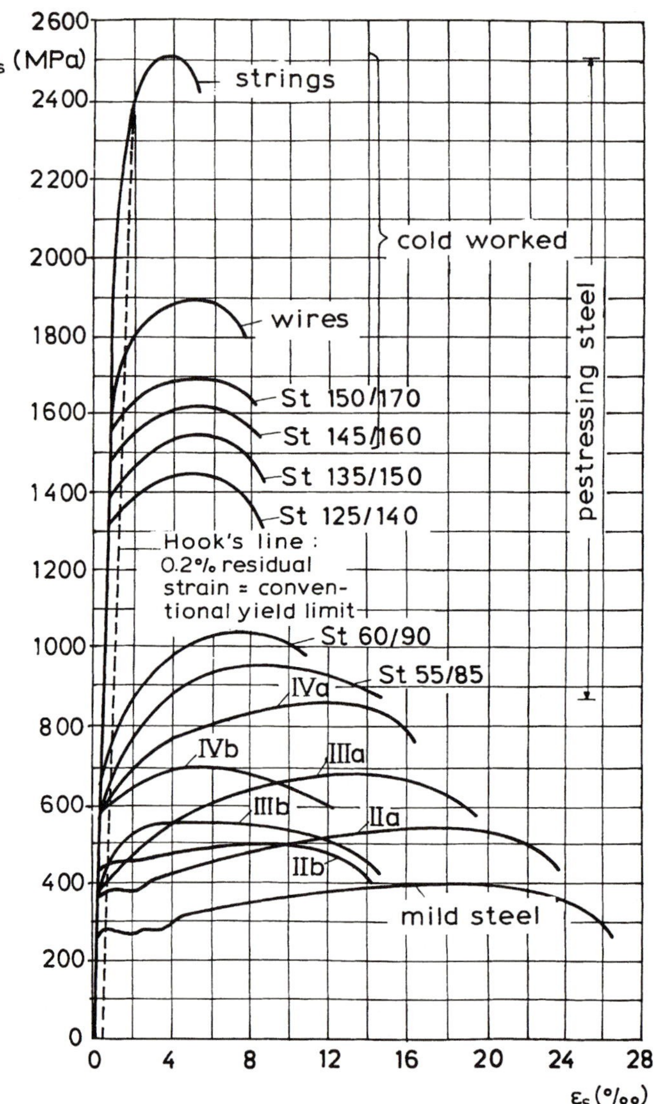

Figure 7.29 Stress–strain diagrams for steel bars of various grades.

that similarly to the case of excessive yield strength, an increased beam strength due to early strain hardening creates a risk of plastic hinge development in adjacent columns. Besides, whenever the moments in a member increase due to strain hardening of steel, shear forces are increased in the same proportion, with a consequent unfavourable effect on member ductility.

4. Steel bars in earthquake-resistant structural elements should be able to develop an efficient composite action with the surrounding concrete, even in regions

where significant inelastic deformations develop as a result of cyclic loading, that is in the plastic hinge regions. Hence the problem of bond between steel and concrete assumes a particular importance in earthquake-resistant construction; this problem will be discussed in section 7.6.

Based on the foregoing remarks it would appear that the most appropriate reinforcement for earthquake-resistant R/C members would consist of mild steel (S220 or similar) deformed bars. Indeed, these reinforcing bars are characterized by very large ultimate strains (Figure 7.29) and relatively low strain hardening which develops well after the first yielding. Besides, ribs or other deformations are necessary for ensuring adequate bond under seismic loading conditions. However, there are a number of practical problems which question the validity of the previous conclusions. For instance, the use of low strength steel (S220) leads, as a rule, to high ratios of reinforcement and often to the selection of large diameter bars. Notwithstanding the economic consequences of such a selection (increased cost of labour, increased time of construction works), high ratios of reinforcement lead to a reduction in member ductility (section 8.2.1(a)), while large diameter bars have a more problematic behaviour than smaller ones, with regard to bond and cracking. Moreover, some quite practical problems should also be taken into account, such as the *in situ* distinction between different steel grades, which is the main reason that in countries like Greece S220 deformed bars are not commercially available. Nevertheless, some design codes (SANZ, 1982) recommended the use of steel grades similar to S220 for horizontal members of the seismic load-resisting system (beams), while steel grades similar to S400 are recommended for vertical members.

7.5.2 Response to monotonic loading

In Figure 7.29 are given stress (σ_s)–strain (ε_s) diagrams for various grades of steel used in Germany, Greece and elsewhere, subjected to monotonic tensile loading. It is clear from these diagrams that as the strength of steel increases, its ultimate deformation decreases, a tendency similar (but more marked) than that found for plain concrete (section 7.3.1). Moreover, the ratio of peak stress (f_u) to yield stress (f_y) increases with the steel grade, that is the influence of strain hardening is larger in high strength steel, for which the threshold of the hardening branch is closer to the yield strain than in low strength steel.

For analysis purposes the σ_s–ε_s diagram for monotonic loading may be idealized as an elastic–perfectly plastic one (Figure 7.30(a)) that is the effect of strain hardening is ignored. This simple diagram which is adopted by most design codes, EC2 among them, facilitates everyday calculations, but is only adequate for design against vertical loads. In the case of seismic design the relative strength of adjacent members is quite important (capacity design), thus it is not permitted to ignore the effect of strain hardening on member strength. A commonly accepted procedure is to use the simple elastoplastic diagram of Figure 7.30(a) and to increase the resistance calculated on this basis by a coefficient (overstrength factor, γ_{Rd}) specified by the code (SANZ, 1982; CEB, 1985; CEN, 1995). Alternatively, a more refined model may be selected for the σ_s–ε_s diagram, such as the one shown in Figure 7.30(b), which has often been used for research purposes (Park and Paulay, 1975; Kappos, 1993).

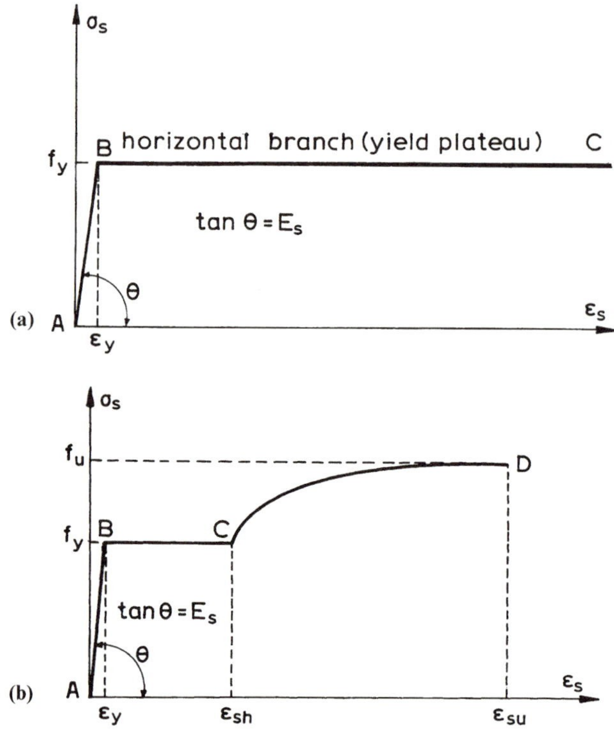

Figure 7.30 Idealized stress-strain diagrams for steel subjected to monotonic loading.

Three regions may be recognized in the diagram of Figure 7.30(b). The first one (AB) is the elastic one and its slope is equal to the modulus of elasticity E_s, which may be taken equal to 200 GPa (2.0×10^5 MPa) for all concrete grades (CEN, 1991). As soon as the stress exceeds the yield limit (f_y), an approximately horizontal branch (BC) follows, which extends up to a strain ε_{sh} whose value decreases as the steel grade increases (ε_{sh} ranges from about 1% up to more than 4%, the second value corresponding to mild steel). The third branch (CD) is the strain-hardening one, for which Park and Sampson (1972) have suggested the following equation:

$$\sigma_s = f_y \left[\frac{m\,(\varepsilon_s - \varepsilon_{sh}) + 2}{60\,(\varepsilon_s - \varepsilon_{sh}) + 2} + \frac{(\varepsilon_s - \varepsilon_{sh})\,(60 - m)}{2\,(30r + 1)^2} \right] \tag{7.56}$$

where

$$m = \frac{f_u/f_y\,(30r+1)^2 - 60r - 1}{15r^2} \tag{7.57(a)}$$

and

$$r = \varepsilon_{su} - \varepsilon_{sh} \tag{7.57(b)}$$

It is pointed out that the diagram is terminated at the 'ultimate' strain ε_{su}, ignoring any subsequent descending branch, along which steel fails in a fast and difficult to control mode. The use of the foregoing model, as well as of similar more recent

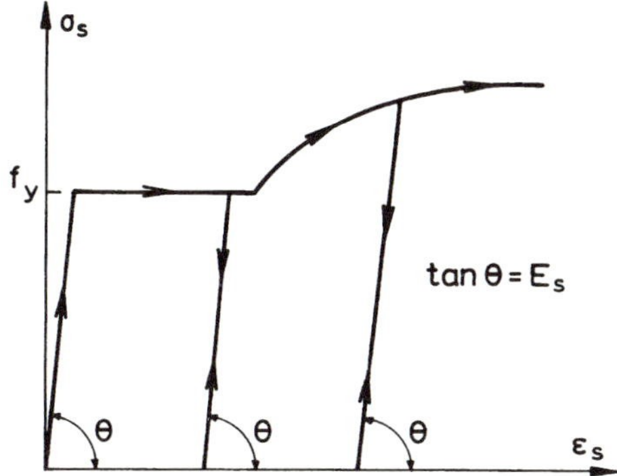

Figure 7.31 Idealized stres-strain diagram for steel subjected to repeated loading.

ones (CEB, 1991), presupposes knowledge of a number of parameters, some of which are easily determined by standard tests (f_y, f_u, ε_{su}) and some that require more sophisticated testing procedures (ε_{sh} and the initial slope of the strain-hardening branch).

For the σ_s–ε_s diagram in compression it is commonly accepted that it coincides with that in tension, provided of course that the surrounding concrete and the transverse reinforcement prevent buckling of the steel bar. Hence, the aforementioned diagrams (Figure 7.30) may be used for both tension and compression (Park and Sampson, 1972).

7.5.3 Response to cyclic loading

In Figure 7.31 is shown an idealization of the stress–strain diagram of steel subjected to repeated loading, with full unloading (but no stress reversal). It is seen in the figure that unloading from the yield branch, as well as from the strain-hardening branch, proceeds along a path parallel to that of the elastic branch (E_s). In experimentally derived curves a narrow hysteresis loop is formed during unloading and subsequent reloading, that is a small energy dissipation takes place. It has to be pointed out that, similarly to the case of concrete (section 7.3.2), the envelope of repeated loading in steel practically coincides with the curve resulting from monotonic application of the loading.

In the general case of seismic loading, the sign of the stress in the steel reinforcement which is in tension under gravity loading may reverse, although significant compressive strains do not usually develop (Popov, 1977). In this case it is important to know the behaviour of steel under cyclic loading conditions. As Figure 7.32 clearly shows, stress reversal in the inelastic range causes a reduction in stiffness (nonlinear behaviour) at stress levels significantly lower than the yield limit. This property of steel, known as the **Bauschinger effect**, is strongly influenced by the loading history, while temperature and time also have some influence (Park and Paulay, 1975; Popov, 1977). On the other hand, unloading subsequent to stress reversals continues to take place almost elastically, as indicated in the idealized diagram of Figure 7.32.

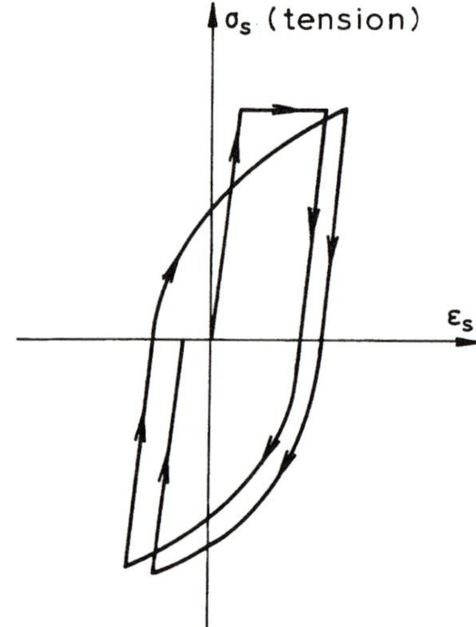

Figure 7.32 Idealized stress–strain diagram for steel subjected to cyclic loading.

Depending on the level of sophistication sought in analysis, the constitutive law used for steel may range from the simple elastoplastic formulation to particularly complex models which take into account the path dependence of the parameters involved in describing the load cycles (Popov, 1977; CEB, 1991). A relatively simple model proposed by Kent and Park (see Park and Paulay, 1975) and shown in Figure 7.33, will be described in the following.

The slope of the unloading cycles is equal to the elastic one, E_s (see also Figure 7.31) for tension, as well as for compression loading. Subsequent to first yielding

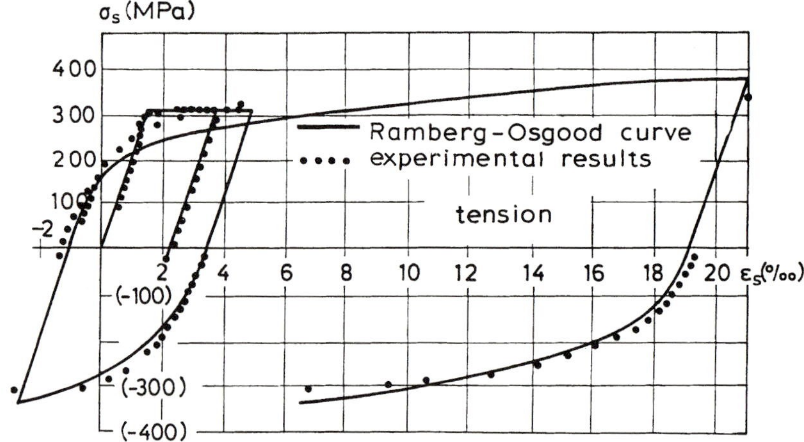

Figure 7.33 Modelling of the cyclic loading behaviour of steel and comparissons with relevant test data (Park and Paulay, 1975).

($\sigma_s > f_y$), reloading in the opposite direction is described by the following Ramberg–Osgood equation:

$$\varepsilon_s - \varepsilon_{si} = \frac{\sigma_s}{E_s}\left\{1 + \left|\frac{\sigma_s}{\sigma_{ch}}\right|^{r-1}\right\} \tag{7.58}$$

where ε_{si} is the strain corresponding to zero stress (residual strain) at the beginning of the ith reloading cycle, σ_{ch} is a characteristic stress which depends on the yield strength and on the plastic strain during the previous loading cycle, and r is a parameter depending on the number of load cycles. The parameters σ_{ch} and r are estimated from empirical formulae (Park and Paulay, 1975) derived from curve fitting to available test data. Equation (7.58) describes the curving of the hysteresis loops due to the Bauschinger effect and, as shown in Figure 7.33, the resulting correlation with the corresponding experimental data is satisfactory.

Ma, Bertero and Popov (see Popov, 1977) have suggested a similar, though more complicated model, where the Ramberg–Osgood equations are used for describing the unloading curves, as well as the reloading ones, so long as the material has entered the strain-hardening range. It is understood that the foregoing models, as well as others, some of which are reviewed in a recent CEB Task Group 22 report (CEB, 1991), presuppose the use of an appropriate software, and are meant primarily for research, rather than for design purposes.

Finally, with regard to the effect of strain rate which is typical in seismic loading, it is noted that, as in the case of concrete (section 7.3.1), fast strain (or stress) rates cause an increase in the (apparent) yield strength, as well as in the slope of the strain-hardening branch, Soroushian and Sim (1986) have suggested the following equation for estimating the 'dynamic' yield stress of steel:

$$f_{y.dyn} = f_y[-6.54 \times 10^{-4} f_y + 1.46 + (-1.33 \times 10^{-4} f_y + 0.0927)\log \dot{\varepsilon}_s] \tag{7.59}$$

where f_y is given in MPa. By applying equation (7.59) it follows that for a strain rate $\dot{\varepsilon}_s = 0.02\ \text{s}^{-1}$ (which is typical for seismic loading), the yield strength of steel increases by 20% with respect to static loading ($\dot{\varepsilon}_s \approx 10^{-5}\ \text{s}^{-1}$) for low grade steel ($f_y = 240$ Mpa), and by 12% for higher grade steel ($f_y = 420$ MPa); the variation of $f_{y.dyn}$ with the strain rate $\dot{\varepsilon}_s$ is given in Figure 7.34 for two typical steel grades. An equation similar to (7.59), that is a linear function of $\log \dot{\varepsilon}_s$, has been suggested

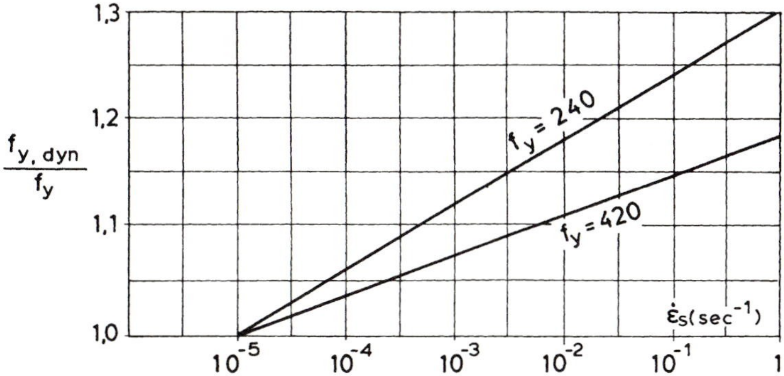

Figure 7.34 Strain rate dependence of the yield strength of steel.

by Soroushian and Sim (1986) for the strain-hardening modulus E_{sh} (the initial slope of the strain-hardening branch).

7.5.4 Relevant code provisions

With regard to reinforcing steel, EC8 prescribes different minimum requirements (additional to those of EC2) for each ductility class; these additional rules are summarized in Table 7.1. The rules strictly apply only for critical regions of R/C members (see sections 8.3.2, 8.5.2, 9.4.2), but it is understood that, for practical reasons, the same steel will be used for the entire member length. The aim of these rules is to ensure adequate plastic hinge lengths (by specifying minimum f_u/f_y ratios, as already discussed in section 7.5.1), and high local ductilities (by specifying minimum ultimate strains ε_{su}). Moreover, the specification of maximum values for the actual to nominal yield stress ratio ($f_{y.act}/f_{y.nom}$) and for the f_u/f_y ratio aims at ensuring a reliable control of the desired strength hierarchy of members, established through the capacity design procedures.

The manufacture of steel bars which simultaneously satisfy all the requirements set forth in Table 7.1, is not a routine problem for the industry. Based on the current status of steel production in Europe, it appears that the requirements which are more difficult to meet are those concerning the maximum $f_{y.act}/f_{y.nom}$ ratio and the maximum f_u/f_y ratio (in combination with the corresponding minimum ratio). Recognizing these practical difficulties, EC8 specifies that the overstrength (γ_{Rd}) factors (see Chapters 8 and 9) should be increased in proportion to the differences between the required and the actual ratios of f_u/f_y and $f_{y.act}/f_{y.nom}$, whenever such differences cannot be avoided. Table 7.1 also applies for prestressing steel at critical regions of beams and columns, unless all tendons are located inside the kernel of the cross-section. It is recalled here that elongations of even 5% are quite difficult to achieve in prestressing steel (Figure 7.29).

In addition to the requirements set forth in Table 7.1, EC8 specifies that only deformed (high bond) bars should be used as longitudinal reinforcement in critical regions of R/C members. Smooth bars are only allowed for hoops or crossties. The importance of adequate bond under seismic loading conditions has already been emphasized in section 7.5.1 and will be discussed in detail in the following section. Moreover, the use of welded meshes is not allowed in critical

Table 7.1 Requirements for reinforcing steel in critical regions, according to EC 8

Properties		DC "L"	DC "M"	DC "H"
1. Uniform elongation at maximum load (characteristic value)	$\varepsilon_{su,k}$	$\geqslant 5\%$	$\geqslant 6\%$	$\geqslant 9\%$
2. Tensile strength to yield strength ratio (mean values of the ratio)	$\frac{f_u}{f_y}$	$\geqslant 1.08$	$\geqslant 1.15$	$\geqslant 1.20$
3.		–	$\leqslant 1.35$	$\leqslant 1.35$
4. Actual to nominal yield strength ratio (mean values)	$\frac{f_{y,act}}{f_{y,nom}}$	–	$\leqslant 1.25$	$\leqslant 1.20$

regions of members, including walls, unless they conform to the requirements of Table 7.1. Their use is permitted outside the critical regions and also as slab reinforcement.

It is worth mentioning here that in the CEB (1985) Seismic Code the use of steel grades higher than S400 was not allowed, unless proof of its adequate ductility was provided. The exclusion of this provision from EC8 is one of the several changes which reflect recent developments in the technology of reinforcing steel.

7.6 BOND BETWEEN CONCRETE AND STEEL

It is well known that the composite action of concrete and steel in reinforced concrete is due to the bond between the two materials; ensuring adequate bond conditions is one of the main goals of design and detailing of R/C structural elements. The satisfactory performance of R/C members with regard to anchorage, splicing, cracking and deflections depends primarily on the adequacy of bond ensured. The role of bond becomes dominant with respect to the seismic behaviour of R/C structures, since in addition to the aforementioned factors, bond also affects stiffness and seismic energy dissipation capacity. In this section the main characteristics of bond between concrete and steel bars will be examined for monotonic, as well as for cyclic, loading conditions, while the consequences of bond deterioration on the seismic behaviour of R/C structural elements will be discussed in Chapter 8.

7.6.1 Constitutive equations of bond

In Figure 7.35(a) are shown the forces acting on an infinitesimal element of a steel bar surrounded by concrete. Equilibrium of forces requires that

$$A_s \mathrm{d}\sigma_s = \tau u \,\mathrm{d}x$$

where τ is the bond stress and u the length of the perimeter of the bar. For the typical case of a circular reinforcing bar, having a diameter d_b, the previous relationship takes the following form:

$$\boxed{\frac{\mathrm{d}\sigma_s}{\mathrm{d}x} = \frac{4}{d_b}\tau} \tag{7.60}$$

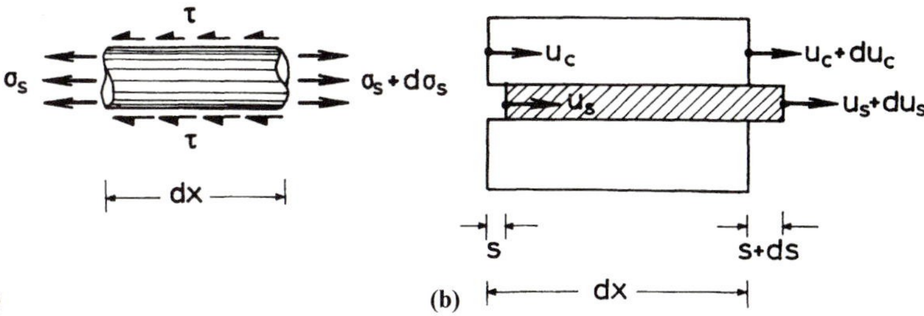

Figure 7.35 Bond between concrete and steel; (a) steel stresses and bond stresses in an infinitesimal element of the bar; (b) displacements and relative slip in an infinitesimal R/C element (tie).

It is clear from equation (7.60) that the bond stress is equal to zero wherever there is no stress gradient in the steel reinforcement (constant moment areas), and assumes its peak values at points of steep gradients of σ_s (for instance in regions where point loads are applied). Figure 7.35(b) shows the displacements of the bar (u_s) and of the surrounding concrete (u_c) in an infinitesimal R/C element, as well as the corresponding relative slip (s) between the two materials. It is seen from the figure that $s = u_s - u_c$ and $ds = du_s - du_c$, hence

$$\frac{ds}{dx} = \frac{du_s}{dx} - \frac{du_c}{dx}$$

Since the two terms on the right-hand side of the equation are, by definition, the strain in the reinforcement (ε_s) and in concrete (ε_c), the foregoing equation becomes

$$\boxed{\frac{ds}{dx} = \varepsilon_s - \varepsilon_c} \tag{7.61}$$

The bond stress τ is related to the relative slip s through the relationship

$$\tau = \tau(s) \tag{7.62}$$

while for the reinforcing steel stress σ_s the standard relationship

$$\sigma_s = \sigma(\varepsilon_s) \tag{7.63}$$

applies, as discussed in detail in the previous section. Equations (7.60)–(7.63) are the constitutive equations of bond. The concrete strain ε_c in equation (7.61) is commonly ignored (as negligible with respect to ε_s), while the problem is solved numerically (section 7.6.2(b)). In the following the basic equation (7.62) will be examined, which describes the relationship between the **local bond stress** (τ), that is the bond stress in an infinitesimal element[2] and the **local slip** (s) with emphasis on the case of seismic loading.

7.6.2 Bond under monotonic loading

(a) Discussion of test data

Various τ–s diagrams, extending up to the range of very large slip values (strain-controlled tests), taken from an extensive experimental study by Eligehausen, Popov and Bertero (1983), are shown in Figure 7.36. A typical diagram can be idealized as a sequence of linear segments, as shown in Figure 7.37.

Up to a certain value of stress (τ_0 in Figure 7.37) bond is due to chemical adhesion of the cement paste on the surface of the steel bar and practically no slip takes place; typical values of τ_0 range from 0.5 to 1.0 MPa (ACI Committee 408, 1991). For $\tau > \tau_0$ adhesion breaks down and bond is provided by friction and wedging action between the cement paste and the microscopic anomalies (pitting) of the bar surface and also, in the case of deformed bars, by mechanical interlock of the deformations and the surrounding concrete. Due to these interlock forces, at

[2]In practice the quantity that can be measured is the mean value of bond stress within a small length, typically between $3d_b$ and $5d_b$ (Eligehausen, Popov and Bertero, 1983).

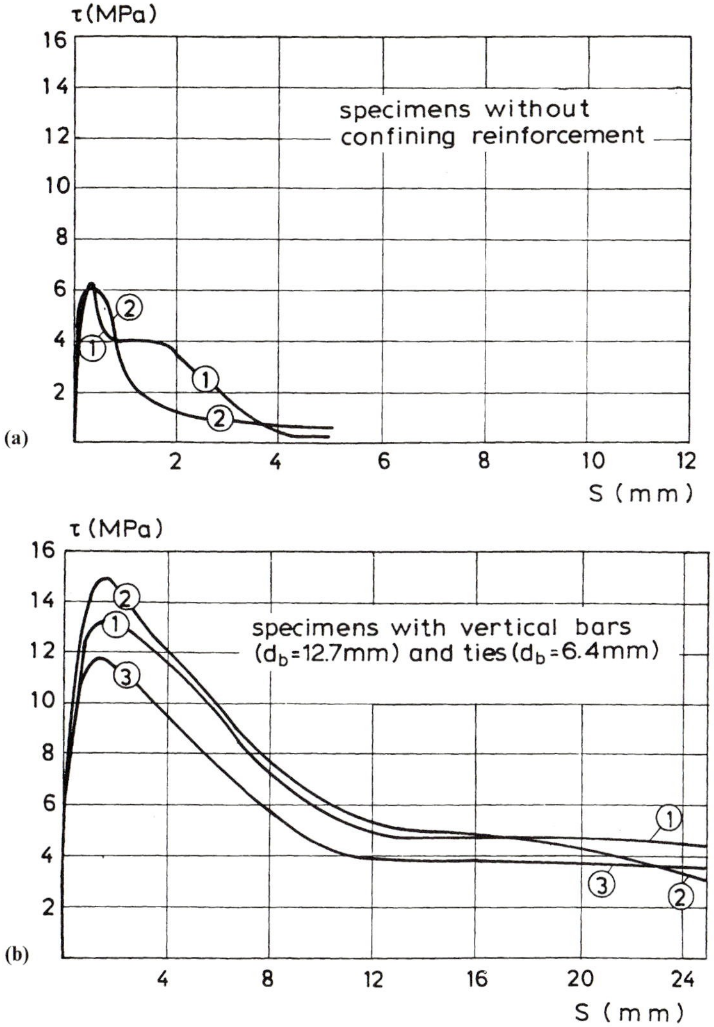

(a)
τ(MPa)
16
14
12
10
8
6
4
2
specimens without
confining reinforcement
2
4
6
8
10
12
S (mm)
(b)
τ (MPa)
16
14
12
10
8
6
4
2
specimens with vertical bars
(d_b=12.7mm) and ties (d_b=6.4mm)
4
8
12
16
20
24
S (mm)

Figure 7.36 Experimentally derived local bond stress (τ)–slip (s) diagrams (Eligehausen, Popov and Bertero, 1983).

τ
τmax
τ2
τ1
τ0
splitting failure
τ3
s
s0
s1
s2
sτmax
s3

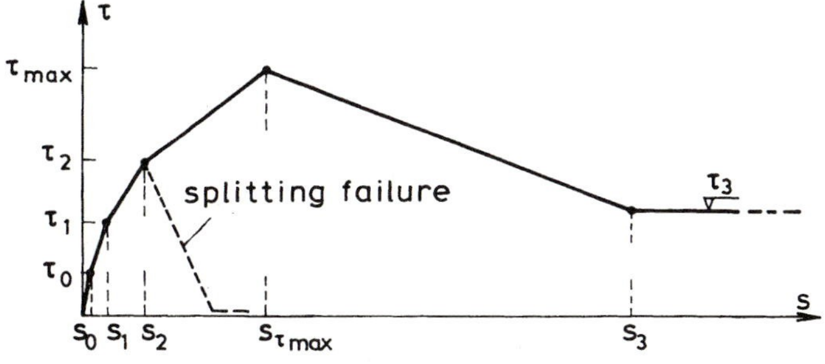

Figure 7.37 Idealized local bond–slip curve for monotonic loading.

a stress level $\tau=\tau_1$ (which is a function of the tensile strength of concrete) bond cracks form, as shown in Figure 7.38 (Popov, 1977; Eligehausen, Popov and Bertero, 1983). At approximately the same time, separation of concrete from the reinforcing bar takes place in the region of primary (flexural) cracks. This separation causes an increase in the circumference of the concrete surface previously in contact with the bar, and as a result circumferential tensile stresses develop (Park and Paulay, 1975). These stresses, in combination with the radial component of the force carried by the ribs or indentations, shown in Figure 7.38, lead to splitting cracks. Whenever at a stress $\tau=\tau_2$ (Figure 7.37) these cracks propagate up to the external face of the member and at the same time there is not enough confinement, bond is destroyed and a splitting failure occurs (Figures 7.36 and 7.37). If, on the other hand, the presence of adequate confining reinforcement inhibits the propagation of splitting cracks, the bond stress can reach substantially higher values (τ_{max} in Figure 7.37).

The slope of the consecutive branches of the τ–s diagram gradually decreases, in other words the value of the relative slip corresponding to a given increment of bond stress increases. Along the branch defined by the stresses τ_2 and τ_{max} a gradual deterioration of the concrete lugs (keys) between adjacent ribs occurs, until for a value $\tau=\tau_{max}$ these lugs fail in shear (Tassios, 1979; Eligehausen, Popov and Bertero, 1983).

The descending branch of the τ–s diagram ($s>s_{\tau_{max}}$) corresponds to a complete deterioration (pulverization) of concrete between adjacent ribs, and for $s>s_3$ the moderate amount of residual bond stress (τ_3) is due exclusively to friction at the cylindrical surface defined by the tips of the ribs. The stress τ_3 can remain practically constant for high values of slip, as shown in the experimental curves (Eligehausen, Popov and Bertero, 1983) of Figure 7.36. The value of the slip s_3 almost coincides with the spacing of ribs, since when a rib is displaced to the position occupied by the adjacent one when loading started, the only remaining mechanism of bond transfer is friction at the cylindrical failure surface.

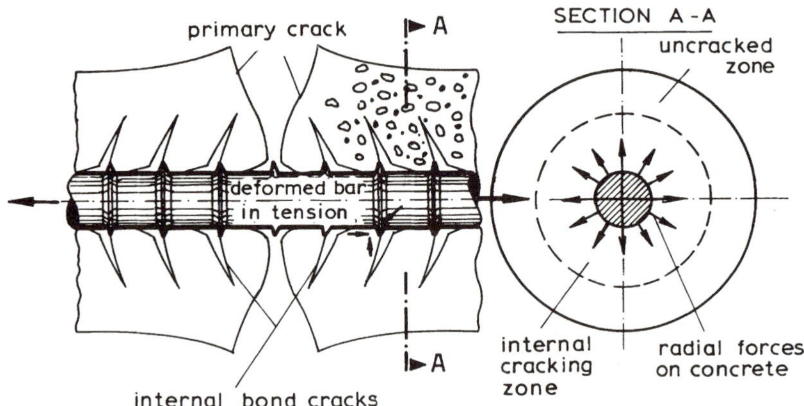

Figure 7.38 Primary (flexural) crack and bond cracks (the width of the cracks is drawn in enlarged scale), and forces acting on the concrete around a bar, in an R/C tie of circular cross-section.

The τ–s curve for monotonic loading remains approximately the same for loading in tension, as well as in compression (Tassios, 1979; Eligehausen, Popov and Bertero, 1983), on condition that an adequate degree of confinement exists. In the case of unconfined zones, such as the areas of a beam–column joint which lie outside the column reinforcement, the τ–s curve is different for tension and compression (Ciampi *et al.*, 1982; Eligehausen, Popov and Bertero, 1983). Besides, when the stress in a bar exceeds its yield strength ($\sigma_s > f_y$), the lateral contraction (for tensile loading) or expansion (for compressive loading) will cause a decrease or increase, respectively, in bond strength. It has been found (Eligehausen, Popov and Bertero, 1983) that this lateral deformation cannot affect the bond strength more than 20–30%, even for very large steel strains.

Figure 7.39 shows the influence of transverse (confining) pressure, resulting either from compressive axial loads (such as those acting on beam–column joints where beam bars are anchored or pass through) or from confinement. As can be seen in the figure, both the maximum bond stress and the residual stress due to friction (τ_3) increase with the confining pressure, but in a non-linear fashion (Figure 7.39(b)). These data indicate the favourable effect of confinement with regard to bond conditions. Indeed, the presence of confining reinforcement inhibits premature, brittle types of bond failure due to splitting, and in addition it increases bond strength.

(b) Analytical modelling

Early models for bond slip typically ignored the portion of the τ–s diagram beyond τ_{max} (descending branch), apparently because test data for this region were not available at that time. A model for the complete τ–s, relationship, including the strain-softening region, suggested by Tassios (1979), is shown in Figure 7.37. The τ-s curve is approximated by a sequence of linear segments, each one corresponding to a specific stage of the response, as explained in the previous section, the only difference being that a single segment is used up to the stress τ_1, which corresponds to the formation of bond cracks. Analytical expressions are proposed by Tassios for estimating the stresses τ_1, τ_2, τ_{max} and τ_3, as well as the slip values s_1, s_2 and $s_{\tau_{max}}$ (Figure 7.37), but on his own admission, the scatter in the experimental results then available was very large, especially with regard to slip values, for which the suggested expressions are only a first approximation.

Eligehausen, Popov and Bertero (1983), based on the findings of their extensive experimental programme described previously, suggested the following non-linear equation for the ascending branch of the τ–s diagram up to the value τ_{max}:

$$\tau = \tau_{max}(s/s_1)^{\alpha} \tag{7.64}$$

where $\alpha = 0.33$–0.45, while, as shown in Figure 7.40, the slip s_1 corresponding to τ_{max} is different from the s_1 of Figure 7.37. A horizontal branch follows ($\tau = \tau_{max}$) up to a value of slip $s = s_2$, and then the bond stress is reduced linearly down to a value of τ_f which corresponds to the residual strength due to friction, assumed to remain constant for $s > s_3$. Also shown in Figure 7.40 are numerical values for the various parameters of the model, based on the tests by Eligehausen, Popov and Bertero on concrete specimens ($f_c = 30$ MPa) with 25 mm diameter deformed bars, having a rib spacing of 10.5 mm ($= s_3$) and a relative rib area of 0.66. The scatter in the experimentally derived values of the various parameters was also pointed

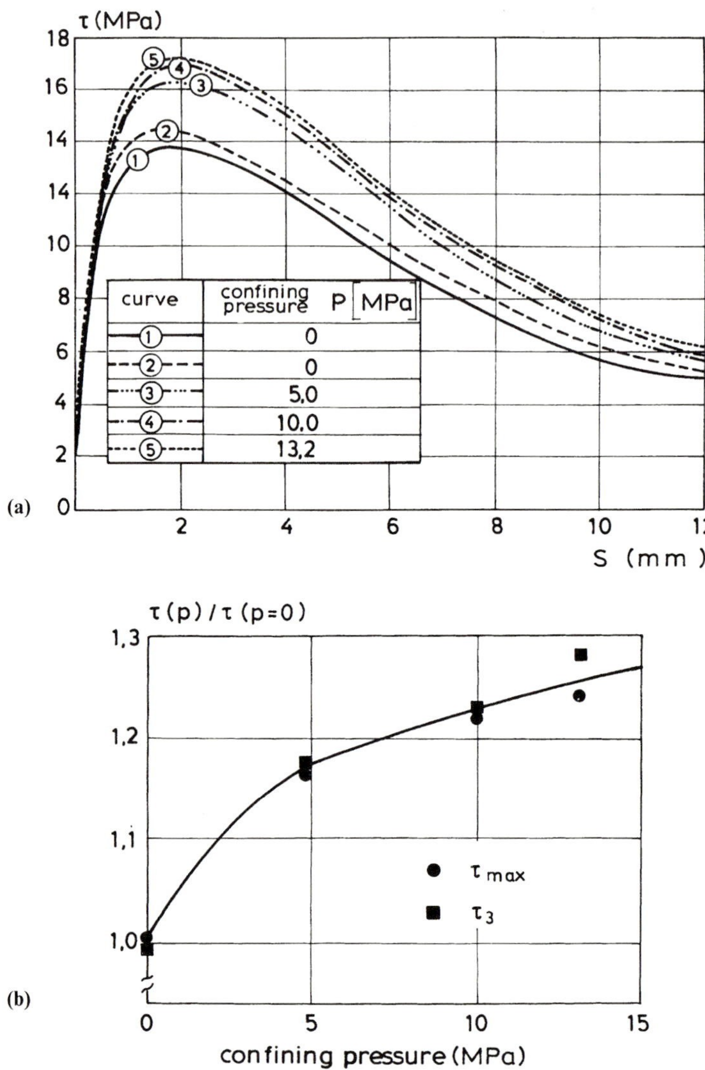

Figure 7.39 Effect of lateral pressure on local bond–slip relationship.

out by these investigators, and as a result they decided to suggest empirical coefficients for correcting the values given in Figure 7.40 whenever the shape of the deformed bar, the concrete strength and the spacing of bars are different from those used in their tests. Values of (local) τ_{max} in the literature vary from about 10 to 21 MPa, while values of $s_{\tau max}$, which show considerable scatter, vary from 0.25 to 2.5 mm (ACI Committee 408, 1991).

With regard to the influence of confining pressure, Eligehausen, Popov and Bertero (1983) are suggesting the relationship shown in Figure 7.39(b), while for the effect of rate of application of slip (s), shown in Figure 7.41 for three different

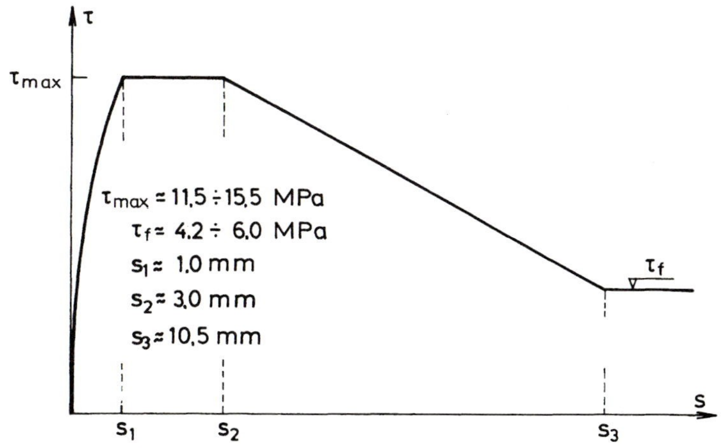

Figure 7.40 Idealized local bond stress–slip relationship for monotonic loading (Eligehausen, Popov and Bertero, 1983). Numerical values of parameters as calculated in the tests.

rates, the approximation of Figure 7.41(b) is suggested. It is noted that an increase in the slip rate s of 100 times results in increases of τ_{max} and τ_3 of about 15%.

The constitutive law of bond (τ–s) can be introduced in a refined analysis of the behaviour of a bonded bar, typically by introducing discrete springs connecting the bar at various points along its length to the surrounding concrete. To construct the stiffness matrix of the bonded bar model, knowledge of the steel (equation (7.63)) and bond constitutive laws is required, as well as the histories and values of displacement at the points where bond springs are connected to the bar. Soroushian, Obasaki and Marikunte (1991) have suggested an efficient algorithm for this problem, which unlike the previous ones (Ciampi *et al.*, 1982; Eligehausen, Popov and Bertero (1983) does not involve iterative solution of nonlinear equations.

7.6.3 Bond under cyclic loading

Bond behaviour under cyclic loading is affected by the following factors (ACI Committee 408, 1991):

1. concrete compression strength
2. cover thickness and bar spacing
3. bar size (diameter)
4. anchorage length
5. geometry of bar deformations (ribs)
6. steel yield strength
7. amount and position of transverse steel
8. casting position, vibration and revibration
9. strain (or stress) range
10. type and rate of loading (strain rate)
11. temperature
12. surface condition – coating.

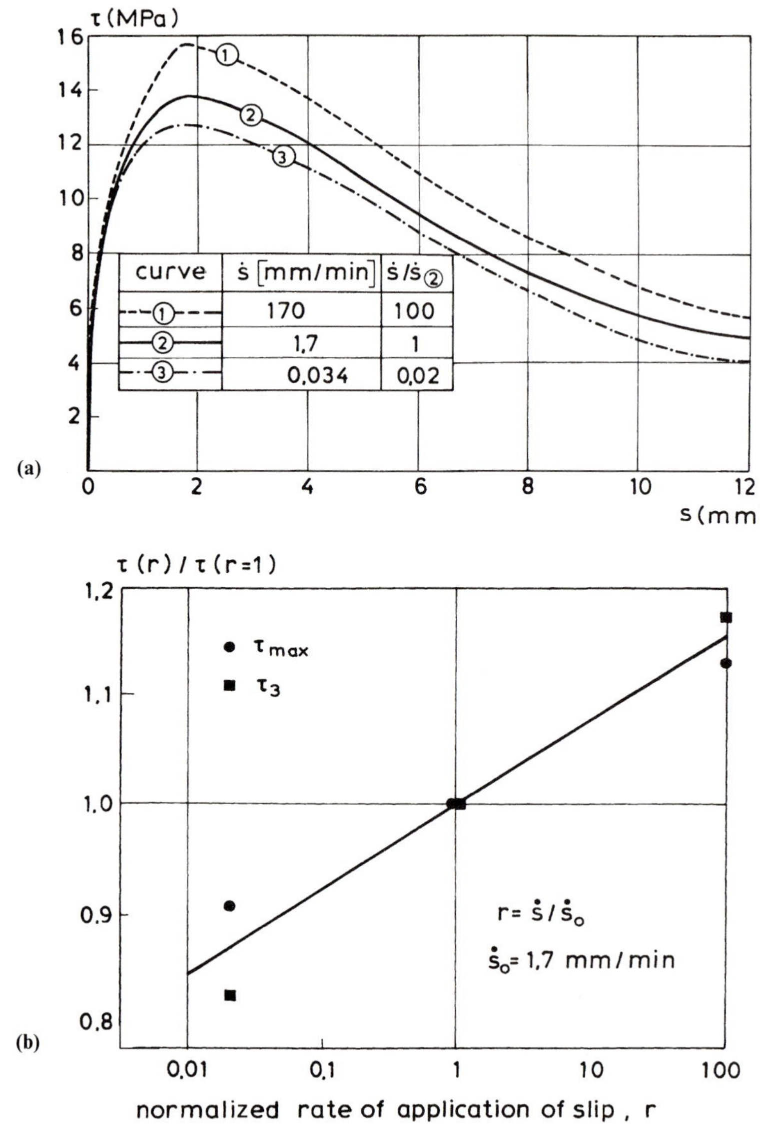

Figure 7.41 Effect of rate of application of slips ($\dot{s}$) on the local bond stress–slip relationship.

It has to be pointed out that the influence of many of the foregoing parameters on bond resistance is only qualitatively understood. Parameters (1)–(3), (6)–(8) and (12) appear to be the ones mostly affecting bond under monotonic loading, while (9), (10) and the value of maximum imposed bond stress, in addition to the previous parameters, are very important under cyclic loading conditions (ACI Committee 408, 1991).

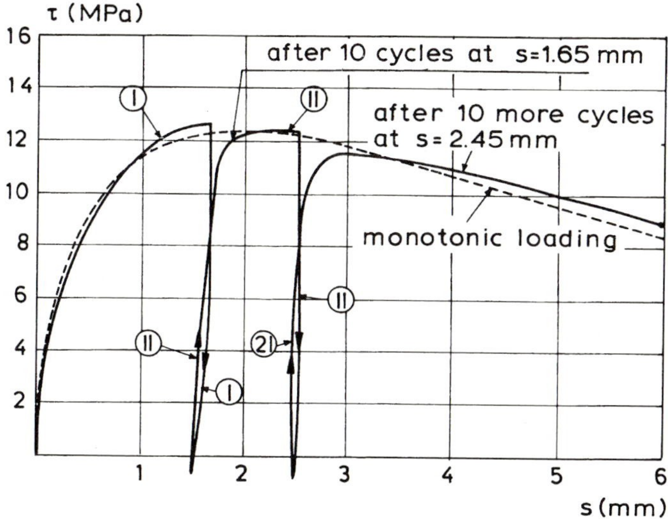

Figure 7.42 Local bond stress–slip relationship for repeated loading.

(a) Discussion of test data

As shown in Figure 7.42, the envelope of τ–s hysteresis loops for repeated loading lies very close to the curve resulting from monotonic loading (Eligehausen, Popov and Bertero, 1983). It is worth pointing out that even after 20 cycles of consecutive loading and unloading, the envelope of the loops remains quite close to the monotonic loading curve, which means that the mechanism of bond deterioration remains the same as that described in the previous section.

Figure 7.43 shows τ–s hysteresis loops derived from cyclic loading tests of specimens with one 25 mm deformed bar and confinement reinforcement (Eligehausen, Popov and Bertero, 1983). It is first pointed out that, as in the case of repeated loading, the residual slip during unloading (branch EF in Figure 7.43(b)) is quite large, which should be attributed to the fact that the elastic part of s consists of the concrete deformation only, which is just a small portion of the total slip. Whenever the sign of the bond stress reverses, the slope of the curve remains significant up to a level of stress (τ_f^-); this increased stiffness is due to friction between the bar and the surrounding concrete. When the frictional resistance is overcome, the bar starts to slip in the opposite direction (with respect to that of initial loading OAE) until the ribs of the bar again come into contact with the surrounding (intact) concrete (point I in Figure 7.43(b)). It is understood that the foregoing apply when the level of loading is such that the concrete lugs between adjacent ribs (see also Figure 7.38) have been ground, thus creating gaps between the side faces of the ribs and the surrounding concrete.

Reloading in the opposite direction (branch IA$'_1$ in Figure 7.43(b)) is now taking place at a significantly increased slope and the path followed is similar to that of monotonic loading. However, if the maximum previously attained (absolute) value of bond stress is higher than 70–80% of τ_{max}, the new envelope (OA$'_1$B$'_1$C$'_1$D$'_1$ in Figure 7.43(b)) has reduced ordinates with respect to the original one

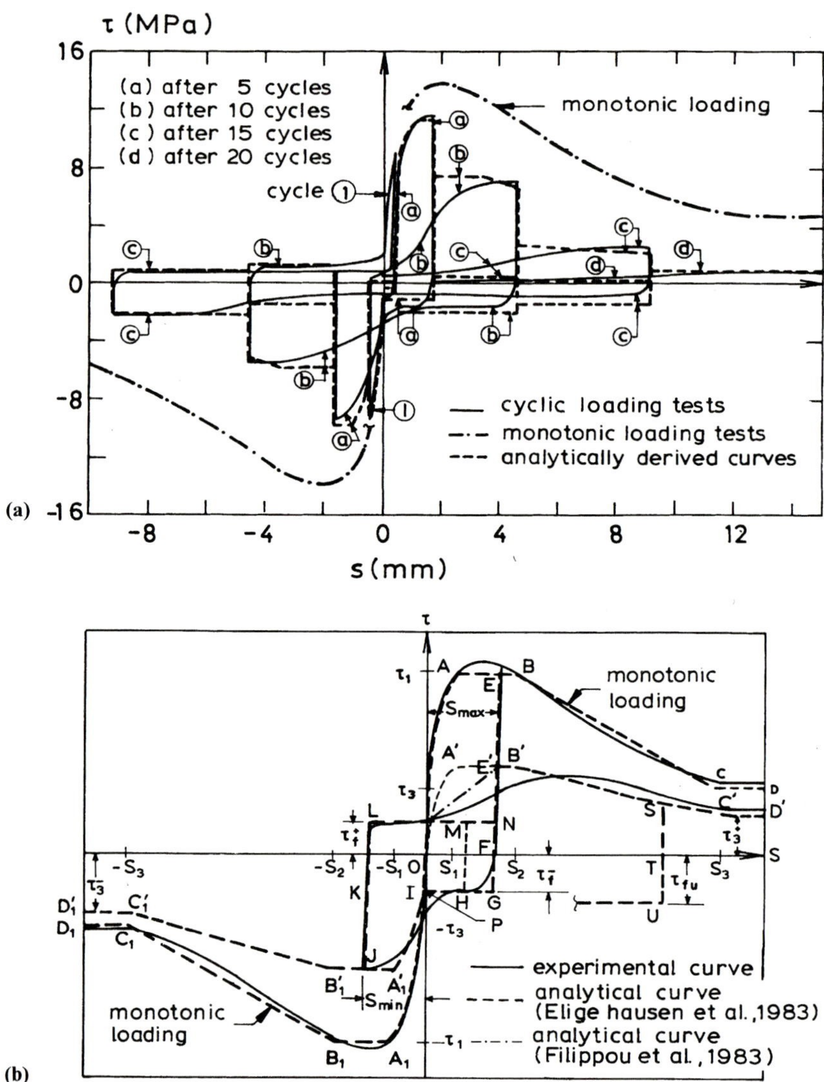

Figure 7.43 Local bond stress–slip curves for cyclic loading of confined concrete specimens (Eligehausen, Popov and Bertero , 1983): (a) comparison of analytical and experimental curves; (b) analytical modelling.

($OA_1B_1C_1D_1$). This reduction in available bond resistance is more pronounced as the values of slip between which cycling takes place increase, and also as the number of cycles increases (Eligehausen, Popov and Bertero, 1983; Tassios, 1979).

Whenever at a certain point (J in Figure 7.43(b)) the sign of loading changes (more precisely, in a deformation–controlled test, when the sign of applied slip changes), the unloading and friction branches (JKLMN) are similar to the previous ones (EFGHI).Further loading is now taking place along a new envelope

(O′A′B′C′D′) whose ordinates are reduced with respect to the initial one (OABCD). If the level of loading is high enough for shear cracks to form in concrete lugs between adjacent ribs, only a portion of these lugs can contribute to the resistance of the system, hence the envelope O′A′B′C′D′ has lower ordinates than the previous one ($OA'_1B'_1C'_1D'_1$). Moreover, if unloading takes place at a point along the descending branch of the τ–s curve which corresponds to a pronounced degradation of the concrete lugs due to shear, the frictional resistance (τ_{fu}) will be higher than its previous value (τ_f^-),since at this stage the interface between the bar and the surrounding concrete is rougher. This characteristic may be verified if the corresponding branches of the loops a and b in Figure 7.43(a) are compared.

It is seen from the foregoing discussion of the bond degradation mechanism under cyclic loading, and also from the evaluation of pertinent test data (Eligehausen, Popov and Bertero, 1983), that most of the damage occurs during the first loading cycle. During subsequent cycles a gradual smoothening of crack interfaces occurs, which causes a reduction of mechanical interlock and friction forces.

(b) Analytical modelling

On the basis of their cyclic loading tests, described in the previous section, Eligehausen, Popov and Bertero (1983) proposed the analytical model shown in Figure 7.43(b) together with corresponding experimental curves. It has to be pointed out that this model applies in the case that adequate confining pressure is present and inhibits premature splitting failure (see also Figure 7.36), and also when deformed bars are used. The envelopes of the hysteresis loops, which are assumed to be identical for both tension and compression (OABCD and $OA_1B_1C_1D_1$) are calculated using equation (7.64) and the rules described in section 7.6.2(b) (Figure 7.40). Unloading is considered to be taking place at a constant slope of 180 MPa mm^{-1}, a value which strictly applies for $f_c = 30$ MPa (for higher concrete grades larger values of the slope were estimated). The same slope is retained during reloading towards the envelope (branch NE in Figure 7.43(b)).

The reduction in available bond resistance with increasing slip and number of loading cycles (reduced envelopes OA′B′C′D′ and $OA'_1B'_1C'_1D'_1$) is calculated using a **damage index** D, which is equal to zero for no damage and to one when bond breaks down completely ($\tau = 0$). The estimation of the index D is shown in Figure 7.44, where the pertinent test data are also plotted. The index D is a function of the ratio E/E_0, where E is the hysteretic energy dissipation at the instant that unloading is taking place, and E_0 is the energy corresponding to the area under the monotonic τ–s curve up to the value s_3 (see also Figure 7.40). In calculating the value of E, only 50% of the energy due to friction is taken into account, as the remaining 50% is assumed to be spent in overcoming the frictional resistance without causing any bond degradation. The index D is used to estimate the reduction in the maximum stress τ_{max}, as shown in Figure 7.44, while a similar index (D_f), calculated on the basis of the energy dissipated through friction only, is used for the estimation of the residual bond stress τ_f(Figure 7.43(b)). The introduction of the residual parameters D and D_f allows the generalization of the model to a random loading history, since no reference has to be made to the number of loading cycles at a constant amplitude (s_{max}), a parameter commonly used in previous models (Tassios, 1979).

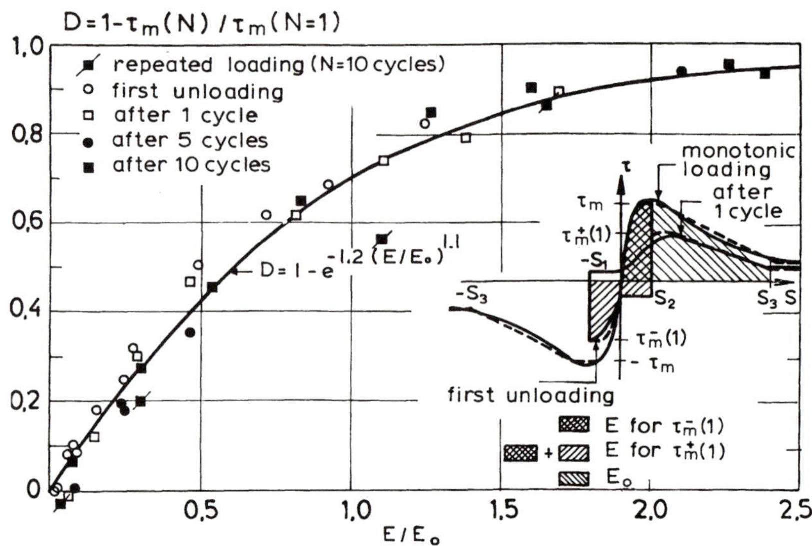

Figure 7.44 Damage index D, as a function of the energy dissipation, in the Eligehausen, Popov and Bertero (1982) model.

The correlation between experimental and analytical results obtained from the Eligehausen, Popov and Bertero (1983) model is satisfactory, as shown in Figure 7.43, with the exception of the region of reloading in the opposite direction (branch LNE′ in Figure 7.43(b)), for which Filippou, Popov and Bertero (1983) have suggested the correction shown in Figure 7.43(b). Nevertheless it has to be pointed out that the aforementioned correlation was carried out only for the test data, on the basis of which the model was developed. The reliability of this (or any other) model can only be established if adequate correlations with data from other tests are carried out; it is also necessary to define the numerous parameters of the model through more general expressions and not through empirical coefficients for correcting specific values of the parameters measured in the tests. The latter has been done for the case of monotonic loading parameters in the model incorporated in the CEB (1993) Model Code, presented in a subsequent section (Figure 7.46).

Another difficult problem consists in adopting the previous model for regions of R/C structural elements which lie outside their confined cores. Suggestions for defining modified τ–s envelopes for these regions (Ciampi *et al.*, 1982; Eligehausen, Popov and Bertero, 1983), while increasing substantially the volume of the required numerical work, appear to lead to moderately satisfactory results, especially in the case that the steel exceeds its yield limit (Ciampi *et al.*, 1982).

It is worth pointing out that more recent models, such as the one suggested by Soroushian, Obasaki and Marikunte (1991) essentially follow the hysteresis laws of Eligehausen, Popov and Bertero (1983), as modified by Filippou, Popov and Bertero (1983). Older, as well as recent models for bond under generalized cyclic loading are evaluated in the report by the CEB Task Group 22 (CEB,1991), where a review of models for concrete-to-concrete and steel-to-concrete interfaces (aggregate interlock, dowel action) may also be found.

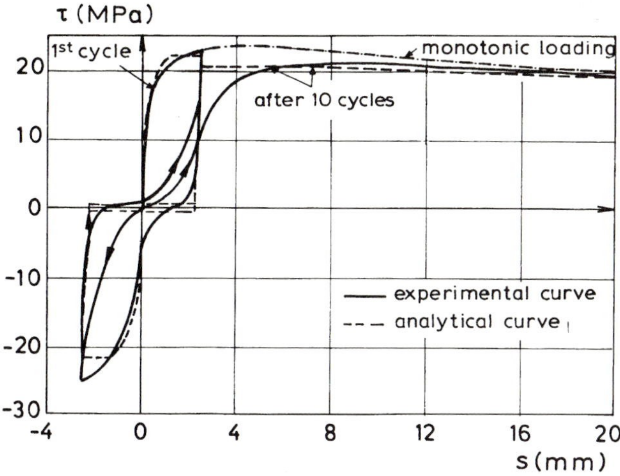

Figure 7.45 Local bond stress–slip curves for deformed bars with hooks (Eligehausen, Bertero and Popov, 1982).

(c) Effect of hooks

Significant improvement of the performance of an anchorage subjected to seismic loading can be achieved when hooks (typically at a 90° angle) are formed at the ends of the bars, a practice which is very common at exterior beam–column joints (section 9.2.4). Eligehausen, Bertero and Popov (1982) have studied experimentally the behaviour under cyclic loading of specimens with 25 mm diameter deformed bars, having a clear length of $5d_b$ and hooks at their ends. Shown in Figure 7.45 are typical local bond stress–slip curves for a hooked bar specimen subjected to monotonic and cyclic loading. It is clearly seen that the available bond resistance under monotonic loading remains almost constant, even for very large values of slip, in contrast to what happens in bars without hooks (Figure 7.36). Furthermore, during successive reversed loading cycles a significant drop in available bond resistance is observed for values of the slip lower than the previously attained peak (s_{max}), but after this value is reached, bond resistance is soon recovered, and the corresponding curve at large values of slip lies quite close to the monotonic loading curve, even after 10 loading cycles. This improved behaviour, compared with that of straight bars (Figure 7.43), should be attributed to the fact that a significant part of the force developed in the bar is carried by the pull-out resistance of the hook angle. In contrast, the bond resistance due to friction was found to be significantly lower in hooked bars than in straight bars. Eligehausen, Bertero and Popov (1982) have adopted the parameters of the bond model of Figure 7.43(b) for the case of hooked bars. The resulting analytical curves are compared with the corresponding experimental curves in Figure 7.45.

7.6.4 Relevant code provisions

The ultimate bond stress to be used for calculating anchorage lengths is given in EC2 (CEN, 1991) by the following equations:

- For high bond bars (with ribs or indentations)

$$f_{bd} = 2.25\, f_{ctk0.05}/\gamma_c \tag{7.65}$$

- For plain bars

$$f_{bd} = 2.25\sqrt{f_{ck}}/\gamma_c \tag{7.66}$$

where $f_{ctk0.05}$ is the characteristic tensile strength of concrete (5% fractile) and $\gamma_c = 1.5$ is the material safety factor. Given that according to EC2

$$f_{ctk,0.05} = 0.7\,(0.30 f_{ck}^{2/3}) = 0.21 f_{ck}^{2/3}$$

it follows from equation (7.65) that $f_{bd} \approx 0.32 f_{ck}^{2/3}$. The values of f_{bd} for deformed bars vary from 1.6 to 4.3 MPa for concrete grades from C12 to C50, while as was mentioned in section 7.6.2(a), measured values of local maximum bond stress vary from 10 to 21 MPa. However, as pointed out in the ACI Committee 408 (1991) report, average bond stresses (along a bar length of at least $15d_b$) are lower than local stresses, hence, for design purposes maximum values of 3.8–5.5 MPa are recommended. Based on the foregoing, EC2 values for f_{bd} appear to be moderately conservative. It has to be pointed out that equations (7.65) and (7.66) apply for good bond conditions (bars inclined 45°–90° to the horizontal during concreting, or bars in members with depths lower than 250 mm or located in the lower half of deeper members); for poor bond conditions the previous values should be reduced by 30%.

With regard to the effect of confining pressure (that is pressure transverse to the potential plane of splitting failure), EC2 provides that the f_{bd} values from equations (7.65) and (7.66) should be multiplied by the factor

$$1/(1 - 0.04p) \not> 1.4 \tag{7.67}$$

where p is the confining pressure in MPa. No further comments on the possible sources of p are given in EC2, hence it is not very clear whether in addition to pressure due to compressive axial loads, the confining pressure provided by closed ties or spirals can be introduced in equation (7.67). Moreover, the problem of calculating f_{bd} is not addressed by EC8, hence some ambiguity exists with regard to the design bond strength in members with confined concrete. A very simple way to deal with this problem is to assume that good bond conditions apply along the whole depth of members (with horizontal bars) if confinement according to EC8 is provided.

A rather complete constitutive law for bond stress–slip has been included in the CEB (1993) Model Code, largely based on the Eligehausen, Popov and Bertero (1983) model described in the preceding sections. The τ–s curve for monotonic loading and deformed bars is shown in Figure 7.46; the ascending branch is described by equation (7.64) with $\alpha = 0.4$, while the following values are prescribed for the rest of the model parameters:

1. For confined concrete:
   $s_1 = 1.0$ mm, $s_2 = 3.0$ mm, $s_3 =$ rib spacing
   $\tau_{max} = 2.5\sqrt{f_{ck}}$ and $1.25\sqrt{f_{ck}}$, for good and poor bond conditions, respectively, $\tau_f = 0.4\tau_{max}$
2. For unconfined concrete:

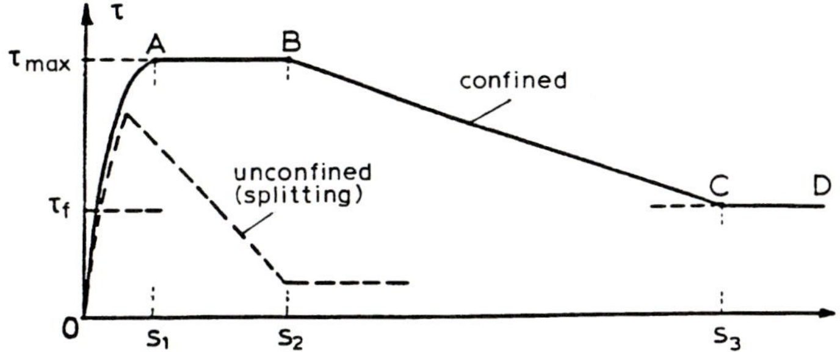

Figure 7.46 Bond stress–slip model adopted by the CEB (1993) Model Code.

$s_1 = 0.6$ mm $= s_2$, $s_3 = 1.0$ and 2.5 mm, for good and poor bond conditions, respectively

$\tau_{max} = 2.0\sqrt{f_{ck}}$ and $1.0\sqrt{f_{ck}}$, for good and poor bond conditions, respectively, $\tau_f = 0.15\tau_{max}$

The parameters for confined concrete are applicable whenever the transverse pressure $p \geqslant 7.5$ MPa (compression) or closely spaced transverse reinforcement is present, satisfying the conditions $\Sigma A_{sw} \geqslant nA_s$, where ΣA_{sw} is the area of stirrups over a length equal to the anchorage length, n is the number of bars enclosed by the stirrups, and A_s is the area of one longitudinal bar. For

$$0.25 \leqslant \Sigma A_{sw}/(nA_s) \leqslant 1.0 \quad \text{or} \quad 0 \leqslant p \leqslant 7.5 \text{ MPa}$$

linear interpolation between the values of confined and unconfined concrete may be used to derive the model parameters. For smooth hot rolled bars the behaviour is substantially different, with $s_1 = s_2 = s_3 = 0.1$ mm, $\alpha = 0.5$ and $\tau_{max} = \tau_f = 0.3f_{ck}^{1/2}$ and $0.15f_{ck}^{1/2}$ for good and poor bond conditions, respectively. It is worth pointing out that in the CEB (1993) Code's commentary it is explicitly recognized that the scatter in bond test results is considerable, the coefficient of variation in the bond stresses being up to about 30%.

For repeated loading the CEB (1993) model accepts a constant value of the unloading branch equal to 200 MPa mm^{-1}, independent of the slip value; this value is essentially the same as that suggested by Eligehausen, Popov and Bertero (1983) (section 7.6.3(b)). No model for reversed cyclic loading has been included in the CEB Model Code, as this type of loading is not explicitly covered by the code, possibly because the amount of data available for this case is clearly smaller than the corresponding amount from monotonic loading tests.

7.7 REFERENCES

ACI Committee 408 (1991) Abstract of state-of-the-art report: bond under cyclic loads. *ACI Materials Journal*, **88** (6), 669–73.

Ahmad, S.H. and Shah, S.P. (1982) Stress–strain curves of concrete confined by spiral reinforcement. *Journal of the ACI*, **79** (6), 484–90.

Aoyama, H. and Noguchi, H. (1979) Mechanical properties of concrete under load cycles idealizing seismic actions, *Bull. d' Inf. CEB*, **131**, 29–63.

Aoyama, H *et al.*, (1992) Development of advanced reinforced concrete buildings with high-strength and high-quality materials. *Proceed. of 10th World Conf. on Earthq. Engng*, July 1992, Madrid, Spain. Balkema, Rotterdam. **6**, 3365–70.

ASCE Task Committee on Finite Element Analysis of Reinforced Concrete Structures (1982) *State-of-the-art Report on Finite Element Analysis of Reinforced Concrete*, ASCE, New York.

Bazant, Z.P. and Bhat, P.D. (1977) Prediction of hysteresis of reinforced concrete members. *Journal of the Struct. Div., ASCE*, **103** (ST1), 153–67.

Bazant, Z.P. and Prat, P.C. (1988) Microplane model for brittle-plastic material–Part I: Theory; Part II: Verification. *Journal of Engng Mech.*, ASCE, **114** (10), 1672–702.

Blakeley, R.W.G. and Park, R. (1973) Prestressed concrete sections with cyclic flexure. *Journal of the Struct. Div., ASCE*, **99** (ST8), 1717–42.

Buyukozturk, O. and Tseng, T.-M. (1984) Concrete in biaxial cyclic compression. *Journal of Struct. Engng, ASCE*, **110** (3), 461–76.

Carreira, D.J. and Chu, K.-H. (1985) Stress–strain relationships for plain concrete in compression. *Journal of the ACI*, **82** (6), 797–804.

CEB (1983) Concrete under multiaxial states of stress. *Bull. d' Inf. CEB*, **156**, Paris.

CEB (1985) Model Code for Seismic Design of Concrete Structures. *Bull. d' Inf. CEB*, **165,** Paris.

CEB (1989) Design aspects of high strength concrete. *Bull. d' Inf. CEB*, **193**, Lausanne.

CEB (1991) Behaviour and analysis of reinforced concrete structures under alternate actions inducing inelastic response–Vol. 1: General models. *Bull. d' Inf. CEB*, **210**, Lausanne.

CEB (1993) CEB-FIP Model Code 1990. *Bull d' Inf. CEB*, **213/214**, Lausanne.

CEN Techn. Comm. 250 SG2 (1991) *Eurocode 2: Design of Concrete Structures–Part 1: General Rules and Rules for Buildings (ENV 1992-1-1)*, CEN, Berlin.

CEN Techn. Comm. 250/SC8 (1995) *Eurocode 8: Earthquake Resistant Design of Structures–Part 1: General rules (ENV 1998-1-1/2/3)*, CEN, Berlin.

Chen, E.-S. and Buyukozturk, O. (1985) Constitutive model for concrete in cyclic compression. *Journal of Engng. Mech. ASCE*, **111** (6), 797–814.

Chen, W.F. (1982) *Plasticity in Reinforced Concrete*, Mc Graw-Hill, New York.

Ciampi, V *et al.* (1982) Analytical model for concrete anchorages of reinforcing bars under generalized excitations. *Report EERC-82/23*, Univ. of California, Berkeley.

Darwin, D. and Pecknold, D.A. (1976) Analysis of RC shear panels under cyclic loading. *Journal of the Struct. Div., ASCE*, **102** (ST2), 355–69.

Dougill, J.W. (1976) On stable progressively fracturing solids. *ZAMP (Jnl. of Applied Mathem. and Physics), Birkhäuser*, **27** (4), 423–37.

Eligehausen, R., Bertero, V.V. and Popov, E.P. (1982) Hysteretic behavior of reinforcing deformed hooked bars in R/C joints. *Proceed. 7th Europ. Conf. Earthq. Engng.*, Athens, Greece, **4**, 171–178.

Eligehausen, R., Popov, E.P. and Bertero, V.V. (1983) Local bond stress-slip relationships of deformed bars under generalized excitations. *Report EERC-83/23*, Univ. of California, Berkeley.

Elwi, A.A. and Murray, D.W. (1979) A 3D hypoelastic concrete constitutive relationship. *Journal of the Engng Mech. Div., ASCE*, **105** (EM4), 623–41.

Filippou, F.C., Popov, E.P. and Bertero, V.V. (1983) Modeling of R/C joints under cyclic excitations. *Journal of Struct. Engng, ASCE*, **109** (11), 2666–84.

Gerstle, K *et al.*, (1980) Behavior of concrete under multiaxial stress states. *Journal of the Engng Mech. Div., ASCE*, **106** (EM6), 1383–403.

Han, D.J. and Chen, W.F. (1985) A nonuniform hardening plasticity model for concrete materials. *Mech. of Materials*, **4**, 283–302.

Hobbs, D.W., Newman, J.B. and Pomeroy, C.D. (1977) Design stresses for concrete structures subjected to multi-axial stresses. *The Struct. Engineer*, **55** (4), 151–64.

Kappos, A.J. (1991) Analytical prediction of the collapse earthquake for R/C buildings: suggested methodology. *Earthq. Engng and Struct. Dynamics*, **20**(2), 167–76.

Kappos, A.J. (1993) *RCCOLA-90: A Microcomputer Program for the Analysis of the Inelastic Response of Reinforced Concrete Sections.* Lab of Concrete Structures, Dept of Civ. Engng, Aristotle Univ. of Thessaloniki, Greece.

Karsan, I.D. and Jirsa, J.O. (1969) Behavior of concrete under compressive loadings. *Journal of the Struct. Div., ASCE*, **95** (ST12), 2543–63.

Kotsovos, M.D. and Newman, J.B. (1979) A mathematical description of the deformational behaviour of concrete under complex loading. *Magazine of Concrete Research*, **31** (107), 77–90.

Krajcinovic, D. and Fonseka, G.U. (1981) The continuous damage theory of brittle materials–Part I: General theory: Part II: Uniaxial and plane response modes. *Journal of Applied Mech.* **48**, 809–24.

Kupfer, H., Hilsdorf, H.K. and Ruesch, H. (1969) Behavior of concrete under biaxial stresses. *Journal of the ACI*, **66** (8), 656–66.

Mc Dermott, J.F. (1978) Mechanical characteristics and performance of reinforcing steel under seismic conditions. *proceed. of Workshop on Earthq. Resistant Reinforced Concrete Building Construction*, Univ. of California, Berkeley, July 1977, II, pp. 629–57.

Madas, P. and Elnashai, A.S. (1992) A new passive confinement model for the analysis of concrete structures subjected to cyclic and transient dynamic loading. *Earthq. Engng and Struct. Dynamics*, **21** (5), 409–31.

Mander, J.B., Priestley, M.J.N. and Park, R. (1988) Theoretical stress–strain model for confined concrete. *Journal of Struct. Engng, ASCE*, **114** (8), 1804-26.

Meyer, C. and Okamura, H.(eds) (1986) *Finite Element Analysis of Reinforced Concrete Structures*. ASCE, New York.

Moehle, J.P. and Cavanagh, T. (1985) Confinement effectiveness of crossties in RC. *Journal of Struct. Engng. ASCE*, **118** (2), 3268–84.

Mueller, H.S. and Hilsdorf, H.K. (1993) Constitutive relations for structural concrete. *Bull. d' Inf. CEB*, **217**, 19–65.

Noguchi, H. (1986) Analytical models for cyclic loading of RC members, in *Finite Element Analysis of Reinforced Concrete Structures*, ASCE, New York, (eds C. Meyer and H. Okamura) pp. 486–506.

Otter, D. and Naaman, A.E. (1989) Model for response of concrete to random compressive loads. *Journal of Struct. Engng, ASCE*, **115** (11), 2794–809.

Ottosen, N.S. (1977) A failure criterion for concrete. *Journal of the Engng Mech. Div., ASCE*, **103** (EM4), 527–35.

Park, R. and Paulay, T. (1975) *Reinforced Concrete Structures*, J. Wiley & Sons, New York.

Park. R., Priestley, M.J.N. and Gill, W.D. (1982) Ductility of square confined concrete columns. *Journal of the Struct. Div., ASCE*, **108** (4), 929–50.

Park. R. and Sampson R.A. (1972) Ductility of reinforced concrete column sections in seismic design. *Journal of the ACI*, **69** (9), 543–51.

Popov, E.P. (1977) Mechanical characteristics and bond of reinforcing steel under seismic conditions. *Proceedings of Workshop on Earthquake Resistant Reinforced Concrete Building Construction*, Univ. of California, Berkeley, **II**, 658–682.

Sixth World Conf. on Earthq. Engng, New Delhi, India, V. **II**, pp 1933–8.

Sakai, K. and Sheikh, S. A. (1989) What do we know about confinement in reinforced concrete columns? *ACI Struct. Journal*, **86** (2), 192–207.

SANZ (Standards Association of New Zealand) (1982)(a) *Code of Practice for the Design of Concrete Structures* (NZS 3101–Part 1: 1982); (b) *Commentary on Code of Practice for the Design of Concrete Structures* (NZS 3101–Part 2: 1982), Wellington.

SANZ (1984) *Code of Practice for General Structural Design and Design Loadings for Buildings* (NZS 4203: 1984), Wellington.

Scott, B.D., Park, R. and Priestley, M.J.N. (1982) Stress–strain behaviour of concrete confined by overlapping hoops at low and high strain rates. *Jounal of the ACI*, **79** (1), 13–27.

Shah, S.P., Fafitis, A. and Arnold, R. (1983) Cyclic loading of spirally reinforced concrete. *Journal of Struct. Engng, ASCE*, **109** (7), 1695–710.

Sheikh, S.A. and Uzumeri, S.M. (1980) strength and ductility of tied concrete columns. *Journal of the Struct. Div., ASCE*, **106** (ST5), 1079–102.

Sheikh, S.A. and Uzumeri, S.M. (1982) Analytical model for concrete confinement in tied columns. *Journal of the Struct. Div., ASCE*, **108** (ST12), 2703–22.

Sheikh, S.A. and Yeh, C.C. (1986) Flexural behaviour of confined concrete columns. *Journal of the ACI*, **83**(3), 389–104.

Soroushian, P., Obasaki, K. and Marikunte, S. (1991) Analytical modelling of bonded bars under cyclic loads. *Journal of Struct. Engng. ASCE*, **117** (1), 48–60.

Soroushian, P. and Sim, J. (1986) Axial behaviour of reinforced concrete columns under dynamic loads. *Journal of the ACI*, **83** (6), 1018–25.

Stankowski, T. and Gerstle, K.H. (1985) Simple formulation of concrete behavior under multiaxial load histories. *Journal of the ACI*, **82** (2), 213–21.

Tassios, T.P. (1979) Properties of bond between concrete and steel under loads idealizing seismic actions. *CEB Bull. d' Inf.*, **131**, Paris, 67–122.

Tassios, T.P. (1989) Specific rules for concrete structures–justification note No. 6; required confinement for columns, in *Background document for Eurocode 8–Part 1, vol. 2–Design Rules*, CEC DG III/8076/89 EN, pp. 23–49.

Tasuji, E., Slate, F.O. and Nilson, A.H. (1978) Stress–strain response and fracture of concrete in biaxial loading. *Journal of the ACI*, **75** (7), 306–12.

Taylor, M.A. (1978) Constitutive relations for concretes under seismic conditions. *Proceed. of Workshop on Earthquake Resistant Reinforced Concrete Building Construction*, Univ. of California, Berkeley, July 1977, II, 569–93.

Vallenas, J. and Bertero, V.V. (1977) Concrete confined by rectangular hoops subjected to axial loads. *Report EERC-77/13*, Univ. of California, Berkeley.

Yang, B., Dafalias, Y.F. and Herrmann, L.R. (1985) A bounding surface plasticity model for concrete. *Journal of Engng Mech., ASCE*, **111** (3), 359–80.

Yankelevsky, D.Z. and Reinhardt, H.W. (1989) Uniaxial behavior of concrete in cyclic tension. *Journal of Struct. Engng, ASCE*, **115** (1), 166–82.

8

Earthquake-resistant design of reinforced concrete linear elements

8.1 INTRODUCTION

The fundamental principles governing the design of R/C structural elements are the following (Park and Paulay, 1975; Dowrick, 1987; CEN 1995):

1. The dissipation of seismic energy should take place mainly in elements which possess adequate ductility and which are relatively easy to repair. For an R/C building this principle leads to the requirement that beam failure should precede column failure, while for an R/C bridge the usual requirement is that energy dissipation take place in the piers rather than in the superstructure or in the foundation (Priestley and Park, 1987).
2. Seismic energy dissipation should take place through flexural yield mechanism as opposed to shear or bond-slip mechanisms. This means that flexural failure of an element should always precede shear failure, as well as anchorage failure.
3. Joints, that is common areas of adjacent elements, should not fail before these elements attain their full strength.
4. The reinforcement required for ensuring a ductile behaviour in structural elements should not be so much as to cause major difficulties during construction.
5. Structural elements which for design purposes are not considered as part of the seismic action resisting system, should be in a position to maintain their load-bearing capacity during a strong earthquake excitation.

At this point it is deemed appropriate to recall two general principles which are often forgotten in everyday practice of R/C design and construction. The first one refers to the need for regularity of the structural system (see also section 5.1). It is well established by now that a careful dimensioning and detailing of structural elements can only marginally reduce the unfavourable consequences of a high degree of irregularity. The second principle refers to the need for a systematic quality control. This control should cover all stages of production and use of a structure, more specifically the design project, the construction and the maintenance during service life.

The present chapter deals with the seismic behaviour and the earthquake-resistant design of R/C linear elements, that is beams and columns, while planar elements are dealt with in Chapter 9. The behaviour of R/C elements is presented with a focus on results from cyclic load testing of members and subassemblages, while occasionally use is made of some simple analytical models. A brief discussion of the response of R/C elements to monotonic loading precedes the treatment of behaviour under cyclic loading, with a view to providing first a clear understanding of the basic characteristics of strength and ductility. For each type of element the design and detailing procedures for flexure and shear according to the Eurocodes (CEN, 1991, 1995) are presented, and the background for the various code provisions is also given. Finally, design examples involving beams and columns of a multi-storey R/C structure are provided to facilitate the understanding of the EC8 design and detailing requirements.

8.2 SEISMIC BEHAVIOUR OF BEAMS

Beams are the structural elements in which the larger portion of seismic energy dissipation takes place, through stable **flexural yield mechanisms**. Given that, the need arises for a design and detailing of beams aiming at a sufficiently ductile behaviour.

8.2.1 Behaviour under monotonic loading

The available ductility of a beam can be conveniently estimated with reference to the moment (M)-curvature (ϕ or $1/r$) diagram (Fig. 8.1) of its critical sections. It is known (Park and Paulay, 1975) that the curvature of a beam section, according to the classical bending theory, is the ratio of strain at a certain fibre of the section to the corresponding distance from the neutral axis. Referring to Figure 8.2 the curvature can be written as

$$\phi = \varepsilon_c / x \tag{8.1(a)}$$

where ε_c is the strain (shortening) of concrete at the top compression fibre and x the neutral axis depth; alternatively

$$\phi = \frac{\varepsilon_{s1}}{d - x} \tag{8.1(b)}$$

where ε_{s1} is the strain (elongation) of tension reinforcement and d the effective depth of the beam. Using equations (8.1) the curvature may be written as

$$\phi = \frac{\varepsilon_c + \varepsilon_{s1}}{d} \tag{8.2}$$

It is seen from equation (8.2) that increasing the ultimate strain of concrete (ε_{cu}) and steel (ε_{su}) leads to an increase in the ultimate curvature of the beam, that is in its available ductility. A useful index for expressing quantitatively the deformation capacity of an element is the **curvature ductility factor**

$$\boxed{\mu_\phi = \frac{\phi_{max}}{\phi_y}} \tag{8.3}$$

where ϕ_{max} is the maximum curvature at the critical section and ϕ_y the corresponding yield curvature, that is the starting-point of the post-elastic branch of

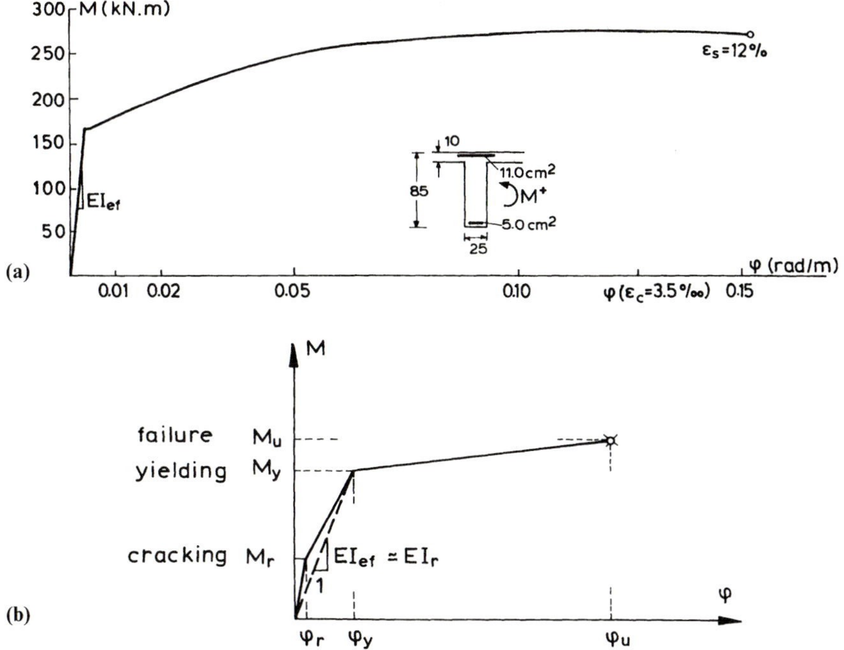

Figure 8.1 Moment–curvature diagram for R/C beams: (a) Exact diagram, using the RCCOLA-90 code (Kappos, 1993); (b) bilinear and trilinear idealisation of the M-ϕ diagram.

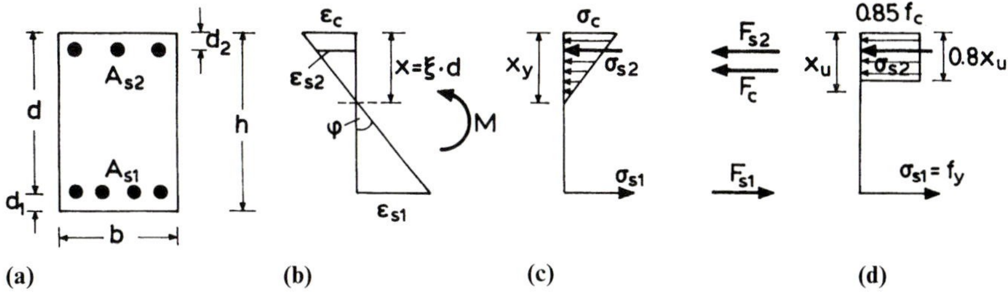

Figure 8.2 Data for the calculation of the curvature ductility factor (μ_ϕ) for a doubly reinforced rectangular section.

a linearized M–ϕ diagram, such as the one in Figure 8.1(b). For an adequately designed R/C beam this point corresponds to yielding of the tension reinforcement. If failure is defined at the material level, the **ultimate curvature** is given as the lower value resulting from the following relationships

$$\phi_u = \frac{\varepsilon_{cu}}{x_u} \tag{8.4(a)}$$

$$\phi_u = \frac{\varepsilon_{su}}{d - x_u} \tag{8.4(b)}$$

where ε_{cu} and ε_{su} are the ultimate strains of concrete and steel, and x_u is the neutral axis depth at failure. Appropriate procedures for defining ε_{cu} and ε_{su} have already been presented in sections 7.4 and 7.5, respectively. Alternatively, the ultimate curvature (under monotonic loading) may be defined with regard to a prescribed drop in strength (ΔM) in the M–ϕ diagram (Park and Paulay, 1975), or the load–displacement (F–δ) diagram of the beam (Priestley and Park, 1987).

(a) Approximate procedure for estimating curvature ductility

The available curvature ductility of a beam section can be calculated using closed-form relationships in the case of sections with simple geometry and arrangement of reinforcement. Thus, for the doubly reinforced rectangular section of Figure 8.2(a), yield curvature may be estimated from equilibrium of normal forces and the assumption of linear behaviour of concrete up to yielding of tension steel, which is valid for $\sigma_c \leqslant 0.7f_c$ (section 7.3). From the strain profile shown in Figure 8.2(b) it is seen that

$$\frac{\varepsilon_c}{\xi d} = \frac{\varepsilon_{s1}}{d-\xi d} = \frac{\varepsilon_{s2}}{\xi d - d_2}$$

Substituting the values $\varepsilon_c = \sigma_c/E_c$, $\varepsilon_{s1} = \sigma_{s1}/E_s$ and $\varepsilon_{s2} = \sigma_{s2}/E_s$ in the above relationships, solving for the stresses in the reinforcements and defining α as the ratio of Young's moduli E_s/E_c, the following equations result:

$$\sigma_{s1} = \frac{1-\xi}{\xi}\alpha\sigma_c \tag{8.5(a)}$$

$$\sigma_{s2} = \frac{\xi d - d_2}{\xi d}\alpha\sigma_c \tag{8.5(b)}$$

Applying the requirement for equilibrium of normal forces in concrete (F_c) and steel (F_{s1}, F_{s2}) results in

$$F_c + F_{s2} - F_{s1} = 0$$

Substituting in the previous relationship the values of the three forces resulting from the stress diagram of Figure 8.2 (c) leads to the following equation:

$$\frac{1}{2}b\xi d\sigma_c + \sigma_{s2}A_{s2} - \sigma_{s1}A_{s1} = 0 \tag{8.6}$$

Now, if the values of σ_{s1}, σ_{s2} from equations (8.5) are substituted in equation (8.6), it turns out that

$$\frac{1}{2}b\xi d + \frac{\xi d - d_2}{\xi d}\alpha A_{s2} - \frac{1-\xi}{\xi}\alpha A_{s1} = 0$$

Setting $A_{s1}/bd = \rho_1$ and $A_{s2}/bd = \rho_2$ and rearranging terms in the previous relationship, the following second-degree (with respect to ξ) equation results:

$$\xi^2 + 2\alpha(\rho_1 + \rho_2)\xi - 2\alpha\left(\rho_1 + \frac{d_2}{d}\rho_2\right) = 0$$

The solution of this equation gives the value of the neutral axis parameter

$$\xi=\left[\alpha^2(\rho_1+\rho_2)^2+2\alpha\left(\rho_1+\frac{d_2}{d}d_2\right)\right]^{1/2}-\alpha(\rho_1+\rho_2) \tag{8.7}$$

Therefore, the yield curvature can be calculated from the relationship

$$\phi_y=\frac{\varepsilon_y}{d-x_y}=\frac{f_y/E_s}{d(1-\xi_y)} \tag{8.8}$$

where f_y is the yield strength of steel and ξ_y is given from equation (8.7).

For the ultimate curvature ϕ_u to be estimated, a certain assumption is required for the shape of the concrete stress-strain diagram. The procedure is simplified if the rectangular stress block of Figure 8.2 (d) is assumed, the use of which is allowed by modern codes for flexural design (CEN, 1991). However, the calculation can also be based on parabolic or other shape of the σ_c–ε_c diagram. Moreover, an assumption has to be made regarding the stress in the compression reinforcement.

For the case where the compression steel yields at failure (i.e. $\sigma_{s2}=f_y$), the equilibrium condition can be written

$$0.85f_c0.8x_ub+A_{s2}f_y-A_{s1}f_y=0$$

and solving for the neutral axis depth

$$x_u=\frac{(A_{s1}-A_{s2})f_y}{0.68\,f_cb} \tag{8.9}$$

The coefficient 0.85 in the concrete stress block accounts for the effect of sustained (long-term) loading (CEN, 1991) and could be omitted in the case of seismic loading which is a short-term one. In this case the multiplier of the denominator of equation (8.9) would be 0.80 instead of 0.68. However, for design purposes it is recommended to retain the coefficient 0.85 to account (roughly) for the reduced capacity of concrete due to cyclic loading (see also section 8.2.2).

The ultimate curvature can now be estimated from equation (8.4(a)), if the value of x_u from equation (8.9) is substituted, resulting in

$$\phi_u=\varepsilon_{cu}\frac{0.68f_cb}{(A_{s1}-A_{s2})f_y} \tag{8.10}$$

Finally, using the definition of μ_ϕ (equation (8.3)) and the values of ϕ_u, ϕ_y from equations (8.8) and (8.10), the curvature ductility factor is

$$\boxed{\mu_\phi=\frac{0.68f_c\varepsilon_{cu}E_s}{f_y^2(\rho_1-\rho_2)}(1-\xi_y)} \tag{8.11}$$

where the value of ξ_y is calculated from equation (8.7). It is apparent from equation (8.11) that the available ductility of a beam increases linearly with the ultimate strain in concrete ε_{cu}, which, as already shown in section 7.4, increases with the amount of transverse reinforcement (hoops or spirals). Moreover ductility increases with increasing compression reinforcement and decreases with increasing neutral axis depth and increasing tension reinforcement.

As mentioned previously, equation (8.11) applies when compression steel yields at failure, that is when the strain

$$\varepsilon_{s2}=\varepsilon_{cu}\left(1-\frac{d_2}{x}\right)$$

equals or exceeds the yield strain $\varepsilon_y = f_y/E_s$. In case that $\varepsilon_{s2} < \varepsilon_y$ it can be shown (Park and Paulay, 1975) that μ_ϕ can be calculated from the relationship

$$\mu_\phi = 0.8E_s\varepsilon_{cu}(1-\xi_y)\left(f_y\left\{\left[\left(\frac{\rho_2\varepsilon_{cu}E_s - \rho_1 f_y}{1.7f_c}\right)^2 + \frac{d_2}{d}\frac{\rho_2\varepsilon_{cu}E_s}{1.06f_c}\right]^{1/2} - \left(\frac{\rho_2\varepsilon_{cu}E_s - \rho_1 f_y}{1.7f_c}\right\}\right)^{-1} \tag{8.12}$$

(b) Refined procedure for estimating curvature ductility

In case either the geometry of the section is more complex, or the reinforcing bars are arranged in multiple layers, as well as whenever an improved accuracy is sought with regard to material models (see sections 7.3–7.5), recourse should be made to a more general methodology for calculating M–ϕ relations, such as the one described here.

As shown in Figure 8.3, the section is divided into n horizontal layers (of depth h/n), while for the determination of the strain profile the 'plane sections remain plain' assumption is made, although the latter is not an indispensable feature of the method. The concrete layers, as well as the various reinforcement layers, are defined by their distances (y_i) from a reference axis which is commonly taken either as the centre of gravity of the section or as the top compression fibre.

For a given axial load N and for every value of strain (ε_c), at the top fibre, it is possible to find the moment, M, and the corresponding curvature, ϕ, of the section. The usual procedure is to assume an initial position of the neutral axis depth (x), thus the strain profile. For a given strain ε_{ci} at the centre of fibre i, the corresponding stress σ_{ci} is defined from the σ_s–ε_c diagram, which can be of any shape. Similarly, from the strain values ε_{si} at the centre of each reinforcement layer the corresponding strresses ε_{si} are determined using the (also arbitrarily shaped) σ_s–ε_s diagram. If the total number of steel layers is m, the following equilibrium equation should apply:

$$M = \sum_{i=1}^{n} \sigma_{ci}A_{ci} + \sum_{i=1}^{m} \sigma_{si}A_{si} \tag{8.13}$$

In general the equilibrium condition is not satisfied, therefore the assumed neutral axis depth is corrected and the procedure repeated until equation (8.13) converges within a specified tolerance. Subsequently the moment at the section can

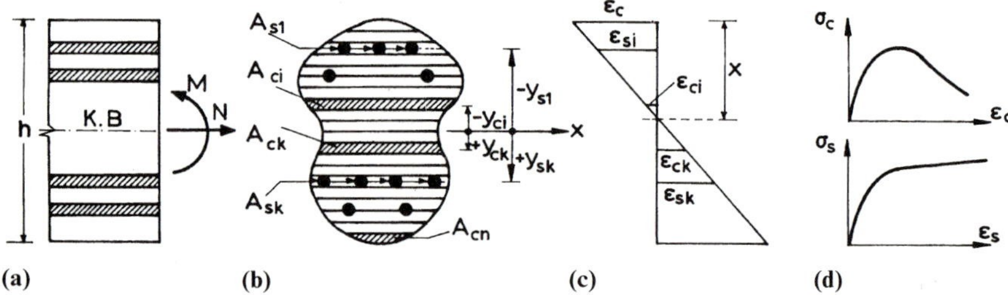

Figure 8.3 Calculation of moment-curvature diagram for an arbitrarily shaped R/C section subjected to uniaxial bending and axial force.

be calculated from the relationship

$$M=\sum_{i=1}^{n} \sigma_{ci}A_{ci}\,y_{ci}+\sum_{i=1}^{m} \sigma_{si}A_{si}\,y_{si} \tag{8.14}$$

while the corresponding curvature is found from equation (8.1(a)). Following the procedure for various ε_c values, the full moment–curvature diagram for the section can be determined, and if the analysis is repeated for various axial loading levels, it is also possible to construct *M–N* interaction diagrams. Such a procedure is followed in the computer code RCCOLA-90 (Kappos, 1993), which analyses arbitrarily shaped R/C sections subjected to monotonic loading, using a variety of models for confined and unconfined concrete, and for steel (section 7.4). An example M–ϕ diagram for a T-beam was given in Figure 8.1. It is pointed out that RCCOLA-90, as well as similar programs (Blakeley and Park, 1973; Kaba and Mahin, 1984) for the analysis of the inelastic response of R/C sections, are based on the assumption that concrete carries no tensile stresses (tension stiffening ignored), and that the σ_s–ε_s curve for steel is identical for tension and compression. It is understood that such a procedure is valid as long as premature buckling of compression bars does not occur, which actually is the case when hoops are closely spaced. Pertinent experimental data show that hoop spacing should not exceed six times the diameter of longitudinal reinforcement, otherwise buckling may occur immediately following the spalling of cover concrete (see section 8.2.2).

(c) Estimation of flexural deformations

Given the distribution of curvature along the span of an R/C member, for the elastic as well as for the inelastic stage of its response, it is then possible to estimate the corresponding **flexural deformations**. Thus the **rotation** between two sections *a* and *b* of an element is given by the moment–area relationship

$$\theta_{ab}=\int_a^b \phi\,\mathrm{d}x \tag{8.15}$$

where d*x* is the length of an infinitesimal element. The corresponding **relative deflection** between *a* and *b* (displacement transverse to the longitudinal axis of the element) is given by the moment–area relationship

$$\delta_{ab}=\int_a^b x\phi\,\mathrm{d}x \tag{8.16}$$

where *x* is the distance of the infinitesimal element with length d*x* from section *a*.

The deformations calculated according to this procedure are not the actual total deformations of an R/C element. In fact, two different types of deformations should be added to the flexural ones, namely **shear deformations** which are significant subsequent to the formation of diagonal cracks in the element, and deformations due to **bond-slip** of longitudinal bars (section 7.6). Ignoring these sources of deformations would result in underestimating actual rotations and deflections, especially in the case of cyclic loading, as explained in section 8.2.2. On the other hand, ignoring the effect of **tension stiffening** (that is the ability of concrete between cracks to carry some tensile stresses) leads to underestimating the actual stiffness of the element and therefore to overestimating the total deflections. This compensates to a certain extent for ignoring shear and bond-slip deformations. It has to be emphasized that tension stiffening decreases as the element approaches yield conditions and is negligible in plastic hinge regions (Park and Paulay, 1975).

Consider now the beam of Figure 8.4 which is subjected to end moments (of the seismic type), having such a magnitude as to force the end regions of the beam to enter the post-elastic range. The length of these regions (commonly called **plastic hinge** regions) is generally larger than the value defined on the basis of the bending moment diagram (sections with $M > M_y$), since the presence of shear leads to an increase in the stress of tension reinforcement and therefore to a spread of yielding in a wider area (Bachmann, 1970). The rotations at the beam ends can be calculated by appropriate integration of the curvature diagram shown in Figure 8.4(c), according to equation (8.15). Such a refined procedure, also including the effect of bond-slip and tension stiffening, has been suggested by Eligehausen and Fabritius (1993).

A simpler procedure, suitable for hand calculations, can be developed if the actual curvature diagram is replaced by the linearized one shown in Figure 8.4 (c). The **equivalent plastic hinge length** l_p, along which the maximum curvature ϕ_u extends, is defined in such a way that the area of the corresponding parallelogram equals the area of the actual plastic curvature ($\phi > \phi_y$) diagram. Thus, the plastic rotation θ_p can be estimated from the simple relationship

$$\theta_p = (\phi_u - \phi_y) l_p \tag{8.17}$$

An example of plastic hinge rotation claculation is shown in Figure 8.5 in the form of a θ_p–x/d diagram (see Figure 8.2 for definitions of x and d). The curves in Figure 8.5 correspond to different amounts of transverse reinforcement and were constructed using the σ_c–ε_c model for confined concrete suggested by Kappos (1991) and the σ_s–ε_s model for steel suggested by Park and Paulay (1975). The ultimate and yield curvatures calculated using these models were introduced in equation (8.17) to estimate the rotational capacity. Also shown in the figure is the

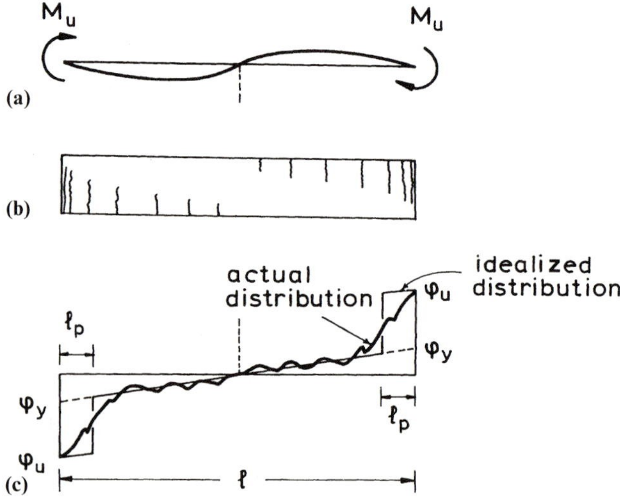

Figure 8.4 Actual and idealized distribution of curvatures in the inelastic range, for a beam subjected to seismic loading.

curve adopted by EC2 for nonlinear analysis purposes. It is clearly seen that the code curve is very conservative, especially in the case of high hoop ratios ρ_w, since it ignores the favourable effect of confinement in increasing rotational capacity.

(d) Estimation of rotational ductility

A useful index for quantitatively expressing the ductility of an element is the **rotational ductility factor**, which is defined (similarly to μ_φ) by the relationship

$$\boxed{\mu_\varphi = \frac{\theta_{max}}{\theta_y}} \tag{8.18}$$

where θ_y is the yield rotation, which in the case of beams usually corresponds to yielding of tension steel (see also Figure 8.6). The μ_φ value corresponding to $\theta = \theta_u$ (rotation at failure) expresses the available ductility of a member in terms of rotation. The main advantage of the μ_θ index is that it can be measured experimentally in an easier and more reliable way than μ_φ, which is a localized quantity. This renders μ_θ a particularly useful quantity in comparing experimental and analytical results. On the other hand, the disadvantage of this index is that it depends on the loading pattern of the element, which influences the value of yield rotation θ_y. For a beam subjected to antisymmetric loading, as shown in Figure 8.4, the rotation at its ends due to the couple of moments M_y is

$$\theta_y = \frac{M_y l}{6EI_{ef}} \tag{8.19}$$

where EI_{ef} is the effective stiffness factor (modulus of elasticity times the effective moment of inertia) which in the pre-yield region ($M < M_y$) may be taken as equal

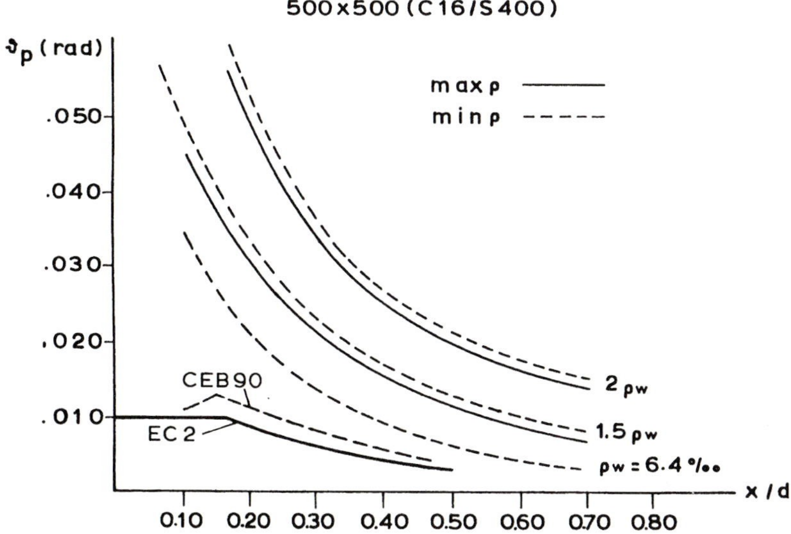

Figure 8.5 Plastic rotation capacities calculated for a 500 mm square column with varying amounts of hoop reinforcement.

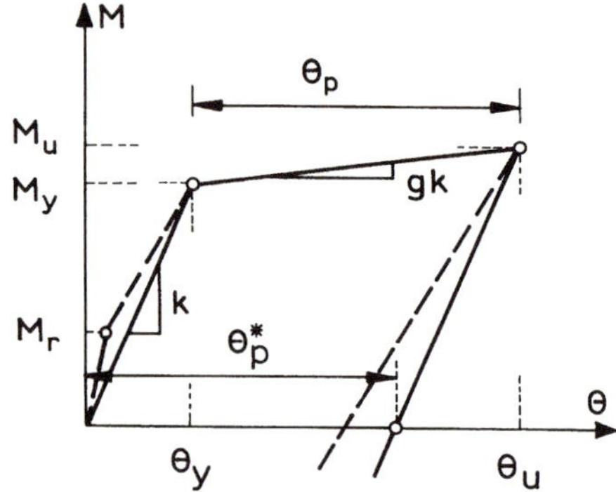

Figure 8.6 Required parameters for the estimation of rotational ductility factor μ_θ.

to the slope of the bilinear idealization of the M–θ diagram, as shown in Figure 8.1. However, for an arbitrary ratio M_i/M_j of the end moments, θ_y is given by the relationship (Kappos, 1986a)

$$\theta_{yk} = \frac{M_{yk} l}{\lambda_k EI_{ef}} \quad (k = i,j) \tag{8.20(a)}$$

where

$$\lambda_k = \frac{6\beta}{2\beta - 1} \quad (\beta = M_i/M_j) \tag{8.20(b)}$$

For the member end subjected to the higher moment, λ_k varies from 2 to 6. It is understood that for a structural element subjected to (time-varying) seismic loading the value of θ_{yk} depends on the ratio M_i/M_j at the instant of yield.

Comparisons with experimental results show that calculation of θ_y using equations (8.19) and (8.20) which are based solely on flexural considerations, underestimates the actual value of yield rotation in an R/C member. A more general expression for θ_y should include the contribution of shear deformations ($\theta,_V$) and fixed end rotations due to bond-slip ($\theta,_S$), that is

$$\theta_y = \theta,_M + \theta,_V + \theta,_s \tag{8.21}$$

where $\theta,_M$ is the flexural contribution (equations (8.19) or (8.20)). Park and Ang (1985) have suggested the following equation for the shear contribution

$$\theta,_V = \frac{0.002}{l_0/d - 0.5} \tag{8.22}$$

based on an evaluation of 244 tests involving R/C beams and columns. In equation (8.22) l_0/d is the shear span ratio (replaced by 1.5 if $l_0/d < 1.5$), where l_0 is the length of the equivalent cantilever ($M = Vl_0$). For squat members ($l_0/d < 4$) and members with high shear (effective shear stress $\tau > 0.4 f_c^{1/2}$ MPa) empirical correction terms are added to equation (8.22). The rotation due to bond slippage at a member end can be estimated from the relationship

$$\theta,_s = \frac{s}{d - d_2} \tag{8.23}$$

where the slip s can be estimated from the assumed local bond stress (τ_b) – slip relationship (section 7.6). Park and Ang (1985) also suggest a correction factor for $\theta_{,M}$ to account for inelasticity of concrete and axial load effects ($\theta'_y = \delta\, \theta_y$)

$$\delta = 1.05 + \left(\frac{0.45}{0.84 + 2\rho_2 - \rho_1} - 0.05 \right) \frac{\nu}{0.3} \tag{8.24}$$

where $\nu = N/(f_c\, b\, d)$ is the normalized axial load and ρ_1, ρ_2 the reinforcement ratios defined previously. It is understood that for relatively deep members, poor bond conditions and high axial loading, the resulting yield rotation is significantly higher than indicated by equation (8.19).

Referring now to Figure 8.6, the rotational ductility factor may be written as

$$\mu_\theta = 1 + \frac{\theta_p}{\theta_y} \tag{8.25}$$

and it is in this form that μ_θ is commonly used in the framework of an inelastic dynamic analysis of structures subjected to seismic excitation (Kappos, 1986a, b) wherein for each member end the plastic hinge rotation θ_p is calculated.

Whenever shear and bond-slip are not significant, the rotational ductility may be estimated analytically using the corresponding curvature ductility. The following relationship between the rotational ductility factor (μ_θ) and curvature ductility factor (μ_ϕ) can be established (Kappos, 1986a)

$$\boxed{\mu_\phi = 1 + \frac{l}{\lambda^* l_p} (\mu_\theta - 1)} \tag{8.26}$$

with $\lambda^* = \lambda\,(EI_{ef}/EI_r)$; λ is given from equation (8.20(b)), while EI_{ef} is the effective stiffness and EI_r the cracked section stiffness ($\approx M_y/\phi_y$) of the member. For antisymmetric bending and $EI_{ef} = EI_r$, $\lambda^* = 6$.

For typical values of the plastic hinge length ($l_p/l \approx 0.05$–0.08) the values of μ_ϕ from equation (8.26) are approximately twice the corresponding μ_ϕ values. By calculating μ_ϕ from equations (8.3) and (8.4) and using (8.26) it is possible to estimate analytically the available rotational ductility at a certain member end in a rather simple and straightforward way, provided of course that flexure dominates.

An alternative, more refined procedure for calculating plastic hinge rotations by estimating crack opening values in the plastic hinge region has been proposed by Bachmann (1970). This method is much more complicated than the one based on the integration of the curvature diagram, and it requires a reliable knowledge of certain parameters which are difficult to estimate, such as the location and the inclination of shear cracks, as well as the distribution of bond stresses between cracks. Apparently because of these difficulties the application of this method has been rather limited.

(e) Estimation of the plastic hinge length

A weak point of the presented straightforward methodology, even in the case that shear and bond-slip can be neglected, is the appropriate determination of the plastic hinge length l_p, which depends on the characteristics of steel and (to a lesser extent) of concrete, on the type of loading (which affects the distance l_0 between the critical section $M = M_{max}$ and the point of contraflexure), on the geometry of the section, and on shear which can increase substantially the length of the inelastic region. During the 1960s, when inelastic analysis methods attracted a great deal of attention, several empirical relations for l_p were proposed, some of the best-known among them shown in Figure 8.7. It is seen that for the typical case $l_0 = l/2$, the value of the plastic hinge length varies from 5 to 10% of the span l, for common values of d/l.

In a more recent work (Priestley and Park, 1987) it is suggested including the longitudinal bar diameter (d_b) in the formula for l_p, along with l_0 and h (or d), thus,

$$l_p = C_1 l_0 + C_2 d_b + C_3 h \tag{8.27}$$

The reason for including d_b in equation (8.27) is to account for penetration of yield in the joint into which the member frames. Based on results from cyclic loading tests on R/C columns, Paulay and Priestley (1992) suggest the following form of equation (8.27)

$$\boxed{l_p = 0.08 l_0 + 0.022 f_y d_b} \tag{8.28}$$

which implies that $C_3 \approx 0$ (no correlation between h and l_p). Values of l_p calculated using equation (8.28) are comparable to those given by the empirical formulae of Figure 8.7 for low d/l ratios, but they may be quite lower for high d/l ratios in the case where small diameter bars ($d_b = 0.02$–$0.03d$) are used. It is worth mentioning that in the tests by Priestley and Park (1987) no dependence of l_p on the axial load level or on the reinforcement ratio was found. It is understood that the plastic hinge length increases with the level of plastification of the member under consideration. Equation (8.28) is valid for displacement ductilities in excess of 4, beyond which l_p tends to stabilize. In contrast for lower ductility levels scatter in the experimental results was significant (Priestley and Park, 1987) and it was difficult to suggest generally applicable formulae for this range. As a first approx-

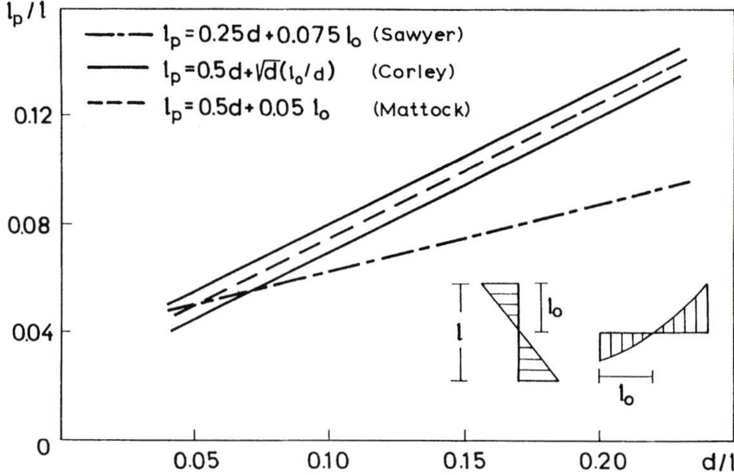

Figure 8.7 Empirical relationships for the estimation of the plastic hinge length, l_p (for $l_0 = l/2$).

imation, Kappos (1991) suggested the relationship

$$l_p = \left(\frac{\mu_\theta - 1}{3}\right)^{1/2} l_{p0} \not> l_{p0} \tag{8.29}$$

where l_{p0} is the plastic hinge length for $\mu_\theta > 4$, given from equation (8.28). For ductility calculations it is necessary to set a lower bound to equation (8.29) of 0.7 l_{po} or 0.3h (whichever is the smaller), otherwise unrealistically low available ductilities may be calculated. The suggested lower bound appears to be reasonably conservative compared with experimental results (Priestley, private communication, 1993).

(f) Empirical estimation of ductility

Given the uncertainities involved in the response of R/C elements to actions inducing a high degree of inelasticity, perhaps the most reliable method for estimating available ductility is through empirical relationships based on the largest possible database. Seen from this perspective, the empirical equation suggested by Park and Ang (1985) for the ultimate ductility factor $\mu_u = \theta_u/\theta_y$, based on 144 monotonic loading tests on beams and columns, deserves special attention. The suggested relationship is

$$\mu_u = \left(\frac{\varepsilon_P}{\varepsilon_{cl}}\right)^{0.218\rho_w - 2.15} \exp[0.654\rho_w + 0.38] \tag{8.30(a)}$$

where

$$\varepsilon_p = 0.5\varepsilon_{s2} + 0.5\sqrt{\varepsilon_{s2}^2 + \theta,_V^2} \tag{8.30(b)}$$

In equation (8.30) ρ_w is the volumetric ratio of confinement reinforcement (section 7.4.2), $\varepsilon_{cl} \geqslant 0.002$ is the concrete strain at maximum stress, ε_{s2} is the concrete strain at the location of the compression reinforcement at yield (Fig. 8.2) corresponding to ϕ_y as calculated from equations (8.8) and (8.24), while $\theta,_V$ is estimated from equation (8.22). The parameter ε_p defined by equation (8.30(b)) is the principal concrete strain at yielding and it was found to correlate well with μ_u (Park and Ang, 1985). The correlation between μ_u and $(\varepsilon_p/\varepsilon_{cl})$ is best when failure is defined as a 10% strength drop in the load–deformation curve. However, Park and Ang defined failure of gradually failing members at the 20% strength drop, because the majority of tests indicated that total repair is generally needed beyond this point. For suddenly failing members failure was easy to identify and it typically corresponded either to fracture of the longitudinal or the transverse bars (hoops) and/or buckling of the compression reinforcement.

(g) Relationship between local and overall ductility

To conclude the treatment of the notion of the ductility factor, it is important to establish a relationship between the **displacement ductility factor** μ_ϕ (already introduced in section 3.3.4) which typically refers to the structure as a whole and with reference to which modern seismic codes define the design seismic loading, and the curvature ductility factor μ_ϕ which quantifies ductility requirements at the section level (local ductility). Notwithstanding its practical usefulness, such a correlation is in general difficult to establish and it takes a different form for each type of structural system.

In case the structure under consideration can be reasonably modelled as a vertical cantilever (such a model may be used for an isolated wall or core, or for a bridge in its transverse direction) of height L with a plastic hinge length L_p at its base, it can be shown (Priestley and Park, 1987) that

$$\mu_\delta = 1 + \frac{3}{C}(\mu_\phi - 1)\frac{L_p}{L}\left(1 - 0.5\frac{L_p}{L}\right) \tag{8.31}$$

where C is a coefficient depending on the flexibility of the foundations: for a rigid support $C=1$, while for a flexible one $C>1$. The yield displacement for the flexibly supported structure is $C\,\delta_y$ where δ_y is the yield displacement for the rigidly supported structure. The factor μ_ϕ refers to the critical section of the structure, that is the base of the cantilever.

From equation (8.31) with the assumption that, $L_p/L \approx 0.10$ it results that for a displacement ductility $\mu_\delta = 4$(SANZ, 1984), the required $\mu_\delta = 11.5$ for $C=1$, while $\mu_\phi = 32.5$ for a flexible foundation with $C=3$. There are two important conclusions that may be drawn from this example. First, that in general the required curvature ductility is significantly higher than the corresponding displacement ductility and this should always be kept in mind when a quantitative estimation of ductility requirements in R/C elements is sought. Second, that ductility factors alone do not always provide a clear picture of the response of a structure. Indeed what the previous discussion does not clarify is that the total displacement δ_{max} is not the same for $C=1$ and $C=3$. This can be better demonstrated with reference to Figure 8.8, where it is seen that the total displacement corresponding to a displacement ductility of 4 is three times larger in the flexible structure compared with the rigid structure (points d and c, respectively). Therefore, the plastic part of δ_{max} has also to increase to match the increased yield deformation, and this, of course, leads to increased plastic curvature demands. On the other hand, it is high-

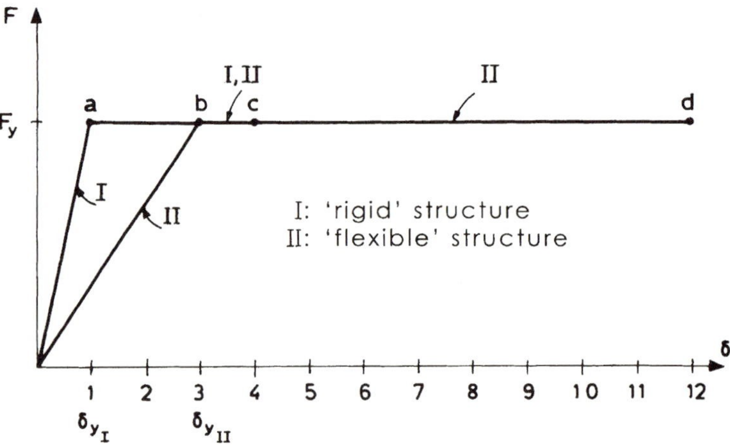

Figure 8.8 Idealized force–displacement diagram for two systems with equal strength but different stiffness.

ly questionable whether the total displacement of an actual flexibly supported structure would indeed be C times the corresponding value for the rigidly supported structure (spectral amplification for higher period structures is typically lower than for lower period ones, as discussed in section 3.2.3). It appears, therefore, that it is is more correct to describe inelastic response in terms of both ductility factor and maximum displacement (or plastic displacement).

8.2.2 Behaviour under cyclic loading

The response of an R/C beam to monotonic loading constitutes an upper bound to its response to a strong seismic excitation. The reason is that the cyclic nature of the seismic loading has certain consequences leading to a decrease in the beam's ductility, conceived in its broader sense, that is ability to sustain large inelastic deformations without a substantial decrease in strength, and at the same time ability to dissipate the seismic energy (section 7.1). The requirement for ductility is particularly imperative in the **critical regions** of structural elements which, in the case of beams subjected to prevailing seismic loading, are their end zones.

(a) Flexure-dominated beams

A clear picture of the behaviour of a critical region under cyclic loading is best given by a force–displacement diagram, which is drawn either in terms of moments (M) and rotations (θ), or of loads (F) and deflections (δ), such as that of Figure 8.9

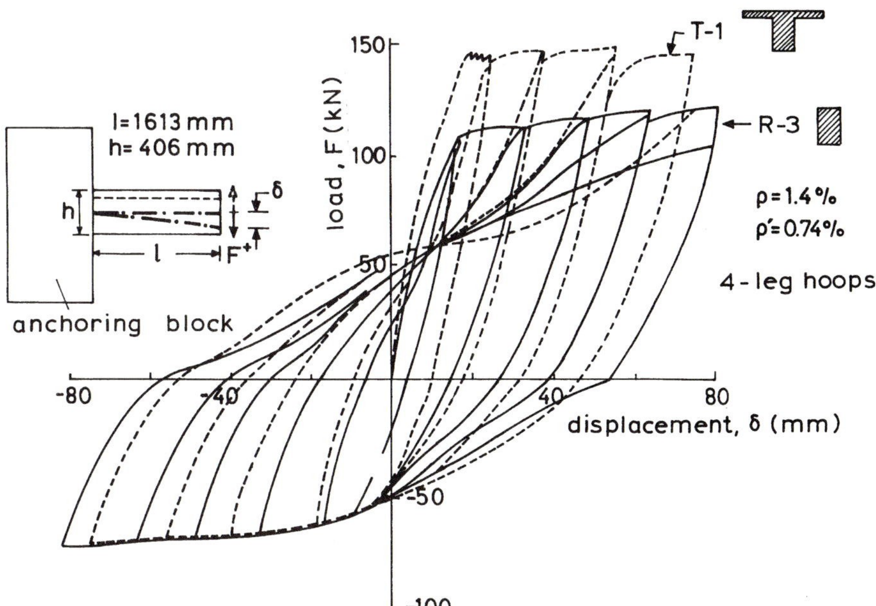

Figure 8.9 Hysteresis loops for R/C members subjected to predominantly flexural cyclic loading (low shear) (Bertero and Popov, 1977).

(Bertero and Popov, 1977). The stress state in the cantilever shown in the figure is similar to that in a beam subjected to seismic moments at its ends (the moment diagram of the cantilever coincides with the corresponding diagram of the beam part extending from its end to the point of contraflexure). The ratio of shear span to the effective depth of the cantilever beam is $M/(Vd)=4.5$ and the maximum (nominal) shear stress $\tau=V/(bd)$ is equal to $0.26\,f_c^{1/2}$, where the cylinder strength of concrete f_c is expressed in MPa. For this beam the percentage of the total deflection attributed to shear deformations did not exceed 10% during all stages of loading.

For the beam of Figure 8.9 a **flexural failure** was observed, caused by crushing of the concrete cover to the reinforcement (called **spalling**) and subsequent inelastic buckling of longitudinal bars. This type of failure is characterized by a high value of the ductility factor (for the beam under consideration $\mu_\delta \approx 5.0$), as well as by significant energy dissipation during cyclic loading. As shown in Figure 8.9, the width of the hysteresis loops does not decrease substantially even at high ductility levels. Nevertheless the shape of the loops is not elastoplastic, especially in the regions of reloading to the opposite direction (second and fourth quadrant in the diagram). Thus reloading (after complete unloading, i.e. $F=0$) takes place at a slope smaller than that of unloading, and it is seen that the reduction in the slope of the reloading branch is more pronounced as the level of inelastic deformation increases. This stiffness degradation under predominantly flexural loading should be attributed to three main factors:

1. The Bauschinger effect in steel (section 7.5.2) which results in a reduction in stiffness at loading levels significantly lower than those corresponding to yielding.
2. The fact that yielding of the reinforcing steel at one side of the beam results in permanent elongations of the bars, which prevent closing of cracks when the sign of loading is reversed. Therefore as long as the elongated bars do not yield in compression, flexural cracks remain open at both sides of the beam and all the moment is carried by a couple of forces in the reinforcement layers (top and bottom). If, according to common practice, the reinforcement at the bottom side of a beam support region is lower than that at the top side, the force that can be developed when the bottom reinforcement is in tension is not large enough to cause yielding of the top reinforcement, thus the flexural cracks at the top of the beam will remain continuously open.
3. The penetration of yielding of reinforcing bars into the (theoretically) fixed end, which destroys bond between concrete and steel and leads to local (concentrated) rotations at the beam–column interface (in the system of Figure 8.9 at the interface between the cantilever and the anchoring block). As presented in detail in section 7.6.3, the local bond stress (τ_b)–slip (s) diagram is characterized by a very small (almost flat) slope in the second and fourth quadrant, as shown in Figure 7.30. Therefore the larger the influence of slip in longitudinal bars, the more pronounced the pinching of the hysteresis loops in an R/C member. During the Berkeley tests (Bertero and Popov, 1977), although special attention was given to anchoring of bars, the portion of the deflection attributed to bond slip was found to be up to 44%, while in an actual R/C structure member failures due to inadequate anchorage of reinforcement often occur during earthquakes.

Also shown in Figure 8.9 is the F–δ diagram for a T-beam (T-1) with the same depth and web width as the rectangular beam (R-3). It was found that the increase in strength was proportional to the increase in top reinforcement due to the contribution of the slab steel, while strength in the other direction practically coincided with that of the rectangular beam. Besides, during the first loading cycle the stiffness of the T-beam was higher than that of R-3, but this increase in stiffness tended to vanish as the maximum displacement increased.

(b) Buckling of longitudinal bars

It has already been pointed out that failure of beams with predominantly flexural response is usually caused by buckling of longitudinal bars in compression which takes place in regions of the beam where spalling has taken place. The determination of the required hoop diameter and spacing to prevent premature buckling of bars is a rather complicated problem. Actually in any analytical approach (Bertero, 1979; Scribner, 1986; Papia, Russo and Zingone, 1988; Mau, 1990; Monti and Nuti, 1992) the need to specify the tangent modulus of elasticity (E_{st}) for the bars arises, which after the yielding of steel is not constant either along the buckling length or along the two faces of the bar, the compression and the tension one if bending of the bar is taken into account, and it is quite difficult to estimate in a reliable way. It has to be pointed out that a simple elastic–perfectly plastic σ_s–ε_s behaviour cannot be assumed since in this case $E_{st}=0$ and the required stirrup spacing would be zero. In fact, as suggested by Mau (1990), two different buckling events may actually take place. The first buckling occurs at the yield load ($E_{st}\approx 0$), but if steel exhibits strain-hardening behaviour, straightening of the initially buckled bar takes place (the bar returns to its straight position). A second event occurs when E_{st} along the strain-hardening branch reduces to a value sufficiently low to lead to buckling again; this second event leads to actual failure of the bars in compression and subsequently of the member as a whole. Another complication arises with respect to the buckling length of longitudinal bars which is difficult to estimate, since buckling can extend along one, two or even more hoop spacings (Figure 8.10). As a rule, the stiffer the hoop, the smaller the buckling length (Papia and Russo, 1989). Furthermore, the possibility of buckling is larger in bars which are not restrained by a hoop angle or hook (Scribner, 1986).

An interesting analytical approach to the bar buckling problem has been suggested by Papia, Russo and Zingone (1988) and Papia and Russo (1989), based on the limit stability condition of longitudinal reinforcing bars and the effective modulus (E_{st}) approach. All the relevant parameters are taken into account (slenderness of longitudinal bar, stiffness of hoop, hardening characteristics of steel), however the method is practically applicable only in cases where the longitudinal bars enter the strain-hardening range, and its calibration against experimental results has been very limited. A useful procedure for estimating the ultimate concrete strain (ε_{cu}) corresponding to buckling of longitudinal steel has been developed by these investigators (Papia and Russo, 1989); the procedure has been incorporated in the RCCOLA-90 code (Kappos, 1993) to define one of the possible failure criteria.

Given the uncertainties and limitations of analytical procedures, it is more appropriate to define the maximum spacing and the minimum diameter of hoops to avoid early buckling, on the basis of experimental results. Thus Bertero (1979)

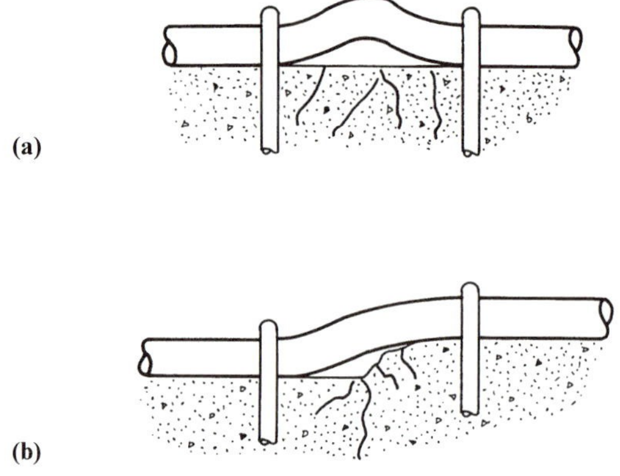

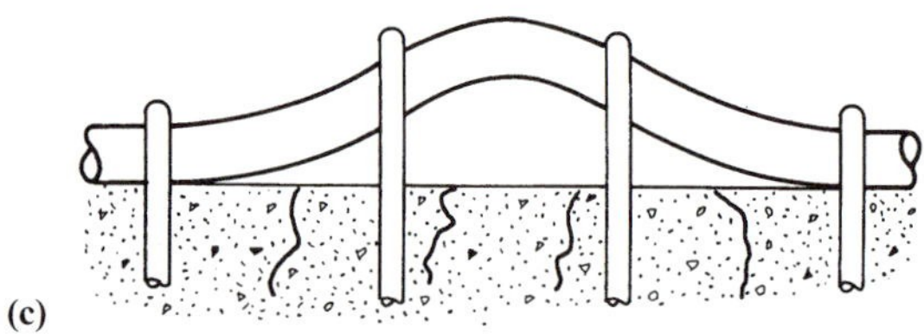

Figure 8.10 Different modes of buckling in reinforcing bars.

found that the required hoop spacing (s) varies from 3.5 to 7 times the diameter (d_b) of the longitudinal bar for a steel grade close to S400, while the New Zealand Code (SANZ, 1982) imposes the requirement

$$s \leqslant 6d_b \leqslant d/4 \tag{8.32}$$

for R/C beam plastic hinge regions. With regard to minimum hoop diameter requirements, Scribner (1986) recommends values not exceeding half the longitudinal bar diameter, while in the New Zealand Code, as well as in the CEB MC/SD (CEB, 1985) for ductility level III beams, the following relationship is included

$$\boxed{A_{sw} \geqslant \frac{\Sigma A_{sl} f_{yl}}{16 f_{yw}} \frac{s}{100}} \tag{8.33}$$

where A_{sw} is the area of one hoop leg, ΣA_{sl} the total area of longitudinal bars (having a yield strength f_{yl}) restrained by the hoop, while the other symbols have their usual meaning. In the case that only one bar is restrained by the hoop, equation (8.33) becomes

$$d_{bw} \geqslant 0.25 d_{bl} \left(\frac{f_{yl}}{f_{yw}}\right)^{1/2} \left(\frac{s}{100}\right)^{1/2} \tag{8.34}$$

where d_{bw} and d_{bl} are the diameters of the hoop and the longitudinal bar, respectively. In the case where two bars are restrained the coefficient 0.25 in equation

(8.34) becomes 0.35. Equation (8.33) is based on the empirical observation that the resistance of a hoop is adequate whenever its tensile strength is not lower than 1/16th of the total yield force in the bars it restrains, when $s = 100$ mm (SANZ, 1982).

In concluding the treatment of beams with predominantly flexural behaviour, it is emphasized that a very favourable response to cyclic loading may be achieved by providing an adequate amount of bottom reinforcement in the support region (independently of requirements resulting from the analysis), adequate anchorage length of longitudinal bars, and adequate diameter (equation (8.34)) and spacing (equation (8.32)) of hoops. It is pointed out that the behaviour of beams such as that of Figure 8.9 is the most favourable one that can be expected in an R/C element detailed to usual construction practices.

(c) Shear-dominated beams

The critical parameter with respect to the type of failure in a well-detailed R/C beam is shear, most effectively expressed in terms of the nominal (or average) shear stress $V/(bd)$ (Bertero, 1979; Paulay and Bull, 1979; French and Schultz, 1991). The beams of Figure 8.11, both from the Berkeley tests (Bertero and Popov, 1977), are identical except for the shear span. For beam R-5 it is $l/d = M/(Vd) = 2.75$, while for beam R-6 it is $l/d = 4.46$. The corresponding shear stresses are $0.44 f_c^{1/2}$ (MPa) for R-5 and $0.29 f_c^{1/2}$ for R-6. It is clearly seen in Figure 8.11 that the hysteresis loops for R-5, the beam with the higher shear stress, have a substantially smaller area, as well as a smaller average slope (secant modulus) than the corresponding loops for the beam with the lower shear stress (R-6). The differences are more significant in the region around the origin of the diagram, where the loops for beam R-5 are characterized by a marked pinching.

The main factor causing the different behaviour of the two beams is the contribution of shear deformations to the total displacement. Plotted in Figure 8.12 is the contribution of each mode of deformation to the deflection of beams R-5 and R-6, estimated through appropriate measuring techniques (Bertero and Popov, 1977), as a function of the displacement ductility factor. The portions corresponding to slippage of longitudinal bars ($\delta_{,s}$) and to shear deformations ($\delta_{,V}$) increase with the level of inelasticity (as expressed by μ_δ) for both beams, however the value of $\delta_{,V}$ is equal to 12% of the total displacement δ in the case of beam R-6, while it reaches 37% of δ for beam R-5. Besides, the maximum plastic rotation (θ_p) was 38% higher in the low-shear beam R-6.

The importance of shear deformations with regard to the behaviour of beams under cyclic loading can also be inferred from Figure 8.13, where the load (F)–shear strain (γ) diagram for beam R-5 is plotted. The deflections (δ) of the beam corresponding to points 38, 42, 46 and 49A in the diagram are all equal, however the corresponding shear deformations increase substantially from cycle to cycle. Moreover, F–γ loops display a characteristic pinching with almost zero stiffness at low levels of loading. This is attributed to the fact that during reloading (after inelastic unloading) cracks crossing reinforcement bars that have already yielded, remain open and shear is carried solely by the reinforcement. It is only when closure of cracks takes place and composite action of concrete and stirrups is restored, that stiffness increases again.

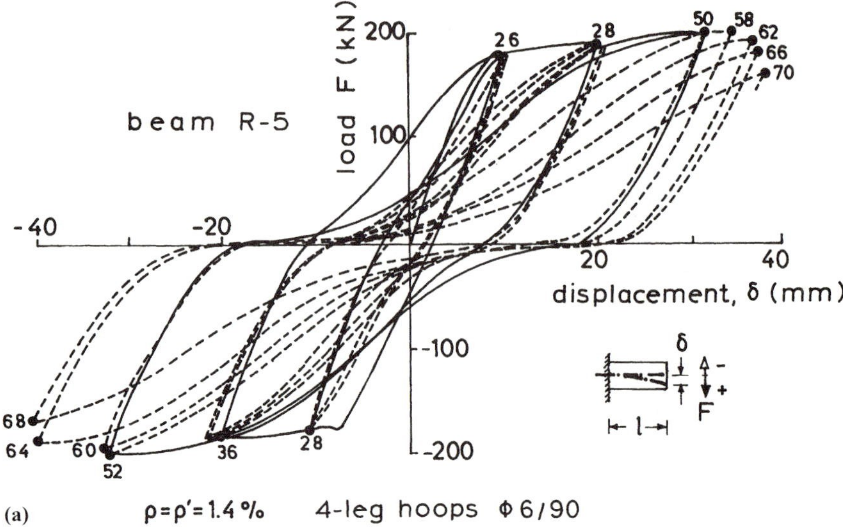

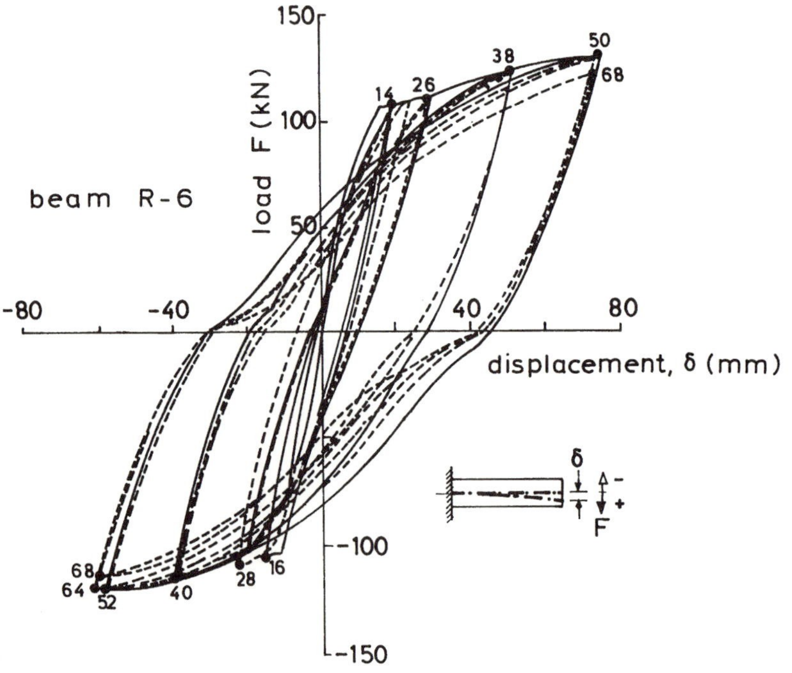

Figure 8.11 Hysteresis behaviour of beams with different levels of shear (Bertero and Popov, 1977).

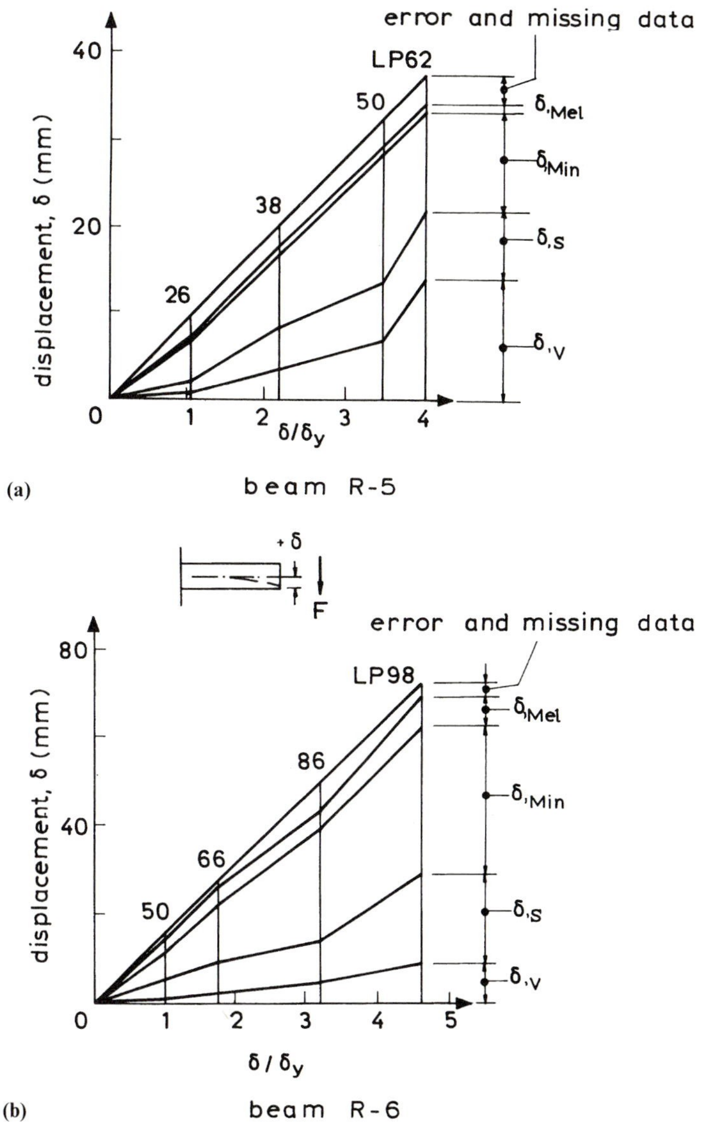

Figure 8.12 Contribution of various deformation modes to the total deflection of the beams of Figure 8.11: (a) beam R-5; (b) beam R-6.

(d) Shear transfer mechanisms and sliding shear failure

The fundamental **shear transfer mechanisms** in an R/C element are the following (Park and Paulay, 1975; Paulay and Priestley, 1992):

1. the action of stirrups at the inclined cracks, through the so-called 'truss' action;
2. the flexural compression zones of concrete;

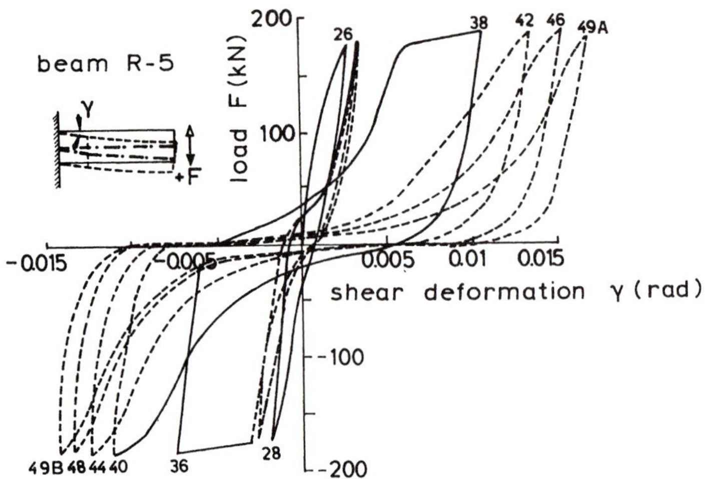

Figure 8.13 Load-shear strain diagram for a beam with high shear stress ($\tau = 0.44\, f_c'^{1/2}$ MPa) (Bertero and Popov, 1977).

3. aggregate interlock along shear crack interfaces;
4. the dowel action of longitudinal reinforcement.

During cyclic loading, flexural cracks in the compression zone remain open unless the reinforcing bars yield in compression (as explained previously). The presence of open cracks in the compression zone of beams renders the shear mechanism (2) ineffective. Moreover, cyclic loading results in a gradual smoothening of initially rough crack interfaces (inclined as well as vertical), thus leading to a degradation of mechanism (3). This degradation increases with the cycles of loading and the level of inelasticity induced. A similar degradation occurs with respect to mechanism (4) as well, as cyclic loading gradually destroys the bond of longitudinal reinforcement (section 7.6), which renders the dowel action ineffective. Particularly unfavourable with regard to this mechanism is spalling of cover concrete which occurs when the strain in the top compression fibre exceeds about 0.4%. From the foregoing discussion it appears that the main shear transfer mechanism under cyclic loading is the one involving the stirrups. However, for high levels of inelasticity and provided the web reinforcement yields, the truss mechanism also tends to lose its efficiency, particularly whenever bond between stirrups and concrete has degraded. Nevertheless, the presence of closely spaced hoops (with end detailing as described in section 8.3.3), allows the development of large inelastic deformations, provided, however, that the shear stress is kept relatively low.

Cyclic loading inducing high levels of inelastic deformation (μ_δ of the order of 3 or more) may lead to the formation of full depth cracks, which remain open during reloading (Bertero, 1979; Paulay and Bull, 1979), as shown in Figure 8.14. At this stage the whole shear force has to be transferred along the interface of the full depth crack, mainly through dowel action of the longitudinal bars. Given that the top, as well as the bottom, reinforcement has yielded, the vertical displacement of the bars required to transfer the shear is significant, leading to a sub-

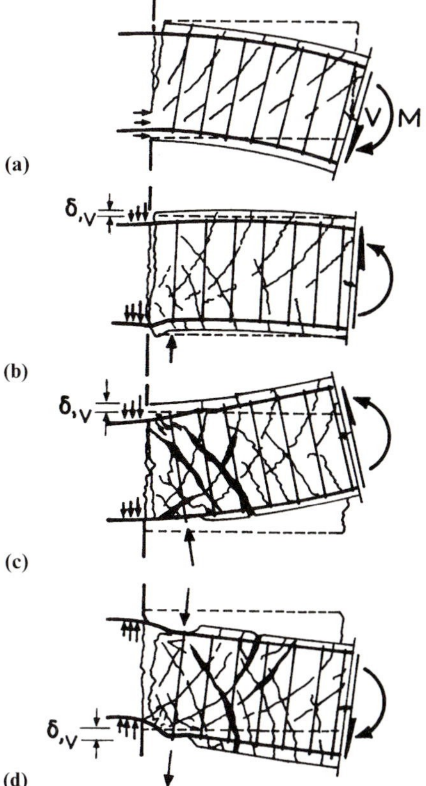

Figure 8.14 Characteristic phases of the responses of an R/C beam to reverse cyclic loading.

stantial decrease in the stiffness of the beam (see also Figures 8.11 and 8.13). If the bottom reinforcement is large enough to develop a compression force in the top reinforcement equal to its yield strength, the crack at the critical section may close (Figure 8.14(c)). However, due to the previous vertical displacement ($\delta_{,V}$) an uneven bearing at the crack interface in the newly formed compression zone has occurred, which leads to grinding of concrete at relatively low compressive stress. Thus, as cyclic loading continues, a gradual smoothening of crack interfaces takes place, which when combined with the elongation of the beam due to accumulations of plastic elongations in the reinforcing bars, leads to the formation of full depth cracks which remain open during the rest of the loading (Figure 8.14(d)). The severe stressing of longitudinal bars crossing such a crack and carrying all the shear by dowel action, while at the same time being subjected to normal stresses due to flexure, leads eventually to their failure, usually in a buckling mode. This type of failure which is common to beams with high level of shear stress, subjected to large inelastic flexural deformations, is known as **sliding shear** failure. A special feature of this type of failure is that it cannot be avoided, even when closely spaced hoops are used, because the critical crack is vertical, that is parallel to the hoops. Of course the probability of failure depends on the value of shear stress (V/bd), as well as on the magnitude of inelastic deformations and on the number of loading cycles. Based on pertinent experimental results (Bertero, 1979; Paulay and Bull, 1979; Scribner and Wight, 1980) the behaviour of beams reinforced conventionally (with longitudinal bars and hoops) can be divided into the following three categories:

1. beams with τ less than 0.25–$0.3f_c^{1/2}$ which sustain a large number of inelastic loading cycles without a significant decrease in their energy dissipation capacity;
2. beams where $0.3 < \tau/f_c^{1/2} \leqslant 0.5$, characterized by a significant decrease in stiffness and energy dissipation capacity during inelastic cyclic loading;
3. beams with $\tau > 0.5f_c^{1/2}$ where a (relatively) premature sliding shear failure is expected.

Given the complexity of flexure–shear interaction in the post-elastic range, boundaries between two successive categories are not easy to draw. In Figure 8.15 are shown the estimated displacement ductility factors for cyclic loading tests on beams with different geometries and longitudinal and web reinforcement ratios (French and Schultz, 1991), as a function of the nominal shear stress. Because of differences in geometries and reinforcement ratios, but also because of the large number of uncertainties involved (including different testing techniques, since the results are from eight different experimental programmes) a large scatter is observed. Nevertheless, some interesting conclusions may be drawn from the figure, for instance that 26 out of the 28 beams that failed in shear were subjected to shear stresses in excess of $0.35f_c^{1/2}$ while all failures that occurred at $\tau > 0.5f_c^{1/2}$ were of the shear type.

An additional important factor that has to be taken into account is whether the shear reversal is full or partial. A beam in a realistic R/C structure is subjected to vertical loading which can prevent the reversal of the sign of shear force at its ends. In such a case the situation is more favourable than that in experiments

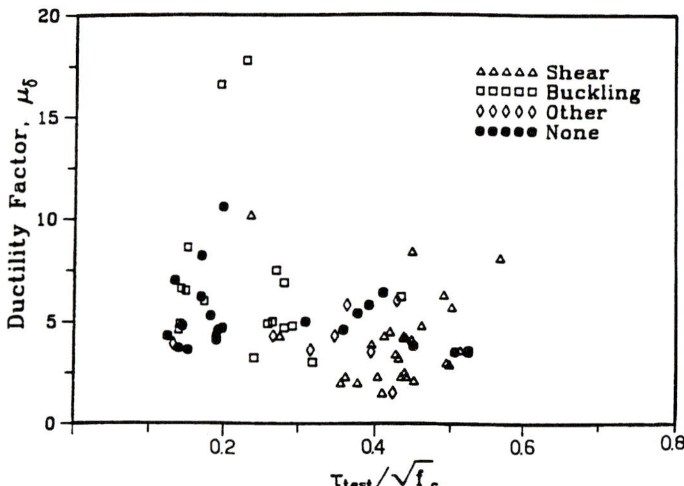

Figure 8.15 Displacement ductility factor as a function of shear stress from cyclic loading tests on beams (French and Schultz, 1991).

with full load reversals, where $V^+ \approx |V^-|$ (Figure 8.9 and 8.11). Based on this observation, Paulay and Bull (1979) proposed the following relationship for the limiting shear stress, beyond which conventional reinforcement with hoops cannot prevent a shear failure:

$$\tau = 0.3(2+\zeta)\, f_c^{1/2} \tag{8.35}$$

The coefficient ζ in equation (8.35) is the ratio V^-/V^+ of the maximum ('negative') shear developed at the beam end section when the bottom reinforcement yields in tension to the maximum ('positive') shear at the same section developed when the top reinforcement yields (see also section 6.1.3). Values of ζ range from -1 (full shear reversal) to 0 (no reversal), thus the corresponding limits of shear stress are $0.3\, f_c^{1/2}$ and $0.6\, f_c^{1/2}$ respectively.

(e) Protecting beams against cyclic shear

The best way to mitigate the adverse consequences of high shear of R/C members is to use the largest possible cross-section dimensions, while keeping reinforcement ratios as low as practicable. As observed, among others, in the experiments by Nmai and Darwin (1986), beams with a longitudinal reinforcement ratio less than 1% show a performance under cyclic loading which is superior to that of similar beams with higher percentages of reinforcement (an appropriate performance criterion based on normalized energy dissipation was used by Nmai and Darwin to compare the behaviour of beams with different strength). The main reason for the superior performance of beams with low reinforcement ratios is the reduced level of shear that develops, in combination with a reduction in compression stresses in concrete, leading to a reduced rate of degradation with respect to the energy dissipation capacity of the beam. It is worth mentioning here that, as verified in the previously mentioned experimental programme, an increase in the bottom reinforcement at beam supports offers not only advantages, but also disadvantages due to the development of larger shear forces.

Whenever architectural or other reasons do not permit the use of relatively large beam sections with low reinforcement ratios, behaviour under cyclic shear can be improved by using intermediate longitudinal bars, located between the top and bottom reinforcement layers, as shown in Figure 8.16(a). The use of these bars leads to the formation of a larger number of full depth cracks with limited width, and it is recommended to use relatively large bar diameters, which are more effective with respect to dowel action (Paulay and Bull, 1979). As shown in the tests by Scribner and Wight (1980), the improvement in the behaviour (increased energy dissipation capacity) due to use of intermediate bars is more marked in beams of the second category ($0.25 < \tau/f_c^{1/2} \leqslant 0.5$).

Finally, for beams with a very high level of shear ($\tau > 0.5 f_c^{1/2}$), the recommended reinforcing pattern consists of **cross-inclined diagonal** (or '**bidiagonal**') bars in the plastic hinge region, as shown in Figure 8.16(b). These bars which offer the advantage of crossing every possible full depth vertical crack, were first proposed by Paulay for coupling beams in coupled walls systems (see Park and Paulay, 1975), which are characterized by low shear spans l/d. In the case of normal slenderness beams, it suffices to put these reinforcements in regions where large inelastic deformations are expected and a sliding shear failure may occur, that is within a distance d from the end support section (Paulay and Bull, 1979).

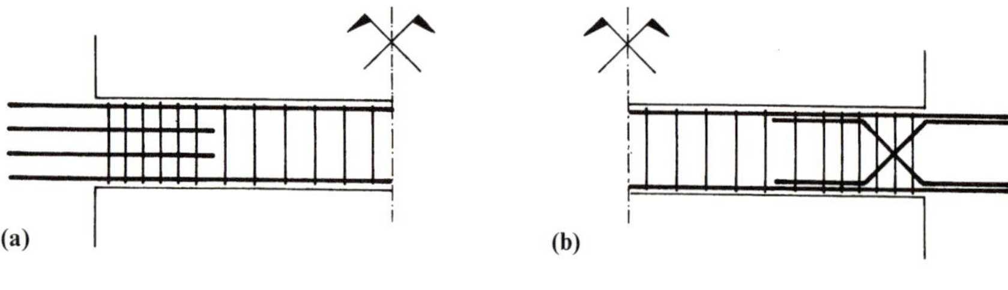

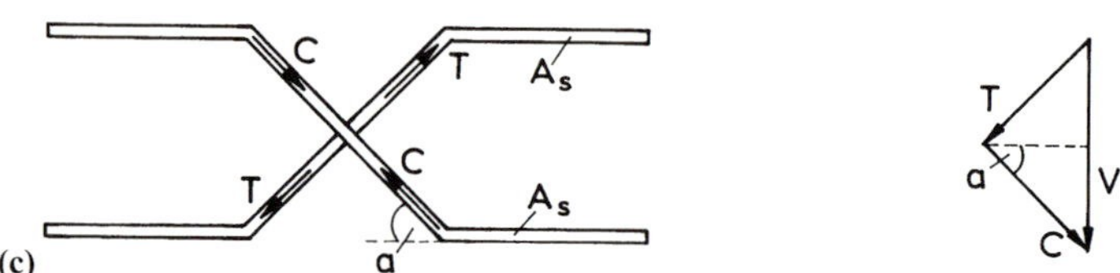

Figure 8.16 Non-conventional types of reinforcement for beams with high shear: (a) inermediate straight bars; (b) diagonal cross-inclined bars; (c) calculation of forces in the diagonal bars.

The required area of diagonal bars in each direction may be calculated assuming that they yield simultaneously in both directions; thus (see Figure 8.16(c)) from force equilibrium it results that,

$$V = 2T \sin \alpha = 2C \sin \alpha \tag{8.36}$$

where α is the angle of the bidiagonal bars relative to the beam axis (usually equal to 45°, resulting $\sin \alpha = 2^{1/2} / 2$). Since $T = C = A_s f_y$, the required area of diagonal bars in each direction is

$$\boxed{A_s = \frac{V}{2f_y \sin \alpha}} \tag{8.37}$$

while the calculation using equation (8.37) is very simple and the behaviour of beams with bidiagonal bars is very satisfactory, practical problems may arise in placing these inclined bars in regions where congestion of reinforcement occurs. A way to avoid the need for bidiagonal reinforcement is to curtail longitudinal bars in such a way as to relocate the plastic hinge at a distance ($\geqslant d$) from the column face (Paulay and Bull, 1979; Abdel-Fattah and Wight, 1987). Such a reinforcing pattern offers the additional important advantage of improving anchorage conditions at the joints and is examined in more detail in the section on design of beam–column joints (see in particular section 9.2.4).

(f) Modelling of beams subjected to cyclic loading

The general methodology for the calculation of moment–curvature diagrams presented in the previous section may also be applied in the case of cyclic loading.

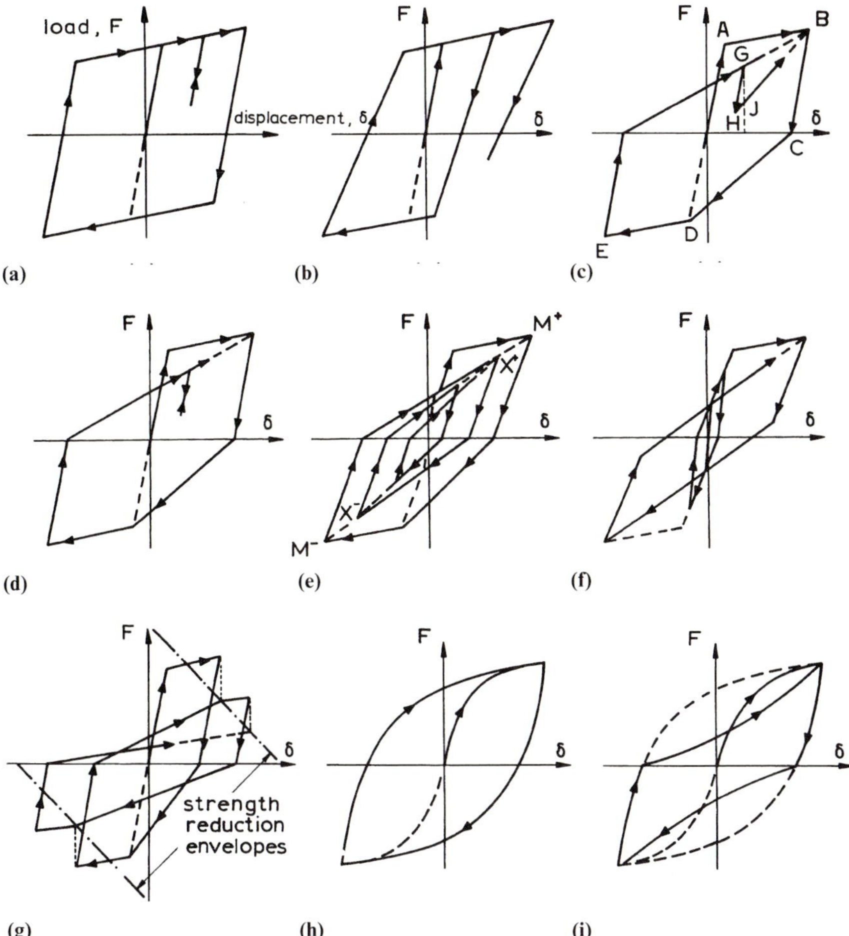

Figure 8.17 Hysteresis models for flexure-dominated R/C elements: (a) elastoplastic (with strain-hardening); (b) bilinear stiffness-degrading; (c) Clough model; (d) modified Clough model (by Riddell and Newmark); (e) Takeda model; (f) trilinear stiffness-degrading; (g) Takayanagi–Schnobrich model; (h) Ramberg–Osgood model as modified by Jennings; (i) Celebi–Penzien model.

The necessary condition for this extension is the use of appropriate σ–ε models describing all stages of material behaviour, namely initial ('elastic') loading, yielding, unloading and reloading in the opposite direction. Such models have already been presented in Chapter 7, both for concrete (sections 7.3.2 and 7.4.3) and for steel (section 7.5.2).

If the force–displacement diagram (in terms of M–θ or F–δ) is calculated by integration of the corresponding moment–curvature diagram, the actual deflections of the beam may be significantly underestimated (beam stiffness overestimated), because, as already explained, deformations due to shear and bond-slip may be substantial, especially in the case of cyclic loading. It follows that the prerequisite for a reliable analytical model for cyclic loading is to take into account all significant sources of deformation. A local bond stress (τ_b)–slip (s) model for cyclic loading has already been presented in section 7.6, while shear force–shear deformation (V–γ) models have also been proposed (see Figure 8.18) on the basis of appropriate idealization of corresponding experimental diagrams such as the one in Figure 8.13. However, the simultaneous use of all these mechanical models (with appropriate account of possible interaction among them) in any type of finite element analysis would inevitably lead to a high degree of complexity, as well as an increased cost of analysis (conceived mainly in terms of required manpower, rather than required CPU time which is not a major consideration with personal computers).

An attractive alternative approach to the problem consists in the use of intermediate mechanical models based on an **element-to-element discretization** (one structural element generally coincides with a finite element of the model), wherein some account is taken of all main parameters. Some of the best-known **phenomenological hysteresis models** for flexure-dominated beams are given in Figure 8.17, and for shear-dominated beams in Figure 8.18 (full references for the models may be found in Kappos, 1986a). These models are appropriate for inelastic dynamic analysis of multi-storey R/C buildings subjected to seismic base accelerations, in which case the hysteresis rules apply for the moment (M_i)–rotation (θ_i) curve at the end i of an element. A detailed treatment of these models is beyond the scope of this book and reference is made to the pertinent specialized literature (Keshavarzian and Schnobrich, 1985; Kappos, 1986a,b).

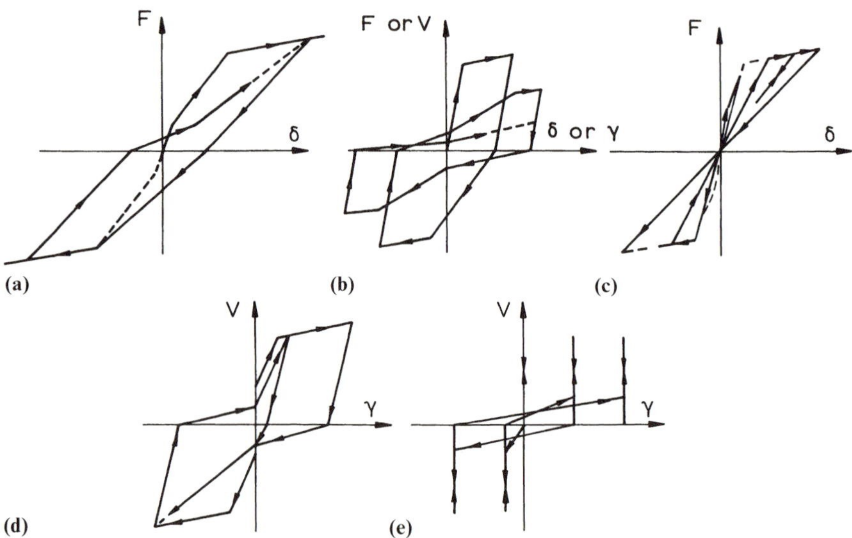

Figure 8.18 Hysteresis models for shear-dominated R/C elements: (a) Saiidi–Sozen Q-model; (b) Takayanagi *et al.* model; (c) Umemura *et al.* model; (d) Kustu–Bouwkamp model; (e) Ma *et al.* model.

8.3 SEISMIC DESIGN OF BEAMS

In the following, design of R/C beams for earthquake resistance according to EC8 (CEN, 1995) is presented; a design example is given in section 8.6.2.

8.3.1 Design for flexure

The resistance of R/C beams subjected to bending moment and axial load (provided $N > -0.1A_c f_{cd}$, otherwise the provisions for columns apply) can be calculated according to EC2 (CEN, 1991), using the same procedure as in the case of vertical loading. The coefficient 0.85 in the σ–ε block of concrete is retained, as explained in section 7.3.4, to take into account the strength reduction due to degradation caused by cyclic loading. The partial safety factors for the materials are those given in EC2 for the fundamental load combination ($\gamma_c = 1.50$ for concrete and $\gamma_s = 1.15$ for steel). The previous assumptions have the practical consequence that section design for flexure due to the seismic combination of actions (section 4.6.5), can be carried out using the same design aids (tables, diagrams, software) as in the case of standard flexural design for permanent and variable actions.

(a) Limitations regarding the longitudinal reinforcement

The **minimum longitudinal reinforcement ratio** $\rho_{mm} = A_s/(b_t d)$ where b_t is the mean width of the tension zone, is

$$\boxed{\rho_{min} = 0.5\,\frac{f_{ctm}}{f_{yk}}} \tag{8.38}$$

The resulting reinforcement ratios for the usual concrete and steel grades are given in Table 8.1. The aim of equation (8.38) is to ensure that the yield moment M_y of a beam section is larger than the corresponding cracking moment M_r (Figure 8.1), since otherwise only one major flexural crack forms and subsequent behaviour of the beam is rather brittle, not excluding the possibility of fracture of longitudinal bars (Paulay and Priestley, 1992).

An upper limit for the cracking moment of a rectangular section (Figure 8.2), ignoring the contribution of steel (a reasonable assumption for lightly reinforced

Table 8.1 Maximum and minimum tension reinforcement ratios (%) for R/C beams (EC8)

Reinforcement ratio	Steel grade	C12	C16	C20	C25	C30	C35	C40	C50
ρ_{min}	S220	0.36	0.43	0.50	0.59	0.66	0.73	0.80	0.93
	S400	0.20	0.24	0.28	0.33	0.36	0.40	0.44	0.51
	S500	0.16	0.19	0.22	0.26	0.29	0.32	0.35	0.41
ρ_{max}	S220	1.51	1.96	2.41	2.98	3.30	3.30	3.30	3.30
	S400	0.89	1.14	1.40	1.70	2.01	2.32	2.63	3.25
	S500	0.75	0.95	1.15	1.40	1.65	1.90	2.15	2.65

Note: ρ_{max} corresponds to $\rho_2/\rho_1 = 0.5$ and to DC 'M'. An upper limit equal to $0.03bh \approx 0.033bd$ was set to ρ_{max}, which is the value allowed for DC 'L' beams.

beams) can be estimated from the relationship

$$M_r = f_{cmk0.95} W_I \tag{8.39(a)}$$

where $f_{cmk0.95}$ is the 5% upper fractile of the modulus of rupture (flexural tensile strength of concrete), defined as in EC2 (CEN, 1991), and W_I is the moment of resistance of the uncracked ('State I') concrete section, equal to $bh^2/6$ for a rectangular section, thus

$$M_r = f_{cmk0.95} \frac{bh^2}{6} \tag{8.39(b)}$$

The yield moment of the cracked ('State II') section is given by the relationship

$$M_y = F_s z \tag{8.40}$$

where F_s is the force in tension steel and $z = \zeta d$ the internal lever arm of the section forces. Putting $F_s = A_s f_{yk}$ and conservatively assuming $\zeta = 0.9$, it follows from equation (8.40) that

$$M_y = 0.9 A_s f_{yk} d \tag{8.41}$$

Equating the values of M_r and M_y from equations (8.39(b)) and (8.41) and assuming $h \approx d/0.9$ results in the following relationship:

$$0.9 A_s f_{yk} d = 0.21 f_{cmk0.95} bd^2 \tag{8.42}$$

Solving equation (8.42) for the reinforcement ratio $\rho = A_s/bd$ results in a relationship for the minimum ratio

$$\rho_{min} = 0.23 f_{cmk0.95}/f_{yk} \tag{8.43}$$

Assuming now an average ratio between the upper 5% fractile and the mean tensile strength $f_{cmk0.95}/f_{cmm} \approx 1.3$ (EC2) and a ratio between flexural tensile strength and direct (uniaxial) tensile strength $f_{cmm}/f_{ctm} \approx 1.2$, it follows that $f_{cmk0.95} \approx 1.56 f_{ctm}$; substituting in equation (8.43) gives

$$\rho_{min} = 0.36 f_{ctm}/f_{yk} \tag{8.44}$$

Equation (8.44) which is based on simple flexural considerations (the effects of phenomena such as stress localization and tension stiffening are ignored), implies that if ρ_{min} is calculated from equation (8.38), a ratio M_y/M_r of about 1.4 is ensured, which appears to be quite acceptable, if not somewhat overconservative (since conservative assumptions regarding the cracking moment were made).

The potential plastic hinge regions are considered as critical regions and they extend up to a distance l_{cr} from the column face, as well as from both sides of any other cross-section where yielding is expected under the design seismic load combination (Figure 8.19). The length l_{cr} is equal to $2.0h_b$, $1.5h_b$ and $1.0h_b$ for DC 'H', 'M' and 'L', respectively.

The **maximum tension reinforcement ratio** ρ_{max} within the potential plastic hinge regions, defined for DC 'M' beams, is

$$\boxed{\rho_{max} = 0.65 \frac{f_{cd}}{f_{yd}} \frac{\rho_2}{\rho_1} + 0.0015} \tag{8.45}$$

where ρ_2 is the compression reinforcement ratio present in the section and ρ_1 the corresponding ratio for the tension reinforcement, that is the one limited by ρ_{max}.

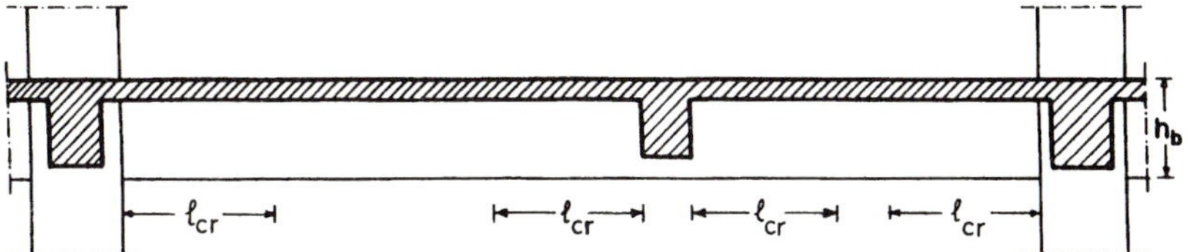

Figure 8.19 Beam critical regions.

The coefficient 0.65 in equation (8.45) is changed to 0.35 for DC 'H' beams, while for DC 'L' ρ_{max} should not exceed 75% of the value allowed by EC2, that is $0.75\times 4.0=3.0\%$ of the concrete section (bh). Values of ρ_{max} corresponding to the common value $\rho_2/\rho_1=0.5$ (at beam support sections) are given in Table 8.1. Equation (8.45) aims at ensuring adequate ductility in the critical beam regions and also at avoiding congestion of reinforcement in those regions where closely spaced web reinforcement is also present (see next section). It appears that equation (8.45) overestimates the effect of concrete grade on ductility; for example going from C20 to C40 and assuming S400 steel and $\rho_2/\rho_1=0.5$, equation (8.45) indicates a 90% increase in ρ_{max}, while an increase of only 66% results from the New Zealand Code (SANZ, 1982).

Within the critical regions the compression reinforcement ratio ρ_2 should not be less than half the amount of the actual tension reinforcement ratio ρ_1, regardless of the requirements arising from the analysis. This provision aims at ensuring adequate ductility in the critical region (section 8.2.2) and also to account for a possible increase in positive moment at the support, beyond the value resulting from the analysis for the seismic action combination. It has to be remembered here that symmetric reinforcement ($\rho_1=\rho_2$) at beam supports, although favourable with regard to flexural ductility requirements, has the disadvantage that it increases the maximum shear and might change the failure mode to a shear-dominated one (Nmai and Darwin, 1986). Therefore, the designer should be reluctant to increase ρ_2 beyond the values required by EC8 and other seismic codes.

EXAMPLE

Consider a rectangular beam section 250 by 600 mm, made of grade 20 concrete (C2025) and grade 400 steel (S400). For the maximum reinforcement ratio $\rho_{max}=1.4\%$ (Table 8.1) and a web reinforcement consisting of 8 mm hoops at 150 mm spacing (volumetric ratio $\rho_w=0.44\%$), resulting in an ultimate concrete strain ε_{cu} of about 0.7% (section 7.4.3), the available curvature ductility factor may be estimated from equations (8.7) and (8.11). Introducing the appropriate value for $\alpha=E_c/E_s=200/29=6.9$ and assuming $d_2=40$ mm and no compression reinforcement ($\rho_2=0$), the neutral axis depth factor from equation (8.7) is

$$\zeta=\left[6.9^2\,(0.014+0)+2\times 6.9\times\left(0.014+\frac{40}{560}\times 0\right)\right]^{1/2}-6.9(0.014+0)=0.353$$

and the curvature ductility from equation (8.11)

$$\mu_\phi=\frac{0.68\times 20.0\times 0.007\times 200\,000}{400^2\,(0.014-0)}\,(1-0.353)=5.5$$

If the code provision $\rho_2 = 0.5\rho_1$ is now taken into account, the resulting μ_ϕ equals 11.5, which means that the use of the required compression reinforcement led to a doubling of the available ductility. It is recalled once more that the foregoing is an approach based solely on flexural considerations for monotonic loading and the actual behaviour under cyclic loading might not be well described using such an approach (refer to section 8.2.2).

The behaviour of the previous beam section could be compared with that of a 350 × 750 mm one, with a longitudinal reinforcement ratio $\rho_1 = 0.56\%$, that is twice the minimum required (Table 8.1). The two sections have approximately the same moment capacity: note, however, that the actual area of reinforcing bars is about 30% lower in the large beam. According to EC8, the compression reinforcement should be equal to $\rho_1/2$, that is 0.28%, the minimum ratio from Table 8.1. For this section the calculated ductility factor from equation (8.11) is $\mu_\phi = 32.7$, almost triple the corresponding value for the section with $\rho_1 = \rho_{max}$. It is clearly seen that even for low degrees of confinement (8 mm hoops at 150 mm) the ductility of beams (at least under monotonic loading) is very high, provided that reinforcement ratios are kept close to the minimum values prescribed by the code. Low tension reinforcement also leads to low values of shear, which is a particularly desirable characteristic for beams subjected to seismic loading.

(b) Arrangement of longitudinal bars

The arrangement of flexural reinforcement in a beam should be carried out in such a way as to provide for an unexpected distribution of bending moments, that is a distribution which at certain points exceeds the moment envelope resulting from the design actions combinations. The pertinent detailing rules, which are common to most modern codes covering seismic design, are summarized in Figure 8.20. The diameter of longitudinal bars passing through interior or exterior beam–column joints should be limited in order to prevent bond degradation during seismic loading; relevant EC8 provisions are given in section 7.2.4. It has to be added here that longitudinal bars passing through interior joints should not be terminated within the critical region (l_{cr}), and also that at least two 14 mm diameter bars (S400) should be provided both at the top and the bottom along the entire length of the beam.

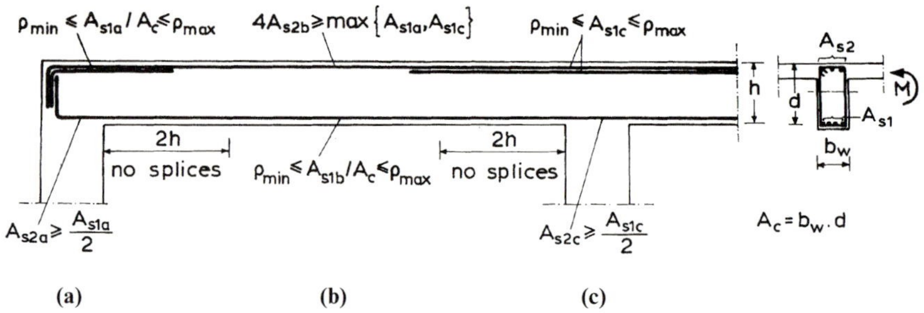

Figure 8.20 Arrangement of longitudinal reinforcement in earthquake-resistant R/C beams.

For beams monolithically cast with slabs (T-beams or L-beams) the top reinforcement at the support regions can include some of the bars placed within parts of the flange, smaller than the effective slab width, up to a certain distance from the column face defined as in Figure 8.21. However, top reinforcement bars should be placed mainly within the width of the beam; since no specific percentages are included in EC8, 75% of total reinforcement, which is prescribed in other codes (CEB, 1985), is suggested. The effective slab widths contributing to flexural strength of T-beams shown in Figure 8.21 appear to be small compared with relevant experimental data. For instance, in the beam of Figure 8.9 where the overhanging portions of the slab on each side of the web were equal to $6h_f$, it was found (Bertero and Popov, 1977) that the total amount of slab reinforcement contributed to the strength of the T-beam, although no transverse beam was present.

It has to be emphasized here that in estimating the design action effects for columns (section 6.1.4), the actual resisting moments of beams must be computed taking into account the effective slab widths shown in Figure 8.21, otherwise the actual beam strength and therefore the corresponding requirement for the column are underestimated.

8.3.2 Design for shear

Shear resistance evaluation and verification according to EC8 are carried out using different provisions for the critical regions than for the rest of the beam. The critical regions have been defined in the previous section (Figure 8.19) and they are typically the end regions which, under prevailing seismic actions, are subjected to the highest values of both bending moment and shear force. According to the capacity design criterion, the design shear forces for DC 'H' beams are determined considering the equilibrium of the beam when the actual flexural strengths are attained at its ends, as described in section 6.1.3.

(a) Evaluation of shear resistance

The shear resistance (strength) of a beam according to modern codes, including

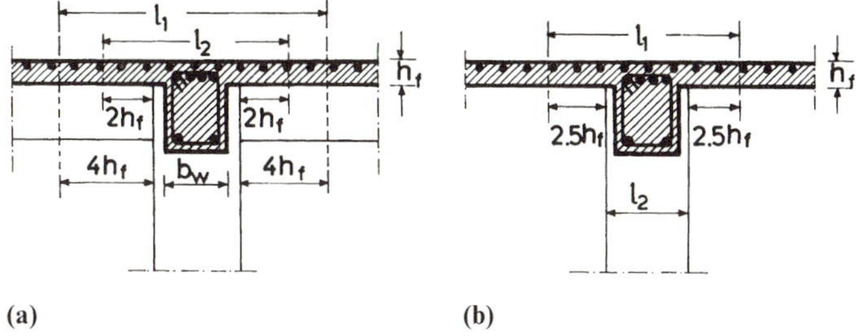

Figure 8.21 Slab reinforcement which can be included in the required reinforcement of T-beams, lies within a width equal to l_1 for interior columns and l_2 for exterior columns: (a) case of transverse beam framing into the joint; (b) case of no transverse beam.

the Eurocodes (CEN, 1991, 1995), is given by the equation

$$\boxed{V_{Rd} = V_{cd} + V_{wd}} \tag{8.46}$$

where V_{wd} is the shear carried by the web reinforcement through the truss mechanism (refer also to section 8.2.2), and V_{cd} is the shear attributed to the rest of the shear transfer mechanisms, which as explained in section 8.2.2, are gradually destroyed during severe cyclic loading.

Outside the potential plastic hinge regions V_{cd} is calculated as in EC2 using the relationship

$$\boxed{V_{cd} = [\tau_{Rd} k(1.2+40\rho_l) + 0.15\sigma_{cp}] b_w d} \tag{8.47}$$

where $\tau_{Rd} = 0.25 f_{ctk0.05}/\gamma_c$ is the basic design shear strength ($\gamma_c = 1.5$); $k = 1.6 - d$ (d in m), not to be taken less than 1;$\rho_l = A_{sl}/(b_w d)$ is the longitudinal reinforcement ratio (b_w is the web width); $\sigma_{cp} = N_{sd}/A_c$, where N_{sd} is the (minimum) design axial force and A_c is the total cross-sectional area of the concrete section.

The use of equation (8.47) implies that EC8 recognizes that outside the critical regions no significant degradation of shear transfer mechanisms takes place, thus design is carried out as for gravity loading. In contrast, within the critical regions degradation of mechanisms associated with concrete (compression zones, aggregate interlock) and dowel action of longitudinal steel is very substantial, especially in members with low axial load (section 8.2.2). Recognizing this, EC8 provides that V_{cd} within beam critical regions be taken as follows:

1. For DC 'H' beams, $V_{cd} = 0$.
2, For DC 'M' beams, V_{cd} is equal to 40% of the value calculated according to EC2 (equation (8.47)).
3. For DC 'L' beams, V_{cd} is calculted from equation (8.47).

The term V_{wd} (shear carried by web reinforcement) is calculated taking into account the algebraic value of the ratio $\zeta = V_{Smin}/V_{Smax}$ between the minimum and the maximum design shear forces. As explained in section 8.2.2, this ratio gives an indication of the degree of cycling (in the sense of shear reversal) which is expected in the beam, hence of the possibility of a sliding shear failure. The following cases are distinguished in EC8:

1. If $\zeta \geqslant -0.5$, i.e. when only a partial shear reversal is expected, V_{wd} is calculated from the well-known equation

$$\boxed{V_{wd} = \frac{A_{sw}}{s} 0.9 d f_{ywd}} \tag{8.48}$$

resulting from the truss model if vertical web reinforcement (stirrups or hoops) is considered. In equation (8.48) A_{sw} is the area of hoop reinforcement (area of one leg times the number of legs) within a spacing s, and f_{ywd} is the design yield strength of the hoop steel.

2. If $\zeta < -0.5$, i.e. when a high degree of shear reversal is expected, then:

 (a) If $|V_{Smax}| \leqslant \beta_1 (2+\zeta)\, \tau_{Rd} b_w d$, equation (8.48) applies and shear is resisted solely by hoop reinforcement. For DC 'H' beams $\beta_1 = 3$, and for DC 'M' beams $\beta_1 = 4$.

(b) If $|V_{Smax}|$ exceeds the above value, bidiagonal reinforcement at $\pm 45°$ shall be provided to mitigate the efffects of sliding shear failure and/or excessive diagonal cracking. The distribution of shear between hoops and bidiagonal reinforcement depends on the value of 'negative' shear (typically the value corresponding to tension at the bottom of the support under consideration) V_{Smin}.

If $|V_{Smin}| \leqslant \beta_2(2+\zeta)\ \tau_{Rd} b_w d$, half of V_{Smax} shall be resisted by hoops and half by bidiagonal bars.

If $|V_{Smin}| > \beta_2(2+\zeta)\ \tau_{Rd} b_w d$, the entire V_{Smax} shall be resisted by bidiagonal bars. In either case the amount of bidiagonal reinforcement required may be estimated using equation (8.37), introducing the appropriate portion of V_{Smax}. For DC 'H' beams $\beta_2 = 6$, and for DC 'M' beams $\beta_2 = 8$.

(c) For DC 'L' beams V_{wd} is always estimated using equation (8.48) and shear is carried solely by hoops.

As already mentioned in section 8.2.2, testing of R/C members under shear reversal conditions has shown that there is a critical value of the shear stress τ (equation (8.35)), beyond which the presence of vertical hoops cannot prevent shear failure by sliding along a full depth vertical crack. Equation (8.35) expresses τ as a multiple of $f_c'^{1/2}$ according to common practice in the USA, New Zealand and elsewhere. To find an equivalent expression in Eurocode terms, $f_c' = f_{ck}$ is assumed and the EC2 equations $f_{ctm} = 0.3 f_{ck}^{2/3}$ and $f_{ctk0.05} = 0.7 f_{ctm}$ are used. It follows that $f_{ck}^{1/2}$ equals 15–18 τ_{Rd}, depending on the concrete grade. Thus, for $f_{ck}^{1/2} = 17\tau_{Rd}$ equation (8.35) may be written in terms of shear force:

$$V_{lim} = 5(2+\zeta)\ \tau_{Rd} b_w d \tag{8.49}$$

Comparison of equation (8.49) with the EC8 provisions (β_1 coefficient) reveals that a 'safety margin' of 1.67 and 1.25 for DC 'H' and 'M' respectively has been adopted in taking shear sliding failure into account. About 17% lower safety margins exist with regard to the β_2 coefficient (which determines the limits beyond which bidiagonal bars must carry the entire shear force) if $V_{Smin} \approx V_{Smax}$, while for ζ close to -0.5 the provision $|V_{Smin}| < \beta_2\ (2+\zeta)\ \tau_{Rd} b_w d$ appears to be unsafe, as the corresponding V_{Smax} is about twice that value ($\tau_{max} = 12(2+\zeta)\ \tau_{Rd}$ for DC 'H' beams) and experimental evidence, as well as the New Zealand code (SANZ, 1982) require that for such a τ value, the entire shear be carried by bidiagonal reinforcement.

To prevent crushing of the concrete diagonal struts which form part of the idealized truss (Park and Paulay, 1975), a limit on the maximum shear force should be set; according to EC8 this limit is that of EC2

$$\boxed{V_{Rd2} = 0.5\left(0.7 - \frac{f_{ck}}{200}\right) f_{cd} b_w 0.9d} \tag{8.50}$$

for all ductility classes. This restriction can be critical in beams with particularly thin webs, as may be found in prestressed concrete structures.

(b) Transverse reinforcement requirements

According to EC8, the transverse reinforcement within the critical regions (outside l_{cr} EC2 applies) should satisfy the following requirements:

1. Hoop diameters $d_{bw} \geqslant 6$ mm shall be used.

2. The spacing of hoops shall be determined by

$$s_w = \min \{h_w/4;\ 24d_{bw};\ 5d_{bl};\ 150\ \text{mm}\} \tag{8.51(a)}$$

for DC 'H' beams and by

$$s_w = \min \{h_w/4;\ 24d_{bw};\ 7d_{bl};\ 200\ \text{mm}]\} \tag{8.51(b)}$$

for DC 'M' beams. The EC2 minimum requirements apply for DC 'L' beams.
3. The first hoop shall be placed within 50 mm from the end section of the beam.

It has already been explained in section 8.2 that hoops in the plastic hinge regions play a triple role, that is confinement of concrete, protection of longitudinal bars against buckling, and tranfer of a substantial portion of the shear force. Estimating the combined effect of all these on the hoop's stress state is a very difficult task, hence the requirements of equation (8.51) are largely based on experience gained from available test data. From the practical design point of view, the restriction which typically governs the detailing of hoops in beams of usual dimensions is the limitation of s_w with regard to the diameter of longitudinal bars (5–$7d_{bl}$, depending on the DC), which is derived from buckling prevention considerations (Mau, 1990) as expalined in section 8.2.2(b); in that section equation (8.34) was given, which may be used for an appropriate selection of the hoop diameter.

8.3.3 Other design requirements

(a) Geometrical constraints

For reasons of ease of construction, but also to reduce the slenderness of beams, their dimensions should exceed certain limits which are defined on the basis of previous experience and to a lesser extent, of relevant research findings. Thus, EC8 requires that the minimum width (b_w) of beams be equal to 200 mm. To avoid the possibility of lateral instability of the web in potential plastic hinge regions, the width to height ratio of the web must satisfy the relationship (included in EC2 for beams susceptible to lateral buckling)

$$b_w/h \geqslant 0.4 \tag{8.52}$$

For DC 'M' beams the limit 0.25 (instead of 0.4) applies. It is noted that limiting the aspect ratio according to equation (8.52) also leads to lower shear stresses and ensures that no crushing of concrete struts (compare equation (8.50)) takes place. With regard to lateral instability considerations, the New Zealand Code (SANZ, 1982) gives the following additional requirements

$$l_n/b_w \leqslant 25 \tag{8.53(a))}$$

for continuous beams, and

$$l_n/b_w \leqslant 15 \tag{8.53(b))}$$

for cantilever beams; a 50% increase in these limits is allowed for T- and L-beams. On the other hand, as already explained in section 8.2.2, nor should the beams have a very low slenderness, to ensure a prevailing flexural mode of failure.

According to EC8 this is considered to be the case when

$$l/h \geqslant 3 \tag{8.54}$$

It will be shown in a later section (9.4.5) that equation (8.54) is not usually satisfied in the case of coupling beams in coupled wall systems.

In order to ensure an efficient transfer of seismic moments at beam–column connections forming part of the main earthquake-resisting structural system, the **eccentricity** of the beam (centroidal) axis relative to that of the column should be less than one-quarter the width of the column ($b_c/4$).

Finally, to take advantage of the favourable effect of the column compressive axial load on bond conditions regarding horizontal beam bars passing through joints, the width of the beam b_w should not exceed the column width b_c by more than half the beam height h_b on each side of the column, that is

$$b_w \leqslant b_c + h_b \tag{8.55}$$

and in addition b_w should always be kept below $2b_c$.

(b) Anchorage of reinforcing bars

With regard to the anchorage of longitudinal reinforcement in beams the following EC8 requirements apply:

1. Longitudinal bars bent in joints for anchorage shall always be placed inside the corresponding column hoops and detailed as shown in Figure 9.6 of section 9.2.4. In that section limitations concerning the diameter of bars passing through joints (to ensure favourable bond conditions) are also presented.
2. The required **anchorage length** is given by the EC2 relationship

$$\boxed{l_{b,net} = \alpha_a \left(\frac{d_{bl}}{4} \frac{f_{yd}}{f_{bd}} \right) \frac{A_{s,req}}{A_{s,prov}} \not< l_{b,min}} \tag{8.56}$$

   where α_a is a coefficient depending on the type of anchorage, given in EC2 ($\alpha_a = 1.0$ for straight bars, $\alpha_a = 0.7$ for curved bars in tension); f_{bd} is the design bond stress (section 7.6.4); $A_{s,req}$ is the reinforcement area required on the basis of the analysis and $A_{s,prov}$ is the area actually provided (taking minimum code requirements and practical considerations into account); $l_{b,min}$ is equal to 30% of the term in parentheses for bars in tension and 60% of the same term for bars in compression.
3. In anchoring beam bars at exterior beam–column joints (Figure 9.8) a portion of the bar length equal to $k_b d_{bl}$ measured from the point the bar enters the joint, shall not be included in $l_{b,net}$, since bond along this portion may have been destroyed due to yield penetration from the adjacent plastic hinge (Paulay, 1986). Since bond deterioration depends on the level of inelasticity (section 7.6.3), $k_b = 10$ for DC 'H', $k_b = 5$ for DC 'M' and $k_b = 0$ for DC 'L'.

With regard to the anchorage of transverse reinforcement, the hoops required in the critical region are closed stirrups with 135° bent-in ends (Figures 8.20 and 8.21) and $10d_{bw}$ long extensions (d_{bw} is the stirrup diameter). As explained in section 7.4.3, confinement to concrete can only be provided if closed stirrups with their hooks projecting inside the concrete core are used. Open stirrups and /or 90°

hook extensions, although satisfactory for normal gravity load design, are completely inefficient with respect to confinement. Their use is therefore restricted to DC 'L' beams only.

(c) Splicing of bars

Lapped splices within critical regions are only permitted in DC 'M' and 'L' beams. As already mentioned, bond conditions in plastic hinge regions are very unfavourable and splicing of bars by overlapping in these regions may endanger beam strength (Gergely, 1977), especially if the ductility demand is high, as in the case of DC 'H' beams. It has to be pointed out that although EC8 does not preclude lapped splices in critical regions of DC 'M' and 'L' beams, it is required that bars passing through interior joints be extended a distance l_{cr} outside the joints; it appears therefore that, practically speaking, lapped splices within l_{cr} can hardly be used for any DC. Typically bar splices in beams are located at midspan.

Splicing by **welding** is not allowed within critical regions of beams, as there are not sufficient data to demonstrate that such systems have adequate ductility under seismic loading conditions.

The necessary **lap length** (according to EC2) is

$$l_s = \alpha_1 l_{b,net} \nless l_{s,min} \qquad (8.57)$$

where $l_{s,min} = 0.3\alpha_a\ \alpha_1(d_{bl}/4)\ (f_{yd}/f_{bd}) \nless 15d_{bl} \nless 200$ mm. The coefficient α_a is the same as in equation (8.56), while α_1 is defined as follows:

- $\alpha_1 = 1.0$ for bars in compression, and for bars in tension where less than 30% of the section reinforcement is lapped, and in addition the clear distance between adjacent spliced bars is not less than $10d_{bl}$ and the distance from the outer face of the spliced bar to the lateral surface of the concrete section is not less than $5d_{bl}$.
- $\alpha_1 = 1.4$ for bars in tension where either: (1) more than 30% of reinforcement is lapped or (2) the clear distance between adjacent spliced bars is less than $10d_{bl}$ or the distance of the outer face of the spliced bar from the concrete surface is less than $5d_{bl}$ (but not both 1 and 2).
- $\alpha_1 = 2.0$ for tension splices if both the foregoing conditions regarding distances apply.

It has to be emphasized that in seismically loaded beams there is always the possibility that a bar in compression (according to the envelope of design actions) may actually be subjected to tension due to an unexpected earthquake-induced moment reversal. Hence, it is recommended always to assume tension conditions when calculating lap lengths.

The transverse reinforcement within a lap length (even outside the critical regions) shall consist of hoops, whose spacing is limited by the following relationship:

$$s = \min\ \{h_w/4;\ 100\ \text{mm}\} \qquad (8.58)$$

It is pointed out that the 100 mm requirement is more critical than the corresponding ones for confinement reinforcement in beam critical regions (equations (8.51)), even for DC 'H' structures. The favourable effect of closely spaced hoops with regard to bond conditions has already been outlined in section 7.6.2.

Regardless of ductility class, the EC2 requirement concerning transverse reinforcement at lap splices should be respected, that is for $d_{bl} \geqslant 16$ mm the sum of all legs of transverse reinforcement parallel to the layer of the spliced reinforcement should not be less than the area (A_{sl}) of the larger lapped longitudinal bar

$$n\,A_{sw} \geq A_{sl} \tag{8.59}$$

where n is the number of transverse bars (typically stirrups) along the lap length l_s.

(d) Beams supporting cut-off vertical elements

According to EC8, beams (and slabs) supporting vertical R/C walls are not permitted.

Although not recommended, beams supporting cut-off columns are allowed; in this case the following requirements apply (in addition to analysis requirements, namely that the vertical component of seismic action has to be taken into account, even through a partrial model; see section 4.4):

1. Beam shall be resting on at least two direct supports, consisting of columns or walls.
2. No eccentricity of the cut-off column axis relative to that of the beam is allowed.
3. A length of the beam equal to $2h_b$ on each side of the cut-off column (compare Figure 8.19) is considered as a critical region and the corresponding requirements for DC 'H' and 'M' beams apply.

8.4 SEISMIC BEHAVIOUR OF COLUMNS

A fundamental principle of capacity design (section 6.1.2) is that in R/C buildings plastic hinge formation in columns should be avoided. To achieve this, column design moments are derived from equilibrium conditions at beam–column joints, taking into account the actual resisting moments of beams framing into the joint, as outlined in section 6.1.4. However, there are a number of reasons why the capacity procedure included in EC8, as well as similar procedures adopted by other codes (CEB, 1985; ICBO, 1994) cannot achieve this goal; these reasons are discussed in the following section.

8.4.1 Uncertainties regarding the capacity design of columns

1. Whenever the degree of inelasticity at the beam ends is high (typically this would be the case with DC 'H' beams), the longitudinal bars enter the strain-hardening range and this may cause an increase in beam strength between 10 and 25%, depending on the steel characteristics and the ductility factor attained.
2. In calculating actual strengths of beams, reinforcing bars in slabs integrally built with the beams are either completely neglected or taken into account considering an effective slab width in tension that is clearly smaller than that observed in relevant tests, as already mentioned in section 8.3.1. The corresponding increase in the actual beam strength may range from 10 to 30% (Paulay, 1986).
3. As shown in Figure 8.22, the flexural strength of a column varies considerably with the axial load level. During a strong earthquake motion the axial load in

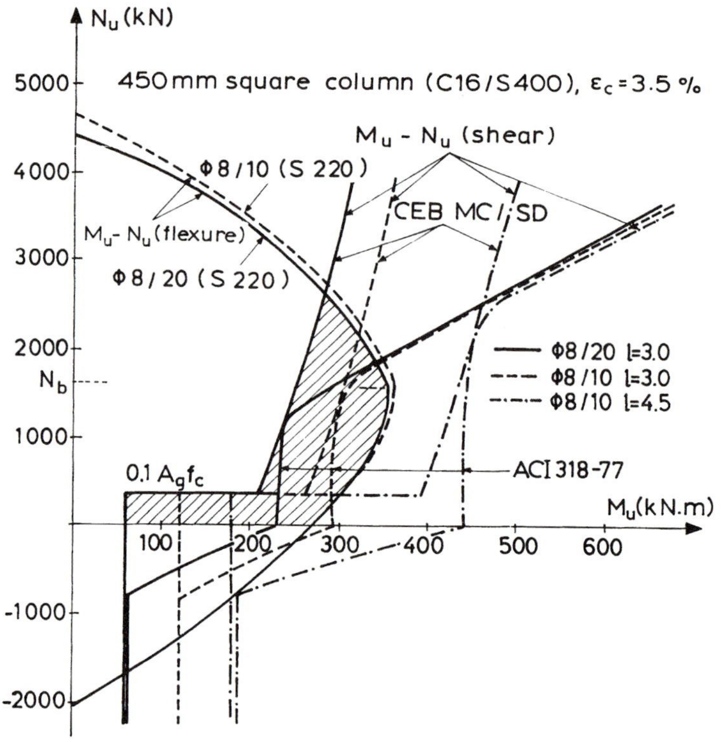

Figure 8.22 *M–N* and $M_{,V}$–*N* interaction diagrams for a 450 mm square column.

a column is continuously changing due to the combined effect of overturning moments and the vertical acceleration of the motion; this effect is more pronounced in columns at the perimeter of the building. The range of variation of *N* may be wider than that predicted by the analysis for the design actions (Park and Paulay, 1975; Kappos, 1986a), particularly when the vertical motion is significant. Therefore, at certain stages of the seismic response, the strength of a column may be substantially lower than that taken into account in the capacity design.

4. Analysis of the inelastic response of multi-storey R/C buildings subjected to earthquake excitation (Park and Paulay, 1975; Kappos, 1986a) have shown that the point of contraflexure in columns shifts considerably during the excitation, leading to a distribution of bending moments substantially different from that resulting from the code-prescribed analysis (especially when the latter is an equivalent static one). In addition to differences between static and dynamic response (influence of higher modes), the shift of the contraflexure point is caused by the formation of hinges in beams adjacent to the column and even by extensive cracking in parts of the column, as all these factors alter the stiffness of the beam–column subassemblage, hence the moment distribution. Therefore, ensuring that the sum of column moments at a joint exceeds the sum of the corresponding beam moments does not necessarily mean that the moment in each

single column always remains lower than the corresponding flexural strength. It is not uncommon that in the course of seismic loading a plastic hinge forms in the column below a certain joint, while the column above the joint remains in the elastic range, as it is subjected to significantly lower moments.

5. The direction of propagation of seismic waves does not in general coincide with a principal axis of the building (if indeed such an axis exists), and this, combined with the effect of eccentricities in plan (section 5.4.5) leads to a biaxial stress state in columns (particularly the corner ones). Checking the relative strength of beams and columns at a joint separately in each direction (as allowed by most codes, including EC8), does not necessarily ensure that a column has adequate capacity to resist an arbitrary biaxial loading history (section 8.4.3), especially when all beams framing into the joint (in two or more directions) form a plastic hinge (Park and Paulay, 1975; Bertero, 1979).

According to Paulay (1986) a beam overstrength factor (γ_{Rd}) in the range of 2.0–2.5 is required in the capacity design of columns to ensure that no plastic hinge will form in any column. This is clearly much higher than the values recommended by EC8 (γ_{Rd} between 1.20 and 1.35) and other seismic codes, even if the fact that EC8 includes some portion of slab reinforcement in estimating beam strength is taken into consideration. According to Paulay, a design based on such γ_{Rd}-values is feasible and does not lead to an excessive increase in cost, provided that it is combined with a substantial decrease in transverse reinforcement of columns (hoops, spirals), as well as with reduced requirements for lap lengths and for joint core reinforcement. It cannot be said that a consensus has been reached on this subject; however, it is clear that EC8 has not adopted this approach of drastically reducing the possibility of hinging in columns.

It follows from the foregoing discussion that design of R/C columns (in building structures) based on current practice cannot preclude that, at least in some columns, plastic hinges will form during a strong earthquake. This is clearly seen in the analytical studies presented in Chapter 10 which concern multi-storey buildings designed to EC8. Therefore, it is concluded that some ductility should be ensured in the case of columns as well. The ductility of columns under seismic loading is examined in the following sections.

8.4.2 Behaviour under monotonic loading

The available ductility of a column can be expressed through the curvature ductility factor (μ_ϕ) defined in equations (8.3) and (8.4). The general method for the calculation of the moment (M)–curvature (ϕ) diagram outlined in section 8.2.1(b) is, of course, applicable in the case of columns as well. Using computer codes such as that presented in that section (Kappos, 1993), it is possible to calculate M–ϕ diagrams for columns having an arbitrary cross-section and reinforcement arrangement (Figure 8.3), subjected to monotonic loading to failure, which in members with relatively high axial load typically occurs in the compression zone (concrete crushing, buckling of bars in compression, hoop fracture). The simplified procedure outlined in the following, although restricted to a relatively simple geometry, gives a clearer picture of the effect of certain cirtical parameters on column ductility than does the general method.

(a) Approximate procedures for estimating column ductility

Consider the doubly reinforced rectangular section shown in Figure 8.23 subjected to a bending moment, M and an axial load, N. Let the normalized axial load be defined as

$$\nu = \frac{N}{f_c bh} \tag{8.60}$$

and the reinforcement ratios as

$$\rho_1 = \frac{A_{s1}}{bh} \quad \rho_2 = \frac{A_{s2}}{bh} \tag{8.61}$$

where for columns with prevailing seismic loading typically $\rho_1 = \rho_2$. From the strain profile shown in Figure 8.23 it follows that the neutral axis depth at yielding of tension steel is

$$x_y = \frac{\varepsilon_c}{\varepsilon_c + \varepsilon_y} d \tag{8.62}$$

For $d \approx 0.9h$ and assuming $\sigma_c \approx f_c$ (which is only an approximation) and also taking into account that $\sigma_{s1} = f_y$ and $\sigma_{s2} = \lambda f_y (\lambda \leqslant 1)$, the equilibrium of axial forces acting on the section yields

$$N_y = F_c + F_{s2} - F_{s1} \tag{8.63}$$

Substituting the values for concrete and steel forces and using definitions (8.60) and (8.61), equation (8.63) becomes

$$\nu f_c bh + \rho_1 bh f_y - \lambda \rho_2 bh f_y = \tfrac{2}{3} f_c b x_y \tag{8.64}$$

Introducing in equation (8.64) the value of x_y from equation (8.62) and eliminating the term bh, the following relationship results:

$$\nu f_c + (\rho_1 - \lambda \rho_2) f_y = 0.6 f_c \frac{\varepsilon_c}{\varepsilon_c + \varepsilon_y} \tag{8.65}$$

The neutral axis depth at failure ($\varepsilon_c = \varepsilon_{cu}$) is

$$x = \frac{\varepsilon_{cu}}{\varepsilon_{cu} + \varepsilon_s} d \tag{8.66}$$

If the assumption is made that the stress in the compression reinforcement remains equal to λf_y (which is just a rough approximation) it follows from the equilibrium

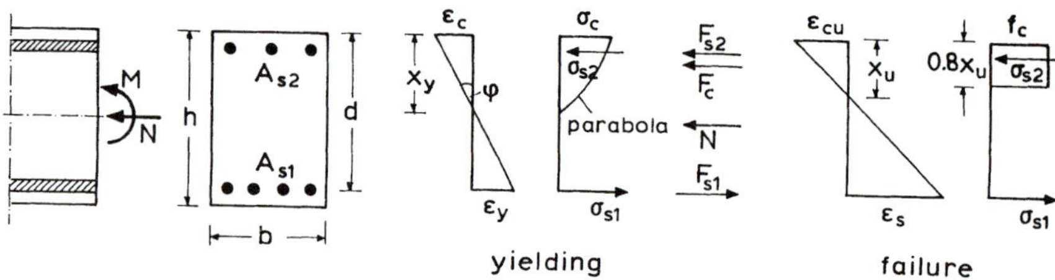

Figure 8.23 Calculation of the curvature ductility factor in a rectangular section subjected to flexure and axial load.

of forces at failure (also taking equation (8.66) into account) that

$$\nu f_c + (\rho_1 - \lambda\rho_2) f_y = 0.72 f_c \frac{\varepsilon_{cu}}{\varepsilon_{cu} + \varepsilon_y} \tag{8.67}$$

The concrete strain at yield may be found from equation (8.65):

$$\varepsilon_c = \frac{\varepsilon_y[\nu + (\rho_1 - \lambda\rho_2) f_y / f_c]}{0.6 - \nu - (\rho_1 - \lambda\rho_2) f_y / f_c} \tag{8.68}$$

The curvature ductility factor $\mu_\phi = \phi_u / \phi_y$ may be written

$$\mu_\phi = \frac{\varepsilon_{cu} / x_u}{\varepsilon_c / x_y} \tag{8.69}$$

If the values of x_y, x_u from equations (8.62) and (8.66) are introduced in equation (8.69) the curvature ductility factor is

$$\mu_\phi = \frac{\varepsilon_{cu} + \varepsilon_s}{\varepsilon_c + \varepsilon_y} \tag{8.70}$$

Equating the second parts of equations (8.65) and (8.67) and rearranging terms yields

$$\frac{\varepsilon_{cu} + \varepsilon_s}{\varepsilon_c + \varepsilon_y} = 1.2 \frac{\varepsilon_{cu}}{\varepsilon_c} \tag{8.71}$$

Using equations (8.68), (8.70) and (8.71) the curvature ductility factor may be expressed by the following relationship (Tassios, 1989a):

$$\boxed{\mu_\phi = 1.2 \frac{E_s}{f_y} \left[\frac{0.6}{\nu + (\rho_1 - \lambda\rho_2) f_y / f_c} - 1 \right] \varepsilon_{cu}} \tag{8.72}$$

Parametric investigations (Tassios, 1989a) have led to the following values of the compressive stress parameter λ:

1. For $\nu < 0.1$ ('beams') $\lambda = 0.5 + 18\rho_1$.
2. For $\nu = 0.1$, $\lambda \approx 2/3$.
3. For $\nu = 0.2$, $\lambda \approx 0.9$.
4. For $\nu > 0.2$, $\lambda \approx 1.0$.

Equation (8.72), although approximate, offers the advantage that it gives a clear picture of the influence of each parameter affecting the ductility of an R/C section. The effect of geometric and material parameters has already been discussed with reference to beams (sections 8.2.1); with regard to the axial load, equation (8.72) shows that the ductility of a column decreases as the (compressive) axial load increases. This is clearly seen in Figure 8.24, where curvature ductility factors for two column sections with varying degrees of volumetric ratio of hoop reinforcement are plotted as a function of axial loading. Also shown in the figure are the EC8 ductility requirements for columns; it is clear that for high axial loads ($\nu > 0.4$) it is not feasible to achieve the target ductilities, even with high degrees of confinement.

The ultimate concrete strain ε_{cu}, which is linearly related to μ_ϕ as indicated by equation (8.72), can be defined in the following ways (see also section 7.3.2).

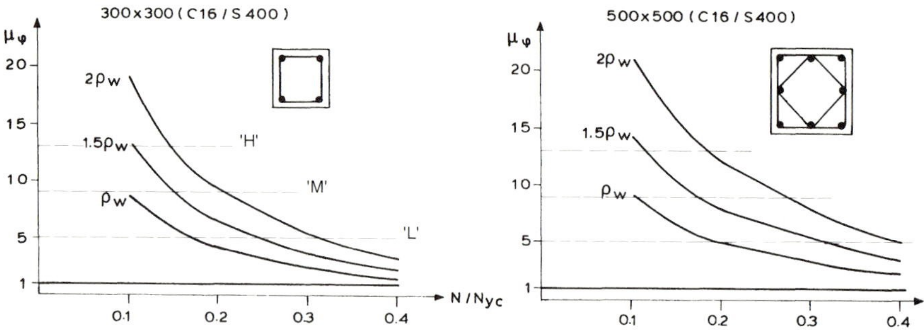

Figure 8.24 Variation of curvature ductility with the axial load and the hoop ratio, for rectangular column sections.

1. On the basis of a stress level of $0.85f_c$ along the descending branch of the σ_c–ε_c curve, which is the design strength for purposes of flexural strength according to many modern codes, including EC2 (CEN, 1991) and the CEB Model Code (CEB, 1993). This definition of ε_{cu} although lacking a physical meaning, is quite convenient for design purposes, thus it was also adopted by EC8 to define the conventional curvature ductility factor (equation (8.4)), which forms the basis of the local ductility criterion. The following expression is suggested by Tassios (1989b) for calculating the ultimate concrete strain.

$$\boxed{\varepsilon_{cu} = \varepsilon_{cu,\,unc} + 0.1\alpha_n\alpha_s\omega_w} \tag{8.73}$$

where $\varepsilon_{cu,\,unc}$ is the ultimate strain for unconfined concrete (to be taken as 0.0035–0.004) and the confinement effectiveness coefficients are given by

$$\alpha_n = 1 - \frac{\Sigma(b_i^2)}{6A_0} \tag{8.74(a)}$$

$$\alpha_n = \left(1 - \frac{s}{2b_0}\right)^2 \tag{8.74(b)}$$

In equation (8.74) the notation introduced in section 7.3.2 is used and it is clearly seen that α_n and α_s are essentially the coefficients defining the effectively confined area in the Sheikh–Uzumeri (1982) model. These coefficients have been adopted in the final version of EC8 (CEN, 1995) to define the confining hoop requirements in columns (section 8.5.2).

2. On the basis of fracture of the first hoop, which according to the empirical equation suggested by Paulay and Priestley (1992) based on the work of Mander, Priestley and Park (1988) occurs at

$$\boxed{\varepsilon_{cu} = \varepsilon_{cu,unc} + 1.4\rho_w\varepsilon_{su}\frac{f_{yw}}{f_{cc}}} \tag{8.75}$$

where the symbols have their usual meaning and the yield stress f_{yw} of the hoop is given in MPa, while ε_{su} is the transverse steel strain at maximum tensile stress (f_u in Figure 7.17). Although empirical, equation (8.75) has the advantage that it has a clear physical meaning, as it corresponds to a phenomenon which for all practial purposes can be considered as failure.

3. On the basis of buckling of compression reinforcement, which again is a situation relatively straightforward to identify and is practically equivalent to failure, although some strength reserves still exist after buckling of longitudinal bars. A comprehensive, though rather complex, analytical procedure for determining the concrete strain corresponding to buckling was suggested by Papia and Russo (1989), as already mentioned in section 8.2.2(b).

EXAMPLE

Consider a 400 mm square column, made of grade 16 concrete and grade 400 steel. For a transverse reinforcement consisting of 8 mm single hoops (two legs) at 200 mm (which is a pattern not allowed for DC 'M' and 'H' columns), and a cover to the reinforcement equal to 20 mm, the resulting volumetric ratio of transverse reinforcement is

$$\rho_w = \frac{4 \times 352 \times 50}{352^2 \times 200} = 0.00284$$

and the corresponding mechanical ratio for $f_{ym} \approx 440$ MPa, $f_{cm} \approx 24$ MPa (estimated mean values)

$$\omega_w = 0.00284 \times \frac{440}{24} = 0.052$$

The confinement effectiveness coefficients can be calculated from equation (8.74) introducing $b_i = b_c = 352$ mm, $A_c = b_c^2$:

$$\alpha_n = 1 - \frac{\Sigma b_i^2}{6A_0} = 1 - \frac{4 \times 352^2}{6 \times 352^2} = 0.333$$

$$\alpha_s = \left(1 - \frac{s}{2b_0}\right)^2 = \left(1 - \frac{200}{2 \times 352}\right)^2 = 0.513$$

The ultimate concrete strain can now be calculated from equation (8.73):

$$\varepsilon_{cu} = 0.004 + 0.1 \times 0.333 \times 0.513 \times 0.052 = 0.0049 (0.49\%)$$

If the transverse reinforcement is increased to three-leg 10 mm hoops at 100 mm, the volumetric ratio $\rho_w = 0.0134$ (which is 4.7 times the previous one) and the corresponding mechanical ratio $\omega_w = 0.246$. The resulting confinement coefficients are $\alpha_n = 0.667$ and $\alpha_s = 0.736$ and the ultimate strain $\varepsilon_{cu} = 0.016$ (1.6%), which means that the ductility of the column has increased by 227%; this is a clear indication of the paramount role of hoop reinforcement with regard to column ductility. Once again, however, it has to be remembered that this is a purely flexural approach to the problem and the whole situation might change if cyclic shear effects are taken into account (section 8.4.3).

For the column under consideration, if ultimate concrete strains are calculated using equation (8.75) with $\varepsilon_{su} \approx 0.09$ and $f_{cc} = Kf_c$ (K from equation (7.28)), a value of 1.0% results in the case of 8 mm hoops at 200 mm, and a value of 2.9%

for 10 mm hoops at 100 mm spacing. In the first case the calculated ε_{cu} is 104% higher than the value resulting from the previous approach (equation (8.73)), while in the second case it is 81% higher. It has to be noted that the empirical equation (8.75) was derived on the basis of experimental data concerning heavily confined columns and piers, and it appears to overestimate ε_{cu} in cases of low hoop reinforcement ratios. On the other hand, the semi-empirical equation (8.73) appears to be generally conservative and possibly appropriate for design purposes (see also section 8.5.2).

With regard to the criterion of buckling (section 8.2.2(b)), assumptions concerning the characteristics of steel have to be made. If only the minimum EC8 specifications for DC 'M' structures are assumed, that is $f_u/f_y=1.15$, $\varepsilon_{su}=0.06$, the slenderness of longitudinal bars is too high in the case where single 8 mm ties at 200 mm are used and early buckling is expected immediately after spalling, hence $\varepsilon_{cu}\approx 0.4\%$ which is even lower than the 0.49% predicted by the generally conservative '$0.85f_c'$' criterion. On the other hand, when three-leg 10 mm hoops at 100 mm spacing are used, buckling is estimated to occur at $\varepsilon_{cu}=2.0\%$, which is 31% lower than that calculated using equation (8.75), but 25% higher than the value resulting from equation (8.73). It is pointed out that if a more realistic value $f_u/f_y=1.45$ is assumed for strain-hardening, together with $\varepsilon_{su}=0.09$, the estimated buckling strain is $\varepsilon_{su}=2.8\%$, which almost coincides with the value resulting from the hoop fracture criterion.

The foregoing example points to the need for taking into account all possible modes of failure when attempting to estimate the available ductility of R/C elements.

(b) Test results and empirical formulae

The high ductility of properly confined columns was confirmed by several tests involving monotonically increasing concentric compression (Sheikh and Uzumeri, 1980; Scott, Park and Priestley, 1982) or monotonic flexure with axial load (Bertero, 1979; Sheikh and Yeh, 1990).

Typical test results for monotonic compression for various hoop patterns were given in section 7.4.3, while a typical M–ϕ diagram for monotonic flexure with axial load is given in Figure 8.25, where M_{code} is the theoretical strength of the column calculated according to the American code (ACI 318–89) using measured material strengths (the procedure is very similar to that used in EC2 and other concrete codes). It is clear from the figure that high axial loading not only leads to reduced ductility (compare specimens D-5 and D-15), but may also lead to reduced strength (for axial loads above the one corresponding to 'balanced' conditions, see also Figure 8.22), especially when the amount of hoop reinforcement is low (specimen D-14 in Figure 8.25). It is worth pointing out that the value $\nu=0.75$ used in two of the specimens is not allowed by EC8 for either DC 'H' or 'M' columns. Another aspect of the unfavourable influence of axial loading is the more pronounced effect of spalling on the strength of columns; as clearly shown in Figure 8.25, spalling of cover concrete in specimens D-14 and D-15 is accompained by a substantial reduction in moment capacity (residual capacity is about 15% lower than the code value in specimen D-14). Spalling is almost immediately followed by yielding of the perimeter hoop and subsequently of the inner (octagonal) hoop; however if (as in the case of specimen D-14) the quantity of transverse steel is low, the resulting con-

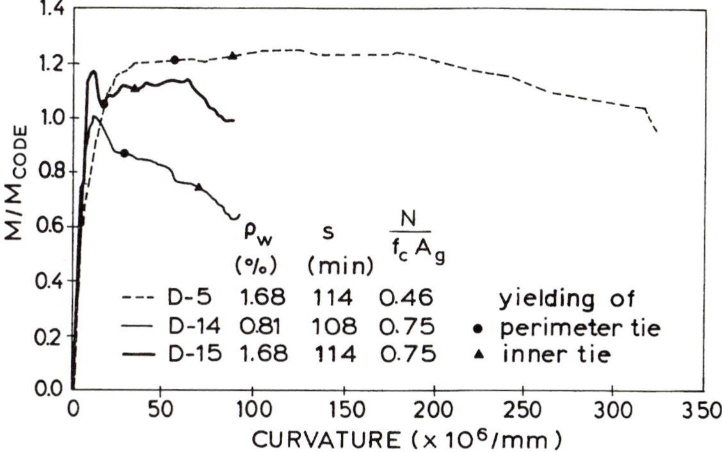

Figure 8.25 Effects of axial load and amount of transverse reinforcement on the behaviour of a 305 mm square column (Sheikh and Yeh, 1990).

finement pressure is not sufficient to maintain the integrity of the column. In the tests by Sheikh and Yeh (1990) involving 15 square columns with various arrangements and quantities of hoop reinforcement, increases in flexural capacity due to confinement up to 26% were recorded, while curvature ductility factors ranged from as low as 3 (for columns with single hoops subjected to $v=0.78$) to as high as 50 (for columns with multiple closely spaced hoops subjected to $v=0.6$).

The behaviour of columns subjected to monotonic flexure and axial loading, inducing high levels of inelasticity, may be estimated analytically using the refined or the approximate procedures presented previously. However, as already pointed out for beams, due to the large number of parameters involved and the uncertainties associated with each of them, probably the most appropriate procedure is using empirical relationships calibrated on the basis of a large amount of experimental data. With regard to yield rotation (θ_y), equations (8.21)–(8.24) suggested by Park and Ang (1985) may be used to supplement the theoretical equation (8.20); the effect of axial loading (v) on the yield rotation of columns is incorporated in the correcting factor of equation (8.24). The rotational capacity can be estimated as $\theta_u = \mu_u \, \theta_y$, where the ductility factor (μ_u) for monotonic loading is given by equation (8.30(a)), which is applicable for $0 \leqslant v \leqslant 0.55$ (Park and Ang, 1985). A scatter of 38% was estimated with regard to 142 beam and column specimens used for deriving equation (8.30(a)).

8.4.3 Behaviour under cyclic loading

The degradation of the response of R/C columns due to reversed cyclic loading takes place in a way similar to that observed for beams (section 8.2.2). The main factor leading to a different response of columns is, of course, axial loading, which may cause favourable, as well as unfavourable effects, which are briefly summarized below.

(a) Effect of compressive axial loading

1. The presence of compressive axial loading contributes to closing of flexural, as well as shear cracks, especially in the region of low moments. As shown in Figure 8.26, an increase in axial loading leads to an increase in column stiffness, which is more marked in the region around the origin of the coordinate system. Thus, the width of the hysteresis loops is larger than in the case of beams ($N = 0$) with similar geometry and reinforcement. The hysteresis loops of Figure 8.26 refer to specimens tested at Rice University (Texas) by Jirsa (1974); the axial loading is expressed as a percentage of the balanced load N_b which corresponds to the point of maximum flexural capacity in an M–N interaction diagram (compare Figure 8.22). The specimens were symmetrically reinforced ($\rho_l = \rho_2$) as is typical in the case of columns, and this was an additional factor contributing to closing of cracks in the compression zone (see also section 8.2.2(a)).

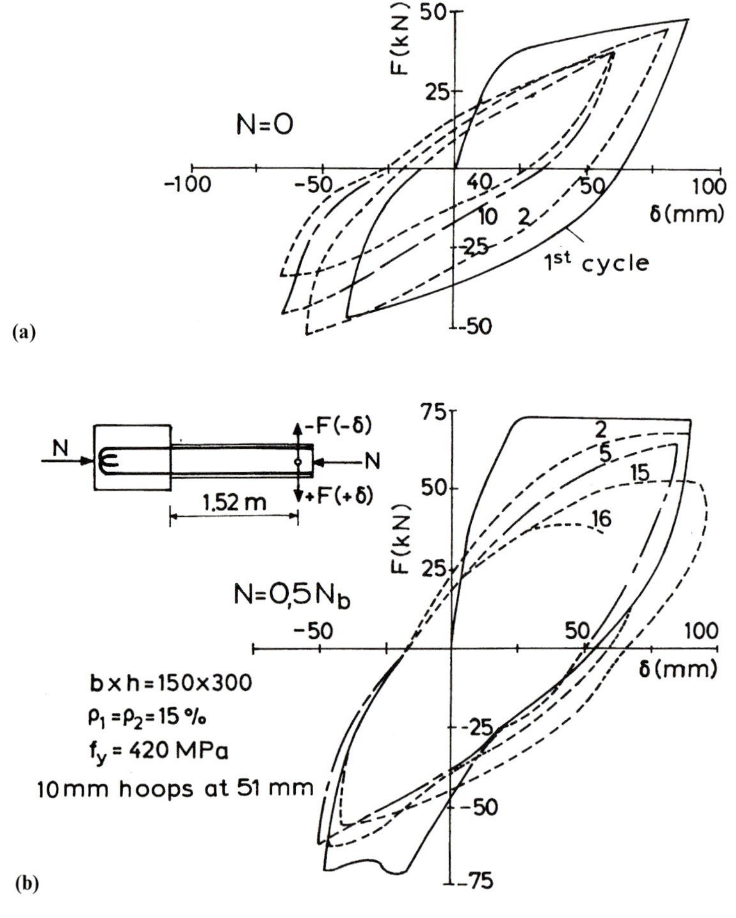

Figure 8.26a,b

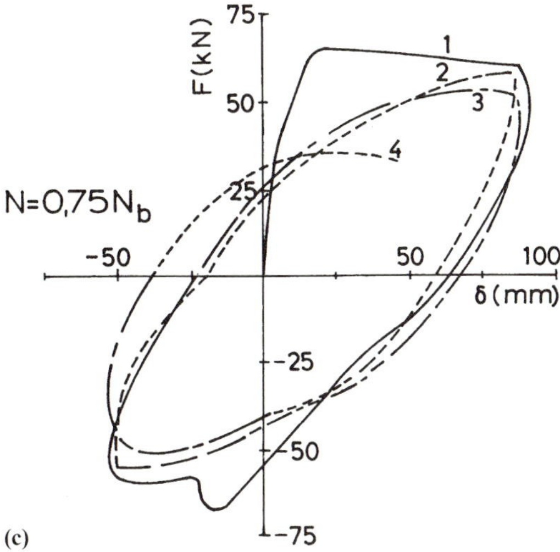

Figure 8.26 Hysteretic behaviour of elements subjected to various levels of axial loading (Jirsa 1974).

The influence of axial loading is also significant with regard to the behaviour of columns under cyclic shear, as it contributes to the closing of cracks perpendicular to the axis of the member, thus mitigating premature failures due to sliding shear. The favourable effect of axial loading on shear strength can be seen in Figure 8.22, where the variation with axial load of the moment corresponding to the shear capacity of the column (estimated using code procedures) is plotted, for various arrangements of transverse reinforcement and column slenderness values. The moment corresponding to shear capacity was estimated assuming (conservatively) antisymmetric flexure ($M = Vl/2$). Figure 8.22 suggests that for the usual ranges of axial loading level ($0 \leqslant N \leqslant N_b$) shear failure may precede flexural failure, if an adequate amount of hoop reinforcement is not present in the column. It is pointed out that the $M_{,V}$–N curves shown in Figure 8.22 were derived using code formulae for shear capacity which are, of course, quite conservative (especially the CEB MC/SD formula).

The previously mentioned tests of Jirsa (1974) have also indicated the favourable effect of compressive axial loading with respect to bond; premature anchorage failures in columns subjected to large inelastic deformations are precluded if a moderate amount of axial loading is present (see also section 7.6.2).

2. The presence of compressive axial loading results in a larger compression zone in the member, thus in higher demands regarding concrete strain, compared with the case of zero axial load. This leads to spalling of cover concrete at relatively low levels of displacement and a subsequent drop in strength, which is more pronounced when the cover concrete constitutes a substantial portion of the section (which is the case when column dimensions close to minimum requirements are used in areas where environmental conditions impose severe demands on minimum cover). This drop in strength due to spalling is clearly seen in Figure 8.26(b) and (c) in the region of 'negative' displacements (ten-

sion at the bottom of the element). Relatively premature spalling increases the risk of buckling of longitudinal column reinforcement. The most common type of failure in columns subjected to high axial loading is buckling of longitudinal bars combined with crushing and degradation of concrete in the regions where spalling has occurred (Jirsa, 1974; Sheikh and Yeh, 1990); the situation is aggravated when shear cracking is also present. As can be clearly seen in Figure 8.26, the number of inelastic cycles to failure (here for a ductility $\mu_\delta \approx 5$) drastically reduces as the level of axial loading is increased.

3. It is well known that the lateral displacement of the head of a column relative to its base gives rise to second-order moments ('P–Δ effect' where P is the axial load on the column and Δ the lateral displacement). It is seen in Figure 8.26(b) and (c) that the post-yield branch has a negative slope which is due to the presence of P–Δ moments and, as expected, is more pronounced in the column with the higher level of axial loading ($N = 0.75N_b$). When the imposed ductility level is high ($\mu_\delta > 4$) second-order moments are significant and they lead eventually to failure of columns (and subsequently of the structure as a whole) due to lateral instability. Such a situation may arise in buildings where column sidesway mechanisms form (Park and Paulay, 1975).

In conclusion, it may be stated that while relatively low levels of axial loading have, as a rule, a favourable effect on the seismic performance of R/C columns, as they increase their energy dissipation capacity and prevent sliding shear failures, high levels of axial loading may drastically reduce the ductility of columns and they induce failure modes that may lead to partial or total collapse of the structure (in the sense of actually falling down).

(b) Effect of tensile and varying axial loading

Although design codes such as EC8 or the CEB MC/SD define columns as elements with a minimum amount of compressive axial loading (typically $N < -0.1A_c f_c$), referring, of course, to design seismic loads, it is not uncommon that during an actual earthquake overturning moments combined with the effect of the vertical component of the seismic motion lead to drastic reduction of column axial loads and even to net tension in some of them (typically lying at the perimeter of a building). The seismic behaviour of such elements is markedly different from that described in the previous section.

Given in Figure 8.27 are hysteresis loops recorded in beam–column sub-assemblages (Townsend and Hanson, 1977) where the axial load in the column was zero or tensile (equal to 65% of the load causing yield under uniaxial tension). The pinching of the loops is more pronounced in the case of the column subjected to tensile axial loading, which also suffered a significant loss of strength during cycling at a constant displacement. The most drastic reduction in the capacity of the column occurs during the second inelastic cycle of loading, when the drop in strength (for a given μ_δ) was of the order of 20% in the column with tensile loading. It appears that the presence of axial tension causes flexural, as well as shear, cracks to remain open in the course of cyclic loading, which leads to a reduction of stiffness, energy dissipation capacity, bond of longitudinal bars, and also to a degradation of shear-resisting mechanisms (section 8.2.2(d)), thus increasing the possibility of sliding shear failure (Bertero, 1979).

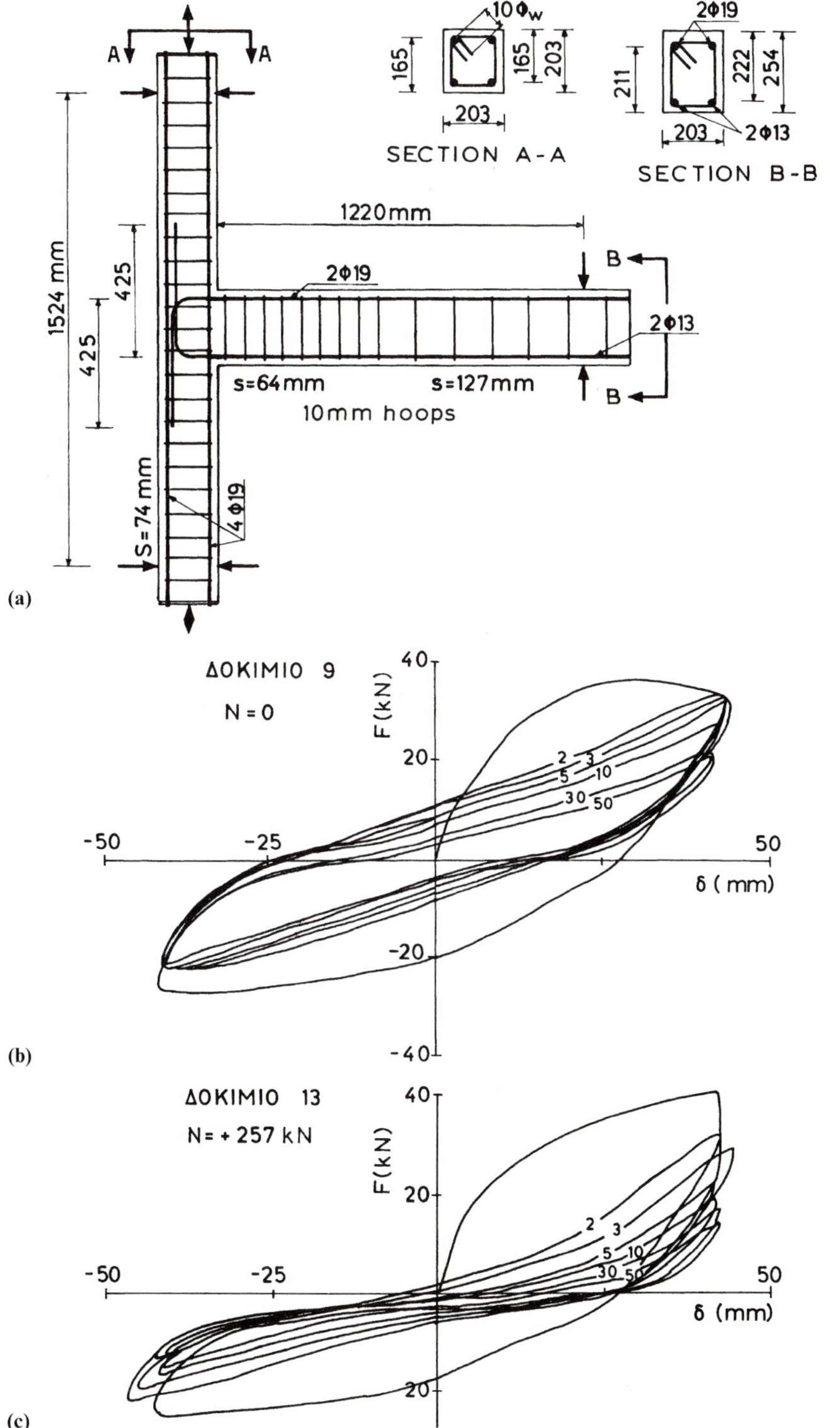

Figure 8.27 Hysteretic behaviour of beam–column subassemblages under various levels of axial loading (Townsend and Hanson, 1977).

It has to be pointed out that in the experimental study presented previously (Townsend and Hanson, 1977), axial loading in columns was kept constant during cyclic loading. It is understood that in an actual R/C building subjected to a seismic motion, the axial load in a column may change (within a fraction of a second) from high compression to net tension, therefore its behaviour is more complex than previously described. A limited number of experimental studies (Jirsa, Maruyama and Ramirez, 1980; Abrams, 1987; Tsonos, Tegos and Penelis, 1995) have addressed the problem of changing axial loading in columns subjected to cyclic shear, wherein the typical loading history was that high shear in the column was accompanied by high compression and low shear by low compression or tension.

The aforementioned studies have shown that the variation in axial loading influences the strength, stiffness and ductility of a column. With regard to flexural strength, decreasing compression and/or tension lead to lower moment capacity (Abrams, 1987), as would be expected from an *M–N* interaction diagram (compare Figure 8.22). With regard to stiffness, the most unfavourable case appears to be when axial force varies, but it tends to be constant after yielding, which is actually the common case when yielding beams frame into columns; in this case displacement increases rapidly with decreasing axial load (previously opened cracks do not close), while progressive accumulation of plastic strain occurs in column bars subjected to tensile strains during cycles of increasing axial compression as the column tries to reach a symmetrical deflection in each direction with an increased stiffness due to the high axial compression (Abrams, 1987). Variation in axial loading is critical with respect to the shear strength of a column; increasing axial compression leads to increased flexural strength and stiffness of the column, thus to increased shear force. The favourable effect of compressive axial loading on shear capacity does not necessarily suffice to compensate for the development of this increased shear force, particularly when open cracks due to accumulated plastic strains are present.

(c) Confinement requirements

The paramount importance of transverse reinforcement in potential plastic hinge regions has already been pointed out in previous sections (7.4,8.2). Specifically with regard to columns, it has to be emphasized that the existence of compressive axial loading reduces to a certain extent the required shear reinforcement, but increases the required confinement reinforcement. The latter is due to the fact that the presence of axial compression leads to larger compression zones and imposes higher requirements for concrete strains if a given curvature limit has to be respected (compare equations (8.4)). It follows that in order to ensure a constant ductility factor in a column, the required confinement reinforcement increases with the level of axial loading (Priestley and Park, 1987; Cheung, Paulay and Park 1992).

The dependence of hoop steel requirements on axial loading is recognized by some design codes (NZS 3101, CEB MC/SD and EC8), while it is ignored by others, the American code (ICBO, 1994) among them. The first equation for the volumetric ratio (ρ_w) of hoops expressed as a function of normalized axial loading was included in the New Zealand Code (SANZ, 1982); the equation can be written in the following form

$$\omega_{wd}=0.78\ (A_c/A_0-1)\ (0.5+0.93\nu_d) \nless 0.31\ (0.5+0.93\nu_d) \tag{8.76}$$

where $\omega_{wd} = \rho_w f_{yd}/f_{cd}$ is the mechanical volumetric ratio of hoops in terms of design material strengths, $\nu_d = N/(A_c f_{cd})$ is the normalized design axial load acting on the column, and A_c/A_0 is the ratio of the gross section area to the area of the confined core. A similar equation was included in the CEB MC/SD (CEB, 1985) for ductility level III (high) columns. A comparison of hoop requirements resulting from equation (8.76) with the corresponding requirements given by the EC8 equations (section 8.5.2), is shown in Figure 8.28, where it is clear that the New Zealand Code places more emphasis on the effect of spalling (expressed through the A_c/A_0 term) and less on the effect of axial loading. Recent research (Cheung, Paulay and

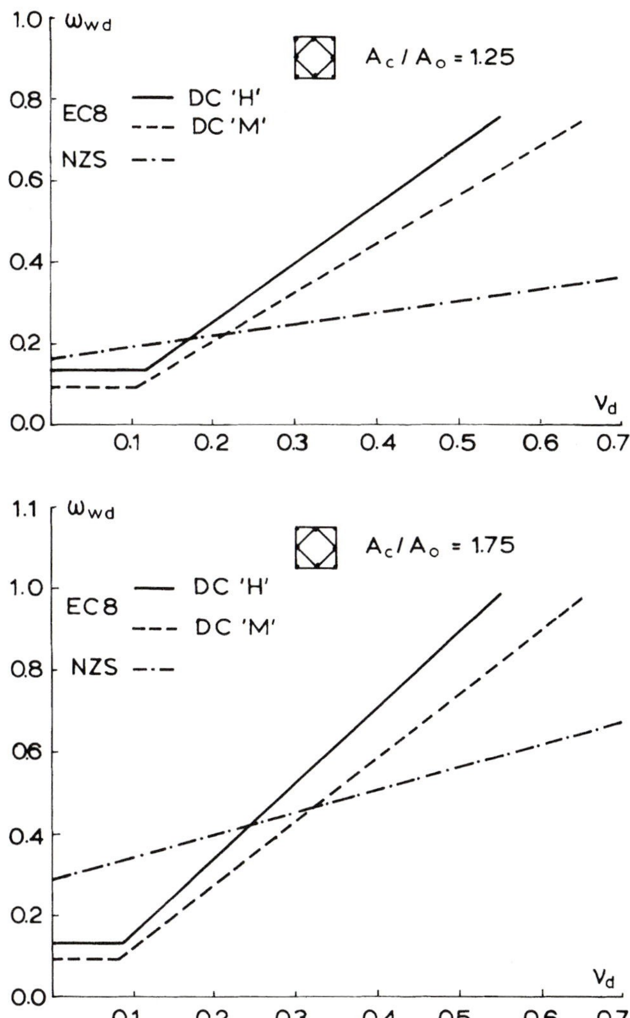

Figure 8.28 Required mechanical volumetric ratios of hoop reinforcement according to EC8 and the New Zealand Code.

Park, 1992) has indicated that the effect of axial loading on confinement requirements is more pronounced than indicated by equation (8.76) and this trend is reflected in the EC8 equations. It is pointed out that for high axial loads the required hoop reinforcement is quite large and difficulties in placing it on site may arise; the situation is aggravated in the case of relatively small colums where A_c/A_0 is quite high. On the other hand, it has been found experimentally (Priestley and Park, 1987) that columns designed according to equation (8.76) can sustain displacement ductility factors $\mu_\delta \geqslant 6$, which is more than adequate for most practical purposes.

(d) Short columns

The information presented in the foregoing sections regarding the influence of axial loading on the seismic behaviour of columns applies to members of normal slenderness, namely for $l/h>4$. For an antisymmetrically loaded column ($V=2M/l$) the previous condition means that the shear span ratio $\alpha=M/Vh$ (where h is the column depth) is greater than 2. Members with $\alpha \leqslant 2$, commonly referred to as **short columns**, have a substantially different behaviour under cyclic shear, characterized by an increased possibility of brittle failure (Minami and Wakabayashi, 1980; Tegos and Penelis, 1988). Short columns are often met in industrial or school buildings, where discontinuous masonry infills are used to create openings extending along a whole span of the R/C frame (see Figure 8.32). These low slenderness elements, when conventionally reinforced with longitudinal and transverse bars, and subjected to relatively high axial loading, fail by splitting of concrete along their diagonals, as shown in Figure 8.29(a) which refers to the short columns ($\alpha=2$) tested by Tegos and Penelis (1988). If the axial loading level is low, the most probable mode of failure is by shear sliding along full depth cracks at the member ends. The latter case is more common in coupling beams of structures with coupled walls (section 9.4.4).

During the 1980s a lot of effort was spent on finding new arrangements of reinforcement which could improve the seismic performance of short R/C columns. Minami and Wakabayashi (1980) have proposed the use of cross-inclined diagonal bars (**bidiagonal** reinforcement), similar to those previously suggested by Paulay (Park and Paulay, 1975) for coupling beams of R/C walls. More recently Tegos and Penelis (1988) have proposed the use of multiple cross-inclined bars, forming a rhombic truss, as shown in Figure 8.29(c). Test results have indicated that the use of either bidiagonal or rhombic reinforcement leads to an increase in shear capacity, as well as in stiffness and energy dissipation of short columns (Minami and Wakabayashi, 1980; Tegos and Penelis, 1988). As shown in Figure 8.29(b) and (c), the use of non-conventional reinforcement patterns prevents brittle modes of failure from diagonal splitting and leads to a more favourable shear-flexure type of failure, similar to that observed in normal slenderness columns. Bidiagonal reinforcement is practically preferable for slendernesses $l/h<1.5$, while rhombic reinforcement is preferable for $1.5 \leqslant l/h \leqslant 4.0$, as in this case a shift of the inflection point away from column mid-height is possible (Tegos and Penelis, 1988).

According to an ultimate strength model developed by Tegos and Penelis (1988), the total shear capacity of a short column ($\alpha \leqslant 2$) reinforced with inclined bars,

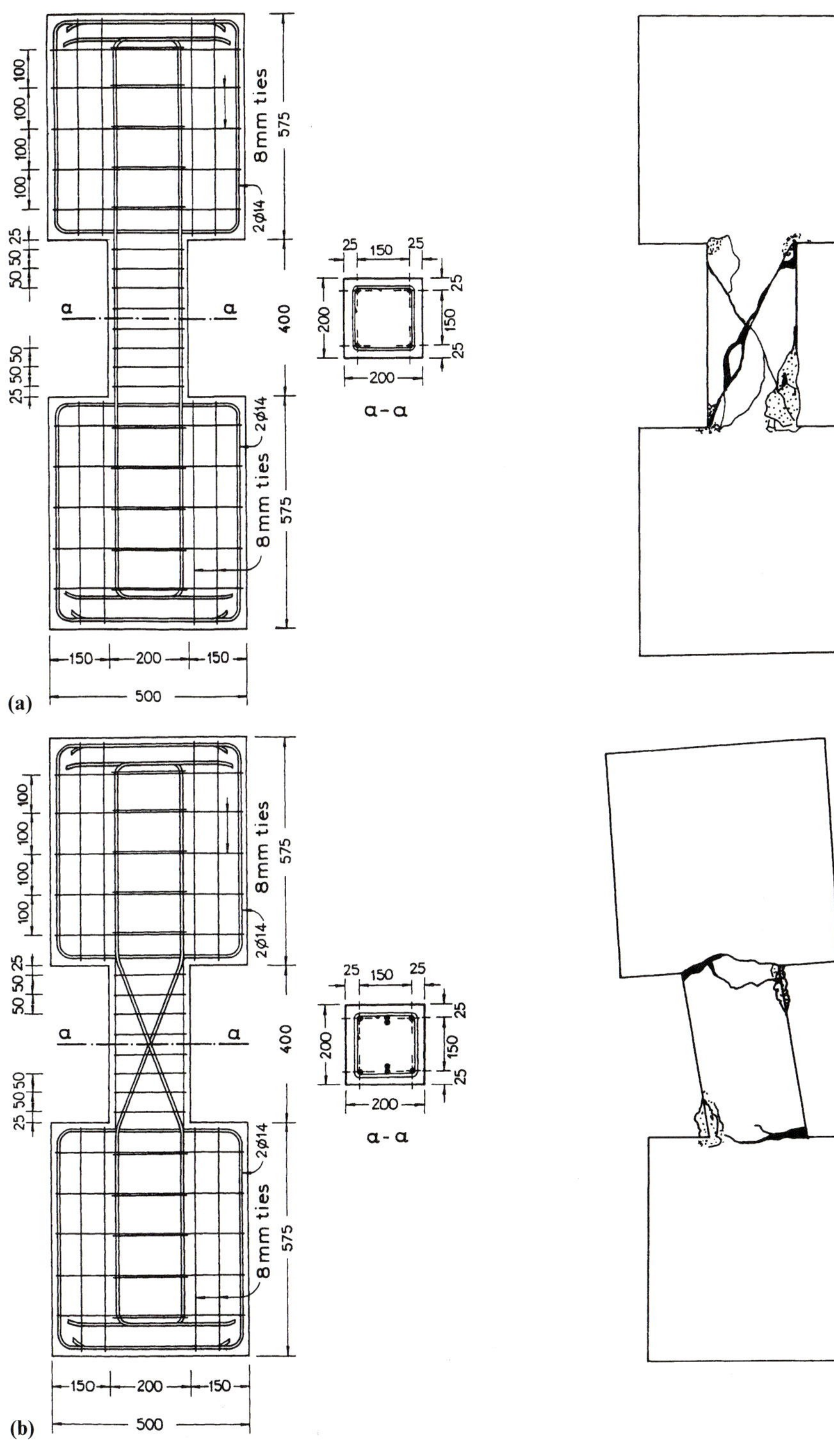

100
100
100
100
50 50 25
50 50 25
25 50 50
a
a
8 mm ties
575
2Ø14
400
575
8 mm ties
2Ø14
25
150
25
200
25
150
25
200
a - a
150
200
150
500
(a)
(b)

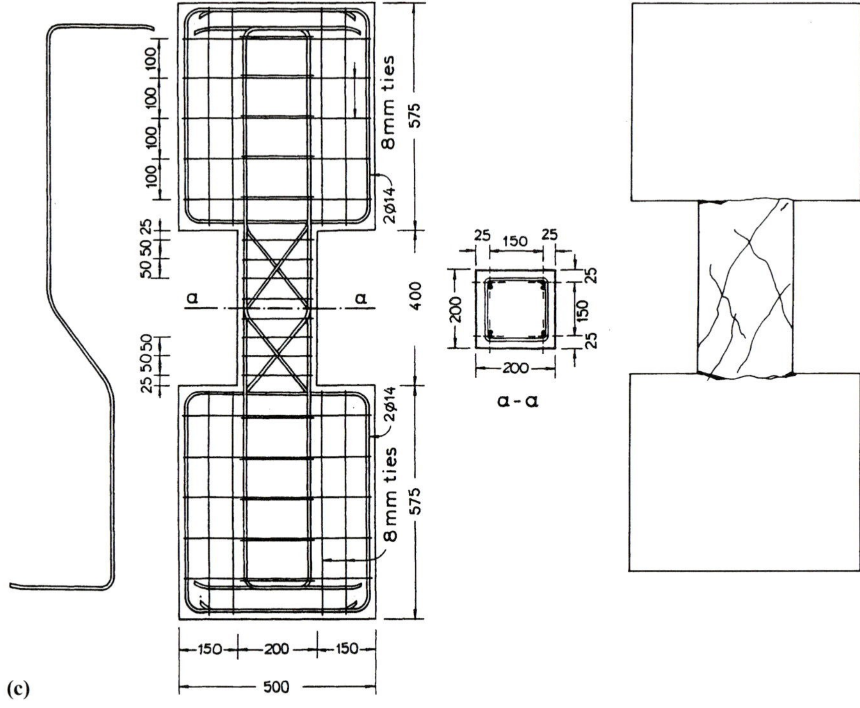

Figure 8.29 Arrangement of reinforcement and corresponding failure modes in short columns: (a) conventional reinforcement (closely spaced ties); (b) bidiagonal reinforcement and ties; (c) rhombic reinforcement and ties.

longitudinal bars and hoops, may be estimated by superposition of the following three partial mechanisms:

1. The well-known truss mechanism, wherein hoops are the web elements in tension and concrete struts between inclined shear cracks are the elements in compression, the longitudinal steel is the tension cord and the compression flexural zone is the compression cord; this mechanism carries a shear force

$$V_{R1} = (\tan\theta - \sin\theta) A_s f_y \tag{8.77}$$

   where A_s is the area of reinforcement in the tension cord and θ the angle of inclined concrete struts with respect to the column axis.
2. The rhombic truss mechanism of the inclined bars, which is able to carry a shear force

$$V_{R2} = 2A_s f_y \sin\theta \tag{8.78}$$

   where A_s is the area of inclined bars, which coincides with the area of tension reinforcement (to be entered in equation (8.77)) if no extra (straight) longitudinal bars are used.
3. A compression parallelogram formed by the compressive stress path in concrete, as compression is tramsmitted from one end of the member to the other

through a double-arch action; this mechanism carries a shear

$$V_{R3} = 0.5N \tan \theta \tag{8.79}$$

where N is the axial load of the column. The total shear capacity is given by the relationship

$$V_R = V_{R1} + V_{R2} + V_{R3} \tag{8.80}$$

Transverse reinforcement (hoops) is required both for the truss mechanism and for balancing the splitting force of the compression parallelogram. The required amount of hoops may be calculated from the relationship (Tegos and Penelis, 1988)

$$\rho_w = (V_R - V_{R2})/2\alpha bhf_{yw} \tag{8.81}$$

which implies that no transverse reinforcement is required for the development of the rhombic truss mechanism. Nevertheless, independent of the amount of shear carried by the inclined bars (V_{R2}), it is essential that a certain amount of hoops be present to provide the necessary confinement to concrete.

(e) Biaxial cyclic loading

It has already been pointed out at the beginning of this section that columns in buildings are subjected to a biaxial stress state during an earthquake. A first consequence of this type of stressing is a reduction in column stiffness, compared with the case of uniaxial stress state (Bertero, 1979; Jirsa, Maruyama and Ramirez, 1980), which can lead to increased lateral displacements of the building, hence to a more pronounced effect of second-order actions (P–Δ effect). Besides, when lateral displacements in one direction induce relatively high levels of inelasticity ($\mu_\delta > 3$), it is possible that certain members yield simultaneously in the orthogonal direction, even if the angle of seismic action with respect to the building axis is small. For example, in an entirely symmetric building if displacements in one directions (i) reach the value $\delta_i = 3\delta_y$ ($\mu_\delta = 3$) it needs only a displacement $\delta_j = \delta_i/3$ in the orthogonal direction (j) to cause yielding; such a displacement may develop for an angle of the seismic action with respect to the i-axis of only 18.4°.

An R/C column with beams framing in both directions, will develop moments in the direction of the seismic action larger than the (uniaxial) strength of the beams. This increase of moments, which may reach up to 42% ($\sqrt{2}$:1) when beams at right angles yield simultaneously, will probably cause a plastic hinge to form in the column. Furthermore, the shear acting on the column will also increase in the same proportion.

Biaxial testing of columns has indicated that the shear capacity in the diagonal direction is hardly any larger than the value corresponding to uniaxial (unidirectional) loading (Bertero, 1979; Jirsa, Maruyama and Ramirez, 1980; Umehara and Jirsa, 1984). This might cause problems regarding the seismic performance of a column, in case it is not designed for a shear larger than that resulting by considering the seismic action separately along each principal axis of the building (if indeed such axes exist). Tests on square columns reinforced with closely spaced transverse reinforcement (according to the New Zealand Code) have shown that their behaviour under biaxial cyclic loading (along the diagonal of the cross-section) is quite satisfactory, characterized by stable hysteresis loops even at ductility levels $\mu_\delta > 8$ (Priestley and Park, 1987; Zahn, Park and Priestley, 1989).

The exact loading history of a column belonging to a building subjected to an actual earthquake is quite complex and difficult to determine; it is pointed out that the earthquake motion not only induces cycling along an axis not coinciding with the principal ones of a column section, but also along other axes which might even be orthogonal to the previous one (bidirectional loading). The influence of cyclic loading history on the behaviour of low-slenderness square and rectangular columns was investigated experimentally at the University of Austin, Texas (Jirsa, Maruyama and Ramirez, 1980; Umehara and Jirsa, 1984). When cyclic loading was applied along the diagonal of the column section (specimen O-D, Figure 8.30), it was found that, although the response in each direction is significantly lower than that corresponding to uniaxial monotonic loading, the resultant of the response (vector sum of the two individual components) almost coincides with the uniaxial loading curve up to the development of the maximum shear, but it is lower than that in the region of large inelastic displacements. This result implies that the biaxial strength is almost the same as the uniaxial one and the shape of the strength interaction curve (M_u vs M_v) is close to a circle for square, symmetrically reinforced columns, and close to an ellipse for rectangular columns (Zahn, Park and Priestley, 1989). A similar conclusion was drawn with regard to **shear capacity**, based on their experimental data on two-thirds scale column specimens, by Umehara und Jirsa (1984). Measured maximum shear forces for various types of biaxial and/or bidiagonal cyclic loading was close to a circle or an ellipse, with a dispersion of only 10%. Moreover, for cycling well into the inelastic range the biaxial response appears to be inferior to the uniaxial one. Even less favourable conditions arise if a **bidirectional** loading history (which may involve cycling either along principal axes of the section, or at an angle to them) such as the square loading history (specimen O-S, Figure 8.31) is applied; in this case the reduction in strength is more pronounced, especially for large inelastic displacements.

In the aforementioned experiments by Jirsa, Maruyama and Ramirez (1980) the effect of varying axial loading (including net tension) on the biaxial behaviour of columns was also studied. It was found that the presence of axial tension significantly reduces the stiffness and the shear strength of biaxially loaded columns. Nevertheless, the total response under alternating tension and compression was similar to that under constant axial loading for the same lateral load history.

8.5 SEISMIC DESIGN OF COLUMNS

In the following, the design of R/C columns for earthquake resistance according to the relevant Eurocodes (CEN, 1991, 1995) is presented.

8.5.1 Design for flexure and axial loading

(a) Dimensioning of column section and design of longitudinal reinforcement

The design of columns for the bending moments (M_d) and axial forces (N_d) resulting from the analysis for the design seismic action (Chapter 5) is carried out using the assumptions and the procedure for flexural design prescribed by EC2 (CEN, 1991) for normal (vertical) loads. This is allowed because, according to EC8, the partial safety factors γ_c for concrete and γ_s for steel retain the same values as in EC2.

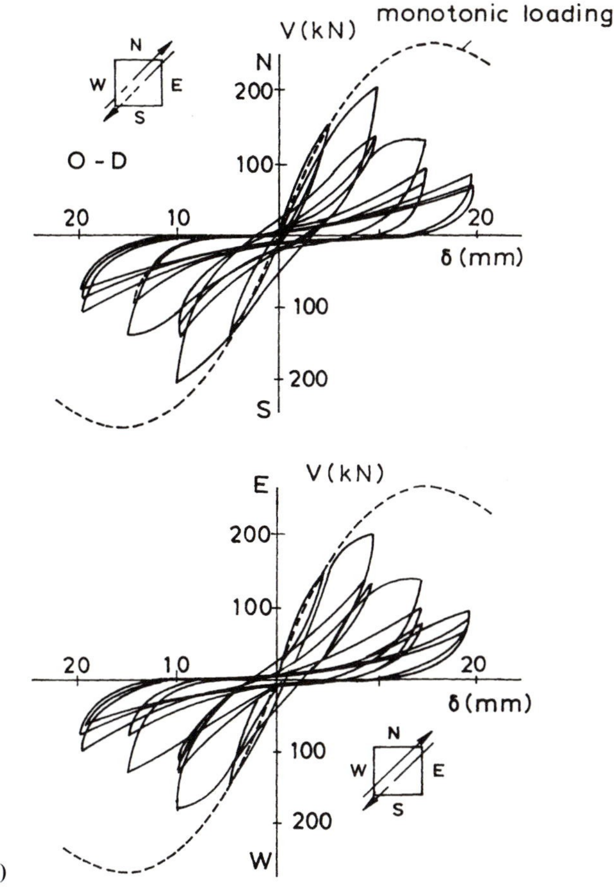

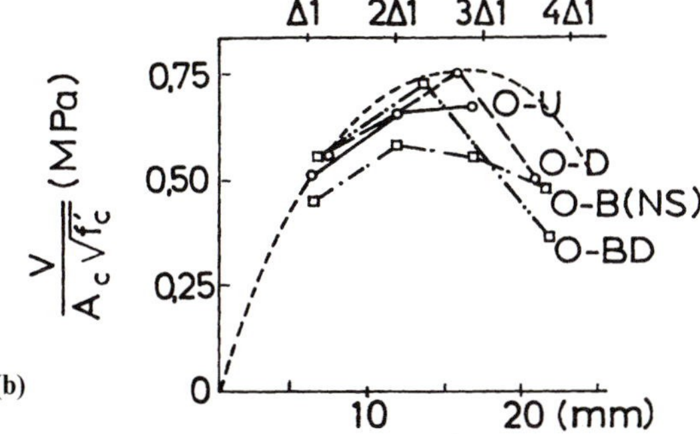

Figure 8.30 Diagonal cyclic loading of columns; (a) hysteresis loops in each direction; (b) envelope of resultant shear vs. displacement (Jirsa, Marujama and Ramirez, 1980).

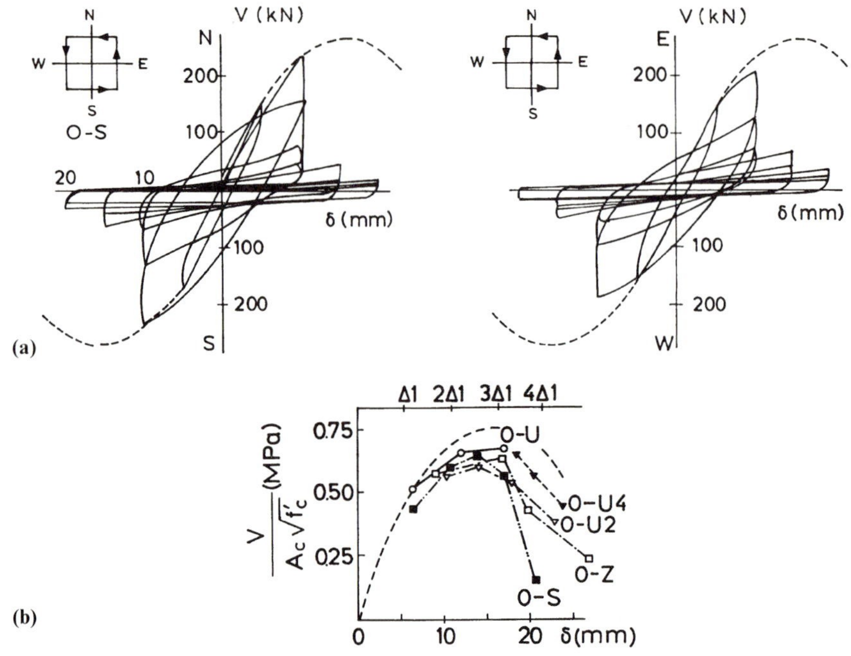

Figure 8.31 Square-like cyclic loading of columns: (a) hysteresis loops in each direction; (b) envelope of resultant shear vs. displacement (first loading cycle) (Jirsa, Marujama and Ramirez, 1980).

The bidirectional character of the seismic action has to be appropriately taken into account in designing column sections. According to EC8, biaxial bending (with axial load) should be considered in DC 'H' columns, while approximate procedures may be used for the other ductility classes. The approximation consists in carrying out the verification separately in each direction, but with a bending resistance reduced by 30%, that is

$$0.7M_{\mathrm{R}id}(N_{\mathrm{Sd}}) \geqslant M_{\mathrm{S}id,\mathrm{CD}} \quad (i=x,y) \tag{8.82}$$

where $M_{\mathrm{S}id,\,\mathrm{CD}}$ is the acting bending moment, appropriately modified according to capacity design procedures and N_{Sd} the most unfavourable axial load resulting from the seismic action.

Design aids (typically charts) for uniaxial and biaxial flexure with axial loading, such as those prepared within the CEB framework (CEB, 1982) can be used for this purpose, since essentially the same assumptions as in EC2 have been made for deriving the design charts. If a fully automated design procedure is sought, analytical approximations to the M–N interaction curves may be derived and appropriately incorporated in relevant computer codes for the design of R/C structures.

On the basis of the column strengths derived from the previous step (taking the most unfavourable value of N_{d} into account) the capacity design criterion (Chapter 6), based on the equilibrium of moments at beam–column joints (section 6.1.4), is applied and, if necessary, column reinforcement (and more rarely section dimensions) are appropriately modified.

It is recalled that if the inter-storey drift sensitivity coefficient θ (equation (5.67)) exceeds the value of 0.1, approximate or explicit account of second-order effects has to be taken in deriving design actions.

A major consideration in dimensioning the column section is the limitation of the normalized axial force $v_d = N_d/(A_c f_{cd})$ imposed by EC8 for ensuring adequate ductility (sections 8.4.2(a), 8.4.3(a)). Depending on the ductility class, the following limits are specified.

- for DC 'H' $v_{d,max} = 0.55$
- for DC 'M' $v_{d,max} = 0.65$
- for DC 'L' $v_{d,max} = 0.75$

These limitations are all stricter than the corresponding one in the New Zealand Code ($v_d = 0.94$), while the CEB MC/SD imposes a limit of 0.75 regardless of ductility class, thus it is less conservative than EC8 with respect to DC 'M' and 'H' structures. It will be shown later that high axial loading leads to excessive demands regarding confinement reinforcement (section 8.5.2), and this is one more reason for selecting appropriately large cross-section dimensions in columns.

(b) Limitations regarding the longitudinal reinforcement

The total longitudinal reinforcement ratio ρ_{tot} in a column, according to EC8, should not be less than 1.0%, to ensure appropriate substitution of cracked concrete without yielding, that is $M_y > M_r$ where M_r is the cracking moment (section 8.3.1(a)). Unlike beams, symmetric column cross-sections are symmetrically reinforced ($\rho_1 = \rho_2$), which is a factor contributing to an increase in ductility. Unsymmetric reinforcement should be used only in large-span, gravity load dominated frames (for reasons of economy).

In order to enhance the rotational capacity at critical regions, the limitation $\rho_{tot} \ngtr 4\%$ is imposed by EC8, regardless of ductility class. It is not clarified in the code whether this limitation also applies in regions of lapped splices; according to the CEB MC/SD, ρ_{tot} should not exceed 6%, including the regions of lapped splices, that is when all column bars are spliced $\rho_{tot} \ngtr 3\%$. It is understood that the situation is quite different if splicing is within the critical region or outside it; in the former case it is recommended that the EC8 limitation should apply for the total reinforcement in the region of lapped splices. Futhermore, it has to be emphasized that high reinforcement ratios also lead to high shear demands in the column, with all the unfavourable consequences described in section 8.4.3.

The influence of the aforementioned limitations regarding longitudinal reinforcement ratio on column ductility is illustrated by the followed example.

EXAMPLE

Consider the 400 mm square column, for which the ultimate concrete strain ε_{cu} was estimated for various patterns of hoop reinforcement in the example of section 8.4.2(a), and assume it is reinforced with eight 16 mm bars symmetrically distributed around the perimeter (three bars on each side), resulting in a total reinforcement ratio $\rho_{tot} = 0.01$, that is the minimum required by EC8.

1. For a moderate amount of axial compression $v_d = 0.20$, which is equivalent to $v = N/(A_c f_{ck}) = 0.133$ (v_d refers to $f_{cd} = 10.67$ MPa, while v refers to $f_{ck} = 16.0$ MPa), the following ductility factors may be calculated from the approximate formulae (8.72) if $\rho_1 = \rho_2 = 0.0038$ is introduced (the two intermediate bars are ignored in estimating ρ_1, ρ_2):

   (a) In the case where single 8 mm ties at 200 mm spacing are used, $\varepsilon_{cu} = 0.4\%$ was found in the example of section 8.4.2(a) based on the buckling criterion, thus

$$\mu_\phi = 1.2\,\frac{200\,000}{400}\left[\frac{0.6}{0.133 + 0.0038(1 - 0.5 - 18 \times 0.0038)\,400/16} - 1\right]0.004 = 5.3$$

   It is pointed out that if design, rather than characteristic, material strengths are used in the above equation, the resulting value $\mu_\phi = 3.8$ is 28% lower than the previous one. It is also worth noting that a refined analytical procedure using appropriate constitutive laws for confined concrete and steel (Kappos, 1991, 1993) yields a value $\mu_\phi = 6.2$, which is quire close to that predicted by the approximate formula.

   (b) In the case where three-leg 10 mm hoops at 100 mm are used, $\varepsilon_{cu} = 1.6\%$ was found, resulting in $\mu_\phi = 21.2$, which is larger than the EC8 minimum requirement even for DC 'H' structures (section 8.5.2(b)).

2. For a compressive load corresponding to $v_d = 0.65$, which is the maximum allowed by EC8 for DC 'M' structures, it is $v = 0.43$. Introducing this value in equation (8.72), with $\lambda = 10$, a value of $\mu_\phi = 0.95$ results for 8 mm single ties at 200 mm, which implies a brittle type of failure. A more refined analysis yields a value of $\mu_\phi = 1.8$ in this case, which is about twice the approximate value, but still points to an extremely low ductility situation. On the other hand, for 10 mm three-leg hoops at 100 mm, equation (8.72) yields $\mu_\phi = 3.8$, which is below the minimum EC8 requirement for DC 'L' columns (section 8.5.2(b)).

Consider now that the column is reinforced with eight 22 mm bars, resulting in $\rho_{tot} = 1.9\%$, which is very close to the ρ_{max} specified by EC8, if splicing of all column bars in the critical region is assumed. Introducing $\rho_1 = \rho_2 = 0.0071$ in equation (8.72) yields the following estimates of curvature ductility:

1. For $v_d = 0.20$ ($v = 0.133$), $\mu_\phi = 5.9$ for 8 mm ties at 200 mm ($\varepsilon_{cu} = 0.0049$), and $\mu_\phi = 19.3$ for 10 mm three-leg ties at 100 mm spacing ($\varepsilon_{cu} = 0.016$).
2. For $v_d = 0.65$ ($v = 0.43$), $\mu_\phi = 1.2$ for 8 mm ties at 200 mm, and $\mu_\phi = 3.8$ for 10 mm three-leg ties at 100 mm spacing. Note that the latter is the same value calculated for $\rho_{tot} = 1.0\%$, since for $v > 0.2$, $\lambda = 1.0$ in equation (8.72) and the term involving ρ_1 and ρ_2 vanishes. This implies that for $v > 0.2$ the longitudinal reinforcement ratio does not affect the ductility of symmertrically reinforced sections, which appears to be a rather rough approximation in the light of more refined analytical procedures.

The foregoing example indicates that the combination of high longitudinal reinforcement ratios with high axial loading (close to the EC8 specified maximum), results in unacceptably low ductilities. while for ρ_{tot} close to the minimum requirement, ductility may not be sufficient in the case where high axial loads are combined with inadequate hoop reinforcement (see also section 8.5.2(b)).

(c) Arrangement of longitudinal bars

A minimum of three longitudinal bars should be used at each side of the column, two at the corner and an intermediate one; this enhances the shear resistance of beam–column joints, as will be explained in section 9.2.2.

8.5.2 Design for shear and local ductility

(a) Column critical regions

Similar to the case of beams, shear and local ductility requirements are different within critical regions of colums and outside them. According to EC8, the potential regions for plastic hinge formation (critical regions) in columns are located between the beam (or slab)–column interface and a section at a distance l_{cr} defined as follows (see also Figure 8.32):

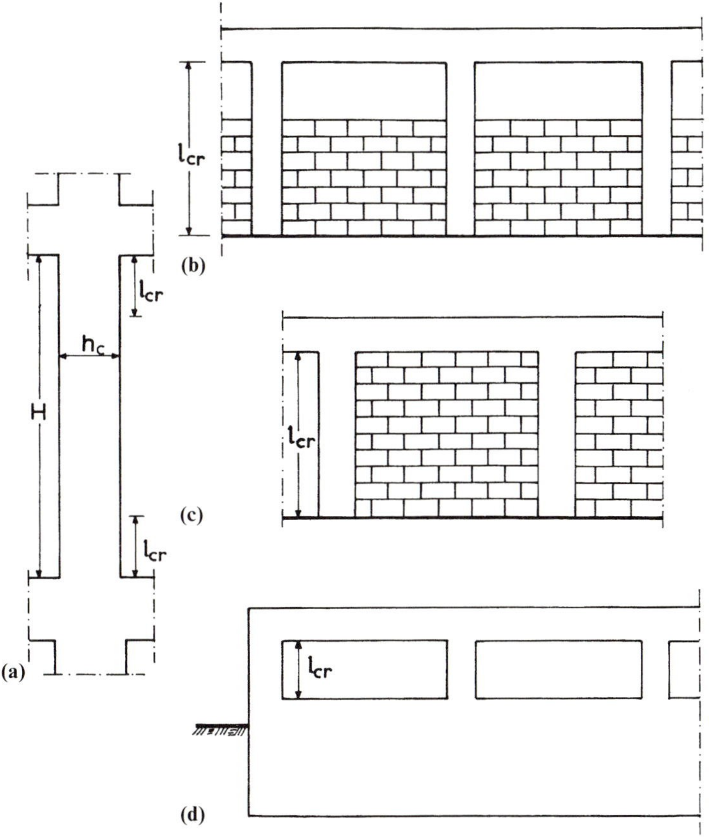

Figure 8.32 column critical regions.

1. In DC 'H' buildings:

$$l_{cr} = \max\{1.5h_c;\ l_c/5;\ 600\ \text{mm}\} \tag{8.83(a)}$$

where h_c is the largest dimension of the column cross-section and l_c the clear height of the column.
2. In DC 'M' buildings:

$$l_{cr} = \max\{1.5h_c;\ l_c/6;\ 450\ \text{mm}\} \tag{8.83(b)}$$

3. In DC 'L' buildings:

$$l_{cr} = \max\{h_c;\ l_c/6;\ 450\ \text{mm}\} \tag{8.83(c)}$$

In addition to the above requirements, the entire height of the column shall be considered as a critical region ($l_{cr} = l_c$), in case $l_c/h_c < 3$, that is whenever a short column situation arises (Figure 8.32(b), (d)).

Furthermore, in R/C buildings with masonry infills, if the infill is located at only one side of the column (Figure 8.32(c)), which is the typical situation with corner columns, $l_{cr} = l_c$ should be considerd.

The critical regions defined by equations (8.83) are those where maximum column moments appear and the possibility of plastic hinge formation exists, although yielding in columns is not permitted by seismic codes; the reasons for this have already been explained in section 8.4.1. The critical lengths (l_{cr}) specified by equations (8.83) are in general adequate in the case of columns located in parts of the building above the ground storey, especially in cases where an explicit capacity design procedure (section 6.1.4(a)) has minimized the inelasticity demand in these columns. However, at the base of the ground storey columns, where the design does permit the formation of plastic hinges, these critical lengths might not be adequate. This is due to factors such as strain-hardening of longitudinal bars, increase in concrete strength due to confinement (with a subsequent increase in column strength), and also the shift of inflection point above the mid-height of the column. Experimental studies have indicated that in such cases it is possible to have a rather brittle type of failure in a column region outside the critical one, wherein hoop reinforcement has been reduced (Paulay, 1986). It is in recognition of such situations that the New Zealand Code (SANZ, 1982) provides for a 50% increase in l_{cr} when high axial loads ($\nu > 0.27$) are present, in which case large quantities of confinement reinforcement are required (section 8.5.2(c)); in addition, a gradual, rather than a sudden increase in hoop spacing is required beyond the critical region. In a similar way, EC8 requires that in the lower two storeys of a building, l_{cr} should be increased by 50% with respect to the values specified by equations (8.83), for DC 'H' and 'M' columns. It is believed that from the practical point of view (ease of construction), it is preferable to consider the entire height of the ground storey columns as the critical region.

With regard to the possibly unfavourable interaction of R/C frames with masonry infill walls, the corresponding code requirements are mainly based on experience from damage caused to columns adjacent to infill walls. In the case where the wall is located at one side of the column only, an adverse effect may be created when the masonry fails in shear along cross-inclined diagonal cracks (see Figure 11.4) or, even more unfavourably, along a nearly horizontal crack close to mid-height (Tassios, 1984); in the latter case the possibility of plastic hinging at column mid-height exists and the least that has to be done is to provide confinement along the entire height ($l_{cr} = l_c$). Furthermore, EC8 points out the increased vulnerability of infill walls at ground storeys and requires the entire height of ground storey columns to be considered as critical. Taking into account the previous point regarding the extent of inelasticity beyond l_{cr} in these columns, and also the EC8 requirement that $l_{cr} = l_c$ when a considerable reduction of infill walls occurs in a ground storey, it appears that in most buildings ground storey columns should be properly confined along their entire height.

The possibility of relatively brittle shear failures in short columns, which may either be created by adjacent masonry infills terminating at a distance from the top of the column (Figure 8.32(b)), or appear in storeys lying partially below the ground (Figure 8.32(d)), has already been discussed in section 8.4.3(d), thus the need for taking $l_{cr} = l_c$ in these elements is rather obvious.

(b) Evaluation of shear resistance

The shear resistance of columns should be calculated according to EC2 (CEN, 1991) and compared with acting shear forces derived using the capacity design procedure described in section 6.1.4(b). The resistance against diagonal cracking is evaluated using equations (8.46)–(8.48). Unlike what is done for beams, the shear attributed to concrete mechanisms (V_{cd}) in columns is non-zero, even within critical regions; note that the axial stress σ_{cp} in equation (8.47) is positive for compressive N_{Sd} and negative for tensile N_{Sd} (in whcih case a reduced value of V_{Sd} is calculated. The favourable effect of compressive axial loading and the unfavourable effect of tensile axial loading on shear capacity of columns has already been discussed in sections 8.4.3(a) and (b).

The resistance against crushing of diagonal concrete struts is checked using equation (8.50), but is hardly ever critical in the case of columns.

(c) Transverse reinforcement requirements

A minimum conventional curvature ductility factor μ_ϕ (or $\mu_{1/r}$ using the EC8 notation) is required in column critical regions, to satisfy the plastic rotation demands compatible with the ductility class and the corresponding q-factor used for design. EC8 defines μ_ϕ as the ratio of the curvature at the post-peak $0.85M_{Rd}$ level, to the curvature at yield, provided the available limiting strains of concrete (ε_{cu}) and steel (ε_{su}) are not exceeded. The notion of curvature ductility factor and the various possible definitions for ε_{cu} and ε_{su} have already been presented in detail in sections 8.2.1 and 8.4.2, where the problems of using strength-based criteria (such as the $0.85\ M_{Rd}$ criterion of EC8) in column sections where spalling of cover concrete does not necessarily coincide with failure have also been discussed.

If for the specified μ_ϕ a concrete strain $\varepsilon_c > 0.0035$ is needed, compensation for the strength loss due to spalling should be achieved by means of adequate confinement of the concrete core. EC8 does not require an explicit evaluation of the available curvature ductility in column critical regions (the designer may, nevertheless, estimate it using one of the approximate or refined procedures presented in the foregoing sections); instead, the code specifies a minimum amount of transverse reinforcement (hoops or spirals), which is to be considered as confinement reinforcement, and is deemed to satisfy the above-mentioned criterion.

The mechanical volumetric ratio of confinement reinforcement prescribed by EC8 for column critical regions is given by the relationship

$$\boxed{\alpha\omega_{wd} \geq k_0\mu_\phi\nu_d\varepsilon_{yd}(0.35A_c/A_0+0.15)-0.035} \quad (8.84)$$

where the symbols ω_{wd}, ν_d, A_c, A_0 have their usual meaning explained previously, $\varepsilon_{yd}=f_{yd}/E_s$ is the yield strain of steel (design value), and α is the coefficient of global effectiveness of confinement, $\alpha=\alpha_n\ \alpha_s$, where the partial coefficients accounting for the hoop pattern (α_n) and the hoop spacing (α_s) are defined by equations (8.74). In the case of circular hoops or spirals $\alpha_n=1$, while in the case of circular spirals the square power in equation (8.74(b)) is dropped. The coefficient k_0 takes into account the different level of confinement (and thus of ductility) required for each ductility class, namely:

1. For DC 'H', the minimum ductility factor $\mu_\phi=13$ and a hoop ratio ω_{wd} calculated by equation (8.84) for $k_0=55$, not to be less than $\omega_{wd,\,min}=0.13$, are deemed to ensure this ductility value.
2. For DC 'M', the minimum $\mu_\phi=9$ and the corresponding values $k_0=60$ and $\omega_{wd,\,min}=0.09$.
3. For DC 'L', the minimum $\mu_\phi=5$ and the corresponding values $k_0=65$ and $\omega_{wd,\,min}=0.05$.

In order to achieve an adequate confinement in column critical regions, EC8 imposes the following additional requirements regarding the transverse reinforcement, whose volumetric ratio is calculated from equation (8.84):

1. The diameter of hoops should not be less than

$$\boxed{d_{bw}=\beta\, d_{bl.\,max}(f_{yld}/f_{ywd})^{1/2}} \quad (8.85)$$

   for DC 'H' and 'M' columns, where $d_{bl,\,max}$ is the maximum diameter of longitudinal bars, f_{yld} and f_{ywd} the design yield strengths of longitudinal and transverse steel, respectively, $\beta=0.40$ for DC 'H' and $\beta=0.35$ for DC 'M' columns. For all ductility classes d_{bw} should not be less than 6 mm.
2. The spacing of hoops should not exceed the following limits:

   (a) for DC 'H' columns

$$s_w=\min\ \{b_0/4;\ 100\text{ mm};\ 5d_{bl}\} \quad (8.86(a))$$

   where b_0 is the minimum dimension of the confined core;
   (b) for DC 'M' columns

$$s_w=\min\ \{b_0/3;\ 150\text{ mm};\ 7d_{bl}\} \quad (8.86(b))$$

   (c) for DC 'L' columns

$$s_w=\min\ \{b_0/2;\ 200\text{ mm};\ 9d_{bl}\} \quad (8.86(c))$$

3. Hoop patterns should be selected so as to maximize the effectively confined area of the concrete core (sections 7.4.1, 7.4.3). For DC 'H' columns only multiple hoop patterns (Figure 8.33) are allowed and the distance between consecutive longitudinal bars restrained by hoop bends or cross-ties should not exceed 150 mm. For DC 'M' columns single hoops are not strictly forbidden,

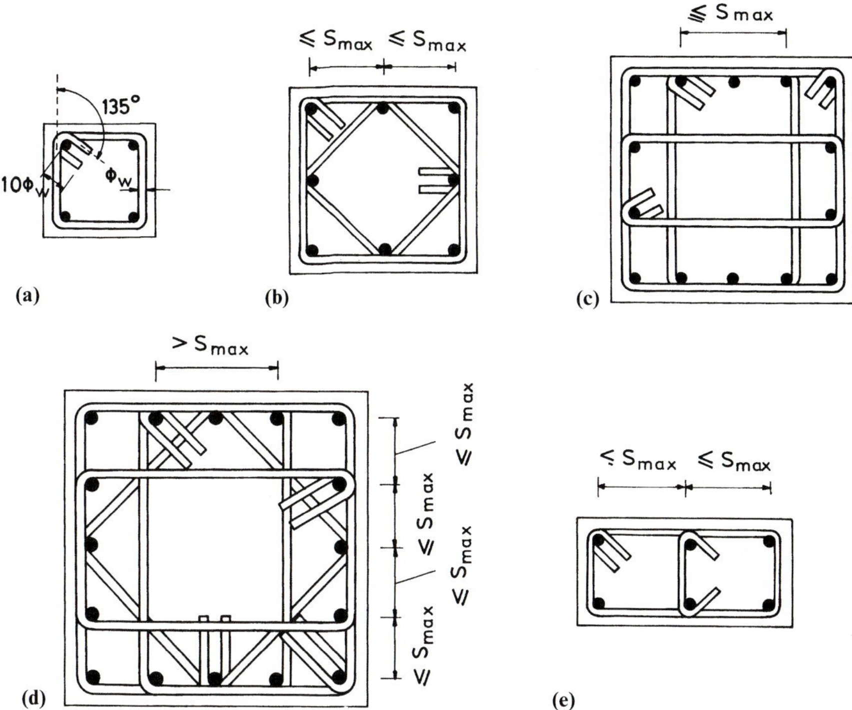

Figure 8.33 Required hoop patterns in critical regions of columns.

but the distance between consecutive restrained bars should not exceed 200 mm, thus the use of single hoops is practically restricted to columns with $b_c \approx 250$ mm (which is the minimum dimension allowed by EC8). Finally, for DC 'L' columns the distance between consecutive restrained longitudinal bars should not exceed 250 mm, which means that single hoops may be used in the common 300 mm square columns. It is pointed out that for DC 'L' columns the transverse reinforcement requirements may be determined according to EC2, rather than EC8, provided that the normalized axial force $v_d \leqslant 0.20$ and the basic value of the behaviour factor $q_0 \leqslant 3.5$; in this case a less strict set of rules applies, the most important differences being perhaps that minimum hoop spacing should not exceed $12d_{bl}$, rather than the $9d_{dl}$ specified in equation (8.86(c)), and of course that the hoop ratio might be less than that required by equation (8.84). Note, however, that DC 'L' buildings in regions of moderate seismicity might be subjected to quite high seismic actions (due to low q-factors), so that shear reinforcement requirements might govern the design of hoops.

The transverse reinforcement requirements resulting from the combined application of all the foregoing rules might be quite high, especially in the case where axial loads close to the permitted $v_{d,\,max}$ are used. The volumetric ratios of hoops resulting from equation (8.84) for DC 'H' and 'M' columns with double hoops are plotted in Figure 8.28 together with the corresponding requirements of the

New Zealand Code (SANZ, 1982). It is pointed out that not only are the EC8 requirements higher than those of NZS3101 (at least for common values of the A_c/A_0 ratio) for most design axial loads, but in addition difficulties in placing the required hoop reinforcement on site may arise. For example in a DC 'H' 300 mm square column the required transverse reinforcement corresponding to the maximum permitted axial loading ($v_d = 0.55$) and to $A_c/A_0 = 1.6$ (25 mm clear cover), is 10 mm double hoops (Figure 8.33(b)) at 65 mm spacing, if S400 steel is used for hoops; for larger cover thickness and/or for lower steel grade the required hoop reinforcement is even more difficult to place on site.

8.5.3 Other design requirements

(a) Geometrical constraints

For reasons of ease of construction, of reducing the slenderness, and also for limiting the adverse effect of spalling of cover concrete, design codes impose certain minimum dimensions for columns. Thus, EC8 requires that the minimum width be 300 mm for DC 'H' columns, 250 mm for DC 'M' columns and 200 mm for DC 'L' columns.

Moreover, unless the inter-storey drift sensitivity coefficient of equation (5.67) $\theta \leqslant 0.1$, the slenderness of columns should be limited by requiring their dimensions not to be smaller than a certain fraction of the larger distance (l_0) between the inflection point ($M = 0$) and the end section of the column (typically the beam or slab–column interface), for bending within a plane parallel to the column dimension considered. Assuming that the critical check is in the direction of the minimum cross-sectional dimension (b):

$b/l_0 \not< 1/8$ for DC 'H' columns
$b/l_0 \not< 1/10$ for DC 'M' and 'L' columns

It is recalled here that very low slenderness is also a problem for columns in seismic regions, as short column type of failure is rather brittle (section 8.4.3(d)).

From the practical point of view, another significant consideration is the reduction in column dimensions due to the placement within their cross-section of various types of tubes (typically gutters or conduits of the plumbing system of the building). Even if in the design of the column account is taken of this reduction in cross-sectional dimensions, the problem still remains of repairing a pipe located inside an R/C column, without causing damage to the R/C element. Good construction practice in earthquake-resistant building does not allow pipes to run along the height of the column, nor to run through the column in potential plastic hinge regions (running through column zones outside the critical one is not forbidden).

(b) Anchorage of reinforcing bars

In calculating the required anchorage length ($l_{b,\ net}$) of longitudinal column bars inside critical regions, the ratio $A_{s,req}/A_{s,prov}$ in EC8 equation (8.56) should be taken as equal to 1, to account for bond degradation during cyclic loading (section 7.6.3).

The anchoring of column bars inside a beam-column joint region (a situation which typically arises at the top storey of buildings), should be done in the way shown in Figure 8.34, by extending the bars as close as practicable to the top of the beam and bending them at 90° angles towards the interior of the joint. The anchorage length ($l_{b,net}$) is measured from a point at a distance equal to $k_b\, d_{bl}$ from the bottom face of the beam, where

$k_b = 10$ for DC 'H' columns
$k_b = 5$ for DC 'M' columns
$k_b = 0$ for DC 'L' columns

The foregoing k_b values are specified for taking into account the different degree of yield penetration in the joint, due to different levels of imposed inelastic deformations anticipated for each ductility class. EC8 is not very clear with regard to anchorage requirements at corner joints; a conservative interpretation of the foregoing requirement for such a joint is shown in Figure 8.34(b).

If the axial load of the column, resulting from the seismic load combination, is tensile, the anchorage length defined by equation (8.56) should be increased by 50% according to EC8; the reason for this increase is the precipitation of bond degradation during cyclic loading when tensile axial loading is present (sections 7.6.3 and 8.4.3(b)).

With regard to the anchorage of transverse reinforcement, the hoops required in column critical regions are either spirals or closed stirrups with 135° hooks as shown in Figure 8.33, with $10d_{bw}$ extensions. For columns with a rectangular cross-section ($b < h$) it is usually more convenient to combine closed stirrups with cross-ties, as shown in Figure 8.34 (e) Experimental studies (Moehle and Cavanagh, 1985) have indicated that the practically convenient detail of using 90° hooks at one end of these cross-ties, does not lead to an inferior seismic performance of the column, provided that the 90° hooks are located in alternating faces of consecutive cross-sections. However, other studies (Sheikh and Yeh, 1990) have shown that for high levels of axial loading these 90° hooks open at large deformations and offer no confinement.

(c) Splicing of bars

Lapped splices within critical regions are permitted by EC8 for DC 'M' and 'L' columns, while it is suggested that they should be avoided in DC 'H' buildings.

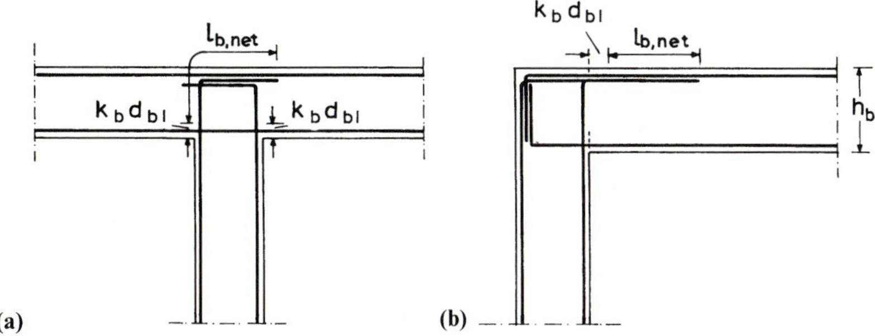

Figure 8.34 Anchoring of column bars in an interior and an exterior beam–column joint.

The required lap length is calculated according to EC2, with the additional rules given in section 8.3.3(c). In the rather common case that all column bars are spliced at the same sections (the top of the floor slab), it is quite probable that $\alpha_l = 2.0$ should be used in equation (8.57). Assuming grade 400 steel and class 20 concrete, the resulting lap length is equal to $75d_{b_l}$; for a moderately large bar diameter $d_{b_l} =$ 22 mm, the required lap length is 1.65 m, which is more than half the clear height of a typical building column.

The foregoing requirements regarding the splicing of column bars appear to be not only excessive, but also an inefficient approach to the problem. Experimental studies (Gergely, 1977; Paulay, 1982) have shown that in the case of cyclic loading, the main factors affecting the behaviour of lapped splices are the diameter and the spacing of transverse reinforcement. As already mentioned in section 7.6.3, bond degrades under cyclic loading conditions, hence yield penetrates from one or both sides of the lapped splice towards its interior. It follows that the length along which transfer of forces between spliced bars takes place is gradually reduced (unzipping effect) and an increase in the lap length does not lead to marked improvement in the behaviour of the splice (Paulay, 1982). For this type of loading the efficient approach to the problem consists in improving the bond conditions, which can be achieved by appropriate transverse reinforcement.

The hoop steel requirements within the lap length of longitudinal bars in columns, spliced at the same location, are calculated by the following EC8 formula:

$$A_{sw} = s\frac{d_{bl}}{50} f_{yld}/f_{ywd} \tag{8.87}$$

where A_{sw} is the area of one leg of the hoop, d_{bl} the diameter of the spliced bar and the other symbols have their usual meaning. It is reminded that the spacing of hoops along the laps is limited by equation (8.58) which, nevertheless, is usually less critical than the requirements of confinement reinforcement resulting from equation (8.84).

The seismic performance of a lapped splice depends on the level of the imposed inelastic deformations, thus on the ductility class. On the basis of pertinent test data (Paulay, 1982), Priestley and Park (1987) suggest that lapped splices could be located within the critical region, when the overall displacement ductility of the building does not exceed a value $\mu_\delta \approx 3$, which may be roughly corresponded to DC 'M' R/C buildings. The presence of closely spaced hoops improves the performance of lapped splices located in plastic hinge regions, but cannot preclude the concentration of plastic deformations in the zone of the splice adjacent to the joint,thus leading to failure of this zone whenever the imposed ductility is relatively high, $\mu_\delta > 3$ (Paulay, 1982).

Splicing by welding is forbidden by EC8 within column critical regions, as there is danger of embrittlement of the steel adjacent to the weld, unless preheating is carried out in a very rigorous manner (Paulay and Priestley, 1992). Splicing by mechanical couplers (annular sleeves) is allowed in columns and walls if they are covered by appropriate testing under cyclic loading compatible with the ductility class selected. It is pointed out that a tension test involving a single splice is not adequate to demonstrate satisfactory seismic performance; instead, realistically sized column specimens with several bars spliced by mechanical couplers should be tested.

8.6 DESIGN EXAMPLE

8.6.1 General data and analysis procedure

(a) Design action

The 10-storey plane frame of Figure 8.35 has been designed according to EC8 for DC 'M' and a design ground acceleration of $0.25g$. The estimated fundamental natural period of the structure using the formula of Annex C of Part 1–2 of the code (see also section 5.4.4) is

$$T_1 = 0.075h^{3/4} = 0.075 \times 30.0^{3/4} = 0.96 \text{ s}$$

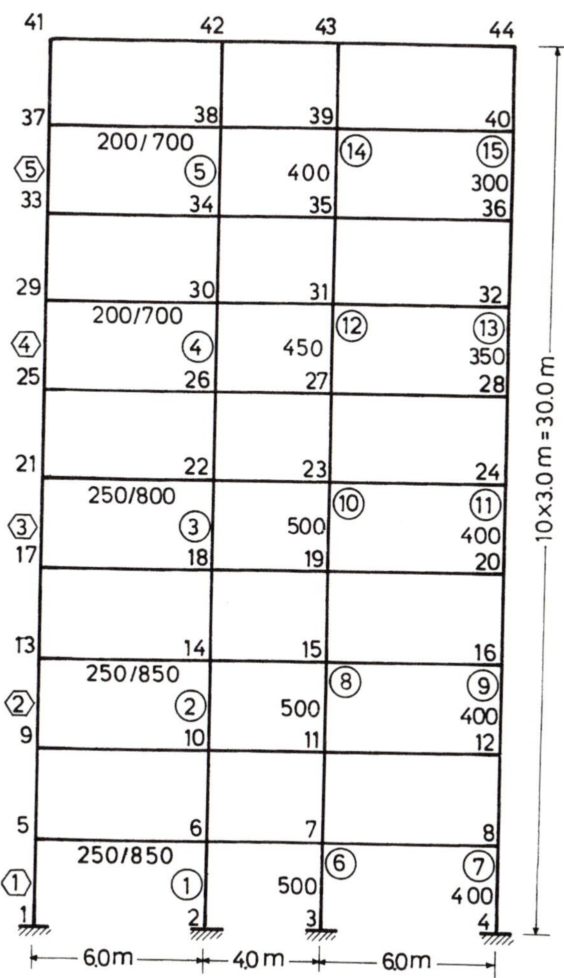

Figure 8.35 Geometric data for the frame structure studied.

while the behaviour factor (section 4.4.6) is

$$q = q_0 k_D k_R k_w = 5.0 \times 0.75 \times 1.0 \times 1.0 = 3.75$$

that is a regular DC 'M' R/C frame is assumed.

The design base shear coefficient, assuming subsoil class A (stiff deposits or rock) is (section 4.4.5)

$$S_d(T_1) = \alpha S \frac{\beta_0}{q} \left(\frac{T_C}{T} \right) k_{d1} = 0.25 \times 1.0 \times \frac{2.5}{3.75} \left(\frac{0.40}{0.96} \right)^{2/3} = 0.093$$

The total gravity loading to be taken into account for determining the seismic action ($\Sigma G_k + \Sigma \psi_E Q_k$, as specified in section 4.5) was found to be 4011 kN, assuming frames are spaced at 3.0 m centres in the direction orthogonal to the one under consideration. Note that the structure is similar to that examined in section 6.3, except that the centrally located wall is replaced here by the interior columns and the central span of the beam. The design seismic action (base shear) is therefore

$$V_b = S_d(T_1) W = 0.093 \times 4011 = 373 \text{ kN}$$

As the conditions for applying the simplified modal response spectrum analysis (section 5.4.1) are obviously satisfied (planar structure with $T_1 < 4T_c = 1.6$ s), the frame is analysed for horizontal forces derived from the equation (section 5.4.3)

$$H_i = V_b \frac{z_i W_i}{\Sigma z_j W_j}$$

where z_i, z_j are the distances of storey masses m_i, m_j from the foundation level (full fixity to the foundation is assumed for simplicity).

(b) Structural analysis

The analysis has been carried out on the basis of R/C member stiffnesses estimated taking the effect of cracking into account as suggested by Kappos (1986b) and by Paulay and Priestley (1992), that is $EI_{ef} \approx 0.4 EI_g$ for beams, $EI_{ef} \approx 0.8 EI_g$ for columns). The calculated displacements should satisfy the requirement (section 6.2.3)

$$q d_{ei} / h_i < 0.004 v = 0.004 \times 2.0 = 0.008$$

where $q d_{ei}/h_i$ are the estimated inter-storey drift ratios for the design earthquake (d_{ei} are the inter-storey displacements resulting from the application of forces F_i), and $v = 2.0$ corresponds to importance category III (ordinary buildings). The calculated drift ratios ranged from 0.003 to 0.006, the largest value corresponding to the seventh storey (where all member cross-sections are reduced with respect to the storey belowm see Figure 8.35).

(c) Material characteristics

The materials used in the design were the following:

1. Strength class C20/25 concrete according to EC2, with a design compressive strength $f_{cd} = f_{ck}/\gamma_c = 20/1.5 = 13.33$ MPa, a Young's modulus $E_{cm} = 29\,000$ MPa and a design shear stress $\tau_{Rd} = 0.26$ MPa.

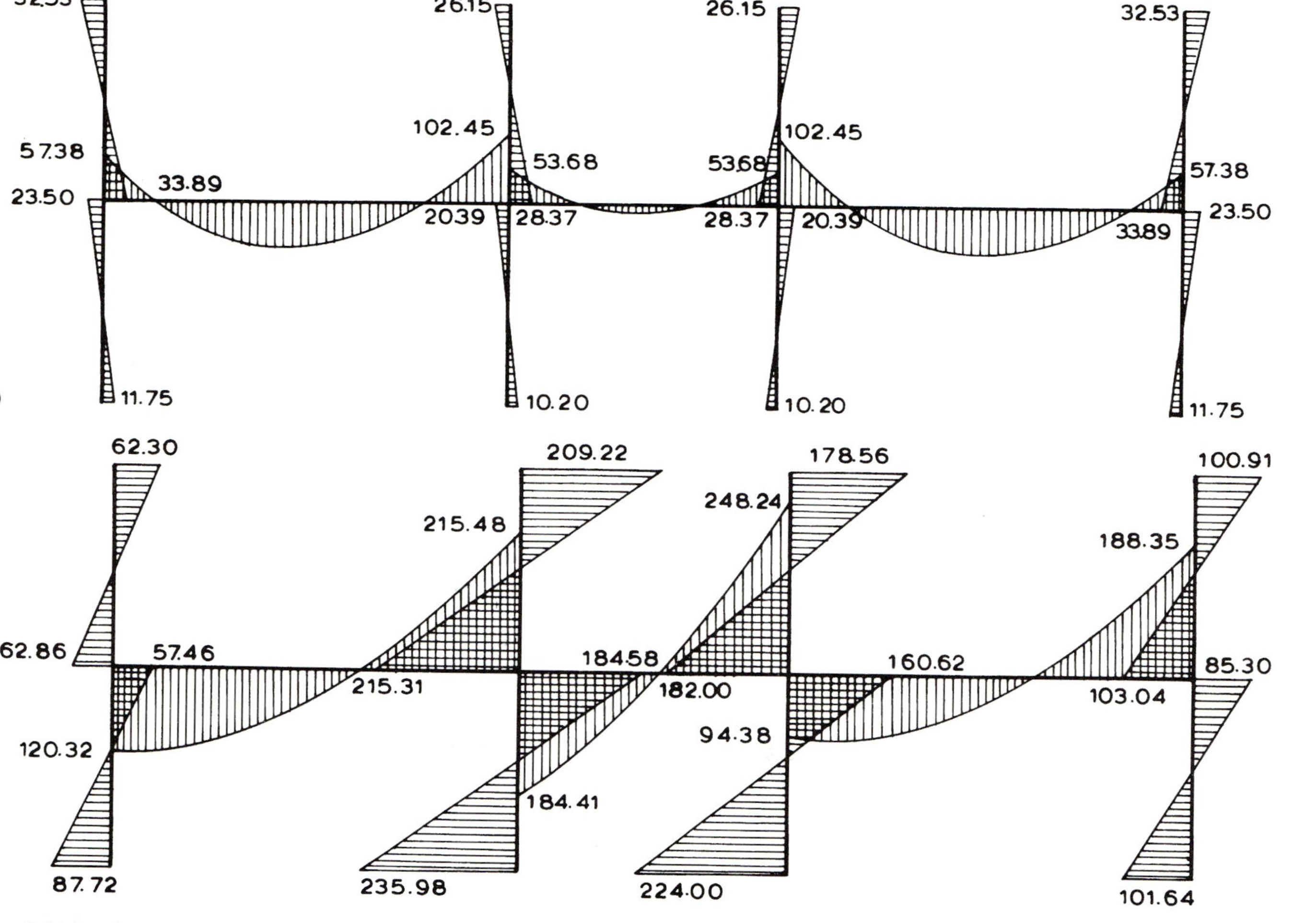

Figure 8.36 Bending moment diagrams at the bottom of the frame for the following load combinations: (a) $E_d = E(1.35G+1.5Q)$; (b) $E_d = E(G+0.3Q+H)$.

2. Grade 400 steel (deformed bars) with a design tensile strength $f_{yd}=f_{yk}/\gamma_s=400/1.15=347.8$MPa. The same grade was used for both the longitudinal and the transverse reinforcement, although grade 220 steel might have resulted in a more economical solution with regard to hoop reinforcement in some of the beams.

In the following, details of the design of beams and columns in the ground storey are given to demonstrate the application of the design rules presented in sections 8.3 and 8.5. It is pointed out that explanations and comments, not normally part of an actual design project, are included to facilitate the understanding of the procedure.

8.6.2 Design of beams

(a) Longitudinal reinforcement requirement

The bending moment diagram resulting from the seismic load combination ($G+0.3Q+H$) for the bottom subframe (ground storey beam with the columns below and above it, which is the common substructure considered for gravity load calculations) is shown in Figure 8.36. This combination results in bending moments for the beam approximately twice those resulting from the gravity load combination $1.35G+1.50Q$ (Figure 8.36(a)) which is ignored in the following; it is pointed out, however, that at the top storey beam the maximum moment from the seismic load combination is only 8% higher than that resulting from the gravity loading.

Exterior beam support The maximum negative and positive moments from the analysis are

$$M^-_{Sd}=188.4 \text{ kNm} \quad M^+_{Sd}=120.3 \text{ kN m}$$

For the negative moment the beam works as a rectangular section, thus the corresponding normalized action (see CEB, 1982) is (for $d\approx h\text{-}40=810$ mm)

$$\mu_{Sd1}=\frac{M^-_{Sd}}{b_w d^2 f_{cd}}=\frac{188.4}{0.25\times 0.81^2\times 13\,333}=0.086$$

From the CEB (1982) design aid for flexural design, reproduced in Table 8.2, the mechanical ratio of steel (ω_1) corresponding to the above value of μ_{Sd} is found to be 0.092, thus the required area of longitudinal steel is

$$A_{s1}=\omega_1 b_w d f_{cd}/f_{yd}=0.092\times 25\times 81\times 13.3/347.8=7.1 \text{ cm}^2$$

For the positive moment the beam works as a T-beam, with an effective width whcih may be estimated according to the EC2 approximate formula

$$b\approx b_w+\frac{1}{5}\, l_0=0.25+\frac{1}{5}\times 0.6\times 6.0=0.97 \text{ m}$$

where the distance l_0 between points of zero moment is estimated for the bending moment diagram corresponding to the seismic load combination (Figure 8.36(b)) and not to the gravity load moments on which the EC2 recommendation is based.

Table 8.2. Design table for rectangular sections subjected to bending moment and axial force, without compression reinforcement (CEB, 1982)

μ_{sd}	ω	$\xi = \frac{x}{d}$	$\zeta = \frac{z}{d}$	ε_c(‰)	ε_s(‰)	σ_{sd} (MPa) S220	S400	S500
0.01	0.0102	0.050	0.983	−0.52	10.00	191	348	435
0.02	0.0205	0.072	0.975	−0.77	10.00			
0.03	0.0310	0.089	0.969	−0.98	10.00			
0.04	0.0415	0.104	0.963	−1.16	10.00			
0.05	0.0522	0.118	0.958	−1.34	10.00			
0.06	0.0630	0.131	0.953	−1.51	10.00			
0.07	0.739	0.144	0.947	−1.68	10.00			
0.08	0.0849	0.156	0.942	−1.85	10.00			
0.09	0.0961	0.168	0.937	−2.03	10.00			
0.10	0.1074	0.181	0.931	−2.21	10.00			
0.11	0.119	0.194	0.925	−2.40	10.00			
0.12	0.131	0.207	0.919	−2.60	10.00			
0.13	0.143	0.220	0.912	−2.82	10.00			
0.14	0.155	0.233	0.905	−3.04	10.00			
0.15	0.167	0.247	0.899	−3.27	10.00			
0.16	0.179	0.261	0.892	−3.50	9.92			
0.17	0.192	0.280	0.884	−3.50	9.02			
0.18	0.206	0.299	0.878	−3.50	8.22			
0.19	0.219	0.318	0.868	−3.50	7.50			
0.20	0.233	0.338	0.859	−3.50	6.85			
0.21	0.247	0.359	0.851	−3.50	6.26			
0.22	0.261	0.380	0.842	−3.50	5.72			
0.23	0.276	0.401	0.833	−3.50	5.22			
0.24	0.291	0.423	0.824	−3.50	4.77			
0.25	0.307	0.446	8.814	−3.50	4.35			
0.26	0.323	0.470	0.805	−3.50	3.95			
0.27	0.340	0.494	0.795	−3.50	3.59			
0.28	0.357	0.519	0.784	−3.50	3.24			
0.29	0.375	0.545	0.773	−3.50	2.92			
0.30	0.394	0.572	0.762	−3.50	2.62			
0.31	0.413	0.600	0.750	−3.50	2.33			435
0.32	0.434	0.630	0.738	−3.50	2.05			410
0.33	0.455	0.662	0.725	−3.50	1.79		348	358
0.34	0.478	0.695	0.711	−3.50	1.54		308	308
0.35	0.503	0.731	0.696	−3.50	1.29		258	258
0.36	0.529	0.770	0.680	−3.40	1.05	191	210	210
0.37	0.559	0.812	0.662	−3.50	0.81	162	162	162
0.38	0.592	0.860	0.642	−3.50	0.57	104	104	104
0.39	0.630	0.915	0.619	−3.50	0.32	64	64	64

More refined procedures are available for estimating the effective width of T-beams, however the subject has not been well studied as far as seismic loading is concerned (Kappos, 1986b). Using the well-known approximate formula for T-beams with relatively large flanges (also included in the 1982 CEB Manual), the required bottom reinforcement is

$$A_{s2}=\frac{1}{f_{yd}}\left(\frac{M^{+}_{Sd}}{d-h_{ef}/2}\right)=\frac{1}{347.8}\left(\frac{120.3\times 10^{3}}{0.81-0.10/2}\right)=455\ \text{mm}^2=4.6\ \text{cm}^2$$

The longitudinal bars to be selected should fulfil the EC8 requirement regarding the limitation of diameter of beam bars anchored along beam column joints (section 9.2.4); the critical check is for the interior joints wherein

$$d_{bl}\leqslant 4.5(f_{ctm}/f_{yd})\ (1+0.8\nu_d)\ b_c=4.5\ (2.2/347.8)\ (1+0.8\nu_d)b_c$$

For the ground storey the normalized column axial loading due to the seismic load combination is $\nu_d=0.298$ for the 500 mm square interior column, thus $d_{bl}\leqslant 17.6$ mm. A similar requirement results for reinforcement passing through the exterior column, hence for the section under consideration.

Given the above, four 16 mm bars (total area of 8.0 cm^2) are selected as top reinforcement and three 16 mm bars (6.0 cm^2) as bottom reinforcement. The corresponding ratios ($\rho=A_s/bd$) are $\rho_1=0.39\%$ and $\rho_2=0.30\%$ which are both higher than the minimum ratio required by EC8, $\rho_{min}=0.28\%$ (Table 8.1, section 8.3.1(a)). Besides, they are lower than the maximum ratio for DC 'M' beams

$$\rho_{max}=0.65\ \frac{f_{cd}}{f_{yd}}\ \frac{\rho_2}{\rho_1}+0.0015=0.65\ \frac{13.3}{347.8}\ \frac{0.30}{0.39}+0.0015=0.0206$$

Interior beam support Maximum moments:

$$M^{-}_{Sd}=248.2\ \text{kN m}\qquad M^{+}_{Sd}=184.4\ \text{kN m}$$

$$\mu_{Sd1}=\frac{248.2}{0.25\times 0.81^2\times 13\,333}=0.113$$

hence $\omega_1=0.123$ and

$$A_{s1}=0.123\times 25\times 81\times\frac{13.3}{347.8}=9.5\ \text{cm}^2$$

Four 16 mm and one 14 mm diameter bars are selected as top reinforement (total area of 9.6 cm^2), resulting in $\rho_1=0.47\%$.

The effective width on the basis of the seismic load combination (Figure 8.36(b)) is

$$b\approx 0.25+(1/5)\times 0.5\times 4.0=0.65\ \text{m}$$

The normalized positive moment is

$$\mu_{Sd2}=\frac{M^{+}_{Sd}}{bd^2f_{cd}}=\frac{184.4}{0.65\times 0.81^2\times 13\,333}=0.032$$

For $h_f/d=100/810=0.12$ and $b/b_w=650/250=2.6$, the mechanical ratio of reinforcement (ω_2) is found from the CEB (1982) design aid for T-beams, reproduced here as Table 8.3, $\omega_2=0.034$, hence the required bottom reinforcement is

$$A_{s2}=\omega_2 bdf_{cd}/f_{yd}=0.034\times 65\times 81\times 13.3/347.8=6.8\ \text{cm}^2$$

Three 16 mm and one 14 mm diameter bars are selected as bottom reinforcement (total area of 7.6 cm^2). It is pointed out that it is not recommended to use Table 8.3 for the design of the exterior support, since the corresponding μ_{Sd2} is 0.014, which is lower than the minimum μ_{Sd2} value included in this table, thus an uneconomical design may arise.

The arrangement of longitudinal bars along the beam is determined on the basis of the modified envelope of bending moments (to account for the effect of shear on flexural reinforcement) and the corresponding envelope of flexural resistances, is shown in Figure 8.37. The horizontal displacement of the M_{Sd} envelope in accordance with EC2 is

$$a_1=z(1-\cot a)/2 \not< 0$$

where $\cot a=0$ for vertical stirrups and $z\approx 0.9d$, thus

$$a_1=0.9\times 810(1–0)/2=365\ \text{mm}$$

(b) Transverse reinforcement requirements

The design shear forces for the beam are derived directly from analysis, as no capacity procedure for shear is required by EC8 for DC 'M' beams (section 6.1.3).

The maximum shear force $V_{Sd}=115.6$ kN at the exterior support is calculated for the seismic load combination (for the gravity load combination $V_{sd}=105.4$, that is 9% lower). The negative shear corresponding to $V_{Sd}=115.6$ is only 0.22 kN, that is

$$\zeta=V_{Smin}/V_{Smax}=-0.22/115.6=-0.002>-0.5$$

hence the EC2 design procedure for shear (section 8.3.2(a)) is followed. The shear resistance of the beam in the critical region is

$$V_{Rd3}=V_{cd}+V_{wd}$$

where

$$V_{cd}=0.4[\tau_{Rd}k(1.2+40\rho_1)+0.15\sigma_{cp}]b_w d$$
$$=0.4[260.0\times 1.0\ (1.2+40\times 0.0039)+0.15\times 0.0]0.25\times 0.81=28.6\ \text{kN}$$

and

$$V_{wd}=\frac{A_{sw}}{s}\,(0.9d)\,f_{ywd}$$

The minimum hoop diameter $d_{bw}=6$ mm and the maximum spacing (in the critical regions)

$$s_w=\min\ \{h_w/4;\ 24d_{bw};\ 7d_{bl};\ 200\ \text{mm}\}$$
$$=\min\ \{850/4;\ 24\times 6;\ 7\times 16;\ 200\}=\min\ \{212,\ 144,\ 112,\ 200\}=112\ \text{mm}$$

Thus 6 mm two-leg hoops at 110 mm are selected, resulting in

$$V_{wd}=2\times\frac{28.3}{110}\times 0.9\times 810\times 347.8\times 10^{-3}=130.5\ \text{kN}$$

Table 8.3 Desgn table for T-beam sections subjected to bending moment and axial force (CEB, 1982)

		$h_f/d=0.05$ 1000ω for $b/b_w=$					$h_f/d=0.10$ 1000ω for $b/b_w=$					$h_f/d=0.15$ 1000ω for $b/b_w=$				
	μ_{Sd}	10	5	3	2	1	10	5	3	2	1	10	5	3	2	1
	0.02	20	20	20	20	21	21	21	21	21	21	21	21	21	21	21
	0.04	41	41	41	41	42	42	42	42	42	42	42	42	42	42	42
	0.06	65	63	63	63	63	63	63	63	63	63	63	63	63	63	63
	0.08		91	87	85	84	84	84	85	85	85	85	85	85	85	85
	0.10			114	110	107	111	108	108	107	107	107	107	107	107	107
	0.12			146	137	131	138		134	132	131	130	130	130	130	131
	0.14				166	155			164	158	155		157	155	155	155
	0.16				199	179			200	188	179		192	184	182	179
	0.18				237	206				220	206			219	211	206
	0.20					233				259	233				244	233
	0.22					261					261				283	261
	0.24					291					291					291
	0.26					323					323					323
	0.28					357					357					357
	0.30					394					394					394
	0.32					434					434					434
S220 }	μ_{lim}	0.070	0.099	0.138	0.186	0.330	0.106	0.131	0.164	0.205	0.330	0.139	0.160	0.189	0.224	0.330
S400 }	$1000\omega_{lim}$	84	125	180	249	455	122	159	208	270	455	160	192	237	291	455
S500	μ_{lim}	0.069	0.096	0.133	0.178	0.316	0.104	0.128	0.159	0.198	0.316	0.138	0.157	0.184	0.217	0.316
	$1000\omega_{lim}$	81	119	170	233	424	119	153	198	254	424	157	187	226	276	424

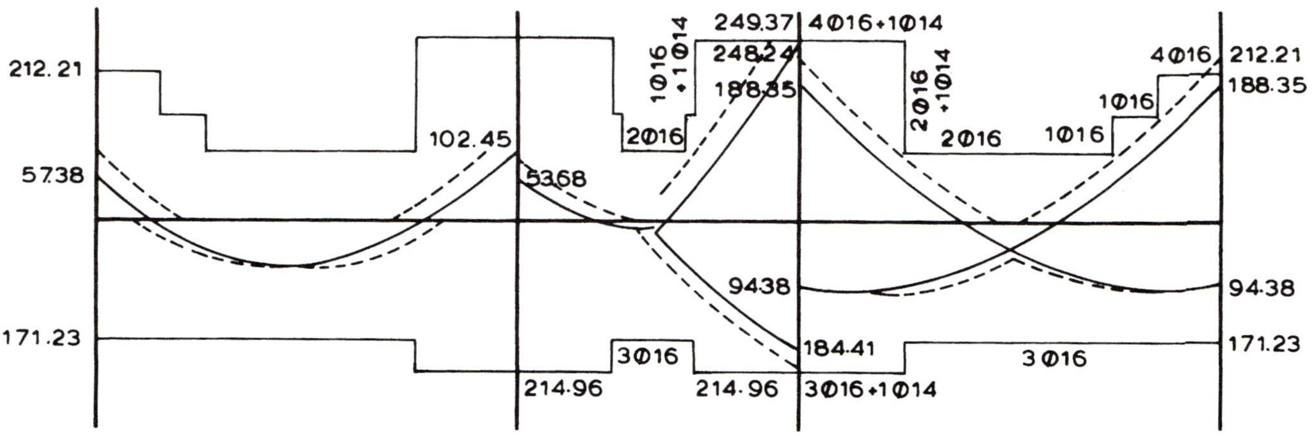

Figure 8.37 Envelope of moments of resistance for the ground storey beam.

hence

$$V_{Rd3} = 28.6 + 130.5 = 159.1 > V_{Sd} = 115.6$$

In addition it has to be checked that V_{Sd} does not exceed

$$V_{Rd2} = 0.5\ (0.7 - f_{ck}/200) f_{cd} b_w 0.9d\ (1 + \cot a)$$
$$= 0.5\ (0.7\text{–}20/200)\ 13\ 333 \times 0.25 \times 0.9 \times 0.81\ (1+0) = 729.0\ \text{kN}$$

As expected (section 8.3.2(a)), V_{Rd2} is much larger than V_{Sd}.

With regard to the interior support, the critical check is at the side of the central span, where the maximum shear from the seismic load combination is $V_{Sd} = 140.3$ kN, which is 156% higher than the shear derived for the 1.35G+1.50Q combination. The substantially different analogy between seismic and gravity loading in the exterior and the interior span of the beam is pointed out; this difference is attributed to the different slendernesses *l*/*h* of the two spans. The ratio ζ in the interior span is

$$\zeta = V_{Smin}/V_{Smax} = -76.0/140.3 = -0.54 < -0.5$$

hence a high degree of shear reversal is expected and the following check has to be carried out:

$$\beta_1(2+\zeta)\tau_{Rd} b_w d = 4(2\text{–}0.54)260.0 \times 0.25 \times 0.81 = 307.5 > |V_{Smax}| = 140.3$$

Therefore shear can be resisted solely by hoops. As already calculated, the minimum required 6 mm hoops at 110 mm carry a shear $V_{wd} = 130.5$ kN; the shear resisted by the other mechanisms is

$$V_{cd} = 0.4[260.0 \times 10(1.2 + 40 \times 0.0047)] \times 0.25 \times 0.81 = 29.2\ \text{kN}$$

thus

$$V_{Rd3} = 29.2 + 130.5 = 159.7 > V_{Sd} = 140.3$$

It is seen that all beam critical regions can develop a shear resistance in excess of the corresponding requirements if the minimum EC8 provisions concerning the diameter and the spacing of hoops are satisfied, provided grade 400 steel is used for the transverse reinforcement; in case S220 steel is preferred (for ease of construction) the diameter has to be increased and/or the spacing be reduced. Outside the critical regions, it is found (using the same procedure as above, but without the 0.4 reduction factor in V_{Sd}) that 6 mm ties at 300 mm (the maximum spacing allowed by EC2) are able to resist the required shear force.

(c) Detailing requirements

The cross-section of the beam satisfies the requirement

$$b_w = 0.25\ \text{m} > 0.25h = 0.25 \times 0.85 = 0.21\ \text{m}$$

The arrangement of the longitudinal and the transverse reinforcement along the beam is shown in Figure 8.38.

The anchorage lengths of the longitudinal bars have been calculated from equation (8.56); as an example, the required length for the 16 mm bars in the interior

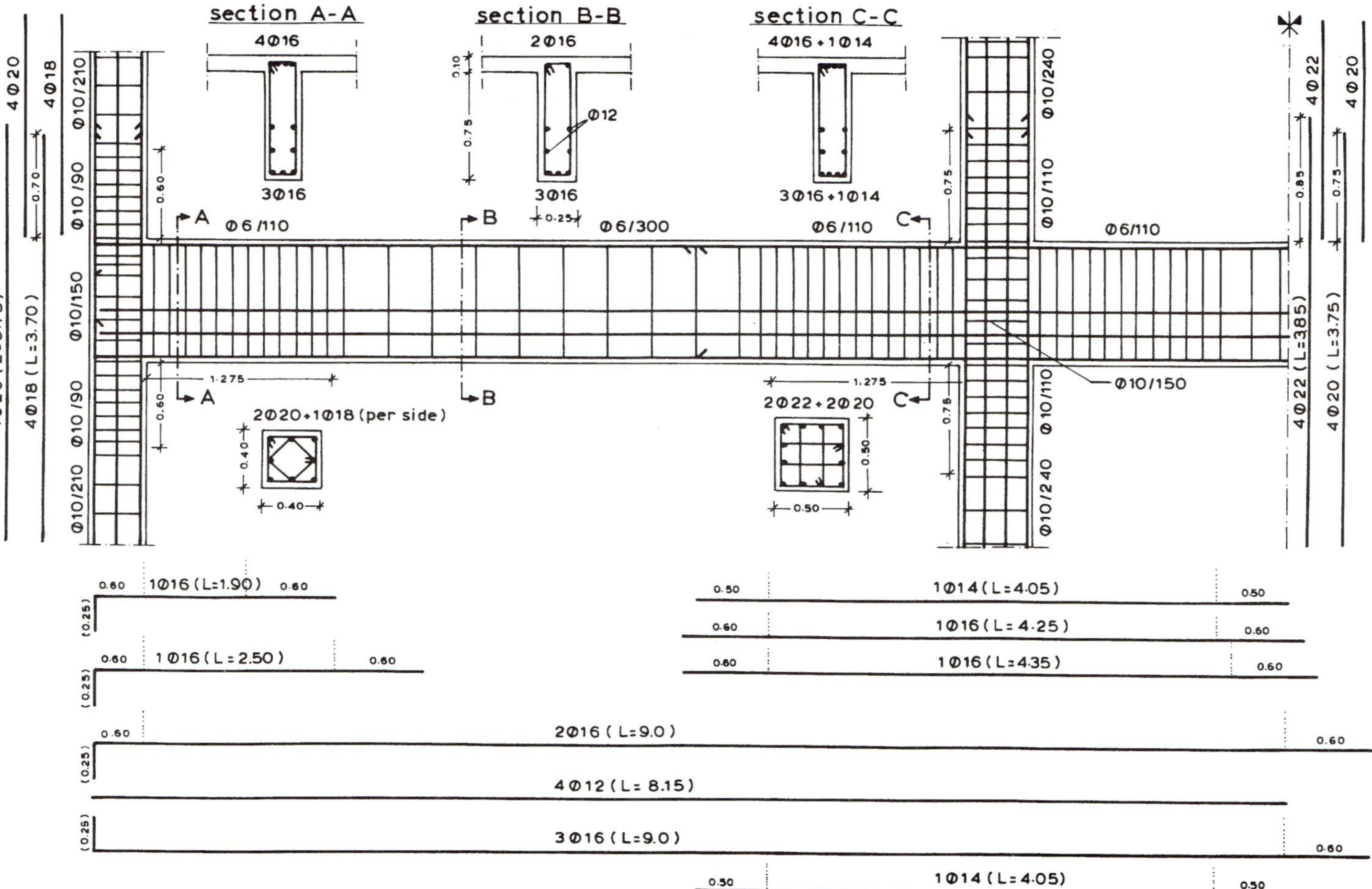

Figure 8.38 Arrangement of longitudinal and transverse reinforcement at the bottom of the frame (small inclined hooks merely indicate the curtailing position of longitudinal bars).

supports is

$$l_{b,net} = \alpha_a \frac{d_{bl}}{4} \frac{f_{yd}}{f_{bd}} \frac{A_{s,req}}{A_{s,prov}} = 1.0 \times \frac{16}{4} \times \frac{347.8}{2.3} \times \frac{9.5}{9.6} \approx 600 \text{ mm} \quad (=37d_{bl})$$

The anchorage lengths of bars at the exterior supports are increased by $5d_{bl}$, as required by EC8 for DC 'M' beams.

Lapped splices are located in the middle of the interior span; the required lap length is equal to twice the anchorage length $l_{b,net}$ calculated previously ($\alpha_l = 2.0$ has to be used in equation (8.57)).

The length of the critical regions is equal to

$$l_{cr} = 1.5h_b = 1.5 \times 0.85 = 1.28 \text{ m}$$

measured from each column face towards the interior of the span. Hoop reinforcement within the critical regions consists of closed stirrups with 135° hooks and an extension of $10d_{bw}$. Outside the critical regions stirrups with 90° hooks are allowed (EC2 applies in this case), while the required extension is the same as for hoops ($10d_{bw}$).

With regard to curtailment of longitudinal bars it is pointed out that EC8 requires that no bars be terminated within distances l_{cr} on both sides of interior joints; this provision has been conservatively interpreted here by measuring the required anchorage length of such bars beyond the end of the critical zone (Figure 8.37). Finally, is is noted that the two 16 mm bars which are continuous along the entire beam length exceed one-quarter of the maximum top reinforcement at supports, as required by EC8.

8.6.3 Design of columns

(a) Longitudinal reinforcement requirements

The critical combination of actions for the columns is obviously the seismic one, as clearly shown in Figure 8.36. Design moments for the seismic combination are more than an order of magnitude higher than those resulting from the gravity load combination in the interior columns at the bottom of the building. Nevertheless the differences decrease with height, and at the top storey the maximum seismic moment is only 65% higher than the gravity load moment. At the exterior columns, design moments from the seismic load combination are 332% higher at the ground storey, but only 13% higher at the top storey. To simplify the construction somewhat, column (as well as beam) sections are kept constant every two storeys.

Exterior columns The critical section of the 400 mm square exterior column (of the subframe shown in Figure 8.35) is at its base, where the following M–N pairs were found:

$$M_{Sd} = 101.6 \text{ kN m} \quad N_{Sd} = -1199.8 \text{ kN ('leeward' column)}$$

$$M_{Sd} = 87.7 \text{ kN m} \quad N_{Sd} = -464.0 \text{ kN ('windward' column)}$$

The normalized actions resulting from the first pair are

$$\mu = \frac{M_{Sd}}{bh^2 f_{cd}} = \frac{101.6}{0.4^3 \times 13\ 333} = 0.119$$

$$\nu = \frac{N_{Sd}}{bh f_{cd}} = \frac{-1199.0}{0.4^2 \times 13\ 333} = -0.562$$

From the CEB (1982) design aid for columns with S400 steel and $d_1/h = 0.10$, reproduced here as Figure 8.39, the total mechanical reinforcement ratio corresponding to the previous normalized actions is $\omega_{tot} = 0.09$. It is pointed out that the normalized axial loading is less than the limit ($\nu_d = 0.65$) allowed by EC8 for DC 'M' columns. For the second pair of *M–N*, the corresponding normalized actions are

$$\mu = 0.103 \quad \nu = -0.217$$

and the corresponding reinforcement ratio (from Figure 8.39) $\omega_{tot} = 0.08$, that is less than the previously found value. The minimum total reinforcement ratio is $\rho_{tot} = 1\%$ (section 8.5.1(b)), corresponding to a mechanical ratio

$$\omega_{tot,min} = \rho_{tot}(f_{yd}/f_{cd}) = 0.01(347.8/13.3) = 0.261$$

which is larger than the previously found values. Note, however, that if an attempt is made to arrive at a more economical solution by reducing the column section to, say, 350 mm, the resulting ν_d is equal to 0.73 which is higher than the maximum limit of 0.65, therefore such a solution is not admissible.

The required reinforcement can now be calculated:

$$A_{s1} = A_{s2} = \tfrac{1}{2}\ \omega_{tot}\ bh\ (f_{cd}/f_{yd}) = 0.5 \times 0.261 \times 40^2\ (13.3/347.8) = 8.0\ \text{cm}^2$$

Two 20 mm bars at the corners and one intermediate 18 mm bar are selected (total area per side equal to 8.8 cm^2) for the exterior column, resulting in a total reinforcement ratio $\rho_{tot} = 1.4\%$. A somewhat more economical solution may be obtained if the minimum ratio ρ_{tot} is referred to the total number of bars available in the section; thus if eight 16 mm bars are used (three on each side of the column), $\rho_{tot} = 1.0\%$ results, which is exactly equal to the EC8 minimum requirement. However, such an economy may not be advisable in the portion of the building studied (bottom storeys) which is the most critical one with respect to seismic performance (sections 8.4.3, 8.5.2 and 10.2).

It is worth pointing out that minimum reinforcement requirements dominate the design of all exterior columns, where cross-sections had to be selected to satisfy axial loading (ν_d), as well as drift (d_{ei}/h_i) limitations imposed by EC8.

Interior columns The bending moment–axial force combinations for the 500 mm square ground storey column are

$$M_{Sd} = 224.0\ \text{kN m} \quad N_{Sd} = -1753.6\ \text{kN}$$

$$M_{Sd} = 236.0\ \text{kN m} \quad N_{Sd} = -1099.7\ \text{kN}$$

The most critical combination is the first one, for which

$$\mu = 0.134 \quad \nu = -0.526$$

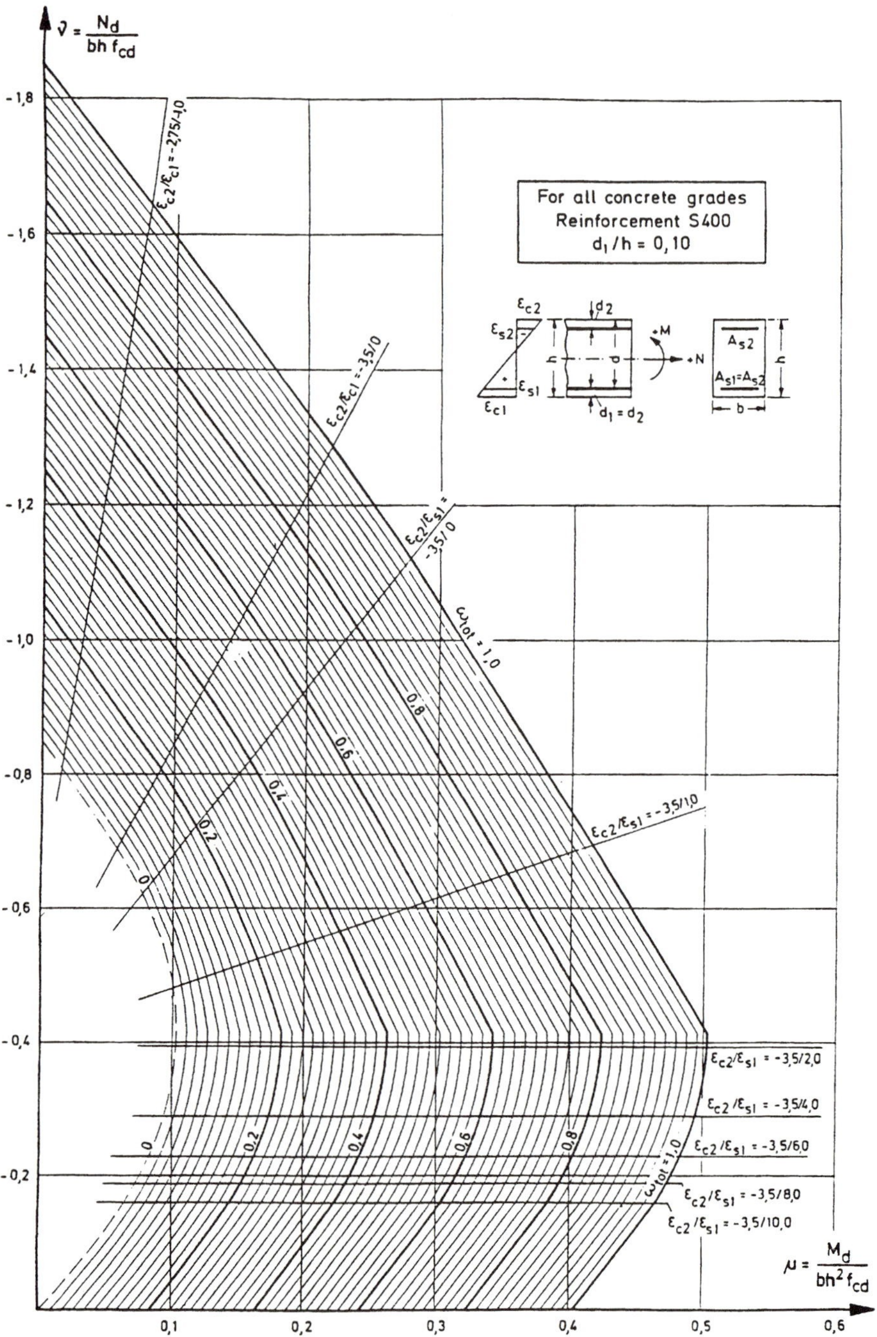

Figure 8.39 CEB (1982) design chart for symmetrically reinforced rectangular sections subjected to flexure and axial loading.

Entering these values in the design chart of Figure 8.39 gives $\omega_{tot} = 0.12$ which is less than the minimum required $\omega_{tot,min} = 0.261$, thus

$$A_s = \tfrac{1}{2} \times 0.261 \times 50.0^2 \,(13.3/347.8) = 12.5 \text{ cm}^2$$

Four 20mm bars (12.6 cm^2) are selected for each side of the column, resulting in a total reinforcement ratio $\rho_{tot} = 1.5\%$. The comments made previously regarding possible more economical solutions apply here as well. Note that if the cross-section dimension is reduced to 450 mm, $v_d = 0.65$ results, which is equal to the EC8 specified limit; using such a cross-section is not advisable from the ductility point of view, and besides problems in satisfying the drift limitations arise.

It is pointed out that minimum reinforcement requirements dominate the design of all the interior columns, however it is not possible to reduce cross-section dimensions without violating the drift control criterion (section 8.6.1(b)).

Capacity design considerations An additional check of the flexural capacity of columns is required by EC8, taking into account the strength of beams framing into beam–column joints, as explained in section 6.1.4(a). Note that one out of the four columns present in each storey may be excluded from this check, but for a symmetric structure such as the one under consideration it is not advisable to make use of this exemption, as it would apparently lead to highly asymmetric yield mechanisms during an earthquake excitation (see also section 10.2.1).

For the exterior beam–column joint (nodes 5 and 8 in Figure 8.35) the actual beam strength (in terms of design material properties) is

$$M_{Rd} = 188.4(8.0/7.1) = 212.3 \text{ kN m}$$

where the term in brackets is the ratio of available to required area of top reinforcement (section 8.6.2(a)). The sum of moments ratio (Figure 8.36(b) and equation (6.5)) is

$$\alpha_{CD1} = \gamma_{Rd} \frac{|M^l_{R1}| + |M^r_{R1}|}{|M^o_{S1}| + |M^u_{S1}|} = 1.2 \times \frac{0 + 212.3}{103.0 + 85.3} = 1.35$$

for direction '1' of seismic action, resulting in tension at the top of beam support and maximum moments at the columns (Figure 8.36(b)). It is pointed out that, for simplicity, the subscript 'd' has been dropped from both the acting and the resisting moments (M_{Si} and M_{Ri}). The α_{CD} factor for the opposite direction '2' of seismic action, whereby tension at the bottom of the beam results ($M^+_{Rd} = 120.3 \times 6.0/4.6 = 156.9$ kN m) is

$$\alpha_{CD2} = \frac{156.9}{62.9 + 57.5} = 1.56$$

The moment reversal factor (section 2.8.1.1.1 of EC8) for direction 1 is

$$\delta_1 = \frac{|M^r_{S1} - M^l_{S1}|}{|M^r_{R1}| + |M^l_{R1}|} = \frac{|0 - 188.4|}{0 + 212.3} = 0.88$$

and for direction 2

$$\delta_2 = \frac{|M^r_{S2} - M^l_{S2}|}{|M^r_{R2}| + |M^l_{R2}|} = \frac{|0 - 120.3|}{0 + 156.9} = 0.77$$

Hence the capacity design requirement for direction 1 is expressed as follows:

$$M_{Sd1,CD} = |1 + (\alpha_{CD} - 1)\delta_1| M_{Sd1} \not> qM_{Sd1}$$

$$M^{0}_{Sd1,CD} = |1 + (1.35 - 1)0.88| \times 103.0 = 1.308 \times 103.0 = 134.7 \text{ kN m}$$

$$M^{u}_{Sd1,CD} = 1.308 \times 85.3 = 111.6 \text{ kN m}$$

The axial load acting on the column above the exterior joint (for direction 1) is $N^0 = -1053.8$ kN, resulting in $\nu = -0.494$. Given that the mechanical ratio is

$$\omega_{tot} \approx 2_{\rho 1}(f_{yd}/f_{cd}) = 0.011(347.8/13.3) = 0.287$$

$\mu \approx 0.21$ results from the chart of Figure 8.39, thus

$$M^{0}_{Rd} = \mu bh^2 f_{cd} = 0.21 \times 0.4^2 \times 13\,333 = 179.2 \text{ kN m} > M^{0}_{CD} = 134.7$$

which means that there is no need to revise the column reinforcement. Note that for estimating the flexural capacity, $2\rho_1 = 2\rho_2$ instead of ρ_{tot} was used to calculate ω_{tot}, that is the contribution of the intermediate (18 mm) bars was ignored; this is a conservative assumption which renders it possible to use the available design charts for symmetrically reinforced rectangular sections.

The axial load acting on the bottom column is $N^u = -1199.8$ kN, resulting in $\nu = -0.562$; for $\omega = 0.287$, $\mu \approx 0.19$ results from Figure 8.39, thus

$$M^{u}_{Rd} = 0.19 \times 0.4^3 \times 13\,333 = 162.1 > M^{u}_{CD} = 111.6 \text{ kN m}$$

Considering now the actions resulting for direction 2 of the earthquake, the capacity criterion is

$$M_{Sd2,CD} = |1 + (\alpha_{CD2} - 1)\delta_2| M_{Sd2} = |1 + (1.56 - 1)0.77| M_{Sd2} = 1.431 M_{Sd2}$$

hence

$$M^{0}_{Sd2,CD} = 1.431 \times 62.9 = 90.0$$

$$M^{u}_{Sd2,CD} = 1.431 \times 57.5 = 82.3$$

The corresponding axial loads are $N^0 = -421.1$ and $N^u = -464.0$ kN, resulting in $\nu^0 = -0.197$ and $\nu^u = -0.218$, respectively. For $\omega = 0.287$ the following flexural resistances are calculated using the chart of Figure 8.39:

$$M^{0}_{Rd} = 0.19 \times 0.4^3 \times 13\,333 = 162.1 > 90.0$$

$$M^{u}_{Rd} = 0.20 \times 0.4^3 \times 13\,333 = 170.7 > 82.3$$

Note that the capacity check is less critical in direction 2; however, the choice of the most critical direction is not an obvious one, since M_{Rd} values depend on many factors whose interrelation is different in each structure, and even in different parts of the same structure.

With regard now to the interior beam–column joint, wherein two beams frame, the beam strengths on the basis of actually provided reinforcement are

$$M^{-}_{Rd} = 248.2(9.6/9.5) = 250.8 \text{ kN m}$$

$$M^{+}_{Rd} = 184.4(7.6/6.8) = 206.1 \text{ kN m}$$

The sum of moments ratio is (Figure 8.36(b))

$$\alpha_{\mathrm{CD}} = 1.2 \times \frac{250.8+206.1}{184.6+215.3} = 1.37$$

and the moment reversal factor is

$$\delta = \frac{|184.4+215.5|}{206.1+250.8} = 0.87$$

Thus the capacity criterion is

$$M_{\mathrm{Sd,CD}} = |1+(1.37-1)0.87|M_{\mathrm{Sd}} = 1.322M_{\mathrm{Sd}}$$

$$M^{0}_{\mathrm{Sd,CD}} = 1.322 \times 215.3 = 284.6$$

$$M^{\mathrm{u}}_{\mathrm{Sd,CD}} = 1.322 \times 184.6 = 244.0$$

The corresponding axial forces are $N^0 = -994.3$ kN, $N^{\mathrm{u}} = -1099.7$ kN, resulting in $\nu^0 = -0.298$ and $\nu^{\mathrm{u}} = -0.330$. For $\omega = 0.262$ the flexural resistances from the chart of Figure 8.39 are

$$M^{0}_{\mathrm{Rd}} = 0.20 \times 0.5^3 \times 13\,333 = 333.3 > 284.6 \text{ kN m}$$

$$M^{\mathrm{u}}_{\mathrm{Rd}} = 0.21 \times 0.5^3 \times 13\,333 = 350.0 > 244.0 \text{ kN m}$$

thus the capacity criterion is satisfied. In a similar manner it is verified that the check in the opposite direction is less critical, hence there is no need to revise the column reinforcement. It is worth pointing out that although the selected cross-section dimensions, as well as the corresponding reinforcement ratios, were well below those required on the basis of design moments and axial forces, they are almost just what is required to satisfy the capacity design criterion at certain locations, such as the bottom of the second storey column. In the light of this observation, it appears that the alternative, more economical, solutions mentioned in the previous section would subsequently have to be modified on the basis of the capacity design criterion.

Checks of the relative resistances of members framing into beam–column joints along the rest of the height of the building, with the exception of the top storey, where no capacity design of columns is required by EC8, showed that there was no need to revise the column reinforcements originally selected. This, of course, is the positive side of the severe limitations regarding maximum allowable axial loading and (especially) maximum inter-storey drift ratio.

Consideration of biaxial bending As mentioned in section 8.5.1(a), biaxial bending effects in DC 'M' columns may be considered in a simplified way, by reducing the flexural resistances in each direction (separately) by 30%. In the example under consideration only a symmetric plane frame is studied and only one direction of earthquake attack has been considered. In a more realistic (spatial) structure biaxial bending would result in all columns, the situation usually being aggravated in the case of members at the perimeter of the building.

Some account of biaxial loading effects has been taken in the frame studied by reinforcing columns symmetrically in both directions. Furthermore, it appears from the discussion presented in the previous section that even if the 30% reduction in the flexural resistances of columns is accounted for, these resistances are often higher than the corresponding design actions; in fact it was found that in 80% of the column sections studied the criterion $0.7M_{Rd} > M_{Sd,CD}$ was satisfied and in the rest of the sections the maximum 'shortage' of strength did not exceed 16%. Hence, it can be said that the design of the structure is a realistic one, conforming in general to EC8, even if biaxial effects are taken into account.

Finally, from the practical point of view, the simplified design for biaxial bending can best be carried out by increasing the normalized moment μ by 43% ($1/0.7 = 1.43$) before entering the design chart, thus ensuring a priori that $0.7M_{Rd} > M_{Sd}$. Note that an increase in μ of 43% may lead to doubling or even tripling of the required reinforcement ratio (Figure 8.39).

The arrangement of longitudinal bars in the columns of the bottom subframe is shown in Figure 8.38.

(b) Transverse reinforcement requirements

The design shear forces for the columns are derived on the basis of the capacity design procedure outlined in section 6.1.4(b). Thus for the exterior ground storey column the design shear is

$$V_{Sd,CD} = \gamma_{Rd}\frac{M^{0}_{Rd} + M^{u}_{Rd}}{3.0} \approx \frac{\gamma_{Rd}(2M^{0}_{Rd})}{l_c} = \frac{1.2(2 \times 162.1)}{l_c} = 129.7 \text{ kN}$$

Note that $M^{0}_{Rd} = 162.1$kN m is the flexural resistance of the column at its top, calculated in the section on capacity design for bending, and assumed, for simplicity, equal to the flexural resistance at the bottom (that is the small variation of M_{Rd} due to the self-weight of the member is ignored). Similarly the design shear for the interior column at ground storey is

$$V_{Sd,CD} = \frac{1.2(2 \times 350.0)}{3.0} = 280.0 \text{ kN}$$

It is worth pointing out that due to the shape of the M–N interaction curve (Figure 8.39), the most adverse axial loading for the calculation of the design shear is not obvious and typically both N_{min} and N_{max} should be tried in deriving $V_{Sd,CD}$. For the column under consideration the value $M_{Rd} = 350.0$ kN m corresponds to minimum compression $N_{min} = -1099.7$kN; for $N_{max} = -1753.6$ kN the flexural resistance is $M_{Rd} = 283 < 350.0$, that is not critical. This is due to the fact that the normalized axial loading ($\nu = -0.526$) corresponding to N_{max} lies above the balanced point of the μ–ν interaction curve (Figure 8.39), which is a rather exceptional case; usually the maximum M_{Rd} (for a given reinforcement) corresponds to the maximum axial load.

The column critical regions, assuming that no unfavourably arranged infill panels are present (section 8.5.2(a)), extend along a distance

$$l_{cr} = \max\{1.5h;\ l_c/6;\ 450 \text{ mm}\}$$

as specified by EC8 for DC 'M' columns. For the exterior column it is

$$l_{cr} = \max\{1.5 \times 400;\ 2150/6;\ 450\} = 600 \text{ mm}$$

while for the interior column

$$l_{cr} = \max\{1.5 \times 500;\ 2150/6;\ 450\} = 750 \text{ mm}$$

It is pointed out that for all columns of the structure the critical requirement with respect to l_{cr} is 1.5h.

Regarding now the selection of transverse reinforcement, the experience of the authors is that within the critical regions hoop requirements are usually dominated by confinement criteria (section 8.5.2(c)) or, less often, by maximum spacing limitations; it is very rare that shear reinforcement is not sufficient if the previous criteria are satisfied. Therefore, the recommended design procedure for hoop reinforcement is:

1. Select the hoop pattern according to the ductility class and the cross-sectional dimensions.
2. Select the hoop spacing so that the EC8 requirements regarding $s_{w,max}$ are satisfied. For axial loads close to the maximum permitted value ($v_{d,max}$) it is recommended to reduce further the spacing allowed by the code.
3. Calculate the required mechanical ratio of hoops (ω_{wd}) using equation (8.84) and revise hoop spacing, if necessary. If very close spacing results (say $s_w < 50$ mm), it is recommended to revise the hoop pattern (use more hoop legs).
4. Calculate the shear capacity of the column (section 8.5.2(b)) corresponding to the hoop reinforcement selected previously and revise it, if necessary.
5. In the typical case that shear reinforcement requirements are less critical than local ductility requirements, the spacing of stirrups outside l_{cr} may be increased to the maximum allowed to satisfy shear resistance requirements.

The foregoing procedure may be applied to the exterior column of the ground storey as follows:

1. Since the maximum distance between adjacent bars restrained by hoop bends is 200 mm, a double hoop pattern is selected for this 400 mm square column (Figure 8.38).
2. The maximum spacing allowed is determined by equation (8.86(b)): $s_w = \min \{b_0/3;\ 7d_{bl};\ 150 \text{ mm}\} = \{330/3;\ 7 \times 20;\ 150\} = 110$ mm
   Note that the diameter of the corner bar has been introduced in the above equation (see pertinent discussion in Papia, Russo and Zingone, 1988). The maximum normalized axial loading is $v_d = -0.56$, quite close to $v_{d,max} = -0.65$, therefore it is advisable to further reduce the spacing by, say 20%, thus to $s_w = 90$ mm.
3. The confinement coefficients (equations (8.74)) are

$$\alpha_n = 1 - \frac{\Sigma b_i^2}{6A_0} = 1 - \frac{8}{3n} = 1 - \frac{8}{3 \times 8} = 0.667$$

   where $n = 8$ is the total number of b_i (the distance between adjacent restrained bars); note that for a square column $nb_i = 4b_0$

$$\alpha_s = \left(1 - \frac{s_w}{2b_0}\right)^2 = \left(1 - \frac{90}{2 \times 330}\right)^2 = 0.746$$

   thus

$$\alpha = \alpha_n\, \alpha_s = 0.667 \times 0.746 = 0.497$$

For DC 'M' columns it is $k_0=60$, hence introducing $A_c/A_0=0.40^2/0.33^2=1.47$ in equation (8.84), the required mechanical ratio of confinement reinforcement is

$$\omega_{wd}=\frac{1}{\alpha}[k_0\mu_\phi\nu_d\varepsilon_{yd}(0.35A_c/A_0+0.15)-0.035]$$

$$=\frac{1}{0.497}[60\times 9\times 0.562\times 0.00174\ (0.35\times 1.47+0.15)-0.035]=0.635$$

For the selected hoop pattern of 10 mm double hoops at 90 mm the volumetric ratio is

$$\rho_w=\frac{(4\times 330+2(2)^{1/2}\times 330)0.79}{330^2\times 90}=0.0181$$

and the corresponding mechanical ratio is

$$\omega_{wd}=\rho_w(f_{yd}/f_{cd})=0.0181\ (347.8/13.3)=0.473$$

which is less than the required value, therefore the selection of hoop reinforcement was not appropirate. Typically a trial and error procedure is required to arrive at a satisfactory solution; it is finally found that for 10 mm hoops at 65 mm spacing the required $\omega_{wd}=0.612$ and the provided $\omega_{wd}=0.655$, thus this solution is admissible.

4. The shear resistance of the column can be estimated from equations (8.47) and (8.48) as follows:

$$V_{cd}=[\tau_{Rd}(1.6-d)\ (1.2+40_{\rho l})+0.15\sigma_{cp}]b_w d$$

$$=\left\{260[1.6-0.36]\left[1.2+40\left(\frac{8.82}{40\times 36}\right)\right]+0.15\left[\frac{464}{0.4^2}\right]\right\}0.4\times 0.36=129.6\text{ kN}$$

$$V_{wd}=\left(\frac{A_{sw}}{s}\right)0.9df_{ywd}=\frac{(2+2^{1/2})\ 0.79}{6.5}\ 0.9\times 36\times 34.78=467.6\text{ kN}$$

Thus

$$V_{Rd}=V_{cd}+V_{wd}=129.6+467.6=597.2>V_{Sd,CD}=129.7$$

It is pointed out that the 'concrete mechanism' (V_{cd}) alone is able to carry the entire design shear force. In fact, for all the columns in the structure the shear resistance was at least twice the required value.

For completeness, the resistance against crushing of the concrete struts should be checked, using equation (8.50):

$$V_{Rd2}=0.5\left(0.7-\frac{f_{ck}}{200}\right)f_{cd}b_w\ 0.9d$$

$$=0.5\left(0.7-\frac{20}{200}\right)13\ 333\times 0.4\times 0.9\times 0.36=518.4>129.7\text{ kN}$$

5. Outside the critical regions the spacing of stirups may be increased to the max-

imum permissible according to EC2:

$$s_w = 12d_{bl} = 12 \times 18 = 216 \text{ mm}$$

If the same hoop diameter $d_{bw} = 10$ mm used inside l_{cr} is retained, single stirrups at 220 mm spacing may be selected. The corresponding V_{wd} is equal to 80.9 kN, only 24% of the value within the critical region; however, $V_{Rd} = 210.5 > 129.7$, which means that there is no need for more transverse reinforcement.

It is pointed out that the diameter of hoops in the critical regions should not be less (equation(8.85)) than

$$d_{bw} = 0.35\ d_{bl,max}\ (f_{yld}/f_{ywd})^{1/2} = 0.35 \times 20 \times 1.0 = 7 \text{ mm}$$

if the same steel grade (S400) is used for both the longitudinal and the transverse reinforcement.

(c) Detailing requirements

The minimum dimension used in the frame under consideration (Figure 8.35) is 300 mm, which is higher than the minimum width specified EC8 for DC 'M' columns ($b_{min} = 250$ mm). Furthermore, the lowest ratio $b/l_0 = 300/1500 = 1/5 > 1/10$, which is the maximum value allowed for this ductility class.

According to usual practice, lapped splices are arranged at the top of the floor slab; it is recalled that splicing within the critical regions is allowed by EC8 for DC 'M' columns. The required lap length may be calculated using equations (8.57) and (8.56), setting $A_{s,req}/A_{s,prov} = 1$(section 8.5.3(b)) and $\alpha_1 = 2.0$, since more than 30% of the bars are spliced and their distance from the exterior surface of the column is less than $5d_{bl}$, thus

$$l_s = \alpha_1\left(\frac{d_{bl}}{4}\ \frac{f_{yd}}{f_{bd}}\right) = 2.0\ \frac{d_{bl}}{4}\ \frac{347.8}{2.3} \approx 75d_{bl}$$

For the 20 mm bars used in the subframe under consideration (Figure 8.38), it is $l_s = 1.50$ m, which is half the total height of the column. In the light of the discussion presented in section 8.5.3(c), a more realistic lap length corresponding to $\alpha_1 = 1.0$ was adopted, as shown in Figure 8.38. The area of one hoop leg within the lap length of the longitudinal bars must also satisfy equation (8.87)

$$A_{sw} = s_w \frac{d_{bl}}{50}\ \frac{f_{yld}}{f_{ywd}} = 65\ \frac{20}{50}\ 1.0 = 26.0 \text{ mm}^2$$

For 10 mm hoops it is $A_{sw} = 78.5$ mm^2, thus there is no problem in satisfying the previous requirement.

The arrangement of longitudinal and transverse bars in the columns of the bottom subframe of the structure studied is shown in Figure 8.38.

8.7 REFERENCES

Abdel-Fattah, B. and Wight, J.K (1987) Study of moving beam plastic hinge zones for earthquake-resistant design of R/C buildings. *ACI Struct. Journal*, **84**(1), 31–9.

Abrams, D.P. (1987) Influence of axial force variations on flexural behavior of reinforced concrete columns. *ACI Struct. Journal*, **84**(3), 246–54.

ACI (1989) *Building Code Requirements, for Reinforced Concrete (ACI 318-89) and Commentary (ACI 318R-89)*, Detroid, Michigan.

Bachmann, H. (1970) Influence of shear and bond on rotational capacity of reinforced concrete beams. *IABSE Publications*, **30** (Part II), 11–28.

Bertero, V.V (1979) Seismic behavior of structural concrete linear elements (beams, columns) and their connections. *Bull. d' Inf. CEB*, **131**, Paris.

Bertero, V.V. and Popov, E.P (1977) Seismic behaviour of ductile moment-resisting reinforced concrete frames, in *ACI SP53: Reinforced Concrete Structures in Seismic Zones*, ACI, Detroit, Michigan, pp 247–91.

Blakeley, R.W.G. and Park, R. (1973) Prestressed concrete sections with cyclic flexure. *Journal of the Struct. Div., ASCE*, **99** (ST8), 1717–42.

CEB (1982) CEB-FIP Manual on Bending and Compression. *Bull., d' Inf. CEB*, **141**, Paris.

CEB (1985) Model Code for Seismic Design of Concrete Structures. *Bull. d' Inf. CEB*, **165**, Paris.

CEB (1993) CEB-FIP Model Code 1990. *Bull.d' Inf. CEB*, **213/214**, Lausanne.

CEN Techn. Comm. 250/SC2 (1991) *Eurocode 2: Design of Concrete Structures–Part 1: General Rule and Rules for Buildings (ENV 1992-1-1)* CEN, Berlin.

CEN Techn. Comm. 250/SC8 (1995) *Eurocode 8: Earthquake Resistant Design of Structures–Part 1: General Rules and Rules for Buildings (ENV 1998-1-1/2/3)*, CEN, Berlin.

Cheung, P.C., Paulay, T. and Park, R. (1992) Some possible revisions to the seismic provisions of the New Zealand Concrete Design Code for moment resisting frames, *Bull. of the New Zealand Nat. Society for Earthq. Engng*, **25**(1), 37–43.

Dowrick, D.J. (1987) *Earthquake Resistant Design for Engineers and Architects*, J. Wiley & Sons, New York.

Eligehausen, R. and Fabritius, E. (1993) Steel quality and static analysis. *Bull. d' Inf. CEB*, **217**, Lausanne, 67–107.

French, C.W. and Schultz, A.E. (1991) Minimum available deformation capacity of reinforced concrete beams, in *ACI SP-127: Earthquake-resistant Concrete Structures–Inelastic Response and Design*, ACI, Detroit, Michigan, 363–419.

Gergely, P. (1977) Experimental and analytical investigations of reinforced concrete frames subjected to earthquake loading. *Proceed. of Workshop on Earthquake Resistant Reinforced Concrete Building Construction*, Univ. of California, Berkeley, July 1977, **III**, pp. 1175–95.

ICBO (Int. Conf. of Building Officials) (1994) *Uniform Building Code–1994 Edition*, Whittier, California.

Jirsa, J.O. (1974) Factors influencing the hinging behaviour of reinforced concrete members under cyclic overloads. *Proceed. of 5th World Conf. on Earthq. Engng*, June 1973, Rome, Italy, **2**, pp. 1198–204.

Jirsa, J.O., Maruyama, K. and Ramirez, H. (1980) The influence of load history on the shear behaviour of short RG columns. *Proceed. 7th World Conf. on Earthq. Engng*, Istanbul, Turkey, **6**, pp. 339–46.

Kaba, S.A. and Mahin, S.A. (1984) Refined modelling of reinforced concrete columns for seismic analysis. *Report EERC-84/03*, Univ. of California, Berkeley.

Kappos, A.J. (1986a) Evaluation of the inelastic seismic behaviour of multistorey reinforced concrete buildings. Doctoral Thesis (in Greek), *Scientific Annual of the Faculty of Engineering*. Aristotle University of Thessaloniki, **10** (Annex 8).

Kappos, A.J. (1986b) Input parameters for inelastic seismic analysis of R/C frame structures. *Proceed. of 8th European Conf. on Earthq. Engng*. Lisbon, Portugal, Sept. 1986, **3**, pp 6.1/33–40.

Kappos, A.J. (1991) Analytical prediction of the collapse earthquake for R/C buildings: suggested methodology. *Earthq. Engng and Struct. Dynamics*, **20**(2), 167–76.

Kappos, A.J. (1993) *RCCOLA-90: a Microcomputer Program for the Analysis of the Inelastic Response of Reinforced Concrete Sections*, Lab. of Concrete Structures, Dept of Civil Engng, Aristotle Univ. of Thessaloniki, Greece.

Keshavarzian, M. and Schnobrich, W.C (1985) Analytical models for the nonlinear seismic analysis of reinforced concrete structures. *Engng Struct.* **7**, April, 131–142.

Mander, J.B., Priestley, M.J.N. and Park, R. (1988) Theoretical stress–strain model for confined concrete. *Journal of Struct. Engng, ASCE,* **114**(8), 1804–26.

Marujama, K., Ramirez, H. and Jirsa, J.O. (1984) Short R/C columns under bilateral load histories. *Journal of the Struct. Div. ASCE,* **110**(1), 120–37.

Mau, S.T. (1990) Effect of tie spacing on inelastic buckling of reinforcing bars, *ACI Struct. Journal,* **87**(6), 671–7.

Minami, K. and Wakabayashi, M. (1980) Seismic resistance of diagonally reinforced concrete columns. *Proceed. of the Seventh World Conf. on Earthq. Engng,* Istanbul, Turkey, **6**, pp. 215–22.

Moehle, J.P. and Cavanagh, T. (1985) Confinement effectiveness of crossties in RC. *Journal of Struct. Engng, ASCE,* **111**(10), 2105–20.

Monti, G. and Nuti, C. (1992) Nonlinear cyclic behavior of reinforcing bars including buckling. *Journal of Struct. Engng, ASCE,* **118** (12), 3268–84.

Nmai, C.K. and Darwin, D. (1986) Lightly reinforced concrete beams under cyclic load. *Journal of the ACI,* **83**(5), 777–83.

Papia, M. and Russo, G. (1989) Compressive concrete strain at buckling of longitudinal reinforcement. *Journal of Struct, Engng, ASCE,* **115**(2), 382–97.

Papia, M., Russo, G. and Zingone, G. (1988) Instability of longitudinal bars in RC columns. *Journal of Struct. Engng, ASCE,* **114**(2), 445–61.

Park, R. and Paulay, T. (1975) *Reinforced Concrete Structures,* J. Wiley & Sons, New York.

Park, R., Priestley, M.J.N. and Gill, W.D. (1982) Ductility of square confined concrete columns. *Journal of the Struct. Div. ASCE,* **108**(4), 929–50.

Park, Y.-J. and Ang, A.H.-S. (1985) Mechanistic seismic damage model for reinforced concrete. *Journal of Struct. Engng, ASCE,* **111**(4), 722–39.

Paulay, T. (1982) Lapped splices in earthquake-resisting columns. *Journal of the ACI,* **79**(6), 458–69.

Paulay, T. (1986) A critique of the special provisions for seismic design of the Building Code Requirements for Reinforced Concrete (ACI 318-83). *Journal of the ACI,* **83**(2), 274–83.

Paulay, T. and Bull, I.N. (1979) Shear effects on plastic hinges of earthquake resisting reinforced concrete frames. *Bull d' Inf. CEB,* **132**, Paris.

Paulay, T. and Priestley, M.J.N. (1992) *Seismic Design of Reinforced Concrete and Masonry Buildings,* J. Wiley & Sons, New York.

Priestley, M.J.N. and Park, R. (1987) Strength and ductility of concrete bridge columns under seismic loading. *ACI Stuct. Journal,* **84**(1), 61–76.

Scott, B.D., Park, R. and Priestley, M.J.N. (1982) Stress–strain behavior of concrete confined by overlapping hoops at low and high strain rates. *Journal of the ACI,* **79**(1), 13–27.

Scribner, C.F. (1986) Reinforcement buckling in reinforced concrete flexural members. *Journal of the ACI,* **83**(6), 966–73.

Scribner, C.F. and Wight, J.K. (1980) Strength decay in R/C beams under load reversals. *Journal of the Struct. Div., ASCE,* **106** (ST4), 861–76.

Sheikh, S.A. and Uzumeri, S.M. (1980) Strength and ductility of tied concrete columns. *Journal of the Struct. Div., ASCE,* **106** (ST5), 1079–102.

Sheikh, S.A. and Uzumeri, S.M. (1982) Analytical model for concrete confinement in tied columns. *Journal of the Struct. Div., ASCE,* **108** (ST12), 2703–22.

Sheikh, S.A. and Yeh, C.-C (1990) Tied concrete columns under axial load and flexure. *Journal of Struct. Engng, ASCE,* **116**(10), 2780–800.

SANZ (Standards Association of New Zealand) (1982) (a) *Code of Practice for the Design of Concrete Structures* (NZS 3101–Part 1:1982); (b) *Commentary on Code of Practice for the Design of Concrete Structures* (NZS 3101–Part 2: 1982), Wellington.

SANZ (1984) *Code of Practice for General Structural Design and Design Loading for Buildings* (NZS 4203:1984), Wellington.

Tassios, T.P. (1984) Masonry, infill and RC walls under cyclic actions. *CIB Symposium on Wall Structures,* Warsaw, Poland, June 1984.

Tassios, T.P. (1989a) Specific rules for concrete structures – justification note no. 4: Ductility of beams–Maximum steel ratio, in *Background Document for Eurocode 8–Part 1, Vol. 2–Design Rules*, CEC DG III/8076/89 EN, pp. 15–16.

Tassios, T.P (1989b) Specific rules for concrete structures–justification note no. 6: Required confinement for columns, in *Background Document for Eurocode 8–Part 1, Vol. 2–Design Rules*, CEC DG III/8076/89 EN, pp 23–49.

Tegos, I. and Penelis, G.G. (1988) Seismic resistance of short columns and coupling beams reinforced with inclined bars. *ACI Struct. Journal*, **85**(1), 82–8.

Townsend, W.H. and Hanson, R.D. (1977) Reinforced concrete connection hysteresis loops, in *ACI SP53: Reinforced Concrete Structures in Seismic Zones*, ACI, Detroit, Michigan, pp. 351–70.

Tsonos, A. G., Tegos, I.A. and Penelis, G.G. (1995) Influence of axial force variations on the seismic behavior of exterior beam–column joints. *European Earthq. Engng*, **IX**(3), 51–63.

Umehara, H. and Jirsa, J.O. (1984) Short rectangular RC columns under bidirectional loading. *Journal of Struct. Engng, ASCE*, **110**(3), 605–18.

Zahn, F.A. Park, R. and Priestley, M.J.N. (1989) Strength and ductility of square reinforced concrete column sections subjected to biaxial bending. *ACI Struct. Journal*, **86**(2), 123–31.

9

Earthquake-resistant design of reinforced concrete planar elements

9.1 INTRODUCTION

A number of rather different elements are treated in the present chapter, which however share at least one common feature: their thickness is small compared with the other two dimensions, thus they behave essentially like disks or plates loaded along their median plane, and their reinforcement is similar in two orthogonal directions.

The elements addressed are:

- beam–column joint cores
- structural walls
- slabs acting as diaphragms.

It is worth pointing out that as far as modelling is concerned, it is not always obligatory to treat all these elements as two-dimensional (typically: plane stress) ones. In fact it is quite common to model R/C walls as linear elements (fat columns), especially if their slenderness is relatively high (sections 5.5.4 and 5.5.5).

As in the case of linear elements dealt with in Chapter 8, the behaviour of each planar element under monotonic loading to failure is first discussed, followed by the case of cyclic loading and subsequently of pertinent design requirements and detailing rules; the latter are largely based on the corresponding EC8 provisions.

9.2 BEAM–COLUMN JOINTS

The common regions of intersecting structural elements are called joints; in general it is advisable to refer to these regions using the term **joint core** rather than the more general term 'joint'. Whenever the area of these regions is limited, as in the case of linear elements (beams and columns) framing into each other, it is essential to verify their **shear** resistance, as well as the **anchorage** conditions of reinforcement passing through the joint region.

A fundamental requirement for an R/C structure is that for the members of the structure to be able to develop their full strength, premature failure of their joints should be precluded. It has to be emphasized that the concept according to which the stress state in joints having dimensions equal to or larger than those of the interconnecting elements is not critical and does not warrant any special verification, is an erroneous one, especially in the case of joints subjected to seismic loading (Park and Paulay, 1975). However, design practice in all countries, including those with problems of seismic risk, was based on this erroneous concept, at least up to the late 1970s.

Research on the behaviour of beam–column joints, with emphasis on cyclic loading conditions was initiated during the late 1960s and has today reached such a level as to render it possible to develop reliable, experimentally verified design methods and detailing rules, which were incorporated in several modern codes (SANZ, 1982; CEB, 1985; ACI, 1989; CEN, 1995). Nevertheless, there are still significant differences in both the design approach and the detailing requirements adopted by different codes, as will be clearly seen from the discussion that follows.

9.2.1 Basic design principles

Notwithstanding the fact that quite different procedures for the design of R/C beam–column joints have been adopted by each country, the main schools of thought being the American (ACI–ASCE, 1985) and the New Zealand (SANZ, 1982) one, the latter having been adopted by the CEB (1985) and to a lesser extent by EC8 (CEN, 1995), the basic principles regarding the seismic design of joints are more or less the same, and can be summarized as follows (see also Paulay, Park and Priestley, 1978):

1. The strength of the joint should not be inferior to that of the weakest member framing into it. This fundamental requirement emerges from the need to avoid seismic energy dissipation through mechanisms characterized by strength and stiffness degradation under cyclic loading conditions, as well as from the fact the joint core region is difficult to repair.
2. The load-bearing capacity of a column should not be jeopardized by possible strength degradation of the joint core.
3. During an earthquake excitation of moderate intensity (on which the serviceability limit state is based, according to the EC8 approach) the joint should preferably remain in the elastic range, so that no repair is required.
4. The reinforcement required for ensuring an adequate seismic performance of the joint should not be such as to cause construction difficulties due to congestion of bars in this region.

The feasibility of satisfying all the foregoing requirements depends first on the type of the joint. More difficulties arise in the case of interior joints (Figure 9.1(a)), somewhat less in the case of exterior joints (Figure 9.1(b)), while in the case of corner joints at the top of a frame (Figure 9.1(c)) the situation is more favourable and typically no additional checks are required.

Shown in Figure 9.2 are schematic representations of different modes of failure at an interior beam–column joint. It is the aim of seismic design to ensure that failure occurs in the mode shown in Figure 9.2(a), which is characterized by hinge formation at the beams framing into the joint. As already explained in sections 8.2 and 8.3, proper detailing of the plastic hinge regions can provide a high ductility to the beam–column subassemblage (and hence to the structure as a whole) and minimize the possibility of collapse during a strong earthquake. In contrast,

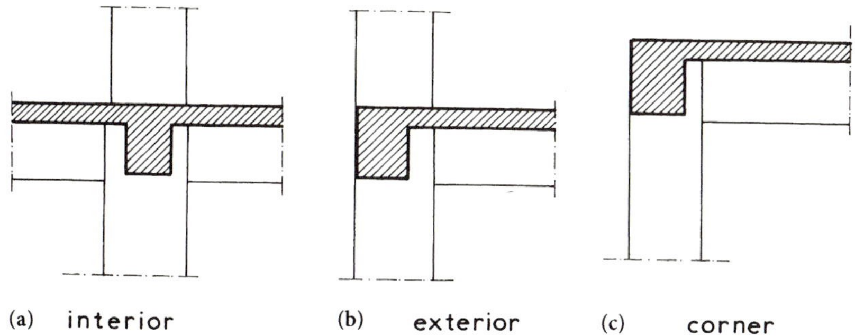

Figure 9.1 Types of joints in monolithic R/C structures: (a) interior; (b) exterior; (c) corner.

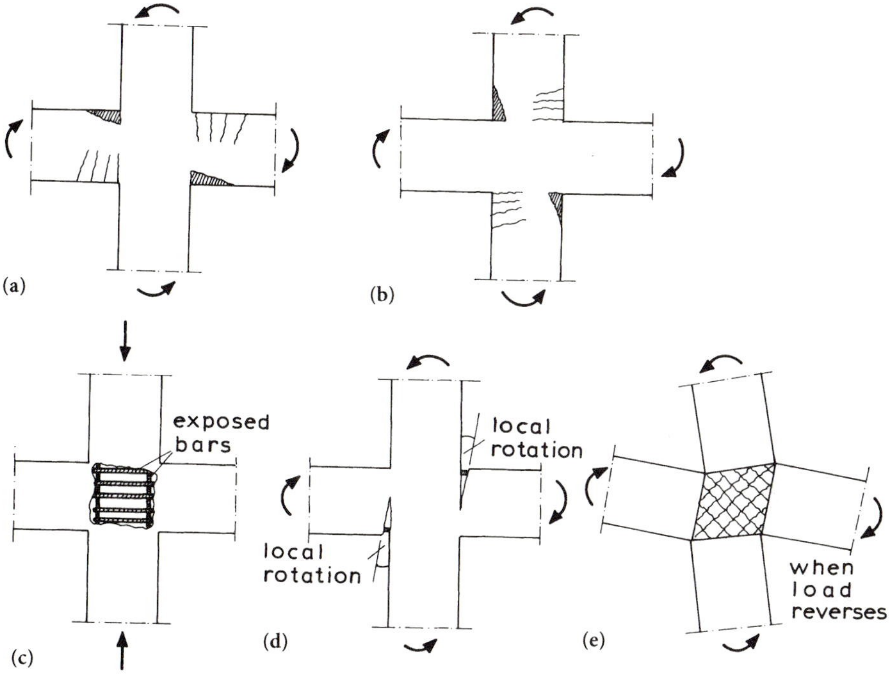

Figure 9.2 Types of failure at beam–column joints: (a) attainment of deformational capacity of the beam; (b) attainment of deformational capacity of the column; (c) spalling of joint core; (d) anchorage failure of beam bars; (e) shear failure of the joint core.

the formation of plastic hinges at the columns should be precluded for the reasons already explained in sections 6.1 and 8.4.

The remaining three types of failure shown in Figure 9.2, refer to damage incurred within the joint core; these are:

1. *Spalling* of cover concrete at the faces of the joint core, which can lead to a significant reduction in the bearing capacity of the column (the amount of

reduction depends, of course, on the ratio of the confined column area to the area of the section prior to spalling).

2. *Anchorage* failure in the longitudinal bars of the beam passing through the joint, which leads to strength deterioration and significant permanent deformations and consequent local rotations (fixed-end rotations) at the beam–column interface, hence to a drastic reduction in the stiffness of the beam–column subassemblage.
3. Failure of the joint core due to diagonal tension caused by shear (Figure 9.2(e)), with consequences on the strength and the stiffness of the subassemblage similar to those caused by the previous failure mode.

In order to ensure that the favourable mechanism of Figure 9.2(a) will form, it is necessary to verify the following:

1. The relative flexural resistances of beams and columns framing into a joint; the pertinent EC8 requirement regarding the capacity design of columns has already been given in section 6.1.4(a).
2. The shear resistance of the joint core, in the horizontal, as well as in the vertical, direction.
3. The required anchorage of longitudinal bars of the beam and the column, passing through the joint or anchored within its core.

The background, as well as the implementation in terms of code provisions, of the verifications (2) and (3) will be presented in the following sections.

9.2.2 Behaviour of joints under cyclic shear

The behaviour of beam–column joints under cyclic loading is characterized by an interaction of shear, bond and confinement mechanisms, with regard to all of which considerable uncertainties exist, especially at the stage of the response well beyond the first yield excursion. This should explain, at least to a certain extent, the significant differences that still exist among seismic codes with regard to not only the shear transfer mechanisms assumed for developing provisions for joint shear reinforcement, but even with regard to the design forces for the joint core. In the following the simple approach for the determination of joint shear forces, adopted by both the New Zealand (SANZ, 1982; Paulay and Priestley, 1992) and the American (ACI–ASCE, 1985) methods, is presented, while the revised relationships incorporated in EC8 are given in the next section (9.2.3).

(a) Shear forces in the joint core

The forces acting on the core of an interior joint, forming part of an R/C frame subjected to predominantly seismic actions, are shown in Figure 9.3. Provided the design of the frame has ensured that plastic hinges will form at the beams (rather than at the columns), it is necessary to take into account in deriving the joint forces the increase in beam moments ($M_1^*$and M_2^*) due to: (1) the possibility of a yield stress in beam steel higher than the design value, and (2) the increased stress of beam longitudinal reinforcement due to strain-hardening. Furthermore, in estimating M_1^* and M_2^*, the actual steel areas A_{s1}, A_{s2} rather than the calculated val-

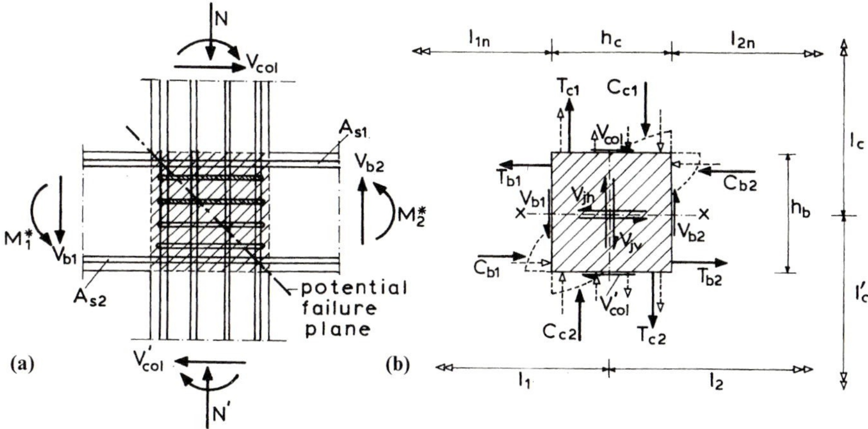

Figure 9.3 (a) Seismic actions in the joint core: (b) internal forces in the joint core (T is the resultant to tensile forces and C the resultant of compressive forces, at each face of the joint).

ues should be used (SANZ, 1982; CEB, 1985; ACI-ASCE, 1985). With regard to the top reinforcement (A_{s1}) due allowance for slab steel in this region should also be made.

The equilibrium of horizontal forces in the joint with respect to the level x–x (see Figure 9.3(b)) gives the following equation for the horizontal shear force acting in the joint

$$V_{jh} = T_{b1} + C_{b2} - V_{col} \qquad (9.1(a))$$

and since at the right face of the beam the resultant (C_{b2}) of the compressive stresses in concrete and steel equals the tensile stress (T_{b2}) at the bottom reinforcement, equation (9.1(a)) may be written as

$$V_{jh} = T_{b1} + T_{b2} - V_{col} \qquad (9.1(b))$$

Assuming that both the bottom and the top reinforcement of the beam yield, the horizontal joint shear may be expressed as

$$\boxed{V_{jh} = \gamma_R f_y (A_{s1} + A_{s2}) - V_{col}} \qquad (9.2)$$

where the overstrength factor (γ_R) accounts for increased yield stress and strain-hardening, as mentioned previously.

During the inelastic response of the frame to a given earthquake the possible distributions of the total moment ($M_1^* + M_2^*$) to the columns above and below the joint are difficult to determine, due to the continuous variation of the stiffness of members framing into the joint, as well as the rest of the factors mentioned in section 8.4.1. Furthermore, the value of the column shear (V_{col}) also depends on the value of moment at the other end of the column, which is also subject to similar uncertainties. Given these, Paulay, Park and Priestley (1978) suggested a capacity relationship for estimating the column shear on the basis of the beam moments at the column faces

$$V_{col} = \frac{(l_1/l_{1n})M_1^* + (l_2/l_{2n})M_2^*}{(l_c + l'_c)/2} \qquad (9.3)$$

where, as shown in Figure 9.3(b), l_1 and l_2 are the beam spans measured from the theoretical support points (column centrelines), l_{1n} and l_{2n} are the corresponding clear spans, and l_c, l_c' are the column heights measured from the beam centrelines. Equation (9.3) was adopted by the New Zealand Code (SANZ, 1982) and the CEB Model Code (1985), but not by EC8 (section 9.2.3).

The vertical joint shear V_{jv} can be derived in a similar fashion using equilibrium considerations (in the vertical direction), or, more simply, by noting that the horizontal shear stress in the joint should equal the corresponding vertical shear stress, $\tau_{jh} = \tau_{jv}$. Assuming for simplicity that τ_{jh} and τ_{jv} are uniformly distributed along each face of the joint, the following equation can be written for the two shear stresses

$$\frac{V_{jh}}{b_j h_c} = \frac{V_{jv}}{b_j h_b}$$

where b_j is the effective width of the joint (section 9.2.3). Hence the vertical joint shear is

$$\boxed{V_{jv} = V_{jh} \frac{h_b}{h_c}} \tag{9.4}$$

The previously described procedure may also be applied to estimate the horizontal and the vertical shear at an exterior joint (Figure 8.27(a) and 9.1(b)) where, due to the fact that there is only one framing beam, the resulting shear forces are lower than in a similar interior joint. In this case equation (9.2) is written as

$$V_{jh} = \gamma_R f_y A_{s1} - V_{col} \tag{9.5}$$

while in equation (9.3) the second term in the numerator (the one involving M_2*) should be disregarded. It is pointed out that the critical situation at an exterior joint typically arises when the top beam reinforcement is in tension, since not only is the top steel at least equal to the bottom steel, but also in a typical monolithic joint, slab steel contributes to the negative resistance (M_1^* in Figure 9.3(a)).

(b) Mechanisms of shear transfer

Shear transfer mechanisms in R/C beam–column joints are quite complex, since an interplay of shear, bond and confinement takes place within a rather limited area. Therefore, it is not surprising that at the present time conflicting views exist with regard to whether joints should be desisgned for horizontal shear, vertical shear or both, and also with regard to the role of the hoop reinforcement (Pantazopoulou and Bonacci, 1992; Cheung, Paulay and Park, 1992). In the present and the following sections the model originally proposed by Paulay, Park and Priestley (1978) and later partially revised by the same investigators will be presented; this model (in its original form) was adopted by the 1982 New Zealand Code, as well as by the 1985 CEB Model Code (only for high ductility frames). The models and/or design approaches followed in the USA and by EC8 will be discussed subsequently.

According to Paulay, Park and Priestley (1978) the total shear within a joint core is carried partly by a **diagonal concrete strut**, formed between the corners of the joint subjected to compression (Figure 9.4(a)), and partly by an idealized **truss**

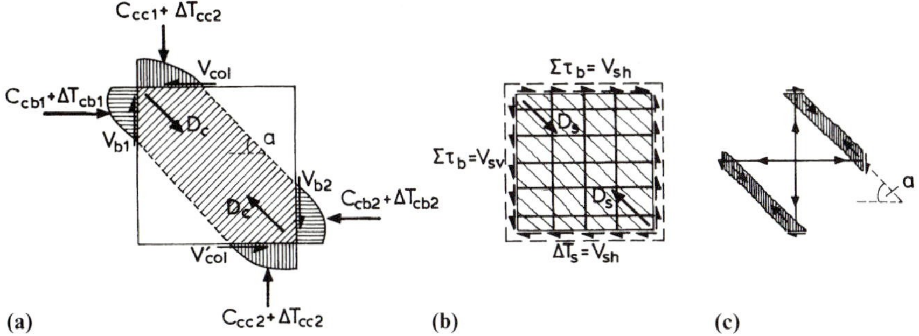

Figure 9.4 Shear transfer mechanisms in a joint core: (a) diagonal concrete strut; (b) truss mechanism; (c) typical components of the truss (steel bars in tension, concrete bars in compression).

consisting of horizontal hoops, intermediate column bars, and inclined concrete bars between shear cracks (Figure 9.4(b)(c)). It is pointed out that in Figure 9.4 the case of a column without axial loading is shown, which is the most unfavourable one in the typical situation of columns in compression. On the other hand the presence of axial tension would undoubtedly lead to a rapid degradation with cyclic loading (Figure 8.27(c)).

As shown in Figure 9.4(a), compression forces in concrete (C_{cbi} on the beam faces and C_{cci} on the column faces), together with beam and column shear forces (V_{bi} and V_{col}) and bond forces (ΔT_{cbi}, ΔT_{cci}) transferred by the reinforcement bars within the compression zones, can form a system of forces in equilibrium. The main component of this mechanism is a diagonal strut, carrying a compression force D_c. A substantial portion of the total joint shear (horizontal, as well as vertical) can be resisted by this mechanism. This portion increases in the presence of (compressive) axial load in the column, as this implies an increased depth of the compression zone, hence a wider concrete strut.

Tests on beam–column joints (Paulay, Park and Priestley, 1978) indicated that the contribution of other common shear transfer mechanisms, such as aggregate interlock and dowel action of longitudinal bars, is very limited, due to the fact that shear deformations of a magnitude sufficient for these mechanisms to be activated cannot develop in the joint core. Paulay, Park and Priestley (1978) suggested that the contribution of concrete mechanisms to the transfer of horizontal shear in a joint core can be estimated from the relationship (Figure 9.4(a))

$$V_{ch} = D_c \cos(\alpha) \tag{9.6(a)}$$

and the corresponding contribution to vertical shear transfer from the relationship

$$V_{cv} = D_c \sin(\alpha) \tag{9.6(b)}$$

As indicated in Figure 9.4, only a part of the bond forces (ΔT_c) is transmitted within the compression zones of the joint, while the rest of these forces (ΔT_s) is introduced along the faces of the joint core, whenever yield penetration from the longitudinal bars has not occurred. These bond forces produce shear stresses within the joint core which, depending on their magnitude, may lead to diagonal tension cracks (Figure 9.4(b)); the main crack is along the diagonal of the core (failure plane), but other cracks, approximately parallel to it, may also form (Paulay, Park and Priestley, 1978; Durrani and Wight, 1985). Under the aforementioned conditions, a truss mechanism may be activated, consisting of appropriately anchored vertical and horizontal bars (hoops), as well as a system of inclined bars in compression formed by concrete between adjacent shear cracks. It is pointed out that for this mechanism to function, horizontal hoops are not sufficient, but intermediate (not corner) vertical column bars are also required (see also Figure 9.4(c)). These intermediate bars are usually in compression, due to the column axial loading, hence they are quite capable of carrying the tension produced by the truss mechanism. It is seen that this is a shear transfer mechanism similar to, and yet distinct from, the one encountered in beams (section 8.2.2(d)).

From the equilibrium of forces shown in Figure 9.4(b) it follows that the truss mechanism can carry a horizontal shear equal to

$$V_{sh} = D_s \cos(a) \tag{9.7(a)}$$

and a vertical shear

$$V_{sv} = D_s \sin(\alpha) \tag{9.7(b)}$$

where a is the angle of the inclined cracks with respect to the horizontal axis of the joint.

According to the approach for shear design presented in previous sections (8.3.2(a), 8.5.2(b)), the total shear resistance of the joint core can be expressed as the sum of the two previously described mechanisms, that is

$$V_{jh} = V_{ch} + V_{sh} \tag{9.8(a)}$$

$$V_{jv} = V_{cv} + V_{sv} \tag{9.8(b)}$$

where the contributions of concrete and of shear reinforcement (through the truss mechanism) are given by equations (9.6) and (9.7), respectively. For practical design purposes the force (D_c) of the concrete strut mechanism has to be evaluated in terms of known quantities rather than in terms of bond forces and concrete forces (for instance it is seen from Figure 9.4(a) that the horizontal component V_{ch} is equal to $C_{cb1} + \Delta T_{cb1} - V_{col}$, but this relationship is hardly appropriate for design purposes). Therefore, practical design is carried out with the aid of semi-empirical relationships, similar to those used for beams, as discussed in section 9.2.3.

(c) Effect of cycling on shear transfer mechanisms

With regard to the effect of earthquake-type (cyclic) loading on the aforementioned mechanisms, it is noted that while during the first loading cycle in the inelastic range the contribution of the concrete strut in carrying the joint shear is significant, it deteriorates with increasing inelastic load cycles. As shown in Figure 9.4(a), the value of the compressive force D_c depends on the compressive forces in concrete (C_{cbi} and C_{cci}) and on the bond forces (ΔT_{cbi} and ΔT_{cci}) induced in the compression zones. Cycling at high levels of inelastic deformation was already shown (section 8.2.2) to cause permanent elongations of beam bars which may eventually lead to full depth open cracks at the beam–column interface.

Under these conditions, compression forces in concrete become negligible, with a subsequent increase in bond forces of beam bars in compression passing through the joint. These bars are now required to transmit the compressive forces previously carried by the compression zone of the beam, which leads to a significant increase in bond stresses within the joint core, especially in the zones away from the beam faces, as yield penetration from the beam to the joint core leads to bond deterioration (Figure 9.7). Particularly high bond stresses may then develop close to the centre of the joint and it is inevitable that slip of beam bars will be significant (see also section 7.6.3); the presence of axial compression in the column can improve bond conditions (section 7.6.2), but cannot prevent yield penetration from the beam. The foregoing lead to a drastic reduction in the contribution of the concrete strut to the transfer of horizontal joint shear and a consequent increase in the contribution of the truss mechanism. It is pointed out that the 1982 New Zealand Code adopted the conservative approach that $V_{ch}=0$ for joints where axial compression in the column is low ($\nu<0.1$); however, more recent studies (Park and Dai, 1988; Cheung, Paulay and Park, 1992) have shown that bond forces in beam bars (ΔT_{cbi} in Figure 9.4(a)) can be quite significant, hence a modified equation for V_{ch} has been proposed by Cheung, Paulay and Park (1992):

$$V_{ch}=0.3(1+3.5\nu)V_{jh} \tag{9.9}$$

It is seen that even for negligible column compression ($\nu\approx0$) at least 30% of the horizontal joint shear may be carried by the concrete mechanism.

With regard now to the vertical joint shear (V_{cv}), no significant modification of the shear transfer mechanisms is expected to occur with cycling. As long as the column reinforcement does not yield (as should be the case for an efficiently designed column), a substantial part of the vertical shear may be carried by compressive forces (C_{cci}) in the concrete, as well as by bond forces (ΔT_{cci}) in the column bars transferred within the compression zone. This implies that lower requirements are imposed on the truss mechanism, which in practical terms means that less vertical shear reinforcement is required. According to Cheung, Paulay and Park (1992) the vertical joint shear resisted by the concrete strut mechanism may be estimated from the relationship

$$V_{cv}=0.5V_{jv}+N \tag{9.10}$$

The mechanisms shown in Figure 9.4 correspond to a monotonic type of loading, while in the case of cyclic loading diagonal cracks in the direction of the other diagonal of the joint are also expected to form. The presence of these cross-inclined shear cracks will lead to an effective compressive strength of the concrete struts between adjacent cracks (the 'bars' of the idealized truss) which is lower than the cylinder strength of concrete. It is thus important to keep the nominal joint shear stress below certain limits to ensure that no premature crushing of concrete struts occurs during cyclic loading; the pertinent EC8 equations are given in section 9.2.3.

The response to cyclic loading of interior beam–column joints designed by different approaches can be seen in Figure 9.5, where hysteresis loops for two specimens tested by Park and Dai (1988) are given. The specimen of Figure 9.5(a) was designed in full conformance with the NZS 3101 (SANZ, 1982) requirements for highly ductile frames, while in the specimen of Figure 9.5(b) the horizontal

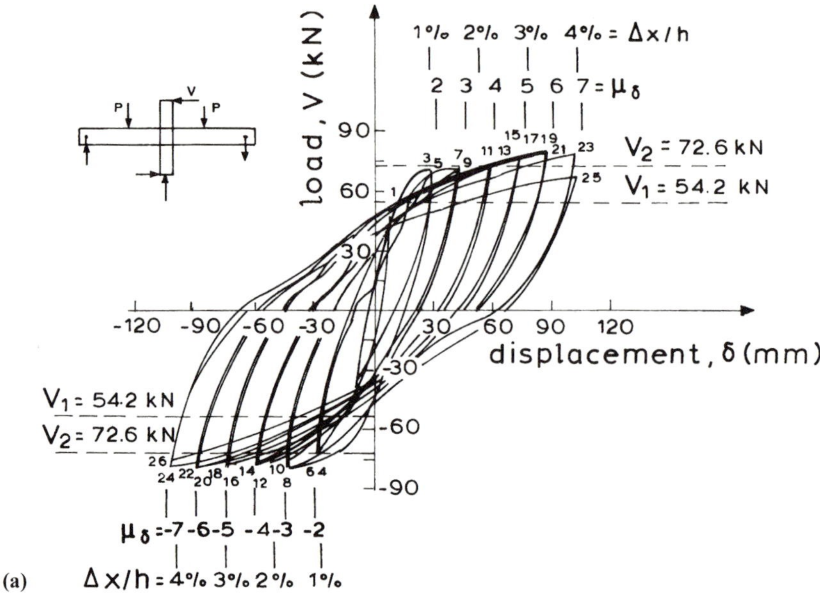

(a)

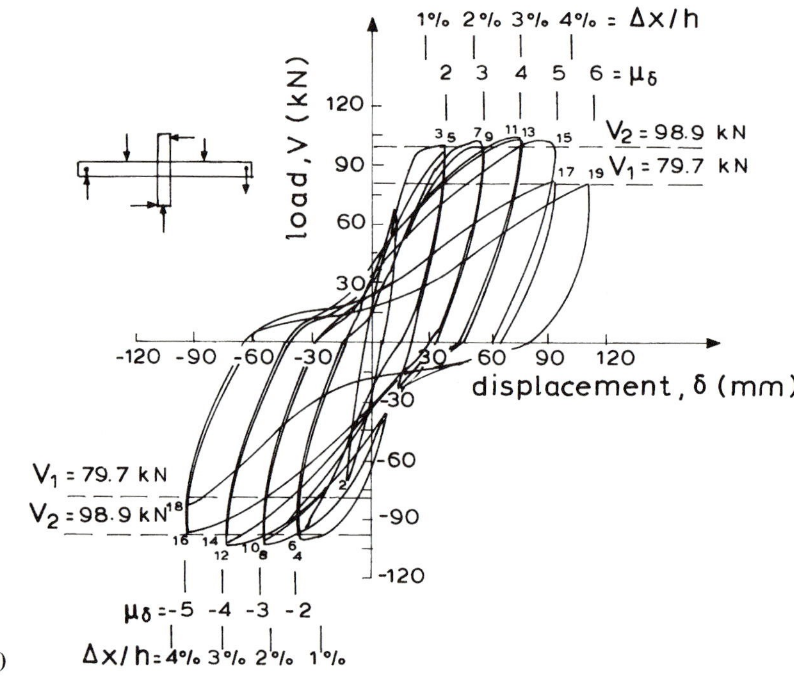

(b)

Figure 9.5 Hysteresis loops for interior beam–column joints: (a) specimen designed according to NZS 3101 (1982); (b) specimen with less horizontal and vertical shear reinforcement, and larger beam bar diameter, with respect to the NZS 3101 requirements.

shear reinforcement in the joint core was 58% of that required by the New Zealand Code and the corresponding vertical shear reinforcement (intermediate column bars) was 82% of the required amount. Furthermore, the diameter of longitudinal beam bars in the specimen of Figure 9.5(b) was 72% greater than that permitted by NZS 3101. Also shown in the figure are the theoretical loads (V_1) when the first plastic hinge formed at the critical positive moment section (tension at the bottom of the beam), and the corresponding loads (V_2) when the second plastic hinge formed at the critical negative moment section.

It is seen that the beam–column subassemblage fully conforming to the New Zealand Code (Figure 9.5(a)) had an excellent response, characterized by high strength (up to 11% higher than the theoretical value V_2) and stiffness, and high energy dissipation capacity. Plastic hinges formed (as intended by the design) at beam ends, while cracking in the column was very limited. More extensive cross-inclined cracking occurred in the joint core, with a measured maximum crack width of 0.6 mm, which is only slightly higher than the values expected under service conditions. At a displacement ductility of 7 the specimen was still maintaining its strength and energy dissipation capacity. Nevertheless, the most interesting finding in this investigation was that the specimen which did not conform to NZS 3101 (Figure 9.5(b)) also showed a quite satisfactory performance, with a maximum strength 8% higher than the theoretical value (V_2) and a residual strength at a ductility of 6 equal to 0.81 V_2. Stiffness degradation and pinching of hysteresis loops were, as expected, more pronounced than in the specimen of Figure 9.5(a), while the maximum crack width measured at the joint core was 1.1 mm, attributed to yielding of the joint shear reinforcement. Although inferior to that of the specimen of Figure 9.5(a), the behaviour of the specimen with less hoops and larger beam bar diameter than permitted by NZS 3101 is certainly quite good (it even satisfies the New Zealand ductility criterion described in section 7.1), which is a strong indication that a less conservative shear design approach can be adopted for beam–column joints (section 9.2.3).

At this stage it is worth pointing out that a number of researchers including the majority of those from North America, have long expressed reservations as well as different opinions regarding the validity of the joint shear transfer models described previously. Relevant tests (Meinheit and Jirsa, 1981) have indicated that the use of hoop reinforcement in the joint core does increase the shear strength, but not to the extent implied by the superposition rule of equation (9.8). Moreover, it was found that an increase in the longitudinal reinforcement of the column leads to an increase of the joint shear strength even less than that caused by the increase of transverse reinforcement, thus it is practically negligible. Finally, it was observed that the value of column axial loading does not influence the joint shear strength, it merely increases the shear force corresponding to first cracking (Meinheit and Jirsa, 1981). Commenting on the foregoing experimental findings, Paulay (1986) pointed out that in the tests under consideration failure was due to anchorage deterioration, rather than shear capacity being exceeded.

In any case, the fact remains that the approach to joint shear design suggested by Paulay and his associates is a mechanical model validated by a number of experimental findings, while the approach adopted by US investigators, which gave rise to the corresponding ACI–ASCE (1985) recommendations, is purely empirical, as it consists in establishing appropriate lower bounds to the largest available number of test results and defining a maximum allowable nominal shear strength ($\tau_{max} = \gamma f_c^{1/2}$) for each type of joint. Finally, as was clearly demonstrated by the example of Figure 9.5, the shear design of joint cores is strongly dependent on the type of performance criteria adopted for these regions.

(d) Effect of transverse beams and floor slabs

In the foregoing discussion no explict account has been taken of the presence of transverse beams in the joint (Figure 9.1), that is of beams orthogonal to the axis of the beam in the direction considered. In tests where these transverse beams were unloaded, an increase in shear strength of interior joints was found; Meinheit and Jirsa (1981) estimated that in the case where the transverse beam area was about 70% the total joint area, the increase in joint shear strength (with respect to a similar joint with no transverse beams) was of the order of 20%. This favourable effect is due to the increased area of the joint working in shear, as well as to the confinement offered by transverse beams, as they prevent the dilation of the joint core after cracking. This favourable effect of transverse beams was also observed in the case of exterior joints (Ehsani and Wight, 1985). However, it has to be pointed out that transverse beams in a real structure also form part of the seismic load-resisting system, hence it is quite possible that plastic hinges form at their ends close to the joint face. The behaviour of a joint with beam plastic hinges at all four faces is not necessarily superior to that of similar joints with only two beams (in fact it is inferior with regard to bond conditions, as will be discussed in section 9.2.4). Nevertheless, if the behaviour of an actual, structurally indeterminate, frame is considered, rather than that of a beam–column subassemblage, it is seen that restraint of joint core expansion is offered by adjacent beams which develop axial forces contributing to the confinement of the joint (Pantazopoulou and Bonacci, 1992).

The presence of a slab monolithically cast with the rest of the members framing into the joint, which is the typical situation in practical construction (Figure 9.1), was also found to affect favourably the behaviour of the joint, by increasing both its strength and its stiffness (Ehsani and Wight, 1985; Durrani and Zerbe, 1987; Kitayama, Otani and Aoyama, 1989). There are, nevertheless, two points of concern that have to be raised with regard to the influence of floor slabs on the seismic behaviour of joints.

The first is that slab reinforcement parallel to the axis of the beam enhances its negative moment flexural resistance; as already mentioned in section 8.4.1, ignoring this increase in beam strength when designing the adjacent columns may lead to hinge formation at the column rather than at the beam ends. During tests on realistic two-way interior beam–column joints with floor slabs, subjected to unidirectional, as well as bidirectional cyclic loading, Cheung, Paulay and Park (1992) noted that the contribution of slab reinforcement is affected by the level of inelasticity induced at the joint, the maximum strength developing at displacement ductilities between 4 and 6. Based on these test results, they suggested that the effective width of the slab, to be taken into account in estimating the beam flexural strength, may be assumed to be the lesser of: (1) one-quarter of the beam span at each side from the beam centreline; (2) one-half or one quarter of the distance between adjacent beams, at each side of the beam centreline, at interior or exterior columns, respectively. Durrani and Zerbe (1987), based on results from tests of exterior joints with floor slabs, have suggested an effective slab width equal to one depth of transverse beam at each side of the column face. In most practical situations all the previously proposed effective widths are larger than the corresponding code values shown in Figure 8.21.

A second point should be made regarding the effect of slabs on transverse beams at exterior beam–column connections, wherein torsion is induced by the floor slab to the edge beam. Durrani and Zerbe (1987) found that transverse beams were initially effective in confining the joint, but once they reached their torsional cracking strength, this effectiveness was drastically reduced. It appears that, at least for this type of joint, transverse beams should not be relied upon to improve the behaviour of the joint core.

9.2.3 Design for shear

In the previous section discrepancies between various approaches to shear design of beam–column joint cores have been pointed out. A critical issue, however, on which a consensus appears to have been reached by investigators dealing with the subject, is that the level of shear at a joint, as expressed by the nominal shear stress, is a crucial factor affecting both the strength and the stiffness of the system. As indicated by test results (Durrani and Wight, 1985; Ehsani and Wight, 1990), the level of shear stress influences shear cracking and deformation of the joint core, as well as bond conditions of longitudinal bars, resulting in a pinching of hysteresis loops and a reduction of the energy dissipation capacity of the beam–column subassemblage. It has been observed that at displacement ductilities in excess of 2, the unfavourable influence of high shear stresses is practically independent of the amount of hoops placed in the joint core (Durrani and Wight, 1985). Thus, if it is also taken into account that for beam reinforcement ratios of 1.5% or more the amount of hoops required in the joint is such as to cause construction difficulties, it is concluded that the combination of low joint shear stress and limited amount of joint reinforcement is more effective, at least from the energy-dissipation point of view, than the combination of high shear stress and heavy joint reinforcement.

(a) Design action effects

A verification of the shear resistance of beam–column joint cores is required by EC8 for both high and medium ductility classes; this goes one step further than the approach adopted by the CEB (1985) Seismic Code, which required joint shear verification only for ductility level III (high) frames, and is a clear indication of the increasing importance the design of joints has assumed in recent years.

According to the capacity design approach adopted by EC8, the horizontal shear force acting on the joint core is given by the expressions (Figure 9.3)

$$\boxed{V_{jh}=\gamma_{Rd}\left[\frac{2}{3}\left(A_{s1}+\frac{9}{5}A_{s2}\right)f_{yd}\right]-V_{col}} \qquad (9.11)$$

for interior beam–column joints, and

$$\boxed{V_{jh}=\gamma_{Rd}(2/3)A_{s1}f_{yd}-V_{col}} \qquad (9.12)$$

for exterior joints, where the symbols have already been explained with reference to Figure 9.3, and $\gamma_{Rd}=1.25$ or 1.15 for DC ‘H’ and ‘M’ respectively.

It is seen that equation (9.11) is similar to equation (9.2), with two differences. First it is assumed that only two–thirds of the beam bar forces result in bond forces acting inside the joint core (Tassios, 1989b), hence the term 2/3 in equation (9.11), and second it is assumed that the force developing at the top bars in compression of the beam support depends on the ductility requirement at the joint, the full yield force ($A_{s2}f_{yd}$) of the bottom bars developing only for $q=5.0$ which is the maximum behaviour factor given by EC8 (section 4.4.5). The factor γ_{Rd} takes account of the different level of strain-hardening in beam bars expected for each ductility class.

It is clear that equations (9.11) and (9.12), although derived from capacity design considerations similar to those of the New Zealand Code (which adopts equation (9.2)), lead nevertheless to substantially lower design shear forces for the joint core. In a typical situation wherein $A_{s2}\approx A_{s1}/2$, the value of $V_{jh}+V_{col}$ from equation (9.11) may be up to 46% lower than the value derived from equation (9.2), assuming the same γ_{Rd} factor for both cases and $q=1.5$ which is the minimum value specified by EC8. Furthermore, the value of the column shear (V_{col}) to be entered in equations (9.11) and (9.12) is not estimated by the capacity-based expression (9.3) given in section 9.2.2(a), but is taken directly from the analysis for the seismic combination. This further contributes to producing lower V_{jh} values, compared with those derived from equations (9.2) and (9.4).

No explicit calculation of the vertical joint shear V_{jv} is required by EC8; a previous version of the code had adopted equation (9.4), while in the final text (CEN, 1995) only detailing requirements for the vertical joint reinforcement are included (see next section).

(b) Design resistance verification

EC8 requires that the diagonal compression induced by the strut mechanism (section 9.2.2(b)) shall not exceed the bearing capacity of concrete. If more refined models are not available, the following application rules are suggested.

1. For interior joints

$$V_{jh}\leqslant 20\tau_{Rd}b_jh_c \tag{9.13(a)}$$

2. For exterior joints

$$V_{jh}\leqslant 15\tau_{Rd}b_jh_c \tag{9.13(b)}$$

In equations (9.13) the horizontal joint shear is derived using equation (9.11), h_c is the column depth and b_j is the **effective joint width** which may be estimated as follows:

1. If the column width exceeds the beam width ($b_c\geqslant b_w$)

$$b_j=\min\{b_c,b_w+0.5h_c\} \tag{9.14(a)}$$

2. If $b_c<b_w$ (which in general is not a recommended practice)

$$b_j=\min\{b_w,b_c+0.5h_c\} \tag{9.14(b)}$$

Equation (9.13) is equivalent to $\tau_{jh}\leqslant 0.25f_c$ or $0.20f_c$ (for interior and exterior joints, respectively); this can be seen if $f_{ct}\approx 0.1f_c$ is assumed and a simplified linear σ_c–σ_{ct}

biaxial failure curve is used to express the failure criterion for the concrete of the diagonal strut (Tassios, 1989b).

Instead of adopting the truss mechanism of shear transfer suggested by Paulay, Park and Priestley (1978), EC8 refers to a confinement mechanism, developing subsequent to the formation of open cracks at the beams framing into the joint, as well as of extensive diagonal cracks within the joint core. Under such conditions the need for adequate confinement (both horizontal and vertical) of the joint is recognized; for design purposes this confinement is deemed to be achieved if the maximum diagonal tensile stress of concrete is kept below the design value of tensile strength

$$\max \sigma_{ct} \leqslant f_{ctm}/\gamma_c \tag{9.15}$$

According to EC8, equation (9.15) is satisfied if horizontal hoops are provided to the joint core according to

$$\boxed{\frac{A_{sh} f_{yd}}{b_j h_{jw}} \geqslant \frac{V_{jh}}{b_j h_{jc}} - \lambda \sqrt{\tau_{Rd}(12\tau_{Rd} + v_d f_{cd})}} \tag{9.16}$$

and, in addition, vertical reinforcement is provided to the joint core according to

$$\boxed{A_{sv,i} \geqslant \frac{2}{3} A_{sh} \frac{h_{jc}}{h_{jw}}} \tag{9.17}$$

In equation (9.16) A_{sh} is the total area of hoop legs in the direction considered, h_{jw} is the clear distance between the axes of the top and bottom reinforcement of the beam and h_{jc} the clear distance between the extreme layers of column reinforcement, v_d is the normalized design axial force of the column (under the combination considered), and $\lambda = 1.0$ or 1.2 for DC 'H' and 'M' structures, respectively; the rest of the symbols have their usual meaning. In equation (9.17) $A_{sv,i}$ is the area of intermediate bars, located between the corners of the column.

The derivation of equation (9.16) using the criterion (9.15) may be found in Tassios (1989b),while equation (9.17) is easily derived if equation (9.4) is taken into account and it is assumed that intermediate column bars are subjected to compression approximately equal to 50% of their yield strength, thus offering a tensile stress margin of 1.5 $A_{sv,i} f_{yd}$ when they are required to act as confining reinforcement in the vertical direction.

From the practical point of view, it has to be emphasized that equation (9.16) recognizes the contribution of concrete mechanisms in carrying part of the joint shear, through the second term of the right-hand side (note that the λ-factor results in a larger concrete shear contribution for the lower ductility class). This is in agreement with the revised New Zealand model (Cheung, Paulay and Park 1992), and in fact equation (9.16) is qualitatively similar to equation (9.9), as both involve a concrete shear term which increases with the compressive axial loading.

Finally, a number of special cases which are not explicitly covered by EC8 should be briefly mentioned here. First, when beams are prestressed through the joint, the contribution of concrete to shear transfer is clearly higher than indicated by equation (9.16); in this case the CEB Model Code (1985) recommends that an additional term

$$V_{ch} = 0.7 P_{cs} \tag{9.18}$$

where P_{cs} is the force due to permanent loads in the prestressing steel located within the central third of the beam depth, should be added to the right-hand side of equation (9.16), with the appropriate (negative) sign. Furthermore, when parts ($A_{a1}$and A_{a2}) of the beam top and bottom reinforcement are bent vertically and anchored in the tensile face of the column, as shown in Figure 9.6, an additional term

$$V_{ch} = A_a f_{yd} \tag{9.19}$$

may be included in equation (9.16), thus reducing the required hoop reinforcement (CEB, 1985). Further 'concrete' terms (V_{ch}) recommended by the CEB Model Code for 'elastic' joints and for joints where beam bars are anchored outside the column core are given in section 9.2.5 (which refers to special types of joints).

(c) Detailing requirements

The horizontal confinement reinforcement in DC 'H' and 'M' beam–column joints should consist of hoops, satisfying the following minimum requirements:

1. The diameter of the hoops shall not be less than 6 mm.
2. The spacing of the hoops is determined by the expression.

$$s_w \leqslant \min\{h_c/4;\ 100\ \text{mm}\} \tag{9.20}$$

3. If beams are framing at all four faces of a joint, the spacing of the hoops may be increased to

$$s_w \leqslant \min\{h_c/2;\ 150\ \text{mm}\} \tag{9.21}$$

The paramount importance of hoops with regard to the shear behaviour of joints, as well as the favourable effect of transverse beams in statically indeterminate structures, have already been discussed in the foregoing sections. It is pointed out

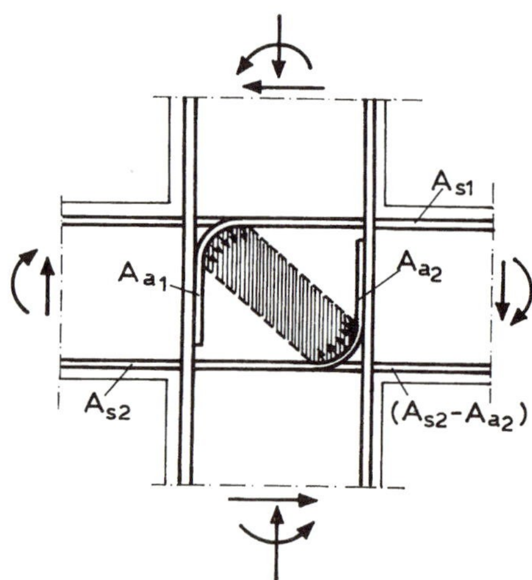

Figure 9.6 Shear transfer in a joint core through radial forces resulting from vertical bending of beam bars.

that for DC 'L' frames the sole requirement with regard to horizontal reinforcement in joint cores is that it should be equal to that provided in column critical regions (section 8.5.2).

With regard to vertical joint reinforcement, EC8 requires that at least one intermediate column bar be provided at each side of the joint, regardless of ductility class. In addition, for DC 'H' frames, the distance between consecutive column bars should not exceed 150 mm. It is worth mentioning here that in cases where the existing intermediate column bars do not satisfy equation (9.17), and /or the previous requirement, additional bars should be provided. extending either along the full column height or at least a full anchorage length ($l_{b,net}$) beyond each side of the joint core.

9.2.4 Anchorage of reinforcement in joints

(a) Bond within the joint core

Bond conditions for bars passing through beam–column joints are adversely affected by the following factors (Paulay, Park and Priestley, 1978):

1. Degradation of concrete due to extensive cross-inclined cracking of the core, caused by cycling at high levels of inelasticity. As mentioned in the previous section, the crack width in the joint core generally increases as the amount of hoop reinforcement decreases.
2. Yield penetration from longitudinal bars located in plastic hinges adjacent to the joint core (typically appearing in the beams). This is usually the main reason for bond degradation in the joint core.
3. Transverse tensile deformations of the concrete core due to the presence of bars orthogonal to the direction considered, that is of longitudinal bars in transverse beams.

With regard to bond stresses, the most critical situation arises in the case of interior joints, such as the one shown in Figure 9.7, where, as a rule, plastic hinges form at the beams. As is clearly seen in Figure 9.7 longitudinal beam bars passing through the joint core, when subjected to seismic loading, develop tension at one end and compression at the other. The tensile force at the one end ($+A_s f_y$ in Figure 9.7) may well exceed the yield capacity of steel, while the corresponding compressive force at the other may reach yield strength if $A_{s1} \approx A_{s2}$ (equal top and bottom reinforcement) and the compression zone in the beam is no longer available due to the presence of nearly full depth cracks. It follows that bond stresses developing at the interior of the joint core are required to carry a force up to approximately $f_y(A_{s1}+A_{s2})$. In general, the concrete surrounding the beam bars is not able to carry such stresses and a certain amount of slippage occurs (Paulay, Park and Priestley, 1978), which leads to significant local rotations ('fixed end rotations') at the beam–column interface (Figures 9.2(d), 9.7).

(b) Limitations regarding bar diameters

The distribution of bond stresses shown in Figure 9.7 is a very adverse one, since it involves yield penetration and subsequent bond degradation at both sides of

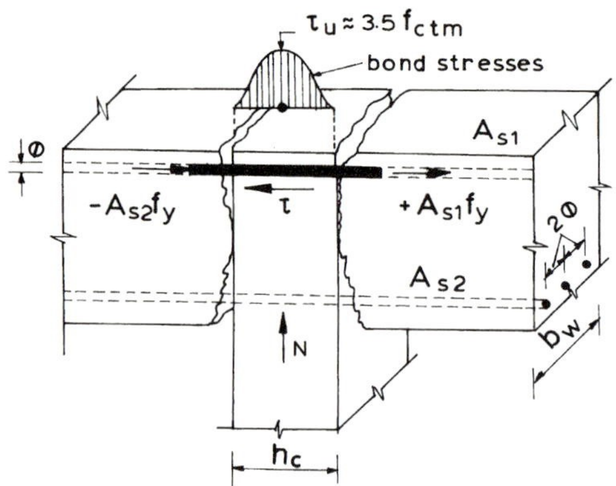

Figure 9.7 Bond stress distribution at an interior beam–column joint, where permanent open cracks have formed at the beam ends.

the joint. For such adverse conditions and for the case that highly ductile behaviour of the beam–column subassemblage is required, Paulay, Park and Priestley (1978) and the New Zealand Code (SANZ, 1982) suggest that the diameter of the beam longitudinal bars should be limited according to the equation

$$d_{bl} \leqslant \frac{11}{f_y} h_c \tag{9.22}$$

in order to minimize slippage within the joint core. Equation (9.22) has also been adopted by the CEB (1985) Seismic Code, with a slight modification (increase in allowable d_{bl} by about 20%) in the case of S400 steel, but only for high-ductility structures. The over-conservatism of this equation has been pointed out by Paulay and Priestley (1992), who suggested that the following factors would allow some relaxation of the severe limitation of equation (9.22):

1. In the typical case that $A_{s2} \leqslant A_{s1}$ (Figure 9.7) the force in the compression top bars will be significantly lower than $A_{s1} f_y$. A simple empirical equation has been suggested by Tassios (1989a) for estimating the force C_s in top bars in compression

$$C_s = A_{s2} f_y \frac{q}{q_{max}} \frac{\rho_1}{\rho_{max}} \tag{9.23}$$

where q_{max} and ρ_{max} are the EC8 maximum values for the behaviour factor and the beam flexural steel ratio. Equation (9.23) is based on the premise that the larger the top reinforcement, the higher the bond resistance with regard to closing of the open crack in the compression zone.
2. When gravity loads are more significant than the seismic ones, a plastic hinge at the joint face with the bottom bars in tension may never develop, and more favourable bond conditions will prevail.
3. Additional bond resistance may result from the presence of axial compression on the column which exerts clamping forces across splitting cracks formed parallel to the top beam bars.

Taking into account all the previous considerations, Tassios (1989a) suggested the following expression for the limiting d_{bl}/h_c ratio for beam bars anchored across interior joints

$$\frac{d_{bl}}{h_c} \leqslant \frac{7.5+8.5\mu_f(b_c/b_w)v_d}{1+(q/q_{max})(\rho_2/\rho_{max})} \frac{f_{ctm}}{\gamma_{Rd}f_{yd}} \tag{9..24}$$

where $\mu_f \approx 0.75$–1.0 is a friction coefficient required for taking into account the contribution of axial loading (expressed in non-dimensional form through v_d) to additional bond resistance, as mentioned previously. It is pointed out that in deriving equation (9.24) it was assumed that $f_{ctm} \approx 0.15 f_{cd}$, and also that equation (9.23) was taken into account. Equation (9.24) includes all the main parameters affecting bond conditions within the joint core, with the exception of the amount of hoop reinforcement, which is not directly incorporated in the formula, nevertheless it is a quite significant factor in providing improved bond conditions through its confining effect (see also section 7.6). An indirect way of including the effect of confinement reinforcement might be the appropriate selection of the μ_f coefficient (higher values for high confinement).

(c) Anchorage at exterior joints

Unfavourable anchorage conditions also appear at exterior joints, despite the fact that the developing bond stresses are not so high as in the case of the interior joints (the maximum force to be anchored through bond cannot exceed $\gamma_{Rd}A_{s1}f_y$ at the top bars). The development of splitting cracks along the beam bars, which should be anchored as far away from the interior column face as practicable, using a 90° hook as shown in Figure 9.8(a), adversely affects the efficiency of the part of the anchorage before the hook (Paulay and Priestley, 1992). The situation is aggravated by yield penetration from the adjacent beam plastic hinge, especially during inelastic cycling.

Particularly unfavourable conditions also develop with regard to the anchorage of column bars at the exterior face of the joint core. These bars are required to carry high bond stresses, due to the fact that they are in compression at one end of the joint and in tension at the other, while they are also affected by radial forces (compare Figure 9.6) developing at the hooks of the beam bars. A very common type of damage at these joints is the development of large splitting cracks parallel to the outer column bars, which lead to extensive spalling at the exterior face of the joint (Figure 11.14(c)). It is not uncommon that this spalling reduces the flexural strength of the columns above and below the joint (Paulay and Priestley, 1992), with a possibility of plastic hinge formation at the column ends. Tests by Ehsani and Wight (1985) have indicated that the presence of transverse beams at the joint improves anchorage conditions of beam bars, but not of outer column bars.

(d) Relevant code provisions

According to EC8, the diameter d_{bl} of the longitudinal beam bars passing through an interior beam–column joint should be limited by the following equation

$$\boxed{\frac{d_{bl}}{h_c} \leqslant \frac{7.5 f_{ctm}}{\gamma_{Rd}f_{yd}} \cdot \frac{1+0.8v_d}{1+k_D\rho_2/\rho_{max}}} \tag{9.25}$$

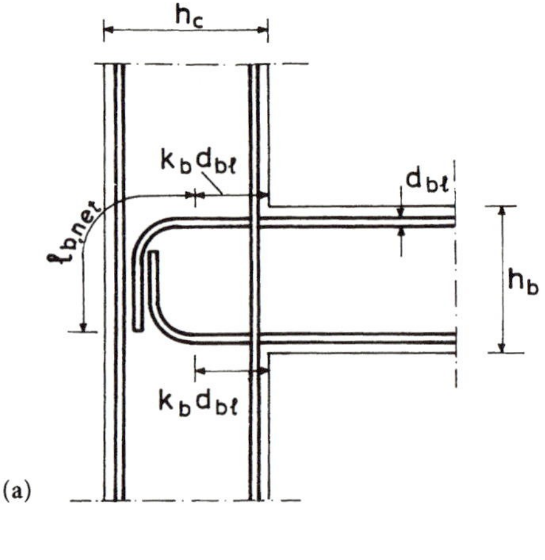

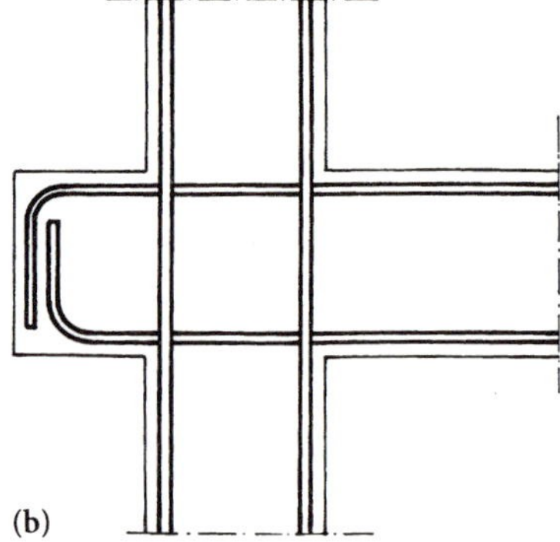

Figure 9.8 Anchorage of beam bars at exterior joint: (a) inside the column; (b) at an exterior anchoring stub.

where h_c is the dimension of the column parallel to the bars (Figure 9.3), $k_D=1.00,0.75,0.50$ and $\gamma_{Rd}=1.25,1.15,1.00$, for DC 'H', 'M' and 'L' respectively. It is seen that equation (9.25) may be derived from equation (9.24), assuming $\mu_f\approx0.75$ and $b_c\approx b_w$, if the coefficient of v_d, as well as the ratio q/q_{max}, are rounded for simplicity. For exterior joints the term $(1+k_D\rho_2/\rho_{max})$ may be taken equal to 1.0, which in general leads to larger allowable bar diameters than those given by equation (9.25). As a further simplification, EC8 suggests the following equations for the beam bars anchored along beam–column joints:

- For interior joints:

$$d_{bl}\leqslant\alpha_1\,(f_{ctm}/f_{yd})\,(1+0.8v_d)h_c \tag{9.26(a)}$$

  where $\alpha_1=4.0$, 4.5, 6.0 for DC 'H', 'M', 'L', respectively.
- For exterior joints:

$$d_{bl}\leqslant\alpha_2\,(f_{ctm}/f_{yd})\,(1+0.8v_d)h_c \tag{9.26(b)}$$

  where $\alpha_2=6.0$, 6.5, 7.5 for DC 'H', 'M', 'L', respectively.

Equations (9.26) are derived from equation (9.25) assuming $\rho_2 \approx 0.5\rho_{max}$, which is a rather conservative assumption (see the examples in section 9.2.6).

It is worth pointing out that no direct limitations are placed by EC8 on column longitudinal bars passing through joint cores, while in the CEB (1985) Seismic Code the limiting value 1/25 was set for the ratio d_{bl}/h_b, where d_{bl} is the diameter of the longitudinal column bars passing through the joint. It is quite clear that if (as an exception) a plastic hinge is expected to form in the column, the previously given equations (9.25) or (9.26) should also be applied to the column bars.

Finally, it is recalled here that in order to account for the possibility of yield penetration from the plastic hinges to the joint as discussed previously, EC8 requires that the anchorage length ($l_{b,net}$) of bars anchored within the joint core be measured from a point at a distance equal to $k_b d_{bl}$ from the face of the beam, where $k_b = 10$, 5, 0 for DC 'H', 'M' and 'L', respectively; this detail is shown in Figure 9.8(a) for beam bars.

9.2.5 Special types of joints

In this section brief reference is made to certain types of beam–column joints, where either the arrangement or the anchorage of reinforcement is carried out in a non-conventional manner. These arrangements, although in general leading to improved seismic performance of the joint, often cause construction difficulties and consequently they have not had wide application as yet.

(a) Elastic joints

The design of beam–column joints for shear and anchorage, as presented in the foregoing sections, poses two major construction problems:

1. Closely spaced hoops are often required in the joint core, resulting in difficulties with regard to placing the bars on site, as well as proper casting and compacting of concrete.
2. A large number of small-diameter bars may be required to satisfy the limitation on beam bar diameter (equations (9.25) or (9.26)).

One way to tackle these problems is to avoid the formation of beam plastic hinges at the beam–column interface, that is, at the face of the joint core. Popov et al. (1977) have long proposed the arrangement of beam reinforcement in such a way that plastic hinges form at the beam, but at such a distance from the column face as to ensure that no yield penetration occurs in the joint core. Various alternative arrangements have been studied experimentally (Popov *et al.*, 1977; Wong, Priestley and Park, 1990) and it was found that very good performance of the joint was achieved when half of the top beam reinforcement was bent downwards (at an angle equal to 60° with respect to the beam longitudinal axis) and half of the bottom reinforcement was bent upwards, the two intersecting at a distance equal to h_b from the column face. This arrangement of cross-inclined reinforcement prevented the slippage of beam bars at the joint, and stable hysteresis loops without significant pinching were obtained up to a displacement ductility of 4.0 (Popov *et al.*, 1977).

From the construction point of view it is easier to use intermediate beam bars (distributed between the top and bottom layers) which are terminated at such a distance from the joint that the beam moment at the column face can only reach yield if the moment at the (theoretical) point of termination of intermediate bars reaches a specified overstrength value (corresponding to a steel stress of $\gamma_{Rd}.f_y$). An improved joint behaviour was found in tests where this arrangement was used (Wong, Priestley and Park, 1990), but it was pointed out that the intermediate beam bars which pass through the joint should not be used as a total replacement for conventional hoop reinforcement in joint cores: a minimum amount of hoops is always necessary to confine the joint and control diagonal tension cracking.

When use of one of the previously mentioned arrangements of reinforcement ensures that the joint core will essentially remain in the elastic range, even for the most adverse case that no compression is present in the column, the required shear reinforcement for the joint core does not have to correspond to more than half of the joint shear, that is $V_{sh} \ngtr 0.5V_{jh}$ (Paulay, Park and Priestley, 1978). No specific mention is made in EC8 to the design of 'elastic' joints, but the CEB (1985) Seismic Code provides for an increased contribution of the concrete mechanism given by the equation

$$V_{ch} = \frac{A_{s2}}{A_{s1}} \frac{V_{jh}}{2} \left(1 + \frac{N_d}{0.4A_c f_{ck}}\right) \tag{9.27}$$

where N_d is positive for compressive column loads; for tensile loads causing a nominal stress $\sigma \geqslant 0.2f_{ck}$, V_{sh} should be taken equal to zero, while for $0.2f_{ck} > \sigma > 0$ linear interpolation between 0 and the value given by equation (9.27) may be used.

Finally, it has to be pointed out that, besides any potential construction problems caused by the use of non-conventional arrangements of reinforcement, the formation of plastic hinges away from the column face means that, for the same inter-storey drift, a larger rotation in the plastic hinge is required to satisfy the kinematics of the yield mechanism (Popov *et al.*, 1977). It follows that in these regions the need arises for confinement reinforcement and, more generally, for a detailing which ensures a fully ductile behaviour.

(b) Non-conventional reinforcing of the joint core

Park and Paulay (1975) have long pointed out that the typical shear reinforcement of joint cores, which consists of hoops and vertical bars, may be replaced by cross-inclined bars resulting from bending part of the longitudinal reinforcement of the beam. It is understood that such a detailing causes construction difficulties and it is infeasible to apply it in two orthogonal directions (when beams are framing into all four faces of a joint), hence it has not been applied in practical situations.

On the other hand, the results of tests at the University of Thessaloniki (Tsonos, Tegos and Penelis, 1992), where use was made of cross-inclined bars resulting from bending of column reinforcement, as shown in Figure 9.9, have shown superior performance of the non-conventionally reinforced beam–column subassemblages compared with similar specimens reinforced only with hoops and vertical straight bars. As shown in Figure 9.10, the hysteresis loops for specimens with cross-inclined column bars were considerably more stable and less pinched than the loops for similar conventionally reinforced specimens. The improvement was attributed mainly to the prevention of slippage of column bars within the joint core, and to the increase in shear strength caused by the cross-inclined bars (see also section 8.4.3(d)). It is also pointed out that for this arrangement of bars the

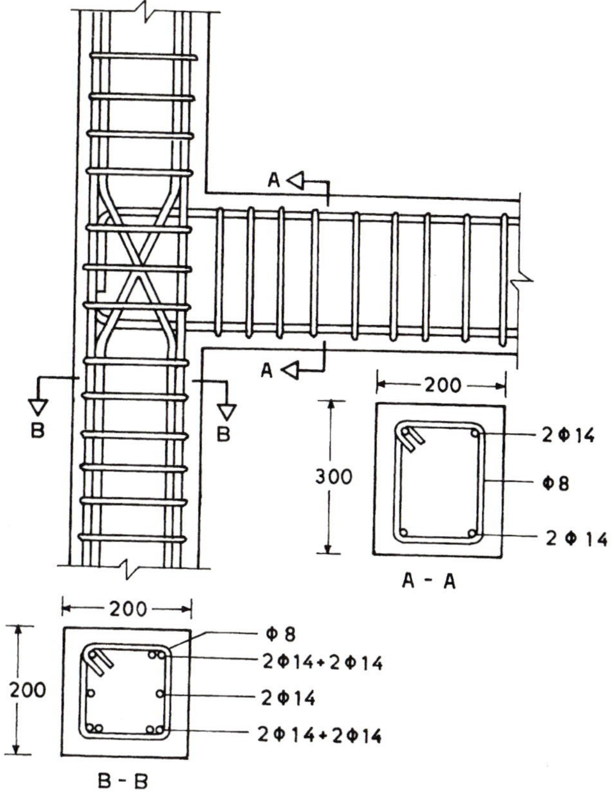

Figure 9.9 Arrangement of cross-inclined bars at an exterior joint core.

required hoop reinforcement is less than in conventionally reinforced joints. However, the use of cross-inclined column bars in two-way frames presents insurmountable difficulties. A possible way of resolving this problem might be the use of a pair of inclined bars placed along the diagonal of the column section; such a novel reinforcing pattern should, of course, first be studied experimentally to evaluate its efficiency as joint shear reinforcement.

(c) Special anchorages at exterior joints

In order to improve anchorage conditions of beam bars, as well as of longitudinal bars at the outer face of the column, in the case of exterior (T-) joints, it is possible to use an exterior stub (corbel), as shown in Figure 9.8(b) (Park and Paulay, 1975; Paulay, Park and Priestley, 1978). In this case radial bearing forces developing at the 90° hook of the beam bar are transferred to a concrete mass which is not subjected to the high joint shear, while at the same time the horizontal anchorage length is increased, which is a very desirable feature, especially in the case of small column sections. Furthermore, bond conditions for column bars at the outer face of the column are significantly improved and the possibility of spalling (section 9.2.4(c)) at this critical region is minimized. According to the CEB (1985) Seismic Code if the arrangement of Figure 9.8(b) is used, the 'concrete contribution' V_{ch} may be estimated from equation (9.27). Notwithstanding its advantages, the use of exterior stubs leads to constructional, as well as architectural problems, and their application has been rather limited, mainly to a few buildings in New Zealand (Paulay, Park and Priestley, 1978).

If, for any reason, longitudinal beam bars are not anchored with a 90° hook, it is possible to use **mechanical anchorages**, such as steel plates which are welded

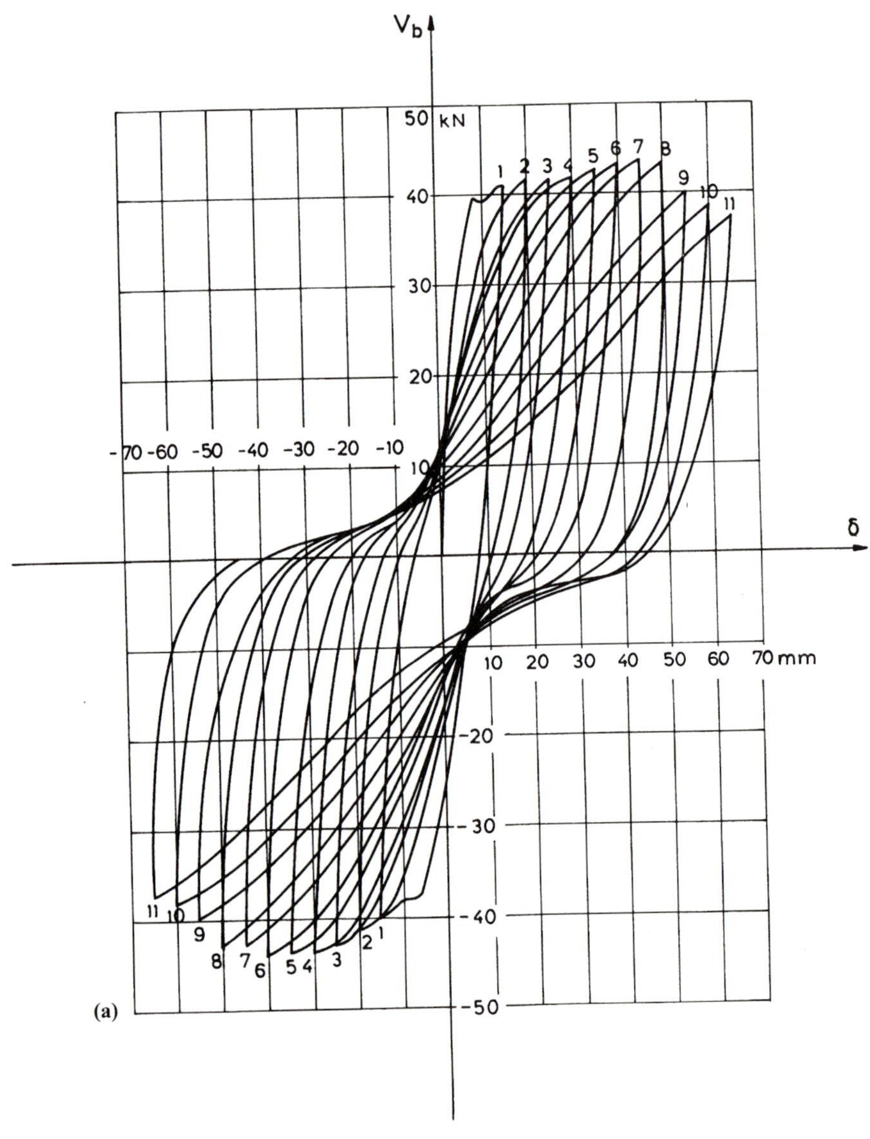

(a)

Figure 9.10a

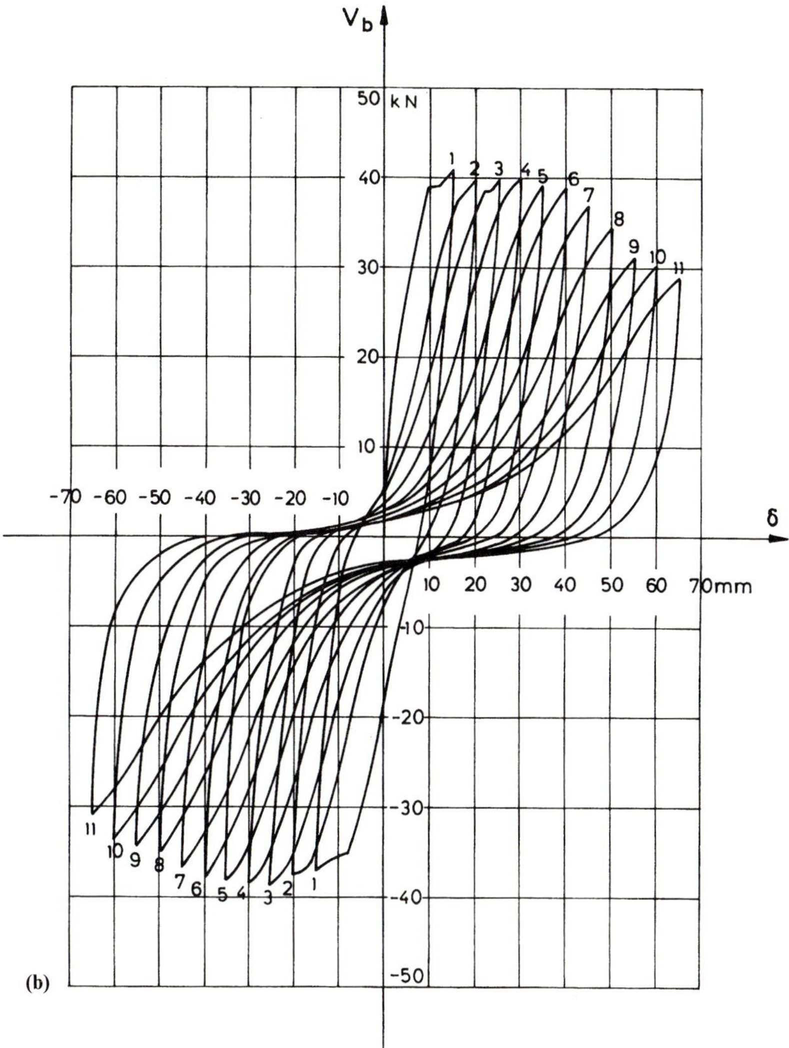

Figure 9.10 Hysteresis loops for exterior beam–column joints: (a) with cross-inclined bars; (b) with conventional reinforcement.

to the ends of beam bars. This solution was found to lead to improved performance both for exterior and interior joints (Paulay and Priestley, 1992).

9.2.6 Design example

In section 8.6 the design of the beams and columns of the 10-storey frame shown in Figure 8.35 was presented and the longitudinal and transverse reinforcement shown in Figure 8.38 was selected for the bottom part of the structure. In the following the design of the beam–column joints of this frame for shear and anchorage will be presented.

(a) Shear design and hoop requirements

The shear foreces expected to develop at the exterior joint of the ground storey will be estimated first. Referring to Figure 8.36(b), where the moment diagram for the seismic combination ($G+0.3Q+H$) is given, and using equation (9.12), the following horizontal shears are calculated:

- When the top beam bars (four 16 mm diam.) are in tension

$$V_{jh}=\gamma_{Rd}\left(\frac{2}{3}A_{s1}f_{yd}\right)-V_{col}=1.15\left(\frac{2}{3}\times 8.04\times 34.78\right)-68.0=146.4\text{ kN}$$

  where $V_{col}=68.0$ is the shear at the column above the joint under consideration (right-hand exterior joint in Figure 8.36(b)), and $\gamma_{Rd}=1.15$ is the overstrength factor for DC 'M' frames.
- When the bottom beam bars (three 16 mm diam.) are in tension

$$V_{jh}=\gamma_{Rd}\left(\frac{2}{3}\frac{q}{5}A_{s2}f_{yd}\right)-V_{col}$$

$$=1.15\left(\frac{2}{3}\times\frac{3.75}{5}\times 6.03\times 34.78\right)-48.4=72.2\text{ kN}$$

  where $V_{col}=48.4$ is the shear at the column below the joint considered.

As expected, the maximum joint shear develops when the top beam reinforcement (A_{s1}) which normally exceeds the bottom reinforcement (A_{s2}), is in tension; although not specifically mentioned in EC8, it is recommended to include in A_{s1} slab reinforcement within the widths specified in Figure 8.21.

The effective width of the joint can now be calculated using equation (9.14(a)), since $b_c=400>b_w=250$ mm.

$$b_j=\min\{b_c,b_w+0.5h_c\}=\min\{400,250+0.5\times 400\}=400\text{ mm}$$

Taking into account that the transverse reinforcement selected (in sections 8.6.2(b) and 8.6.3(b)) for the beam and the columns is 6 mm bars at 100 mm, and 10 mm bars at 90 mm, respectively, and also that the concrete cover is 30 mm, the rest of the joint core dimensions can be calculated:

$$h_{jc}=400-2\times 30-2\times 10-2\times 20=280\text{ mm}$$

$$h_{jw}=850-2\times 30-2\times 6-2\times 16=746\text{ mm}$$

The normalized axial loading, required for calculating the hoop requirements for the joint core, may be estimated using the value of the top column axial load developing simultaneously with the shear force ($V_{col}=68.0$) used in deriving V_{jh}, hence

$$\nu_d=\frac{N_d}{A_c f_{cd}}=\frac{1053.8}{0.4^2\times 13\,333}=0.494\text{ (compression)}$$

This value should be entered in equation (9.16) with a positive sign, to estimate the required area of hoops A_{sh} in the core

$$\frac{A_{sh}f_{yd}}{b_j h_{jw}} \geqslant \frac{V_{jh}}{b_j h_{jc}} - 1.2\sqrt{\tau_{Rd}(12\tau_{Rd} + v_d f_{cd})}$$

$$\frac{A_{sh} \times 347.8}{400 \times 746} \geqslant \frac{146.4 \times 10^3}{400 \times 280} - 1.2\sqrt{0.26(12 \times 0.26 + 0.494 \times 13.33)}$$

$$1.16 \times 10^{-3} A_{sh} > 1.307 - 1.906 = -0.599$$

It is seen that the concrete contribution (τ_{ch}, the second term on the right-hand side of equation (9.16)) exceeds the horizontal joint shear ($\tau_{jh} = V_{jh}/b_j h_{jc}$), thus no horizontal reinforcement is required. In such cases the minimum hoops according to equation (9.21) should be placed within the joint core (assuming that transverse beams are indeed framing into the joint)

$$s_w = \min\{h_c/2, 150 \text{ mm}\} = \min\{400/2, 150\} = 150 \text{ mm}$$

As 10 mm hoops at 90 mm centres are used for the column critical regions, it is in practice more convenient to select 10 mm hoops at 150 mm for the joint core; note that, in contrast to column hoops (Figure 8.38), the joint hoops do not have to be multiple ones.

The integrity of the diagonal strut mechanism is checked using equation (9.13(b)):

$$V_{jh} \leqslant 15\tau_{Rd} b_j h_c \Rightarrow 146.4 \leqslant 15 \times 260 \times 0.4 \times 0.4 = 624.0 \text{ kN}$$

With regard now to vertical shear requirements, equation (9.17) may be directly used to calculate the required vertical joint reinforcement ($A_{sv,i}$) in terms of the corresponding horizontal reinforcement (A_{sh}). Since the required $A_{sh} = 0$ in the joint under consideration, the minimum requirement consists in providing at least one intermediate column bar at each face of the column; the existing 18 mm bars are sufficient for this purpose.

The remaining exterior joints are designed for shear in a similar way to the one presented previously. It is pointed out that up to the sixth storey $\tau_{ch} > \tau_{jh}$ and no hoops are (theoretically) required. Column hoops are continued through the joint core at the reduced spacing of 150 mm up to the eighth storey, and only for the top two storeys, where the axial loadings are very low, has the spacing of hoops within the joints to be reduced (8 mm hoops at 120 mm and 6 mm hoops at 80 mm, for the ninth and the tenth storey, respectively). The existing intermediate column bars suffice as vertical reinforcement for all joints.

Considering now the interior joints, the horizontal joint shear acting at the interior joint of the first storey (Figure 8.38) may be estimated from equation (9.11):

$$V_{jh} = 1.15\left[\frac{2}{3}\left(9.6 + \frac{3.75}{5}7.6\right)34.78\right] - 120.2 = 287.8 \text{ kN}$$

The effective width of the joint is found from equation (9.14(a))

$$b_j = \min\{500, 250 + 0.5 \times 500\} = 500 \text{ mm}$$

The axial loading compatible with $V_{col} = 120.2$ is $N = 1535.0$ kN (compression) thus

$$\nu_d = \frac{1535.0}{0.5^2 \times 13\,333} = 0.460$$

The required hoop reinforcement can now be calculated from (9.16)

$$\frac{A_{sh} \times 347.8}{500 \times 746} \geqslant \frac{287\,800}{500 \times 376} - 1.2\sqrt{0.26(12 \times 0.26 + 0.46 \times 13.33)} = 1.53 - 1.86 < 0$$

that is $\tau_{jh} < \tau_{ch}$ and minimum reinforcement requirements prevail. Using the same diameter as in the column, 10 mm hoops at 150 mm spacing are selected for the joint core (Figure 8.38).

The integrity of the concrete strut is assured, since from equation (9.13(a))

$$V_{jh} = 287.8 < 20\tau_{Rd} b_j h_c = 20 \times 260 \times 0.5 \times 0.5 = 1300 \text{ kN}$$

Given that $A_{sh.req} = 0$, $A_{sv.i}$ is also zero from equation (9.17) and the existing 20 mm intermediate column bars are sufficient with regard to vertical joint reinforcement.

Similar checks are carried out for the remaining interior joints of the frame and it is found that, with the exception of the top storey joints, the column hoops spaced at the maximum allowable distance of 150 mm constitute an appropriate joint shear reinforcement, together with the already existing intermediate column bars.

It is concluded that, at least for the type of structure studied, the EC8 method for shear design of joints leads to very moderate requirements, which do not cause any construction difficulties.

(b) Anchorage requirements

For the exterior joints the EC8 limitation on beam bar diameter to ensure appropriate anchorage through the joint is given by equation (9.25)

$$d_{bl} \leqslant \frac{7.5}{\gamma_{Rd}} \left(\frac{f_{ctm}}{f_{yd}}\right) (1 + 0.8\nu_d) h_c$$

Introducing $\gamma_{Rd} = 1.15$ for DC 'M', $f_{ctm} = 2.2$ MPa for C20 concrete and $f_{yd} =$ 347.8 MPa, the equation becomes

$$d_{bl} \leqslant 0.041(1 + 0.8\nu_d) h_c$$

Using the minimum axial loading calculated from the analysis $\nu_d = 0.197$ for the bottom two storeys, where $b_c = 400$ mm for the exterior column, the previous equation gives

$$d_{bl} \leqslant 0.041(1 + 0.8 \times 0.197)\ 400 = 19.0 \text{ mm}$$

It is seen in Figure 8.38 that the maximum diameter used for the beam bars is 16 mm < 19, thus the requirement is satisfied. However, the limitation becomes more critical in the upper storeys, as both ν_d and h_c decrease. At the top of the building the maximum allowable diameter is only 13 mm, which might be a serious limitation; nevertheless for the frame considered four 12 mm bars are sufficient to satisfy flexural reinforcement requirements at the top storey beams.

For the interior joints the corresponding limitation regarding beam bar diameters is given from equation (9.25)

$$d_{bl} \leqslant \frac{7.5}{\gamma_{Rd}}\left(\frac{f_{ctm}}{f_{yd}}\right)\left(\frac{1+0.8\nu_d}{1+0.75\rho_2/\rho_{max}}\right)b_c$$

Introducing $\gamma_{Rd}=1.15$, $f_{ctm}=2.2$ and $f_{yd}=347.8$ the equation becomes

$$d_{bl} \leqslant 0.041 b_c \frac{1+0.8\nu_d}{1+0.75\rho_2/\rho_{max}}$$

At the ground storey interior support $A_{s2}=7.6$, thus $\rho_2=0.38\%$ while $\rho_{max}= 0.65\,(f_{cd}/f_{yd})(\rho_2/\rho_1)+0.0015=0.65\,(13.33/347.8)\,(0.38/0.47)=2.0\%$

For the minimum $\nu_d=0.298$ the allowable diameter for the two lower storeys is

$$d_{bl} \leqslant 0.041 \times 500 \times \frac{1+0.8 \times 0.298}{1+0.75 \times 0.38/2.0} = 22.2 \text{ mm}$$

It is worth pointing out that if the simplified expression (9.26(a)) is used instead of equation (9.25) the resulting limitation is

$$d_{bl}=4.5\frac{2.2}{347.8}(1+0.8 \times 0.298)500=17.6 \text{ mm}$$

which is considerably more stringent than the previous one. It is clear that for lightly reinforced beams, such as the one under consideration, it is preferable to use the basic equation (9.25).

As expected, the allowable beam bar diameter for the interior joints decreases at the upper storeys, down to a value of about 15 mm at the top of the building. Again, for the structure considered this limitation is not difficult to satisfy.

9.3 SEISMIC BEHAVIOUR OF WALLS

The role of **structural walls** (or 'shear walls') in cast *in situ* concrete structures is mainly the transfer of seismic actions. In many cases walls carry a major part of the seismic base shear, while the existing frames are designed primarily to act as a second line of defence against earthquakes, after extensive cracking and /or failure of walls. It is also worth pointing out that in countries where the construction of highly ductile R/C frames is not feasible practically or presents major difficulties, the use of walls is clearly the recommended choice with regard to earthquake resistance, in particular for medium and high-rise buildings.

9.3.1 Advantage of structural walls

The main advantage offered by earthquake-resisting R/C walls is the significant increase in the stiffness of the building, which leads to a reduction of second-order effects and a subsequent increase of safety against collapse, as well as a reduced degree of damage to non-structural elements, whose cost is often higher than that of the structural elements. Furthermore, the significant reduction of psychological effects on the inhabitants of high-rise buildings subjected to earthquake-induced displacements, should be pointed out.

Another advantage of structural walls is that, even after their extensive cracking, they are able to maintain most of their vertical load–bearing capacity, which is not always the case with columns.

A further advantage is that the behaviour of buildings with structural walls is generally more reliable than that of buildings consisting exclusively of frames. This is due to the fact that plastic hinges form at the beams and not at the walls (particularly when the latter have been designed according to the capacity procedures prescribed by modern codes), and also that the uncertainties resulting from the presence of masonry infills (with regard to structural regularity) are typically less significant in buildings with walls. The latter is a major advantage in the common case of asymmetric arrangement of masonry infills; it is pointed out that in the case of symmetrically arranged infills, frame systems benefit more, in the sense that the relative increase in strength (and stiffness) against lateral loading is higher than in similar structures with walls.

Despite the paramount importance of the aforementioned advantages, seismic codes until recently used to prescribe lower behaviour factors for buildings with structural walls, than for buildings with frame systems. For instance in the CEB (1985) Seismic Code q-factors for frame structures vary from 2.0 to 5.0, for coupled shear walls from 2.0 to 4.0 and for isolated walls from 1.4 to 2.8; it is seen that the values for structures with walls are up to 44% lower than for frames. In the American UBC (ICBO, 1994) the behaviour factor (or 'structural response modification factor') is 50% lower for buildings with R/C shear walls, compared with ductile frame systems; intermediate values are specified for medium-ductility frames and for dual (frame and wall) systems.

The main reason for penalizing structures with walls has been the possibility of non-ductile seismic behaviour, in particular of brittle type shear failure. However, the current trends consist in increasing behaviour factors for dual structures with R/C walls to values similar to those used for frame structures; this is supported by the results of revelant analytical studies (see, for example, Kappos, 1991) and is reflected in the most recent codes, such as EC8, wherein behaviour factors for dual systems are practically the same as for frame systems (Table 4.3). The adequate seismic performance of properly designed R/C walls will be seen from the discussion that follows.

9.3.2 Behaviour under monotonic loading

The ductility under monotonic loading conditions of a wall detailed to the provisions of modern design codes is high, as can be seen in Figure 9.11. The walls referred to in the figure are one-third scale models of the first three storeys of a 10-storey wall designed to the American code (UBC), including portions of the floor slabs monolithically cast with the wall (Bertero, 1980). The section of wall 3, which was subjected to predominantly monotonic loading, was rectangular with enlarged boundary elements (barbell section), that is, it corresponds to the common case of a wall extending a whole bay (from column to column). Web reinforcement consisted of two orthogonal grids, while the main flexural reinforcement was concentrated at the wall boundary elements surrounded by closely spaced hoops. This wall has shown a very large ductility ($\mu_\delta \approx 10$ with reference to the top displacement), without any noticeable drop in load-bearing capacity.

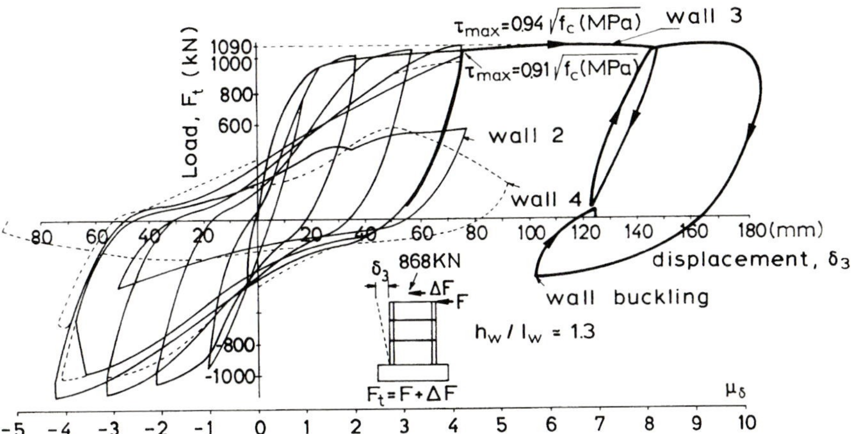

Figure 9.11 Load–displacement curves for R/C walls with barbell cross section subjected to monotonic (wall 3) and to cyclic (walls 2,4) loading.

However, when the sign of horizontal loading was reversed, the wall failed due to out-of-plane buckling for a shear force equal to only one-third of its maximum strength in the previous direction.

It is understood that the seismic design of R/C walls could only be based on their behaviour under reversed cyclic loading, which will be presented in section 9.3.3. Nevertheless, some fundamental characteristics of their inelastic response can be revealed from monotonic loading tests, such as that shown in Figure 9.11. One of these fundamental characteristics is that, when failure is caused by buckling, barbell or flanged sections offer an advantage over rectangular sections. Indeed, the small width of typical wall sections (usually ranging from 200 to 300 mm) leads to a drastic reduction of their stiffness in the weak (out-of-plane) direction after the spalling of parts of the cover concrete. Another characteristic regards the flexural strength of walls, which was found to be very little influenced by web reinforcement (Bertero, 1980). Nevertheless, a closer spacing of the bars in the grids leads to an increase in the ductility of the web of the wall, which, however, is not proportional to the grid reinforcement ratio. The flexural strength of a wall subjected to monotonic loading was found to be similar or slightly higher (2–11%, according to Endo, Adachi and Nakanishi, 1980) than the maximum strength under cylic loading. The monotonic load–displacement curve lies reasonably close to the envelope of the hysteresis loops up to failure, as shown in Figure 9.11.

The analytical determination of the flexural strength of a wall may be carried out using the general procedure (fibre modelling) and the corresponding computer codes mentioned in section 8.2.1(b). It has to be pointed out that the assumption that plane sections remain plane after bending (Bernoulli) does not strictly apply in the case of walls, especially those of low aspect ratio ($h_w/l_w < 2$), which behave like planar rather than linear elements. However, results from the application of the methodology originally developed for beams and columns are satisfactory for design purposes and in reasonable agreement with corresponding test measurements (Paulay, Priestley and Synge, 1982) on low slenderness wall specimens.

For the common case of walls with low axial loading, the flexural strength may be estimated using one of the following relationships (Tassios, 1984):

1. For walls with rectangular cross-section, reinforced with vertical grids corresponding to a reinforcement ratio $\rho v = A_{sv}/(b_w l_w)$, where b_w is the width and l_w the depth of the cross-section, and with concentrated reinforcement at the ends with an area $A_{s1} = A_{s2} = A_s$,

$$M_u = \left[(1 - \frac{\xi}{2})\frac{A_s}{b_w l_w} f_y + \frac{1}{2}(1 - \xi)(\rho_v f_y + \sigma_0)\right] b l_w^2 \tag{9.28}$$

where $\sigma_0 = N/(b_w l_w)$ is the average stress due to axial loading alone and

$$\xi = \left(\frac{A_s f_y}{b_w l_w f_c} + \rho_v \frac{f_y}{f_c} + \frac{\sigma_0}{f_c}\right)\frac{1}{1 + \rho_v f_y / f_c} \tag{9.29}$$

is the neutral axis depth ratio (x/l_w) at the ultimate limit state, while the rest of the symbols have their usual (previously defined) meaning.

2. For walls with barbell (or dumb-bell) section (boundary columns having a width l_c, as shown in Figure 9.20), with the main reinforcement ($A_{s1} = A_{s2} = A_s$) concentrated at the boundary elements

$$M_u \approx (A_s f_y + \frac{N}{2})(l_w - l_c') \tag{9.30}$$

that is the contribution of the vertical grid may be ignored, since, as already mentioned, it is negligible (with regard to flexural resistance only).

The application of equations (9.28) and (9.30) is limited to cases that wall failure is due to flexure and not to shear or buckling.

The analytical estimation of the **curvature ductility**, μ_ϕ (see equation (8.3)) for a wall section can be carried out with the aid of the general methodology described in section 8.2.1. However, it has to be pointed out that for such an estimation to be reliable, appropriate stress–strain (σ_c–ε_c, see section 7.4) relationships should be used for each part of the wall section, as the degree of confinement is different at the ends (where, as a rule, closely spaced hoops are present) and at the web, where confinement is provided only by the grid reinforcement. In any case such a type of analysis may only render a rough estimate of the available wall ductility which, as will be discussed in the following section, is usually significantly affected by the level of shear. An evaluation of more refined models for estimating the inelastic behaviour of wall elements may be found in Vulcano and Bertero (1987).

With regard to the definition of ductility factor for a wall, it should be noted that due to the presence of multiple layers of reinforcement, moment–curvature or moment–rotation curves are strongly curvilinear and their transformation into bilinear ones (such as that shown in Figure 8.16) is not carried out in the same way by all researchers, hence particular caution should be exercised when comparisons between measured and /or calculated ductility factors are made. Paulay, Priestley and Synge (1982) have suggested the determination of the yield displacement δ_y, required for the calculation of the displacement ductility μ_δ, on the basis of the point defined by the intersection of the line connecting the origin of the axes to the theoretical point of first yield and a line parallel to the horizontal axis (displacement) at a distance equal to the theoretical (analytical) maximum horizontal force, as shown in Figure 9.16(a).

Finally, the problem of estimating an appropriate value of equivalent plastic hinge length for ductility calculations in walls, is even more complicated than in the case of linear elements (section 8.2.1(e)). The region of flexural yielding ($M \geqslant M_y$), even when accurately defined, is only an indication of the actual plastic hinge length which is further influenced by shear and the (variable) axial loading, while the typically different extent of tensile and compression strains during the inelastic stage of the response should also be appropriately taken into account. The following empirical expression is suggested by Paulay and Priestley (1992):

$$l_p = 0.2 l_w + 0.044 h_w \not> 0.8 l_w \tag{9.31}$$

The value of the plastic hinge length calculated from equation (9.31) need not be taken less than $0.3l_w$, while in cases that the wall length l_w exceeds 1.6 times the first storey height (h_1), the value $1.6h_1$ should be used instead of l_w in calculating l_p.

9.3.3 Behaviour under cyclic loading

Probably the single most decisive factor affecting the seismic behaviour of a wall is its slenderness, commonly expressed in terms of the **aspect ratio** (height to length ratio) h_w/l_w. High slenderness walls ($h_w/l_w \geqslant 2$), when appropriately designed and constructed (section 9.4), are characterized by a ductile behaviour, failing in a predominantly flexural mode, similar to that of beams. On the other hand, in low slenderness or squat walls ($h_w/l_w \leqslant 1$) the factor dominating the seismic performance is shear, especially the possibility of sliding shear failure, which will be discussed in section 9.3.3(b).

(a) Slender walls

Characteristic load–displacement diagrams for high slenderness walls subjected to reversed cyclic loading are shown in Figure 9.12, which refers to a cantilever wall specimen with $h_w/l_w = 2.4$, a thickness $b = 102$ mm (one-third scale model) and a length $l_w = 1910$ mm; the boundary elements of the barbell cross-section consisted of 305 square columns (Oesterle *et al.*, 1980). The wall in the figure was reinforced with orthogonal grids ($\rho_v = 0.29\%$ and $\rho_h = 0.63\%$) and concentrated reinforcement in the boundary elements ($\rho_1 = \rho_2 = 3.67\%$), while during the cyclic loading a constant axial load corresponding to a nominal stress $\sigma_0 = 3.8$ MPa $= 0.08f_c$ was applied to the specimen. Due to the high reinforcement ratio at the edge columns (the specimen was designed to simulate a multi-storey R/C wall), the flexural strength of the wall was quite high and the corresponding shear resulted in $\tau_{max} = 0.9f_c^{1/2}$ MPa. Despite this, the confinement of the boundary columns with multi-leg hoops (having a ratio $\rho_w = 1.35\%$) led to rather ductile behaviour, similar to that of beams (compare Figure 8.9).

With reference to Figure 9.11, it is pointed out that the behaviour of the wall specimens subjected to cyclic loading was satisfactory; despite their relatively low slenderness ($h_w/l_w = 1.3$), they attained displacement ductility factors of the order of 4, under fully reversed cyclic shear. It is important that in the walls of Figure

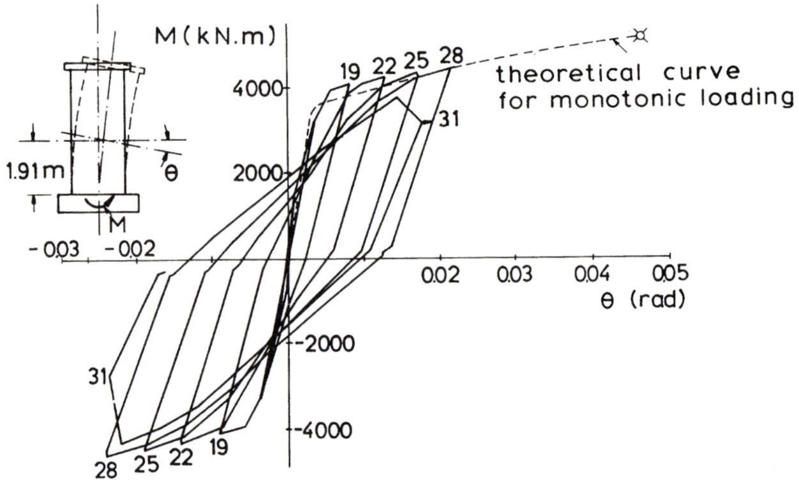

Figure 9.12 Moment–rotation hysteresis loops for slender R/C wall with barbell cross-section, subjected to cyclic loading.

9.11 it was observed that subsequent to the failure of their web, the edge columns remained in a good condition, maintaining their gravity load-carrying capacity.

Collected in Figure 9.13 are rotational ductility factors (μ_θ) measured in a series of tests involving R/C walls with different cross-sections (rectangular, barbell and flanged), plotted as a function of the corresponding maximum nominal shear stress, $\tau = V/(bd)$,where $d \approx 0.8\, l_w$ (Oesterle *et al.*, 1980). It is apparent from the figure that, as in the case of beams (section 8.2.2), the ductility of walls subjected to cyclic loading decreases with increasing values of shear stress. All the specimens of Figure 9.13 where shear stresses $\tau \geqslant 0.6 f_c^{1/2}$ MPa have developed,failed in a shear mode. More specifically it was observed that concrete regions between inclined shear cracks (the compression struts of the idealized truss mechanism) have crushed, with the final failure surface forming either at the wall base or at the interface between the web and the edge column (compare Figure 9.15(c)). An increase in the horizontal grid reinforcement was found to have a negligible influence on the ductility of columns failing due to web crushing, as can be seen by comparing specimens B7 and B8 in Figure 9.13, and taking into account that 2.2 times more horizontal grid reinforcement was provided in specimen B8.

Notwithstanding the previous remarks, the aforementioned shear failure mechanism cannot be classified as a brittle one, given that web crushing only occurred subsequent to the development of significant flexural and shear deformations.

The contribution of each mode of deformation (flexural, shear, fixed end rotation due to slippage of bars at the wall base) to the total displacement of the wall of Figure 9.12, can be seen in Figure 9.14 for two different sections of the specimen, one close to the base and one at a greater distance. It is quite clear that the fraction of the horizontal displacement resulting from shear deformations is quite large, especially in the region close to the base of the wall.

The level of axial loading acting on a typical wall in a building is in general small. The influence of axial load on the ductility of walls with predominantly

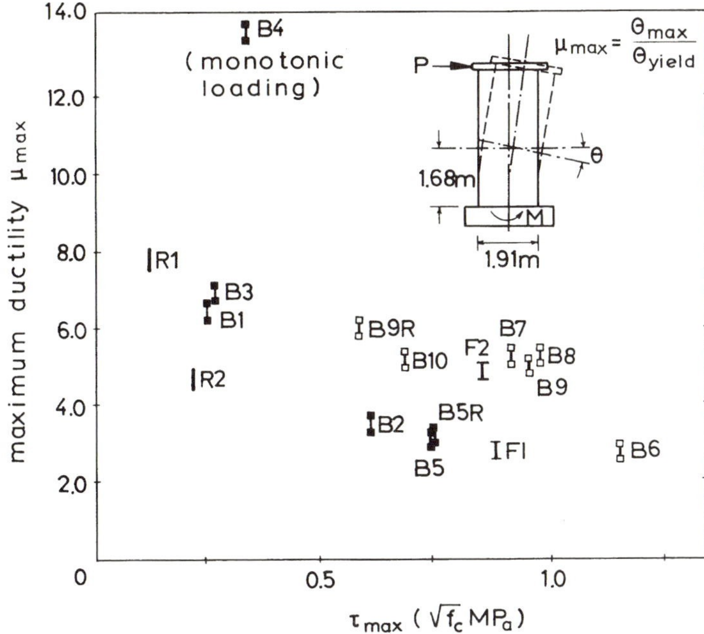

Figure 9.13 Rotational ductility factors, as a function of maximum shear stress for various types of walls (solid shapes indicate walls without axial loading).

flexural response (such as high slenderness walls with rectangular cross-section) is unfavourable, as it leads to an increase in the compression zone depth (section 8.4.2(a)). In contrast, in walls with high level of shear (such as those of Figures 9.11 and 9.12), the presence of compressive axial loading has a favourable effect, since it increases the effectiveness of the aggregate interlock mechanism, thus increasing the stiffness of the wall (Oesterle *et al.*, 1980). This can be seen by comparing the ductility of specimens B5 and B7 in Figure 9.13, which had the same reinforcement, but the second was also subjected to axial compression ($\sigma_0 \approx 0.08f_c$).

With regard to the possibility of out-of-plane buckling in thin-walled sections, it is pointed out that during cycling well into the inelastic range, previously wide open, nearly horizontal cracks at the extreme edge of the wall section may remain open when the sign of loading is reversed and compression develops in the edge previously subjected to tension. At this stage local distress (dislocated concrete particles) combined with the effect of even small amounts of bending in the weak direction of the wall section, may cause an eccentricity of the compression force within the thickness of the section and subsequently lead to significant out-of-plane curvatures. Second-order actions developing at this stage may cause out-of-plane instability of the part of section subjected to compression, especially if this part has a small thickness (no boundary columns or flanges). Paulay and Priestley (1992) suggest the following expression for the critical value of the wall thickness to avoid out-of-plane buckling:

$$b_c = 0.017 l_w \sqrt{\mu_\phi} \qquad (9.32)$$

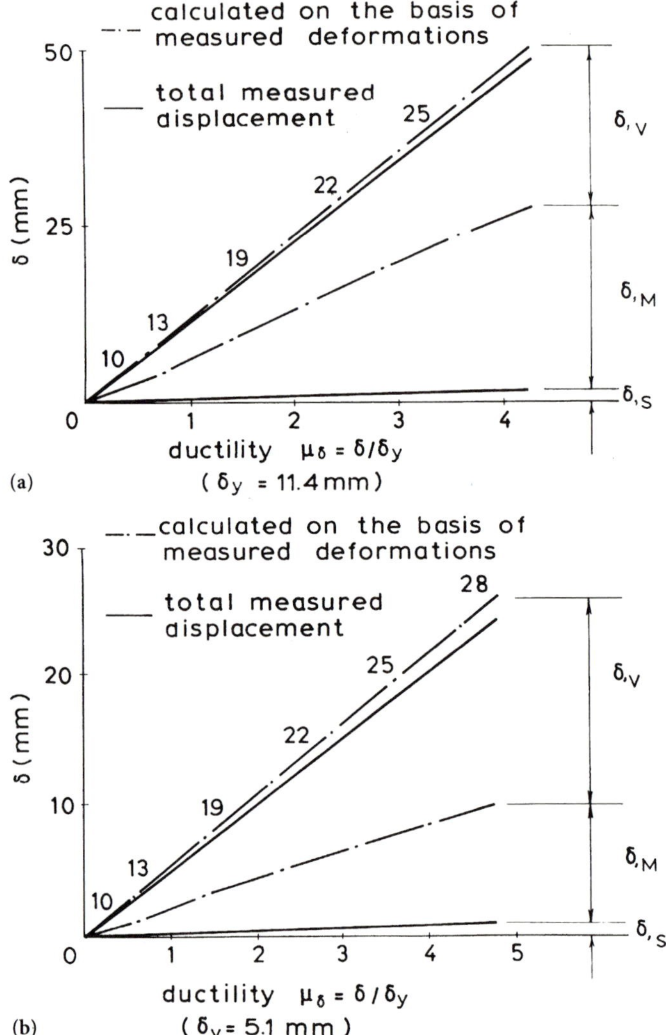

Figure 9.14 Contribution of different modes of deformation to the total displacement of the wall of Figure 9.12: (a) at distance of 1.8 m from the base; (b) at a distance of 0.9 m from the base.

This equation implies that the minimum thickness of a wall subjected to a curvature ductility $\mu_\phi \approx 10$ (which corresponds to a displacement ductility between 3 and 4 in typical situations) is 160 mm when $l_w = 3.0$ mm and 320 mm when $l_w = 6.0$; the example points to the need to increase the stiffness of free edges in elongated walls with rectangular cross-section (see also section 9.4.3).

Summarizing the discussion regarding slender R/C walls, it is pointed out that they behave in a ductile manner, similar to that of beams, on condition that their boundary elements are appropriately designed and detailed for a high degree of confinement (section 9.4.2). Whenever the shear reinforcement is adequate, the

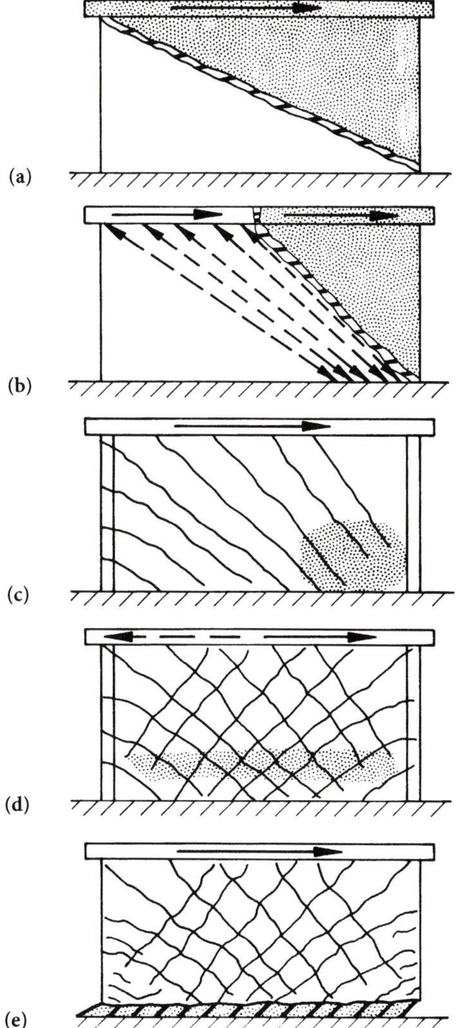

Figure 9.15 Shear failure modes in squat structural walls.

typical mode of failure in walls with rectangular cross-section is out-of-plane buckling and in walls with barbell or flanged cross-section it is web crushing; if no adequate shear reinforcement is present, walls typically fail due to diagonal tension (Figure 9.15(a)). It has to be further mentioned that in high slenderness walls, which as a rule form part of multi-storey R/C buildings, large bending moments develop at their base, which may lead to uplift of their foundation and a subsequent **rocking** (rotation) of the wall about a point close to the compression edge. This is almost inevitable in cases where the stiffness of the foundation system is relatively low (for instance when isolated footings are used instead of a raft). Rocking at the wall base has a favourable effect with regard to flexure (as it decreases the value of bending moment with respect to fully fixed conditions), but not with regard to shear, hence a shear mode of failure may eventually be induced. In the common case that in addition to walls, frames are also present in the building (dual systems), the rigid-body rotation caused by rocking of the wall may increase the rotational ductility demand in beams of frames lying on the tension side of the wall.

(b) Squat walls

In structures such as low-rise buildings it is possible to have walls with a height not exceeding their length in the horizontal direction ($h_w/l_w \leqslant 1$). If these structures are designed for the same behaviour factor as those with slender walls, they are expected to suffer a higher degree of damage. This is due to the fact that because of the lower natural period (of the order of 0.2–0.5, even after inelasticity has occurred) of these structures, the number of load cycles after yielding of critical regions is larger than that in corresponding regions of structures with higher natural periods (Hiraishi *et al.*, 1989), hence the seismic damage (expressed as a cumulative ductility factor) is expected to be higher. For this reason, and in addition because squat walls usually fail in a shear mode, as will be discussed subsequently, it is recommended that these structures are designed for elastic behaviour ($q = 1$). Given that the resulting design actions for low-rise buildings are usually not very high, the resulting reinforcement requirements for walls are in general not excessive (Park and Paulay, 1975; Paulay, 1980). The New Zealand Code (SNZ, 1992) provides for a linear increase of design base shear up to 60%, when the ratio h_w/l_w in a wall reduces from 2.0 to 1.0. The corresponding EC8 requirement (k_w-factor) was given in section 6.1.5; note that according to this requirement the q-factor for wall systems with $h_w/l_w \geqslant 3.0$ is 100% higher than for walls with $h_w/l_w = 1.0$.

The possible failure modes of a squat shear wall are schematically shown in Figure 9.15 (Paulay, Priestley and Synge, 1982). When no adequate horizontal reinforcement is present, failure of the wall due to diagonal tension occurs, along a diagonal (Figure 9.15(a)) crack. This mode of failure may be inhibited if in the design of the wall it is ensured that the grid reinforcement (horizontal and vertical bars) is able to carry a shear higher than that developing at flexural overstrength (capacity design procedure). The relevant EC8 design equations, which recognize both the contribution of 'concrete mechanisms' (V_{cd}) to shear resistance, and the effect of aspect ratio, are given in section 9.4.2. Here it is pointed out that while for slender walls ($h_w/l_w \geqslant 2.0$) only horizontal shear reinforcement is required (as for beams and columns), thus only nominal vertical grid reinforcement need be provided, in squat walls vertical shear reinforcement (vertical bars of the grid) is necessary for the development of the truss mechanism. As seen in Figure 9.15(b) the equilibrium of inclined concrete struts not crossing the boundary elements (where vertical bars are present) cannot be achieved if vertical forces are not present; these are offered by vertical bars and/or axial compression.

Whenever the shear reinforcement is adequate, a **diagonal compression failure** may occur, such as shown in Figure 9.15(c), consisting in crushing of the concrete compression struts in the web of the wall; the compressive strength of these struts is drastically reduced under cyclic loading conditions, since inclined cracks in two directions (Figure 9.15(d)) develop. As already mentioned with reference to slender walls, this web crushing failure may occur in walls with boundary elements (columns or flanges), subjected to a high level of shear stress. This mode of failure can only be inhibited if the average shear stress in the wall critical section does not exceed a certain limit ranging from $0.5f_c'^{1/2}$ to $0.9f_c'^{1/2}$, depending on the ductility requirements imposed on the wall (Park and Paulay, 1975; Oesterle *et al.*, 1980).

A typical failure mode in squat shear walls is due to **sliding shear**, commonly associated with low levels of axial loading and high levels of shear stress. This mode of shear failure, shown in Figure 9.15(e), is similar to that observed in beams subjected to high levels of cyclic shear, discussed in detail in section 8.2.2(d). Its main characteristic is that excessive displacements along a horizontal open flexural crack take place, which implies a drastic reduction in the stiffness of the wall and in its energy dissipation capacity.

Notwithstanding the risk of sliding shear failure, a squat shear wall designed according to modern design codes may show a quite satisfactory seismic behaviour. An example of such a wall is shown in Figure 9.16(a), which refers to a rectangular wall specimen with $h_w/l_w = 0.5$, whose reinforcement was designed to carry a shear 25% higher than that corresponding to the development of flexural

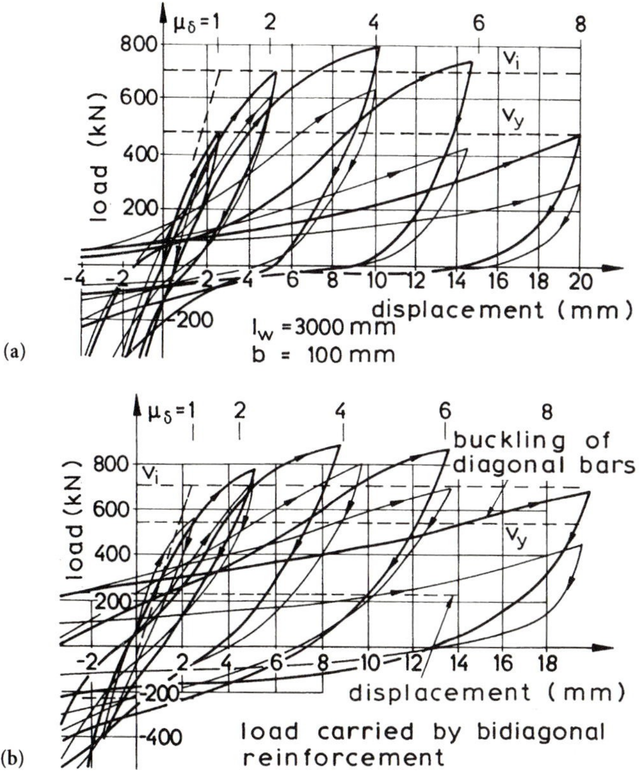

Figure 9.16 Hysteresis loops for squat shear walls ($h_w/l_w = 0.5$) subjected to cyclic loading: (a) conventionally reinforced wall; (b) wall with bidiagonal bars.

strength, with due allowance for strain-hardening (Paulay, Priestley and Synge, 1982). This wall was found to maintain a strength in excess of the theoretical one (V_i in Figure 9.16), up to a displacement ductility of 4.0 under reversed loading conditions. At higher levels of imposed ductility a gradual deterioration of strength was detected, and the wall finally failed in sliding shear at its base. It is worth mentioning that during the second cycle at a displacement ductility $\mu_\delta = 6.0$ the sliding (i.e. the horizontal displacement) measured at the wall base was equal to 65% of the total deflection at the top of the wall.

As was the case with beams, the adverse effects of sliding shear may be limited by using cross-inclined bars, crossing the open failure crack. The wall of Figure 9.16(b) was different from that of Figure 9.16(a) in that cross-inclined (bidiagonal) bars were added (Figure 9.19) designed to provide 30% of the theoretical resistance to shear. In this wall. despite the fact that the level of shear stress was slightly higher than in the wall of Figure 9.16(a) ($0.70f_c^{1/2}$, as opposed to $0.65f_c^{1/2}$), the shear strength under cyclic loading conditions was maintained up to a displacement ductility of 6.0. Moreover, the amount of energy dissipation was larger than in the wall without bidiagonal bars, as can be seen from the width of the hysteresis loops, particularly in the region around the origin of the axes. In the wall of Figure 9.16(b) the sliding at the base at $\mu_\delta = 6.0$ represented only 40% of the total horizontal displacement at the top. An even more favourable behaviour would be expected if the bidiagonal reinforcement had been designed to carry a larger fraction of the shear. On the other hand, this solution presents more construction difficulties than the conventional one of two grids, and it has not met wide application so far. The recognition of its effectiveness by EC8 (section 9.4.2) is expected to change this situation in the near future.

Analytical expressions for the verification of walls against sliding shear may be found in Paulay, Priestly and Synge (1982); the relevant EC8 equations are given in section 9.4.2.

9.3.4 Walls with openings

In practical construction R/C walls are usually pierced with openings to accommodate doors, windows, or even utility ducts. Depending on the size of the opening, the resulting element may be defined as a perforated (or pierced) wall or as a coupled wall. The seismic behaviour of these elements is outlined in the following subsections.

(a) Perforated walls

When the size of the opening is small (which is typically the case with perforations accommodating ducts or pipes or windows of normal size), and/or when the arrangement of openings does not follow a regular pattern, a perforated or pierced wall results. No clear definition of a 'small' opening is usually included in design codes (EC8 among them) but a reasonable criterion might be to limit the opening to wall area ratio to 10 or 15% (at each single storey, not only in an average sense along the height). With regard to the arrangement of openings, EC8 recommends avoiding random openings in walls, unless their influence is either insignificant (see the previous criterion), or taken into account by means of appropriate analysis, dimensioning and detailing. However, no guidance is given in EC8 with respect to the treatment of all these aspects (analytical model, detailing rules).

The behaviour of perforated walls in the inelastic range has been studied to a much lesser extent than solid walls (section 9.3). A comprehensive study on the response of walls with openings to monotonic loading to failure was presented by Yamada, Kawamura and Katagihara (1974), where various sizes of openings in single-storey walls with boundary elements were studied and two failure modes were identified: a rather brittle mode due to diagonal compression in side walls (vertical piers), and a ductile mode due to flexural yielding of the horizontal segments consisting of the beams and the parts of the wall above or below the opening. The thickness of the wall and the amount of reinforcement around the opening were identified as the main parameters influencing the type of failure mode, and a procedure for relating these parameters and the relative area of the opening to the failure mode was suggested.

With regard to the behaviour of walls with openings subjected to cyclic loading the available experimental data are even more limited. Daniel, Shiu and Corley (1986) have tested a one-third scale six-storey wall with centrally located openings corresponding to an opening to wall area ratio of 8.3%, as well as a similar solid wall specimen. Both specimens exhibited large deformation and energy dissipation capacities and their shear strength was reached after the development of large elastic shear deformations in the wall boundary elements. The construction practice (common in the USA) of placing around the opening the vertical bars corresponding to the area of the opening, was judged to function well and no specially designed boundary elements at the sides of the opening were deemed to be necessary.

The inelastic seismic behaviour of walls with staggered openings was studied by Ali and Wight (1991), who tested three one-fifth scale models of a five-storey wall with staggered door openings (Figure 9.17(a)) and a similar solid wall (Figure 9.17(b)). The opening to wall area ratio was equal to 13.4%, while various positions of the staggered openings (various distances from the boundary elements) were studied. As shown in Figure 9.17, both the solid and the pierced walls developed a similar maximum strength and sustained an inter-storey drift of at least 1% without major damage. However, at higher drifts (1.25–1.50%) the walls with openings experienced a shear-compression failure in the end pier (the segment between the opening and the compression edge of the wall), similar to that reported by Yamada, Kawamura and Katagihara (1974) for monotonic loading. When a flexural-shear crack penetrated to the upper corner on the compression side of the opening, it initiated a diagonal splitting of the compression zone, accompanied by compression crushing of the boundary element. The tests showed that door openings located too close to the edge columns (the wall boundary elements) remove the in-plane confinement present in similar solid walls and may cause an early shear-compression failure. It is worth pointing out that at least one multistorey building in Valparaiso, Chile, which had staggered door openings similar to those of the aforementioned specimens (but a relative opening area of only 7.6%) did not suffer significant damage during the $M = 7.8$ earthquake of 1985, which caused considerable damage and casualties in central Chile (Ali and Wight, 1991).

It has long been recognized that it is quite undesirable from the seismic behaviour point of view to discontinue massive structural walls at the top of the ground storey, replacing them with columns (a practice which offers certain architectur-

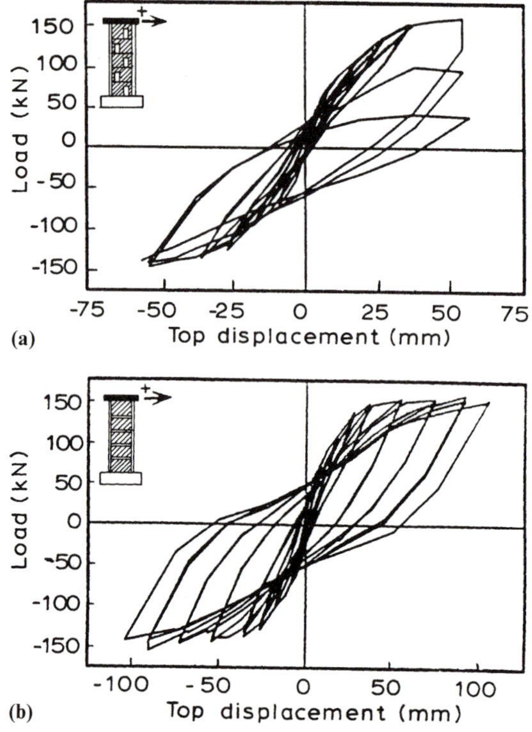

Figure 9.17 Hysteresis loops for **(a)** a perforated wall with staggered openings and **(b)** a corresponding solid wall.

al advantages). The seismic overturning moment is expected to impose very large axial forces (compression, as well as tension) on the columns supporting the discontinued wall, which combined with the high ductility requirement, may lead to failure of the columns in shear and/or compression; this was observed in actual structures hit by strong earthquakes (Park and Paulay, 1975). Shaking table tests on small-scale models of buildings with discontinued walls (Shen and Huang, 1990) (have shown better seismic performance of buildings with full walls at the ground storey (in addition to the discontinued ones), compared with that of similar systems where only columns were present at the ground storey.

(b) Coupled walls

Whenever the relative area of the openings is not small and their arrangement is rather uniform along the height of the wall, the resulting structural system is neither a solid wall nor a real frame, and is referred to as a coupled wall. To avoid confusion, EC8 defines the coupled wall as a 'structural element composed of two or more single walls, connected in a regular pattern by adequately ductile reinforced beams, called coupling beams, able to reduce by at least 25% the base bending moment of each single wall working as a separate cantilever'.

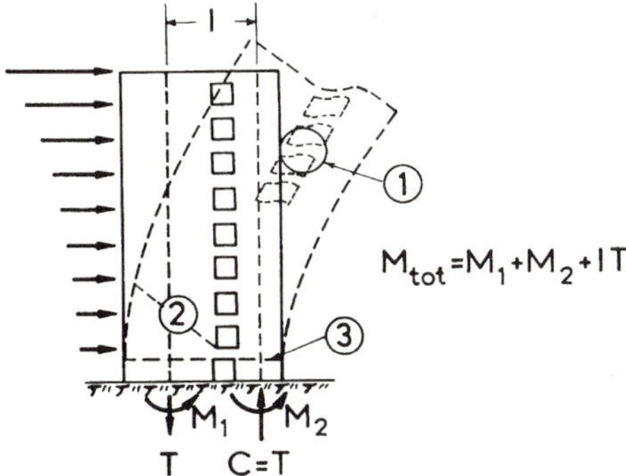

Figure 9.18 Critical regions of coupled walls subjected to horizontal loading: (1) coupling beams; (2) locations of main diagonal tension crack; (3) construction joints.

As shown in Figure 9.18, the total base moment carried by a coupled shear wall is the sum of the bending moments acting at the base of each single wall, and the product $l \cdot T$ (equal and opposite axial forces $C = T$), whose magnitude depends on the distance (l) between the centroids of the single walls and the level of axial loading which can be sustained by each wall. For a given geometry it is understood that the larger the strength of the coupling beams the higher the axial loading on the walls resulting from the beam shears, in other words coupling through strong beams increases the fraction of the seismic overturning moment carried by the couple of wall axial loads and reduces the moments in the walls.

Coupled walls when appropriately designed and detailed, may be a highly ductile structural system, able to dissipate significant amounts of hysteretic energy through flexural yielding of the coupling beams and possibly of the lower part of the walls (which is detailed for a ductile response). The fundamental difference between a frame and a coupled wall is that the relative flexural strengths and stiffnesses of the (coupling) beams is one or even two orders of magnitude lower than that of the adjoining walls, thus it is inevitable that the beams will yield and the walls will remain in the elastic range, except perhaps at their base. The latter depends on the strength of the beams and the intensity of the input motion; ideally the walls will remain elastic for an earthquake corresponding to the serviceability limit state, and will yield at their base when subjected to the design earthquake (ultimate limit state). Whenever the system is subjected to high levels of loading (possibly higher than the one they were designed for), the coupling beams function as 'fuses' by preventing the walls from being seriously damaged; subsequently, the beams may be repaired or even replaced without significant loss of function of the building (Abrams, 1991).

On the basis of the foregoing discussion, it is clear that coupled walls lend themselves to a capacity design procedure, as only the beams need to be reinforced for their full flexural strength, which can be estimated simply and reliably in order to determine subsequently the forces acting on the walls. The uncertainties regarding the mitigation of hinge formation in vertical elements (section 8.4.1) which were significant in frame structures, are minimized if not completely removed in the case of coupled walls.

It has to be emphasized that all the aforementioned advantages of this system are subject to the condition that coupling beams do not fail prematurely due to lack of ductility. These beams are characterized by their low slenderness (low values of the ratio l/h) and their seismic behaviour is similar to that of short columns (section 8.4.3(d)) and of beams subjected to high levels of shear (section 8.2.2(c)). In discussing the behaviour of these elements the risk of sliding shear failure was pointed out and the necessity of using bidiagonal reinforcement was made clear; the corresponding detailing requirements for coupling beams are outlined in section 9.4.5.

Following the pioneering work of Paulay and Santhakumar (1976) on the seismic behaviour of coupled walls, a number of interesting experimental studies have appeared, some involving small-scale (one-tenth or less) models tested on the shaking table (Aristizabal-Ochoa, 1982), and other large-scale models (about one-third) tested under static cyclic loading (Shiu, Takayanagi and Corley, 1984); a complete list of relevant references may be found in Abrams (1991). The main objective of most of these studies was the determination of the **degree of coupling** (which can be expressed as the ratio lT/M_{tot}, see Figure 9.18) that leads to an optimum seismic performance; it is understood that in lightly coupled systems (relatively weak beams) the governing parameter is the ductility of the beams, while in heavily coupled walls (strong beams) the performance of the system depends primarily on the strength and especially the ductility of the walls. Although no generally accepted quantitative criteria on the optimum degree of coupling are available (nevertheless the 25% lower bound of EC8 is recalled here), it can be said that for optimum performance, coupling beams should not be made so strong that the system behaves as a cantilever element (Shiu, Takayanagi and Corley,1984); redistribution of shear from the tension to the compression wall pier should be allowed to take place through the coupling beams, but the amount of axial loading developing on the walls (compression, as well as tension) should not be such as to cause serious problems of flexural ductility or shear degradation (see also sections 8.4.3(a)(b)).

Some difference of opinion appears to exist with regard to the detailing of coupling beams for optimum seismic performance. The New Zealand school of thought (Park and Paulay, 1975; Paulay and Santhakumar, 1976) clearly favours the use of bidiagonal reinforcement, properly anchored and confined, as described in section 9.4.5. In the aforementioned tests by Paulay and Santhakumar (1976), it was found that the conventionally reinforced coupling beams of the tested seven-storey coupled wall failed in sliding shear after several cycles of reversed cyclic loading well into the inelastic range, which indeed is not a poor seismic performance. On the other hand, the beams in a similar coupled wall, which were diagonally reinforced, have shown no significant damage when subjected to the same loading history. In the tests by Shiu, Takayanagi and Corley (1984), it was found that conventionally reinforced beams with a span to depth ratio $l/h \approx 1.25$ were quite effective in coupling walls together and did not fail in a brittle manner. Moreover, in the study by Daniel, Shiu and Corley (1986) mentioned in the previous section (on perforated walls), it was also confirmed that the lintels (coupling beams) effectively coupled the wall piers without the use of bidiagonal reinforcement. It is believed that, similar to the case of beam–column joints discussed in section 9.2, the difference of opinion is mainly a difference in performance criteria adopted by each group of investigators. In general, it is quite clear that carefully detailed bidiagonal reinforcement in coupling beams can ensure a highly ductile behaviour (such as that required in DC 'H' structures, or possibly even higher), while it may also be maintained that conventionally reinforced coupling beams may satisfy DC 'M' and 'L' performance criteria, provided their slenderness does not fall below certain limits, depending on the level of shear stress (section 9.4.5).

9.4 SEISMIC DESIGN OF WALLS

9.4.1 Design for flexure and axial loading

The design of a wall for the bending moments derived using the capacity design procedure outlined in section 6.1.5(a) (see design envelope in Figure 6.6) and the corresponding axial loads, may be carried out in a simplified way by considering an (elongated) rectangular section with concentrated reinforcements at its ends, that is by ignoring the contribution of vertical bars in the grids to the flexural strength. This approach which is conservative with regard to the reinforcement required in the boundary elements, allows the design of walls to be carried out using the same design aids as for rectangular columns (see for instance the design chart of Figure 8.39). However, the approach is not conservative with regard to the shear design of the wall, as it underestimates the actual flexural resistance (M_{Rd}) of the wall required for calculating the magnification factor (equation(6.10)) for shear forces.

A more accurate assessment of the strength of a wall under combined flexure and axial loading can be made either by using equation (9.28) with design material properties, or with the aid of design charts (if available) or software packages. A number of commercial computer codes include modules for the design of polygonal cross-sections with a large number of reinforcement layers, which can be readily used for walls.

In the critical regions of walls (Figure 6.7), where the boundary elements have to be properly confined, the limitations on minimum and maximum longitudinal reinforcement ratio specified by EC8 for columns (section 8.5.1(b)) also apply to the wall boundary elements.

9.4.2 Design for shear and local ductility

In section 6.1.5(b) the EC8 capacity design procedure for deriving the design shear forces for walls, using the magnification factor ε (equation (6.10)) has been outlined. In the following sections the verification procedures against the various possible shear failure modes are presented and discussed.

(a) Verification against diagonal tension failure

The standard EC8 expression for diagonal tension verifications $V_{Sd} \leqslant V_{Rd3}$, where from equation (8.46)

$$V_{Rd3} = V_{cd} + V_{wd}$$

may also be applied to walls, by taking properly into account the effect of the **shear ratio**

$$a_s = \frac{M_{Sd}}{V_{Sd} l_w} \tag{9.33}$$

which depends on the aspect ratio (h_w/l_w) of the wall, although in general it does not coincide with it; for a single-storey (cantilever) wall it is obvious that $a_s = h_w/l_w$ at the base, while for a multi-storey wall subjected to an inverted triangular distribution of horizontal forces, the moment (M_{Sd}) at its base is equal to (2/3) $h_w V_{Sd}$ (where V_{Sd} is the base shear), thus $a_s = (2/3)h_w/l_w$. Caution should be exercised in applying equation (9.33), to introduce M_{Sd} and V_{Sd} values calculated for a given lateral load distribution (or from a dynamic analysis) and not the capacity design values specified in section 6.1.5, since the purpose of a_s is to provide an indication of the effective slenderness of the wall and the actual moment to shear ratio, which influences the mode of failure. Moreover, when stiff horizontal members such as relatively deep beams or continuous perimeter walls at building basements, significantly affect the deformation patterns of the wall, the shear ratio should be calculated at every storey where a discontinuity exists, rather than at the base only.

The following three cases are distinguished by EC8, for DC 'H' and 'M' walls, depending on the value of the shear ratio:

1. If $a_s \geqslant 2.0$, the provisions for columns apply (section 8.5.2(b)), which means that the amount of horizontal bars in the grids is calculated from equation (8.48), which in the case of walls is usually written in the form

   $$V_{wd} = \rho_h f_{yd} b_w z_e \tag{9.34}$$

   where ρ_h is the geometric ratio of horizontal web reinforcement (grid bars), equal to $A_{sh}/(b_w s_h)$, s_h being the spacing of horizontal web bars, and A_{sh} their area (Figure 9.20); f_{yd} is the yield strength of the horizontal web reinforcement; b_w is the web thickness of the wall; z_e is the effective internal lever-arm (distance between the resultants of tension and compression in the section); if a more accurate calculation is not performed, $z_e \approx 0.8 l_w$ may be assumed. It is pointed out that, although not required theoretically, a minimum amount of vertical web bars should be provided (section 9.4.2(d)).
2. If $2.0 > a_s > 1.3$, a simplified truss model may be used, wherein the horizontal web bars are designed on the basis of equation (9.34), but, in addition, vertical web bars, appropriately anchored and spliced along the height of the wall, are required according to the expression

   $$\boxed{V_{wd} = \rho_v f_{yd} b_w z_e + \min N_{Sd}} \tag{9.35}$$

   where ρ_v is the geometric ratio of vertical grid reinforcement $A_{sv}/(b_w s_v)$, s_v being the spacing of vertical web bars, while the rest of the symbols have their usual meaning; it is pointed out that the axial load to be introduced in equation (9.35) is the minimum absolute value of compression (taken with positive sign) or the maximum tensile load (taken as negative). As explained in section 9.3.3(b), vertical reinforcement is required in squat walls to allow the development of the truss mechanism, its quantity decreasing with increasing axial compression;

note that even when $\rho_v \leqslant 0$ is calculated from equation (9.35), a minimum amount of vertical web reinforcement is required.

3. If $a_s \leqslant 1.3$, which is the typical case of squat walls, the fraction of the total shear carried by the web reinforcement is assumed to be distributed between the vertical and the horizontal bars of the grid, according to the empirical equation

$$\boxed{V_{wd} = [\rho_h(a_s - 0.3) + \rho_v(1.3 - a_s)] f_{yd} b_w z_e} \tag{9.36}$$

based on work by Hernandez and Zermeno (1980), where in the case the shear ratio $a_s < 0.3$, it is taken as equal to 0.3. This equation implies that for $a_s = 0.3$ only horizontal reinforcement is (theoretically) required, for $a_s < 0.3$ only vertical reinforcement is required, and for intermediate values of a_s both ρ_h and ρ_v should be provided as web reinforcement. Equation (9.36) reflects the increasing importance of vertical grid bars as the slenderness decreases, as well as the relatively higher shear strength of squat walls, which is recognized by other codes as well (ACI, 1989).

With regard to the 'concrete contribution' (term V_{cd} in equations (8.46)), different values have to be adopted, depending on the sign of the axial load and the region where the shear verification is carried out:

1. Inside the critical regions:

(a) If $N_d \geqslant 0$ (tensile axial force), $V_{cd} = 0$.
(b) If $N_d < 0$ (compression)

$$\boxed{V_{cd} = \tau_{Rd}(1.2 + 40\,\rho_l) b_w z_e} \tag{9.37}$$

where, in accordance with EC2 (CEN, 1991)
$\tau_{Rd} = 0.25 f_{ctk0.05}/\gamma_c$
$\rho_l \;= A_{sl}/(b_w z_e)$ (A_{sl} is the tension reinforcement)

By comparing equations (9.37) and (8.47), it is seen that equation (9.37) is a modified version of the EC2 equation (8.47), to account for the deterioration of shear transfer mechanisms within the plastic hinge regions of walls.

2. Outside the critical regions the EC2 equation (8.47) applies, where the sign of the axial stress σ_{cp} is taken as positive for compression.

For DC 'L' walls the standard EC2 procedure for diagonal tension (equations (8.47)and (8.48)) applies.

Minimum requirements for horizontal and vertical web reinforcement, for all ductility classes, are given in section 9.4.2(d).

(b) Verification against diagonal compression failure

The possibility of web crushing, discussed in section 9.3.3, is checked using the standard expression $V_{Sd} \leqslant V_{Rd2}$, where

1. In critical regions

$$\boxed{V_{Rd2} = 0.4(0.7 - f_{ck}/200) f_{cd} b_w z_e} \tag{9.38}$$

The internal lever arm z_e may be taken equal to $0.8l_w$ if not calculated using a more refined procedure, while f_{ck} should not be taken as > 40 MPa. Assuming $z_e \approx 0.8l_w$ and substituting the relevant values of $f_{ck}, f_{ck}^{1/2}, f_{cd}$, it is seen that V_{Rd2} varies from $0.53\, f_{ck}^{1/2}\, b_w l_w$ to $0.67\, f_{ck}^{1/2}\, b_w l_w$, for f_{ck} varying from 16 to 40 MPa, that is $V_{Rd2}/(b_w l_w) = \tau_{nom} \approx 0.6\, f_{ck}^{1/2}$. As outlined in section 9.3.3(a), walls with $\tau \geqslant 0.6\, f_c^{1/2}$ are susceptible to web crushing failure.

2. Outside the critical regions

$$V_{Rd2} = 0.5(0.7 - f_{ck}/200) f_{cd} b_w z_e \tag{9.39}$$

Equation (9.39) is the same as equation (8.50) suggested by EC8 for diagonal compression checks of beams. The higher degree of vulnerability of walls with respect to web crushing is reflected on the coefficient 0.4 used in equation (9.38) instead of the 0.5 used in equations (8.50) and (9.39).

With regard to diagonal compression failure, the effect of compressive axial loads is unfavourable, EC8 specifies that in walls under compression a reduced value of V_{Rd2} should be calculated using the EC2 equation

$$V_{Rd2,red} = 1.67 V_{Rd2}(1 - \sigma_{cp,ef}/f_{cd}) \ngtr V_{Rd2} \tag{9.40}$$

where V_{Rd2} is given from equation (9.38) or (9.39), and the effective average stress due to axial loading is

$$\sigma_{cp,ef} = \frac{N_{Sd} - f_{yd} A_{s2}}{A_c} \tag{9.41}$$

N_{Sd} being the (compressive) axial load, $A_c = b_w l_w$ the gross concrete section area and A_{s2} the area of compression reinforcement. Trial calculations have shown that equation (9.40) may only become critical in the case of walls at the perimeter of buildings where seismic overturning moments cause high axial compression. Transferring the value of $\sigma_{cp,ef}$ from equation (9.41) to equation (9.40) and rearranging terms, it may be shown that equation (9.40) is critical for axial loads in excess of

$$N_{Sd,lim} = 0.4 f_{cd} A_c - f_{yd} A_{s2} \tag{9.42}$$

(c) Verification against sliding shear failure

From the discussion of sliding shear failure under cyclic loading conditions presented in section 9.3.3 for walls (and, in more detail, in section 8.2.2(d) for beams), it is clear that different shear transfer mechanisms are activated, such as dowel resistance of bars orthogonal to the sliding shear plane, aggregate interlock (friction), and tension in bidiagonal bars, whenever they are present.

The EC8 design equation for sliding shear includes expressions for each of the above mechanisms, in the following form:

$$\boxed{V_{Rd,s} = V_{dd} + V_{id} + V_{fd}} \tag{9.43}$$

where $V_{Rd,s}$ is the total resistance of the wall against sliding shear and V_{dd} is the dowel resistance of vertical bars of the web, or purposely arranged bars in the boundary elements (additional to those resulting from flexural design); the following empirical equation is suggested for this mechanism

$$V_{\mathrm{dd}} = 1.3\ \Sigma\ A_{\mathrm{sj}}\ \sqrt{f_{\mathrm{cd}} f_{\mathrm{yd}}} \not> 0.25 f_{\mathrm{yd}} \Sigma\ A_{\mathrm{sj}} \tag{9.44}$$

where ΣA_{sj} is the total area of the aforementioned vertical bars. The upper bound to equation (9.44) is suggested by Paulay and Priestley (1992) as a reasonable limit for dowel action in squat walls under cyclic loading conditions. The shear V_{id}, transferred by cross-inclined (bidiagonal) bars, is arranged at an angle θ to the horizontal (Figure 9.19) is given by

$$V_{\mathrm{id}} = \Sigma\ A_{\mathrm{si}} f_{\mathrm{yd}} \cos\theta \tag{9.45}$$

where ΣA_{si} is the sum of all inclined bars in both directions. Equation (9.45) is the same as equation (8.37) given for beams (section 8.2.2(e)). The shear carried through friction at the sliding shear crack (aggregate interlock) V_{fd}, is given by the expression

$$V_{\mathrm{fd}} = \mu_{\mathrm{f}}[(\Sigma A_{\mathrm{sj}} f_{\mathrm{yd}} + N_{\mathrm{Sd}})\xi + M_{\mathrm{Sd}}/z] \not> 0.25 f_{\mathrm{cd}} \xi l_{\mathrm{w}} b_{\mathrm{w}} \tag{9.46}$$

where ΣA_{sj} has already been defined with reference to equation (9.44), N_{Sd} is the axial load on the wall (taken as positive for compression), ξ the normalized neutral axis depth and z the corresponding lever arm; the 'concrete-to-concrete friction' coefficient may be taken as equal to 1.0 (cyclic loading conditions) for rough crack interfaces free of laitance. Equation (9.46) is based on the assumption that friction is developed along the compression zone only and the axial forces contributing to friction are the following (Tassios, private communication, 1994):

1. The fraction of the wall axial load corresponding to the compression zone, ξN_{Sd}.
2. The fraction of the total reinforcement ΣA_{sj} (see equation (9.44)) which contributes to friction through damping action, assumed for simplicity equal to $\xi \Sigma A_{\mathrm{sj}}$, although a value $\xi A_{\mathrm{sv}} + A_{\mathrm{s0}}$ (see Figure 9.20) would be more appropriate.
3. The compression force resulting from bending, equal to M_{Sd}/z.

It is pointed out that in earlier versions of EC8 the following simple expression (based on fundamental principles of friction) was used for V_{fd}, instead of equation (9.46)

$$V_{\mathrm{fd}} = \mu_{\mathrm{f}} N_{\mathrm{Sd,min}} \tag{9.47}$$

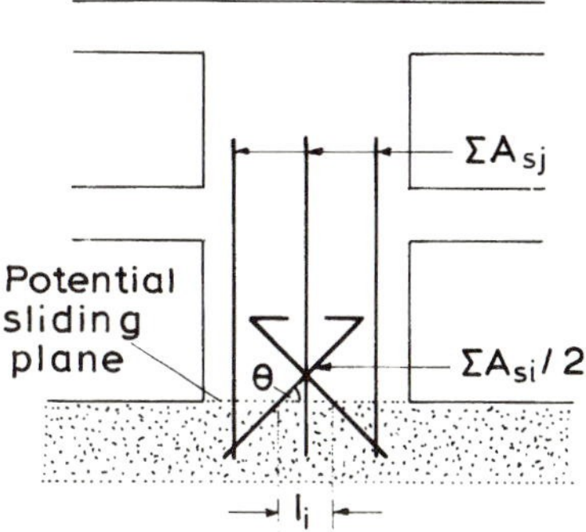

Figure 9.19 Bidiagonal reinforcement in structural walls.

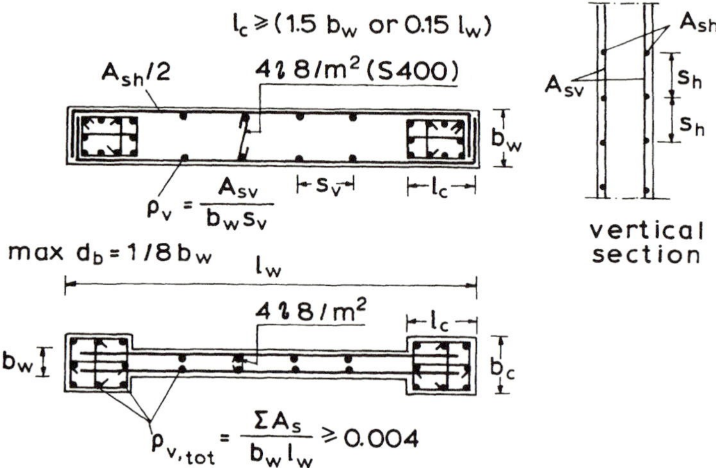

Figure 9.20 Arrangement of horizontal and vertical reinforcement in walls with rectangular or barbell cross-section.

where $N_{Sd,min}$ is the minimum value of axial compression; this simple approach was also adopted by the CEB (1985) Code. To the best of the authors' knowledge (9.46) has not been included in codes other than EC8; comparisons of equations (9.46) and (9.47) for specific cases will be presented later (see the example in section 9.4.4).

For squat walls, EC8 suggests that the following should be ensured:

1. At the base of the wall the shear carried by bidiagonal reinforcement (V_{id}) is greater than $V_{Sd}/2$.
2. At the storeys above the base V_{id} is greater than $V_{Sd}/4$.

The inclined bars (Figure 9.19) lead to an increase of the flexural strength of the wall, the amount of increase depending on the point at which the bidiagonal bars intersect (if the point of intersection lies on the potential sliding plane, the increase is negligible). This increased flexural strength has to be taken into account in estimating design shear forces for walls (it affects the ratio M_{Rd}/M_{Sd} in equations (6.10) and (6.11)); if a more refined procedure is not used, the following simplified methods may be applied:

1. The increase in flexural strength may be estimated as

$$\Delta M_{Rd} = 0.5\ \Sigma\ A_{si} f_{yd} \sin(\theta)\ l_i \tag{9.48}$$

where l_i is the distance between inclined bars at the base of the wall (Figure 9.19). Equation (9.48) is easily derived from equilibrium conditions at the sliding shear plane, assuming that bidiagonal bars reach yield in tension and in compression (the signs depending on the direction of the applied shear).
2. Alternatively, the effective shear resistance of inclined bars against sliding may be estimated as

$$V_{id} = \Sigma\ A_{si} f_{yd}[\cos\theta - 0.5\ l_i \sin\theta/(a_s l_w)] \tag{9.49}$$

This equation is derived from equations (9.45) and (9.48) by taking $\Delta M_{Rd} = \Delta V l_w a_s$ (from equation (9.33)) and subtracting ΔV from V_{id} in equation (9.45). By using equation(9.49) instead of equation (9.48) it is to be verified that the corresponding total shear resistance $V_{Rd,s} > V_{Sd}$, where V_{Sd} must not be revised on the basis of the flexural contribution of the bidiagonal bars.

(d) Web reinforcement requirements

In order to secure walls against premature web shear cracking, EC8 requires a minimum amount of web reinforcement, in the form of orthogonal grids, whose ratio within the critical regions should not be less than

$$\rho_{h,min} = \rho_{v,min} = 0.002$$

Referring to Figure 9.20, $\rho_h = A_h/b_w s_h$ and $\rho_v = A_v/b_w s_v$ are the geometric ratios of horizontal and vertical grid reinforcement, respectively. The diameter of the grid bars should be at least equal to $b_w/8 \nless 8$ mm, and the spacing (s_v and s_h) not greater than 20 bar diameters or 200 mm for DC 'H' walls, and 25 bar diameters or 250 mm for DC 'M' walls; thus at least one grid of 8 mm bars at 160 or 200 mm respectively, is required at each face of the wall cross-section. These grids should be connected by properly spaced cross-ties (typically S-shaped), as shown in Figure 9.20; a minimum of four 10 mm cross-ties per unit area is required if S220 steel is used, or four 8 mm cross-ties if S400 steel is used. Outside the critical region, it is also recommended, but not explicity required, to use the double grid pattern of reinforcement; nevertheless, it is pointed out that even the very demanding New Zealand Code (SANZ, 1982) allows the use of single grids in the noncritical regions of walls with $b_w \ngtr 200$ mm, provided that the corresponding shear stresses do not exceed $0.3 f_c^{1/2}$ MPa.

For DC 'L' walls the EC2 provisions for structural walls apply. The ratio of vertical reinforcement should be between 0.4 and 4.0%; note that reinforcement in the boundary elements is included in these ratios. The spacing of vertical bars (s_v) should not exceed the smaller of $2b_w$ or 300 mm. The horizontal reinforcement should not be less than half the vertical reinforcement, with a spacing (s_h) not exceeding 300 mm.

With regard to bidiagonal reinforcement in the web (Figure 9.19), no specific minimum requirements are specified by EC2; nevertheless, it is recommended to use large-diameter bars ($d_b \geqslant 18$ mm), to reduce the possibility of early buckling. Whenever a highly ductile performance is sought, four bidiagonal bars can be used in each direction,surrounded by closely spaced hoops, thus forming a pair of inclined pseudo-columns, similar to those required in coupling beams (Figure 9.26). To minimize the possibility of plastic hinge formation above the base of the wall, bidiagonal bars should cross all sections of the wall within a distance of $0.5l_w$ or $0.5h_w$ (whichever is smaller) above the critical section (which typically is the base); for the common case that $\theta = 45°$, the previous requirement implies that the intersection point of the cross-inclined bars can be at the critical section, in which case no increase in flexural capacity occurs (equation (9.48)). Inclined bars at $\theta < 45°$ are more efficient in resisting sliding shear (equation (9.45)), but their intersection point should be located above the critical section to satisfy the previous requirement, in which case a non-zero ΔM_{Rd} results.

(e) Confinement requirements

The performance criterion specified by EC8 for walls with prevailing flexural failure mode is that a minimum conventional curvature ductility factor should be ensured within their critical regions (Figure 6.7), depending on the type of wall, as follows: for uncoupled (isolated) walls

$$\boxed{\mu_\phi = q^2} \tag{9.50(a)}$$

and for coupled walls (section 9.4.5)

$$\boxed{\mu_\phi = 0.8q^2} \tag{9.50(b)}$$

If a more refined procedure is not applied, equations (9.50) are deemed to be satisfied if confinement reinforcement is provided in the wall, as follows:

1. For walls with free edges or with barbell (dumb-bell) cross-sections, the mechanical volumetric ratio of confinement reinforcement ($\omega_{wd} = \rho_w f_{yd}/f_{cd}$) in the boundary elements, as well as other required specific measures, shall be the same as those applicable for columns (section 8.5.2(c)), with the μ_ϕ values specified by equations (9.50).
2. In determining the required ratio of hoop or spiral reinforcement (equation (8.84)) the following effective axial load shall be taken into account

$$\boxed{N_{Sd,ef} = 0.5\left(\frac{N_{Sd}}{2} + \frac{M_{Sd}}{z}\right)} \tag{9.51}$$

   where N_{Sd} is the maximum axial load from the analysis (taken as positive for compression) and z is the internal lever arm, to be taken equal to the distance of the centroids of the two confined boundary elements (columns). It is recalled that M_{Sd}/z is the compression force resulting from bending alone (see also equation (9.46)).
3. For flanged or complex cross-sections, the refined procedure for estimating curvature ductility outlined in section 8.2.1(b) may be used; problems is using the EC8 ductility criterion based on the post-peak $0.85M_{Rd}$ strength level have already been discussed in sections 8.2.1, 8.4.2 and 8.5.2(c). A set of empirical conservative equations, included in the informative Annex C of EC8 Part 1.3, may also be used for a preliminary estimation of the ductility-related quantities, before a final analytical verification is carried out.

Multiple hoop patterns are recommended for the boundary elements of walls with free edges, but the use of intermediate cross-ties is also allowed (Figure 9.20). Whenever $b_c > 250$ mm, multiple hoops (or cross-ties) should also be placed between the exterior faces in the direction of the wall longitudinal axis.

At the cross-section level, confinement should extend up to the point where (under cyclic loading) unconfined concrete may spall ($\varepsilon_c > 0.35\%$). For design purposes the length (l_c) of the confined edge may be defined on the basis of a 0.2% compressive strain (Figure 9.21), assuming that the actions M_{Sd}, N_{Sd} are applied to the section. In all cases the value of l_c should not be smaller than $0.15\,l_w$ or $1.5\,b_w$ (Figure 9.20).

The confined regions of the wall should extend vertically along the critical length (Figure 6.7). Above this length and for at least one more storey height, boundary elements shall also be provided, with at least half of the confinement reinforce-

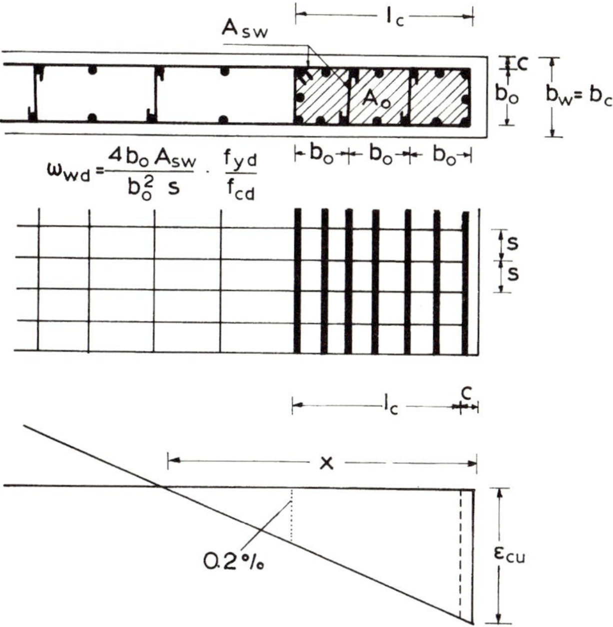

Figure 9.21 Confinement requirements in walls with free edges.

ment calculated for the critical region. In the rest of the wall height a minimum vertical reinforcement equal to $0.005b_w l_c$ shall be provided, enclosed by stirrups, whose spacing is defined according to EC2 (the smaller of b_w, $12d_{bl}$ and 300 mm), where l_c is calculated as previously but the minimum requirements do not apply; in most cases a practical solution consists in constructing a small square column with a dimension equal to the thickness of the wall at its edge (typically $b = b_w$).

In the case of walls with flanged sections, it is quite common to have $\varepsilon_c < 0.2\%$, which means that $l_c = 0$; in this case at least the minimum longitudinal and transverse reinforcement indicated in Figure 9.22 is recommended (this case is not explicitly covered by EC8).

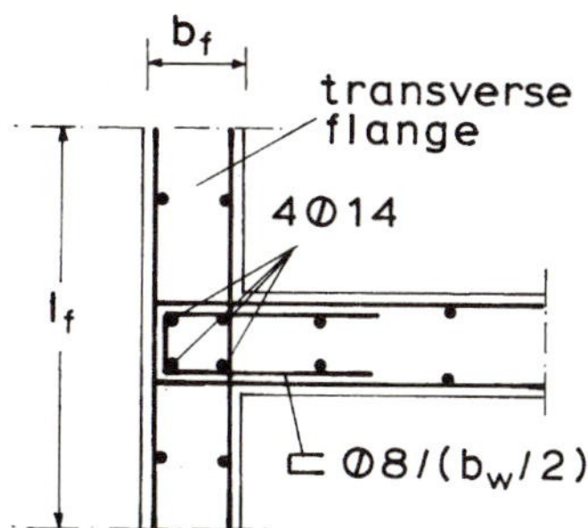

Figure 9.22 Detailing of the web–flange connection in flanged walls.

With regard to the longitudinal reinforcement requirements in boundary elements of DC 'H' and 'M' walls, the corresponding provisions for columns apply, that is, $\rho_{min}=A_s/(b_c l_c)=1.0\%$, $\rho_{max}=4.0\%$ (see further discussion in section 8.5.1(b)).

The total vertical reinforcement ratio $\rho_{v,tot}=\Sigma A_s/A_w$,(where A_w is the total area of the horizontal cross-section of the wall) which includes all longitudinal bars in the web and in the boundary elements, should be at least equal to 0.4%, as required by EC2.

For DC 'L' walls with free edges or barbell cross-section, confined boundary elements with l_c defined as for DC 'M' and 'H' should be provided over the critical region of the wall (Figure 6.7). If the most compressed edge of the wall is connected to a transverse flange with a thickness $b_f \geqslant h_s/15$ and a length $l_f \geqslant h_s/5$ (Figure 9.22), then no confined boundary element is required.

Reinforcement in the boundary elements is determined according to the following rules:

1. The minimum requirements for DC 'L' columns (section 8.5.2(c)) apply, with μ_ϕ and $N_{Sd,ef}$ calculated from equations (9.50) and (9.51) respectively.
2. Whenever

$$\nu_{d,ef}=0.5\,\frac{(N_{Sd}/2+M_{Sd}/z)}{b_w l_c f_{cd}}\leqslant 0.15 \tag{9.52}$$

   only the nominal transverse reinforcement required by EC2 shall be provided, with a minimum vertical reinforcement $A_s=0.005 b_w l_c$ (as for DC 'H' and 'M' walls in the non-critical regions).
3. If $0.15<\nu_{d,ef}\leqslant 0.20$ the previous rules are applicable, provided the q-factor is reduced by 30%; this is a quite inconvenient procedure if it actually implies that an extra analysis with a reduced q-factor has to be carried out.

9.4.3 Other design requirements

(a) Geometrical constraints

The possibility of buckling (usually out-of-plane) of thin-walled sections has already been discussed in section 9.3.3(a). The following EC8 provisions aim at minimizing this possibility, especially at low levels of induced inelasticity.

To avoid unpredictable lateral instability the web thickness of a structural wall should be at least equal to

$$b_w=\min\{q l_w/60;\ h_s/20;\ 150\text{ mm}\} \tag{9.53}$$

where h_s is the storey height.

Additional constraints are provided by EC8 for the confined lengths of a wall (see section 9.4.2(e)), that is:

1. If $l_c \geqslant \max\ \{2b_w, 0.2l_w\}$ the minimum web thickness is

$$b_w \geqslant \{h_s/10,\ 200\text{ mm}\} \tag{9.54}$$

2. If $l_c < \max\ \{2b_w, 0.2l_w\}$ then

$$b_w \geqslant \{h_s/15,\ 200\text{ mm}\} \tag{9.55}$$

3. When the most compressed edge of the wall is connected to a transverse flange (Figure 9.22), having a thickness of at least $h_s/15$ and a length of at least $h_s/5$, and if $l_c \leqslant 3b_w$, the minimum thickness is determined from equation (9.53), otherwise equations (9.54) or (9.55) apply.

The above equations (9.53)–(9.55) are applicable for all ductility classes.

(b) Anchorage and splicing of reinforcement

The detailing of anchorages and lap splices of bars located in the boundary elements of walls with free edges or barbell cross-section is carried out in accordance with the relevant provisions for columns, already given in sections 8.5.3(b) and (c). An example of such a detailing is shown in Figure 9.24 (section 9.4.4(d)). The only difference with respect to columns is that in DC 'H' walls splicing by overlapping must be avoided within the critical regions.

Anchorage and splicing requirements of the web bars in walls are essentially the same as in EC2, with the restrictions already given in sections 8.3.3(b) and (c). It is worth pointing out that while EC8 makes it quite clear that splicing by welding is not allowed within the critical regions of walls, it is not clear whether welded mesh fabrics can be used as grid reinforcement of walls within these regions. To the best of the authors' knowledge no conclusive results are available with regard to the effectiveness of welded mesh fabrics as reinforcement of walls subjected to inelastic cyclic loading.

(c) Construction joints

Construction joints (cold joints) in R/C walls are susceptible to cracking, especially if their preparation is not properly carried out during construction (which is actually the case in many practical situations). These joints can then fail in a sliding shear mode, similar to that already described in sections 9.3.3(b) and 9.4.2(c) for squat walls and/or walls subjected to high levels of shear stress. It is pointed out that failure by sliding can occur in poorly designed cold joints of walls which are neither squat nor subject to high shear forces.

It is therefore, understood that a minimum amount of properly anchored reinforcement crossing the cold joint has to be provided in order to re-establish the shear resistance of uncracked concrete. The EC8 equation for the foregoing requirement is

$$1.5\rho_{min}\sqrt{f_{cd}f_{yd}}+\mu_f\left(\rho_{min}f_{yd}+\frac{N_{Sd,min}}{A_w}\right)\geqslant 1.3f_{ctk0.05} \qquad (9.56)$$

where $\rho_{min}=A_{s,tot}/A_w$ is the geometric ratio of all vertical bars in the wall, including both the web reinforcement and the longitudinal bars in the boundary elements. For the friction coefficient the value $\mu_f=1.0$ may be adopted. Equation (9.56) is based on the cosideration of a friction mechanism of shear transfer, wherein both the axial loading (positive for compression) and the force contributed by vertical bars at yield are contributing, corrected by the empirical term $1.5\rho_{min}f_{cd}^{1/2}f_{yd}^{1/2}$ to account for dowel action and obtain better agreement with existing experimental evidence.

If equation (9.56) is solved for ρ_{min}, the following expression (appropriate for design) results:

$$\boxed{\rho_{min} \geqslant \frac{1.3f_{ctk,0.05} - N_{Sd}/A_w}{f_{yd}[1+1.5(f_{cd}/f_{yd})^{1/2}]}} \tag{9.57}$$

In any case a minimum value $\rho_{min} = 0.25\%$ should be provided at cold joints. Equation (9.57) does not need to be satisfied in cold joints of walls in DC 'L' structures.

It is worth pointing out that equation (9.57) is usually easier to satisfy within the wall critical regions, where large amounts of reinforcement are typically available in the boundary elements, rather than at the upper storeys, where low amounts of wall reinforcement combine with low values of axial loading.

9.4.4 Design example

In section 6.3 analysis of the dual system shown in Figure 6.12 was carried out for a seismic action corresponding to $\alpha = 0.25$ ($A_d = 0.25g$) and DC 'M'. Bending moment diagrams for the seismic combination are shown in Figure 6.14. In the following, the design of the wall located in the middle of this structure is presented in some detail.

(a) Longitudinal reinforcement in the boundary elements

The design bending moments for the wall, using the capacity procedure of the shifted linear envelope (section 6.1.5(a)), is shown in Figure 6.15, while Table 6.6 lists the corresponding moment (M_{Sd}) and axial load (N_{Sd}) values. The M_{Sd} values were calculated using the following formula which is based on the geometry of the diagrams in Figure 6.15, where the pertinent notation is explained

$$M_{Sd}(z_i) = (M_0 - M_n)\left(1 - \frac{z_i - h_{cr}}{H}\right) + M_n$$

for $z_i \geqslant h_{cr}$, otherwise $M_{Sd} = M_0$. The length of the critical region was already found to be equal to 5.0 m ($= H/6$).

Whenever no appropriate software is available, a simple – as well as conservative – procedure for the flexural design of the wall consists in considering it as a rectangular section with concentrated reinforcement layers at the ends. Thus, assuming $d_1/h \approx 0.10$, which is practically equivalent to assuming an $l_c = 0.15 l_w$ to $0.20 l_w$ (a rather common case in the light of the requirements set forth in section 9.4.2(e)), the normalized actions at the wall base are

$$\mu = \frac{4083.06}{0.25 \times 4.0^2 \times 13\,333} = 0.076, \qquad \nu = \frac{-2638.60}{0.25 \times 4.0 \times 13\,333} = -0.198$$

From the design chart of Figure 8.39 it is found that $\omega_{tot} = 0.01$ ($\rho_{tot} \approx 0$), and $\varepsilon_{c2}/\varepsilon_{s1} = -0.35/0.80\%$. The length of the confined boundary elements can now be calculated, based on a neutral axis depth

$$x = \frac{\varepsilon_{c2}}{\varepsilon_{c2} + \varepsilon_{s1}}\, d \approx \frac{0.35}{0.35 + 0.80}\, 3.6 = 1.09 \text{ m} \quad (x/l_w = 1.09/4.00 = 0.27)$$

Using the $\varepsilon_c = 0.2\%$ criterion (section 9.4.2(e), Figure 9.21) the length of the confined boundary element should not be less than

$$l_c = x\left(1 - \frac{0.20}{0.35}\right) = 0.47 \text{ m}$$

Since $0.15l_w = 0.15 \times 4.00 = 0.60 > 0.47$ m, $l_c = 0.60$ m is finally selected. Using the same procedure, it is found that at the second storey the required $\omega_{tot} = 0.015$ ($\rho_{tot} \approx 0$) and the corresponding $l_c = 0.43 < 0.60$ m. Hence along the whole height of the critical region $h_{cr} = 5.0$ (or, preferably from the practical point of view, $h_{cr} = 6.0$ which is equal to two storey heights) the confined boundary elements at the edges of the wall cross-section have a length $l_c = 0.60$ m.

The required reinforcement ratios (ρ_{tot}) are negligible (this is an indication that a smaller l_w might have been used in this structure, had architectural and other considerations allowed it), hence minimum reinforcement requirements prevail at the boundary elements, namely

$$\rho_{min} = \frac{A_s}{b_w l_c} = 0.01 \Rightarrow A_s = 0.01 \times 25 \times 60 = 15.0 \text{ cm}^2$$

Eight 16 mm bars (16.1 cm^2) are an appropriate choice of reinforcement (Figure 9.24).

At this stage it is also required to check the adequacy of the selected width (b_w) of the wall. According to equation (9.53)

$$b_w \geqslant (q/60)l_w = (3.375/60) \times 4.0 = 0.225 \text{ m} < 0.25$$

Furthermore, in the confined boundary elements, since $l_c < 0.2l_w$, equation (9.55) applies

$$b_w \geqslant \{h_s/15, 200 \text{ mm}\} = \{3000/15, 200\} = 200 \text{ mm} < 250$$

Therefore the selected $b_w = 250$ mm may be used along the whole height of the wall, for the entire wall cross-section.

Minimum reinforcement requirements also prevail in the boundary elements of the other storeys. As outlined in section 9.4.2(e), for at least one more storey above the critical region (for the structure under consideration, this is the third storey and, optionally, the fourth storey as well) the 250 × 600 boundary element is continued with a reduced confinement reinforcement (section 9.4.4(c)). The longitudinal reinforcement in these regions, as well as in the upper storeys where a 250 × 250 boundary element at each end is sufficient, should be at least equal to $0.005b_w l_c$; hence 7.5cm^2 are required for the 250 × 600 element and 3.1cm^2 for the 250 × 250 element.

A further requirement concerns the total vertical reinforcement ratio $\rho_{tot} = A_{s,tot}/(b_w l_w)$ which should not be less than 0.4%, that is min $A_{s,tot} = 40.0$ cm^2. Taking into account that the web vertical steel consists of 8 mm bars at 200 mm centres (see next section), it is estimated that at least 13.0 cm^2 are required in each boundary element; thus two 16 mm and six 14 mm bars (13.3 cm^2) are used in all storeys above the critical region. It is worth pointing out that all these reinforcements are well above the values required from flexural design considerations, especially if it is remembered that the (moderate) contribution of the web vertical bars is ignored with respect to the flexural resistance. Nevertheless it will subsequently be shown that most of the extra steel can be used for other purposes, particularly as sliding shear reinforcement (see next section).

(b) Shear design and web reinforcement

For the determination of the design shear forces (V_{Sd}), the magnification factor ε (section 6.1.5(b)) has to be calculated first. The flexural resistance of the base cross-section, reinforced with eight 16 mm bars at each edge, resulting in an $\omega_{tot}=0.084$, and subjected to a normalized axial loading $\nu=0.198$, as calculated previously, can be estimated from Figure 8.39, which gives $\mu\approx0.11$, thus $M_{Rd}=0.11\times0.25\times4.00^2\times13\,333=5866.52$ kN m. The ordinates of the elastic spectrum corresponding to the fundamental period of the structure $T_1\approx0.64$ s (section 6.3.6) and to the upper limit of the constant spectral acceleration branch $T_c=0.40$ s are

$$S_e(T_1)=0.25\times1.0\times1.0\times2.5(0.40/0.64)^{1.0}=0.39$$

$$S_e(T_c)=0.25\times1.0\times1.0\times2.5=0.62$$

Finally, for DC 'M' walls the factor $\gamma_{Rd}=1.15$. Introducing all the relevant quantities in equation (6.10) the magnification factor is

$$\varepsilon=3.375\left[\left(\frac{1.15}{3.375}\times\frac{5866.52}{4083.06}\right)^2+0.1\left(\frac{0.62}{0.39}\right)^2\right]^{1/2}=2.38<q=3.375$$

According to the capacity procedure illustrated in Figure 6.8, the calculated shear forces are multiplied by ε in the lower third of the building height, while in the upper thirds a linear decrease down to $0.5V_{Sd,base}$ at the top is carried out; the resulting diagrams are shown in Figure 9.23. The capacity design shear forces are 138% higher than the values resulting from analysis in the lower third of the wall height, and 156–1964% (!) higher in the upper two-thirds of the height. It will be seen subsequently that these dramatic increases in the theoretically calculated shears do not lead to excessive demands of shear reinforcement.

The required quantity of horizontal web reinforcement may be determined from the diagonal tension verification (equations (8.46), (9.34) and (9.37)). As the axial loading at the base $N_d=-2638.6$ kN<0, the concrete contribution from equation (9.37) is

$$V_{cd}=\tau_{Rd}(1.2+40\rho_l)b_wz_e=260(1.2+40\times0.002)0.25(0.8\times4.0)=266.3\text{ kN}$$

Note that for the calculation of $\rho_l=A_{sl}/b_wz_e$ only the eight 16 mm diameter bars in the boundary element have been considered as tension steel.

The required amount of the horizontal web reinforcement can be estimated from equations (9.34) and (8.46), solving for ρ_h, that is

$$\rho_h=\frac{V_{Sd}-V_{cd}}{f_{yd}b_wz_e}=\frac{1218.1-266.3}{34.78\times25\times0.8\times400}=0.0034$$

Hence 8 mm bars (minimum allowed diameter) at 120 mm centres are selected ($\rho_h=0.0034$).

The shear ratio at the base of the wall is

$$\alpha_s=\frac{M_{S,cal}}{V_{S,cal}l_w}=\frac{4083.1}{512.3\times4.0}=1.99$$

It is pointed out that, as discussed in section 9.4.2(a), calculated rather than capacity derived values of M_S and V_S have to be introduced in equation (9.33) to esti-

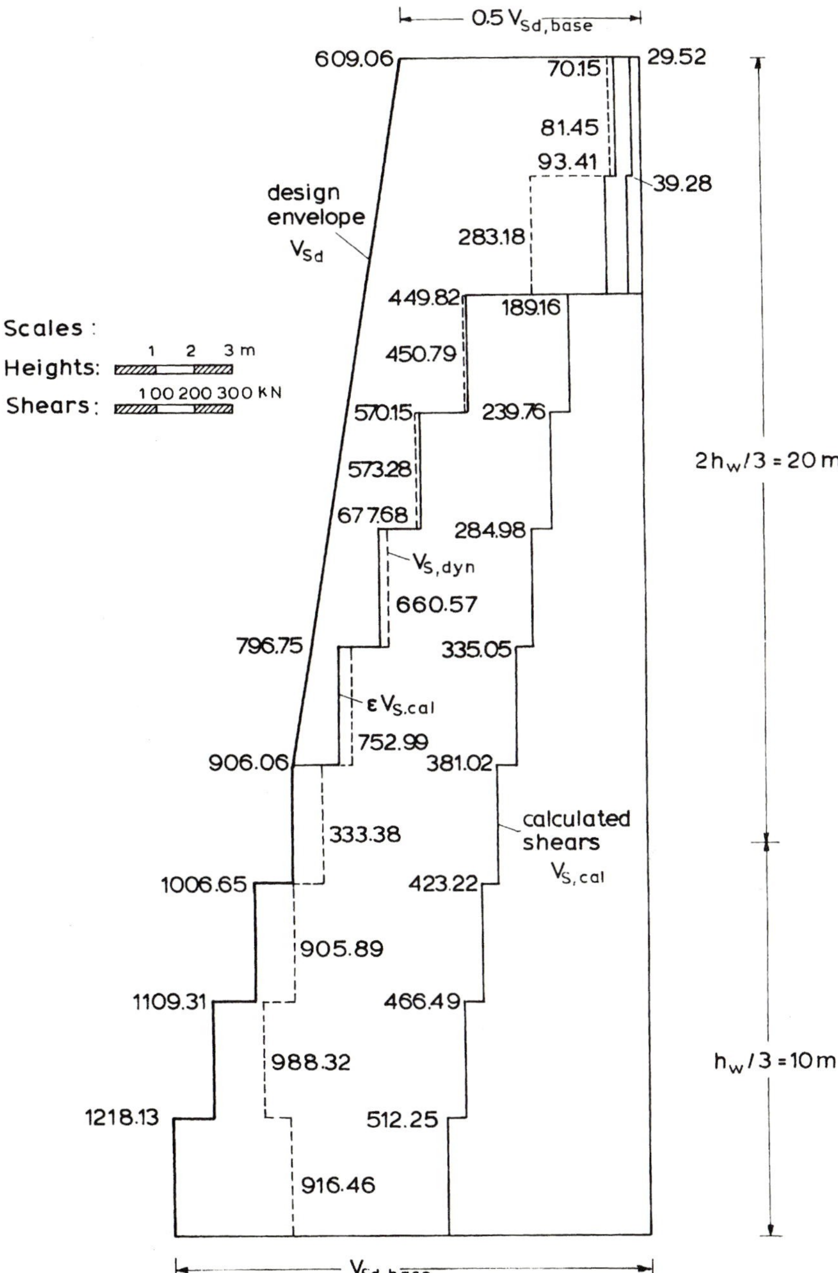

Figure 9.23 Determination of design shear forces for the wall of a dual system (shown in Figure 6.12), and maximum shears from inelastic dynamic analysis.

mate α_s. As $2.0 > \alpha_s > 1.3$ it is also required to check the adequacy of the vertical web reinforcement; assuming the minimum $\rho_v = 0.2\%$, that is 8 mm bars at 200 mm (which means that a rectangular rather than a square grid is selected), V_{wd} is calculated from equation (9.35):

$$V_{wd} = \rho_v f_{yd} b_w z_e + \min N_{Sd} = 0.002 \times 34.78 \times 25 \times 320 + 2638.6 = 3195.1 \text{ kN}$$

Thus, from equation (8.46)

$$V_{Rd3} = V_{cd} + V_{wd} = 266.3 + 3195.1 = 3461.4 > V_{Sd} = 1218.1 \text{ kN}$$

The maximum nominal shear stress in the wall is

$$\tau_{max} = \frac{V_{max}}{b_w l_w} = \frac{1218.1}{0.25 \times 4.00} = 1218.1 \text{ kPa} = 1.2 \text{ MPa} = 0.27 \sqrt{f_{ck}}$$

As pointed out in section 9.4.2(b) a diagonal compression failure may only be possible if τ_{max} exceeds $0.6 f_{ck}^{1/2}$; it is therefore clear that the lightly reinforced wall under consideration is not susceptible to such a type of failure (a value of $V_{Rd2} = 2560$ kN can be calculated from equation (9.38), which is more than twice the capacity-derived shear).

With regard now to the sliding shear verification (section 9.4.2(c)), the resistance of the wall at its base may be calculated using equations (9.43)–(9.46).

The dowel resistance of vertical web reinforcement can be found from equation (9.44), taking into account that the bars available for resisting sliding shear are those of the web (a total of 26 bars with 8 mm diameter, having an area of 13.1 cm^2) and the fraction of the boundary element reinforcement not required for flexural resistance; as the latter quantity is only

$$\omega_{tot}(f_{cd}/f_{yd}) b_w l_w = 0.01(13.3/347.8)25 \times 400 = 3.8 \text{ cm}^2$$

and the existing eight 16 mm bars have an area of 16.1 cm^2, it is seen that $16.1 - 3.8 = 12.3$ cm^2 are available for resisting sliding shear (in each edge element), thus

$$V_{dd} = 1.3 \, \Sigma A_{sj} \sqrt{f_{cd} f_{yd}} = 1.3(13.1 + 2 \times 12.3) \sqrt{1.33 \times 34.78} = 333.3 \text{ kN}$$

This value is slightly larger than the EC8 limit of

$$0.25 f_{yd} \, \Sigma A_{sj} = 0.25 \times 34.78(13.1 + 2 \times 12.3) = 327.8 \text{ kN}$$

therefore $V_{dd} = 327.8$ will be used for the verification.

The shear resistance due to friction at the sliding surface may be estimated from equation (9.46)

$$V_{fd} = \mu_f[(\Sigma A_{sj} f_{yd} + N_{Sd})\xi + M_{Sd}/z]$$

by introducing $\Sigma A_{sj} = 13.1 + 2 \times 12.3 = 37.7 \text{cm}^2$, $\zeta = x/l_w \approx 1.09/4.00 = 0.273$ and $\mu_f = 1.0$, hence

$$V_{fd} = 1.0[(37.7 \times 34.78 + 2638.6)0.273 + 4083.1/3.20] = 1078.3 + 1276.0 = 2354.3 \text{ kN}$$

In this case the EC8 upper bound of

$$0.25 f_{cd} \xi b_w l_w = 0.25 \times 1.33 \times 0.273 \times 25 \times 400 = 907.7 \text{ kN}$$

is much lower than the previous value, thus $V_{fd}=907.7$kN. Note that equation (9.47) with $\mu_f=1.0$, gives $V_{fd}=2638.6$ kN, which is quite close to the value of 2354.3 calculated previously on the basis of equation (9.46). The sum of the two foregoing mechanisms is $327.8+907.7=1235.5>V_{Sd}=1218.1$kN therefore no bidiagonal reinforcement is necessary.

Shear verifications and the design of web reinforcement are carried out in a similar manner for the rest of the wall sections. It is pointed out that significant variations in the shear ratio value take place along the height of the wall, as the bending moment distribution changes from trapezoidal to column-like antisymmetric, as shown in Figure 6.15. For instance, at the fifth storey $\alpha_s=0.53<1.3$ is calculated, hence equation (9.36) has to be used for the diagonal tension verification, that is,

$$V_{wd}=[\rho_h(\alpha_s-0.3)+\rho_v(1.3-\alpha_s)]f_{yd}b_w z_e=\rho_h f_{yd}b_w z_e$$

since $\rho_v=\rho_h=0.002$ has been selected, thus

$$V_{wd}=0.002\times 347.8\times 0.25\times 3.20=0.556\ \text{MN}=556\ \text{kN}$$

Given that the fifth-storey wall is outside the critical section, the concrete contribution is calculated using equation (8.47):

$$V_{cd}=[\tau_{Rd}k(1.2+40\rho_l)+0.15\ \sigma_{cp}]b_w z_e$$

Introducing

$$k=1.0 \quad (\text{as } 1.6-\text{d}=1.6-3.6<1)$$

$$\sigma_{cp}=N_{Sd}/b_w l_w=1578.0/1.0=1578.0\ \text{kPa}$$

$$\rho_l=13.3/(25\times 320)=0.0017$$

thus

$$V_{cd}=[260.0\times 1.0(1.2+4.0\times 0.0017)+0.15\times 1578.0]\ 0.25\times 3.20=453.1\ \text{kN}$$

Therefore

$$V_{Rd3}=556.0+453.1=1009.1\approx V_{Sd}=1056.6$$

Since the difference between V_{Rd3} and V_{Sd} is less than 5%, there is no need to change the spacing of the grid bars (say from 200 to 190 mm). A typical grid of 8 mm bars at 200 mm centres is found to be adequate as web reinforcement along the entire non-critical height of the wall.

(c) Confinement reinforcement

For the isolated wall with free edges under consideration, confinement reinforcement in the form of hoops has to be provided in the boundary elements of the first two storeys ($h_{cr}\approx 6.0$ m). The quantity of hoop reinforcement will be determined using equation (8.84) regarding confinement of columns, on the basis of an effective axial load given by equation (9.51). At the base of the wall, where $l_c=0.60$ m, this axial load is

$$N_{Sd,ef}=0.5(N_{Sd}/2+M_{Sd}/z)=0.5(2638.6/2+4083.1/3.4)=1260.1\ \text{kN}$$

For DC 'M' structures the basic equation (8.84) takes the following form:

$$\omega_{wd} = \frac{60}{\alpha_n \alpha_s}(9\nu_d \varepsilon_{yd})(0.35A_c/A_0+0.15)-0.035$$

Using the relationships given in section 8.5.2(c), the confinement coefficients are

$$\alpha_n = 1 - \frac{\Sigma b_i^2}{6A_0} = 1 - \frac{8 \times 19^2}{6 \times 60 \times 19} = 0.578$$

$$\alpha_s = \left(1 - \frac{s_w}{2b_0}\right)^2 = \left(1 - \frac{6.5}{2 \times 19}\right)^2 = 0.687$$

In the foregoing equations a 30 mm cover to the reinforcement was assumed and two overlapping 10 mm hoops (Figure 9.24) were selected at a spacing determined by

$$s_w = \min\{b_0/3, 150 \text{ mm}, 7d_{bl}\} = \min\{190/3, 150, 7 \times 16\} = 63.3 \Rightarrow s_w = 65 \text{ mm}$$

Introducing the previously calculated values of α_n, α_s, s_w and

$$\nu_d = N_{Sd.ef}/(A_c f_{cd}) = 1260.1/(0.25 \times 0.63 \times 13\,333) = 0.60$$

in the equation for the hoop ratio, results in

$$\omega_{wd} = \frac{60}{0.578 \times 0.687}(9 \times 0.60 \times 0.00174)(0.35 \times 0.157/0.114+0.15)-0.035 = 0.864$$

For the selected hoop pattern of 10 mm bars at 65 mm centres, the available hoop ratio is

$$\omega_{wd} = \frac{(2 \times 60+4 \times 19+2 \times 20)0.79}{19 \times 60 \times 6.5} \times \frac{347.8}{13.33} = 0.656 < \omega_{wd,req} = 0.864$$

As a second trial 12 mm hoops at 65 mm centres are selected resulting in an $\omega_{wd} = 0.942$ which is larger than the required values $\omega_{wd,req} = 0.864$. It is seen that the confinement required at the wall boundary element is rather heavy; if the value of μ_ϕ for isolated walls from equation (9.50(a)) had been introduced in the equation for calculating ω_{wd}, even heavier confinement would have resulted ($q^2 = 3.375^2 = 11.4 > 9$).

According to EC8, half the hoop reinforcement of the critical region has to be continued for at least one more storey; thus two 10 mm overlapping hoops at 90 mm centres ($\omega_{wd} = 0.470$) will be provided in the boundary elements of the third storey. For the remaining storeys 6mm stirrups at

$$s_w = \min\{b_w, 12d_{bl}, 300 \text{ mm}\} = \{250, 12 \times 14, 300\} = 168 \Rightarrow s_w = 170 \text{ mm}$$

are appropriate for the boundary elements.

(d) Detailing requirements

The required minimum thickness of the wall was already found to be less than $b_w = 250$, thus a 250 × 4000 rectangular cross-section is used along the entire height of the wall (a small taper in the direction of the thickness, say from 250 to 200 mm, above the first two storeys could also have been selected).

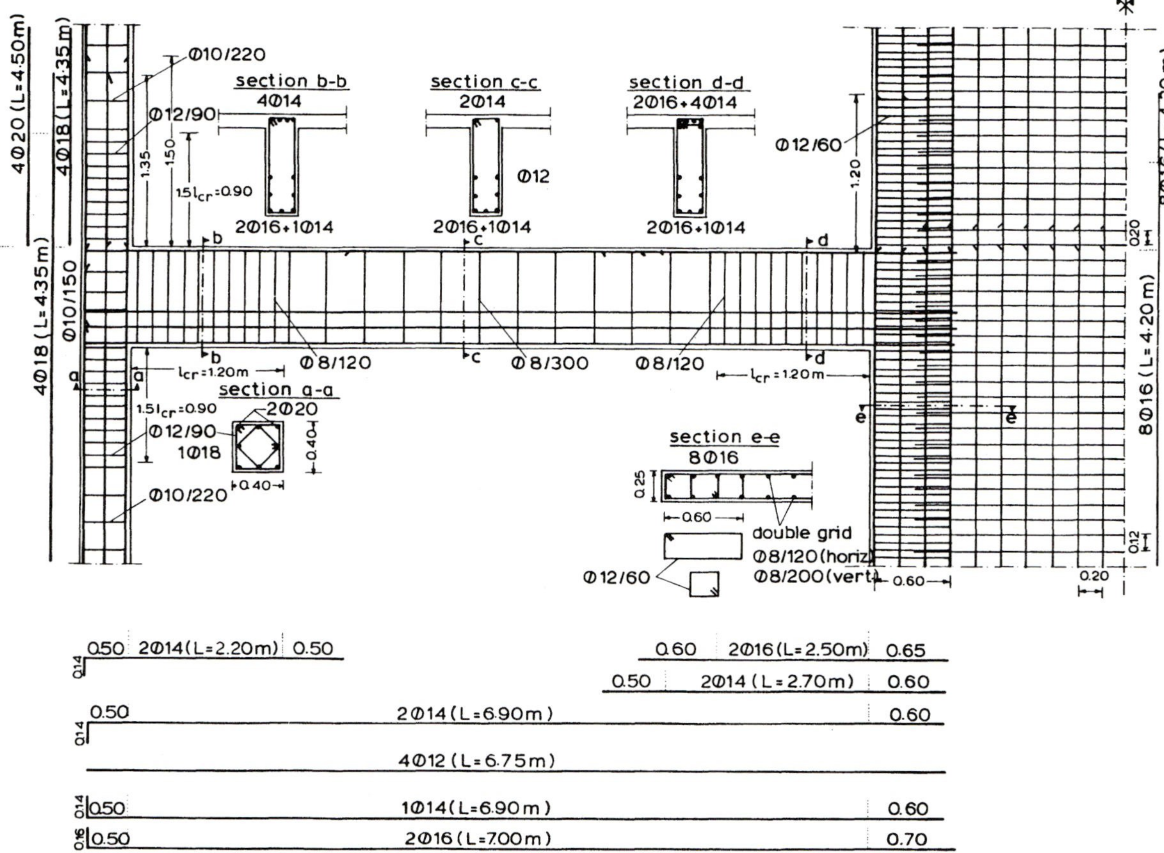

Figure 9.24 Arrangement of longitudinal and transverse reinforcement at the bottom of the dual system (short inclined hooks merely indicate the curtailing positions of bars).

The required anchorage length of the wall longitudinal bars is given by equation (8.56); for the bars in the bottom edge members

$$l_{b,net} = \alpha_a \left(\frac{d_{bl}}{4} \frac{f_{yd}}{f_{bd}} \right) \frac{A_{s,req}}{A_{s,prov}} = 1.0 \left(\frac{d_{bl}}{4} \frac{347.8}{2.3} \right) 1.0 = 38 d_{bl} \not< l_{b,min}$$

where $l_{b,min} = 0.6\ (d_{bl}/4 f_{yd}/f_{bd}) \approx 23 d_{bl} \Rightarrow l_{b,net} = 38 d_{bl}$.

Splicing by overlapping is not forbidden for DC 'M' walls, thus, for ease of construction, the first splices are located above the ground storey as shown in Figure 9.24, where the detailing of the wall reinforcement is presented for the bottom part of the dual structure. The required length of the lap splice is found from equaion (8.57)

$$l_s = \alpha_l l_{b,net} \not< 0.3 \alpha_a \alpha_l (d_{bl}/4)(f_{yd}/f_{bd})$$

where $\alpha_l = 2.0$ (section 8.3.3(c)). Therefore the length of the lapped splices of the edge bars above the ground storey should be $2 l_{b,net} = 76 d_{bl}$. For the bars of the grid it is not required to apply the provisions for columns, hence the term $A_{s,req}/A_{s,prov}$ does not need to be taken as equal to 1.0; indeed for the vertical bars (ρ_v), $A_{s,req} = 0$ and $l_{s,min}$ is used ($23 d_{bl}$).

Finally, for the construction joints, which are typically located at the top of each floor slab, the minimum reinforcement required by equation (9.57) has to be checked. The most critical check is not at the base, but rather at the top storey, where the minimum axial loading is acting on the wall; thus, at the top of the ninth storey slab equation (9.57) leads to the following requirement:

$$\rho_{min} = \frac{1.3 f_{ctk,0.05} - N_{Sd}/(b_w l_w)}{f_{yd}[1 + 1.5(f_{cd}/f_{yd})^{1/2}]} = \frac{1.3 \times 1600 - 261.5/(0.25 \times 4.00)}{347\,800\ [1 + 1.5(13.33/347.8)^{1/2}]} = 0.004$$

This coincides with the minimum total reinforcement required $\rho_{tot} > 0.4\%$, and is indeed provided by the 8 mm vertical bars of the grid (spaced at 200 mm) and the two 16 mm plus six 14 mm bars at each edge of the wall section.

9.4.5 Design of walls with openings

(a) Perforated walls

As already mentioned in section 9.3.4(a), the only EC8 provision regarding pierced (or perforated) walls is to avoid random (unsymmetrical) arrangements, unless the area of the openings is insignificant with respect to that of the wall (a ratio of 0.10–0.15 is usually a reasonably conservative limit, although not included in EC8). It is therefore recommended to use some relevant design provisions from other codes or from the literature, provided they are compatible with the EC8 wall design procedure.

Whenever the openings are not arranged in a symmetrical pattern, it is generally not possible (nor advisable) to use an equivalent frame model for the lateral load analysis. Instead, a plane stress finite element solution may be used, while **'strut and tie' models** are also good candidates, especially in the case of squat R/C walls (Paulay and Priestley, 1992).

In designing perforated walls for shear it is generally not advisable to consider them as separate segments (piers); this can be clearly understood with respect to a squat wall with $h_w/l_w = 1.0$ with openings in the middle having a width equal to $0.1l_w$, whose behaviour is not the same as that of two slender walls with $h_w/l_w = 1.0/0.45 = 2.2$. A simple practice is to design web reinforcement (using equations (9.34)–(9.36) as if the wall were solid, and place the bars corresponding to the area of the perforations around them; improved performance may be obtained if 'pseudo-beams' are formed around the openings, as shown in Figure 9.25. Furthermore, in assessing the concrete contribution, V_{cd} (equation (9.37)) and the resistance to diagonal compression, V_{Rd2} (equations (9.38)–(9.40)) a reduction equal to the opening to wall area ratio should be carried out.

(b) Coupled walls

According to EC8, coupling of walls through slabs alone should not be considered effective; in any case it is quite uncommon that such a system satisfies the 25% criterion described in section 9.3.4(b). Nevertheless, the seismic performance of this system may be improved if rolled steel sections are placed within the floor slab, at the toes of the walls, thus controlling punching shear in these heavily stressed regions (Paulay and Priestley, 1992).

Each vertical segment (pier) of a coupled wall may be designed for flexure and shear as an isolated (typically slender) wall, using the procedures already described in sections 9.4.1–9.4.3. It is pointed out that in contrast to isolated wall cross-sections which typically have a symmetric arrangement of reinforcement, piers of coupled walls are characterized by strongly asymmetric reinforcement patterns, the largest number of bars being concentrated at the edge which is distant from the opening (that is, at the exterior of the entire coupled wall section).

Coupling beams, which are the elements where the largest amount of seismic energy is to be dissipated (section 9.3.4(b)) have to be reinforced with bidiagonal bars, as shown in Figure 9.26, unless one of the following conditions is met:

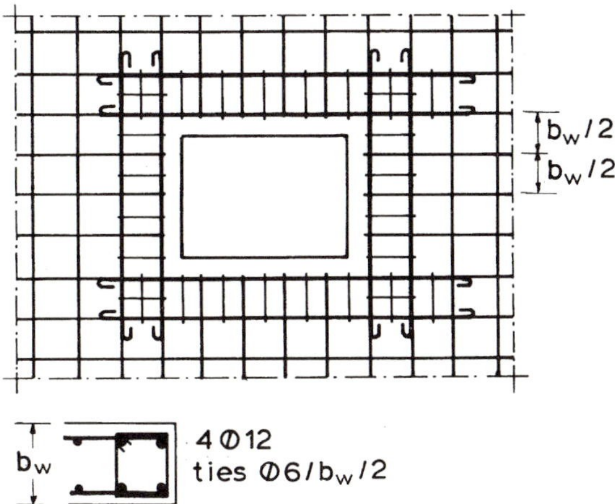

Figure 9.25 Arrangement of wall reinforcement around a small opening.

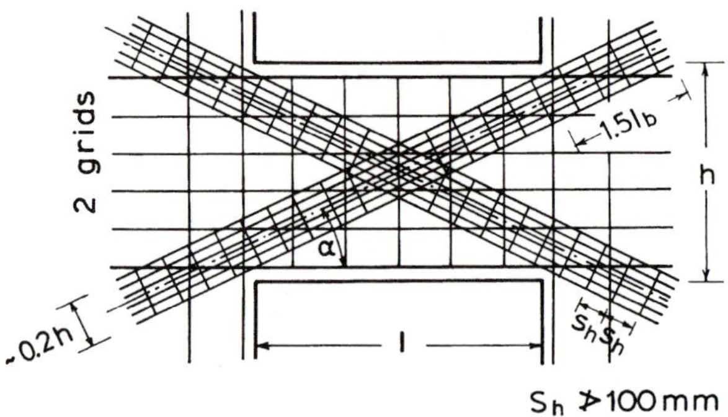

Figure 9.26 Arrangement of reinforcement in a coupling beam.

1. The probability of a bidiagonal cracking failure is low, which according to EC8 is the case when

$$\boxed{V_{Sd} \leqslant 4b_w d\tau_{Rd}} \tag{9.58}$$

Equation (9.58) is equivalent to restricting the nominal shear stress in the coupling beam to about $0.25f_{ck}^{1/2}$ (section 8.3.2(a)), in which case the effect of shear on ductility is minor. It is worth pointing out that for typical coupling beams, in multi-storey buildings whose width does not exceed the thickness of the wall (usually 200–250 mm), equation (9.58) is quite difficult to satisfy.
2. The slenderness of the beam is such as to ensure that the failure mode is predominantly flexural; this is deemed to be the case when $l/h \geqslant 3$ (l is the clear span of the coupling beam, as shown in Figure 9.26).

If neither of the foregoing conditions is satisfied, it has to be verified that

$$\boxed{V_{Sd} \leqslant 2A_{si} f_{yd} \sin\alpha} \tag{9.59}$$

where $V_{Sd}=2M_{Sd}/l$ is the shear force in the coupling beam, A_{si} the total area of diagonal reinforcement in each direction and α the angle between the diagonal bars and the centroid of the coupling beam (Figure 9.26).

Equation (9.59) is the same as equation (8.37), derived in section 8.2.2(e) for beams susceptible to sliding shear failure.

As shown in Figure 9.26, the bidiagonal reinforcement is arranged in column-like elements, with closely spaced hoops in order to prevent premature buckling of the inclined bars. The provisions for DC 'H' and 'M' columns apply with regard to the amount and detailing of these hoops (section 8.5.2(c)) and, in addition, the spacing of hoops in the pseudo-columns should not exceed 100 mm. The anchorage length of the diagonal bars should be 1.5 times the value prescribed by EC2 (equation (8.56)). In practice this length is not difficult to provide, as the length of the wall piers is adequate to accommodate it. On the other hand, the most tedious detail, from the construction point of view, is the intersection of the diagonal pseudo-columns; at least one hoop should engage all intersecting bars, so that a stable reinforcement pattern is ensured.

In addition to the bidiagonal reinforcement, the longitudinal and transverse steel required for beams outside the critical regions (sections 8.3.1 and 8.3.2) apply to coupling beams as well. Although not specifically mentioned in EC8, it is recommended to use intermediate longitudinal bars, that is to form a grid of reinforcement, as shown in Figure 9.26, which provides adequate basketing of the member.

Finally, for DC 'L' coupling beams, EC8 permits avoiding the use of bidiagonal pseudo-columns and providing other arrangements for reinforcement which exhibit a comparable level of energy dissipation capacity. This rather vague statement might possibly be interpreted as using large-diameter bidiagonal bars not surrounded by hoops, whenever the shear force is clearly higher than the limit specified by equation (9.58).

9.5 SEISMIC DESIGN OF DIAPHRAGMS

9.5.1 Requirements regarding configuration and design actions

Floor slabs in R/C buildings are typically designed for vertical loading only; however, during an earthquake they are required to connect all vertical elements (columns, walls) together and distribute the seismic forces to the lateral load-resisting system. This behaviour is called diaphragm action and its implications for the analysis of R/C buildings have already been discussed in sections 5.5 and 5.6. According to EC8 the rigid body condition of a diaphragm is deemed to be satisfied if the in-plane deviations of all points of the diaphragm from their rigid body positions (compare equation (5.34)) are less than 5% of their respective absolute displacements under the actions corresponding to the seismic combination.

It is now well recognized that connections of slabs and vertical elements, especially structural walls, may constitute potential weak links in the path of seismic forces and prevent the vertical elements from developing their strength and energy dissipation capacity. The problem of diaphragm action is particularly relevant in the following cases:

1. Irregular or divided shapes in plan, such as floor slabs with large re-entrant corners, very high aspect ratios (elongated shapes) or large openings (Figures 4.4 and 4.5).
2. Buildings with complex and/or non-uniform layouts of the lateral load-resisting system, such as basements with R/C walls located only in parts of their perimeter or only in part of the ground floor.
3. Whenever structural systems having different characteristics with regard to horizontal deformation are tied (coupled) together through the floor slab, such as dual systems consisting of walls and frames. Major problems are expected in this case only if the distribution of walls is not the appropriate one (Figure 4.5).
4. Structures where significant changes in the stiffness of vertical elements above and underneath the diaphragm take place, such as buildings with setbacks or offsets (Figure 4.6), or with discontinued structural walls (section 9.3.4(a)).

For the aforementioned cases EC8 requires the verification of the R/C slab diaphragm in DC 'H' and 'M' structures. In such cases the action effects in the diaphragms can be estimated by modelling them as plane trusses or as deep beams resting on deformable supports. The action effects derived on the basis of such models must be multiplied by an overstrength factor $\gamma_f = 1.30$ (see also section 6.2.2).

9.5.2 Behaviour under cyclic loading

Experimental data on the behaviour of diaphragms subjected to in-plane cyclic loading are very limited (Chen *et al.*, 1988; Pantazopoulou and Imran, 1992), possibly because of the large size of specimens required. For the critical (but common) case that a lightly reinforced slab is connected to a much stronger R/C wall, it was found (Chen, Huang and Lu, 1988) that the in-plane strength of the slab was dominated by the flexural strength at a major crack, roughly parallel to the wall axis, which formed at the location where negative reinforcement of the slab was terminated (according to standard practice of gravity load design of slabs); this type of failure was quite similar to sliding shear modes encountered in walls (section 9.3.3). An improved behaviour of the diaphragm was found (Pantazopoulou and Imran, 1992) when continuous slab reinforcement (without cut-offs) was used in the slab–wall connection region. In this latter study it was also found that the shear resistance of the lightly reinforced slab diaphragm in the region of the slab–wall connection was well below the values obtained using the ACI-318 (1989) equation, which applies to both walls and diaphragms. Based on available test data, as well as on the results of an inelastic plane-stress analysis of the slab diaphragm, Pantazopoulou and Imran suggested the following equation for estimating the nominal shear resistance of slab–wall connections, where the slab reinforcement ratio $\rho_l \leqslant 1.5\,\%$ (which covers most practical situations)

$$\tau_u = f_y \sqrt{\rho_{lx} \rho_{ly}} \tag{9.60}$$

where ρ_{lx}, ρ_{ly} are the longitudinal reinforcement ratios of the slab in two orthogonal directions.

With regard to the type of slab used as a diaphragm, Chen, Huang and Lu (1988) found that waffle slabs exhibited higher ductility than flat plates or slab-on-beam systems designed for the same live gravity load. Furthermore, the reduction in strength due to cyclic loading (compared with the monotonic case) was less than 5% for the waffle slab, but up to 30% for the other two systems.

9.5.3 Resistance verification

Special mention is made in EC8 for the evaluation of the resistance of diaphragms in buildings with core or wall structural systems; the following requirements are set forth:

1. The nominal shear stress at the interfaces between diaphragms and walls or cores shall be limited to $6\tau_{Rd}$, to ensure that crack widths will not be excessive.
2. In assessing the shear resistance of the slab diaphragm, the contribution of concrete V_{cd} will be disregarded, to ensure adequate strength against shear sliding failure. Additional bars will be provided in the slab, according to the shear demand at the wall–slab connection; the anchorage of these bars will be made

in the same way as for the members of the lateral load-resisting system (sections 8.3.3(b), 8.5.3(b) and 9.4.3(b)).

These EC8 provisions appear to be in fairly good agreement with the limited available test data discussed previously. In the tests by Chen, Huang and Lu (1988) the experimentally estimated shear strengths of the slab diaphragms ranged from $0.17f_c^{1/2}$ to $0.24f_c^{1/2}$ while in the tests of Pantazopoulou and Imran (1992), where no cut-off of slab reinforcement in the connection region was made, τ_u varied from $0.20f_c^{1/2}$ to $0.25f_c^{1/2}$. These values are well below the limit of $6\tau_{Rd}\approx 0.35f_c^{1/2}$ set by EC8, hence it appears that the emphasis of design should be on the selection and detailing of the slab reinforcement in the regions around the walls or cores.

9.6 REFERENCES

Abrams, D.P. (1991) Laboratory definitions of behavior for structural components and building systems, in ACI SP-127: *Earthquake-resistant Concrete Structures – Inelastic Response and Design*, ACI, Detroit, pp. 91–152.

ACI (1989) *Building Code Requirements for Reinforced Concrete (ACI 318-89) and Commentary (ACI 318R-89)*, Detroit, Michigan.

ACI–ASCE Committee 352 (1985) Recommendations for design of beam–column joints in monolithic reinforced concrete structures (ACI 352 R-85). *Journal of the ACI*, **82**(3), 266–83.

Ali, A. and Wight, J.K. (1991) RC structural walls with staggered door openings. *Journal of Struct. Engng, ASCE*, **117**(5), 1514–31.

Aristizabal-Ochoa, J.D. (1982) Dynamic response of coupled wall systems. *Journal of the Struct. Div., ASCE*, **108** (ST8), 1846–57.

Bertero, V.V. (1980) Seismic behavior of R/C wall structural systems. *Proceed. of 7th World Conf. on Earthq. Engng*, Istanbul, Turkey, Sept, 1980, **6**, pp. 323–30.

CEB (1985) Model code of seismic design of concrete structures. *Bull. d' Inf. CEB*. **165**, Paris.

CEN Techn. Comm. 250/SC2 (1991) *Eurocode 2: Design of Concrete Structures–Part 1: General Rules and Rules for Buildings (ENV 1992-1-1/2/3)*, CEN, Berlin.

CEN Techn. Comm. 250/SC8 (1995) *Eurocode 8: Earthquake-Resistant Design of Structures–Part 1: General Rules and Rules for Buildings (ENV 1998-1-1)*, CEN, Berlin.

Chen, S.-J., Huang, T., and Lu, L.-W. (1988) Diaphragm behavior of reinforced concrete floor slabs. *Proceed of 9th World Conf. on Earthq. Engng*, Tokyo–Kyoto, Japan, Aug. 1988, Maruzen, Tokyo, **IV**, pp. 565–70.

Cheung, P.C., Paulay, T. and Park, R. (1992) Some possible revisions to the seismic provision of the New Zealand Concrete Design Code for moment resisting frames. *Bull. of the New Zealand Nat. Society for Earthq. Engng*, **25**(1), 37-43.

Daniel, J.I., Shiu, K.N. and Corley, W.G. (1986) Openings in earthquake-resistant structural walls. *Journal of Struct. Engng, ASCE*, **112**(7), 1660–76.

Durrani, A.J. and Wight, J.K. (1985) Behavior of interior beam-to-column connections under earthquake-type loading. *Journal of the ACI*, **82**(3), 343–9.

Durrani, A.J. and Zerbe, H.E. (1987) Seismic resistance of R/C exterior connections with floor slab. *Journal of Struct. Engng, ASCE*, **113**(8), 1850–64.

Ehsani, M.R. and Wight, J.K. (1985) Effect of transverse beams and slab on behavior of reinforced concrete beam-to-column connections. *Journal of the ACI*, **82**(2). 188–95.

Ehsani, M.R. and Wight, J.K. (1990) Confinement steel requirements for connections in ductile frames. *Journal of Struct. Engng, ASCE*. **116**(3), 751–67.

Endo, T., Adachi, H. and Nakanishi, M. (1980) Force deformation hysteresis curves of reinforced concrete shear walls. *Proceed. of 7th World Conf. on Earthq. Engng*, Istanbul, Turkey, Sept. 1980, **6**, pp. 315–22.

Hernandez, O. and Zermeno, M. (1980) Strength and behaviour of structural walls with shear failure. *Proceed. of 7th World Conf. on Earthq. Engng*, Istanbul, Turkey, Sept. 1980, **4**, pp 121–4.

Hiraishi, H. *et al.* (1989) Experimental study on seismic performance of multistory shear walls with flanged cross-sections. *Proceed. of 9th World Conf. on Earthq. Engng*, (Tokyo–Kyoto, Japan, Aug. 1988, Maruzen, Tokyo, **IV**, pp 553–8.

ICBO (Int. Conf. of Building Officials) (1994) *Uniform Building Code–1994 edition*, Whittier, California.

Kappos, A.J. (1991) Analytical prediction of the collapse earthquake for R/C buildings: case studies. *Earthq. Engng and Stuct. Dynamics*, **20**(2), 177–90.

Kitayama, K., Otani, S. and Aoyama, H. (1989) Behavior of reinforced concrete beam-column–slab subassemblages subjected to bi-directional load reversals. *Proceed. of 9th World Conf. on Earthq. Engng*, Tokyo–Kyoto, Japan, Aug. 1988, Maruzen, Tokyo, **VIII**, pp 581–6.

Meinheit, D.F. and Jirsa, J.O. (1981) Shear strength of R/C beam-column connections. *Journal of the Struct. Div. ASCE*, **107**(11), 2227–44.

Oesterle, R.G., Fiorato, A.E, Aristizabal-Ochoa, J.D. and Corley, W.G. (1980) Hysteretic response of reinforced concrete structural walls, in *ACI SP-63: Reinforced Concrete Structures-Subjected to Wind and Earthquake Forces*, American Concrete Institute, Detroit, pp 243–73.

Pantazopoulou, S. and Bonacci, J. (1992) Consideration of questions about beam-column joints. *ACI Struct. Journal*, 89(1), 27–36.

Pantazopoulou, S. and Imran, I. (1992) Slab–wall connections under lateral forces. *ACI Struct. Journal*, **89**(5), 515–27.

Park, R. and Dai, R. (1988) A comparison of the behaviour of reinforced concrete beam–column joints designed for ductility and limited ductility. *Bull. of the New Zealand Nat. Society for Earthq. Engng*, **21**(4), 255–78.

Park, R. and Paulay, T. (1975) *Reinforced Concrete Structures*, J. Wiley & Sons, New York.

Paulay, T. (1980) Earthquake-resisting shearwalls–New Zealand design trends. *Journal of the ACI*. **77** (3), 144–52.

Paulay, T. (1986) A critique of the special provisions for seismic design of the Building Code Requirements for Reinforced Concrete (ACI 318-83). *Journal of the ACI*, **83**(2), 274–83.

Paulay, T., Park, R. and Priestley, M.J.N. (1978) Reinforced concrete beam–column joints under seismic actions. *Journal of the ACI*, **75**(11), 585–93; also: Discussion by D.F. Meinheit and Authors' closure, *Journal of the ACI*, **76**(5), May 1979, 662–7.

Paulay, T. and Priestley, M.J.N. (1992) *Seismic Design of Reinforced Concrete and Masonry Buildings*, J. Wiley & Sons, New York.

Paulay, T., Priestley, M.J.N. and Synge, A.J. (1982) Ductility in earthquake resisting squat shear-walls. *Journal of the ACI*, **79**(4), 257–69.

Paulay, T. and Santhakumar, A.R. (1976) Ductile behavior of coupled shear walls. *Journal of the Struct. Div., ASCE*, **102** (ST1), 93–108.

Popov, E.P. *et al.* (1977) On seismic design of R/C interior joints of frames. *Proceed. of the Sixth World Conf. on Earthq. Engng*, New Delhi, India, II, pp 1933–8.

Shen, J. and Huang, Z. (1990) Inelastic seismic response of structure with discontinuous shear walls supported on frames. *Proceed. of Int. Conf. on Earthq. Resistant Construction and Design*, Berlin, June 1988, Balkema, Rotterdam, pp. 515–24.

Shiu, K.N., Takayanagi, T. and Corley, W.G. (1984) Seismic behavior of coupled wall systems. *Journal of the Struct. Div. ASCE*, **110**(5), 1051–66.

SANZ (Standards Association of New Zealand) (1982) (a) Code of Practice for the Design of Concrete Structures (NZS 3101–Part 1: 1982); (b) Commentary on Code of Practice for the Design of Concrete Structures (NZS 3101–Part 2: 1982), Wellington.

Standards New Zealand (1992) *Code of Practice for General Structural Design and Design Loadings for Buildings* (NZS 4203: 1992), Wellington.

Tassios, T.P. (1984) Masonry infill and RC walls under cyclic actions. *CIB Symposium on Wall Structures*, Warsaw, June 1984.

Tassios, T.P. (1989a) Specific rules for concrete structures–justification note no. 5: Anchorage of beam's reinforcements across a column, in *Background Document for Eurocode 8–Part 1, Vol. 2–Design Rules*. CEC DG III/8076/89 EN, Brussels, pp. 18–22.

Tassios, T.P. (1989b) Specific rules for concrete structures–justification note no. 8: A rational simplified model for beam–column joint behaviour, in *Background Document for Eurocode 8–Part 1, Vol. 2–Design Rules*, CEC DG III/8076/89 EN, Brussels, pp. 55–64.

Tsonos, A.G., Tegos, I.A and Penelis, G.G. (1992) Seismic resistance of type 2 exterior beam–column joints reinforced with inclined bars. *ACI Struct. Journal*, **89**(1), 3–12.

Vulcano, A. and Bertero, V.V. (1987) 'Analytical models for predicting the lateral response of RC shear walls: evaluation of their reliability. *Rep. No. UCB/EERC-87/19*, Univ. of Calif. at Berkeley.

Wong, P.K.C., Priestley, M.J.N. and Park, R. (1990) Seismic resistance of frames with vertically distributed longitudinal reinforcement in beams. *ACI Struct. Journal*, **87**(4), 488–98.

Yamada, M., Kawamura, H. and Katagihara, K. (1974) Reinforced concrete shear walls with openings: test and analysis, in *ACI SP-42: Shear in Reinforced Concrete*. American Concrete Institute, Detroit, **2**, pp. 559–78.

10

Seismic performance of buildings designed to modern seismic codes

10.1 METHODS FOR ASSESSING THE SEISMIC PERFORMANCE

10.1.1 Introductory remarks

The most common structural systems used in R/C buildings and other structures have already been given in section 4.3, where the problem of irregular configurations has also been discussed. Furthermore, criteria for regularity in plan and elevation have been given in section 5.1 for buildings, and analysis methods appropriate for each type of building have been presented subsequently (sections 5.3–5.7). On the other hand, the seismic performance of individual members (beams, columns, walls, floor slabs), as well as of structural subassemblages ('substructures'), such as beam–column joints (including the joint core region), has been discussed in detail in Chapters 8 and 9.

The purpose of this chapter is to address briefly the problem of evaluating the seismic performance of entire buildings, being composed of the aforementioned subassemblages of beams and columns, with or without structural walls, tied together by floor slabs. An extensive treatment of all structural systems used in R/C building construction falls outside the scope of this book, wherein emphasis will be placed on the two most commonly used systems, namely frame and dual systems, designed to the provisions of modern seismic codes, in particular EC8 and EC2, whose design and detailing requirements have already been given in Chapters 5–9 (including design examples). Some reference to buildings designed to the previous 'generation' of seismic codes, such as the CEB (1985) Model Code will also be made here. On the other hand, only a very brief mention of the seismic performance of existing buildings, most of which have been designed to 'old-fashioned' codes, or even without taking seismic regulations into account, will be included in the present chapter. Nevertheless, the seismic pathology of such structures will be discussed in Chapter 11, while the methods for assessing and repairing seismic damage in existing structures will be explained in the remaining three chapters (12–14) of the book.

10.1.2 Performance assessment through testing of models and inspection of actual structures

From a first view it appears that the ideal procedure for assessing the seismic performance of structures is the careful inspection of the behaviour of actual structures that have been damaged or even collapsed due to earthquakes. However, relatively few buildings have been subjected to extreme seismic forces and for many of them no sufficient data from post-earthquake damage inspection are available. Moreover, in the rare event of collapse, it is usually difficult to develop a reliable 'scenario' of the events that led to collapse, as it is almost impossible to re-create the different stages of the response of the building from the heap of rubble on the ground. On the other hand, very few buildings are equipped with instrumentation to such an extent as to permit a complete description of their inelastic seismic response; most of these buildings are situated in the United States, mainly in California. Finally, as will be made quite clear in the following chapter, damage patterns in existing buildings are quite varied and they often contradict predictions based on previous experience and/or engineering judgement.

Given the above, testing of appropriately scaled models appears to offer an attractive alternative (or, better, complementary) procedure for evaluating the seismic performance of structures. Nevertheless, testing of realistic buildings is expensive, and it is impossible to test even low-rise buildings with large-amplitude dynamic forces, since shaking tables or eccentric mass generators of this calibre are simply not available (Abrams, 1991). It is beyond the scope of this book to discuss the limitations of testing small-scale models on the shaking table (a geometric scale of about 1/10 is commonly used for medium-rise buildings not exceeding 10 storeys); valuable information regarding this problem may be found, among others, in the comprehensive reports by the ACI Committee 444 (ACI, 1982) and by the CEB Task Group III/6 (CEB, 1994).

It is, of course, possible to test larger-scale models of medium-rise buildings (up to seven or eight storeys high) statically, but since inertia and damping forces cannot be reproduced in such a procedure, careful selection of the applied displacement history is required, as well as careful interpretation of the results. During the late 1970s the **pseudodynamic testing** technique evolved, whereby the dynamic inertia forces are simulated by statically applied loads (displacements) whose magnitude is controlled by the dynamic equilibrium equations. The elastic and inelastic restoring forces are measured experimentally, while inertia and viscous damping forces are derived analytically by solving numerically a differential equation such as (3.21). The economic and practical advantages of this method, in comparison with the shaking table procedure, as well as its drawbacks (the major one being possibly its inability to capture the strain-rate effects, dealt with in section 7.3.1), are discussed, among others, by Shing and Mahin (1984) and by Carvahlo and Coelho in the CEB Task Group III/6 report (CEB, 1994). An interesting correlation study involving pseudo-dynamic and shaking table tests on two-storey half-scale models of R/C frames has been presented by Kitagawa *et al.* (1984).

10.1.3 Performance assessment using inelastic dynamic analysis

In section 3.5 the procedure for analysing the inelastic response of multi-storey buildings subjected to base accelerograms was outlined, and the main output parameters from such an analysis have been presented (Figure 3.25). If these parameters are appropriately evaluated, in particular if a meaningful comparison of demanded and supplied quantities (for instance, ductilities) is carried out, inelastic time-history analysis may become a powerful and cost-effective tool in assessing the seismic performance of complicated structures, such as multi-storey R/C buildings. However, a high degree of uncertainty exists with regard to both sides of the equation, that is supply and demand. The study of the uncertainties involved in evaluating the demands, in particular of displacement and local deformation requirements (in the plastic hinge regions) has been the subject of numerous studies, a recent one being that by the CEB Task Group III/6 (CEB, 1994), while the other source of uncertainty, the supplies, in particular the available deformation capacity of R/C members, has also been the subject of a number of studies, the most comprehensive being possibly that by the ACI–ASCE Committee 442 (Abrams, 1991; French and Schultz, 1991; Saatcioglu, 1991; Wood, 1991).

(a) Evaluation of demands

Research carried out at the University of Thessaloniki (Kappos, 1986, 1990) on the influence of various assumptions on the calculated inelastic seismic response of R/C buildings, has clearly indicated the following:

(1) Depending on the assumption made for the stiffness of the elastic part of lumped plasticity member models, which are elements with concentrated plastic hinges at their ends (see section 3.5 and also CEB, 1994), that is whether moderately or fully cracked members are assumed, the calculated inter-storey drifts may differ by more than 100% (Kappos, 1986).
(2) Although a rather refined technique for normalizing all the input motions used to the same spectrum intensity was applied, differences in main response quantities up to about 100% were recorded for typical medium-rise R/C buildings (Kappos, 1990); on the other hand, the same study has indicated that in statistical terms, the variation of response quantities was of the order of 30%, while its distribution along the height of the buildings was quite uniform.
(3) The rest of the input parameters investigated, which included variation of material strengths within the limits usually expected in practice, and various assumptions regarding the effective shear and axial stiffness due to cracking, were found to have a relatively minor effect on the calculated response of R/C frames.

Within the framework of the CEB Task Group III/6, two typical R/C structures (an eight-storey and a four-storey frame), designed to EC2 and EC8 were analysed in the inelastic range of their response by various groups of investigators, each using its own computer code. The geometry of each frame, as well as the material strengths, were the same in all analyses, but the rest of the modelling assumptions made by each group were different. From the results presented in the previously mentioned report (CEB, 1994), it appears that marked differences existed in the calculated displacement time histories, attributed mainly to the different assumptions made regarding the member stiffnesses, and secondly due to different amounts of viscous damping assumed by each group. It is worth pointing out that maximum storey displacements ranged from as low as 38 mm to as high as 165 mm for the eight-storey frame subjected to an artificial accelerogram compatible with the EC8 design spectrum. As expected, maximum forces calculated from each group were much closer to each other than displacements, since the former are controlled by the yield strength of members, which was estimated using fixed values of material strengths. It is also interesting to note that comparisons between the various analyses in terms of rotational ductility demands in the critical regions of R/C members were not even attempted, due to the different modelling of these regions adopted by each group of investigators. Notwithstanding the aforementioned uncertainties, all the individual groups agreed that the performance of both frames subjected to the design earthquake was satisfactory.

(b) Evaluation of supplies

With regard to the estimation of supplies, that is of the deformational capacity of members, it was pointed out in section 8.2.1 that a high degree of uncertainty exists, even for the case of monotonic loading. In that section both the conventional, purely analytical, procedure based on integration of plastic curvatures, and the empirical equations by Park and Ang (1985) were given; it is recalled that the latter were based on a very large database of test results concerning R/C beams and columns. In the previously mentioned report by the ACI–ASCE Committee 442 the available rotational capacity of beams (French and Schultz, 1991), columns (Saatcioglu, 1991) and walls (Wood, 1991), subjected to typical cyclic loading histories were evaluated on the basis of large experimental databases and the following were pointed out:

(1) Significant scatter exists with regard to experimentally measured deformational capacities, expressed in the form of either ductility or drift ratios (Figure 8.15); this scatter reflects both uncertainties in the load transfer mechanisms of R/C members under cyclic loading (see also sections 8.2.2 and 8.3.2), and differences in testing techiques.
(2) The single most important parameter affecting the rotational capacity of members was the level of shear force, better expressed in the form of nominal shear stress, the general trend being that ductility decreases with increasing shear stress.

It has long been recognized that R/C members subjected to cyclic loading fail due to a combination of large deformation, usually expressed as a ductility ratio or as a plastic hinge rotation, and of low-cycle fatigue, usually expressed by the hysteretic energy dissipated by the member. This is reflected in the **seismic damage index** suggested by Park and Ang (1985), which is given by the following equation

$$D = \frac{\theta_{\mathrm{max}}}{\theta_{\mathrm{u}}} + \frac{\beta}{M_{\mathrm{y}}\theta_{\mathrm{u}}} \int \mathrm{E} \qquad (10.1)$$

where θ_{max} and θ_{u} are the maximum required rotation and the rotational capacity (demand and supply respectively) of the member, with θ_{u} referring to monotonic loading conditions; M_{y} is the member yield moment, E is the amount of energy dissipated, and β is an empirical constant depending on the member geometry and reinforcement. It is pointed out that equation (10.1) is expressed in the more convenient (for dynamic analysis purposes) form of rotations, while the original Park–Ang index was based on displacements (δ), which are more convenient to use when test results are evaluated; for isolated members subjected to loading cycles inducing substantial inelastic deformations, the approximation $\theta = \delta / l$ (where l is the shear span) is a reasonable one. A member is assumed to have failed when $D = 1$, which can result either when $\theta_{max} = \theta_u$ (monotonic loading failure), or when $\theta_{max} < \theta_u$ and the second term (based on energy) reaches an appropriate (non-zero) value (cyclic loading failure).

(c) Concluding remarks

Based on the foregoing considerations, it is concluded that, notwithstanding the significant uncertainties involved, it is possible to assess the seismic performance of relatively complicated structures on the basis of an inelastic time-history analysis using different, appropriately normalized, input accelerograms and at least two different structural models, based on upper and lower bounds of member stiffnesses (if more refined procedures are not used) followed by a correlation of the estimated demands in terms of member (and/or global) deformations and the corresponding capacities, estimated empirically and/or analytically. Such a procedure is particularly useful if comparisons between different (possibly alternative) structural solutions are sought, while it is not necessarily adequate for reproducing available test data, for instance data derived from pseudo-dynamic or shaking table studies.

In the following a quantitative assessment of the seismic performance of EC8-designed buildings, in particular a 10-storey frame and a corresponding dual system, will be presented, using mainly the analytical procedure described previously, complemented by data and observations from pertinent experimental studies.

10.2 SEISMIC PERFORMANCE OF FRAMES

In Chapter 8 the design procedure specified in EC8 for R/C linear elements (beams, columns) was outlined and a design example involving a 10-storey frame (Figure 10.2) was given in section 8.6. In the following it will be attempted to evaluate the seismic performance of this frame, when it is subjected to input motions of various intensities corresponding to the serviceability, the design (ultimate) and the survival (collapse) limit state (see also section 4.2.1). It is pointed out that a wealth of data from both experimental and analytical studies involving R/C frames designed to various seismic codes can be found in the report by the CEB Task Group III/6 (CEB, 1994).

10.2.1 Selection of input motions

As already mentioned in the previous section, the calculated inelastic dynamic response is quite sensitive to the characteristics of the input motions, thus for the frame under consideration it was decided to use seven accelerograms, given in Table 10.1, six of which were recorded in Greece during the most damaging earth-

Table 10.1. Basic data for the input motions used

No.	Earthquake	Magnitude (M)	Site	Component	Peck acceleration(g)	Peak velocity (mm s^{-1})	Normalization parameter	Effective duration (s)
1	Volvi, Greece 20.6.1978	6.5	Thessaloniki (city centre)	N30E	0.142	127	2.17	12.0
2	Volvi, Greece 20.6.1978	6.5	Thessaloniki (city centre)	N60W	0.154	167	2.66	12.0
3	Alkyonides, Greece 24.2.1981	6.7	Corinth (city centre)	N35E	0.239	225	1.09	19.4
4	Alkyonides, Greece 24.2.1981	6.7	Cornith (city centre)	N55W	0.286	246	1.32	17.8
5	Kalamata, Greece 13.9.1986	6.2	Kalamata (city centre)	N80E	0.240	323	1.17	7.5
6	Kalamaa, Greece 13.9.1986	6.2	Kalamata (city centre)	N10W	0.273	237	1.24	8.5
7	Imperial Valley, California 18.5.1940	7.1	E1 Centro Site	S00E	0.348	335	0.93	26.5

quakes occurring in the last 15 years, while the seventh is the well-known El Centro 1940, N–S component, which facilitates comparisons with similar studies carried out by other investigators. With regard to seismological data not included in Table 10.1, it has to be mentioned that all six motions from Greece were recorded at sites quite close to the earthquake epicentre, at distances varying from 9.0 km (Kalamata) to 31.5 km (Corinth). Indeed, surface earthquakes at small epicentral distances are the ones that typically cause the most serious damage in Greek cities.

All the input motions were normalized to the same spectrum intensity, using a modified Housner technique suggested by Kappos (1991a), according to which the scaling factor for each motion (n) is equal to the ratio SI_o/SI_n of the shaded areas of the velocity spectra shown in Figure 10.1; it is pointed out that SI_o is the area under the velocity response spectrum derived from the acceleration design spectrum specified by EC8. Instead of using the standard limits of 0.1 and 2.5 suggested by Housner, it was deemed more appropriate (Kappos, 1990) to condense them, taking into account the natural periods of the structures studied (the aforementioned frame as well as the dual system given in section 10.3), thus the area under the spectra was calculated between periods of 0.6 and 1.9 s. The normalization factors corresponding to the EC8 design spectrum for $A=0.25\,g$ (the design seismic action) are given for each motion in Table 10.1, and it can be easily seen that in general they are higher than the ratios of the design acceleration to the peak acceleration of each record (also given in Table 10.1).

It is well established that the destructiveness of a seismic motion depends not only on its peak values but also on its duration. Trifunac and Brady (1975) have defined the duration of strong motion acceleration to be that time interval during which the central 90% of the contribution to the integral of the square of the acceleration $\int A^2\,dt$, takes place.

The values listed in the last column of Table 10.1 as 'effective' durations, also include the intervals from the beginning of the excitation to the beginning of the Trifunac–Brady duration, therefore they are slightly larger than the latter; for instance 26.5 s have been calculated for El Centro, instead of the 24.0 Trifunac–Brady duration, while the entire duration of this record is 53 s. These effective durations have been used in the inelastic time-history analysis, in order to obtain a meaningful comparison between the different records, whose energy characteristics as expressed by $\int A^2 dt$, are quite different.

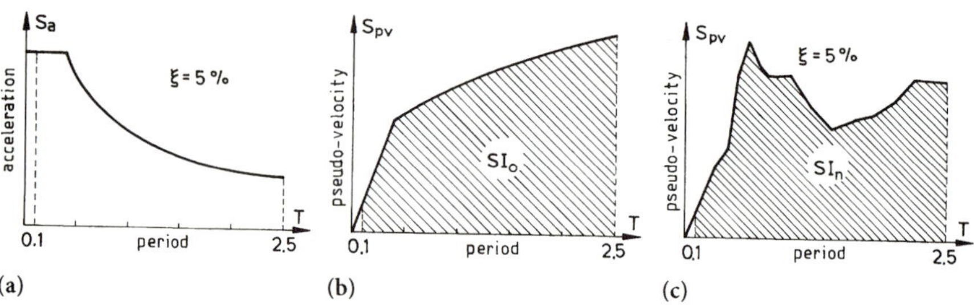

Figure 10.1 Normalization of input motions to the design earthquake intensity: (a) code-prescribed acceleration spectrum; (b) corresponding pseudo-velocity spectrum; (c) pseudo-velocity spectrum for n-motion.

10.2.2 Modelling assumptions and failure criteria

(a) Modelling assumptions

For multi-storey structures, such as the 10-storey frame (Figure 10.2) under consideration, the **member-by-member modelling** approach is typically adopted. Microscopic approaches (**finite element modelling**) are typically restricted to the analysis of isolated members under a few cycles of loading, or of entire structures but under monotonic loading; efforts to extend their application to the time-history analysis of moderately sized structures have not been so successful in establishing overall numerical stability and superiority over less sophisticated models in reproducing pertinent experimental results (CEB, 1994).

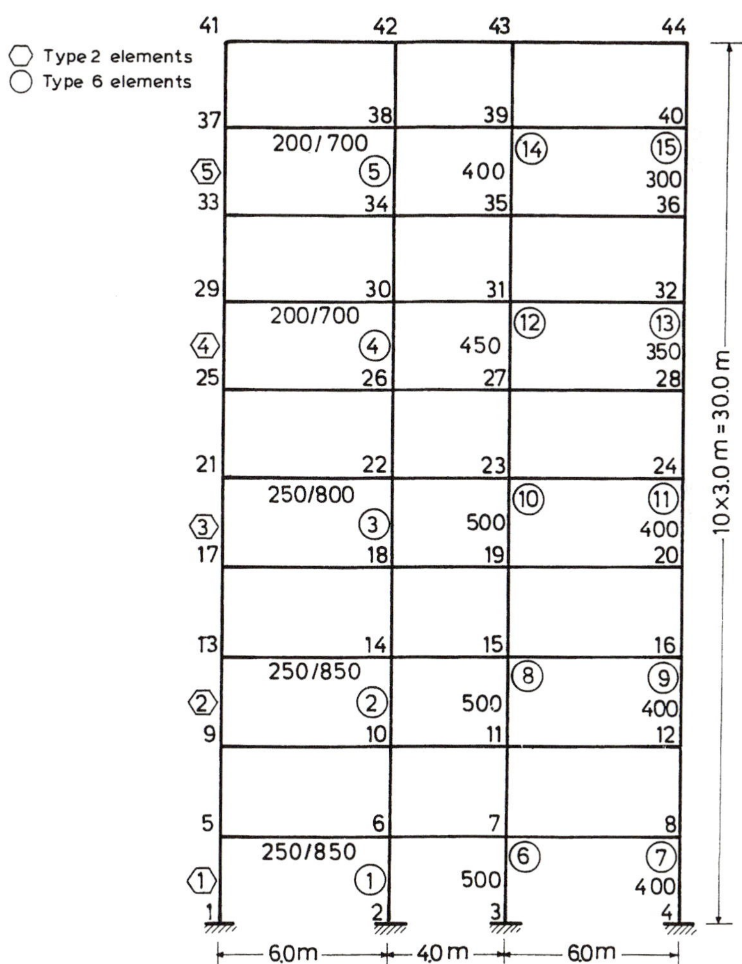

Figure 10.2 Ten-storey frame structure designed to EC8.

The analysis of the 10-storey frame for the aforementioned input motions is carried out using the DRAIN-2D/90 code (Kappos, 1992), whose basic features were outlined in section 3.5, while more detailed information regarding the selection of appropriate model parameters can be found in Kappos (1986). The modified Takeda hysteresis model proposed by Otani and Sozen (1972) was used for all R/C members except the exterior columns, for which it was preferred to use the (less refined) element with bilinear behaviour available in DRAIN-2D/90; the reason for this choice is that the latter has the advantage of taking into account the variation of column yield moment with the axial loading, which is a very important consideration in members with significant variation of axial load due to the seismic overturning moment. Sections with Takeda hysteresis are indicated by circles in Figure 10.2, while sections with bilinear hysteresis are indicated by hexagons.

(b) Local failure criteria

It was pointed out in section 10.1 that a meaningful evaluation of the seismic performance of a structure presupposes a comparison between the demands imposed by the earthquake and the corresponding capacities of the members, or of the structure as a whole. In section 8.2.1 a detailed presentation was made of both the purely theoretical and the empirical procedures for estimating the available rotational capacity of R/C members subjected to essentially monotonic loading. On the other hand, the importance of taking into account the effect of cyclic shear on the seismic capacity of structures was emphasized in sections 8.2.2, 8.4.3 and 9.3.3. The simplest way to account for the effect of shear on the available rotational capacity of R/C members appears to be that suggested by Kappos (1991a), whereby the classical equation (8.17)

$$\theta_{p,av} = (\phi_u - \phi_y)l_p$$

is modified by including a corrector for the effect of shear, $k_V \leqslant 1$. that is

$$\boxed{\theta_{p,av} = k_V(\phi_u - \phi_y)l_p} \tag{10.2}$$

The following expression for the shear corrector k_V has been proposed by Kappos, Antoniades and Konstantinides (1994)

$$k_V = 1.67 - V_{max}/V_{R1} \ngtr 1 \tag{10.3}$$

where in the light of the discussion in section 8.2.2(d), the threshold value of shear is

$$V_{R1} = 3(2+\zeta)\tau_R b_w d \tag{10.4}$$

where $\tau_R \approx 0.1 f_{cm}^{1/2}$, b_w is the width and d the effective depth of the section where the maximum shear, V_{max}, develops (typically the member end). According to equation (10.3) whenever $V_{max} < V_{R1}$ the effect of shear on the rotational capacity is negligible ($k_v = 1$), while for

$$V_{max} > 1.67\ V_{R1} = 5(2+\zeta)\ \tau_R b_w d$$

a brittle failure ($\theta_{p.av} = 0$) is expected, since $k_V = 0$. From the discussion in section 8.2.2(d) it is easily seen that the first assumption (lower limit) is a very reasonable one, being in agreement with available test data, while the second assumption (upper limit) is slightly conservative, since even for $\tau > 0.5(2+\zeta)f_{cm}^{1/2}$, some ductility is available in the member. It is emphasized that the procedure involving equations (10.2)–(10.4) applies only to members with conventional shear reinforcement, and not to those with bidiagonal bars (Figure 8.16); for the latter substantially higher k_v values should be adopted.

In addition to exceeding the available rotational capacity (equation(10.2)), member failure may also result from inadequate shear reinforcement, a factor which is not included in the aforementioned procedure. Ideally a criterion based on shear deformations should be used for this purpose, in a way analogous to that used for estimating the available plastic rotation in flexure-dominated members. However, the available experimental data are very limited since shear deformations (γ) are quite difficult to measure and also subject to more uncertainties than total deformations (Figure 8.12), hence it is not easy to define an ultimate shear strain (γ_u) analogous to the ultimate normal strain (ε_u) used for calculating the ultimate curvature (ϕ_u) of R/C sections (sections 8.2.2(b) and 8.4.2(a)). Nevertheless, some τ–γ (or V–γ) hysteresis models have been proposed, the most recent (Fardis, 1991) including a falling branch after the development of maximum shear strength, but their experimental verification has been very limited and their use very restricted. It is worth pointing out that even if very reliable τ–γ models were available, their incorporation into computer codes based on member-by-member discretization, would at the least be very laborious (see also CEB, 1994).

For the purposes of the present study a strength criterion was used for evaluating the possibility of shear failure, which was assumed to take place whenever the maximum shear force (V_{max}) calculated from the dynamic analysis exceeded the estimated shear resistance (V_R) of a member. The resistance V_R depends, of course, on the type of shear failure and, in general, all three types should be considered, namely diagonal tension failure, diagonal compression failure (web crushing), and sliding shear failure. The corresponding EC8 equations, which are deemed to be consistent with the current state of the art, were used in all cases, without introducing any material safety factors, to avoid, at least to a certain extent, the conservatism inherent in design equations. However, a rather significant additional modification was carried out with regard to the so-called 'concrete mechanism' for which the value $V_{cd}=0$ is suggested by EC8 for plastic hinge regions of beams, as well as of walls. On the basis of available experimental data, Kappos, Antoniades and Konstantinides (1994) suggested the following equation

$$\boxed{V_c = \left[1-\left(\frac{\mu_\theta - 1}{3}\right)^2\right] V_{c0}} \tag{10.5}$$

where V_{c0} is the code specified value of shear resistance for regions outside the critical one (equation (8.47)), again using τ_R instead of τ_{RD}. The first term of equation (10.5) reflects the trend, clearly observed in relevant test results, that shear capacity decreases with the level of inelasticity. The assumption that $V_c=0$ for $\mu_\theta > 4$, is quite reasonable for beams, but somewhat conservative for columns with significant axial loading, for which Priestley and Seible (1992) suggest a residual shear capacity of about $0.33V_{c0}$ for $\mu_\theta > 4$. It is understood that assuming failure of a member whenever the estimated shear capacity is exceeded is usually a conservative approach, which nevertheless offers the great advantage of simplicity.

(c) Global failure criteria

Local failure, however reliably predicted by analysis, does not necessarily indicate collapse of a storey and even more of the building as a whole. Predictions of 'failure' based on member ductility criteria could be either conservative or unconservative. In fact, if failure is defined as the state of a building where the cost of repair is greater than the cost of reconstruction, the local failure criterion may well not be conservative, in particular when the cost of repairing the 'non-structural' members dominates, which is usually the case. On the other hand, if a member not essential to the stability of a building (for instance a beam) fails prematurely and redistribution of actions is possible, the local failure criterion may yield conservative results. Based on the above remarks, it is considered essential to include a global failure criterion in the analytical procedure for assessing the seismic performance of R/C buildings. In the following, it is conservatively assumed that the failure of a single storey is equivalent to the overall failure of the building, although post-earthquake inspections have revealed that this is not always the case, especially in structures with a 'soft' first storey.

Perhaps the single most important response parameter to characterize the seismic behaviour of a storey or a building is the relative inter-storey drift (Sozen, 1981), defined as

$$\Delta x_i/h_i = (x_i - x_{i-1})/h_i \tag{10.6}$$

where x_i,x_{i-1} denote the horizontal displacements of two adjacent floors and h_i the corresponding storey height. This quantity is easy to measure in tests or in actual buildings struck by earthquakes and can be correlated with available data on damage. However, it is very difficult to define a single value of drift corresponding to collapse, to apply for all buildings. Taking into consideration factors such as the vulnerability and importance of the building contents, as well as the cost of repair, it was suggested (Sozen, 1981) that an inter-storey drift of 2% might be set as the collapse limit for about three-quarters of R/C buildings. This value is included in a combined criterion of storey failure in the suggested procedure. It has to be emphasized that for this criterion to be valid, it is essential that the stiffness of the building under consideration be adequately modelled in the analysis, otherwise the procedure may not be conservative.

The inter-storey drift serves also as a measure of the effect of second-order shears and moments, which are not treated in an explicit way by most computer codes for inelastic time-history analysis, including that used in the studies in the following sections. At values of inter-storey drift in excess of 2% the P–δ effect is significant, reducing the lateral force resistance and the stiffness of the vertical structural members and precipitating failure.

Another criterion for storey failure, commonly used in the cases of inelastic static analysis, as well as limit analysis, is the formation of a 'sidesway' mechanism, involving plastic hinges at both the top and the bottom of all vertical members. Previous studies (Kappos, 1991a) indicated that the above criterion is conservative, apparently because at the time a hinge forms at a certain member end, another, already yielding, member may enter the unloading stage and respond with a stiffness equal to, or slightly lower than, the elastic one. Therefore, a combined criterion is adopted involving both the formation of a sidesway mechanism and the occurrence of an inter-storey drift in excess of 2%. However, even in the case that a collapse mechanism does not form, a building is assumed to have failed whenever the maximum inter-storey drift exceeds 3%, since at this stage all non-structural elements have been severely damaged and repair of the building is no longer cost-effective.

10.2.3 Performance under the design earthquake

The variability of the calculated inelastic seismic response of the 10-storey frame with the input motion characteristics (Table 10.1) was found to be quite significant, although all motions were normalized to the spectrum intensity of the design earthquake ($A_{ef}=A_d=0.25g$) using the refined technique described in section 10.2.1. In Figure 10.3 are shown the calculated beam rotational ductility factors (equation(8.25)) for positive and negative bending, for the most critical (with regard to ductility) and the less critical input motion, the criterion always being the maximum calculated μ_θ factor in the beams (at the interior supports, where

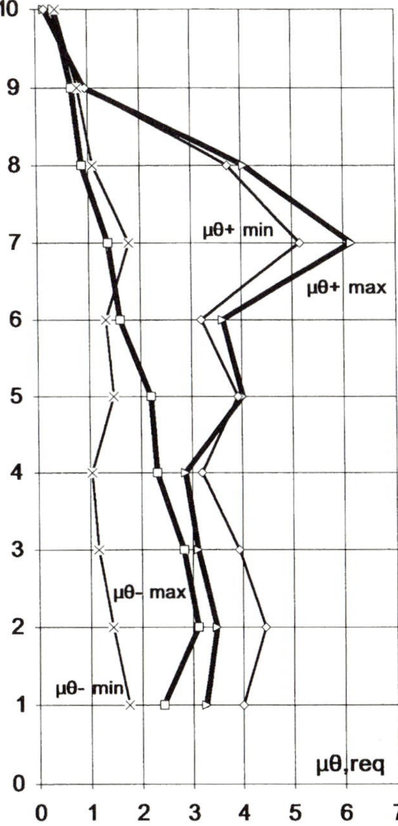

Figure 10.3 Beam ductility factors for the frame structure of Figure 10.2 subjected to the most critical and the least critical motion (for $A_{ef}=0.25g$).

the max μ_θ are recorded). It is seen that although the differences at the peak values are not dramatic (the ratio of maximum to minimum value of peak μ_θ^- factors is about 1.5), the calculated values at individual storeys may differ by up to about 100%. It is also worth pointing out that the same input motion (Thessaloniki N30E) produced the maximum peak μ_θ^- and the minimum peak μ_θ^+ value. Similar trends were found with regard to the other main response quantities; for example, the maximum inter-storey drift ratio ranged from 0.49 to 0.60%, the differences at some storeys being much more marked. Moreover, the variability tended to increase with the level of the excitation, as more inelasticity was induced in the R/C members.

The foregoing are in agreement with the conclusions of previous studies (Kappos, 1990), and suggest that a statistical approach might be more suitable for assessing the seismic performance of structures on the basis of inelastic time-history analyses. However, in the following the more conservative approach of comparing available capacity with the demands resulting from the most critical input motion has been adopted, to compensate roughly for ignoring the effect of bond-slip and of possible geometric errors in the analytical model.

According to the methodology described in section 10.2.2 the possibility of member failure due to inadequate rotational capacity can be assessed using equation (10.2), where an appropriate shear corrector k_V (equation (10.3)) is introduced. The comparison between required ($\theta_{p,req}$) and available ($\theta_{p,av}$) plastic hinge rotations for the beams of the frame structure under consideration is shown in Figure 10.4, where for each storey the point drawn corresponds to the plastic hinge exhibiting the lowest ratio $\theta_{p,av}/\theta_{p,eq}$; typically this corresponds to negative bending, since for positive bending the beams work as T-sections and the neutral axis depth is very small, resulting in very high ductility, as explained in section 8.2.1. It is clearly seen in Figure 10.4 that ample safety margins (expressed as the ratio $\theta_{p,av}/\theta_{p,eq}$) exist at both the exterior and the interior beams of the frame, the 'safety factor' being considerably higher than 6 along the whole height of the building. This is a clear indication that the beams of the structure are in a position to develop an adequate mechanism for dissipating the energy induced by the design earthquake.

The rotational capacity of columns is compared with the corresponding requirements in Figure 10.5, where again for each storey the peak plastic rotation calculated for any of the seven records used is considered. The negligible values of $\theta_{p,req}$, which are lower than 0.002 rad at all storeys and zero in most of the upper half of the frame, are of course much lower than the corresponding capacities, the smallest safety factor (ratio $\theta_{p,av} / \theta_{p,req}$) being equal to 15 (at the ground storey interior columns). This is a clear indication that the capacity design procedure prescribed in EC8 for columns is an efficient one, as inelasticity is not likely to be induced in these elements. On the other hand, it could be argued that the procedure might be overconservative, but for such a conclusion to be substantiated, the performance of the structure under the maximum credible earthquake should also be assessed; this is done in section 10.2.4.

In all the rotational capacity checks shown in Figures 10.4 and 10.5 the shear corrector k_V (equation (10.2)) was found to be equal to 1.0, that is, the slenderness of the members combined with the relatively low reinforcement ratios resulted in shear stresses lower than the limit defined by equation (10.4). As discussed in section 10.2.2 this does not necessarily mean that shear reinforcement in R/C members is adequate, and extra checks should be carried out. In Figures 10.6 and

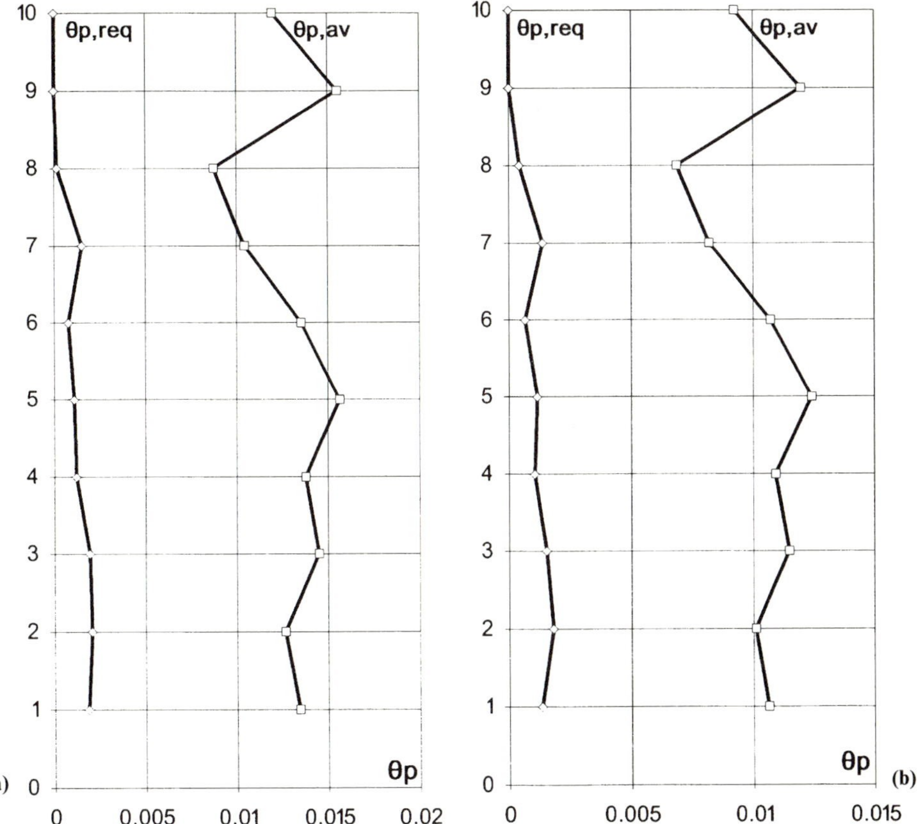

Figure 10.4 Required and available plastic rotations (in rad) in the beams of the frame structure, for the most critical motion: (a) exterior supports; (b) interior supports.

10.7 comparisons are shown of the maximum shear forces (V_{max}) recorded at each storey under the most critical motion normalized to $A_d = 0.25g$) with the corresponding capacities, for the beams (Figure 10.6) and the columns (Figure 10.7) of the frame structure. The most critical shear capacity check for all members was found to be that against diagonal tension failure, hence V_c in the figures is the shear carried by the 'concrete mechanism', estimated using the EC8 formula (equations (8.47), with τ_R instead of τ_{Rd}) and the correction for the induced level of inelasticity suggested by equation (10.5). The shear carried by the transverse reinforcement (V_w) through the well-known truss mechanism is estimated again from the EC8 formula (equation (8.48)) without material safety factors, and it is assumed to remain constant with increasing inelastic deformation ($\mu_\theta > 1$).

As shown in Figure 10.6(a) the shear capacities of the beams at the exterior supports are well above the maximum shear forces calculated in the inelastic analyses, the 'safety factor' $(V_w + V_c) / V_{max}$ ranging from 3.3 to 5.1; it is worth pointing out that in the upper half of the frame the hoops alone (V_w term) are sufficient to carry the entire dynamic shear. On the other hand the shear capacity of the

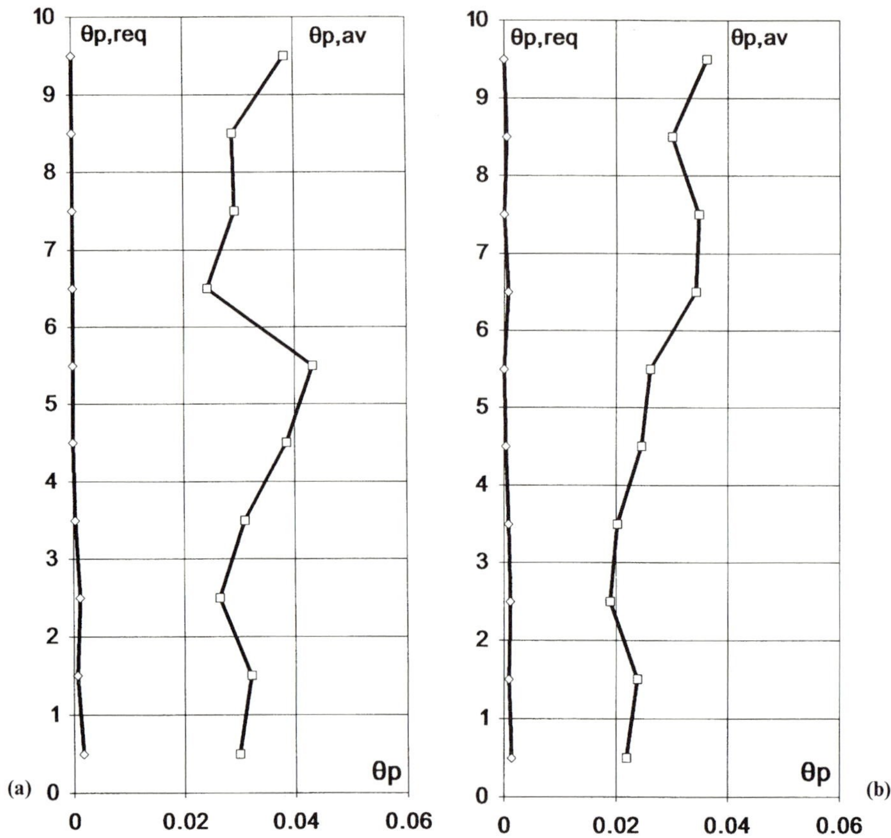

Figure 10.5 Required and available plastic rotations in the columns of the frame structure, for the most critical motion: (a) exterior columns; (b) interior columns.

beams at the interior supports (Figure 10.6(b)) is again higher than the corresponding demands, but the corresponding safety factors are only slightly larger than 1.0 in the lower part of the frame; the value of 1.2 recorded at the second storey is the lowest safety factor estimated for the structure subjected to the design earthquake. The main problem here is that a maximum shear of 210 kN was recorded in the interior beam during the most critical motion (Thessaloniki N30E), while the shear calculated using the design procedure suggested in EC8 (section 8.3.2) was only 141 kN. A first reason for this discrepancy is the difference between design and actual material strengths; the design yield strength of steel was equal to 400/1.15 = 348 MPa, while f_y = 440 MPa was used to estimate strengths for the inelastic analysis, the latter value still being lower than strengths recorded during actual tests of steel bars. Furthermore, the combination of slab reinforcement in the negative moment capacity of the beams was ignored during design according to standard practice (also accepted by EC8), but was taken into account in the inelastic analysis. Finally the strain-hardening of beam reinforcement also contributed to the discrepancy between actually developed and design shear force.

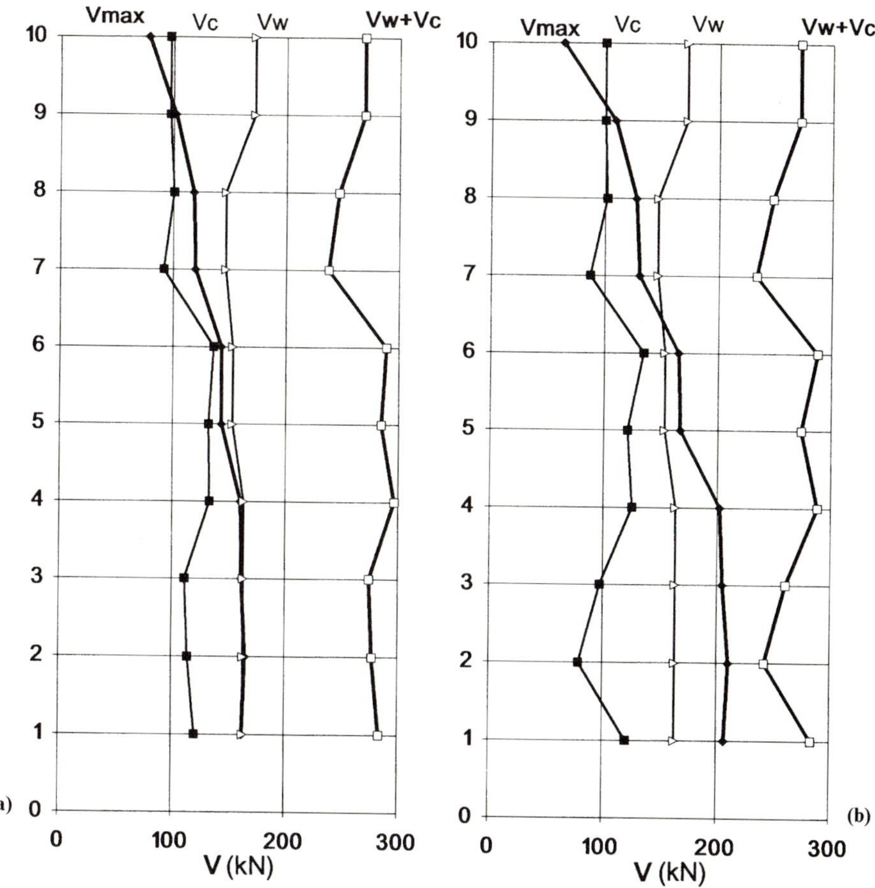

Figure 10.6 Required and available shear capacities of the beams of the frame structure, for the most critical motion: (a) exterior supports; (b) interior supports.

As shown in Figure 10.7, the maximum shears estimated from the dynamic inelastic analysis are substantially lower than the corresponding capacities (diagonal tension) for both the interior and the exterior columns of the frame; along the entire height of the structure hoops alone are in a position to carry the shears induced by the most critical seismic motion. This is attributed to the fact that the amount of hoops required for confinement of column critical regions (section 8.6.3 (b)) was consistently larger than that resulting from shear resistance requirements, in other words all columns are confinement-critical rather than shear-critical. The foregoing, in combination with the fact that ample ductility is available in all columns (Figure 10.5), further point to the conclusion that the design of columns according to EC8 might be overconservative.

Finally, with regard to the global failure criterion, the safety margins available in the frame structure are quite high since, as shown in Figure 10.8, the maximum inter-storey drift ratio calculated for the design earthquake intensity ($0.25g$) does

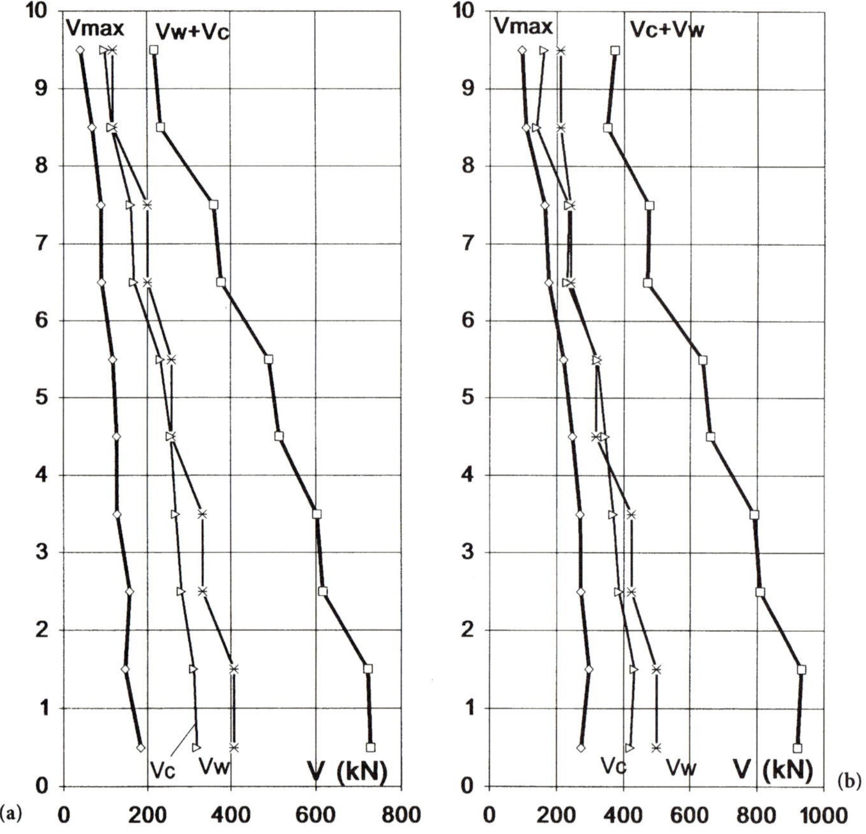

Figure 10.7 Required and available shear capacities of the columns of the frame structure for the most critical motion: (a) exterior columns; (b) interior columns.

not exceed 0.6%, which is much lower than the adopted limit of 2%. Moreover, no column sidesway mechanism was found to form anywhere in the structure.

10.2.4 Serviceability and survival earthquake

Since the results of the analyses presented in section 10.2.3 suggest that the performance of the frame structure at the (design) ultimate limit state is satisfactory, it was deemed appropriate to investigate the seismic performance for two different earthquake intensities, a lower one corresponding to the 'serviceability' limit state, and a higher one corresponding to the 'maximum credible' earthquake, that is, to the 'survival' or 'collapse' limit state (see also section 4.2.1). For the former an intensity equal to 1/2.5 the design one was selected ($A_{ef}=0.10g$), in conformance with the $\nu=2.5$ factor adopted by EC8 for damage control verifications (section 6.2.3), while for the 'survival' earthquake an intensity equal to twice that of the design earthquake was selected ($0.50g$), using data from the literature (see, for

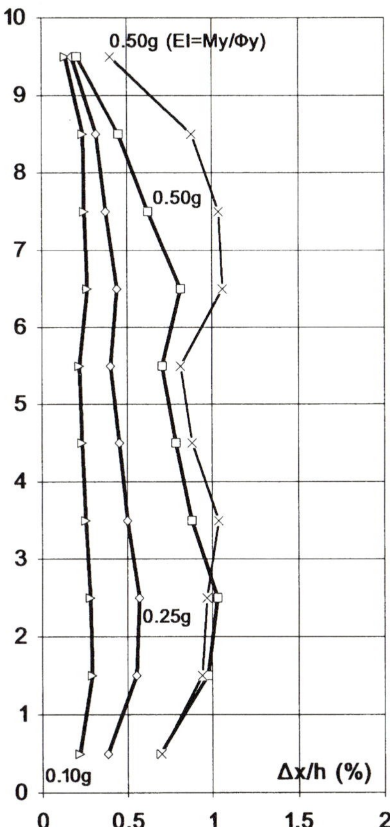

Figure 10.8 Maximum inter-storey drifts calculated for the frame structure under the most critical seismic motion normalized to various intensities.

instance, Paulay and Priestley, 1992), since no explicit or even implicit definition of this earthquake is included in EC8.

As shown in Figure 10.8, the inter-storey drift ratios calculated for the serviceability earthquake (0.10g) are 0.3% along the entire height of the structure, even for the most critical input motion. This is an exceptionally good performance for a bare frame structure which is the most flexible system that could be used for a multi-storey building, and it implies that damage under the serviceability earthquake is expected to be only minor, as intended by the code. This is also confirmed by the fact that no hinges form at the columns, while the maximum ductility factor in the beams does not exceed 1.0 in negative bending and 1.8 in positive bending. The latter implies that under the most critical motion, cracking requiring repair might appear at the bottom of some beam supports; however, the presence of 'non-structural' elements such as infill walls would typically preclude the formation of even those moderate cracks, as will be seen in section 10.5.

Under the maximum credible earthquake (0.5g) significantly more inelasticity than for the design earthquake is expected, hence two models were analysed:

(1) One based on the stiffness assumptions corresponding to moderate levels of inelasticity, that is $0.4EI_g$ for the beams and $0.8EI_g$ for columns (EI_g is the stiffness calculated on the basis of gross section properties), as suggested by Kappos (1986).

(2) A second model, where the secant stiffness of the fully cracked section ($EI = M_y / \phi_y$) was used for all R/C members; the fundamental period of this model is equal to 1.57 s, which is 59% higher than that of the 'standard' model. As the spectral peaks of the Greek records typically appear at relatively low period values, it is expected that the response of the second, more flexible, model will be more favourable than that of the first (stiffer) model, with the possible exception of displacements.

As shown in Figure 10.8, the calculated inter-storey drift ratio under the 'survival' earthquake does not exceed 1.1%, the influence of the stiffness assumption being relatively minor (the peak value recorded for the stiffer model is equal to 1.0%, appearing in the lower part of the structure, while the 1.1% drift recorded for the more flexible model appears at the seventh storey). With regard to local failure criteria, the most unfavourable response was calculated for the stiffer model due to the reasons mentioned above. The safety margins with respect to the rotational capacities of the members were estimated to be 4.7 or more for the beams, and 7.4 or more for the columns. The latter, in combination with the conclusions previously drawn for the columns subjected to the design earthquake, strongly suggest that the EC8 provisions for column confinement are overconservative. On the other hand, the safety margins with respect to the shear capacity of the beams were not adequate in the case of the interior supports, where values as low as 0.7 were calculated. Notwithstanding that the shear criterion is based on strength rather than on deformability (hence it might be too conservative), and that the most critical, rather than the average, calculated response was considered, some concern might be raised with regard to the performance of these beams with respect to shear, during an earthquake stronger than the design one. In the columns of the frame structure a minimum safety factor of 2.1 was estimated with regard to shear failure, which is one more indication of overconservatism in the design of the transverse reinforcement of these members.

10.3 SEISMIC PERFORMANCE OF DUAL SYSTEMS

In Chapter 9 the design procedure specified by EC8 for R/C structural walls was presented and a design example given in section 9.4.4, involving a slender wall forming part of a dual system, whose design actions were derived in section 6.3. In the following an attempt will be made to evaluate the seismic performance of this dual system (Figure 10.9), when it is subjected to the input motions given in Table 10.1, normalized to the intensities of the earthquakes corresponding to the serviceability, the design and the survival limit state (see also section 4.2.1). The assessment of the seismic performance will be made using the procedure outlined in section 10.1.3, which is based on inelastic time-history analysis. Experimental data concerning the seismic response of dual systems may be found in Paulay and Spurr (1977) for the case of cyclic static loading, in Okamoto *et al.* (1985) for the case of pseudo-dynamic loading (of a full-scale seven-storey building), and in

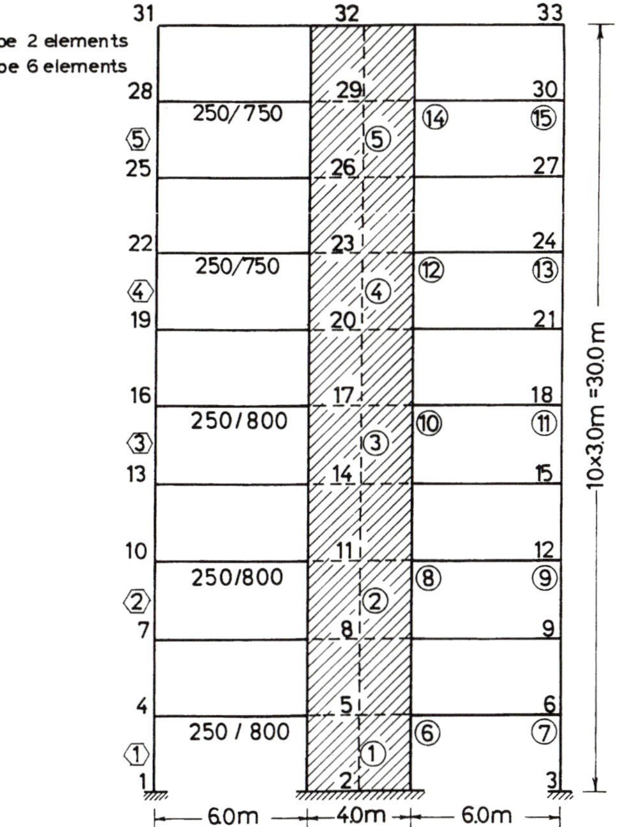

Figure 10.9 Ten-storey dual system designed to EC8.

Bertero *et al.* (1985) for the case of shaking table testing (of a one-fifth scale model of the same building studied by Okamoto *et al.*) A wealth of cyclic test data concerning the behaviour of (isolated) slender R/C walls may be found in Wood (1991).

10.3.1 Modelling assumptions and failure criteria

The beams and columns of the dual system shown in Figure 10.9 were modelled using the member-by-member discretization approach described in section 10.2.2. The same type of modelling was used for the centrally located wall, since its high slenderness ($h_w/l_w = 7.5$) was deemed to justify the selection of a beam–column model. The degrading stiffness Takeda hysteresis element of DRAIN-2D/90 (Kappos, 1992) was used for modelling wall segments extending one storey in height, as previous analyses (Kappos, 1991c) have indicated that further discretization (more elements per storey) leads to only marginal differences in the calculated response of the wall, so long as the same type of element is used in the two alternative models. In order to account for the finite width of the wall elements, fully rigid zones (of 2.0 m width) were introduced at the ends of the beams framing into the wall (Figure 10.9). For the design earthquake intensity, stiffness values between 50 and 60% the gross stiffness EI_g were assumed for the wall elements, the higher value corresponding to the lower part of the structure where the axial loading is higher. Moreover, as in the case of the frame structure, a second model, where fully cracked sections ($EI_{ef} = M_y / \phi_y$) were used for the wall, as well as for the other elements, was analysed in the case of the 'survival' earthquake (section 10.3.3).

In calculating the rotational capacity of the wall critical regions, essentially the same procedure used for columns (sections 8.2.1 and 10.2.2) was followed, with due consideration to the different degree of confinement existing in the different parts of a wall section (heavily confined edge columns, web, cover concrete). For the equivalent plastic hinge length, the empirical equation (9.31) suggested by Paulay and Priestley (1992) was used; trial calculations have shown that the values resulting from equation (9.31) are similar to those calculated on the basis of the extent of flexural yielding in the wall. Introducing $l_w = 4.0$ m and $h_w = 30.0$ m, $l_p = 2.12$ m results from equation (9.31), which is less than the upper bound of $0.8l_w = 3.2$ m suggested by Paulay and Priestley (1992). Due to lack of pertinent test data no attempt was made to relate the wall plastic hinge length to the ductility level (as was done with equation (8.29) for columns and beams); it has, however, to be noted that in all critical ductility checks for the wall at its base (where the plastic hinge formed) the rotational ductility factor μ_θ was of the order of 4.0 or more, hence the use of the empirical equation (9.31) appears to be justified, as it refers to walls responding well into the inelastic range.

With regard to the effect of shear stress on the rotational capacity (θ_p from equation (10.2)) of the wall, the following modified form of equation (10.3) was used.

$$k_V = 2.0 - V_{max}/V_{R1} \ngtr 1.0 \tag{10.7}$$

to reflect the fact that shear dominates the response of R/C walls beyond a limit value slightly higher than that corresponding to beams ($\tau \approx 0.6 f_c^{1/2}$ as opposed to $\tau \approx 0.5 f_c^{1/2}$), as already mentioned in section 9.3.3(a); the value of V_{R1} is again calculated from equation (10.4) .

Finally, with respect to the adequacy of shear reinforcement, a total of three different checks has to be carried out, corresponding to the three shear failure mechanisms identified in sections 9.3.3 and 9.4.2. It is worth recalling here the discussion regarding the adequacy of strength criteria in assessing seismic performance (section 10.2.2), which is perhaps even more pertinent in the case of walls, wherein the estimation of ultimate shear deformations is subject to more uncertainties than in the case of linear elements.

The resistance of a wall to diagonal tension failure was estimated using equations (9.37) for V_c and (9.34)–(9.36) for V_w, and introducing 'actual' rather than design values for material strengths (f_y, τ_R). Moreover, according to the discussion in section 10.2.2, the value of V_c was adjusted to account for the effect of inelasticity, on the basis of equation (10.5), with a lower bound of 0.33 V_{co} whenever the normalized axial loading ν (calculated using f_{cm}) exceeded 0.05 (compression).

The resistance against sliding shear failure was checked using equation (9.44) for the dowel resistance V_d (the lower limit of $0.25 f_y \Sigma A_{sj}$ always dominated V_d), and equation (9.46) for the frictional resistance V_f (the lower limit of $0.25f_c\,\xi l_w b_w$ always dominated V_f); again actual, rather than design, values were used for concrete and steel strength. The resistance against diagonal compression failure, estimated using the unfactored equation (9.38), was found to be well in excess of the shear values developing at the wall under consideration.

10.3.2 Performance under the design earthquake

Figure 10.10 shows the maximum ductility factors calculated for the beams of the dual structure (typically at the interior support, where the beam frames into the wall), for the most critical and the least critical among the seven input motions of Table 10.1, normalized to the spectrum intensity of the design earthquake ($A_{ef} = 0.25g$). As in the case of the frame structure, the variability of the calculated inelastic response is quite significant, although a refined technique was used for normalizing the various motions (section 10.2.1). For instance, the maximum

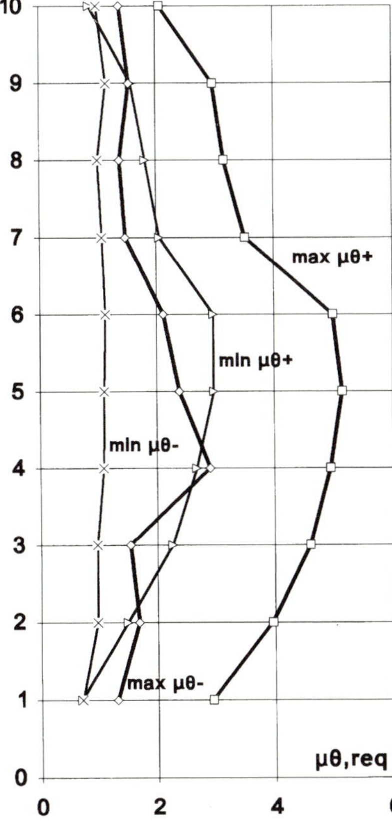

Figure 10.10 Beam ductility factors for the dual structure of Figure 10.9 subjected to the most critical and the least critical motion (for $A_{ef}=0.25g$).

peak value of $\mu_{\bar{\theta}}=2.90$, calculated for the Kalamata N80E record, is 154% higher than the peak value $\mu_{\bar{\theta}}=1.14$ calculated for the Thessaloniki N30E record (which gave the most critical $\mu_{\bar{\theta}}$ value in the case of the frame structure, as discussed in section 10.3.1). Similar trends were found with regard to other response quantities; for instance, the maximum inter-storey drift ratio ranged from 0.24% (for the Thessaloniki N30E record) to 0.37%, that is 56% higher (for the Kalamata N10W record); differences at individual storeys were in general much more significant. It is worth pointing out that the variability in calculated inter-storey drifts of the dual system was higher the in the case of the frame (section 10.2.3). These trends confirm once more the conclusions derived in previous studies (Kappos, 1990), and point to the importance of using a large number of input motions for assessing the seismic performance of structures.

Required and available plastic rotations at the beam critical regions of the dual system are compared in Figure 10.11 for the most critical motion, which is the Kalamata N10W record for the lower six storeys and the El Centro record for the four upper storeys. It is clearly seen in the figure that ample safety margins

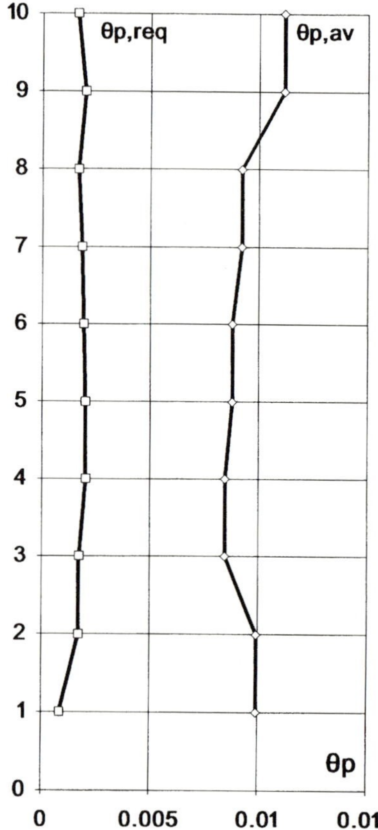

Figure 10.11 Required and available plastic rotations (in rad) in the beams of the dual system, for the most critical motion.

with regard to exceeding the deformational capacity exist, the safety factor (ratio $\theta_{p,\,av}/\theta_{p,\,req}$) being at least equal to 4.2, while higher values were recorded in other storeys, in particular the lower ones. It has to be pointed out that the distribution of safety margins along the height of the building is reasonably uniform, which in combination with the foregoing remarks leads to the conclusion that the design of the beams according to EC8 results in a satisfactory seismic performance of these elements.

The rotational capacity of the vertical members of the dual systems is compared with the requirements calculated for the most critical motion in Figure 10.12 for the columns, and Figure 10.13 for the wall. It is quite clear from both figures that the required plastic rotations in the vertical members are minimal (zero, or nearly so) with the exception of the first storey segment of the wall and the top storey column; note that the appearance of maximum requirements at the bottom of the wall and the top of the frame is typical in dual (wall+frame) systems. As can be seen in Figure 10.12, the available plastic rotation of the top storey column is significantly higher than the corresponding demand (calculated 'safety

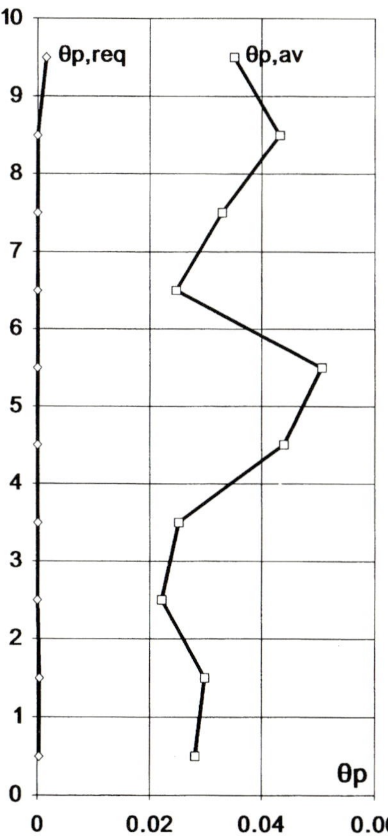

Figure 10.12 Required and available plastic rotations in the columns of the dual system, for the most critical motion.

margin' of 21.4), which is expected, since the required θ_p is only 1.6×10^{-3} rad and the axial load is very low ($\nu = 0.03$). Figure 10.12 suggests an overdesign of the columns with respect to ductility, but again such a conclusion can only be validated if the survival earthquake is also considered (section 10.3.3).

As shown in Figure 10.13, the rotational capacity of the wall at its base is substantially larger than the corresponding demand, the calculated safety margin being equal to 27.8; this is again a strong indication of overconservatism in designing the edge columns of the wall (it is recalled that 12 mm ties at a spacing of 65 mm are required in these parts), and it remains to be seen whether such excessive ductility is indeed required for the survival earthquake. On the other hand, the capacity design of the wall appears to work exactly as intended, that is, inelasticity is only required to develop at the wall base (essentially at the ground storey segment), while the rest of the wall remains below the yield level (μ_θ values between 0.4 and 0.9 were recorded from the third storey upwards).

The resistance of the wall against the two potentially critical shear mechanisms is shown in Figure 10.14 (diagonal tension check) and Figure 10.15 (sliding shear

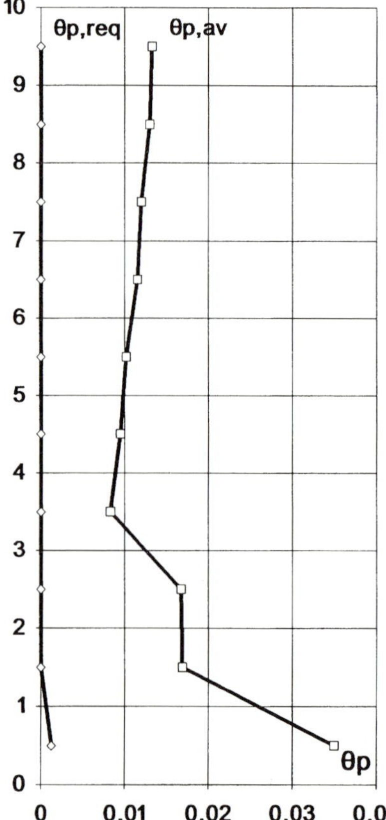

Figure 10.13 Required and available plastic rotations in the wall of the dual system, for the most critical motion.

check); both figures refer to the motions that produce critical response at each storey. Prior to discussing the safety margins against shear failure, it is worth pointing out that the maximum shear forces calculated during the inelastic dynamic analyses are very close to the values resulting from the capacity design of the wall (Figure 9.23), with the exception of the upper third of the structure, where the design appears to be overconservative.

The two different components of shear resistance against diagonal tension V_c ('concrete' contribution) and V_w (truss action), are shown in Figure 10.14, and it is seen that at least in some storeys both components are required to provide a resistance at least equal to the maximum dynamic shear ($V_{R3} \geqslant V_{max}$). The calculated safety margins against this type of shear failure range from 1.5 at the ground storey to 4.2 at the ninth storey and 16.8 at the tenth. Hence it is concluded that the design of the wall against diagonal tension is generally adequate, with a tendency to overconservatism in the upper storeys.

In Figure 10.15 is shown the comparison of the wall's resistance (V_{Rs}) against sliding shear failure, consisting of the dowel mechanism (V_d) and the 'friction'

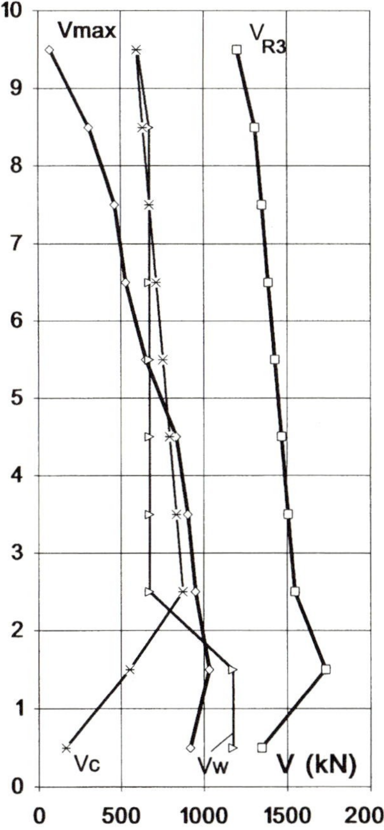

Figure 10.14 Required and available shear capacity of the wall, with respect to diagonal tension failure, for the most critical motion.

mechanism (V_f), and the corresponding demands (V_{max}) estimated for the most critical input motions. It is pointed out that the friction mechanism contributes more than 80% of the total shear capacity at all storeys; this is mainly due to the fact that the area of the wall is quite large for the isolated dual system considered (Figure 10.9), having a transverse spacing of the frames equal to 3.0 m and minimum reinforcement requirements govern the design of its longitudinal reinforcement (in the web, as well as in the edge columns). It is seen from Figure 10.15 that the safety margin against sliding shear failure is quite low at the lower part of the structure, the calculated safety factor at the ground storey amounting to only 1.5, which is the lowest calculated anywhere in the dual system. It is worth pointing out here a possible ambiguity regarding the EC8 provision that the reinforcement contributing to the dowel resistance (ΣA_{sj} in equation (9.44)) is 'additional' to that required from flexural considerations. For the wall under consideration ΣA_{sj} was assumed to include the difference between the actually placed (due to ϱ_{min} requirements) and the theoretically required amount of reinforcement at the edge columns, in addition to the total amount of web vertical bars, at the design stage.

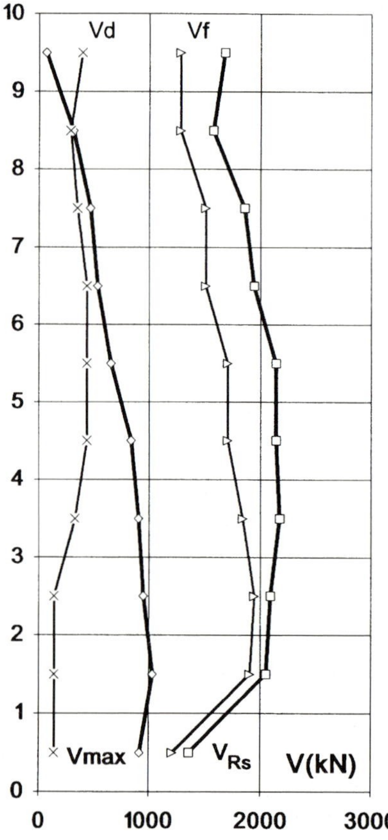

Figure 10.15 Required and available shear capacity of the wall with respect to sliding shear failure, for the most critical motion.

However, during the inelasitc analysis stage since the wall yielded at its base it was assumed (perhaps conservatively) that the edge column bars do not contribute to dowel resistance. It is felt that a clearer definition of the 'additional' dowel reinforcement should be included in the final (EN) version of EC8.

The shear capacity of the beams and the columns of the dual system were also checked, using the same procedure as in the frame structure (compare Figures 10.6 and 10.7). Adequate margins of safety against diagonal tension failure (the critical shear check in these members) were found, having a value of 1.7 or more for the beams, and 3.4 or more for the columns. It appears, therefore, that from the shear capacity point of view, the relatively more critical member is the wall, while the columns exhibit a trend of overdesign with regard to shear as well (in addition to the possible overdesign for ductility).

Finally, as expected for a structure with a large amount of wall area (and stiffness), the global failure criterion, based on inter-storey drifts, was not critical. As shown in Figure 10.16, the calculated drift ratios for the design earthquake ($A_{ef}=0.25g$) do not exceed 0.4%, which is much lower than the limit of 2%. Moreover, no sign of sideway mechanism was detected anywhere in the structure, since column yielding was essentially confined to the top storey and wall yielding to the ground storey.

10.3.3 Serviceability and survival earthquake

From the inter-storey drift ratios plotted in Figure 10.16, it is seen that for the service ability earthquake ($A_{ef}=0.10g$) the maximum value (under the most critical motion) does not exceed 0.19%, which is less than one-tenth the adopted limit of 2%. For such low drift values it is clear that no damage to the non-structural elements is expected, hence the goal of the design appears to be fully satisfied. Moreover, all columns remain in the elastic (pre-yield) stage, with required ductility factors ranging from 0.2 to 0.7, which is a clear indication that only minor cracking of these members is expected, and no repair will be required. The only members where some inelasticity has been induced under the most critical motion normalized to $A_{ef}=0.10g$ are the beams at the bottom of their supports (positive bending) and the wall at its base. Rotational ductility factors up to about 2 were recorded for the beams, while the wall at its base required $\mu_\theta=1.3$ under the most critical motion. These values imply that cracking requiring repair might appear in the aforementioned regions, unless the presence of other members such as masonry infill walls precludes the formation of even these moderate cracks (section 10.5).

The inter-storey drift ratios calculated for the survival earthquake ($A_{ef}=0.50$ g) are shown in Figure 10.16 for both models analysed (two different stiffness assumptions, see section 10.3.1). It is seen that even for the more flexible model, the calculated drift ratios do not exceed 1.3%, that is, they are well below the limit of 2%. This confirms the superior performance of dual systems with regard to drift control, also pointed out in previous studies (Kappos, 1991b). Moreover, no indication of sidesway mechanism was detected, since columns remained elastic in the right half of the structure, while the wall yielded only in the three lower storeys, as shown in Figure 10.17. It is worth pointing out that, with the exception of the top storey, column yielding was very moderate, with required ductil-

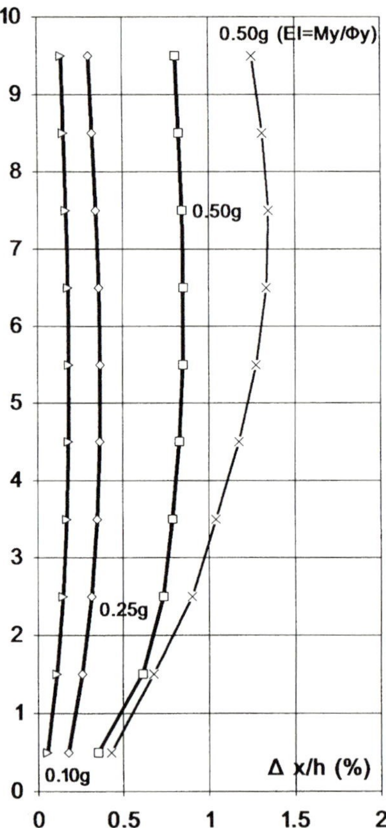

Figure 10.16 Maximum inter-storey drifts calculated for the dual structure under the most critical seismic motion normalized to various intensities.

ity factors just exceeding 3. This is a clear indication that the EC8 capacity design of columns did succeed in preventing significant yielding in unfavourable locations, but on the other hand it also indicates that the very heavy confinement reinforcement required by EC8 for column critical regions does not appear to be actually required, even under the most critical motion the structure might possibly experience. In fact the calculated safety factors with respect to flexural deformation were equal to 6.5 at the top storey and to 13.6 or more at the other storeys. These considerations clearly confirm the conclusion already drawn with respect to the design earthquake, that EC8 leads to an overconservative design of columns.

Plastic rotation demands were also well within the corresponding capacities in the wall; the estimated safety factor at the wall base, where $\theta_p = 0.003$ rad was required under the most critical motion (Kalamata N10W), was found to be equal to 12.2, as the ample confinement provided in the edge columns resulted in an available plastic rotation of 0.036 rad. As in the case of columns, the present analysis indicated a significant degree of overconservatism in the design of the wall boundary elements.

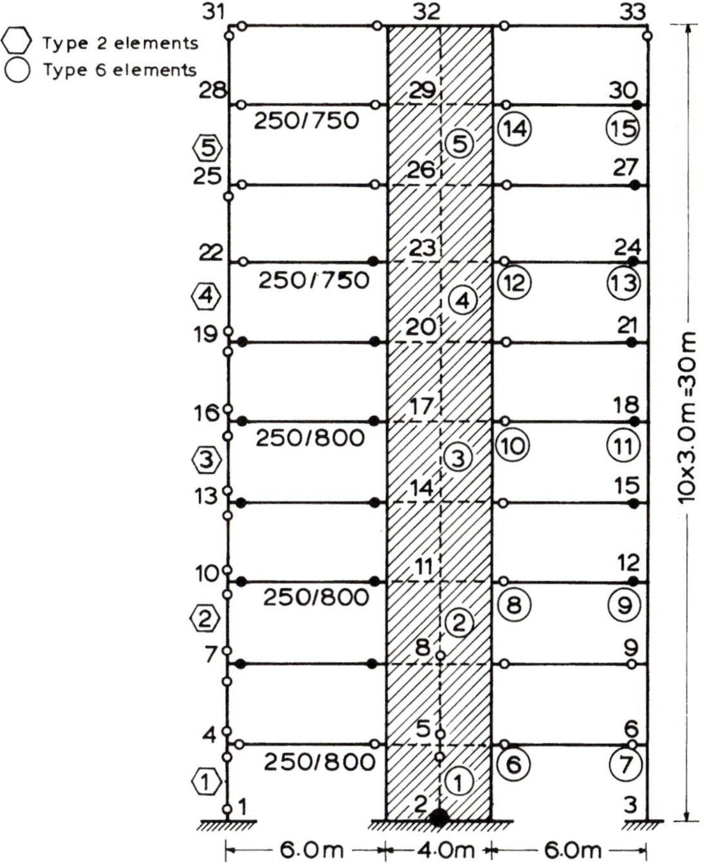

Figure 10.17 Distribution of plastic hinges along the height of the dual structure subjected to the survival earthquake ($A_{ef} = 0.50g$). Full circles indicate yielding at both faces of the member, open circles yielding at one face only.

With regard to shear capacity, the most critical region identified from the analysis for the design earthquake was the wall base (sliding shear check); it was found that the corresponding safety factor under the survival earthquake was 1.1 (both against sliding shear and diagonal tension). Given the already discussed conservatism of the shear force approach, the foregoing value may be considered satisfactory if it refers to the survival, that is, the maximum credible, earthquake. On the other hand, a safety factor less than unity was calculated for some of the beams in the upper four storeys, where a very light transverse reinforcement (6 mm ties at 120 mm centres) is required by EC8. At the most critical location eighth storey beam, interior support) the available shear capacity was only equal to 89% of the corresponding demand; the significant difference with respect to the design earthquake was that due to the much higher value of required ductility ($\mu_\theta = 6.1$) the concrete contribution (V_c) was zero for the survival earthquake, while for the design earthquake the estimated V_c (for $\mu_\theta = 1.7$) was of the same order as V_w (the contribution of hoops). Notwithstanding the conservatism of the approach and the relatively minor importance of beams with respect to the overall seismic capacity of the structure, the shear behaviour of the beams of the upper storeys cannot be considered satisfactory under the survival earthquake. Finally, with respect to columns, the calculated safety margins against shear failure (diagonal tension) were at least equal to 3.5, clearly indicating that the hoop reinforcement in these elements (resulting primarily from confinement considerations) provides a shear capacity well in excess of that required even under the most unfavourable seismic motion.

10.4 INFLUENCE OF DESIGN DUCTILITY CLASS

The aim of EC8 is to ensure a similar level of seismic reliability to the structures designed for the three different ductility classes, the main idea being that the higher design forces specified for the lower ductility classes compensate for the reduced detailing requirements. However sound this concept is, it remains to be seen whether the seismic performance of structures designed for different ductility classes is indeed equally satisfactory, at least for the design earthquake. Moreover, it is quite interesting to see whether the cost of these structures is dependent on the ductility class or remains practically constant. Since no studies addressing these matters in a systematic way have been reported as yet, brief reference will be made hereafter to a study involving frames and dual structures having the same geometry as those addressed in the previous sections (Figures 10.2 and 10.9), which were designed according to the CEB (1985) Model Code for three different ductility levels (classes) DL I, II, and III, generally corresponding to the EC8 DC 'L', 'M' and 'H' (Kappos and Papadopoulos, 1994).

10.4.1 Influence on cost

In Figure 10.18 are shown the estimated quantities of concrete and steel required for the construction of each structural system, including slabs, beams, columns and walls (when present), but not the foundation; note that the values shown refer to an isolated frame or dual system, not to an entire building. It is seen that for the frame structure (Figure 10.18(a)) the volume of concrete, as well as the total weight of reinforcement, reduces with the ductility level. Given that the required transverse steel (ties or hoops) increases with the ductility level (up to 56%), it is clear that the reduction in the total reinforcement is due to the reduced demands of longitudinal steel, caused by corresponding reductions in the design seismic actions.

Similar trends are detected in the dual structures, with regard to the volume of concrete. However, regarding the weight of steel, it is observed that it increases (by about 4%) on going from DL I to DL II, while it reduces (by 8%) on going from DL II to DL III. This trend is mainly due to the fact that, although the seismic action for DL II is 33% lower than for DL I, the design of the wall in the former case has to be carried out according to capacity procedures (a displaced linear envelope of the calculated moments, similar to that of Figure 6.15) which increases the flexural reinforcement demands to twice or even three times the values

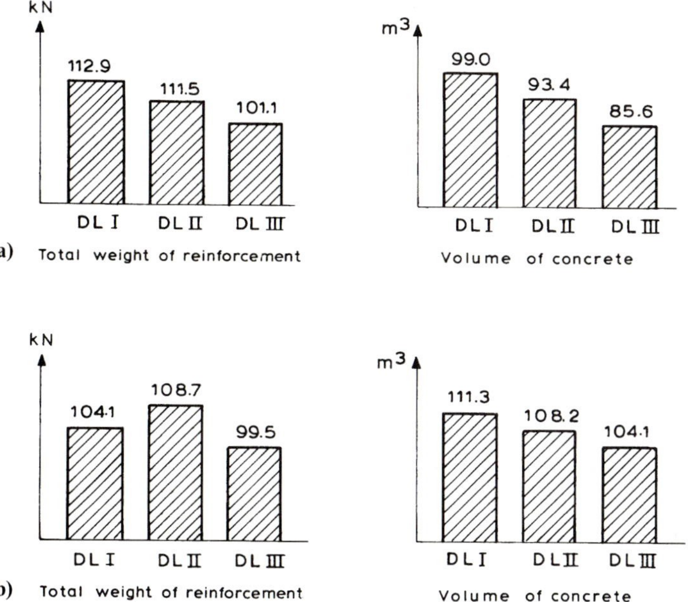

Figure 10.18 Required quantity of materials for 10-storey structures designed for different ductility levels (DL): (a) frames; (b) dual systems.

resulting from the analysis. At the same time, the required longitudinal steel in the beams and columns decreases, in a way similar to that of the frame structure.

The foregoing remarks lead to the conclusion that the main effect of the ductility class consists in the differentiation of the distribution of total reinforcement into longitudinal and transverse, which is a factor of paramount importance with regard to the strength–ductility balance in an R/C structure.

Further, consideration of the cost of labour, which is expected to be higher in the case of DL III buildings, based on limited data from Greek construction practice, indicates that it is not expected to increase by more than 1% of the cost of the structural system (for DL III buildings), while it was shown previously that the total reinforcement is 9–10% lower than in the DL I and II structures and the volume of concrete is also less. Therefore, it is clear that, overall, the most cost-effective solution corresponds to DL III design.

10.4.2 Influence on seismic performance

In Figure 10.19 are shown the plastic hinge rotation requirements calculated for the frame structures designed for three ductility levels, when the most critical input motion is considered for each structure; the input motions considered were those of Table 10.1 in addition to four more records from recent Greek earthquakes. It is seen that the maximum requirement is similar for all ductility levels, the maxθ_p being slightly higher in the DL I structure ($\theta_p = 0.0055$ compared with the

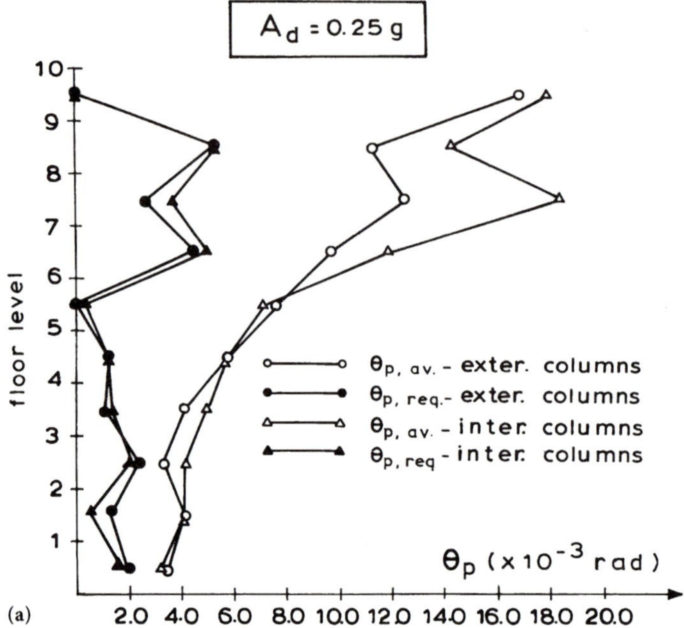
$A_d = 0.25\,g$
floor level
$\theta_{p,\,av.}$ - exter. columns
$\theta_{p,\,req.}$ - exter. columns
$\theta_{p,\,av.}$ - inter. columns
$\theta_{p,\,req}$ - inter. columns
θ_p ($\times 10^{-3}$ rad)
(a)

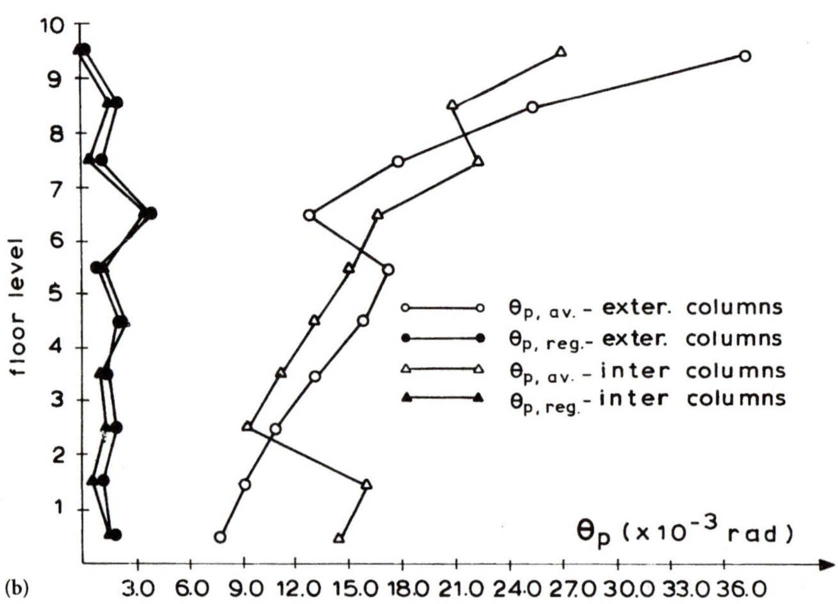
floor level
$\theta_{p,\,av.}$ - exter. columns
$\theta_{p,\,reg.}$ - exter. columns
$\theta_{p,\,av.}$ - inter columns
$\theta_{p,\,reg.}$ - inter columns
θ_p ($\times 10^{-3}$ rad)
(b)

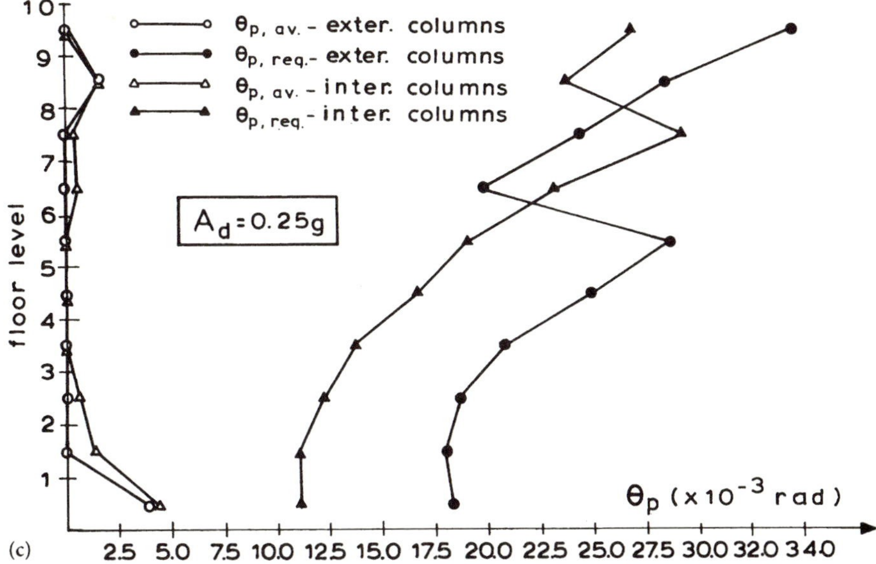

Figure 10.19 Required and available rotational capacities in 10-storey frame designed for different ductility levels: (a) DL I; (b) DL II; (c) DL III.

$\theta_p = 0.0045$ of the DL III frame). However, the distribution of ductility requirements is distinctly different in the low and high ductility structures; while inelasticity in the columns of the DL III frame is essentially confined to the ground storey (in particular at the bottom of its columns) and θ_p is either zero or negligibly small in the rest of the structure, a different picture is seen in the DL I structure where non-zero plastic rotations are required in most storeys, the peak values appearing at the seventh and the ninth storeys, where a rather significant taper of column dimensions takes place. These trends are readily explained if it is recalled that while the relative strength of beams and columns at the joints has been checked at all storeys of the DL III frame, with the exception of the base and the top storey (according to the CEB MC provisions), no such checks were carried out for the DL I frame. Inspection of the diagrams of Figure 10.19 readily leads to the conclusion that the safety margins are significantly higher in the DL III frame. As already mentioned, a minimum safety factor of 1.4 was estimated for the DL I structure, while for the other two ductility levels the governing criterion was the global one, resulting in safety factors of 2.2 and 2.3 for the DL II and III frames respectively. Thus, it is concluded that the corresponding safety margins are 57–64% higher than in the DL I structure, which is clear indication of the increased seismic reliability of the frames designed for medium and high ductility levels.

With regard to the behaviour of dual systems, whose rotational capacities and demands are shown in Figure 10.20, an important point has to be raised regarding the DL I structure, which appears to be the only one among those studied that might not be able to survive the design earthquake. As can be seen in Figure 10.20(a), at the second storey column the estimated rotational capacity is well

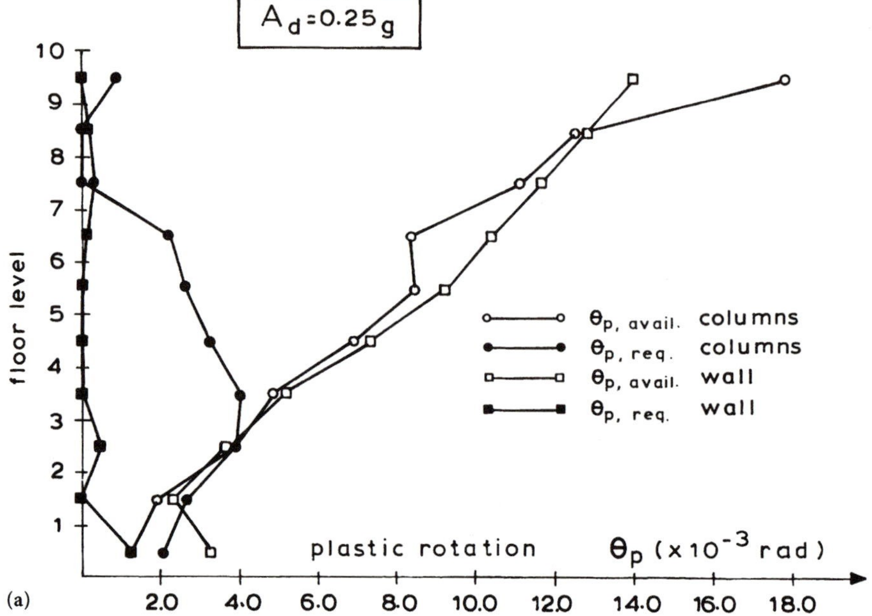
A d=0.25g
floor level
10
9
8
7
6
5
4
3
2
1
θp, avail. columns
θp, req. columns
θp, avail. wall
θp, req. wall
plastic rotation
θp (x10-3 rad)
2.0
4.0
6.0
8.0
10.0
12.0
14.0
16.0
18.0
(a)

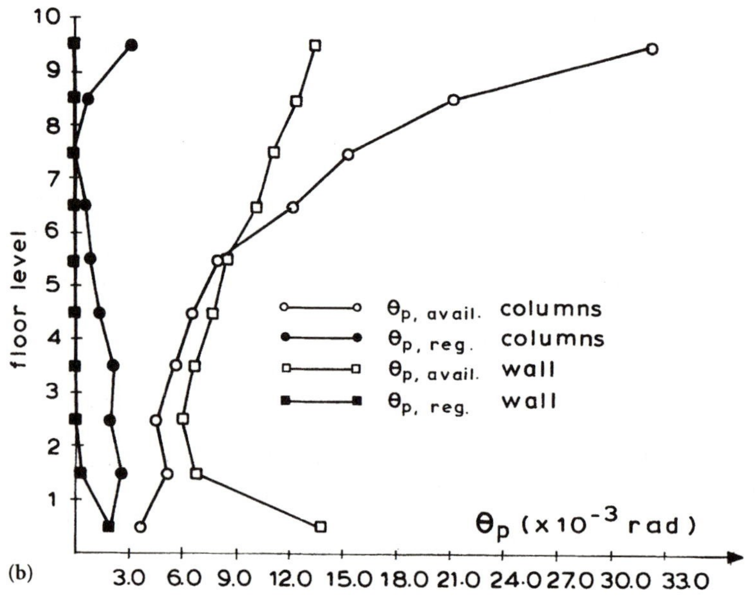
floor level
10
9
8
7
6
5
4
3
2
1
θp, avail. columns
θp, reg. columns
θp, avail. wall
θp, reg. wall
θp (x10-3 rad)
3.0
6.0
9.0
12.0
15.0
18.0
21.0
24.0
27.0
30.0
33.0
(b)

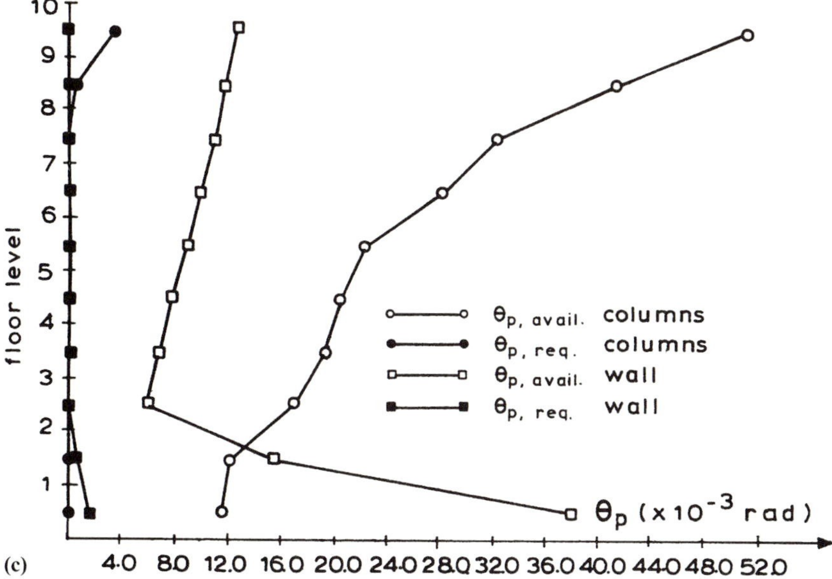

(c)

Figure 10.20 Required and available rotational capacities in 10-storey dual structure designed for different ductility levels: (a) DL I; (b) DL II; (c) DL III.

below the corresponding demand (safety margin of only 0.6) and the same holds for the ground storey column. Additional analyses of this structure for lower earthquake intensities have shown that failure is anticipated at an effective peak acceleration of $0.19g$ which is 24% lower than the design value.

In contrast to the foregoing, the behaviour of the other two dual systems was very satisfactory, with minimum safety factors of 2.0 and 4.8, for the DL II and DL III structures respectively. In the first case the critical criterion was the rotational capacity of columns (at the ground storey), while in the second the minimum safety factor was calculated with respect to the drift criterion, since with respect to local ductility, values of 15 or more were calculated for all elements (Figure 10.20(c)). The behaviour of the DL III dual system was by far the most satisfactory among all those investigated, as this structure combined the well-known high stiffness of buildings with adequate structural walls, with the very high ductilities resulting from the strict detailing provisions of the CEB MC for the critical regions of R/C members.

Having verified that all structures studied were in a position to survive the design earthquake without local and/or storey failures (with the aforementioned exception of the DL I dual system), it is of particular interest to make an estimate of the intensity of the earthquake that would induce some type of failure in each structure, in code terms the earthquake inducing the **collapse limit state**. To this purpose, iterative analyses of the six structures were carried out for the motion which gave the most critical response for $A_d = 0.25g$, until at a certain intensity at least one of the specified criteria was violated. Then all the motions were scaled to this intensity and additional analyses were run to check whether the motion initially selected was indeed the critical one (otherwise more iterations were carried out).

Summarized in Table 10.2 are the effective accelerations of the 'collapse' earthquake for each of the structures studied. It is seen that the three frame structures and the DL II dual system are in a position to survive earthquakes moderately stronger (16–28%) than the design one, while the DL I dual structure is expected to fail at an intensity lower than $A_d = 0.25g$, as mentioned previously. On the other hand, the highly ductile dual system appears to be able to survive earthquakes of an intensity equal to 3.8 times the design earthquake; however unrealistic such a figure might seem, it has to be recalled that test results (Wood, 1992) have been reported, wherein structures were able to survive earthquakes having an intensity up to three times the design value (and peak ground accelerations in excess of $1.0g$).

Based on the limited data of the study presented here, which refers to multistorey structures designed to the CEB (1985) Seismic Code, it is concluded that R/C buildings designed for the highest ductility level prescribed by modern codes provide the highest seismic reliability at a cost lower than that corresponding to the lower ductility levels. Preliminary results (Fardis and Panagiotakos, 1995) from a major co-operative project involving several European research groups, indicate that in multistorey R/C frames designed to EC8, ductility class does not appear to have a systematic effect on the seismic performance in terms of displacement and energy dissipation.

10.5 INFLUENCE OF MASONRY INFILLS

It has been pointed out several times in previous chapters of this book that the presence of infill walls, which in many parts of the world (southern Europe, Central and South America among them) are made of brick masonry, changes significantly the seismic response of R/C buildings, by increasing their strength, stiffness and energy dissipation capacity. In section 5.9 the effect of masonry infills on seismic action was discussed and the corresponding EC8 provisions were presented; it is recalled here that EC8, as well as many national codes, place the emphasis on the effect of irregularities in plan and elevation due to the presence of masonry infills and they generally ignore the favourable effects such as the increased strength and energy dissipation of infilled frames.

Since bare frames and dual structures are far less common than the corresponding infilled structures (at least in most countries), it is quite interesting to supplement the seismic assessment studies of R/C bare structures presented in the

Table 10.2 Earthquake intensities causing failure

Failure criterion	Frame structure			Dual structure		
	DLI	DLII	DLII	DLI	DLII	DLIII
Local						
Ductility				$0.19g$		
Shear capacity	$0.29g$				$0.31g$	
GLOBAL		$0.29g$	$0.32g$			$0.95g$

previous sections (10.2–10.4) with similar studies where the presence of masonry infills is taken into account. To this purpose selective results from a recent study by Michailidis, Stylianidis and Kappos (1995) will be discussed; the study involved the 10-storey frame of Figure 10.2, with different arrangements and quality (strength and stiffness) of brick masonry infill walls. A recently developed refined phenomenological model was used to describe the hysteretic behaviour of the masonry infills; the model is based on an extensive test programme of single-storey infilled frames carried out at the University of Thesaloniki (Valiasis and Stylianidis, 1989). The input motions used in the study were those listed in Table 10.1, and they were normalized to the intensity of the design earthquake ($0.25g$) using the technique already described in section 10.2.1.

Figure 10.21 shows the inter-storey drift ratios calculated for the five models analysed, subjected to the seven motions of Table 10.1; the following abbreviations are used in the diagrams of Figure 10.21.

BF: bare frame (Figure 10.2); fundamental period $T_1 = 0.98$ s
IF1: fully infilled frame, low strength masonry ($\tau_u = 0.27$ MPa); $T_1 = 0.51$ s
IF2: fully infilled frame, high strength masonry ($\tau_u = 0.38$ MPa); $T_1 = 0.42$ s
IF1P: Infilled frame with open ground storey ('*pilotis*'), low-strength masonry; $T_1 = 0.58$ s
IF2P: Infilled frame with *pilotis*, high-strength masonry; $T_1 = 0.51$ s

As shown in Figure 10.21(b) and (c), although a refined normalization technique was used, the variability of the response (difference between maximum values) is quite significant; this appears to confirm that the conclusions regarding the sensitivity of bare R/C structures (section 10.2 and 10.3) also apply to infilled frames.

From the diagram of Figure 10.21(a), where mean values calculated for the seven motions of Table 10.1 are plotted, the significant differences in the response of each type of structure can be seen.

The bare frame is characterized by a rather uniform distribution of inter-storey drifts, the maximum value appearing at the seventh storey, where a stiffness taper at both exterior and interior columns takes place (Figure 10.2). On the other hand the fully infilled frames exhibit a tendency of decreasing drifts (and consequently of decreasing damage) with the height of the building, which is much closer to the picture obtained by inspection of real buildings damaged by earthquakes (Penelis *et al.*, 1988). Finally, the structures with an open ground storey (*pilotis*) are characterized by a very large drift at the ground storey (mean value of about 1.3%) and a drastic reduction of relative displacements in the upper storeys; it is worth pointing out that in this case the quality of the masonry does not play an important role.

Similar trends may be recognized with respect to the distribution of the rotational ductility factors, shown in Figure 10.22. While in the bare frame an approximately uniform distribution of damage (as expressed by μ_θ) along the height is expected, damage in the infilled structures tends to concentrate in the lower storeys. The *pilotis* buildings are characterized by high ductility demands in the ground storey columns, while in all the other storeys the columns remain elastic. It is worth pointing out the difference in the calculated ductilities of exterior and interior columns in the case of fully infilled frames. From Figure 10.22 it is clear that the required ductility factors at the lower storeys are much larger for the exterior columns; a careful examination of the results reveals that these differences

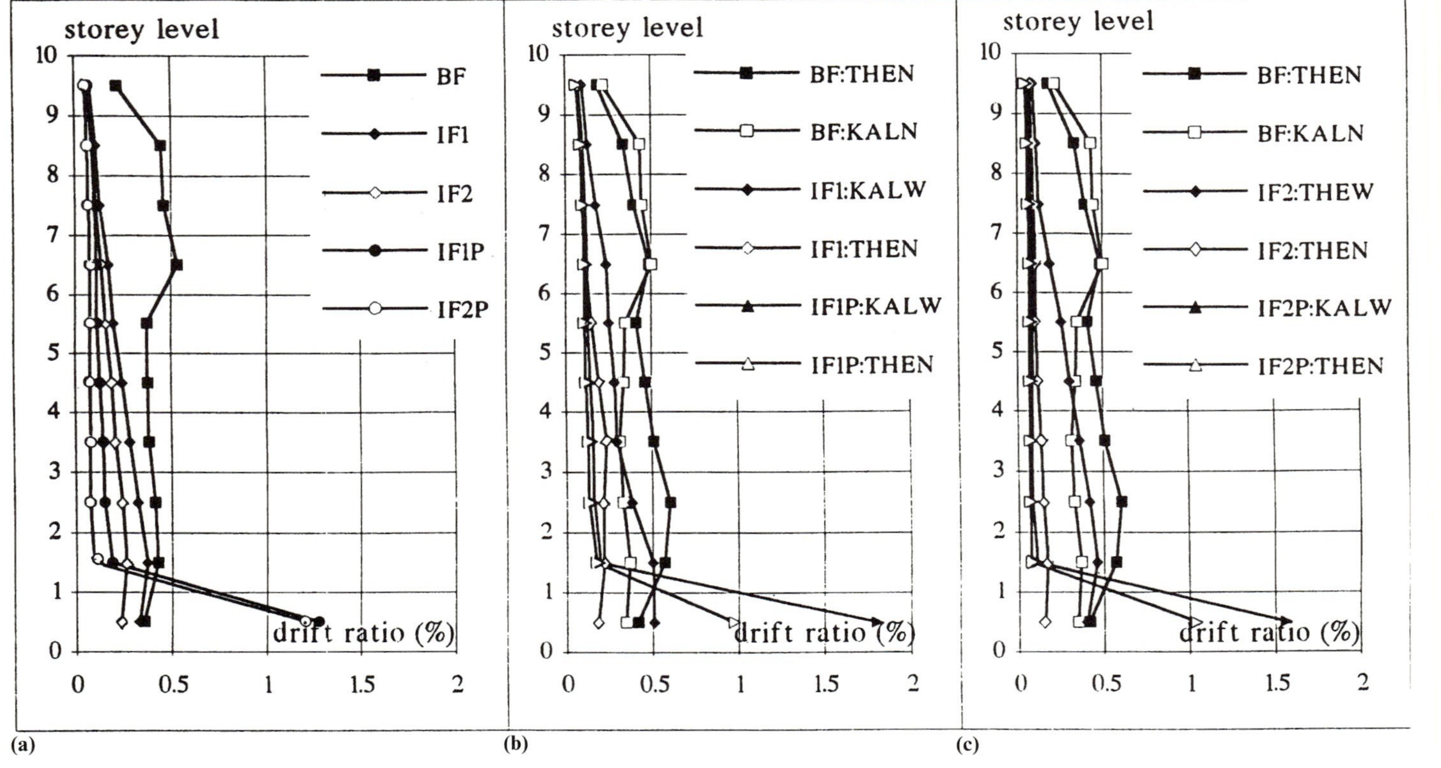

Figure 10.21 Inter-storey drift ratios for the 10-storey structures analysed: (a) mean values for each structure; (b) maximum and minimum values (weak masonry infills); (c) maximum and minimum values (strong masonry infills).

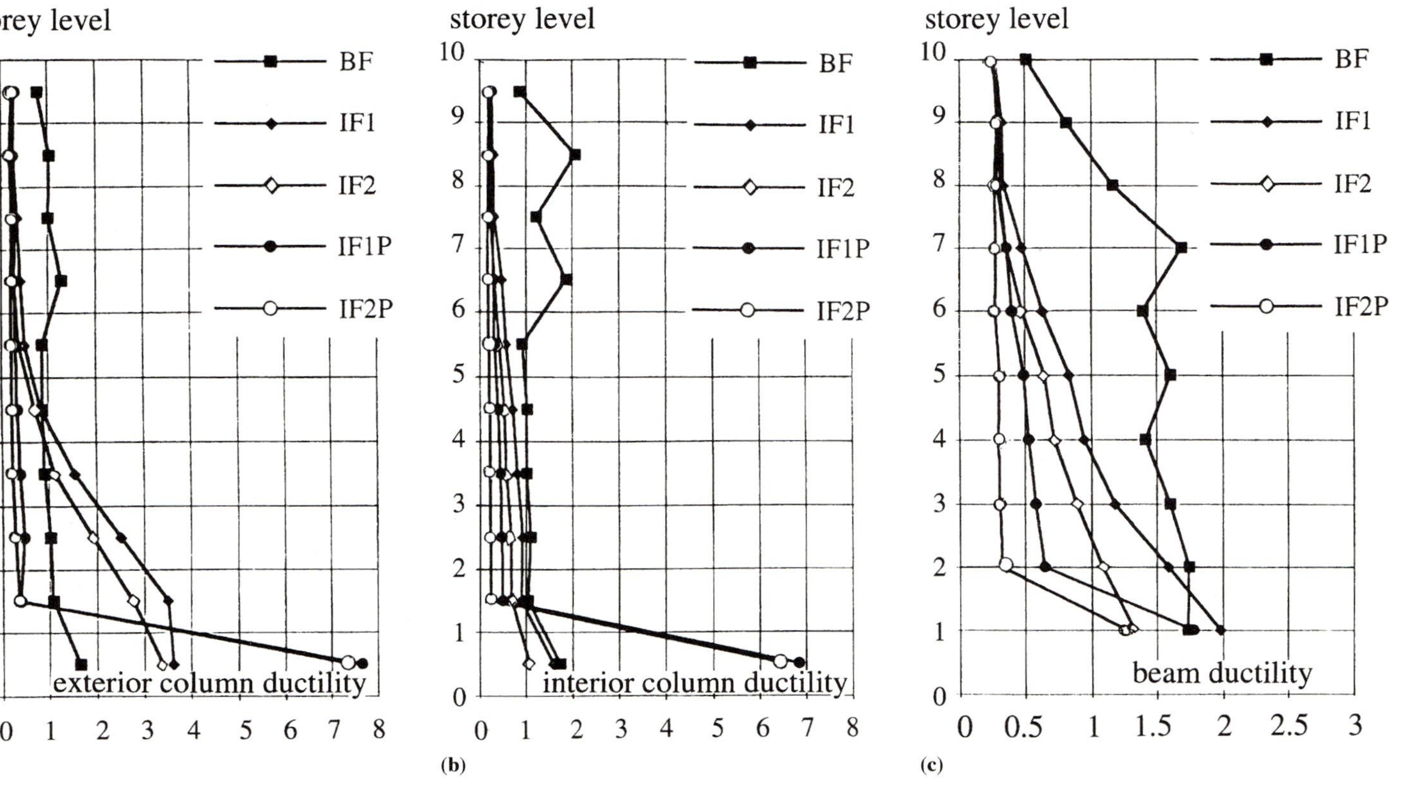

Figure 10.22 Maximum rotational ductility factors calculated for the R/C members of the bare and infilled frames (mean values for the seven motions).

are not always proportional to the plastic hinge rotations in the columns, but they are influenced by the fact that yield rotations θ_p are much lower in the exterior columns due to the influence of varying axial loading. It was found that tensile axial loads up to about 90% of the tensile yield strength developed at the exterior columns and the corresponding θ_y (which are proportional to the yield moment M_y) were quite low, resulting in increased rotational ductility factors. Nevertheless, it was found that even for the most adverse seismic motions, the required θ_p at the exterior columns of the *pilotis* buildings were lower than the rotational capacity of the members. This was the direct result of the heavy confinement of the critical regions (double 10 mm hoops at 90 mm centres), required by EC8, in combination with the relatively moderate axial loading levels (the ratio N/N_u did not exceed 0.27 in the columns of *pilotis* buildings, while balanced conditions correspond to a ratio of 0.35). It is pointed out that in the continuously infilled frames the ratio N/N_u for the ground storey columns was found to be close to, or even slightly above, the value corresponding to balanced conditions; nevertheless, due to the combination of moderate demands (θ_p not exceeding 0.004 rad) and heavy confinement, the available rotational capacity was well above the corresponding requirements. It is pointed out, though, that the model used did not account for the decrease in column stiffness due to the presence of axial tension, nor was the member capacity check extended to shear strength capacity under axial tension. Thus, the previously described picture might be rather optimistic compared with the actual situation.

The foregoing are an indication that the energy dissipation mechanism is different in the two types of structures. Energy calculations, summarized in Figure 10.23, revealed that while in the fully infilled structures about 90% of the total energy dissipation takes place in the infill panels, in the *pilotis* buildings the largest fraction of the energy is dissipated in the R/C elements, mostly in the columns of

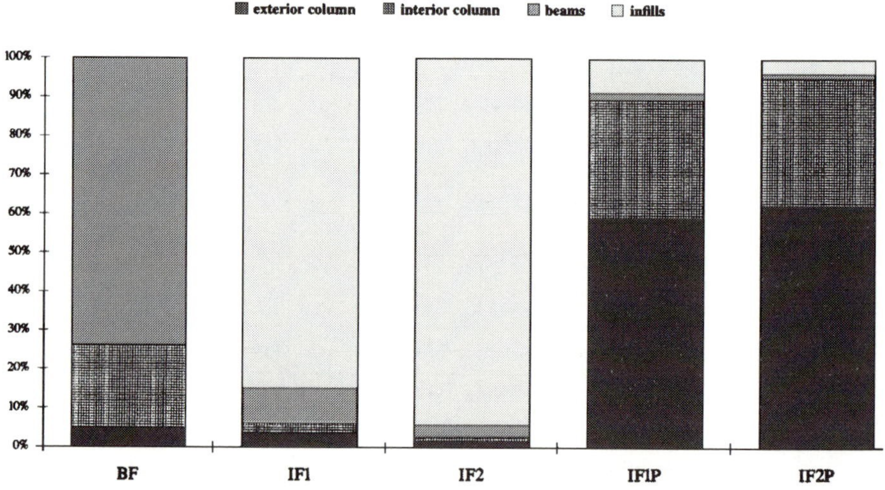

Figure 10.23 Relative contribution of each member type to the energy dissipation of bare and infilled frames.

the ground storey. Although the present study suggests that these columns do have the ability to dissipate these large amounts of energy (due to the careful detailing for ductility according to the EC8 provisions), previous experience from actual earthquakes indicates that it is not advisable, nor recommended, to rely on soft-storey mechanisms of energy dissipation, at least not for strong earthquakes exceeding the design intensity.

The study presented here indicated the superior performance of R/C frames with continuously arranged masonry infills with respect to corresponding bare frames, as well as the effectiveness of the EC8 design procedure (even when no account is taken of irregular arrangement of infills) in providing sufficient ductility and energy dissipation capacity in columns; these properties are particularly needed in the case of infilled structures with an open ground storey.

10.6 CONCLUDING REMARKS

The main purpose of the studies presented in sections 10.2–10.5 was to assess the seismic performance of multi-storey R/C buildings designed to modern codes, in particular EC8 (CEN, 1994) which represents current trends in seismic design, at least in Europe, and also the CEB (1985) Model Code which paved the way for EC8. Although only 10-storey buildings were considered and the methodology used for assessment was subject to numerous limitations and uncertainties, discussed in section 10.1.3, the conclusions drawn are interesting and useful in practice, especially in view of the fact that actual structures designed to these modern codes have not yet been built, and no indication of their seismic performance is available. Useful conclusions regarding the behaviour of medium-rise R/C frame structures designed to the 1988 version of EC8 were drawn based on the pseudo-dynamic testing of a four-storey R/C frame, carried out at the European Laboratory for Structural Assessment, in JRC (Ispra, Italy); preliminary results from this test and of associated analytical studies have been reported by Negro and Pinto (1994).

The main conclusions derived from the studies presented in the preceding sections and also based on the results of experimental studies of similar structures (Abrams, 1991; Bertero *et al.*, 1985; CEB, 1994; Negro and Pinto, 1994; Okamoto *et al.*, 1985) are summarized in the following:

1. The seismic performance of both frames and dual systems, forming parts of multi-storey R/C buildings, designed to modern seismic codes is very satisfactory. Both analyses and tests have shown that these structures are in a position to survive an earthquake having an intensity at least double that of the design earthquake; moreover damage control under the serviceability earthquake appears to be quite effective.
2. Multi-storey structures with R/C walls (dual systems) are characterized by increased stiffness and better drift control than frame structures with similar overall geometry and mass. Nevertheless the analyses presented in section 10.2 have shown that EC8-designed frames develop reasonably low displacements, even when subjected to an earthquake corresponding to the 'survival' limit state (no-collapse requirement). The main reason for this appears to be the limited amount of yielding occurring in columns.
3. The only apparent weakness of the EC8-designed structures was the non-uniform level of seismic reliability characterizing the various members. Columns,

as well as boundary elements of walls, appear to be over-designed with regard to their transverse reinforcement (hoops), as their flexural ductility capacity was estimated to be well in excess of the corresponding requirements, even under the survival earthquake; besides, the shear capacity of columns is also much larger than that required for all seismic intensities considered. On the other hand, the demand to supply ratio was much lower in beams, especially with regard to shear capacity (the possibility of shear failure under the survival earthquake was pointed out in both the frame and the dual system), while low safety margins were also estimated in the walls with respect to shear failure (diagonal tension and sliding shear modes).

4. Possible improvements in EC8 which is now a prestandard (ENV), intended for a trial application of at least three years, might include a revision of confinement reinforcement requirements in columns and wall edge members (to avoid overconservatism), an extension of capacity design requirements for shear to DC 'M' beams (with a reduced γ_{Rd} factor), and probably a revision of the diagonal tension and sliding shear design procedures for walls.
5. Structures designed for the highest ductility level appear to be more cost-effective than those designed for lower ductility levels. With regard to seismic reliability, the poorest design appears to be that for the lowest ductility level, and the best that for the highest ductility level, in particular the dual system which favourably combines adequate ductility with effective drift control. These conclusions, however, refer to CEB (1985) designed structures, while results from recently completed studies indicate that a more uniform seismic reliability characterizes the structures that are designed according to the three ductility classes specified by EC8.
6. The response of multi-storey R/C frames with continuously arranged brick masonry infills was found to be superior to that of similar bare frames; this conclusion is in agreement with observations of seismic behaviour of actual structures of this type in southern Europe and Central and South America. In infilled frames with irregularities, such as an open ground storey, damage was found to concentrate in the levels where the discontinuity occurs; the EC8 design procedure was found to be effective in providing sufficient ductility and energy dissipation capacity in the columns located in the discontinuity areas.

The seismic performance of R/C buildings typical of the existing stock in most countries, that is, structures not designed according to modern seismic codes,will be discussed in Chapter 11.

10.7 REFERENCES

Abrams, D.P (1991) Laboratory definitions of behavior for structural components and building systems, in *ACI SP-127: Earthquake-resistant Concrete Structures–Inelastic Response and Design*, ACI, Detroit, Michigan, pp. 91–152.

Aci (1982) *Dynamic Modeling of Concrete Structures*, Publ. SP-73, ACI, Detroit, Michigan.

Bertero, V.V. *et al.* (1985) Earthquake simulator tests and associated experimental, analytical and correlation studies of one-fifth scale model, in *ACI SP-84: Earthquake Effects on Reinforced Concrete Structures – U.S.–Japan Research*, Detroit, Michigan, pp. 375–424.

CEB (1985) Model Code for seismic Design of Concrete Structures. *Bull. d'Inf. CEB*, **165**, Paris.

CEB (1994) Behavior and analysis of reinforced concrete structures under alternate actions inducing inelastic response–Vol. 2: Frame members. *Bull. d' Inf. CEB*, **220**, Lausanne.

Fardis, M.N. (1991) Member-type models for the nonlinear seismic response analysis of reinforced concrete structres, in *Experimental and Numerical Methods in Earthquake Engineering* (eds. J. Donea and P.M. Jones), Kluwer Academic Publ., Dordrecht, pp. 247–80.

Fardis, M.N. and Panagiotakos, T. (1995) Earthquake response of R/C structures, *Proceed. 5th SECED Conf. on European Seismic Design Practice*, Chester, UK, pp. 11–18.

French, C.W. and Schultz, A.E. (1991) Minimum available deformation capacity of reinforced concrete beams, in *ACI SP-127: Earthquake-resistant Concrete Structures–Inelastic Response and Design*, ACI, Detroit, Michigan, pp. 363–419.

Kappos, A.J. (1986) Input parameters for inelastic seismic analysis of R/C frame structures. *Proceed. of 8th European Conf. on Earthq. Engng*, Lisbon, Portugal, Sept. 1986, **3**, pp. 6. 1/33–40.

Kappos, A.J. (1990) Sensitivity of calculated inelastic seismic response to input motion characteristics. *Proceed. 4th US Nat. Conf. on Earthq. Engng*. Palm Springs, California, May 1990, **2**, pp. 25–34.

Kappos, A.J. (1991a) Analytical prediction of the collapse earthquake for R/C buildings: Suggested methodology. *Earthquake Engng and Struct. Dynamics*, **20**(2), 167–76.

Kappos, A.J. (1991b) Analytical prediction of the collapse earthquake for R/C buildings: Case studies. *Earthquake Engng and Struct. Dynamics*, **20**(2), 177–90.

Kappos, A.J. (1991c) Analytical estimation of behaviour factors for R/C wall-frame structures. *Conf. Intern. sur les Bâtiments à Murs Porteurs en Béton en Zone Sismique*. Paris, France, June 1991, pp. 440–51.

Kappos, A.J. (1992) DRAIN-2D/90: a microcomputer program for the inelastic dynamic analysis of plane structures subjected to seismic loading. *Rep. of the Lab. of Concrete Structures*, Dept of Civil Engng, Aristotle Univ. of Thessaloniki, Greece.

Kappos, A.J., Antoniades, K. and Konstantinides, D. (1994) Seismic behaviour evaluation of R/C buildings designed to the Eurocode 8. *Proceed. of the 2nd Int. Conf. on Earthq. Resistant Construction and Design*. Tech. Univ. of Berlin, Germany, June 1994, Balkema, **2**, pp. 881–8.

Kappos, A.J. and Papadopoulos, D. (1995) Influence of design ductility level on the seismic behaviour of R/C frames. *Proceed. 10th European Conf. on Earthq. Engng,* Vienna, Austria, Aug.–Sept. 1994, Balkema, Rotterdam, 2, pp. 953–8.

Kitagawa, Y. *et al.* (1984) Correlation study on shaking table tests and pseudo-dynamic tests by R.C. models. *Proceed. of 8th World Conf. on Earthq. Engng*, San Francisco, California, July 1984, Prentice-Hall, New Jersey, **VI**, pp 667–74.

Michailidis, C.N., Stylianidis, K.C. and Kappos, A.J. (1995) Analytical modelling of masonry infilled R/C frames subjected to seismic loading. *Proceed 10th European Conf. on Earthq. Engng*, Vienna, Austria, Aug.–Sept. 1994, Balkema, Rotterdam, **3**, pp. 1519–24

Negro, P. and Pinto, A.V. (1994) Analysis and PSD testing of a full-scale R/C frame designed according to EC8. *Proceed. of the 2nd Int. Conf. on Earthq. Resistant Construction and Design*, Tech. Univ. of Berlin, Germany, June 1994, Balkema, **2**, pp. 907–14.

Okamoto, S. *et al.* (1985) Testing, repair and strengthening, and testing of a full-scale seven story reinforced concrete building, in *ACI SP-84: Earthquake Effects on Reinforced Concrete Structures–U.S.–Japan Research*. Detroit, Michigan, pp. 133–61.

Otani, S. and Sozen, M.A. (1972) Behavior of multistory reinforced concrete frames during earthquakes. *Civil Engng. Studies–Struct. Research Series*, **392**, Univ. of Illinois, Urbana.

Park, Y.-J. and Ang, A.H.-S. (1985) Mechanistic seismic damage model for reinforced concrete. *Journal of Struct. Engng, ASCE*, **111**(4), 722–39.

Paulay, T. and Priestley, M.J.N. (1992) *Seismic Design of Reinforced Concrete and Masonry Buildings*, J. Wiley & Sons, New York.

Paulay, T. and Spurr, D.D. (1977) Frame–shear wall assemblies subjected to simulated seismic loading. *Proceed. of 9th World Conf. on Earthq. Engng*, New Delhi, India, **V**. II, pp. 1195–200.

Penelis, G.G., Sarigiannis, D., Stavrakakis, E. and Stylianidis, K.C. (1988) A statistical evaluation of damage to buildings in the Thessaloniki, Greece earthquake of June 20, 1978. *Proceed.*

of 9th World Conf. on Earthq. Engng, Tokyo–Kyoto, Japan, Aug. 1988, Tokyo: Maruzen, **VII** pp. 187–92.

Priestley, M.J.N. and Seible, F. (1992) Performance assessment of damaged bridge bents after the Loma Prieta earthquake. *Bull of the New Zealand Nat. Society for Earthq. Engng,* **25**(1), 44–51.

Saatcioglu, M. (1991) Deformability of reinforced concrete columns, in *ACI SP-127: Earthquake-Resistant Concrete Structures: Inelastic Response and Design*, ACI, Detroit, Mich., pp. 421–52.

Shing, P.S.B. and Mahin, S.A. (1984) pseudodynamic test method for seismic performance evaluation: theory and implementation. *Rep. EERC-84/01*, Univ. of California, Berkeley.

Sozen, M.A. (1981) Review of earthquake response of R.C. buildings with a view to drift control, in *State-of-the-art in Earthquake Engineering*, Kelaynak Press, Ankara, Turkey, pp. 383–418.

Trifunac, M.D. and Brady, A.G. (1975) A study on the duration of strong earthquake ground motion. *Bull. of the Seismol. Society of America*, **65**(3), 581–626.

Valiasis, T.N. and Stylianidis, K.C. (1989) Masonry infilled R/C frames under horizontal loading–Experimental resuls. *European Earthq. Engng*, **III**(3), 10–20.

Wood, S.L. (1991) Observed behavior of slender reinforced concrete walls subjected to seismic loading, in *ACI SP-127: Earthquake-Resistant Concrete Structures: Inelastic Response and Design*, ACI, Detroit, Michigan, pp. 453–77.

Wood, S.L. (1992) Seismic response of R/C frames with irregular profiles. *Journal of Struct. Engng, ASCE*, **118**(2), 545–66.

11

Seismic pathology

11.1 CLASSIFICATION OF DAMAGE IN R/C STRUCTURAL MEMBERS

11.1.1 Introduction

A strong earthquake puts the whole structure through a hard test. As a result, all the weaknesses of the structure, due to either code imperfections or analysis and design errors, or even bad construction, are readily apparent. It is not unusual that a strong earthquake leads to improvements or even drastic changes to the design codes, modifications to the design methods and rejuvenation of the sense of responsibility for the design and execution of construction works.

It is difficult to classify the damage caused by an earthquake, and even more difficult to relate it in a quantitative manner to the cause of the damage. This is because the dynamic character of the seismic action and the inelastic response of the structure while damage is being induced, render questionable every attempt to explain the phenomenon by a simplified static model.

Despite all the difficulties inherent in a damage classification scheme, an attempt will be made in this chapter to divide the damage into categories, and to identify the cause of the damage in each case, according to current concepts on the behaviour of structural elements under cyclic inelastic loading, which simulates sufficiently closely the response of structural members to a strong earthquake.

In this section the damage classification will refer to the individual structural elements (Tegos, 1979), while in the following section reference will be made to the main causes of damage to R/C buildings. In both sections the qualitative analysis will be supplemented with statistical data for the behaviour of structural elements and buildings during strong earthquakes.

The basic source of the statistical data is the research project 'A statistical evaluation of the damage caused by the earthquake of June 20, 1978 to the buildings of Thessaloniki, Greece', which was carried out at the Laboratory of Reinforced Concrete of the University of Thessaloniki, in collaboration with the Ministry of Environment and Public Works (Penelis *et al.*, 1987, 1988), and also the report, *The Earthquake of 19 September 1985–Effects in Mexico City*, of the Committee for Reconstruction of Mexico City (Rosenblueth and Meli, 1985).

In the classification that follows there is no reference to damage due to analysis errors, bad concrete quality, improper reinforcement detailing and so on.

Weaknesses of this type are always present in structures and their frequency and severity depend on the level of technological development of a country. Of course, these weaknesses contribute to the degree of damage caused by an earthquake, and occasionally they become fatal for the stability of buildings.

11.1.2 Damage to columns

Damage to columns caused by an earthquake is mainly of two types:

- damage due to cyclic flexure and low shear under strong axial compression;
- damage due to cyclic shear and low flexure under strong axial compression.

The first type of damage manifests itself with failure at the top and bottom of the column (Figures 11.1 and 11.2). It occurs in columns of moderate to high slenderness ratio, that is

$$\alpha = \frac{M}{Vh} = \frac{L}{2h} > 3.5$$

The high bending moment at these points combined with the axial force, leads to the crushing of the compression zone of concrete, successively on both faces of the column. The smaller the number of ties in these areas, the higher their vulnerability to this type of damage. The crushing of the compression zone is manifested first by spalling of the concrete cover to the reinforcement. Subsequently the concrete core expands and crushes. This phenomenon is usually accompanied by buckling of bars in compression and by hoop fracture. The fracture of the ties and the disintegration of concrete lead to shortening of the column under the action of the axial force. Therefore this type of damage is very serious because the column not only loses its stiffness, it also loses its ability to carry vertical loads. As a result, there is a redistribution of stress in the structure, since the column has shortened due to the disintegration of concrete in the above-mentioned areas. This type of damage is very common; 23.2% of the buildings damaged in their

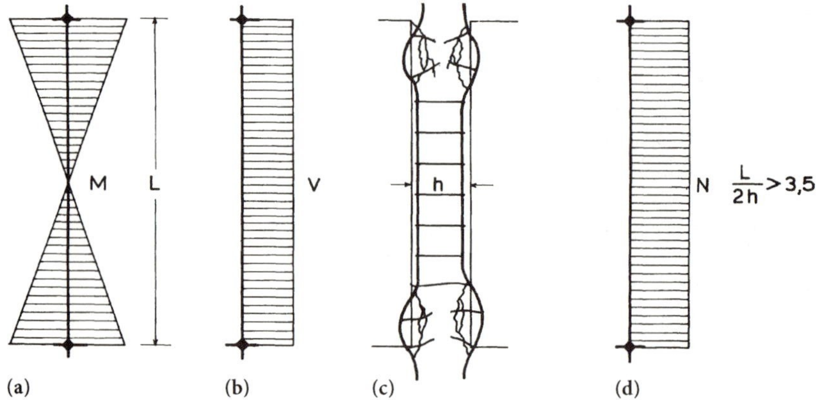

Figure 11.1 Column damage due to strong axial compression and cyclic bending moment: (a) bending moment diagram; (b) shear force diagram; (c) sketch of damage; (d) axial force diagram.

R/C structural systems by the Thessaloniki earthquake of 20 June 1978 showed damage of this type (Penelis *et al.*, 1987; 1988). The great majority of failures of buildings with frame systems during the Mexico City earthquake of 19 September 1985 were caused by column damage (Rosenblueth and Meli, 1985). As the main reasons for this brittle type of failure one should consider the low quality of con-

(a)

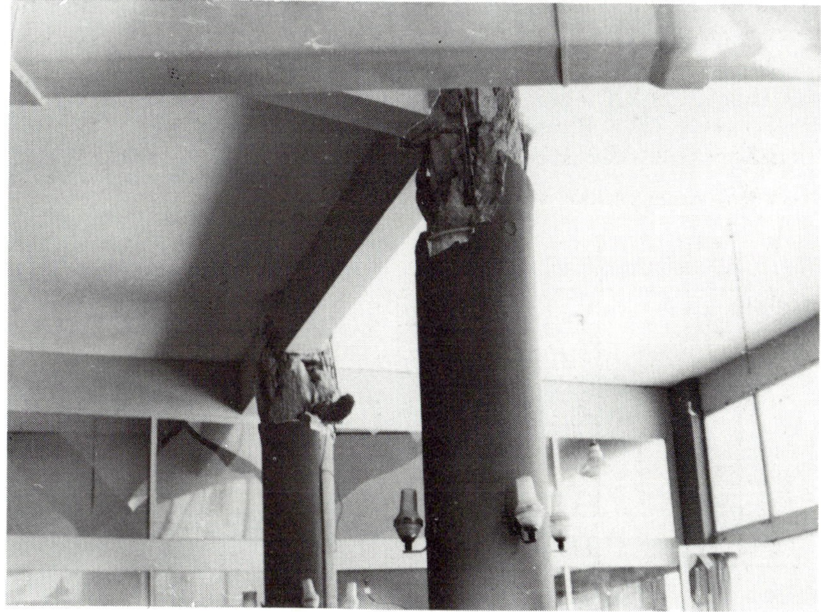

(b)

Figure 11.2 Column damage due to strong axial compression and cyclic bending moment: (a) Bucharest, Romania, 1977; (b) Loutraki, Greece, 1981.

crete, the inadequate number of ties in the critical areas, the presence of strong beams which leads to columns failing first, and finally, of course, the strong seismic excitation inducing many loading cycles in the inelastic range.

The second type of damage is of the shear type and is manifested in the form of X-shaped cracks in the weakest zone of the column (Figures 11.3 and 11.4). It occurs in columns with moderate to small slenderness ratios, that is

$$\alpha = \frac{M}{Vh} = \frac{L}{2h} < 3.5$$

The ultimate form of this type of damage is the explosive cleavage failure of short columns (Figures 11.5 and 11.6), which usually leads to a spectacular collapse of the building. The main reason for this type of damage is that the flexural capacity of columns with moderate to small slenderness ratio is higher than their shear capacity, and as a result shear failure prevails. The frequency of this type of damage is lower than the failure at the top and bottom of the column. It usually occurs in columns of the ground floor, where, because of the large dimensions of the cross-section of the columns, the slenderness ratio is low. It also occurs in short columns which have either been designed as short, or have been reduced to short because of adjacent masonry construction which was not accounted for in the design (Figures 4.8 and 4.9).

Finally, sometimes in the case of one-sided masonry-infilled frames, masonry failure is followed by shear failure of the adjacent columns (Figures 11.7 and 11.8) (Stylianidis and Sariyiannis, 1992).

In conclusion, it has to be stressed that column damage is very dangerous for the structure, because it alters or even destroys the vertical elements of the structural system. Thus, when damage of this type is detected, means of temporary support should be provided immediately.

11.1.3 Damage to R/C walls

The damage which is caused by earthquake to R/C walls is of the following types:

- X-shaped shear cracks;

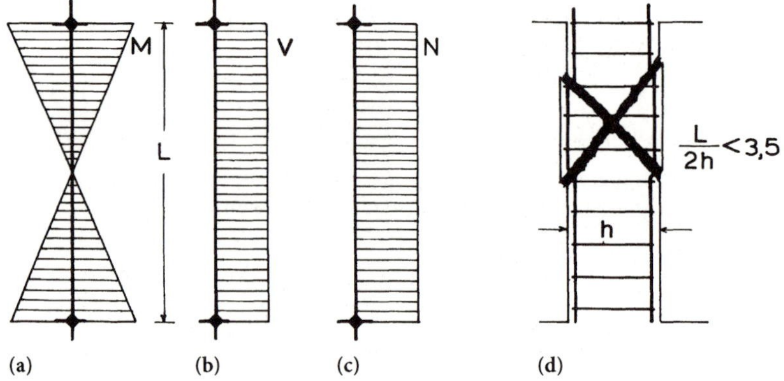

Figure 11.3 Column damage due to strong axial compression and shear: (a) bending moment diagram; (b) shear force diagram; (c) axial force diagram; (d) sketch of damage.

(a)

(b)

Figure 11.4 Column damage due to strong axial compression and shear: (a) Kalamata, Greece, 1986; (b) Mexico City, 1985.

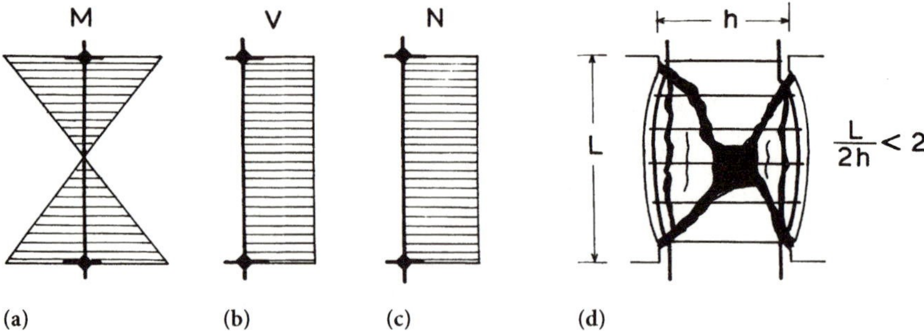

Figure 11.5 Explosive cleavage failure of a short column: (a) bending moment diagram; (b) shear force diagram; (c) axial force diagram; (d) sketch of damage.

- sliding at the construction joint;
- damage of flexural character (horizontal cracks – crushing of the compression zone).

During the Thessaloniki earthquake of 20 June 1978, 28.6% of the buildings which suffered damage in the structural system had damage in the R/C walls (Penelis *et al.*, 1987).

The most frequent type of damage is the appearance of cracks at the construction joint (Figures 11.9 and 11.10). Damage of this type occurred in 88% of the buildings with wall damage caused by the Thessaloniki earthquake of 20 June 1978 (Penelis *et al.*, 1987, 1988). This damage is mainly due to the fact that old concrete is not properly bonded with fresh concrete. All seismic codes in effect today require that extra care should be taken when construction work is discontinued in order to ensure proper bonding of concrete (rough surface, cleaning, soaking, pouring of strong cement first and then concrete). In addition, placement of connecting reinforcement is also required in the form of dowels. The introduction of these requirements is the result of the high frequency of this type of damage. However, it has to be mentioned that this type of damage does not pose a threat to the stability of the building, because, with the horizontal arrangement of the cracks, the wall can still carry vertical loads. Also, from the stiffness point of view, this type of damage has only a slight effect on the entire structural system.

The appearance of X-shaped cracks in R/C walls is the next most frequent damage (Figures 11.11 and 11.12). During the above-mentioned Thessaloniki earthquake the frequency of this damage reached 30% of the buildings with wall damage. This is a shear type of brittle failure. Because of the arrangement of the cracks, under the action of vertical loads, the isosceles triangles which are formed on the two sides tend to separate from the structure and therefore cause its collapse (Figure 11.11). In order to protect the structure from this type of failure, all the current codes require the formation of a column at each side of the wall, which will carry the vertical loads after the shear failure of the web. These columns can either be thicker than the wall and visible, or they can be incorporated in the wall (Chapter 9).

Damage of flexural type occurs very rarely (Figure 11.13). It is the authors' belief that this is due to the fact that the bending moments developing at the base

(a)

(b)

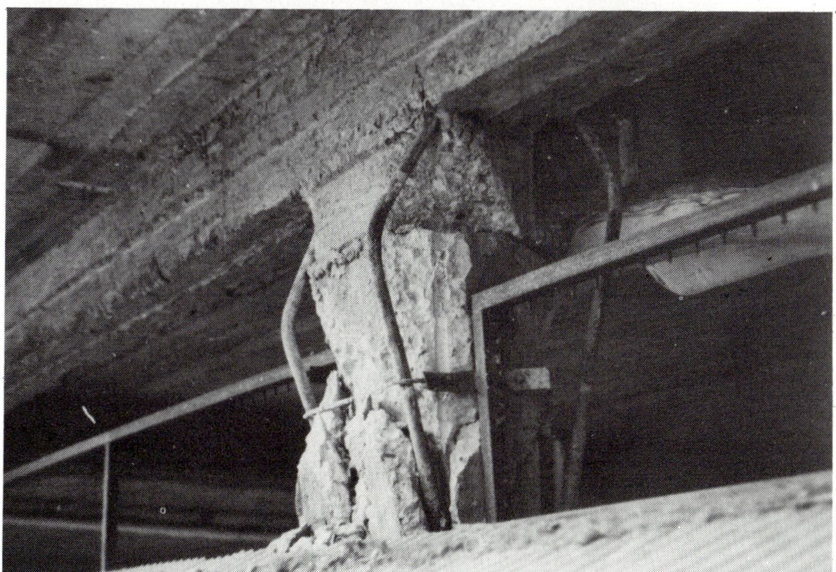

Figure 11.6 Explosive cleavage failure of a short column: (a) general view of a collapsed building, Kalamata, Greece, 1986; (b) detail of a short column of the building in (a).

of the wall are much smaller than those calculated for the design, because the footing rotates as the soil deforms during the earthquake (Figure 5.10). On the other hand, this soil deformation does not much alter the shear force which is carried by the wall, and as a result, shear failure prevails.

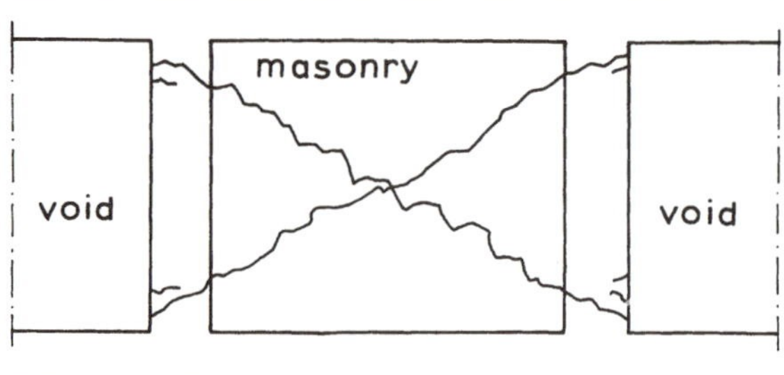

Figure 11.7 Damage in columns in contact with masonry on one side only.

Figure 11.8 Damage in column in contact with masonry on one side only, Loutraki, Greece, 1981.

11.1.4 Damage to beams

The damage which occurs in R/C beams due to an earthquake is as follows:

- cracks orthogonal to the beam axis along the tension zone of the span;
- shear failure near the supports;
- flexural cracks on the upper or lower face of the beam at the supports;
- shear or flexural failure at the points where secondary beams or cut-off columns are supported by the beam under consideration;
- X-shaped shear cracks in short beams which connect shear walls.

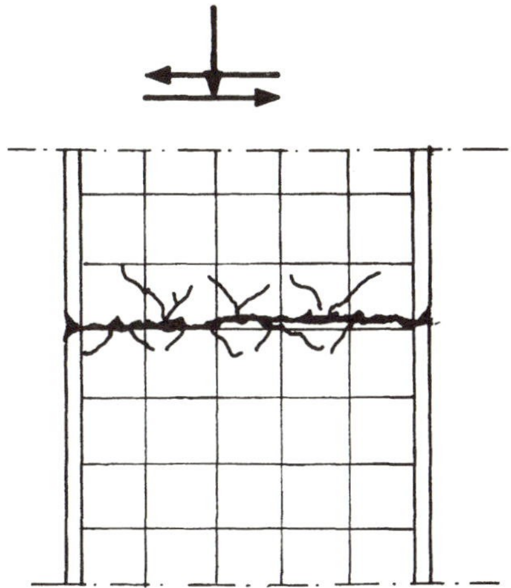

Figure 11.9 Shear wall damage at a construction joint.

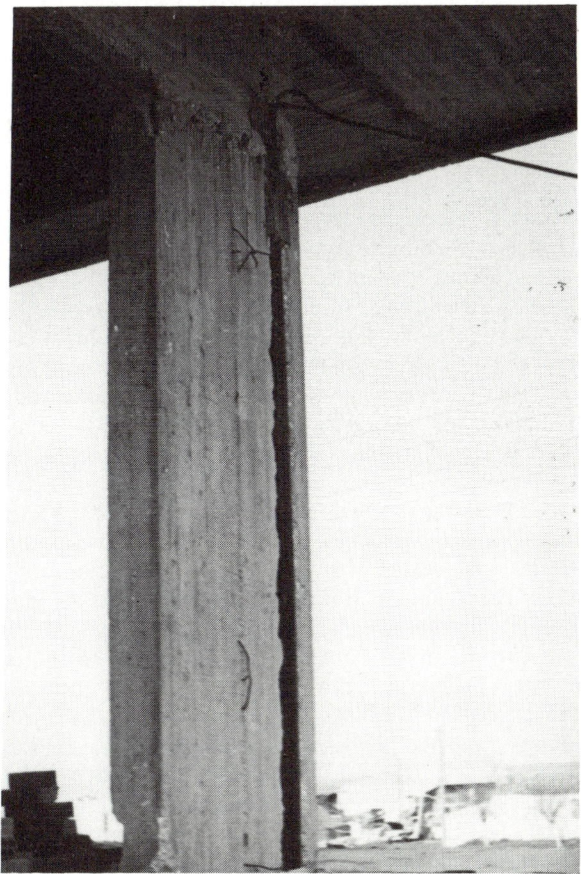

Figure 11.10 Shear wall damage at a construction joint, Kalamata, Greece, 1986.

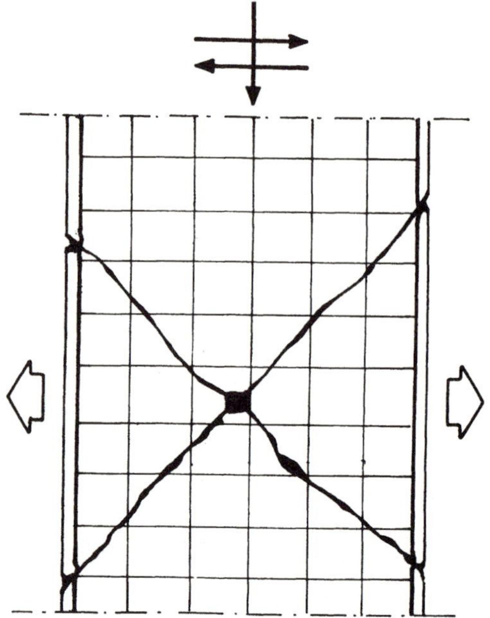

Figure 11.11 Shear wall damage due to shear (X-shaped cracks).

Damage to beams, although, fortunately, it does not jeopardize the safety of the structure, is the most common type of damage in R/C buildings; 32.6% of the buildings whose structural system was damaged during the Thessaloniki earthquake of 20 June 1978, exhibited some type of beam damage.

Cracks in the tension zone of the span constitute the most common type of damage – 83% of the structures with beam damage in Thessaloniki due to the June 1978 earthquake had damage of this type. This type of damage (Figure 11.14) cannot be explained using analytical evidence, given the fact that the action of the seismic forces does not increase the bending moment in the span. However, the vertical component of the seismic action, due to its cyclic character, simply makes visible the microcracks which are due to bending of the tension zone, thus creating the impression of earthquake damage. This is the reason why the large majority of the cases of beams with this type of damage do not jeopardize the overall stability of the structure. It is also understood that the high frequency of damage of this type is rather misleading, since in most cases it is just a manifestation of already existing normal cracking rather than of earthquake damage.

The bending-shear failure near the supports (Figure 11.15) is the second most frequent type of damage (43%) in beams. Undoubtedly it constitutes a more serious type of damage than the previous one, given its brittle character. However, only in very few cases does it jeopardize the overall stability of the structure.

The flexural cracks on the upper and lower face of the beam at the supports (Figure 11.16), can be fully explained if the earthquake phenomenon is statically approximated with horizontal forces. From the frequency point of view, this type of damage is rarer than the shear type (28%). Most of the time cracking of the

(a)

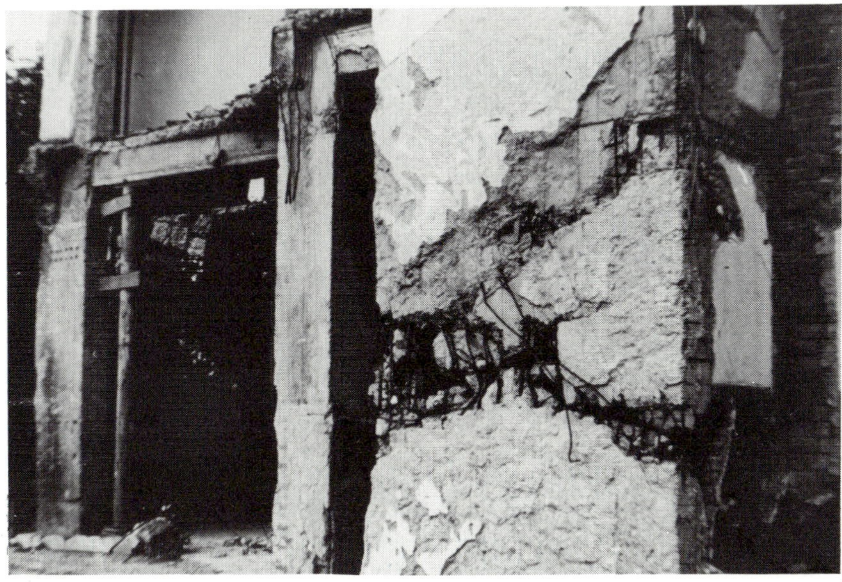

(b)

Figure 11.12 Shear wall damage due to shear (X-shaped cracks): (a) Kalamata, Greece, 1986; (b) Bucharest, Romania, 1977.

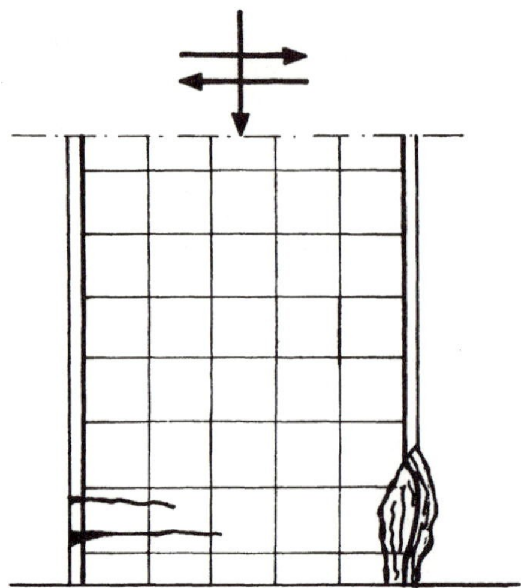

Figure 11.13 Shear wall damage due to flexure and compression.

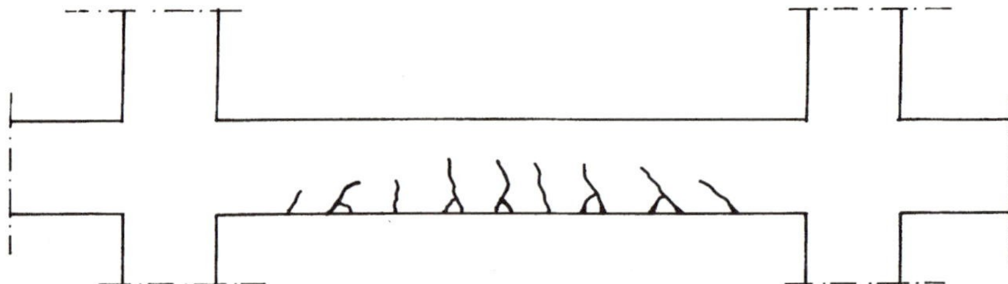

Figure 11.14 Flexural cracks on a beam span.

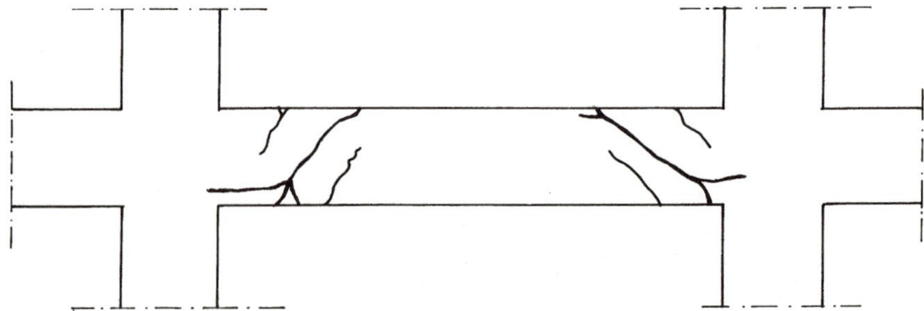

Figure 11.15 Bending-shear cracks near the supports of a beam.

lower face is due to bad anchorage of the bottom reinforcement into the supports, in which case one or two wide cracks form close to the support.

The shear or flexural failure at the points where secondary beams or cut-off columns are supported (Figure 11.17) appears quite frequently. It is due to the vertical component of the earthquake which amplifies the concentrated load.

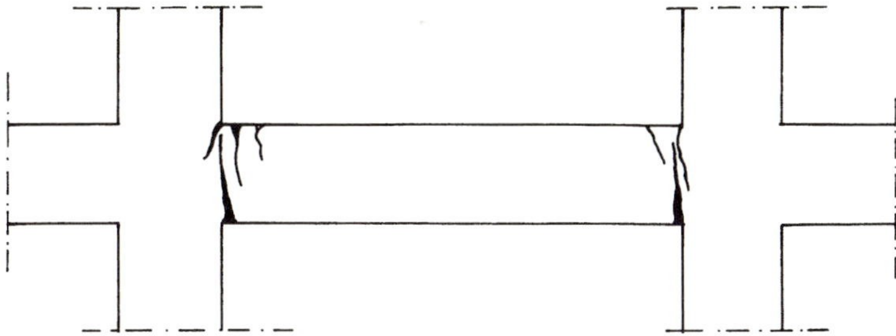

Figure 11.16 Flexural cracks on the lower face of the beam at the support.

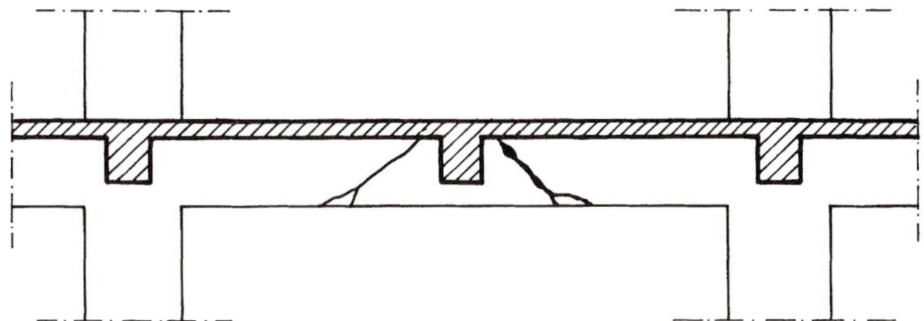

Figure 11.17 Shear failure at the location of an indirect support.

X-shaped shear cracks in short beams coupling shear walls also appear quite often. It is a shear failure similar to that which occurs in short columns (Figure 11.18) but not so dangerous for the stability of the building (see also section 9.3.4).

11.1.5 Damage to beam–column joints

Damage in beam–column joints, even at the early stages of cracking, must be considered extremely dangerous for the structure and be treated accordingly. Damage

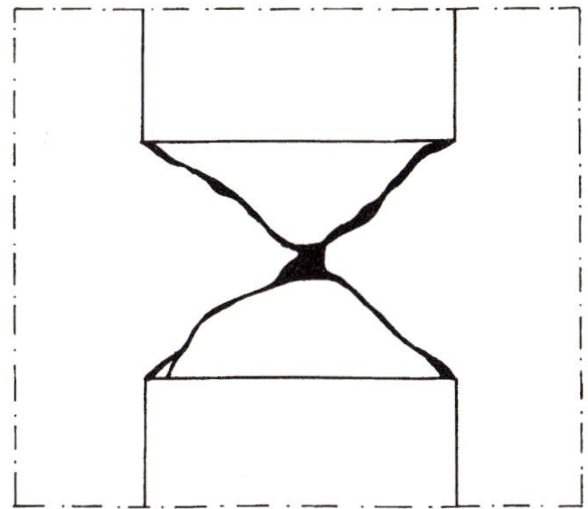

Figure 11.18 Shear failure of a shear wall coupling beam.

of this type reduces the stiffness of the structural element and leads to uncontrollable redistribution of load effects. Common failures of beam–column joints (corner joint, exterior joint of multi-storey structure, and interior joint) are shown in Figures 11.19–11.22.

The flow of internal forces in the reinforcement and the concrete during the successive phases of cyclic loading has already been explained in section 9.2 and will not be discussed here.

11.1.6 Damage to slabs

The most common types of damage which occur in slabs are the following:

- cracks parallel or transverse to the reinforcement at random locations;
- cracks at critical sections of large spans or large cantilevers, transverse to the main reinforcement;
- cracks at locations of floor discontinuities, such as the corners of large openings accommodating internal stairways, light shafts and so on;

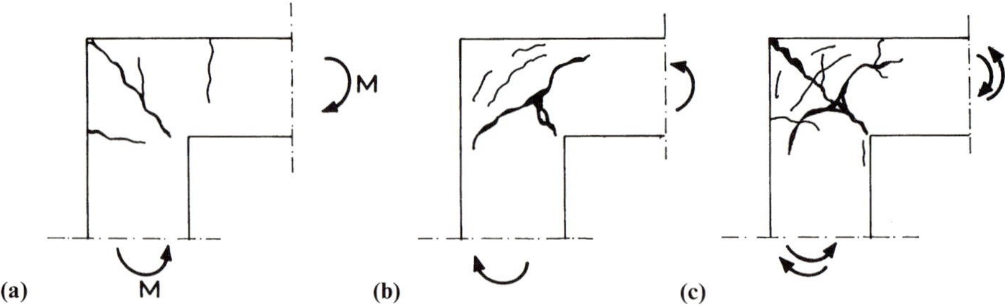

Figure 11.19 Failure of a corner joint: (a) moments subjecting the inner fibre to compression; (b) moments subjecting the inner fibre to tension; (c) cyclic bending moment loading.

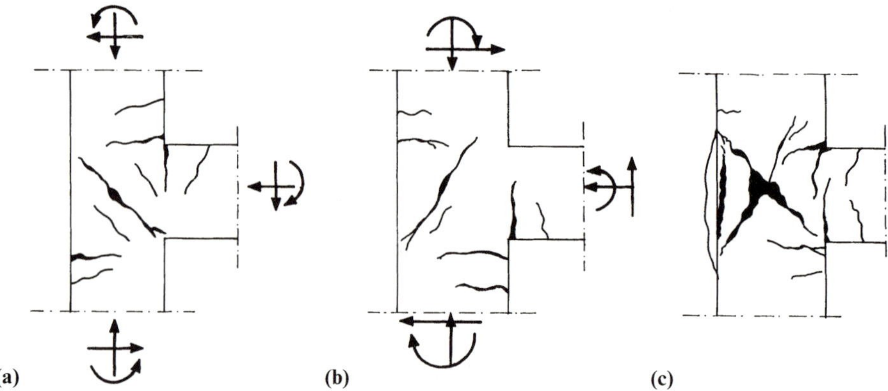

Figure 11.20 Failure of exterior joint in a multi-storey building: (a) moments inducing compression at the lower fibre of the beam; (b) moments inducing compression at the upper fibre of the beam; (c) cyclic bending moment loading.

- cracks in areas of concentration of large seismic load effects, particularly in the connection zones of slabs to shear walls or to columns in flat plate systems.

With the exception of the last type, damage in slabs generally cannot be considered as dangerous for the stability of the structure. However, they create serious aesthetic and functional problems, so they must be repaired. Moreover, the creation of such damage leads to the reduction of the available strength, stiffness and energy dissipation capacity of the structure in case of a future earthquake, and this is an additional reason for their repair.

The first type of damage is the most frequent. Most of the time it is due to the widening of already existing microcracks which are formed either because of bending action or temperature changes or shrinkage and they become visible after the dynamic seismic excitation. Rarely is it due to differential settlement of columns. In such cases, however, the phenomenon is accompanied by extensive cracking of the adjacent beams and masonry infills.

The second and third types of damage are typically due to the vertical component of the earthquake action (Figures 11.23 and 11.24).

The fourth type of damage is usually related to punching shear failure, aggravated by the cyclic bending caused by the earthquake (Figure 11.25). It has already been stressed in Chapter 4 that slabs on columns are seismically vulnerable structures, and they must be avoided as they are not covered by the codes in effect if they are not combined with other seismic-resistant systems (i.e. shear walls or ductile frames).

11.1.7 Damage to the infill panels

As discussed earlier, almost all the infill walls in southern Europe are constructed with masonry, in contact with the surrounding structural members of the frame. Since these infills are constructed with materials (bricks, mortar, plaster) of lower strength and deformability than the structural members, they are the first to fail. Thus, the failure of the infills starts before damage to the frame occurs, and therefore if it is not accompanied by damage in the structural members, the infills can-

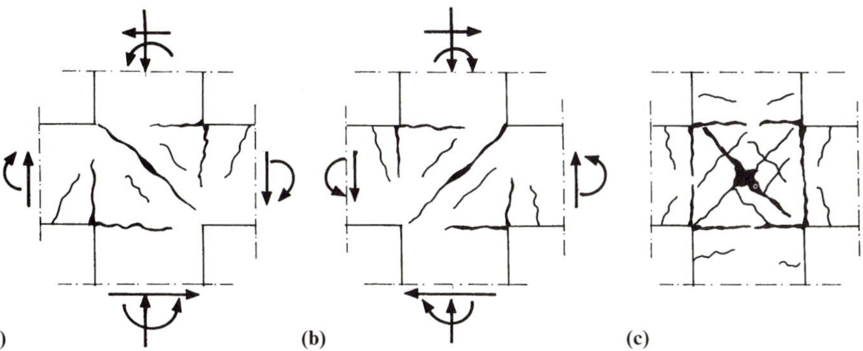

Figure 11.21 Failure of a cross-shaped interior joint: (a) seismic action in the right to left direction; (b) seismic action in the left to right direction; (c) cyclic seismic action.

(a)

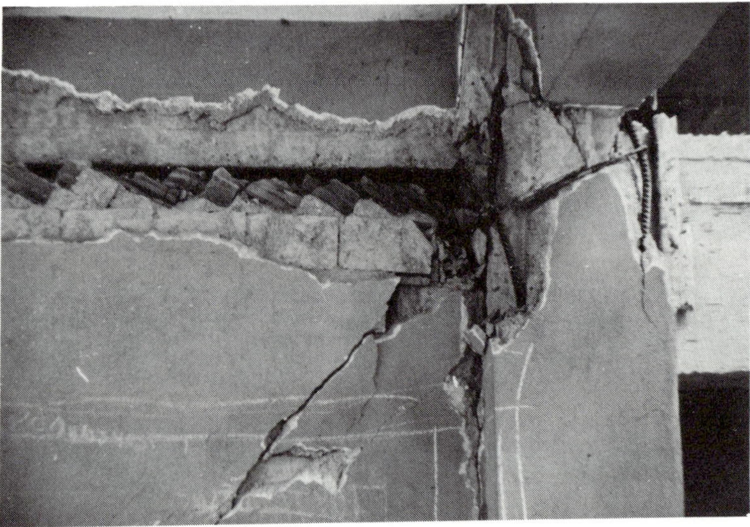

(b)

Figure 11.22 Failure of exterior joint in a multi-storey buildings: (a) Loutraki, Greece, 1981; (b) Kalamata, Greece, 1986.

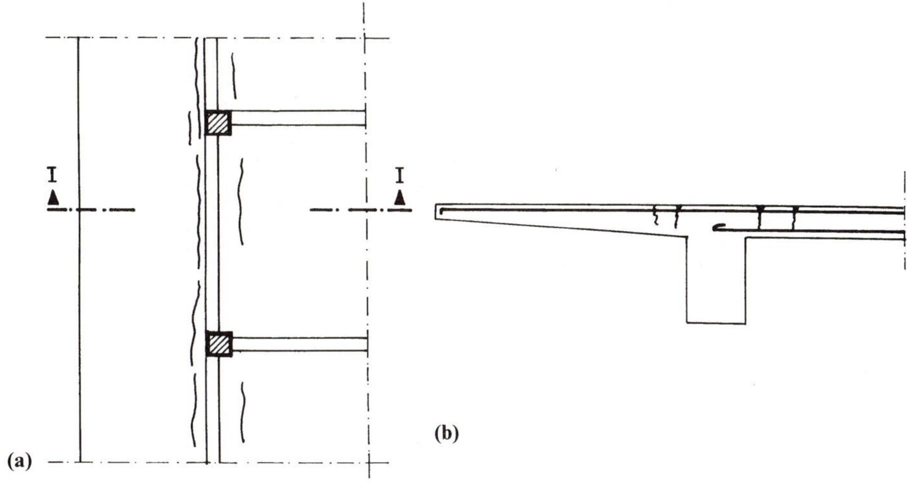

Figure 11.23 Slab damage at the critical area of a cantilever: (a) floor plan of the slab (upper side); (b) section I–I.

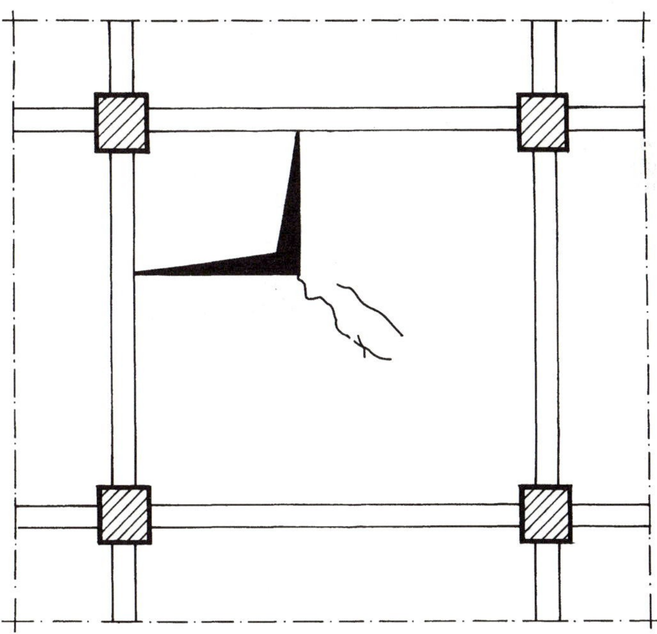

Figure 11.24 Slab damage at the corner of a large opening (lower side of the slab).

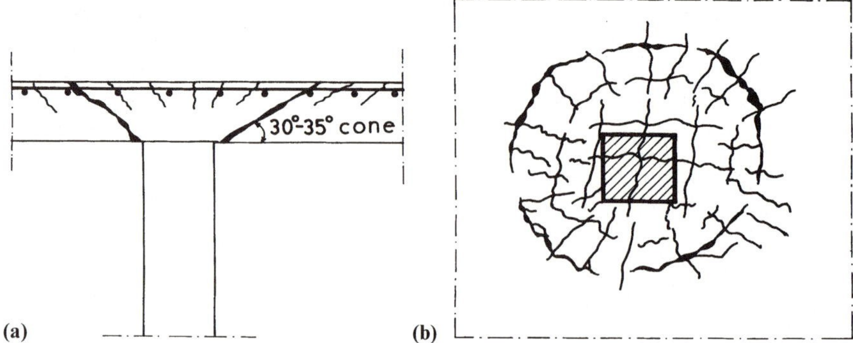

Figure 11.25 Damage at a slab to column connection: (a) section; (b) top side of the slab.

not be considered dangerous for the stability of the structure. However, the largest portion of the repair costs is usually attributed to damage in the infills, because they involve extensive repair of installations and finishing, such as plastering, painting, tiling, plumbing, electric installations and so on.

The damage in the infills occurs as follows: during the excitation of the structure due to the earthquake, the R/C frame starts to deform, and at this stage the first cracks appear on the plastering along the lines of contact of the masonry with the frame. As the deformation of the frame becomes larger, the cracks penetrate into the masonry, and this is manifested by the detachment of the masonry from the frame (Figures 11.26 (a) and 11.27). Subsequently, X-shaped cracks appear, small at the beginning which become larger later, in the masonry itself, in a stepwise pattern following the joint lines (Figures 11.26 (b) and 11.27). When the cracks do not penetrate the whole thickness of the wall the damage is characterized as 'light', otherwise it is a 'serious' damage (X-shaped cracks).

From the above discussion one can conclude that the damage in the infill panels must be the first in frequency of occurrence since they usually precede damage in the R/C structural system (Tiedemann, 1980; Penelis *et al.*, 1987, 1988).

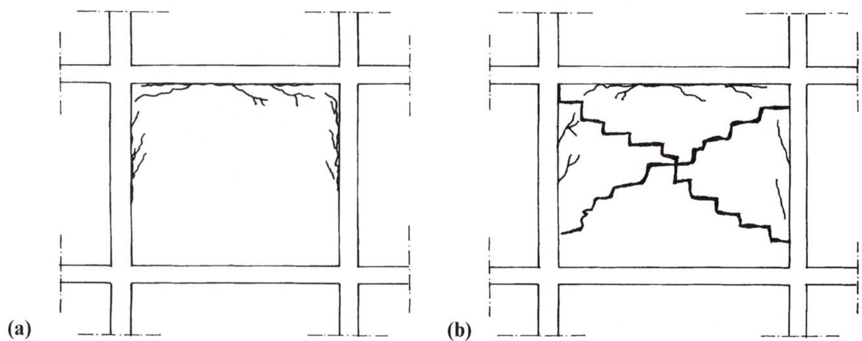

Figure 11.26 Damage in infill panels: (a) Detachment from the frame; (b) X-shaped through cracks.

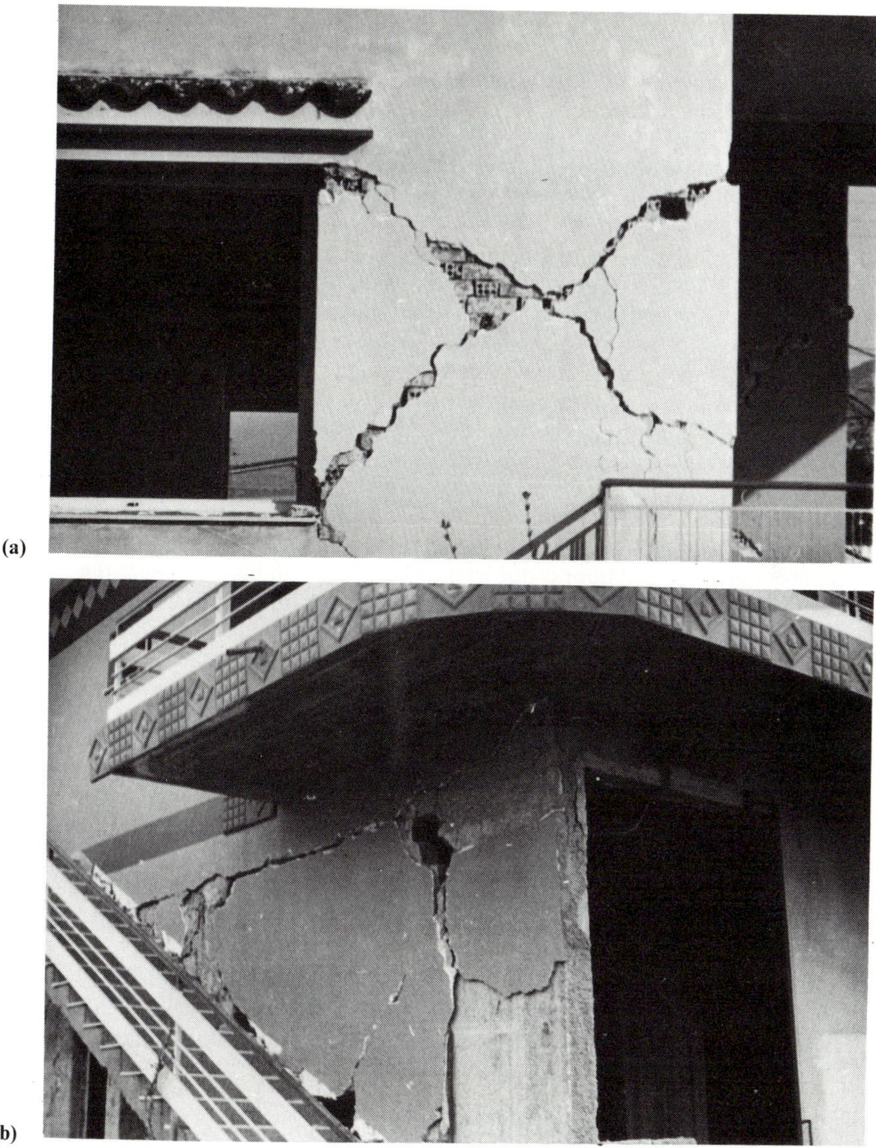

Figure 11.27 Damage in masonry walls: (a) Loutraki, Greece, 1981; (b) Loutraki, Greece, 1981.

During the Thessaloniki earthquake of 20 June 1978 while damage in beams occurred in 7.4% of the buildings, in 5.3% of the columns, and in 6.5% of the shear walls, damage in infill panels occurred in 22.9% of the buildings with an R/C structural system. Also, from the structures that suffered damage in the infill panels, 96% exhibited detachment from the surrounding frame, 79% exhibited X-shaped full-depth cracks and 12% exhibited out-of-plane collapse of the masonry wall.

11.1.8 Spatial distribution of damage in buildings

At this point it would be useful to discuss the distribution of damage in buildings.

Along the vertical direction, the most serious damage occurs in the ground floor. Their frequency and intensity are reduced gradually in the upper floors. This distribution was observed in most recent earthquakes, the Bucharest, Romania, earthquake of 1977, the Thessaloniki, Greece, earthquake of 1978, the Alkyonides, Greece, earthquake of 1981, the Montenegro, former Yugoslavia, earthquake of 1980, the Kalamata, Greece, earthquake of 1986. The Mexico City earthquake of 1985 constitutes an exception to the above observation and will be discussed later. The methodology for analysis and design of earthquake-resistant structures cannot be used to explain this phenomenon. In fact, the lower storeys and particularly the ground storey, due to the higher inertial forces, are subjected to larger seismic effects. However, their structural elements are designed according to rules which apply to the whole building and therefore they conform to common partial safety factors. Therefore, damage is expected to be uniformly distributed throughout the building. The dynamic inelastic analysis of multi-storey buildings (Figure 11.28) (Kappos and Penelis, 1987) also supports the notion of uniform distribution of damage. It is the authors' opinion that the higher degree of damage in the ground floor is due to the fact that the infills contribute the same amount of additional strength to all floors (given the fact that the masonry layout is the same on every floor) which is not usually taken into account when analysing the structure. Indeed, if the masonry did not exist and the required strength for the earthquake were higher than that available by a given percentage, the same for all storeys, there would be a uniform vertical distribution of damage. The addition of the strength of masonry to that of the R/C structural system exceeds the required strength in the upper storeys but not in the lower ones, and that is where damage occurs (Figure 11.29).

In the case of a flexible ground floor (when it is used for stores, in which case the masonry walls are replaced by glass panels, or when the *pilotis* system – open ground storey – is used) the damage in this floor is much more severe and usually occurs only there (Figure 11.30). This subject will be discussed in detail in section 11.2.5.

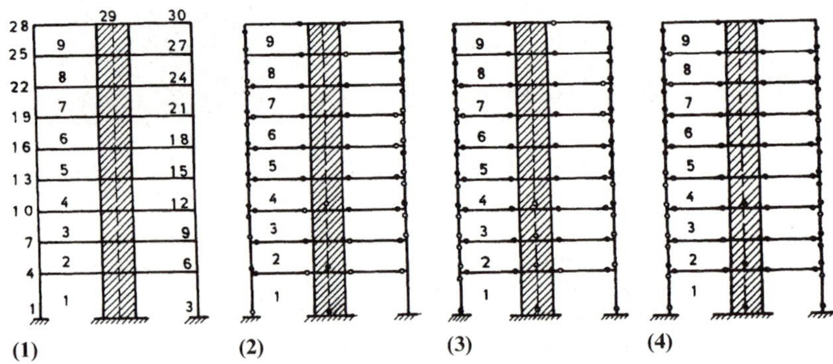

Figure 11.28 Distribution of plastic hinges in a dual nine-storey building, subjected to an El Centro (1940) excitation scaled by 0.75, 1.0 and 1.5, respectively. ● = Yielding of reinforcement on both ends (top and bottom); ○ = Yielding of reinforcement on one end (top and bottom).

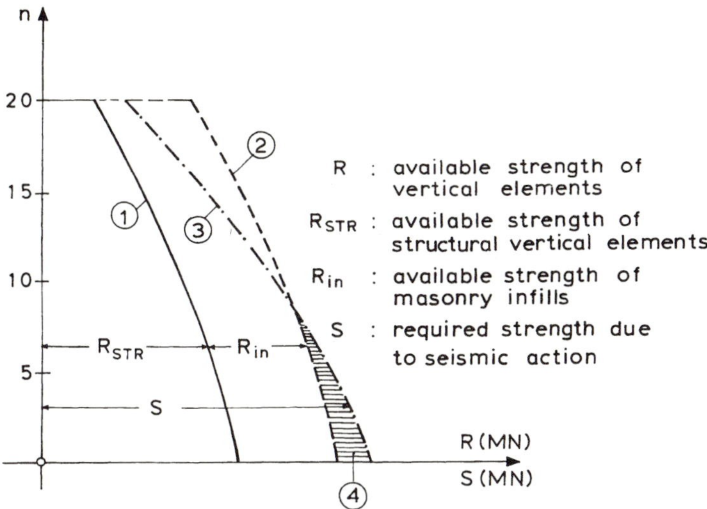

Figure 11.29 Explanation of the higher vulnerability of the lower storeys of a building: (1) shear-strength curve of the vertical structural elements of a building; (2) additional shear strength due to masonry, constant for all storeys; (3) required strength (seismic action) per storey, almost proportional to R_{STR} (e.g. $1.5R_{STR}$); (4) shaded area – storeys with damage.

Figure 11.30 Column damage at the open ground storey (*pilotis* system) Kalamata, Greece, 1986.

As far as the horizontal distribution is concerned, most of the damage occurs in areas which are far from the stiffness centre of the building and mainly on the perimeter of the building.

The recent Mexico City earthquake (1985) was the first during which a large percentage of collapses and large-scale damage occurred in the upper floors of buildings (38%) (Figure 11.31). This can be fully explained considering the fact that the damaged buildings were very tall (with more than 12–15 storeys), with very flexible structural systems (flat plates), wherein higher modes generate large seismic effects in the upper floors. Furthermore, the same types of damage occur in masts or tower-like structures (e.g. bell-towers, minarets, chimneys).

11.1.9 Stiffness degradation

Strong earthquakes induce inelastic deformations to buildings accompanied in most cases by visual damage. As a result, the buildings sustain a stiffness degradation which is displayed by an increase in their fundamental period. Site investigations on actual buildings before and after an earthquake (Ogawa and Abe, 1980) have shown that there is a strong correlation between the extent of damage and the value of the ratio of the fundamental period of the building after the earthquake to that before the event. Shear cracks can be found by visual observation of buildings in which the value of the fundamental periods ratio is more than 1.3, that is

$$\frac{T_2}{T_1} \geqslant 1.30$$

where T_1 is the fundamental period before the earthquake and T_2 the fundamental period after the earthquake.

Figure 11.31 Large-scale damage in the upper floors of a flexible building, Mexico City, 1985.

Taking into account the fact that

$$\frac{T_2}{T_1} = \left(\frac{K_1}{K_2}\right)^{1/2}$$

where K_1 is the equivalent stiffness of the building before the earthquake and K_2 the equivalent stiffness of the building after the earthquake, it is concluded that where visual damage is observed in the building the stiffness degradation is of the order of 40%

$$\frac{K_2}{K_1} \geqq \simeq 0.60$$

11.2 FACTORS AFFECTING THE DEGREE OF DAMAGE TO BUILDINGS

11.2.1 Introduction

In the subsections that follow, there will be an attempt to systematically present the most important factors that seem to affect the degree of damage to buildings. The presence of one of these factors does not necessarily mean that it is the only reason for the damage. Most of the time there is more than one adverse factor in a structure, therefore the determination of how much each of the factors contributes to the damage is not feasible, even after a systematic statistical analysis of the damage.

11.2.2 Divergence between design and response spectrum

The first and most important reason for damage to structures is the inaccurate estimation of the characteristics of the expected earthquake excitation during the design of the structure. As is well known, in every city there are still in use buildings up to 100 years old – besides the monuments which have survived millenniums. Thus, a strong earthquake acts upon a variety of structures, some of which were built with no structural design at all, some were designed only for gravity loads, some were designed for static earthquake horizontal loads with no consideration of ductility requirements and some of them, the most recent ones, were designed according to the current knowledge of seismic design. It is therefore reasonable to except that this spectrum of structures, the great majority of which does not conform to design specifications based on the current state of knowledge, will experience some damage. Furthermore, it is not impossible for damage to occur in engineered structures, designed according to the current codes, mainly for the following three reasons:

1. Even though there has been significant progress in the design of earthquake-resistant structures during the last few decades, this does not mean that the seismic protection problem is solved. Every generation believes that it has taken important steps towards the advancement of an area of interest; future developments, though, usually come to prove this belief wrong. Thus, it is not impossible that structures which are built today according to the most recent

advancement in earthquake engineering do not conform with the specifications in effect in a few years, time.

2. Contemporary structures are designed in such a way that when the design earthquake occurs, they should respond inelastically, that is, they are expected to sustain a controllable degree of damage.
3. Quite often, the design spectrum, scaled according to the behaviour coefficient and the safety factor, does not correlate with the actual response spectrum. The Mexico City (1985) earthquake and the Kalamata (1986) earthquake can be cited as examples (Figures 11.32 and 11.33). Therefore, before rushing into conclusions about the contribution of each damage factor, one should first carefully study the response spectrum of the earthquake which has caused the damage, in relation to the provisions of the code according to which most of the structures in the area were built (Rosenblueth and Meli, 1985; Anagnostopoulos *et al.*, 1986; Penelis *et al.*, 1986).

Independent of the characteristics of the exciting force, a number of the structure's own characteristics which will be discussed subsequently, are factors which contribute to the vulnerability of the structures.

11.2.3 Brittle columns

In the section which refers to the typology of damage in structural elements, the types of column failure have been discussed in detail. The vast majority of fail-

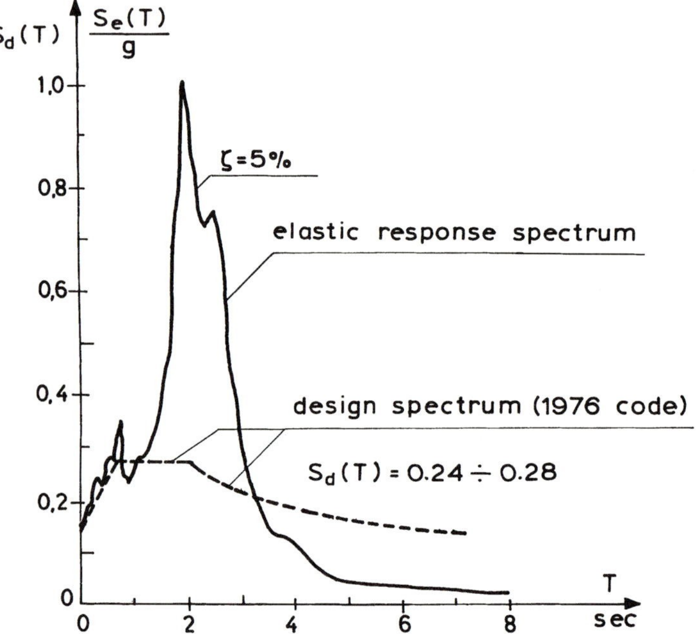

Figure 11.32 The 1985 Mexico City earthquake: comparison of the response spectrum with the design spectrum.

ures in buildings with R/C structural systems is due to column failure, caused by bending and axial load, or by shear under strong axial compression. There are clear indications that in buildings designed during recent decades, due to high axial loading level, most of the time the column reinforcement does not reach the yield point. Column failures must be attributed to the degradation of the mechanical properties of the material due to the high number of loading cycles in the inelastic region (low cycle fatigue). Quite often, the main reason for failure is the large spacing of ties at the critical regions of the column.

11.2.4 Asymmetric arrangement of stiffness elements on the floor plan

It is well known that the core of the staircase is the basic stiffness element in the structural system of a building; therefore, according to what has been discussed in the chapter on the analysis of structures, its central or eccentric position should be of major importance for the behaviour of the building during an earthquake (Figure 11.34). However, a statistical evaluation of the damage which the 1978 earthquake caused to the buildings of Thessaloniki (Penelis *et al.*, 1987, 1988) shows that this factor affected only by 6% the mean value of the percentage of the damaged structures (Figure 11.35). This phenomenon must be mainly attributed to the fact that the infills drastically change the stiffness distribution in the building, and as a result, the effect of eccentricities due to asymmetric arrangement of R/C stiffness elements is reduced. In contrast, asymmetric arrangement of masonry causes markedly inferior behaviour. This asymmetric arrangement of

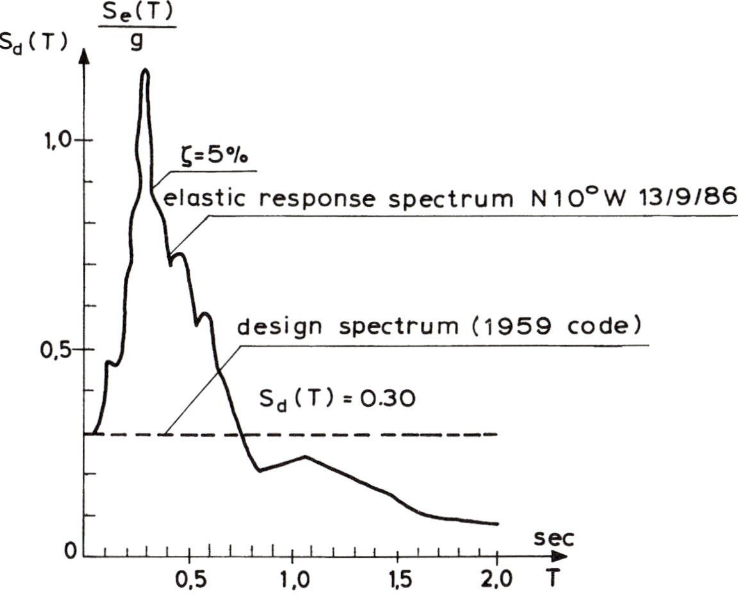

Figure 11.33 The Kalamata, Greece, earthquake of 13 September 1986: comparison of the elastic response spectrum with the design spectrum.

Figure 11.34 Torsional collapse of a building in Mexico City, 1985.

masonry is usually observed in the ground floors of structures located at the corners of building blocks, where the two sides of the perimeter are not filled with masonry because of their usage as shops (Figure 11.36). This is one of the topics which will be discussed next.

11.2.5 Flexible ground floor

The sudden reduction of stiffness at a certain level of the building, typically at one of the bottom floors, results in a concentration of stresses in the structural elements of the flexible floor, which causes damage to those elements. An illustrative example of this fact is the distribution of shear forces which are developed on the R/C staircase core of a 20-storey building with no masonry on the four lower floors (Figure 11.37) (Dowrick, 1987). The shear force distribution has been determined using dynamic inelastic analysis. This example makes obvious the fact that, for the floors with masonry, the shear force acting on the staircase core is much smaller if the infills are taken into account for the analysis, while for the four lower floors without masonry, the resulting shear force is much higher. For this reason the aseismic codes in effect today require an increase in the design

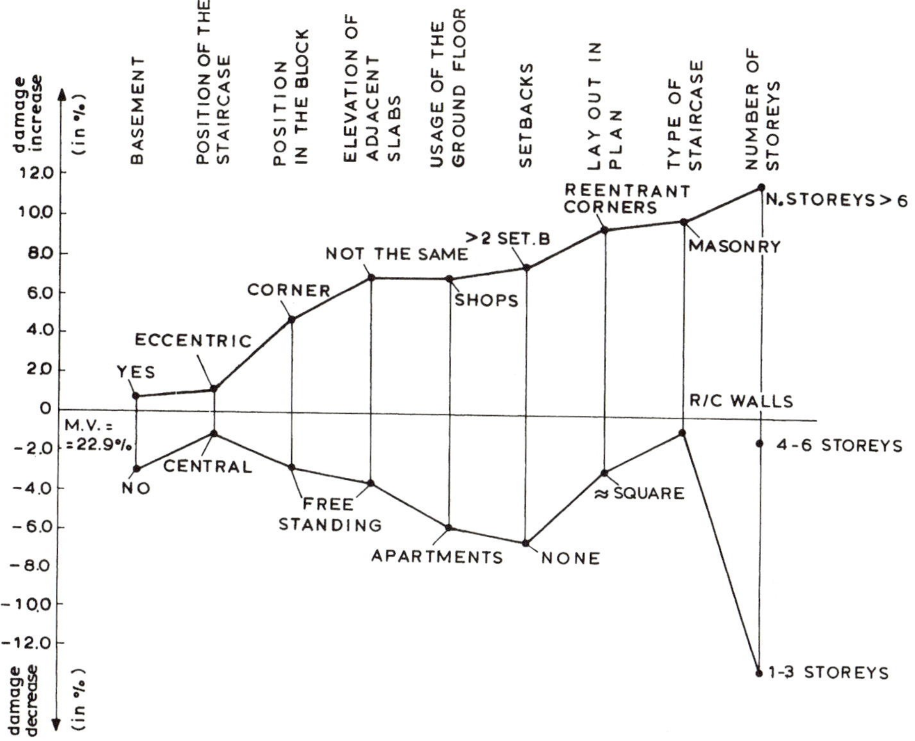

Figure 11.35 Statistical evaluation of the damage in Thessaloniki, Greece, 1978.

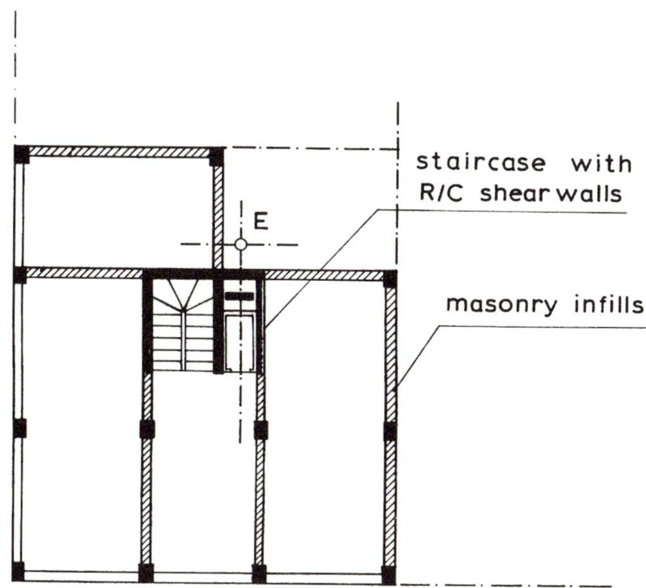

Figure 11.36 Stiffness centre location of a corner structure, when masonry infills are taken into account (approximately).

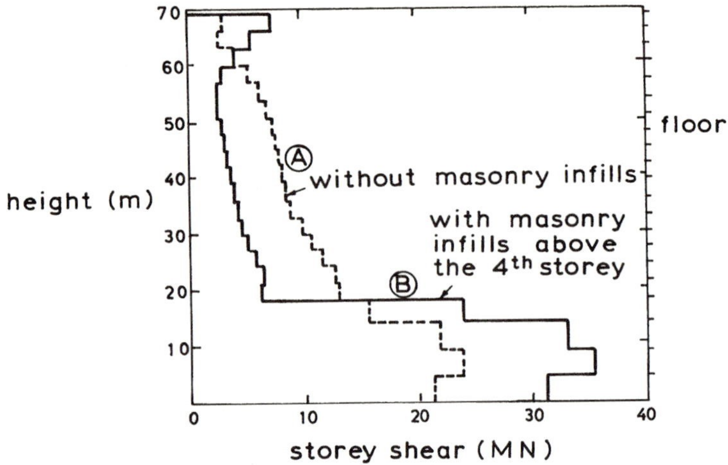

Figure 11.37 Shear force diagram of the staircase core of a 20-storey building, which is indicative of the effect of masonry infills on the storeys above the fourth floor.

shear for the storey with reduced stiffness compared to that of upper floors (sections 4.4.5 and 5.9.4). They also require a high degree of confinement through closely spaced ties or in the form of spirals, throughout the height of the columns of the weak floor, in order to increase their ductility.

The most common case of a flexible floor is the open ground floor (*pilotis* system) or the ground floor used as a commercial area. In such a case, while the upper floors have high stiffness due to the presence of masonry infills, the ground floor has a drastically reduced stiffness because the vertical structural members contribute almost exclusively to it. In these buildings almost all the damage occurs in the vertical structural elements of the ground floor, while the rest of the building remains almost unaffected (Figure 11.38). In contrast in buildings with masonry infills in the ground floor, the damage spreads throughout the structure with usually decreasing intensity from the ground to the upper floors. The 1978 Thessaloniki earthquake caused damage to only 16.4% of the buildings with masonry infills in the ground floor, while damaged buildings having a *pilotis* system or shops in the ground floor, reached 29.8% of the total number of this type of building. During the Mexico City earthquake of 1985, 8% of the buildings which collapsed or exhibited severe damage had a flexible ground floor.

11.2.6 Short columns

It has already been mentioned (see also section 8.4.3(d)) that short columns can experience an explosive shear failure which can lead to a spectacular collapse of the building. This phenomenon, however, appears to be rarer than the failure of regular columns.

Figure 11.38 Collapse of a building with flexible ground floor, Bucharest, Romania, 1977.

11.2.7 Shape of the floor plan

Buildings with a square-shaped floor plan have the best behaviour during an earthquake, while buildings with divided shapes such as +, I, X or with re-entrant corners have the worst. During the 1978 Thessaloniki earthquake, among the damaged buildings (with damage either in the R/C system or in the masonry walls), 19.5% had a square-shaped floor plan, while 32.5% had non-convex shapes of floor plan. This is the reason why the EC8 and the CEB/MC-SD/85 do not allow simplified methods of analysis for earthquake actions when the building under consideration does not have a regularly shaped floor plan.

11.2.8 Shape of the building in elevation

Buildings with upper storeys in the form of setbacks have markedly inferior behaviour than buildings with regular form in elevation. During the 1978 Thessaloniki earthquake, among the total number of damaged buildings, 15.9% were buildings regular in elevation, while 29.9% were buildings with three or more successive setbacks (see also Figure 11.35).

11.2.9 Slabs supported by columns without beams (flat plate systems)

This type of failure has been discussed in the section on slab failures. In the seismically active southern Europe this type of structure is rather recent and therefore there are no statistical data regarding this failure mode. Experimental data, however, as well as statistical data from the 1985 Mexico City earthquake, suggest that this is a very vulnerable type of structure. Indeed, in Mexico City, where this kind of structural system is widely used, 41% of collapse or serious damage occurred in buildings of this type. Structures with such slabs are very flexible and with low ductility. Most of the failures in Mexico City occurred in columns. However, in more than 10% of the cases the columns punched through the slab, under the action of a combination of both vertical and horizontal seismic loads. Besides, the small thickness of the slab does not allow the development of the required bond stresses around the longitudinal reinforcement of the columns and therefore, after a few loading cycles, the joint fails due to the failure of bond mechanisms along the thickness of the slabs (Figure 11.39). For this reason EC8 and CEB/MC-SD/85 do not cover this type of structural system if it is not combined with other seismic-resistant systems (i.e. shear walls, ductile frames).

11.2.10 Damage from previous earthquakes

Buildings which had sustained damage during a previous earthquake and were repaired, usually exhibit the same type of damage in the next earthquake, to a larger extent. This phenomenon was observed in Bucharest during the 1977 earthquake, where many of the buildings repaired after the 1940 earthquake collapsed (Figure 11.40) and also more recently in Mexico City. The main reason for this phenomenon is that the repair works were not carried out carefully enough, and also the fact that 40–50 years ago the repair technology of earthquake damage was in its early stages of development.

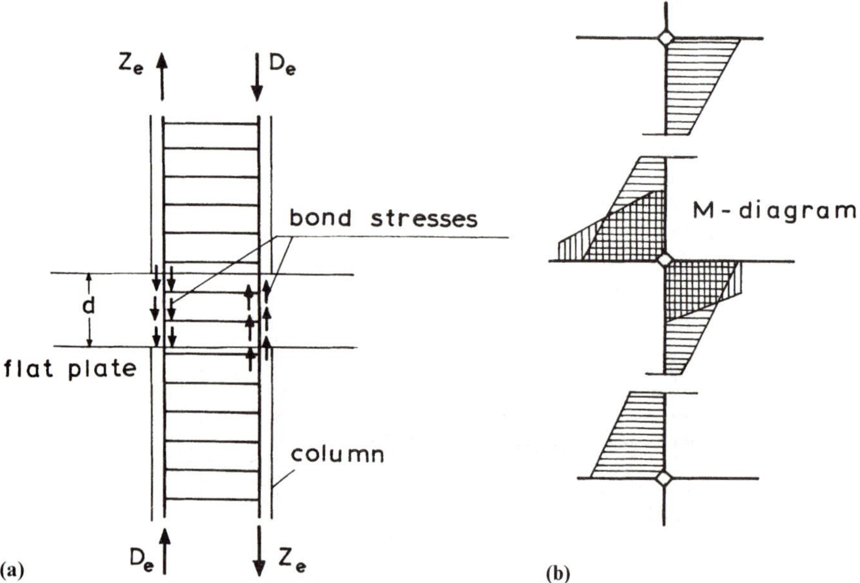

Figure 11.39 Degradation of bond between concrete and column reinforcement within the small thickness h_f, of the slab: (a) sketch of the flow of forces; (b) corresponding bending moment diagram.

11.2.11 Pure frame systems

Frame systems, not inferior to dual systems as far as strength is concerned but superior with regard to available ductility, have lower stiffness than dual systems. As a result, during a seismic excitation, large inter-storey drifts develop, which cause extensive damage in the infill system. Given that the repair of this damage is a very costly procedure, it is understood why the 'frame system' constitutes a source of vulnerability for the building. Therefore, although with frame systems ductile behaviour can be achieved more easily than with dual systems, and that led to extensive use of the frame systems in the 1960s, since about 1975 the idea that shear-wall systems are more suitable for R/C buildings is becoming more widely accepted. Comparative studies of building behaviour during the earthquakes of Managua (1972), San Fernando (1971), Caracas (1967) and Skopje (1963) support the above opinion (Fintel, 1974). From the Thessaloniki earthquake (1978), among the damaged buildings (damage in the structural system or in the infills), 22% were buildings with shear walls and 32.9% were buildings without shear walls. Finally, one of the main observations of the research team from the University of Thessaloniki which visited and studied the earthquake damage

Figure 11.40 Collapse of a building in Bucharest during the 1977 earthquake. The building had been repaired after the 1940 earthquake.

in Kalamata, Greece, in 1986, was the large extent of masonry damage in most multi-storey buildings with a frame structural system (Penelis *et al.*, 1986).

11.2.12 Number of storeys

The number of storeys is directly related to the fundamental period T of the structure, as discussed in previous chapters. Therefore, at least theoretically, the vulnerability of the structure to an earthquake depends on the ordinate of the acceleration spectrum of that specific earthquake corresponding to T, in relation to that of the design response spectrum of the building. In this context the vulnerability of the building should be independent of the number of storeys. However, the existing statistical data from earthquakes show that the vulnerability increases with the height of the buildings. As typical examples, one can cite Bucharest (1977), where damage and collapse were located mainly in buildings with more than six storeys, Mexico City (Table 11.1) (Rosenblueth and Meli, 1985) and Thessaloniki (1978), where among the damaged buildings, 10.9% were low-rise buildings (one to three storeys), while 34.9% were high-rise buildings (over six storeys). In Bucharest and Mexico City the concentration of damage in high-rise buildings is compatible with the response spectrum of the corresponding earthquake, since large acceleration values correspond to high natural periods, corresponding to high-rise buildings. In the case of Thessaloniki, however, this correlation is not possible. The authors' opinion on this issue is that the infill system drastically increases the stiffness as well as the strength of the structure. Given the fact that the masonry layout is more or less the same in every floor and independent of the height of the building, the percentage of additional stiffness and strength due to the presence of masonry is higher in low-rise buildings than in high-rise buildings. As a result, the behaviour of low-rise buildings is better (Figure 11.41).

11.2.13 The type of foundations

The form of the foundation of the structure has two types of effects on the extent of damage in the building: direct and indirect.

Table 11.1 Percentages of collapses and serious damage in Mexico City (1985)

Number of storeys	Percentages of collapses and serious damage (% of every building category)
1–2	0.9
3–5	1.3
6–8	8.4
9–12	13.6
>12	10.5
Total	1.4

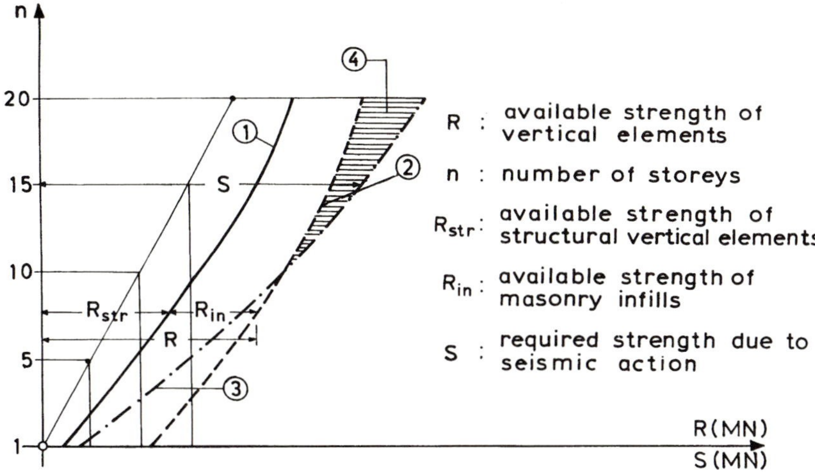

Figure 11.41 Explanation of the seismic vulnerability of high-rise buildings: (a) shear-strength curve of vertical structural elements; (2) additional strength due to masonry, constant for all floors; (3) required shear strength (seismic action), almost proportional to R_{STR}; (4) shaded area – buildings with damage.

Direct effects are manifested in the following ways:

- failure of the foundation members (e.g. fracture of a foundation beam);
- fracture of the foundation soil;
- soil liquefaction;
- differential settlement of the ground;
- partial or general landslide of the foundation soil.

The most usual form of the effects listed above is the differential settlement of the ground, especially in soft soils.

The **indirect effects** are related to the out-of-phase motion of the bases of the individual columns, when their footings are not interconnected (Figure 4.11), or when the existing connection is too flexible. These differential movements in both the horizontal and the vertical directions subject the structure to additional strains. As a result buildings with isolated footings suffer more under the seismic action than others. A characteristic example of the above is the fact that in the 1986 Kalamata, Greece, earthquake the damage to the buildings in the sea-front avenue, where the foundations had large stiffnesses, was limited. The same did not happen in Thessaloniki, Greece, however: although the foundations of the buildings in the coastal zone were either mat foundations or grids of foundation beams, the percentage of damage was high. The interference of other factors irrelevant to the type of the foundation, such as the amplification factor of the seismic excitation which is applied for soft soil deposits and the shift of the prevailing period of the exciting force towards higher values when such soils are present, does not allow a clear statistical evaluation of how the presence of a good foundation indirectly affects the vulnerability of structures.

11.2.14 The location of adjacent structures in the block

The location of adjacent buildings in the block has a great effect on the behaviour of the structure during an earthquake. More specifically, corner buildings are much more sensitive to earthquakes than free-standing ones. In the 1977 Bucharest earthquake, 35 out of the 37 buildings which collapsed were located at the corner of the block (Figure 11.38). In the 1985 Mexico City earthquake, 42% of the buildings which suffered serious damage or collapsed were corner structures (Rosenblueth and Meli, 1985) (Table 11.2). In the 1978 Thessaloniki earthquake (Figure 11.42) among the damaged buildings (with damage in the R/C structural systems or in the infills) only 19.9% were free-standing buildings, while 27.9% were corner buildings. As sources of the higher vulnerability of corner structures the following can be mentioned:

- The asymmetric distribution of stiffness elements on the floor plan due to the lack of masonry on two sides of the perimeter of the ground floor, where the space is usually occupied by stores.
- The transfer of kinetic energy to the end structures of the block (corner structures) during the seismic interaction of adjacent buildings (Figure 11.43). This transfer of energy causes a substantial increase of the inertia forces acting on the end structures (Anagnostopoulos, 1988; Athanassiadou *et al.*, 1994).

11.2.15 Slab levels of adjacent structures

The impulse loading that a building receives from an adjacent structure during an earthquake is a major source of damage. The problem becomes even more serious when the floor slab levels of adjacent buildings do not coincide. In that case, the slabs of one structure during the oscillation pound on the columns of the adjacent building and this results in fracture of the columns. In the 1978 Thessaloniki earthquake, among the damaged structures, the percentage of free-standing buildings or buildings which had the same floor slab levels as adjacent ones was only 19%, while the percentage of buildings with slab levels different from those of adjacent ones was 30.5%, In the 1985 Mexico City earthquake in more than 40% of the structures which collapsed or suffered serious damage, pounding of adjacent structures took place (Table 11.2).

Table 11.2 Causes of failure in Mexico City (1985)

Reason for failure	Percentage
Asymmetric stiffness	15
Cornor structure	42
Weak ground floor	8
Short columns	3
Exceeded vertical design load	9
Pre-existing ground settlement	2
Pounding of adjacent structures	15
Damage from previous earthquakes	5
Punching failure of flat slabs	4
Failure of upper floors	38
Failure of lower floors	40

Figure 11.42 Collapse of an eight-storey corner building in Thessaloniki, Greece, 1978.

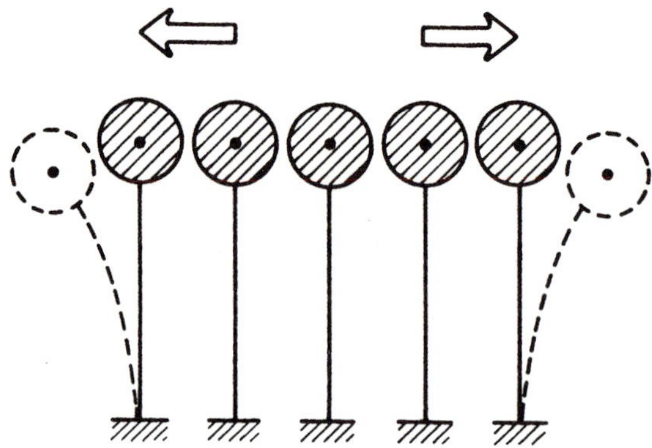

Figure 11.43 Transfer of energy at the end SDOF oscillators in a series of adjacent systems.

11.3 REFERENCES

Anagnostopoulos, S. (1988) Pounding of buildings in series during an earthquake. *Earthq. Engng Struct. Dyn.*, **16**, 443–56.

Anagostopoulos, S. *et al.* (1986) The Kalamata earthquake of September 1986. *Report of the Institute of Eng. Seism. and Earthquake Engineering*, Thessaloniki, Greece (in Greek).

Athanassiadou, C.J., Penelis, G.G. and Kappos, A.J. (1994) Seismic response of adjacent buildings with similar or different dynamic characteristics. *Earthq. Spectra*, **10**(2), 293–317.

Dowrick, D.J. (1987) *Earthquake Resistant Design for Engineers and Architects.*, John Wiley & Sons, New York.

Fintel, M. (1974) Ductile shear walls in earthquake-resistant multistory buildings. *Journal of the ACI*, **71**(6), 296–304.

Kappos, A. and Penelis, G. (1987) Investigation of the inelastic behaviour of existing R/C buildings in Greece. *Technica Chronica*, TCG, **7**(3), 53–86.

Ogawa, J. and Abe, Y. (1980) Structural damage and stiffness degradation of buildings caused by severe earthquakes. *Proceed. of the 7th World Conf. on Earthq. Engng*, Istanbul, Turkey, **VII**, pp. 527–34.

Penelis, G. *et al.* (1986) *Technical Report on the September 13 and 15, 1986 Kalamata Earthquake*, Lab of R/C, Dept of Struct. Engng, University of Thessaloniki, Thessaloniki (in Greek).

Penelis, G. *et al.* (1987) A statistical evaluation of damage caused by the 1978 earthquake to the buildings of Thessaloniki. *Technical Report of the Laboratory of R/C*, Aristotle University of Thessaloniki (in Greek).

Penelis, G. *et al.* (1988) A statistical evaluation of damage to buildings in the Thessaloniki, Greece, earthquake of June 20, 1978. *Proceed. of the 9th World Conf. on Earthq. Engng*, Tokyo–Kyoto, Japan, **VII**, pp. 187–92.

Rosenblueth, E. and Meli, R. (1985) The 1985 earthquake: causes and effects in Mexico City. *Concr. Intern. (ACI)*, **8**(5), 23–36.

Stylianidis, K. and Sariyiannis, D. (1992) Design criterion to avoid column shear failure in infilled frames due to seismic action. *Proceed. of the 1st Conf. on Engng Seismol. and Earthq. Engng*, Athens, Greece, **2**, pp. 220–30 (in Greek).

Tegos, I.A. (1979) *A Systematic Analysis of Damage in R/C Buildings*, a Seminar for the Repair of Buildings with Earthquake Damage, Ministry of Public Works, Thessaloniki (in Greek).

Tiedemann, H. (1980) Statistical evaluation of the importance of non-structural damage to buildings. *Proceed. of the 7th World Conf. on Earthq, Engng*, Istanbul, Turkey, **VI**, pp. 617–24.

12

Emergency post-earthquake damage inspection and evaluation

12.1 INTRODUCTION

The aim of this chapter is to present a reliable procedure which should be followed for the emergency inspection of structures after an earthquake, so that all the structures of the affected area will be inspected in a uniform way. The data which are collected by such inspections assist the state in achieving the following goals (UNIDO/UNDP, 1985):

- to reduce the number of deaths and injuries of occupants of damaged buildings, which might be caused by collapses due to subsequent aftershocks;
- to help the people of the affected area gradually to return to a normal way of life, which presupposes a reliable characterization of the dangerous buildings and full knowledge of the extent of hazardous structures;
- to develop a database for a uniform assessment of risk in economic, social, political and other terms;
- to record and classify earthquake damage, so that the repair of the damaged buildings will follow a priority order;
- to improve earthquake-resistant design, based on the recorded damage.

It should be obvious from this brief introduction that after a destructive earthquake, two levels of building inspection follow. The first level of inspection is performed by the state, during which there is a recording of damage, characterization of hazardous (for usage) structures, demolition of buildings close to collapse and support for those that need it. It is an operation that needs to be carried out quickly, in order gradually to restore the normal way of life in the affected area. During the second level of inspection, which will be discussed in detail in Chapter 13, the residual strength of every affected structure is estimated and the degree of intervention is decided. This is a laborious procedure which starts as soon as the first level of inspection is completed and the frequency and intensity of the aftershocks are diminished. It is also a procedure which is directly related to the decision about the repair and/or strengthening of the structure (EC8/93 Part 1.4–CEN, 1993).

12.2 INSPECTIONS AND DAMAGE ASSESSMENT

12.2.1 Introductory remarks

The purpose of this section is the brief treatment of the problems related to the evaluation of damage to structures after an earthquake. A strong earthquake, like every other hazard, puts on trial not only the citizens, but also the state. The authorities have to face chaotic situations due to lack of information, delays in locating the affected areas, possible interruption of communications and multiple requests for assistance and for inspections of damaged buildings. The first tragic hours, even days sometimes, the affected area stands almost alone, and it is during this initial period when good construction, good communications, good decision-making and good planning pay off in terms of lives and properties saved. The credibility of the state in its citizens' eyes depends on what the state can or cannot do during this early period. These are the views of government officials in charge of disaster relief, who have experienced the situations and problems which arise after a strong earthquake (Office of Emergency Preparedness, 1972; Penelis, 1984).

The foregoing remarks aim at depicting the environment in which the structural engineer is called upon to do an assessment; this should be the prevailing element for the design of the entire operation. Indeed, since damage evaluation sometimes refers to thousands of structures, which have to be assessed in a short period of time in order for the affected area to return to a normal way of life, a special procedure has to be followed, completely different from that used for the assessment of the structural resistance of a single structure.

In the subsections which follow there is a systematic reference to organizational matters of an operation for damage assessment, and also to matters of inspection, as well as problems that the structural engineers face during evaluation of individual cases. This presentation is based on experience from the organization and the realization of such an operation in the Thessaloniki (Greece) metropolitan area in 1978, for which one of the co-authors was the person in charge. It is also based on publications of several international organizations and national committees on the same subject (ATC, 1978; Yugoslav National Report, 1982; Greek National Report, 1982).

12.2.2 Purpose of the inspections

The main purpose of the inspection procedure for the structures after a destructive earthquake, is to minimize the probability of death or injury for the occupants. The danger of this happening in buildings where damage occurred in the main earthquake is serious enough, because it is possible for some of them to partly or completely collapse due to repeated aftershocks, as happened with the Alkyonides, Greece, earthquake (1981), and also with the Kalamata, Greece, earthquake (1986).

There are also other reasons for the inspections, beyond the above-mentioned, which are also of great importance. Thus, after the classification of those damaged buildings which are hazardous to use, life gradually returns to normal, given the fact that the rest of the buildings – as soon as the first psychological reactions begin to disappear – gradually return to their normal usage. Also, based on the first damage evaluation, an approximate idea of the magnitude of the disaster in economic terms may be obtained. These data are needed by all levels of administration almost immediately for them to be able to start a proper planning of aid.

Finally, the statistical data from such an operation are very useful, not only for short-term decision-making regarding temporary housing but also in the long run on matters of evaluation of the construction procedures followed in the past and the factors that could affect them positively.

In closing, it should be stressed that in case the affected area includes a large city, the whole procedure should be managed very carefully, as this becomes a large-scale operation with very high cost and organizing requirements, accompanied by long-term implications. Indeed, the classification of a building as 'damaged' by the state leads to a long-term depreciation of its market value, even though strengthening interventions after the earthquake could have made it stronger than other non-damaged buildings.

12.2.3 Damage assessment

(a) Introduction

It has already been mentioned that the main consideration during damage assessment in structures after an earthquake, is to minimize the possibilities of accidents to occupants, caused by partial or total post-earthquake collapses. Therefore, the problem that the structural engineer has to face in every case is to estimate the residual strength, ductility and stiffness of the structure, and decide whether or not they are sufficient to allow use of the building at an acceptable level of risk. It is understood that this evaluation, based on the existing evidence, is probably the most difficult problem for the structural engineer, much more difficult than the design of a new building. Extensive site inspections are required, first of all for damage survey, and then to check the geometry of the structural system, the quality of the construction materials, the placement of reinforcement at critical structural elements compared to the original drawings, the vertical loads of the structure and the quality of the foundation soil. Subsequently, extensive calculations are needed, using the information collected from site observations, in order to determine the residual strength, stiffness and ductility of the structure. Finally, it has to be estimated whether or not the seismic excitation under consideration did not exceed the level of seismic hazard adopted by the code for the particular zone, as expressed by the design response spectrum. Such a procedure is time-consuming and requires the full involvement of specialized personnel, with a variety of technical means at their disposal. It is therefore obvious that such a procedure cannot be applied during the phase of emergency classification of structures as usable or not, since it is not feasible due to time constraints. It is worth pointing out that such a procedure has not been attempted anywhere in the world up to now.

Thus, engineers come face to face with the building struck by the earthquake, without being able to use for the quantitative evaluation of the structure the scientific tools that they possess, which are the *in situ* measurements, the tests and the analysis. They are compelled by the circumstances to restrict themselves to qualitative evaluations and make decisions which are based only on visual observation of damage, using of course their knowledge and experience on the subject. This last statment is extremely important because it shows that these evalutions are very subjective. Notwithstanding its weaknesses, according to international practice, damage assessment is based on the above procedure (UNDP/UNIDO, 1985; ATC 3-06,1978; Yugoslav National Report, 1982; Greek National Report, 1982).

(b) General Principles of Damage Assessment

Although damage assessment and decision-making regarding the degree of usability of a structure are very subjective, there are some general principles that the structural engineers must keep in mind when they have to make a decision concerning a building damaged by an earthquake; these are as follows:

1. They must have clearly in mind that their judgement must be limited to the evaluation of the risk of partial or total collapse in case of an aftershock, which is an earthquake of smaller magnitude than the main earthquake that comes from the same tectonic fault, therefore it has similar characteristics as the first one (similar acceleration spectrum with smaller maxima). Thus, if the building does not exhibit damage in the structural system from the main shock, it means that if has not exceeded the elastic range, therefore the probability of damage and, even more, of collapse caused by aftershocks, is statistically insignificant.
2. The risk of partial or total collapse of a structure damaged by the main earthquake comes from failure of vertical structural members (columns, structural walls, load-bearing masonry) under the action of vertical loads in combination with the horizontal seismic loads from the aftershocks, which are expected to be smaller than the loads from the main event. The engineer should keep in mind that if damage appears in the structural system of a building, it means that the elastic range has been exceeded and therefore the resistance of the structure to seismic loading has been reduced by the main seismic event.
3. According to the above, engineers who perform the inspection first of all clearly have to find out the layout of the structural system of the building, at least in the cases when it is damaged. If necessary, in the absence of drawings, they should use hammer and chisel in order to determine the location of the vertical structural elements of the building. No reliable damage assessment is possible without a clear understanding of the structural system of the building.
4. In order to estimate the residual strength, stiffness and ductility of a structure, the engineer has to trace out the damage in the structural system as well as in the infill panels. Particularly hazardous is damage in the vertical elements, especially at the ground floor. Crushing of concrete at the top or the bottom of a column accompanied by buckling of the longitudinal reinforcement, X-shaped cracks in shear walls with significant axial loading, X-shaped cracks in short columns, are some of the types of damage which should seriously worry the engineer who performs the inspection. Extensive X-shaped cracks in the infills accompanied by permanent deviation of the structure from the vertical, are also alarming indications. In contrast, cracks in horizontal structural elements, caused by either flexure or shear, are not particularly alarming. The same holds for limited spalling of vertical elements or flexural cracks. However, it has to be stressed that familiarity of the engineer with the various types of earthquake damage is very useful, if not indispensable, for damage evaluation.

5. Before concluding the final evaluation, the engineer should pay particular attention to the following:

   (a) The configuration of the structure. Buildings with symmetric or approximately symmetric floor plans have a better seismic behaviour than asymmetric ones.
   (b) The location of the vertical stiffness members on the floor plan. Symmetrically placed stiffness members drastically reduce the consequences of eccentric loading.
   (c) The existence of a flexible storey. The open ground floor (*pilotis* system) or ground floor occupied by stores make the building extremely vulnerable to seismic actions.
   (d) The quality of the construction material. *In situ* tests with hammer and chisel, or better with special equipment, constitute a very good relative indicator for the engineer, who can draw very useful conclusions after a few repetitions.
   (e) The location of the structure in the block. It should not be forgotten that the vast majority of total collapses worldwide occurred in corner structures.

6. The conclusion of the foregoing discussion is that the engineer can make one of the following decisions:

   (a) to allow use of the building without any restriction, provided it does not exhibit any visible damage in the structural system (classification with, say, **green** colour);
   (b) to classify it as temporarily unusable and limit the access to it (shifting the responsibility to the occupant) because of limited damage, until it is repaired (classification with, say, **yellow** colour);
   (c) to classify it as out of use because of extensive damage, until based on a detailed study it is decided either to repair or demolish it (classification with say, **red** colour).

It is obvious that buildings which fall into the first category can easily be distinguished from buildings of the other two categories, whose capacity to resist an earthquake has been reduced because of damage. In contrast, the distinction between the last two categories is not always easy; therefore, in case of doubt, the more conservative decision should be made.

12.3 ORGANIZATIONAL SCHEME FOR INSPECTIONS

12.3.1 Introduction

The earthquake damage assessment – a job greatly affected by subjective judgement – requires hundreds and sometimes even thousands of engineers, each with a different level of knowledge and experience. Therefore, prior to the earthquake, an appropriate organizational scheme should be developed, which would ensure the following:

- immediate start of the inspections just after the earthquake and the main aftershocks, and the completion of the entire operation in a short time;

- damage assessment that is uniform and as objective as possible, so that mistakes would be statistically minimized;
- detection of possible mistakes made during the first inspection and damage evaluation;
- timely notification of the authorities about buildings which need support or demolition.

The basic features of such a scheme are discussed in the sections that follow.

12.3.2 Usability classification–inspection forms

For an evaluation of the degree of damage that is uniform and as objective as possible, the inspection must be performed by teams consisting of at least two engineers, so that an exchange of ideas can take place. Also, it is necessary to prepare ahead of time special forms for damage description, strength evaluation and usability classification of structures, which should be based on the following principles:

- Easy completion with the data of the structure and the degree of damage, based on visual inspection as already mentioned.
- Assignment of the damage degree into a few clearly defined categories.
- Assessment of the usability of the building on the basis of clearly defined categories. International practice (UNIDO/UNDP, 1985; ATC 3-06, 1978; FEMA, 1986; Yugoslav National Report, 1982) has adopted the three levels of usability mentioned in the preceding section, each one of which is characterized by the colour of the sticker pasted on the buildings, the already mentioned 'green', 'yellow' and 'red' stickers, which were also used after the 1978 Thessaloniki, Greece, earthquake.
- Codification of data for future statistical processing.

It has to be metioned that most of the countries today have standard inspection forms (ATC 3-06, 1978; Yugoslav National Report, 1982). In the framework of the UNIDO/UNDP-funded programme 'Earthquake Resistant Structures in the Balkan Region' such an inspection form was developed (Anagnostopoulos, 1984) as a result of cöoperation of the Balkan countries (Figure 12.1).

In Greece, between 1978 and 1981, simple inspection forms were used, which did not fulfil the requirements given above. An attempt was made to use a form similar to that used by UNIDO (Constantinea and Zisiadis, 1984) in Kalamata (1986). However, the attempt failed for the following reasons:

1. In most of the buildings some apartments were locked up, hence detailed recording of damage was not possible.
2. The engineers were not trained to complete the forms in a uniform way.
3. The requirement for quick completion of the inspections did not allow a systematic and detailed description of the situation.

The authors' opinion is that the forms for the preliminary (emergency) inspection must be as simple as possible, while for the statistical evaluation of the consequences of the earthquake, the engineer responsible for the repair should submit a detailed form, along with the repair project, after performing a second detailed inspection aiming at designing the repair of the building.

EARTHQUAKE DAMAGE INSPECTION FORM

SKETCH OF THE BUILDING

PLAN CROSS SECTION

ADDRESS:
OWNER:

1. TOWN (NAME) 1 □□□□□ 5
2. BUILDING IDENTIFICATION:
   2.1 SECTION NUMBER OF CONSIDERED TOWN AREA OR SETTLEMENT: 6 □□ 7
   2.2 WORKING TEAM NUMBER: 8 □□ 9
   2.3 NUMBER OF THE BUILDING: 10 □□□ 12
3. ORIENTATION OF THE BUILDING PRINCIPAL AXIS (X):
   1.NS, 2.EW, 3. N45E, 4. N45W 13 □
4. POSITION OF BUILDING IN THE BLOCK:
   1. CORNER, 2. MIDDLE, 3. FREE 14 □
5. NUMBER OF STOREYS
   5.1 STOREYS 15 □□ 16
   5.2 APPENDAGES 17 □
   5.3 MEZZANINES 18 □
   5.4 BASEMENTS 19 □
6. GROSS AREA OF THE BUILDING (m^2): 20 □□□ 22
7. USAGE (SEE DESCRIPTION ON BACK PAGE):
   7.1 BUILDING 23 □□□ 25
   7.2 GROUND FLOOR: 26 □□□ 28
8. NUMBER OF APARTMENTS: 29 □□ 30
9. CONSTRUCTION PERIOD (TO BE DEFINED BY EACH COUNTRY):
   1. 2. 3. 31 □

---

10. TYPE OF STRUCTURE (SEE DESCRIPTION ON BACK PAGE): 33 □□□ 35
11. FLOORS:
    1.R.C., 2. STEEL, 3. WOOD, 4. OTHER 36 □
12. ROOF: 1.R.C., 2. STEEL, 3., WOOD, 4. OTHER 37 □
13. ROOF COVERING: 1. TILES, 2. METAL SHEETS, 3. LIGHTWEIGHT ASBESTOS CEMENT, 4. ASPHALT PAPER, 5. HEAVY INSULATION, 6. LIGHT INSULATION, 7. OTHER 38 □
14. QUALITY OF WORKMANSHIP:
    1. GOOD, 2. AVERAGE, 3. POOR 39 □
15. TYPE OF LOAD CARRYING SYSTEM (SEE DESCRIPTION ON BACK PAGE):
    1. BEARING WALLS, 2. FRAMES, 3. FRAMES WITH INFILL WALLS, 6. MIXED, 7. OTHER (SPECIFY) 40 □
16. FIRST FLOOR-STIFFNESS RELATIVE TO OTHERS:
    1. LARGER, 2. ABOUT EQUAL, 3. SMALLER 41 □
17. REPAIRS FROM PREVIOUS EARTHQUAKES:
    1. NO, 2. YES, 3. UNKNOWN 42 □

---

18. DEGREE OF DAMAGE
    STRUCTURAL ELEMENTS (SEE DESCRIPTION ON BACK PAGE):
    1. NONE, 2. SLIGHT, 3. MODERATE, 4. HEAVY, 5. SEVERE
    18.1. BEARING WALLS: 44 □
    18.2. COLUMNS: 45 □
    18.3. BEAMS: 46 □
    18.4. FRAME JOINTS: 47 □
    18.5. SHEAR WALLS: 48 □
    18.6. STAIRS: 49 □
    18.7. FLOORS: 50 □
    18.8. ROOF: 51 □

    NONSTRUCTURAL ELEMENTS AND INSTALLATIONS (SEE DESCRIPTION IN THE MANUAL):
    1. NONE, 2. SLIGHT, 3, MODERATE, 4. HEAVY, 5. SEVERE
    18.11. INTERIOR WALLS: 52 □
    18.12. PARTITIONS: 53 □
    18.13. EXTERIOR WALLS (FACADE): 54 □
    18.14. ELECTRICAL INSTALLATIONS: 55 □
    18.15. PLUMBING: 56 □

---

19. DAMAGE OF ENTIRE BUILDING:
    1.1 NONE, 1.2 SLIGHT, 2.1 MODERATE, 2.2 HEAVY 3.1 SEVERE, 3.2 TOTAL 57 □□ 58
20. INDIRECT DAMAGE (FIRE, SLAMMING, ETC.)
    1. NO, 2.YES 59 □
21. OBSERVED SOIL INSTABILITIES AND GEOLOGICAL PROBLEMS
    1. NONE, 2. SLIGHT SETTLEMENTS, 3. INTENSIVE SETTLEMENTS, 4. LIQUEFACTION, 5. LANDSLIDE, 6. ROCKFALLS, 7. FAULTING, 8. OTHER (SPECIFY): 60 □

---

22. USABILITY CLASSIFICATION AND POSTING:
    POSTED: 1. GREEN, 2. YELLOW, 3. RED
    NOT POSTED: 4 TO BE POSTED GREEN AFTER REMOVAL OF LOCAL HAZARD, 5. SOIL AND GEOLOGICAL PROBLEMS, REINSPECTION REQUIRED, 6. UNABLE TO CLASSIFY, REINSPECTION NECESSARY, 7. BUILDING INACCESSIBLE 62 □

GREEN 1 ORIGINAL SEISMIC CAPACITY HAS NOT BEEN DECREASED	UNLIMITED USAGE
YELLOW 2 ORIGINAL SEISMIC CAPACITY HAS BEEN DECREASED	TEMPORARILY UNUSABLE LIMITED ENTRY
RED 3 BUILDING DANGEROUS AS SUBJECT TO SUDDEN COLLAPSE	ENTRY PROHIBITED

MAIN REASONS FOR YOUR CLASSIFICATION AND POSTING:

23. RECOMMENDATIONS FOR EMERGENCY MEASURES:
    1. NONE, 2. REMOVE LOCAL HAZARD, 3. PROTECT BUILDING FROM FAILURE, 4. URGENT DEMOLITION 64 □
24. ADDITIONAL DATA (PHOT/SKETCHES AND COMMENTS):
    1. NONE, 2. PHOTOS ONLY 3. SKETCH AND COMM. ONLY 4. PHOTOS AND SKETCH AND COMM. 65 □
25. ESTIMATED PRESENT VALUE OF BUILDING (MILLIONS OF ....) 66 □□□ 68
26. ESTIMATED LOSS (% OF ESTIMATED VALUE) 69 □□□ 71
27. HUMAN LOSSES (DEATHS AND INJURIES
    (1) NO; (2) POSSIBLY; (3) YES (4) IF INFORMATION AVAILABLE PLEASE INDICATE: 72 □
    OPTIONAL NO. OF DEATHS 73 □□ 74
    NO. OF INJURIES 75 □□ 76
28. DATE OF INSPECTION: MONTH/DAY 77 □□□□ 80
    NAMES OF INSPECTION ENGINEERS: SIGNATURES
    1.
    2.
    3.

Fig. 12.1 Inspection form of the first-level committee proposed by UNIDO/UNDP, Project RER/79/015/Volume 4.

12.3.3 Inspection levels

In order to locate possible mistakes made during the first (emergency) inspection, a second-degree inspection must be carried out, performed by two-member committees of engineers with high qualifications and experience. In both ATC 3-06 (1978) and the final manual of UNIDO/UNDP (1985) such an inspection is suggested for buildings which were classified as 'red', as well as for buildings whose owners or occupants hindered the (evaluation of the) first inspection. This procedure was also followed in Thessaloniki in 1978.

The timely notification of the authorities about buildings which need immediate support or demolition is one of the first priorities of the inspection committees, during both the first and the second-degree inspection.

Finally, it has to be emphasized that for the inspection of buildings which are of vital importance for post-earthquake life in the affected area, committees of highly qualified engineers must be formed to inspect these buildings first. In Figure 12.2 the chart of the successive inspection levels of structures in the Thessaloniki earthquake is shown, the application of which did not exhibit any particular problems.

12.4 ACTION PLAN

12.4.1 Introduction

It has become clear, from what has been said so far, that such an operation constitutes a large-scale mobilization, from which the state expects a lot. Obviously, there must be a detailed action plan developed before an earthquake occurs, in order to have an immediate and effective mobilization of such a mechanism. This action plan must be a part of the nation-wide general emergency plan for earthquakes. Some basic features of such a plan are presented in the following sections.

12.4.2 State agency responsible for the operation

The existence of a public agency with well-defined responsibilities for the operation in every prefecture (or country) from the pre-earthquake period, is extremely important for the immediate start and implementation of the programme of inspecting the structures after an earthquake. The selection of the public agency must be done in advance, so that its staff will have the time to review the action procedure and prepare all the orders and decisions that must be implemented, as well as the materials and equipment needed for the operation. Obviously, this responsibility must be assigned to an already existing technical service with a good organization and the necessary personnel and means to become the core of the entire operation. Given the fact that in large urban areas there are several such services, belonging to several ministries, the coordination and the duties of each one of them must be clearly defined from the pre-earthquake period in the action plan.

12.4.3 Inspection personnel

In principle, the inspection personnel should consist of civil engineers, organized in two-member teams so that they can exchange ideas during the inspections. The

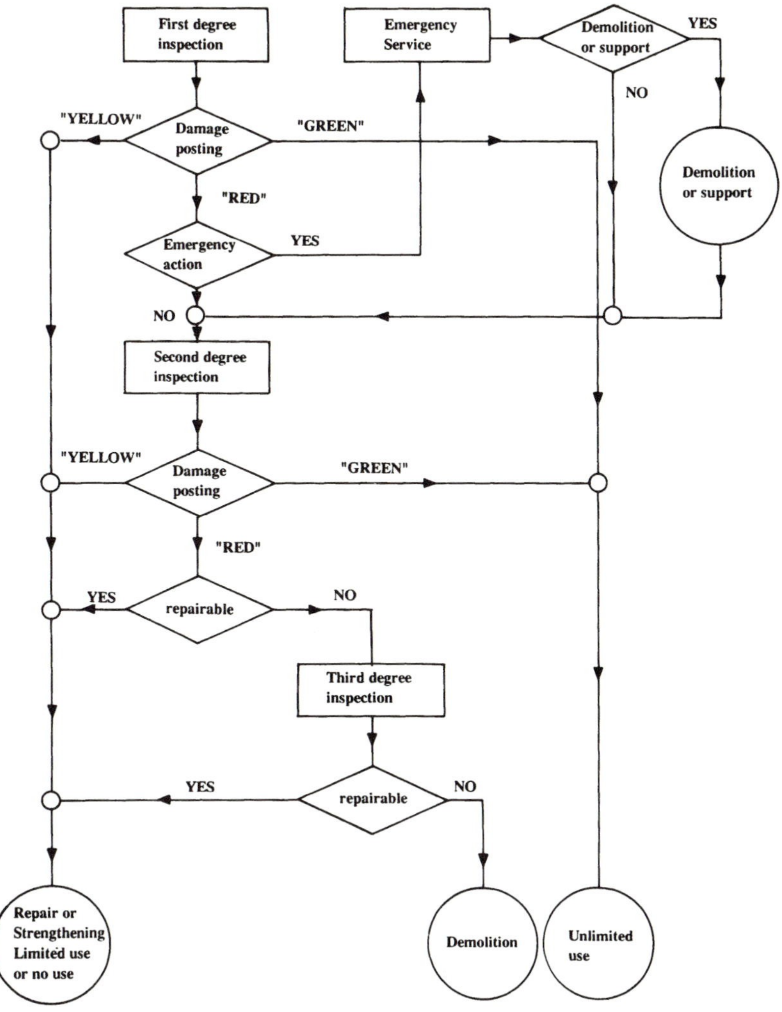

Fig. 12.2 Chart of inspection procedure in Thessaloniki in 1978.

high degree of subjectivity in the decision about the usability of structures does not permit the use of scientists of other specialities at this phase. It has already been mentioned that a large number of engineers will be needed to carry out the inspection. For example, in the case of Thessaloniki in 1978, about 1000 engineers worked for about 45 days for the first and second-degree inspection of about 250 000 apartments located in about 60 000 buildings, while in California in 1971 after the San Fernando earthquake, about 250 engineers were involved for a month with the inspection of 12 000 buildings. These two examples give an idea of the scale of mobilization that would probably be needed. Given the fact that the required number of personnel cannot be covered by the state agency in charge of the operation, there must be a recruiting plan for additional personnel in every area of the country, which must provide successive levels of reserves. These personnel, with a proper annual training, will know in advance the meeting points in case of an emergency. Recourse to civil conscription should not be ruled out in case that after a call the number of the personnel gathered is not adequate. In Thessaloniki the entire operation started on a voluntary basis, but no more than 150–200 engineers from the private sector volunteered. Therefore, after the first week the government was forced to conscript all civil engineers in the region.

12.4.4 Pre-earthquake organizing procedures

The public agency in charge of the inspections should also have the responsibility of organizing the whole operation during the pre-earthquake period. This organizing procedure should cover the following;

- Preparation of inspection personnel lists, providing for several levels of reserves, on the basis of experience and qualifications.
- Determination of local mobilization centres (e.g. police stations).
- Distribution of inspection personnel depending on qualifications to first-level committees, second-level committees, committees for checking buildings of special usage, etc.
- Listing of buildings of vital importance for the survival of the area after an earthquake, such as hospitals, schools, communications buildings, police stations and public buildings, which must be checked first, and formation of inspection teams from highly qualified engineers for inspection of these buildings immediately after the earthquake.
- Acquisition of all materials and equipment needed for the inspections, such as inspection forms, note books, measuring tapes, hammers, chisels, helmets, flashlights, batteries.
- Acquisition of means of communication, given the fact that it is not unusual for telecommunications to be interrupted after an earthquake. It is interesting at this point to mention the report on the San Fernando earthquake, which estimates that if the telecommunication system was down the effectiveness of the entire operation would have been limited to 10%. This is similar to the authors' assessments for the recent earthquakes in Greece.
- Organizing short courses for the inspection personnel on the subject of damage assessment.

Figure 12.3 shows the organizing diagram which was used in the case of Thessaloniki, while Figure 12.4 shows the organizing diagram suggested by the ATC 3–06 for the USA at the state level (ATC 3–06, 1978).

Finally, it is the authors' opinion that the inspection personnel should not render themselves personally liable for any damage that may accrue to persons or property as a result of any act or any omission in carrying out their duties. This is because these inspection procedures are conducted under the pressure of time in emergency situations, and therefore, since engineers are not able to use their main scientific tools, i.e. measurements, test and analysis, they should not be held liable for any mistakes that could occur during the first emergency evaluation. A similar reasoning is followed in ATC 3–06 (1978).

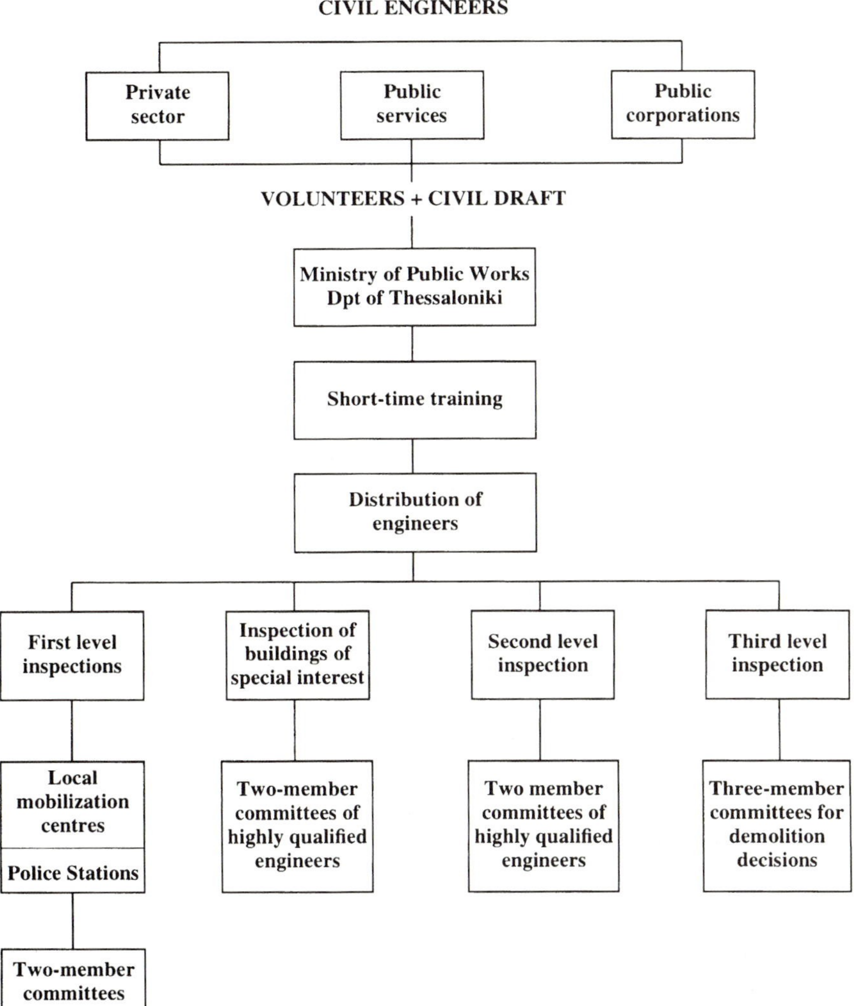

Fig. 12.3 Organization chart for the inspection service in Thessaloniki after the earthquake of 20 June 1978.

12.4.5 Post-earthquake organizing procedures

Right after the occurrence of an earthquake, in order for the inspection mechanism to be set in motion a series of actions are necessary, the most important of which are the following:

1. Determination of the affected areas in order to estimate the scale of the operation. Characteristic of the difficulty of this task is the fact that for the San Fernando, USA, earthquake of 1971, it took 12 hours to locate the boundaries

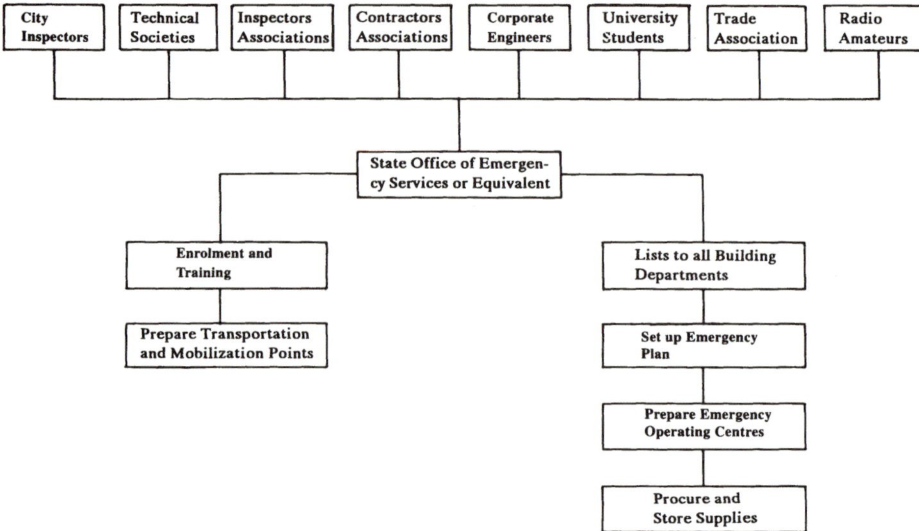

Fig. 12.4 Organization chart for the pre-earthquake preparation period (ATC 3-06 proposal, USA).

of the affected area. In the 1978 Thessaloniki, Greece, earthquake it took two days to locate the boundaries of the affected area.

2. Establishment of the operation headquarters. This centre should be housed if possible in the same building as the disaster relief services, so that easy communication and coordination will be possible. In Thessaloniki the whole operation was housed in the Ministry of Northern Greece, while in Kalamata (1986) the operation headquarters was housed together with the disaster relief headquarters in the telecommunications building. The operation headquarters should perform the following duties:

   (a) coordinate its actions with the activities of the other services for disaster relief;
   (b) mobilize the personnel of inspections and local centres;
   (c) establish top priority for inspections of critical facilities such as hospitals, police and fire stations, etc.;
   (d) set up central and local offices;
   (e) establish and maintain communications;
   (f) provide transportation means for the inspection personnel;
   (g) distribute equipment and supplies to the local mobilization centres;
   (h) provide food and housing for the personnel;
   (i) process inspection reports;
   (j) provide statistical data where necessary;
   (k) coordinate the local mobilization centres.

3. Establishment of local mobilization centres. In Greece, for example, the most appropriate places have proved to be the police stations, where both wireless communications and a lot of data about the area around the station are available.

4. Distribution of the inspection personnel to the local centres with written orders (authorizations).

12.5 FINAL REMARKS

The main conclusion of this chapter is that emergency damage assessment after an earthquake is a completely different procedure from that followed for the evaluation of residual strength of a single structure. This is because in the case of an earthquake very often thousands of structures must be evaluated in a short period of time, so that the affected area can return to normal life.

Therefore, given the fact that hundreds and sometimes even thousands of engineers must be engaged in damage evaluation, each with a different level of knowledge and experience, the main task is the creation of the appropriate organizing scheme, before the earthquake, which will ensure the following:

- damage evaluation that is uniform and as objective as possible, so that statistically mistakes will be minimized.
- quick detection of any serious mistakes in evaluation from the first inspection;
- timely notification about buildings which need immediate support or demolition.

It is understood that such operations which require a high degree of staff organization and are directly associated with public security, should be undertaken by the state.

For the uniform and objective evaluation of the degree of hazard, it is necessary to prepare in advance special forms for damage description, residual strength evaluation and usability classification of structures, which must be based on the following principles:

- easy completion with the data of the structure and the damage level, based on visual inspection;
- assignment of the damage degree into a few clearly defined categories;
- assessment of the usability of the building, on the basis of clearly defined categories;
- organization of data for future statistical processing.

For the effective completion of the engineers' mission, there is a need to organize short training courses, accompanied by visual aids (slides, videotapes, etc.) on the expected damage, classified in categories, and on how to complete the inspection forms.

For the detection of serious errors and for inspection of special buildings, there is a need for a second-level check by experienced engineers.

For timely notification for emergency action on buildings needing supports or demolitions, development of an appropriate communication mechanism is required.

12.6 REFERENCES

Anagnostopoulos, S.A. (1984) Listing of earthquake damage of buildings and building characterization in emergency situations. *Proceed. of the conf. on Earthq. and Structures*, EPPO, Athens, **I**, pp. 550–61 (in Greek).

ATC 3-06 (1978) *Tentative Provisions for the Development of Seismic Regulations for Buildings*, US Goverment Printing Office, Washington D.C.

CEN Techn. Comm. 250/SC8 (1993) Eurocode 8: *Earthquake-resistant Design of Structures– Part 1.4 (Draft): Repair and Strengthening*, CEN, Berlin.

Constantinea, A.M. and Zisiadis, N.A. (1984) A proposal for inspection procedure after an earthquake. *Proceed. of the Conf. on Earthq. and Structures*, EPPO, Athens, **I**, pp. 574–94 (in Greek).

FEMA (Federal Emergency Management Agency) (1986) *NEHRP: Recommended Provisions for the Development of Seismic Regulations for New Buildings*, Part III (App.: Existing Buildings), Washington, DC.

Greek National Report (1982) *On Damage Evaluation and Assessment of Earthquake Resistance of Existing Buildings*, Pr. Rep. 13/05 UNIDO/UNDP, Thessaloniki.

Office of Emergency Preparedness (1972) *Disaster Preparedness*, Report to the Congress, Vols 1–3, Washington, DC.

Penelis, G.G. (1984) Problems after an earthquake – inspections, expert appraisals. *Proceed. of EPPO the Conf. on Earthq. and Structures*, OASP, Athens, **I**, pp. 521–37 (in Greek).

UNIDO/UNDP PR. RER/79/015 (1985) *Post-Earthquake Damage Evaluation and Strength Assessment of Buildings under Seismic Conditions*, Vol. 4, Chapter 2, Vienna.

Yugoslav National Report (1982) *On damage Evaluation and Assessment of Earthquake Resistance of Existing Buildings*. Pr. Rep. 13/05 UNIDO/UNDP, Skopje.

13

Design of repair and strengthening

13.1 GENERAL

The purpose of this chapter is to present the general principles which should govern the structural rehabilitation of a building damaged by an earthquake. It is well known that every intervention constitutes a special case, with its own peculiarities for every building. Despite that, some general principles must be followed in order to obtain a scientifically sound result (EC8/Part 1.4/Draft – CEN, 1993).

Losses due to earthquakes are usually significant, but they can become even more significant because of ignorance or lack of willingness to implement an integrated rehabilitation scheme. Thus, hasty or erroneous design and/or bad execution of the repairs may lead to increased damage and even loss of human life in future earthquakes. Therefore, there is a need to provide the engineer with all the necessary knowledge for the rational design of repair or strengthening, which includes the proper assessment of structural characteristics (including dynamic properties), knowledge of modern techniques and materials for repair and strengthening, design methodology and the appropriate procedure for the execution of the structural rehabilitation. This kind of information is exactly what this chapter intends to provide.

Two tendencies usually appear after every catastrophic earthquake. The first one is known worldwide as 'cover-up' of damage that the structural system and the infills exhibit. It is a quick solution of low cost, but extremely dangerous. A commonly expressed opinion is that 'since the building withstood the earthquake only with some cracks, there is no problem', which leads into ignoring the extent and the pattern of damage, as well as its influence on a possible future seismic loading, which could be fatal for the structurally degraded building and its inhabitants.

The second tendency, which is just the opposite of the first one, comes from people who are terrified by the destruction and suggest a large-scale strengthening of all buildings, damaged or not, disregarding the economic consequences of such a decision.

It is obvious that the best solution can be reached if the problem is approached in a cool and scientifically rational manner, at the same time using international practice on the subject.

The engineer must approach the rehabilitation problem of a damaged building in four successive steps:

1. Examination of the damaged building;

2. Development of alternative rehabilitation schemes;
3. Examination of the technical feasibility of implementing each alternative, as well as its cost estimate and selection of the optimum solution;
4. Final rehabilitation design.

This chapter deals with the first three steps, while the fourth is dealt with in Chapter 14.

13.2 DEFINITIONS

Before proceeding to the objectives and the principles of structural rehabilitation, it is necessary to give some definitions which will be used subsequently.

Required seismic resistance The required seismic resistance V_B, as already explained in Chapter 5, is expressed quantitatively by the seismic base shear force (equation (5.7))

$$V_B = \gamma_I S_d(T_1) W \tag{13.1}$$

where $S_d(T_1)$ is the ordinate of the design spectrum which corresponds to the fundamental period T_1 (equation (4.3)) of the building, γ_I the importance factor of the building and W, the total weight of the building (section 4.5), is

$$W = \Sigma G_{kj} \text{ `+' } \Sigma \psi_{Ei} Q_{ki}$$

Given the fact that damaged structures have been built in different time-periods,they have a different behaviour factor q which depends to a certain degree on the ductility of the building (section 4.4.6) . Thus, in Greece for example, R/C structures which were built before 1985 may roughly be classified in DC 'L' while those built after 1985 are classified in DC 'M' (Figure 13.1).

The required seismic resistance is a value which is defined by the authorities after every catastrophic earthquake. The decision about this value is a function of economic and social parameters on the one hand and the safety of the buildings on the other. The basic criterion of this decision is the elastic response spectrum of the earthquake that caused the damage, as well as the degree of damage.Thus, in the Thessaloniki, Greece, earthquake (1978) the required seismic resistance was kept at the same value as that specified by the code applicable at the time of the earthquake. In the Bucharest, Romania, earthquake (1977) the same response spectrum was used for the repairs as the one in effect before the earthquake. In the Mexico City earthquake (1985) (Jirsa,1994) the ordinates of the design spectrum were increased by 60% in the area mostly affected. In the Kalamata, Greece, earthquake (1986) the required seismic resistance was kept at the same levels as before the earthquake.

Available seismic resistance The available seismic resistance of a structure is expressed quantitatively by the base shear V_C for which the first of the columns or R/C walls of the ground floor level reaches its ultimate limit strength, provided that the structure is in the elastic range. The available seismic resistance refers to the condition of the building prior to earthquake damage, while for the determination of V_C the concrete quality and the reinforcement of the vertical structural elements must be known. It should be pointed out that this definition does

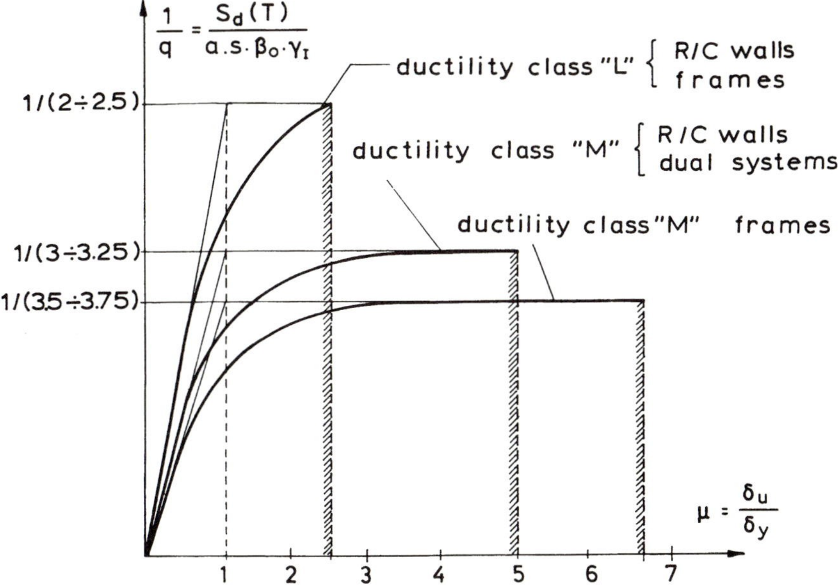

Figure 13.1 Required seismic resistance V_B for ductility classes 'L' and 'M'.

not cover the case of very flexible structures where damage may first appear at the top rather than the bottom of the structure (e.g. the Mexico City case).

Usually V_C is less than V_B for several reasons, such as:

1. The design provisions have not been fully implemented during the construction of the building as far as the quantity of reinforcement, the quality of concrete and detailing are concerned.
2. The structure was designed and built on the basis of design seismic actions different from those specified by the authorities after the earthquake.
3. The usage of the building has been changed, and therefore the gravity loads have been increased.
4. Environmental attacks, such as corrosion of the reinforcing steel bars, have caused a decrease in the load-carrying capacity of members.

Residual seismic resistance The residual seismic resistance V_D of a damaged structure is expressed quantitatively by the base shear under which, assuming an elastic behaviour of the structure, at least one of the (undamaged) columns or walls of the ground floor reaches its ultimate strength. For the estimation of V_D the decrease in the stiffness of the damaged structural elements is taken into account. If the structure exhibits damage due to the earthquake, V_D is always less than V_C (Figure 13.2)(Anagnostopoulos, 1986).

Loss of seismic resistance The difference

$$V_C - V_D \tag{13.2}$$

is defined as loss of seismic resistance.

Repair The term 'repair' means that the damaged structural or non-structural members again reach the minimum strength, stiffness and ductility they ought to have before the earthquake. This means that 'repair' is limited only to the damaged elements and in this sense 'repair' must be considered as a local intervention. V_D is increased with the repair at least up to the value of V_C (Figure 13.2).

Strengthening The term 'strengthening' means the increase of the seismic resistance of the structure with interventions beyond repair, so that the available seismic resistance becomes equal to V_B or to a predefined percentage of it (Figure 13.2). This means that in addition to the local interventions to the damaged elements, interventions of global type will be carried out, so that the overall structural behaviour of the building will be improved.

Strength Index With this term two different quantities can be determined:

1. In the literature (ATC 3-06,1978; UNIDO/UNDP,1985), the strength index is determined as

$$R_C = \frac{V_C}{V_B} \tag{13.3}$$

2. In practice, for the post-earthquake intervention this index is usually replaced (Tassios, 1984;Chronopoulos,1984;EC 8/Part 1.4/Draft – CEN, 1993) by:

$$R_C = \frac{V_D}{V_C} \tag{13.4}$$

As we will see later, R_C and R_D are strongly interrelated; however, R_D can be determined much more easily and reliably than R_C. These two indices constitute a decisive criterion for the level of intervention, that is, whether simple repair or strengthening of the structure is required.

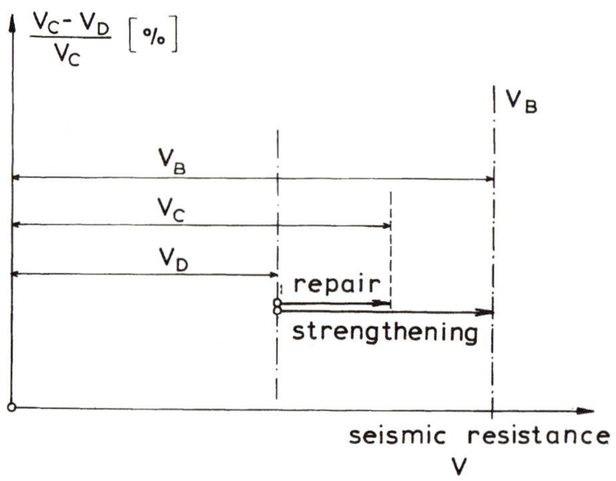

Figure 13.2 Schematic presentation of V_B, V_C, V_D.

13.3 OBJECTIVES AND PRINCIPLES OF INTERVENTION

The main objectives of intervention to an earthquake-damaged building are to protect the structure from collapse in a future strong earthquake, to keep damage at tolerable levels in earthquakes of moderate intensity and to eliminate damage in earthquakes with relatively short return period. In other words, the objectives of an intervention coincide with those set for the design of a new structure.

In this sense the only intervention that can guarantee the above objectives is strengthening, that is, the repair of the damaged structural elements and the increase of the seismic resistance of the structure up to the value of the required seismic resistance, using additional strengthening measures. However, such an approach must be considered as not realistic, because another problem then emerges, that of the periodical strengthening of all buildings in a city (or even a whole country) according to the new codes of the newly acquired seismological data. Indeed, given that the vast majority of buildings in a city, regardless of whether or not they have exhibited damage, have less available seismic resistance than that required, strengthening of just the damaged structures does not appear to be a reasonable decision, if a balanced approach is desired. An operation of intervention to all the buildings, however , is unfeasible due to economic restraints. It should be mentioned that not even countries with strong economies have ever made such an attempt. In California, where an effort was made in this direction, it was limited to school buildings only. A similar effort which started in Thessaloniki, Greece, in 1980 for buildings with very high occupancy (ATC 3-06,1978), never proceeded beyond the design stage. It is the authors' opinion, however, that in high-seismicity areas a cost–benefit analysis would probably prove that it would be much more advantageous to attempt a preventive seismic rehabilitation of buildings with high seismic risk, rather than a post-earthquake intervention, when damage has already occurred.

In view of the foregoing ,the most realistic approach appears to be repair, that is, restoration of the damaged building to pre-earthquake condition. This is based on the notion that if the damaged elements, structural and non-structural, are repaired, the structure more or less regains its pre-earthquake seismic resistance and therefore will behave similarly in a future earthquake with the same characteristics. However, this approach has a weak point: the extent and the seriousness of damage caused by an earthquake constitute the most reliable criterion regarding the difference of the available seismic resistance V_C and the required one V_B. Therefore, in the case of serious damage whose existence proves that the structure came near to collapse, we cannot consider that repairing is enough. The structure should be strengthened to a degree so that in a future earthquake it will behave as the buildings with no damage or with light damage only (Holmes,1994).

Based on the above, the recommended approach to the intervention procedure may be stated as follows:

1. In buildings with light damage, of local nature, intervention should be limited to repair.
2. In buildings with extended or heavy damage, of the global type, intervention should include strengthening of the structure.

However, extended discussion accompanied by divergent opinions on this issue has been found in the literature during the last few years (Anagnostopoulos, Petrovski and Bouwkamp 1989; Freeman,1993).

13.4 CRITERIA FOR REPAIR OR STRENGTHENING

13.4.1 Basic principles

In order to comment on the criteria specified by several agencies and organizations for the choice between repair and strengthening, let us first consider Figure 13.3. The x-axis represents the seismic resistance expressed according to the definitions of section 13.2 as the base shear V. The y-axis represents the loss of seismic resistance, (section 13.2), expressed as the percentage of the available seismic resistance. Curve V_C represents the available seismic resistance and V_D the residual one after the earthquake .

Theoretically speaking, if V_C equals V_B no damage should be expected and V_C should intersect the x-axis at the same point as V_D and V_B. As the ordinate increases, that is, as damage becomes more severe, the distance between curves V_C, V_D and V_B increases, and finally for $(V_C - V_D)/V_C$ =100%, that is, in the case of collapse, curve V_D intersects the y-axis.

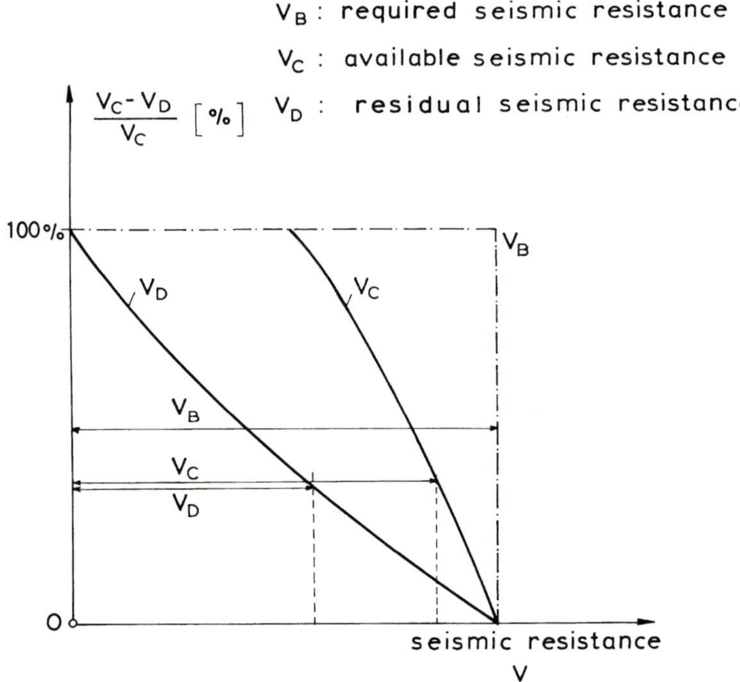

Figure 13.3 Correlation between the indices V_C/V_B and V_D/V_C.

If a reliable value of V_C could be determined analytically, the ratio V_C/V_B would be a safe indication of the expected damage, or in other words, there would be a quantitative correlation between the two ratios V_C/V_B and V_D/V_C and therefore it would not be significant which of the two ratios was to serve as 'index' for the kind of intervention required. However,the reliability of the value of V_C is questionable because of the dynamic character of the problem, the inelastic behaviour of the structure, the materials, the infills and so on. For this reason although the structure does not exhibit serious damage the ratio V_C/V_B often has a very small value or, in contrast, in structures with serious damage this ratio assumes a very large value. Therefore the most reliable index for the choice of intervention level seems to be the ratio V_D/V_C, residual over available seismic resistance, because it expresses undeniable facts, that is, the degree of damage to the structure.

Of course, in the case of pre-earthquake assessment of important buildings, such as schools or hospitals, where there is no damage indication, the available index is the ratio V_C/V_B and this is the one which is usually used. In the manual *Post-Earthquake Damage Evaluation and Strength Assessment of Buildings* of UNIDO/UNDP(1985) the ratio V_C/V_B is used as an index of seismic resistance for both pre-earthquake and post-earthquake assessment.

Greek practice for post-earthquake interventions since 1978 and up to the recent Kalamata (1986) and Pyrgos (1993) earthquakes has used the ratio V_D/V_C as the index of seismic resistance (Tassios, 1984; Chronopoulos, 1984; Penelis, 1979). The values of this index are approximately estimated in practice (EC8/Part 1.4/Draft). Of course in the case of very important structures, highly sophisticated approaches are used (Penelis *et al.*, 1984, 1992).

13.4.2 The UNIDO/UNDP procedure

The procedure which is suggested in the UNIDO/UNDP/PR.RER/79/015 (1985) manual takes into account four factors in order to determine the type of intervention, that is:

- the arrangement of the structural elements
- the strength of the structure
- the flexibility of the structure
- the ductility.

A brief summary of the method will follow.

(a) Arrangement of the structural elements

The structural system of the building, depending on the layout of the structural members, can be classified as:

- *Good*–the arrangement of the structural members is clear, without any irregularities in plan or elevation, and the horizontal forces are carried by clearly defined structural systems of frames or walls in both main directions (Figure 13.4).
- *Acceptable*–the building in general has a good structural system except for some weaknesses, such as large stiffness eccentricity, discontinuity of stiffness in elevation (e.g. open ground floor) and so on (Figure 13.5).

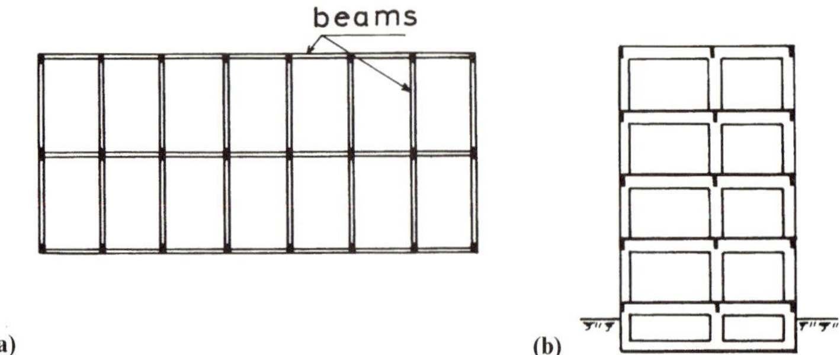

Figure 13.4 Examples of 'good' structural layout.

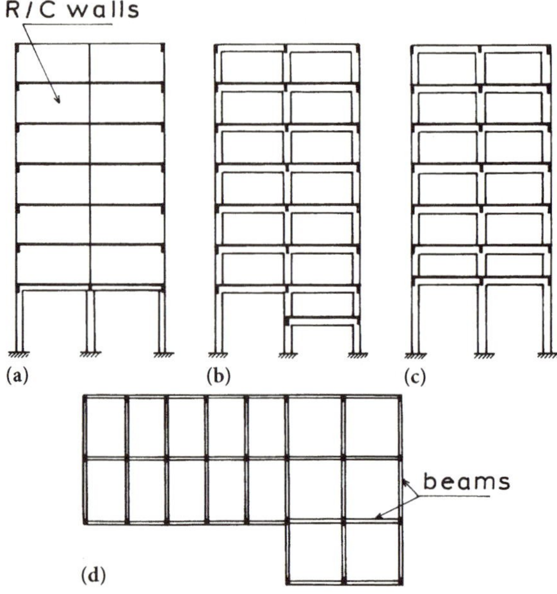

Figure 13.5 Examples of 'acceptable' structural layout.

- *Unclear*–the horizontal forces are carried by systems of structural elements which are not clearly defined (Fig. 13.6).

(b) Strength of the structure

Three levels of the index of seismic resistance $R_C = V_C / V_B$ are adopted for decision-making needs. Of course, the limit values adopted are to be considered as guidelines and not as strict limit conditions.

- $R_C > 0.8$: seismic resistance is considered satisfactory with the probability of somewhat deeper incursions into the inelastic range, without approaching the failure limits. Therefore, repair is adequate.

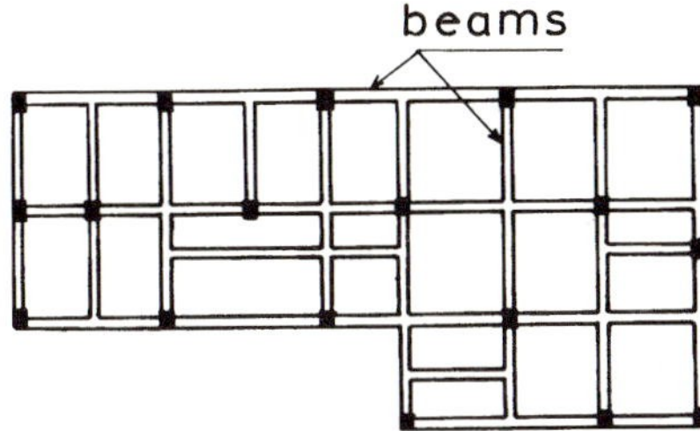

Figure 13.6 Example of 'unclear' structural layout.

- $0.8 > R_C > 0.5$: despite the diminished strength, if enough ductility exists the building can be secured against collapse in a strong earthquake, although this type of structure can reach the failure limits. Therefore, the structure must be strengthened.
- $0.5 > R_C$: The safety of the structure is clearly unsatisfactory.

Combining the estimation for the layout of the structural system with that for the strength index, five categories of the actual structural quality are defined from A to E, given in Table 13.1.

(c) Flexibility of the structure

The flexibility of the structure expressed as the inter-storey drift δ_B/h for loading equal to V_B is compared with two quantities:

- with the deformation limit of the structure itself:

$$\frac{\delta_C}{h} = \frac{0.01 \text{ to } 0.015}{q} \tag{13.5}$$

- with the deformation limit of the infill system:

$$\frac{\delta_D}{h} = \frac{0.007 \text{ to } 0.0075}{q} \tag{13.6}$$

where q is the behaviour factor of the building (section 4.4).

Table 13.1 Five building catagories based on the structural layout and the strength index

Strength index	Structural layout		
	Good	Acceptable	Unclear
$R_C > 0.8$	A	B	C
$0.8 > R_C > 0.5$	B	C	D
$0.5 > R_C$	C	D	E

(d) Ductility of the structure

The ductility requirements specified by modern codes are met only by a very small number of recently built structures. Indeed

- strong columns–weak beams;
- adequate shear reinforcement (ties) so that bending mode of failure be secured;
- confined compression zones with closely spaced hoops or ties

are requirements which were set for the first time in the late 1960s. Thus, in most cases ductility requirements for DC 'M' or 'H' are not met, and the engineer has to choose between large-scale interventions to increase ductility, something which is unrealistic in most cases, or to accept DC 'L', or possibly lower; in this case V_B must be increased, since q corresponding to DC 'L' is lower, and the structure must be strengthened.

(e) Decision for the degree and the type of intervention

As has already been mentioned, four factors are considered for the degree and the type of intervention, that is:

- the layout of the structural system
- the strength of the structure
- the flexibility of the structure
- the ductility.

The type of intervention can either be repair or strengthening.

The type of strengthening, depending on the seriousness of the situation, may be one of following:

- *Type I*: improvement of the ductility and the energy-dissipation mechanism (Figure 13.7) through upgrading of existing structural elements (e.g. using thin jackets on columns with closely spaced ties);
- *Type II*: increase of the strength and stiffness through strengthening of existing structural elements (Figure 13.7) (e.g. increase of the thickness of walls);
- *Type III*: increase of the strength, stiffness and ductility through strengthening of existing structural elements (e.g. increase of the thickness of walls and jackets on columns) (Figure 13.7);
- *Type IV*: increase of the strength, stiffness and ductility through addition of new structural elements (e.g. addition of new walls, jackets on columns, jackets or one-sided strengthening of walls) (Figure 13.7).

The type of intervention in every case can be chosen with the aid of Table 13.2.

The strengthening level (R_{req}) can be determined through probabilistic relationships of seismic risk, which take into account the remaining life of the building in relation with the design life as determined by the code (UNIDO/UNDP, 1985). Thus the design base shear for the strengthening V_{str} can be derived from the relationship

$$V_{str} = \left(\frac{T_{rem}}{T_{des}}\right)^{0.50 \div 0.67} V_B \qquad (13.7)$$

where V_{str} is the base shear for the reanalysis and redesign of the building under strengthening, T_{rem} the remaining lifetime of the building, T_{des} the design lifetime

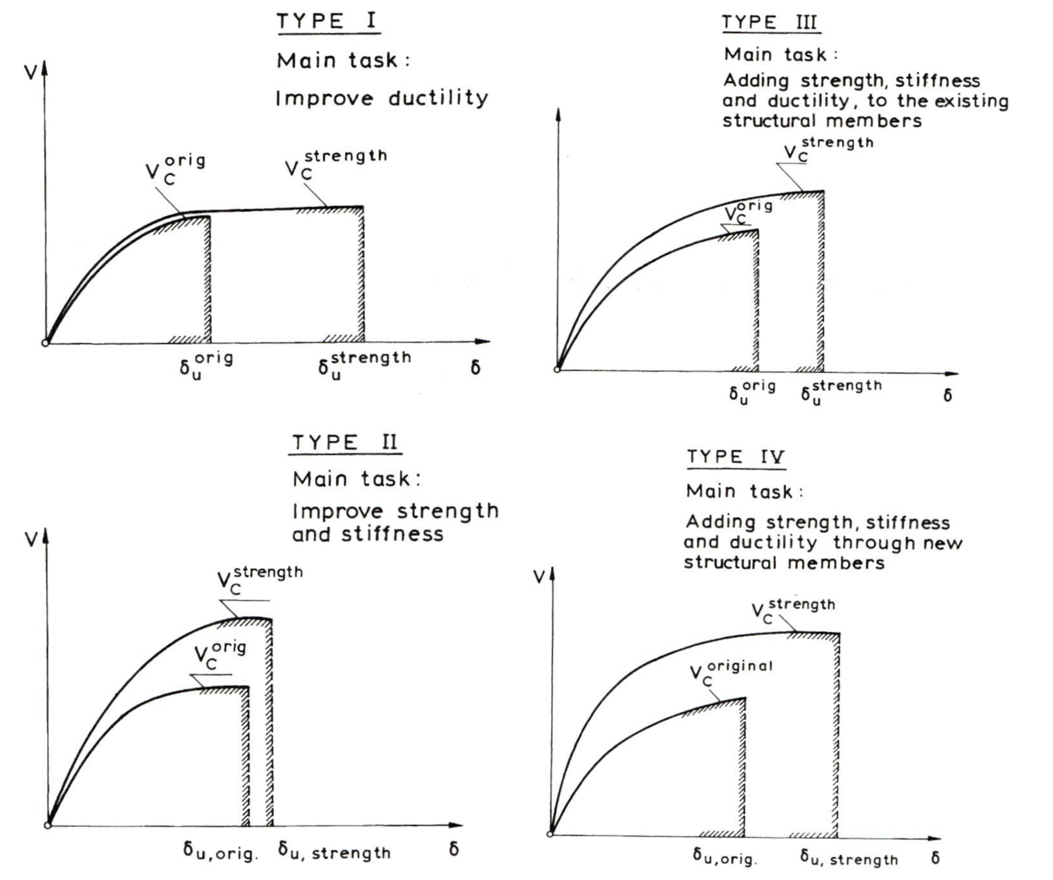

Figure 13.7 Diagrams illustrating the four strengthening types.

of the building and V_B the code-specified base shear for new buildings of the same ductility class.

For some cases in Table 13.2 two or three alternatives appear. In these cases the choice is made based on the cost of intervention.

(f) Critique of the method

The authors' opinion is that the weak point of the method is the use of the ratio V_C/V_B as an index for the choice of degree of intervention, instead of the ratio V_D/V_C. The reasons for that have been explained in the previous subsection.

Another weakness of the method is the attempt to introduce a standard algorithm through Tables 13.1 and 13.2 for the choice of the type of intervention, something which cannot be standarized due to the special problems arising in each case.

While the method gives the impression that the procedure for the choice of the degree and type of intervention is quantitative and therefore objective, it is still highly subjective with regard to the following points:

- modelling of the structural system for the determination of V_C;
- classification of the layout of the structural system;
- choice of the type of strengthening.

Finally, as far as the computational work is concerned, analysis of the system along both main directions is required, in order to determine V_C based on the available sections and reinforcement. This analysis may be avoided with the application of the simplified approximate approaches which are given in UNIDO/UNDP

Table 13.2 Selection of the degree and type of intervention

Structural category	Satisfactory ductility (DC 'M' or 'H')		
	$\delta_R<\delta_D$	$\delta_D<\delta_R<\delta_C$	$\delta_C<\delta_R$
A	R[a]	R or II	III
B	R or II[b]	R or II or IV	III or IV
C	R or II [c]or IV	R or II or IV	IV
D	IV[d]	IV	IV
E	IV[d]	IV	IV
	Unsatisfactory ductility (DC 'L')		
	$\delta_R<\delta_D$	$\delta_D<\delta_R<\delta_C$	$\delta_C<\delta_R$
A	R or I	R or III	III
B	I or II or III	III or IV	III or IV
C	IV	IV	IV
D	IV	IV	IV
E	IV	IV	IV

[a]R = Repair
[b]Category B is strengthened in order to be upgraded to A.
[c]Category C is strengthened in order to be upgraded to A or B, depending on the means available.
[d]Categories D and E are strengthened in order to be upgraded to A or B depending on the means available.

(1985). Then, after the level and the locations of strengthening have been chosen, reanalysis and redesign of the system for the required V_B have to be carried out.

13.5 DESIGN STEPS OF INTERVENTION

13.5.1 General

After the presentation of the procedure which is followed for the decision-making about the degree of intervention, that is, repair or strengthening, the procedure which is followed for the design of the intervention will be given.

This procedure starts with surveying of damage. The recordings must be detailed, with sketches of the pattern of the cracks and other damage, as well as illustrations where possible or necessary.

The engineer in charge of the operation must visit the building to be rehabilitated and inspect it carefully, locating the points which are needed to be recorded in detail after the plaster is removed. After the recording of damage, the engineer must study the original drawings of the structure and revisit the building in order to form a clear idea of the project. Conclusions from the engineer's office, based only on the data of the recording, should not be drawn. One of the most important factors contributing to the decision-making about the intervention type, is the engineering judgement on the extent of damage and the failure mechanism.

Based on the inspection of the structure, the recorded data, the study of the project file and the possible checks for the quality of the materials, the soil and the quality of construction, the engineer has to decide upon the type of intervention, that is, repair or strengthening. The general rules which could help towards this have already been given in a previous section. This decision about repair or strengthening is crucial from the responsibility as well as the rehabilitation cost points of view.

After this decision the actual design starts. It is obvious that the design follows different directions, depending on the type of intervention, that is, repair or strengthening.

13.5.2 Strengthening

In the case of strengthening, that is in the case of an intervention of global type, the following steps are generally followed. At this point, however, it has to be stressed again that the intervention in the structural system must be very careful, so that the structural ability of the building will be actually improved in strength, stiffness and ductility. It is not at all impossible, after much effort and expense, to arrive at the opposite result or to have only a very slight improvement (Newmark and Rosenblueth, 1971).

(a) Information for structural assessment

The information necessary for structural assessment consists of the following items:

1. The original structural drawings and the original structural analysis. If this is not possible there should be detailed sketches of the structural system and the reinforcement arrangement at the points crucial for the safety of the building.

2. A check of how the design was implemented during the construction. This check starts from the most crucial points of the structure and it is extended if necessary to other areas.
3. A check of the quality of the materials used. In reinforced concrete this check is done by combining destructive testing(DT), that is core taking, with non-destructive tests (NDT).
4. A detailed presentation of damage patterns in plan and in elevation, including illustrations if possible.
5. A detailed check of the original structural analysis and design.
6. A structural check of the building as it was constructed, that is, of the adequacy of R/C cross-sections and their reinforcement as found during the structural survey. This check is limited to the minimum if the geometry provided by the design was followed during construction (walls, columns, etc.).

(b) Analysis and verification of the damaged structure

This step is crucial for the evaluation of the residual seismic resistance of the damaged structure and for the decision-making about the degree and extent of strengthening. It includes the following items:

1. *Estimation of the actual structural data.* It is obvious that due to damage the stiffness characteristics of the affected structural elements should be modified. As a result, the overall stiffness of the structural system should be reduced (section 11.2.9). Finally, depending on the detailing of the original structure, a behaviour factor should be estimated. The decision regarding these issues should be based on good engineering judgement, so that a reliable analysis and verification may follow.
2. *Determination of seismic actions.* The peak ground acceleration (PGA) defined for general purposes in the various seismic zones may be modified by national authorities for reasons related to the degree and extent of damage (section 13.2). In this respect, the authorities are responsible for determining the PGA in an area hit by an earthquake. However, the determination of the design spectrum, where various factors are taken into account, is the designer's responsibility.
3. *Analysis of the damaged structural system.* For this step, simplified modal response spectrum analysis, or multimodal response spectrum analysis, is used depending on the characteristics of the structural system. It is understood that the system must also be analysed for vertical loads. It should be noted that the structural data introduced in the analysis are those of the damaged structure.
4. *Seismic resistance evaluation.* The computational evaluation of the seismic resistance of the damaged structure is based on the safety verifications of all cross-sections, by means of the inequality

$$S_d \leq R_d \tag{13.8}$$

where S_d is the design action effects on the structural element and R_d the corresponding design resistance of the same element. For undamaged elements R_d is determined as prescribed in the relevant codes for new structures. For damaged elements the residual R_d may be estimated using the same procedure, introducing at the same time a reduction factor γ_{Rd} in order to take into account the resistance reduction of the damaged element. This factor γ_{Rd} may be estimated on the basis of good engineering judgement.

5. *Final decision on repair or strengthening.* It is obvious that inequality (13.8) will not be fulfilled for the damaged elements, nor for many undamaged ones.The strength index

$$R = \frac{R_d}{S_d}$$

(section 13.2) constitutes the final criterion for the degree and extent of the intervention. In the case where the undamaged structural elements have a value of this index exceeding 0.80 (section 13.4.2), that is

$$R > 0.80$$

the intervention may be limited to the repair of the damaged elements only (repair), otherwise the intervention should be of global type, that is strengthening. It should again be noted here that the evaluation described above will not always lead to a clear explanation of the causes of damage. Sometimes for such an explanation to be obtained, more sophisticated analytical procedures must be considered, such as dynamic analysis in the inelastic range, consideration of the masonry infills and so on. Of course, these cases are exceptional. The type and extent of the strengthening scheme are based on the series of criteria given in section 13.6.

(c) Redesign of strengthening

After the final decision for strengthening, the procedure which is followed for the redesign is the one outlined below (EC8/Part 1.4/Draft):

1. Conceptual design, i.e.-
   (a) selection of techniques and materials, as well as the type and configuration of the intervention;
   (b) preliminary estimation of dimensions of additional structural parts;
   (c) preliminary estimation of the modified stiffness of the strengthened elements;
   (d) preliminary estimation of the appropriate behaviour factor in relation to the local and global ductility of the modified structural system.
2. Reanalysis, i.e.
   (a) identification of the non-seismic actions;
   (b) selection of the seismic actions;
   (c) determination of the action effects taking into account the modified stiffnesses, and possible unfavourable redistribution of action effects due to heavy damage (i.e. column failure, deviations from the vertical axis;)
   (d) implementation of either the simplified modal response spectrum method of analysis or the multimodal one.
3. Safety verifications, i.e.
   (a) selection of the behaviour model of repaired/strengthened elements;
   (b) selection of material partial safety factors (γ_m);
   (c) calculation of design resistances;
   (d) verification of the safety inequalities regarding seismic and non-seismic actions, for both the ultimate limit state (ULS) and the serviceability limit state (SLS).

It should be noted that the level and the extent of the selected reanalysis procedure depends mainly on the importance of the building and the degree of damage. On the other hand, according to EC8/Part 1.4/Draft, national authorities under well-defined conditions may allow a specified level of violation of the safety inequalities in a small number of building elements. For example, in the case of the 1986 Kalamata (Greece) earthquake the ULS inequality was allowed to be violated even up to 30% in the case of undamaged elements, in ordinary buildings for which a global type of intervention was decided (strengthening).

The determination of the stiffness, resistance and ductility of the repaired or strengthened elements must be considered as an issue of major importance in order to ensure the reliability of the procedures described in this section. A detailed reference to these problems will be made in Chapter 14, since these issues are strongly related to the materials and techniques used for repair or strengthening.

(d) Strengthening drawings

The drawings of the strengthening project consist of the general drawings of the new structural elements and those which need repair or strengthening, as well as detailed drawings of the damaged members under repair or strengthening showing the actual condition of the member, the precise position of the new reinforcement and the procedure of execution of the works.

(e) Quantities and cost estimate

For repair or strengthening this is a very difficult job and sometimes of limited reliability, since it is almost impossible to predict all the supplementary works which are required and their cost under the special circumstances met in each case.

13.5.3 Repair

In the case of repair, that is in an intervention of local type, the procedure followed is considerably simpler than that described above, and consists of the following steps.

(a) Information for the structural assessment

1. The original structural drawings and the original structural analysis. If this is not possible, sketches of the structural system should be prepared.
2. A detailed presentation of damage pattern in plan and elevation, including illustrations if possible.

(b) Damage evaluation

1. In the case of repair, damage evaluation is made using the foregoing information and engineering judgement, based on a thorough *in situ* examination of the damaged building.

2. The 'strength index' is determined approximately, without any detailed analysis of the damaged building, and it is used as a criterion for the decision-making regarding repair.
3. The causes of damage in this case are sought mainly in analysis or design inadequacies affecting the damaged structural elements or in their poor construction.

(c) Redesign of repair

As has often been repeated so far, repair is limited only to the restoration of damaged elements.

1. The load effects for this intervention may be taken from the original analysis if there are no obvious mistakes in it. These values should be properly modified to take into account stiffness changes due to the additional material (jacketing).
2. The materials and techniques are then selected, depending on the availability in the market.
3. Based on the load effects derived as described previously and on the type of repair, the design resistances are calculated and used for the safety verifications.

(d) Repair drawings and cost estimate

The material presented in the strengthening subsection regarding drawings, quantities and cost estimate, is also valid here.

13.5.4 Repair of the masonry infills

At this point it is necessary to make a reference to the repair of the infill system. It is obvious that if the infill system is not taken into account in the reanalysis and redesign of the building it is not possible to carry out analytical checks on strength and ductility of the masonry infill, in a way similar to that of the structural system. However, the importance of the infill system with respect to the increase in stiffness and strength of the structure has already been adequately stressed, as well as the importance of the energy that it can dissipate. Therefore, it has to be understood that the restoration of masonry walls, especially if they have suffered serious damage, is of vital importance for the safety of the building, probably of greater importance than the strengthening of a single column or wall. Masonry walls with full depth cracks should be carefully restored, so that they can again become the first line of defence against a new earthquake, not only through the strength they contribute and the energy they dissipate, but also to restore the initial stiffness of the structure. Maybe at this point it has to be stressed that the foregoing refer not only to buildings with structural damage, but also those with no structural damage at all, but with extended damage to their masonry infills.

13.6 CRITERIA GOVERNING STRUCTURAL INTERVENTIONS

It is useful, at this stage, to focus on the intervention problem by giving a set of criteria as they are stated in EC8/Part 1.4/Draft.

13.6.1 General criteria

The intervention scheme should consider the following aspects of the problem:

- costs, both initial and future;
- durability of original and new elements and particularly of their compatibility (chemical, physical, or mechanical compatibility);
- available workmanship, equipment and materials;
- possibilities for appropriate quality control;
- occupancy (impact on the use of the building) both during and after the works;
- aesthetics;
- preservation of the architectural identity of historical buildings;
- duration of the works.

13.6.2 Technical criteria

The following technical criteria may be used as guidelines for the choice of the intervention scheme:

- In the case of highly irregular buildings (both in terms of stiffness and overstrength distributions) their structural regularity should be improved as much as possible.
- If the low damageability requirements regarding the non-structural elements are not fullfilled, appropriate intervention measures for stiffening should be taken.
- All strength requirements of the relevant codes should be fulfiled after the intervention, taking into account the provisions of Chapter 14 for the element redesign.
- The minimum possible modification of local stiffness should be sought, unless otherwise required by the first two criteria.
- Possible increase of local ductility should be sought in critical regions.
- Spreading the areas of potential inelastic behaviour as much as possible across the entire structure should be one of the tasks of the intervention.

13.6.3 Type of intervention

Bearing in mind the above general and technical criteria, an intervention may be selected from the following indicative types individually or in combination:

- no intervention at all;
- restriction or change of use of the building;
- local or global modification (repair or strengthening) of damaged or undamaged elements;
- possible upgrading of existing non-structural elements into structural ones;
- modification of the structural system aiming at stiffness regularity, elimination of vulnerable elements, or a beneficial change of the natural period of the structure;
- mass reduction;

- addition of new structural elements (e.g. bracings, infill walls);
- full replacement of inadequate or heavily damaged elements;
- redistribution of action effects, e.g. by means of relevelling (bringing columns back to their original position) of supports, or by adding external prestressing;
- addition of a new structural system to carry the seismic action;
- addition of damping devices at appropriate parts of the structure;
- base isolation (Chapter 4);
- partial demolition.

In the diagram of Figure 13.8 typical strengthening methods used in Japan are given in a schematic form (Sugano, 1981; Rodriguez and Park, 1991).

13.6.4 Examples of repair and strengthening techniques

Finally, to give a clearer picture of the frequency of implementation of the various techniques, some statistics on the R/C building repair and strengthening techniques in Japan and Mexico are given.

In the case of the 1966 Tokachi-Oki earthquake, the strengthening methods used for the rehabilitation of 157 R/C buildings are listed in Figure 13.9 (Endo *et al.*, 1984; Rodriguez and Park, 1991). In general, more than one method was used for a building, and the most common method of strengthening (85% of cases) was the addition of shear walls cast into existing frames. Column jacketing was used in 35% of the cases.

In the case of the 1985 Mexico City earthquake, the various strengthening methods used for 114 R/C buildings are listed in Table 13.3, in relation to the number of floors of the structures. According to these data, jacketing of columns (designated as concrete JC in Table 13.3) was the most commonly used technique for buildings with 12 storeys or less (Aguilar *et al.*, 1989; Rodriguez and Park, 1991).

13.7 FINAL REMARKS

From what has been presented so far, it can be concluded that very few structural problems are as challenging for the engineer as the confrontation with the consequences of an earthquake.

From the scientific point of view the main tool available to the engineer, i.e. the analysis, has often proved to be inadequate to explain the damage patterns, possibly because the assumptions on which it is based are over-simplified (static loading, elastic response of the system, not taking into account the infill system etc.). Thus, there is always doubt regarding the effectiveness of whatever intervention was decided.

From the practical point of view the determination of the ‘available’ and the ‘residual’ seismic resistance involve a high degree of uncertainty because of the subjectivity involved in the determination of the seismic resistance of the structural elements.

From the technological point of view the various types of intervention which are decided are not always feasible. Therefore, for example, structures which were

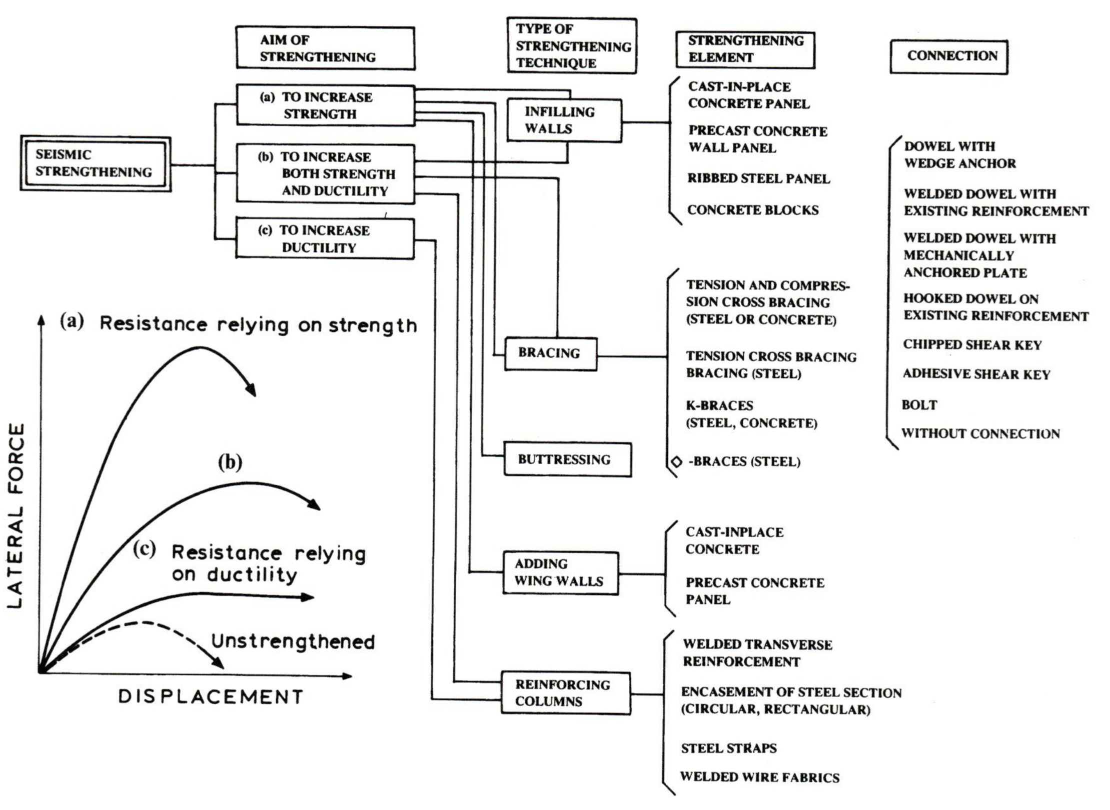

Figure 13.8 Typical strengthening methods used in Japan.

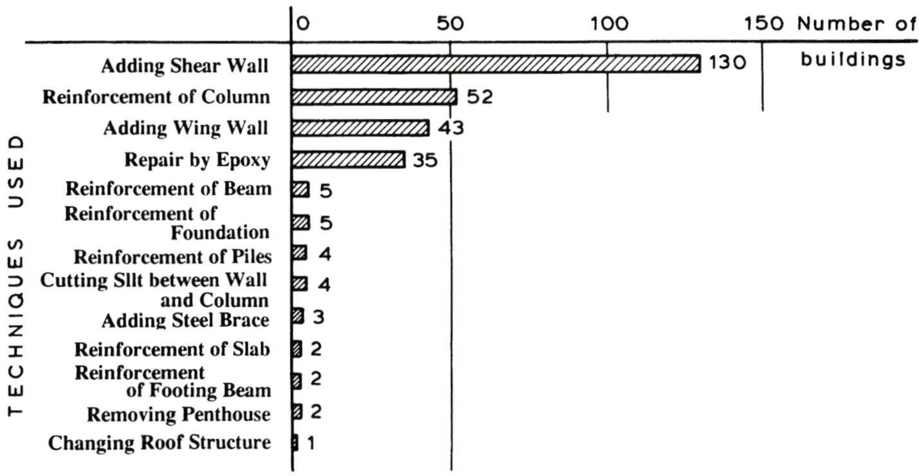

Figure 13.9 Repair and strengthening techniques used for 157 buildings in Japan.

Table 13.3 Repair and strengthening techniques for 114 reinforced concrete buildings in Mexico versus number of floors

Repair and strengthening techniques	Number of floors			
	<5	6–8	9–12	>12
Sealing	1	1	0	0
Resins	2	2	3	2
Replacement	7	8	5	6
Hydraulic jacks	1	1	1	0
Concrete JC	11	18	26	5
Steel JC	2	7	10	2
Concrete JB	4	7	14	2
Steel JB	1	0	3	1
Shear wall	8	12	16	9
Infill wall	4	9	2	2
Steel diagonals	0	7	7	2
Concrete frames	1	3	3	3
Additional elements	3	3	4	2
Straightening	0	1	2	2
New Piles	2	4	8	3

a. JC = Column jacketing.
b. JB = Beam jacketing.

built without respecting the provisions of modern codes (most of the existing structures fall into this category) cannot meet ductility DC 'M' requirements, and possibly not even ductility DC 'L' requirements. On the other hand, a large increase in V_B creates the need for additional strength and stiffness elements which lead to foundation problems, as well as functional problems when the structure is in use again.

Based on the above , the legal framework which is set every time after a destructive earthquake for the restoration of damage cannot withstand strictly scientific criticism. This is because this framework attempts to strike a balance between the desirable and the feasible. In other words, it is a political decision within the broader meaning of the term, which tries to optimize the combination of the scientific knowledge with the technological and financial possibilities in order to confront the acute social problem of the safe restoration of damaged structures.

Independently of the previous general remarks, in summarizing reference should be made to the following special points:

1. The restoration of a seismically damaged building is a much more difficult task than the original design and construction of the building.
2. The difficulties arise during the design, as well as during the inspection and execution of the intervention.
3. A basic factor for the successful outcome of the whole operation is the correct diagnosis of the causes of damage. It is on this diagnosis that the level of intervention depends, that is, repair or strengthening of the structure.
4. The design of the restoration must aim at:
   (a) providing the structure with the stiffness, strength and ductility that it had before the earthquake in the case of local damage;
   (b) providing the structure with the strength, stiffness and ductility required by the current codes in the case of damage of global character (strengthening).
5. Independently of the local or global character of the damage the structural elements must be repaired in such a way that they regain the strength and ductility required by the current codes.
6. For the choice of the repair technique, the market conditions and the feasibility of application of the chosen technique in every particular case must be taken into account.
7. The repair is usually accompanied by the removal of many structural members and therefore special care should be taken with the temporary support of the structure.
8. The outcome of the repair depends to a large degree on the quality control of the design and construction. Therefore very careful inspection is necessary during the execution of the rehabilitation works.
9. The restoration of the heavily damaged infills is very important to the structure and has to be given the appropriate care.
10. Finally, it has to be stressed once again that the structural rehabilitation must have as a reference point the proper combination of strength, stiffness and ductility.

13.8 REFERENCES

Aguilar, J., Juarez, H., Ortega, R. and Iglesias, J. (1989) The Mexico earthquake of September 19, 1985: statistics of damage and retrofitting techniques in R/C buildings affected by the 1985 earthquake. *Earthq. Spectra*, EERI, **5**(1), 145–52.

Anagnostopoulos, S.A. (1986) Proposals on the framework for rehabilitation of Kalamata, Greece. *Report of Inst. Engng Seism. and Earthq. Engng*, Thessaloniki, Greece (in Greek).

Anagnostopoulos, S.A., Petrovski, J. and Bouwkamp, J.G. (1989) Emergency earthquake damage and usability assessment of buildings. *Earthq. Spectra*, **5**(3), 461–76.

ATC 3-06 (1978) *Tentative Provisions for the Development of Seismic Regualtions for Buildings* (Ch. 13), US Government Printing Office, Washington D.C.

CEN Techn. Comm. 250/SC8 (1993) *Eurocode 8: Earthquake-Resistant Design of Structures–Part 1.4 (Draft): Repair and Strengthening*, CEN, Berlin.

Chronopoulos, M. (1984) Damage and rehabilitation cost. *Proceed. of the Conf. on Earthq. and Struct.*, EPPO, Athens, I, pp. 459–69 (in Greek).

Endo, T. *et al.* (1984) Practices of seismic retrofit of existing concrete structures in Japan. *Proceed. of the 8th World Conf. on Earthq. Engng*, San Francisco, USA, I, pp 469–76.

Freeman, S.A. (1993) Assessment of structural damage and criteria for repair. *Proceed. of the 3rd US /Japan Workshop on Urban Earthq. Hazard Reduction*, Publication No. 93-B, EERI, Oakland, California.

Holmes, T.W. (1994) Policies and standards for reoccupancy repair of earthquake-damaged buildings. *Earthq. Spectra*, **10**(1), 197–208.

Jirsa, J.O. (1994) Divergent issues in rehabilitation of existing buildings. *Earthq. Spectra*, **10**(1), 95–112.

Newmark, N.M. and Rosenblueth, E. (1971) *Fundamentals of Earthquake Engineering*, Prentice-Hall, Englewood Cliffs, NJ.

Penelis, G.G. (1979) *Rehabilitation of Buildings Damaged by Earthquakes,* Ministry of Publ. Works, Thessaloniki (in Greek).

Penelis, G.G. *et al.* (1984) The Rotunda of Thessaloniki, in *Repair and Strengthening of Historical Monuments and Buildings in Urban Nuclei*, UNIDO/UNDP PR.RER/79/015, **6**, Vienna, pp. 165–88.

Penelis, G.G. *et al.* (1992) The Rotunda of Thessaloniki: seismic behaviour of Roman and Byzantine structures, in *Hagia Sophia, from the Age of Justinian to the Present*, (eds R Mark and A. Cakmak), Cambridge University Press, USA, pp. 132–57.

Rodriguez, M. and Park, R. (1991) Repair and strengthening of reinforced concrete buildings for seismic resistance. *Earthq. Spectra*, EERI, **7**(3), 439–60.

Sugano, S. (1981) Seismic strengthening of existing reinforced concrete buildings in Japan. *Bulletin of New Zealand Nat. Soc. of Earthq. Engng*, **14**(4), 209–22.

Tassios, T. (1984) Post-earthquake interventions. *Proceed. of the Conf. on Earthq. and Structures*, EPPO, Athens, I, pp. 595–636 (in Greek).

UNIDO/UNDP PR. RER/79/015 (1985) *Post-earthquake Damage Evaluation and Strength Assessment of Buildings under Seismic Conditions*, **4**, Ch.2, Vienna.

14

Technology of shoring, repair and strengthening

14.1 GENERAL

The purpose of this chapter is to briefly present the technological problems associated with the interventions to structures damaged by earthquakes.

In the preceding chapters detailed reference has been made to the procedure followed for decision-making about the extent and type of interventions. At the same time the successive steps for the design of the interventions were discussed in detail. In this chapter there will be reference to the emergency measures for shoring (temporary support), to the materials and techniques of interventions and to the dimensioning of the structural elements for various types of intervention. However, given the fact that several manuals, specifications and codes have been published (UNDP, 1977; NTU, 1977; AUT, 1978, 1979; GMPW, 1978; UNIDO / UNDP, 1983; CEN, 1993), where numerous technical details are given, the focus here will mainly be on some typical repair and strengthening techniques and on the dimensioning of the relevant structural elements.

Specifically on the subject of dimensioning, there are many reservations with regard to the reliability of the proposed methods, for the following reasons:

- There is no adequate experimental verification of these methods.
- Most of them are based on rough and/or simplified models, since analytical models based on experimental and theoretical knowledge have not yet been developed to a degree suitable for practical use.
- The quality of execution of the repair and strengthening works on site drastically influences the results.
- The evaluation of the redistribution of stresses from the old element to its strengthening presents reliability problems.

It should be stressed here that the main issues concerning repair and strengthening, which are materials, techniques and redesign considerations, exhibit different degrees of development regarding research, implementation and codification level. Table 14.1 gives a qualitative picture of this development (Zavliaris, 1994).

Before the individual topics of this chapter are addressed, it is useful to summarize the intervention procedure as given in the previous chapters.

Table 14.1 Development in materials, techniques and redesign considerations

	Materials	Techniques	Redesign considerations
Research and development (R&D)	◯	◯	◯
Implementation	○	○	○
Codification	○	○	○

The diameter of the circles represents the degree of development (qualitatively).

After a destructive earthquake, an inspection operation is usually organized by the state, aimed at locating the buildings which are unsuitable for use on the one hand, and on the other to carry out the necessary demolitions or shorings. After this first phase and once the aftershocks have been attenuated, the procedure for the design of the intervention to every individual damaged building to be retrofitted starts. This second phase is much more systematic than the first, more laborious and more effective, and it requires much time and expense. The preceding chapter as well as the present one cover the approach to problems associated with this second phase.

In closing this brief introduction, it has to be mentioned that the main reference for the intervention techniques for individual structural members presented here, was the UNIDO/UNDP manual, *Repair and Strengthening of Reinforced Concrete, Stone and Brick-Masonry Buildings* (UNIDO/UNDP, 1983), which represents a synthesis of experience and expert knowledge at an international level. With this choice it is felt that some contribution is made to the realization of one of the UNIDO/UNDP goals, that is, the dissemination of this widely accepted up-to-date knowledge on intervention techniques, to the international scientific community with special interest in the subject.

14.2 EMERGENCY MEASURES FOR TEMPORARY SUPPORTS

14.2.1 General

Immediate shoring (temporary support) is recommended for buildings with serious damage in the vertical structural elements (columns, walls). By using shoring the damaged elements are relieved of their loads by temporary additional structures and therefore the danger of collapse due to aftershocks is diminished.

The support must take place initially at the floor where the damage of the vertical element occurred. It is necessary, however, to estimate the ability of adjacent beams to carry the vertical load of the damaged element, and if this is not adequate, support must be extended to other floors as well (Figure 14.1).

The support system must be placed at a certain distance from the damaged element so that enough room is left for the repair work which will follow.

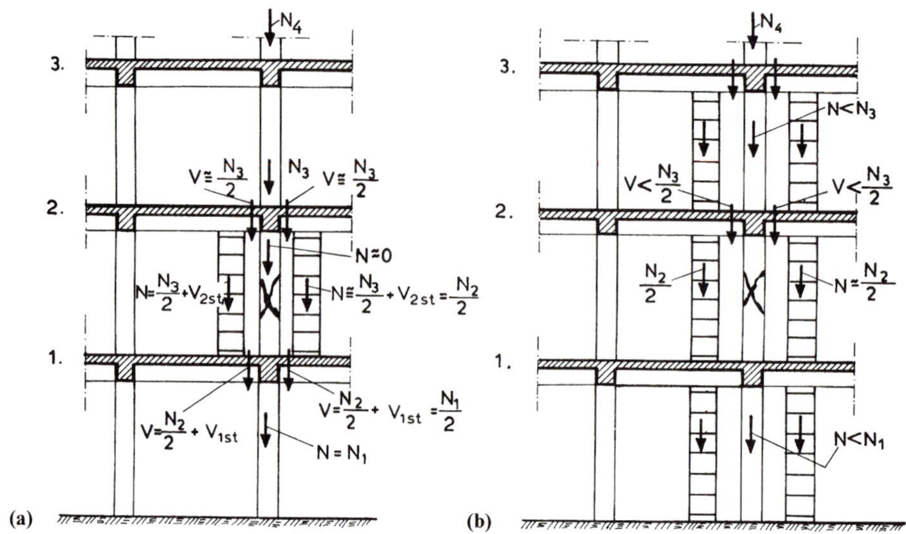

Figure 14.1 Shoring (temporary supporting) of a multistorey building with a damaged column: (a) shoring of only one floor; (b) shoring of more floors.

When there are problems of lateral instability in a structure, lateral support is provided either in the form of ribs or diagonal braces between the frames formed by beams and columns, and even internal tension ties can be used for supporting buildings close to collapse.

The design of temporary supports must be done promptly, with the aid of approximate analysis and design, performed to determine only the order of magnitude of actions and action effects (stresses). The materials and techniques foreseen must be readily available, for instance metal scaffolds, timber, steel profiles, timber grillage, etc.

Given the fact that the shoring of damaged structures is a very hazardous work for the people involved, the time which these people spend in the structure must be kept to a minimum. Therefore, it is recommended that the preparation of all the supporting elements is done away from the damaged structure (based on the dimensions measured on site), so that the work of shoring will be limited to the installation of these elements in the damaged building.

14.2.2 Techniques for supporting vertical loads

(a) Industrial-type metal scaffolds

In the case of small loads, independent industrial-type metal tube shores are used (Figure 14.2) having a load-bearing capacity of 20 kN and a height of about 3.00 m.

For the shoring of beams or slabs, dismountable metal towers are used (Figure 14.3) which are wedged to the surface to be shored with the aid of special screw-type bolts with which all industrial-type scaffolds are equipped.

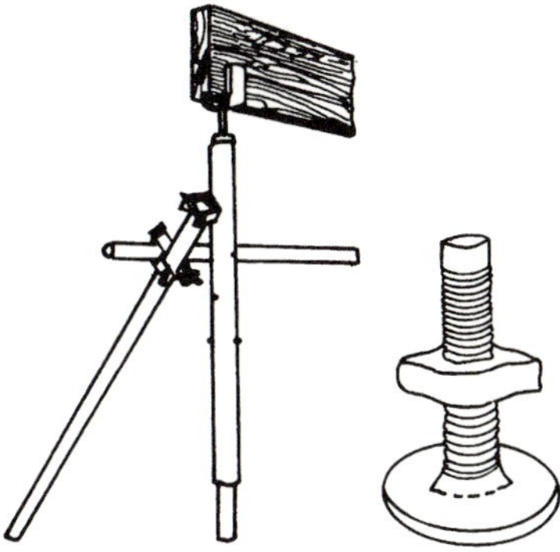

Figure 14.2 Independent industrial–type metal supports.

(b) Timber

Timber elements can also be used for carrying vertical loads, either in the form of logs or telephone poles, or in the form of timber grillages. For every damaged column at least one 250 mm diameter log should be used on each side of the column. The **allowable load** for this diameter and for floor heights up to 3.00 m is estimated at 300 kN per pole for timber of good quality. If the height is greater than 3.00 m or the diameter smaller than 250 mm, the pole must be checked for buckling (DIN 1052, 1981).

In the case where two or more supporting elements are used on each side, they must be connected to each other with X-shaped braces.

If no logs are available, shoring can be achieved with timber grillage (Figure 14.4).

(c) Steel profiles

Steel profiles can be used either in the same manner as timber or as an immediate strengthening means of the damaged column (Figure 14.5). In this case they can be incorporated later into the concrete jacket. In the first case there should always be a buckling check. In the second case the key to success is the tightening of the vertical steel angles to the column with the aid of transverse angles and prestressed ties before the transverse connecting straps are welded to the vertical angles.

1.4.2.3 Techniques for resisting lateral forces

(a) Bracing with buttresses

Bracing with buttresses is the most common way of resisting lateral forces. These forces are due to the deviation of the building from the vertical axis either because of failure of vertical structural elements or because of settlement of the foundations (Figure 14.6). Some critical points of such a bracing system are the following:

- the anchoring of the bracings to the ground, so that they may resist horizontal thrusts;

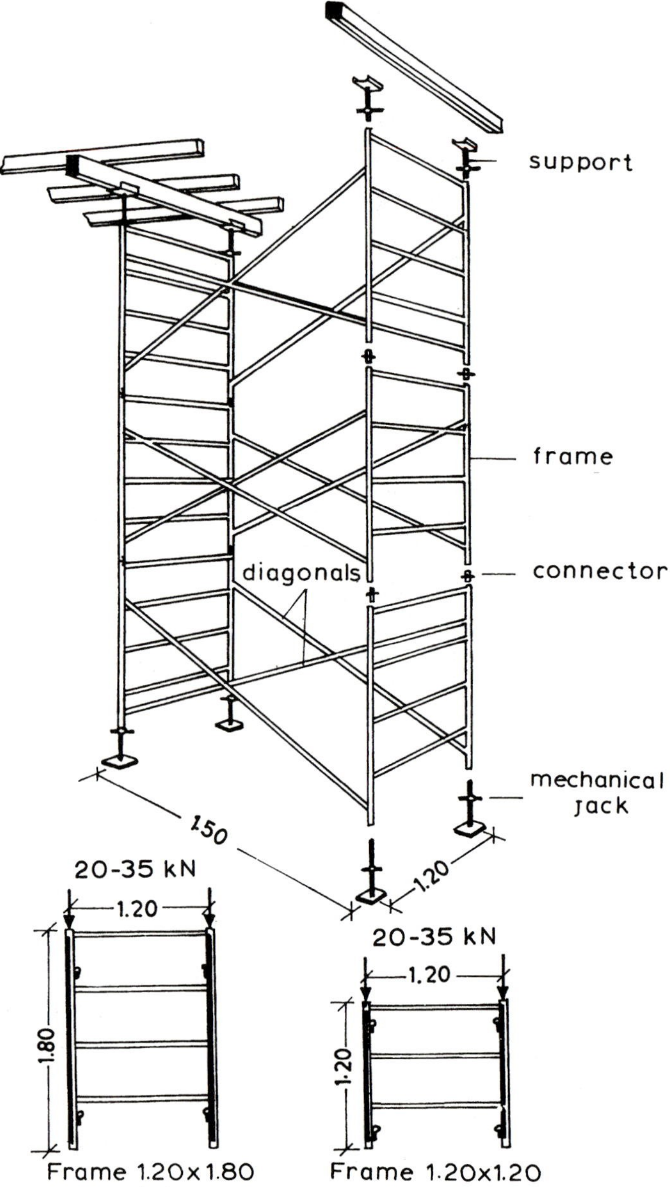

Figure 14.3 Industrial–type metal towers.

- the attachment of the vertical member to the building so that it prevents relative slipping;
- keeping the unbraced length of the inclined member of the lateral bracing below limiting values, to avoid in-plane or out-of-plane buckling.

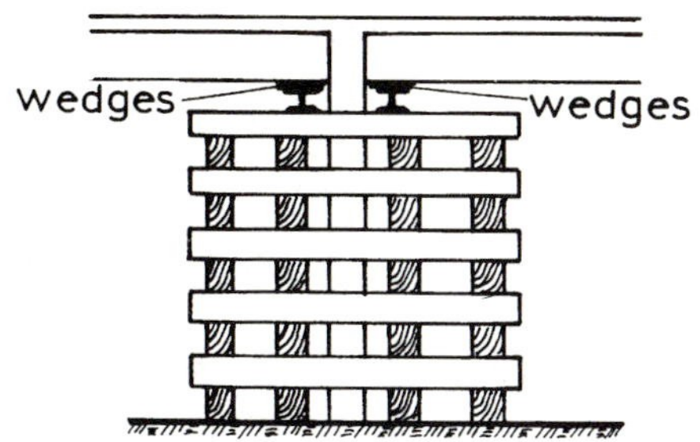

Figure 14.4 Shoring with timber grillage.

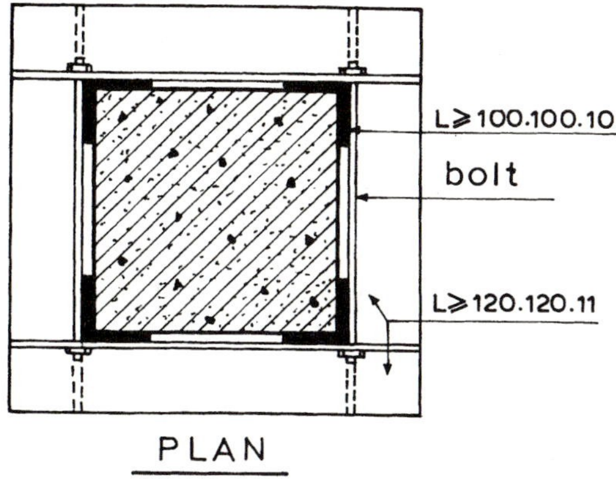

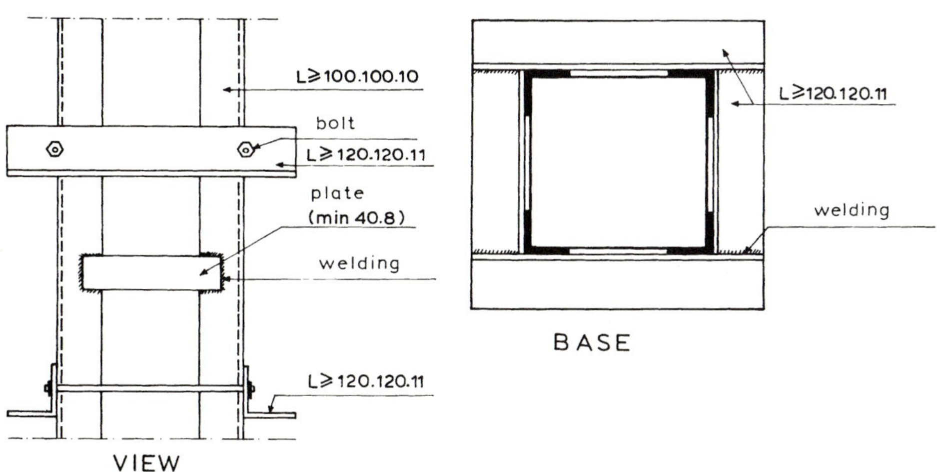

Figure 14.5 Immediate tying of a column with steel profiles.

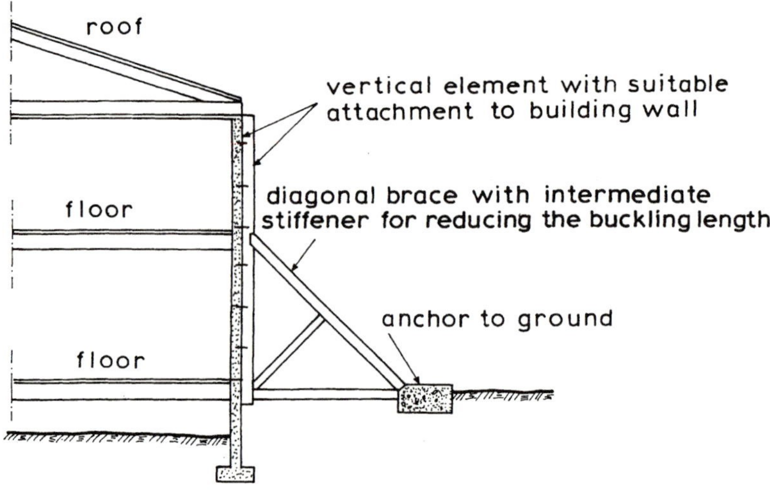

Figure 14.6 Bracing with buttresses.

For this type of shoring usually timber and rarely steel members are used. However, it should be mentioned that the horizontal forces that such a system is assumed to resist, for small deviations from the vertical axis, are not very large and can be easily estimated approximately, using the relationship (Figure 14.7)

$$\Delta H = \frac{\delta}{h} G \tag{14.1}$$

where δ/h is the deviation from the vertical axis and G the total vertical load of the structure which for normal buildings is estimated to be 10.0–12.0 kN per m^2 of floor area.

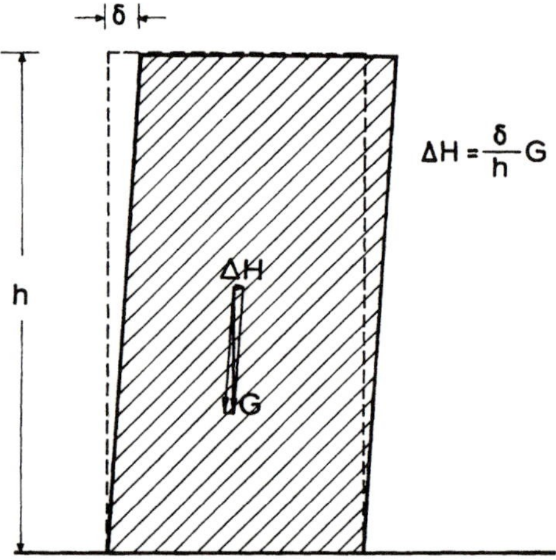

Figure 14.7 Estimation of the horizontal forces due to deviation from the vertical axis.

(b) Bracing with diagonal X-braces

The use of diagonal timber or steel members in the plane of the R/C frames on the one hand allows the partial transfer of gravity loads to undamaged vertical elements, and on the other prevents lateral deformation (Figure 14.8).

Frame bracing can consist of timber, tree logs or steel profiles of sufficient strength considering their potential for buckling. This method is used when external bracing with buttresses cannot be easily carried out.

(c) Bracing with interior anchoring

In the case of hybrid structures consisting of R/C slabs supported by masonry, in order to retain external walls which have been detached and deviate from the vertical axis, metal tensioners are often used which are prestressed with the aid of tensioner couplers (Figure 14.9).

(d) Bracing with tension rods or rings

In the case of deviation from the vertical axis due to arch thrusts, prestressed metal rings or prestressed rods are used, depending on whether the structure is a dome or an arch (UNIDO/UNDP, 1984).

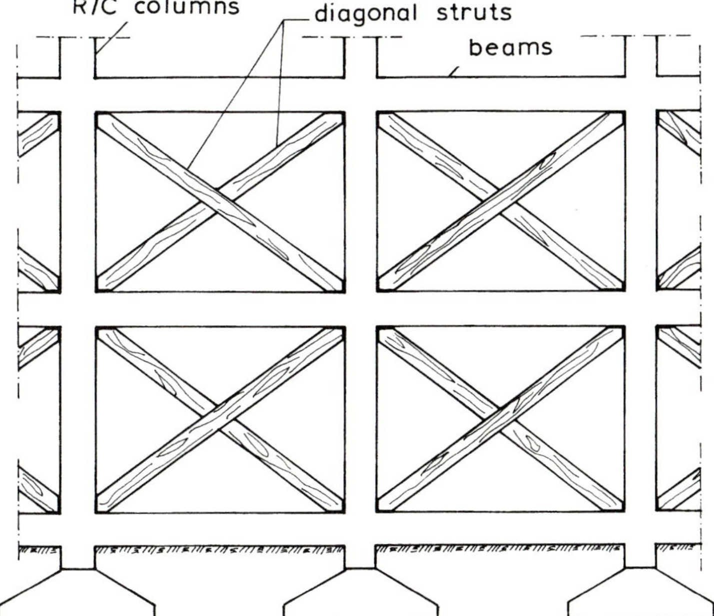

Figure 14.8 Bracing with diagonal struts.

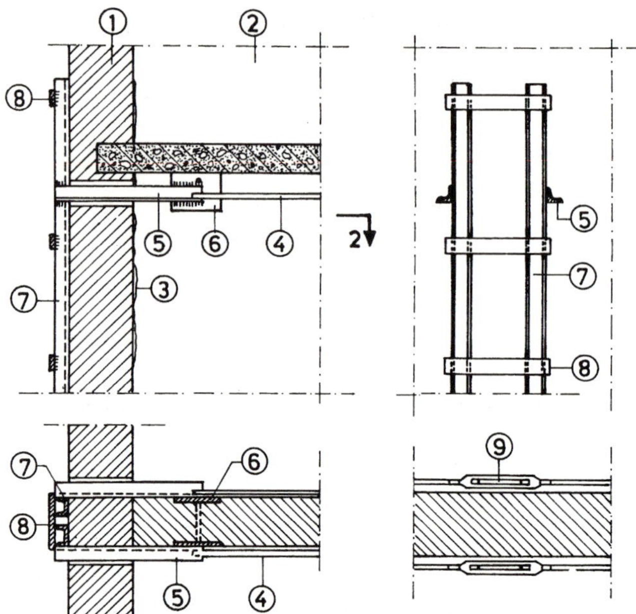

Figure 14.9 Bracing with internal tension ties: 1 = exterior wall; 2 = interior wall; 3 = crack; 4 = steel tensioner; 5 = angle 50.50.5mm; 6 = steel plates; 7 = steel profiles; 8 = steel plates; 9 = tensioner coupler.

14.2.4 Wedging techniques

The wedging procedure is a crucial part of every supporting or bracing procedure, because the transfer of the loads of a damaged element to the shoring or bracing system is accomplished through wedging. Wedging can be achieved by the following means:

- wooden twin wedges (Figure 14.10)
- mechanical jacks (screws) (Figure 14.3)
- hydraulic jacks (Figure 14.11)
- couplers (Figure 14.9).

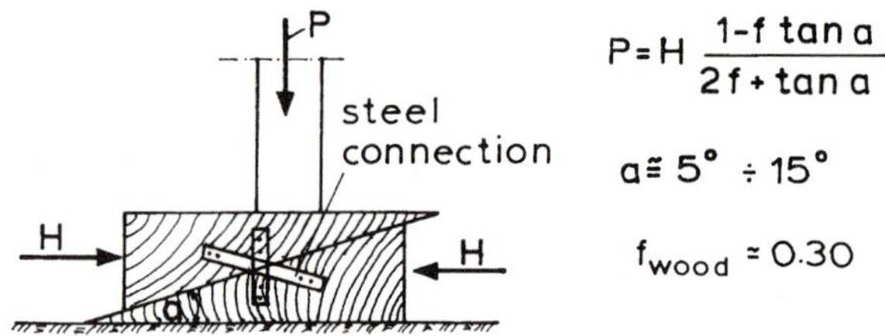

Figure 14.10 Wedging with twin wooden wedges (*f* is the friction coefficient).

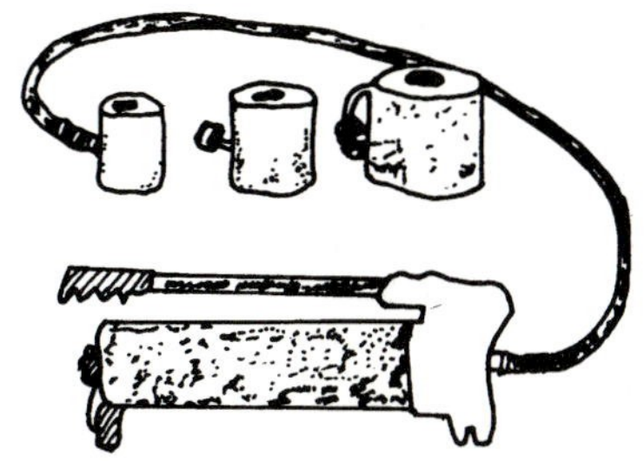

Figure 14.11 Wedging with hydraulic jacks.

14.3 MATERIALS AND INTERVENTION TECHNIQUES

In this section reference will be made to the materials and intervention techniques which are frequently encountered in the repair or strengthening of structures after an earthquake. Given the fact that these special materials, as well as their application techniques, are governed by detailed specifications which are typically related to the know-how that accompanies them, the designer, before considering including any of these materials in a rehabilitation project, must be fully informed about it. In the following, a general presentation of the materials and techniques is given and some critical points related to their advantages, disadvantages and their successful application are discussed.

14.3.1 Conventional cast-in-place concrete

Conventional concrete is very often used in repairs as a cast-in-place material.

In many cases the results are not satisfactory because of the shrinkage of conventional cement, which causes reduced bond between old and new concrete. In order to improve bond conditions and cover additional variabilities in construction operations, the use of concrete having a strength higher than that of the element to be repaired is recommended ($f_{c_{rep}} \geqslant f_{c_{exist}} + 5$ MPa), as well as low slump and water/cement ratio. Such a choice, however renders compaction very difficult, especially when thin jackets are visualized, thus making necessary the use of superplasticizers to increase slump up to 200 mm with the standard method of Abram's cone. The maximum size of aggregates should not exceed 20 mm, so that the mix will be able to pour through the narrow space between the old concrete and the forms.

The procedure of casting the concrete is critical for the success of the intervention. Old surfaces should be made as rough as possible and cleaned in order to increase the adhesion between old and new concrete. After the placement of the reinforcement the forms are placed, which have special lateral openings for casting of concrete. Before concreting there should be a final dusting of the surfaces with compressed air, as well as extensive wetting of the old concrete and the forms. Concrete should be thoroughly vibrated to ensure a high degree of compaction.

14.3.2 High-strength concrete using shrinkage compensating admixtures

For the construction of cast-in-place concrete jackets, very often special dry packed mortar is used which is available in the market under several commercial names. This mortar consists of cement, fine sand (up to 2.0 mm), superplasticizers and expansive admixtures in the appropriate proportions, so that mixing with water of about 15% of weight produces fluid mortar which attains high strength in a very short time (e.g. 30 MPa in a 24-hour period, 70 MPa in 28 days), while at the same time it does not shrink. The attainment of high strength in a short period of time is due to the formation of a special silica calcium hydrate from the reaction between the expansive admixture and the cement. Therefore, very satisfactory repairs are accomplished, without voids and shrinkage cracks, using very thin jackets, e.g. 40 mm. In order for these products to be used, they must be accompanied by a quality control certificate. As far as the rest of the procedure is concerned, it is the same as for conventional concrete.

14.3.3 Shotcrete (gunite)

If the appropriate equipment and trained personnel are available, shotcrete is considered as a very good repair solution. Indeed, due to the fact that forms are not needed, it can be applied on surfaces of any inclination, even on ceilings. Its use is more common extended surfaces such as R/C and masonry walls, but it can also be used for the construction of jackets around columns.

As far as strength is concerned, a strength higher than that of the repaired element is always specified ($f_{c_{rep}} \geq f_{c_{exist}} + 5$ MPa).

The main advantages of the method are the absence of forms, the very good adhesion between old and fresh concrete due to the high degree of compaction energy during shotcreting, and the high strength due to the low water/cement ratio.

As disadvantages of the method, one can consider the fact that the water/cement ratio cannot be quantitatively controlled, given the fact that the fluidity of the mix is controlled only visually by the operator, the high shrinkage which makes necessary the use of wire mesh as additional reinforcement, and finally, the waste of a large fraction of the material due to reflection on the surface of application.

The required equipment for the production and application of the shotcrete includes (NTU, 1978):

- a concrete mixer for dry mixing
- a water tank
- a centrifugal water pump
- a high-capacity compressor
- a gun with one or two chambers
- high-pressure hoses
- a nozzle.

The production procedure is as follows (Figure 14.12):

1. A mixture of 0.5 kN of cement and about 2.0 kN of aggregates with maximum grain size of 7, 12 or 16 mm, depending on the case, is dry-mixed in the concrete mixer.

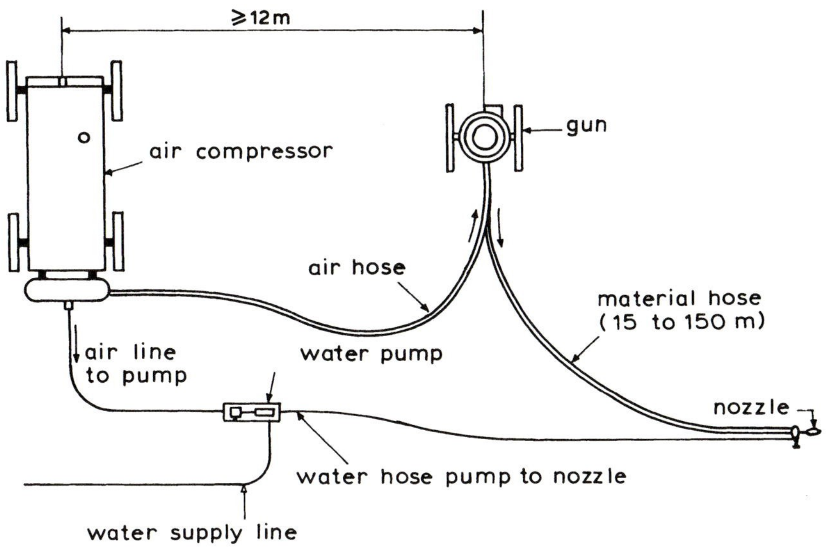

Figure 14.12 Typical arrangement of equipment for shotcreting.

2. The mix is fed into the gun (Figure 14.13) and still in dry form in suspension, reaches the nozzle through a hose with the aid of compressed air.
3. At the nozzle, water is injected into the material. From there the mix is forcefully shot on to the surface to be repaired, which has been previously roughened, wetted and appropriately reinforced. Every layer has a maximum thickness of 30–40 mm. If a larger thickness is required, a second layer should be applied.
4. The resulting surface is very rough, therefore after hardening it must be covered with plain plaster or mortar.

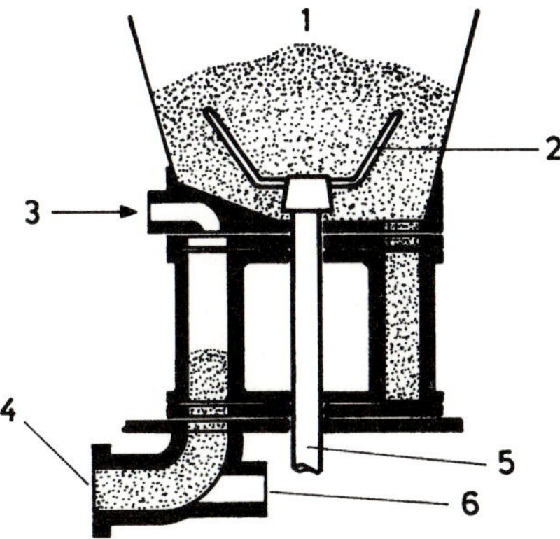

Figure 14.13 The principle of functioning of a shotcreting machine: 1 = dry material supply; 2 = mixer; 3 = compressed air; 4 = material exit under compression; 5 = rotor; 6 = compressed air;

14.3.4 Polymer concrete

Polymer-modified concrete is produced by replacing part of the conventional cement with certain polymers which are used as cementitious modifiers. The polymers, which are normally supplied as water dispersions, act in several ways. They function as water-reducing plasticizers, they improve the bond between old and new elements, they improve the strength of the hardened concrete and so on. However, it should be mentioned that polymer concrete also has several disadvantages. It is vulnerable to fire conditions and due to its lower alkalinity presents inferior resistance against carbonation compared to conventional concrete.

14.3.5 Resins

Resins are usually used for grouting injections into cracks in order to glue together cracked concrete or for gluing thin metal sheets on concrete surfaces. These are materials made of two components which react and harden after they are mixed together. More specifically, one component is the resin in fluid form (epoxy, polyester polyurethane, acrylic, etc), while the second is the hardener (NTU, 1978; AUT, 1978, 1979). There is a great variety of such products with different properties depending on the chemical composition of the components, the mixing ratios, the possible additives such as fillers or sand. Therefore the engineer must have a good knowledge of the properties of such a material before selecting the proper one for a specific use. **Epoxy resins** are the most common type of these materials in use today.

Resins must have an adequate pot life so that a usual dosage can be used before it hardens. Curing requirements should be compatible with the temperature and moisture conditions of the structure. The resin must have excellent bond and adhesion to concrete and steel and present a small to negligible shrinkage. Also, its modulus of elasticity must be generally compatible with that of the concrete to be glued. Resins lose their strength in temperatures higher than 100 °C and therefore such repairs are not fireproof without fire protection (e.g. plaster). Resins which are used in the form of injections must have a viscosity appropriate for the crack width to which the injection is applied. Resins which are used for gluing metal sheets usually have high viscosity. Table 14.2 shows comparative data for strength and deformability of conventional concrete and of epoxy resins (AUT, 1978).

There are several techniques for the application of resin injections. In the simplest case the resin is mixed with the hardener in a separate receptacle and a gun with an injection nozzle is filled with the mixture (Figure 14.14). Sometimes the mixing is done within the gun with separately controlled supply of the two components. The injection is applied with low pressure (up to 1MPa) in which case it is done by hand, or with high pressure (up to 20 MPa) in which case it is done with a pump. The gun is equipped with a pressure gauge. Since epoxy resins are materials which cause irritation to skin, eyes and lungs, the appropriate means of personnel protection are required when working with them (gloves, protective eyeglasses, masks). When the crack width is small (0.1–0.5 mm) pure resin is used. In the case of wider cracks it is useful to mix the resin with filler having grain

Table 14.2 Comparison between mechanical properties of concrete and epoxy resins

Property	Concrete	Epoxy resin
Compressive strength (MPa)	20–70	Up to 250
Tensile strength (MPa)	2–5	3.5–35
Flexural strength (MPa)	3.5–7.0	10–35
Elongation (%)	0.01	0.2–50

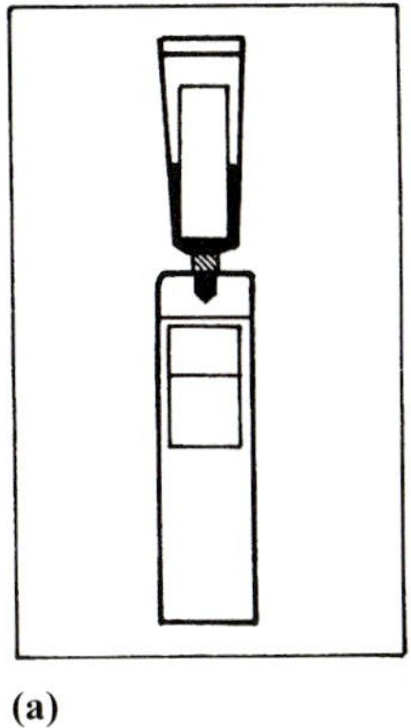

(a)

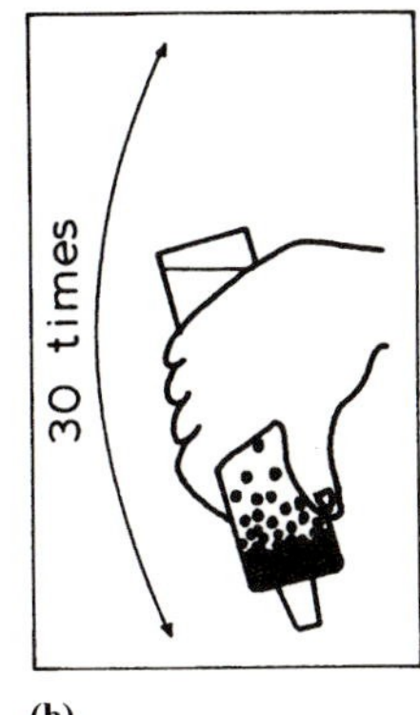

(b)

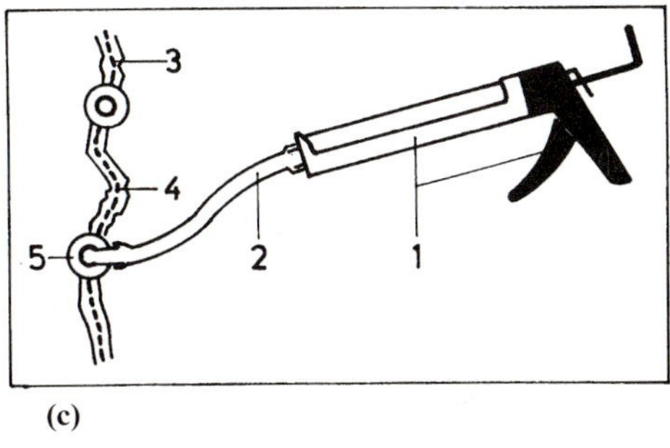

(c)

Figure 14.14 Procedure for the application of resin injections; (a) mixing of resin with the hardening agent; (b) shaking of the mixture to become homogeneous; (c) application of the resin injection; 1 = injection gun; 2 = plastic hose; 3 = crack; 4 = sealer; 5 = nipples.

diameter not larger than 50% of the crack width or 1.0 mm, whichever is smaller. The ratio of resin to filler is usually about 1:1 in weight.

Before the application of resin injections the crack is cleaned with compressed air. Then holes of 5–10 mm in diameter are opened with a drill at certain distances along the length of the crack and nipples or ports of the appropriate diameter are placed on the mouths of the holes to facilitate the execution of the resin injections. The crack is then sealed on the surface with a quick-hardening resin paste and the injections are applied. On vertical surfaces the procedure starts from the lowest nipple or port and as soon as the resin leaks from the mouth of the next nipple the procedure is discontinued, the mouth is sealed, and the same process is repeated for the next nipple. The next day, when the epoxy resin hardens, the resin paste is removed from the surface with an emery wheel.

14.3.6 Resin-concretes

Resin-concretes are concretes in which the cement has been replaced by resin. They are mainly used for substituting pieces of concrete which have been cut off. In order to make sure that there will be enough bonding between the old and the new parts, it is recommended that the old concrete is well cleaned and its surface coated with pure resin before the new resin-concrete is cast in the place of the cut-off piece. Resin-concretes require not only a special aggregate mix to produce the desired properties but also special working conditions, since all two-component systems are sensitive to humidity and temperature.

14.3.7 Grouts

Grouts are often used for the filling of voids or cracks with large openings on masonry or concrete. The usual grouts consist of cement, water, sand, plasticizers and expansive admixtures in order to obtain high strength and minimum shrinkage during hardening (Leonhardt, 1962). Details on the composition of conventional grouts can be found in all prestressed concrete manuals. Grouts are mainly used for the repair of structural masonry. In the case of traditional or monumental buildings the grouts which are used must be compatible with the original construction materials as far as strength and deformability are concerned, therefore a large percentage of the cement is replaced by pozzolans or fly ash and calcium hydroxide (UNIDO/UNDP, 1984; Penelis *et al.*, 1984).

For application, the same procedure is followed as in the case of resin injections. Figure 14.15 shows the general set-up for the application of grouts.

14.3.8 Gluing metal sheets on concrete

This is a relatively new method of intervention (Figure 14.16). The gluing is carried out with epoxy resin on the lower face of beams, on the vertical faces of beams or on the joints. The sheets are made of stainless steel (usually 1.00–1.50 mm thick) so that they can be fitted well and glued on the surface of the element to be strengthened (NTU, 1978; AUT, 1978, 1979).

The intervention procedure includes the following phases: careful smoothing of the concrete surface with an emery wheel or emery paper; washing and drying of the concrete surface; roughening up the sheet surface using the process of sandblasting, coating of the concrete surface with an epoxy resin of high viscosity; covering the steel sheet with an epoxy resin layer, putting it up and keeping it in place with tightening screws for 24 hours, so that it will be glued on the concrete. Repeat this procedure if a second sheet is necessary. Finally cover up of the sheets with wire mesh and cement plaster or shotcrete.

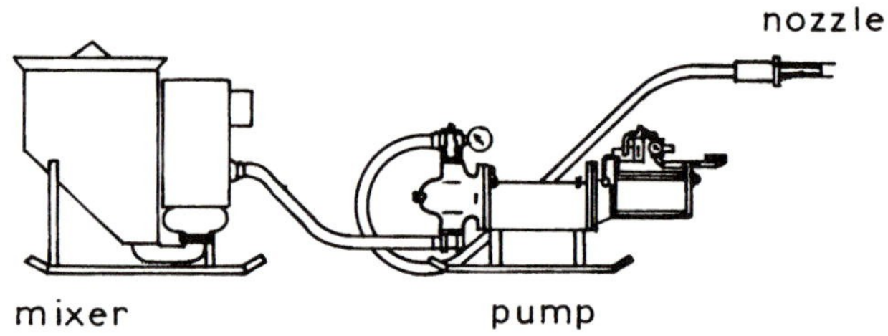

Figure 14.15 Arrangement for application of cement grouts.

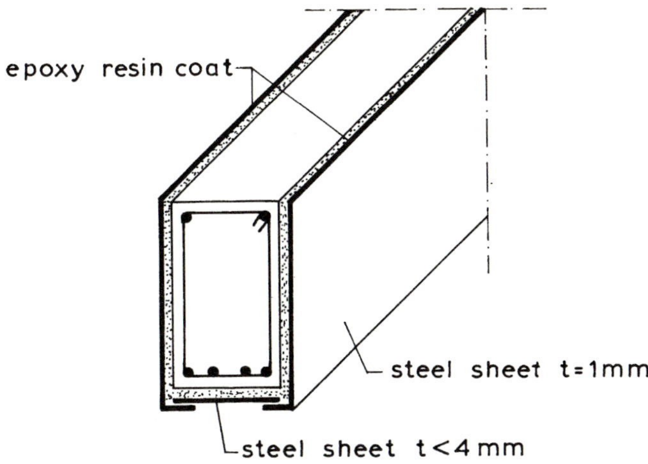

Figure 14.16 Strengthening of a beam with glued metal sheets.

14.3.9 Welding of new reinforcement

The most usual way to strengthen tension regions is the use of new reinforcement, the force transfer of which with the old reinforcement is accomplished through welding (Figure 14.17). Low-alloy steel is preferred to deformed reinforcing bars as new reinforcement because it welds more easily. New bars are welded on the old ones with the aid of connecting bars (bar pieces of the same diameter but not smaller than ø16 and of at least 5ø length, spaced about 500 mm apart).

14.3.10 Gluing Fibre-Reinforced Plastic (FRP) sheets on concrete

The strengthening of R/C structural members (e.g. beams, slabs, columns) with glued fibre-reinforced plastic sheets is a relatively new method of intervention (Balinger, Maeda and Hoshijima 1993; Plevris and Triantafillou, 1994; Saadat-

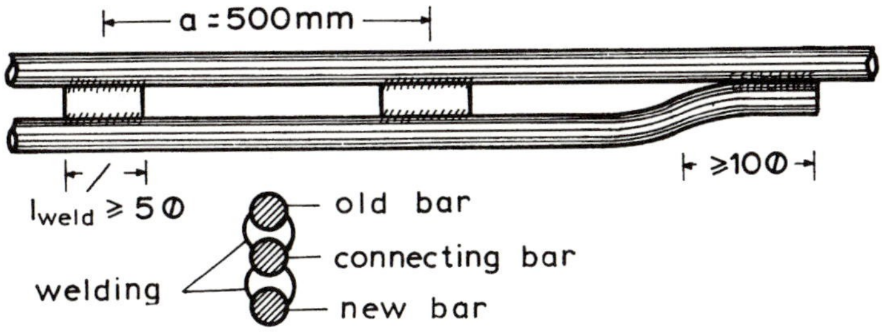

Figure 14.17 Welding of a new reinforcement bar.

manesh and Ehsani, 1991; Triantafillou, 1994) similar to that of gluing metal sheets. The sheets of FRP exhibit the following advantages compared to the steel sheets:

- They are light in weight
- They do not corrode.
- They are available in large dimensions.
- They have low modulus of elasticity accompanied by large elastic deformations up to failure, which are particularly useful properties for prestressing.

At the same time, they also have some disadvantages:

- Due to their elastic behaviour up to failure they must be considered as materials of low ductility.
- They lose a large percentage of their initial strength under permanent loading. This loss ranges from 15 to 60% (Table 14.3).

In practice, FRP sheets are used with fibres in one direction only and with a proportion of reinforcement 50–70% in volume. The mechanical properties of these materials vary and depend on the reinforcement material. For glass, aramid and carbon fibres, these properties are given in Table 14.3, while in Figure 14.18 the constitutive law of these materials is presented schematically.

Thanks to the above advantages, FRP sheets have been used in the development of new techniques for repair and strengthening of R/C structural elements, in which they replace steel sheets. It should be stressed that while these materials can be successfully used to increase strength in bending, shear and compression, they cannot affect stiffness positively. At the same time, in most cases they influence local ductility negatively.

14.4 REDIMENSIONING AND SAFETY VERIFICATION OF STRUCTURAL ELEMENTS

14.4.1 General

Repair and strengthening have to do with several interfaces, which are due to the damage itself or are created by the intervention. New materials are added to the

Table 14.3 Mechanical properties of FRP

Material	Elastic modulus (GPa)	Tensile strength (MPa)	Failure deformation (%)	Loss of strength under permanent load (%)
Glass-FRP	50	1700–2100	3	60
Aramid-FRP	65–120	1700–2100	2–3	50
Carbon-FRP	135–190	1700–2100	1–1.5	15
Steel	200	220–400	0.2[a]	–

[a]Yield deformation.

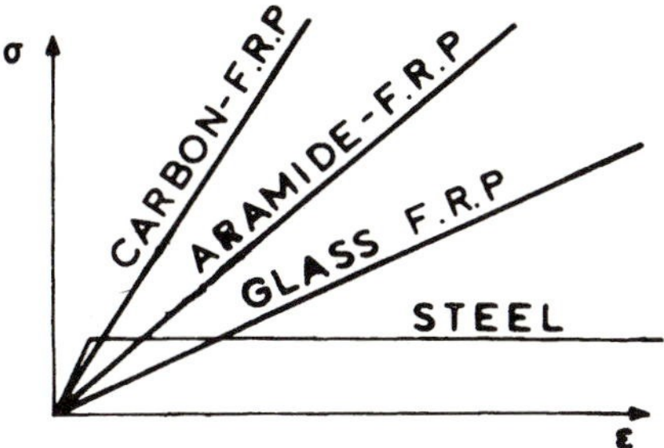

Figure 14.18 Constitutive laws of FRP.

existing structural elements, e.g. concrete to concrete, epoxy resin to concrete, steel to concrete, steel to steel acting through welding and so on. Consequently, load transfer from the original element to the additional 'reinforcing' materials is carried out through discontinuities, by means of unconventional mechanisms like friction, dowel action, large pull-out action and so on. The systematic study of these mechanisms constituting a kind of new mechanics for the non-continuum appears to be a fundamental prerequisite for the rational design of repaired and strengthened structural elements (Tassios, 1983; Tassios and Vintzèleou, 1987; CEB, 1991). However, besides the independent study of these force transfer mechanisms, the proper combination of several of them in integrated physical and mathematical models is needed for the safety verification of the structural elements, since the various repair or strengthening techniques may activate several force transfer mechanisms simultaneously. In this context extensive research is needed to bridge the existing gaps in knowledge in this area until this process is applicable to practical problems.

Therefore, at present the redimensioning and safety verification follow in practice a semi-empirical procedure based on practical rules supported by experimental evidence.

In the subsequent subsections these two methods will be presented in detail.

14.4.2 Revised γ_m-factors

No matter which one of the two methods mentioned above are followed for the redimensioning and safety verification, it should be stressed that special attention should be given to γ_m-factors introduced in the calculation.

Original materials will be factored as foreseen in EC2 (section 6.2.2). The strengths of additional materials attached to the original structural elements must be divided by increased γ_m-factors in recognition of the additional variabilities in reconstruction operations. Particularly in the case of cast-in-place new concrete, having in mind the above considerations, for the construction of new elements the use of concrete with a strength of 5 MPa higher than that of the original elements has been recommended (section 14.3). Thus, the designer may retain the same γ_m-factor for both the original and the new element, with the condition that the strength introduced in the redesign calculations will be that of the original concrete.

14.4.3 Load transfer mechanisms through interfaces

In the subsequent paragraphs the most common transfer mechanisms along the several discontinuities or interfaces between existing and additional material will be presented as they are grouped in EC8/Part 1.4/Draft (CEN, 1993).

(a) Compression against precracked interfaces

During reloading after cracking due to tension, compressive forces may be carried prior to full recovery of the previous extensional deformation, since the protruding elements constituting the rough surface at both faces of a crack may come into earlier contact due to their transversal microdisplacement (uneven bearing). Consequently, it is allowed to account for this phenomenon by means of an appropriate model (Figure 14.19) (Tassios, 1983; Gylltoft, 1984). The quantitative evaluation of such a model needs extensive experimental support.

(b) Adhesion between non-metallic materials

Local adhesion versus local slip between old and new materials may be accounted for by means of appropriate models, but taking into account their sensitivity to curing conditions and the characteristics of possible bonding agents. Taking into account that the value of the slip needed to mobilize adhesion is very low, it is permissible to consider that the entire edhesion resistance is developed under almost zero displacement (Figure 14.20) (Hanson, 1960; Ladner and Weber, 1981; Tassios, 1983).

(c) Friction between non-metallic materials

In several cases, friction resistance may be accounted for as a function of relative displacement (slip) along the discontinuity or along the interface. A consti-

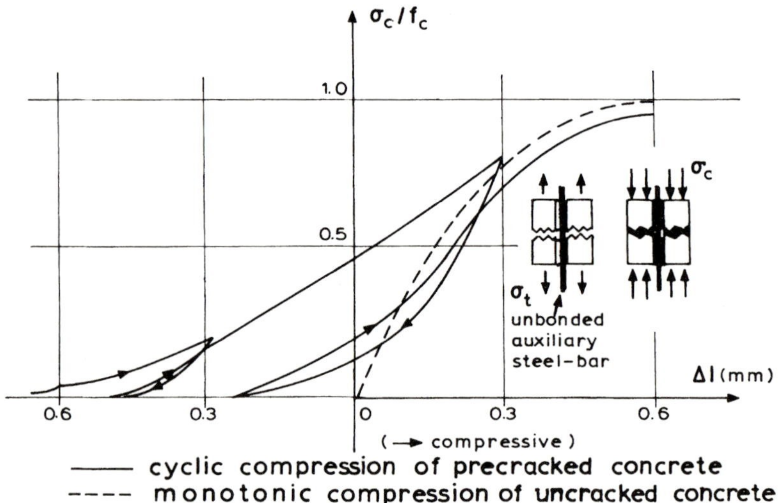

Figure 14.19 Monotonic and cyclic compression of cracked concrete.

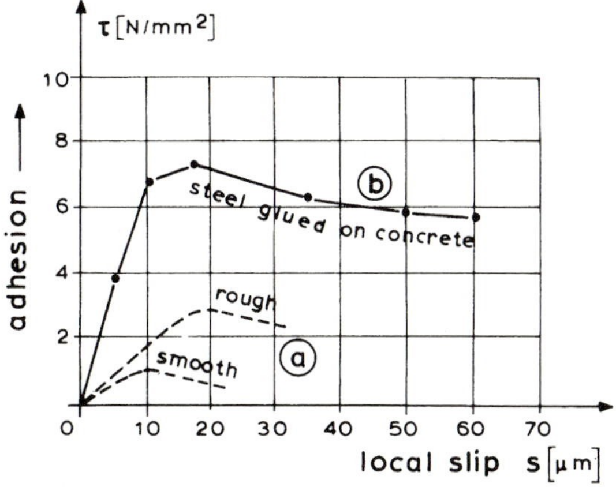

Figure 14.20 Constitutive law of adhesion: (a) concrete to concrete through bonding agent (Hanson, 1960); (b) steel sheets glued to concrete by means of epoxy resin (Ladner and Weber, 1981).

tutive law must be formulated for this purpose based on experimental data (Figure 14.21).

In some cases, when the slip needed to activate the maximum friction resistance (τ_u) is relatively low, the concept of 'friction coefficient' that is

$$\tau_u = \mu\sigma_u \tag{14.2}$$

may be used. However, for relatively low σ-values, the strong relation between 'μ' and 'σ' values must be taken into account (Figure 14.22) (Tassios, 1983).

(d) Load transfer through resin layers

The tensile strength of the contact interface between a resin layer and a given material (e.g. concrete) may be taken as equal to the tensile stength of the weak-

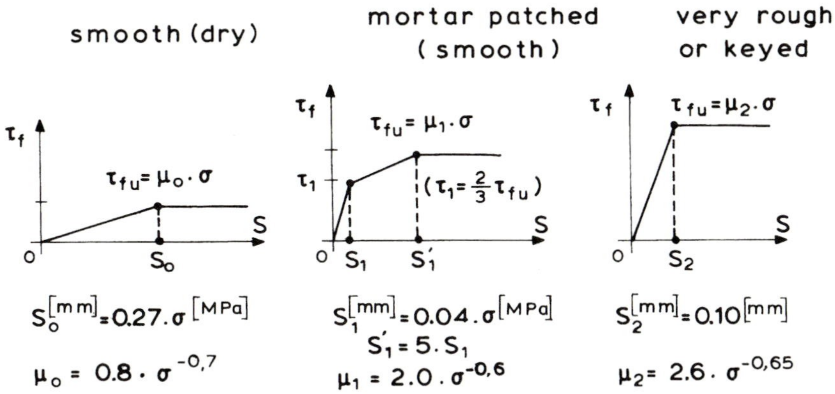

Figure 14.21 Formalistic models for concrete-to-concrete friction, as a function of normal compressive stress, σ.

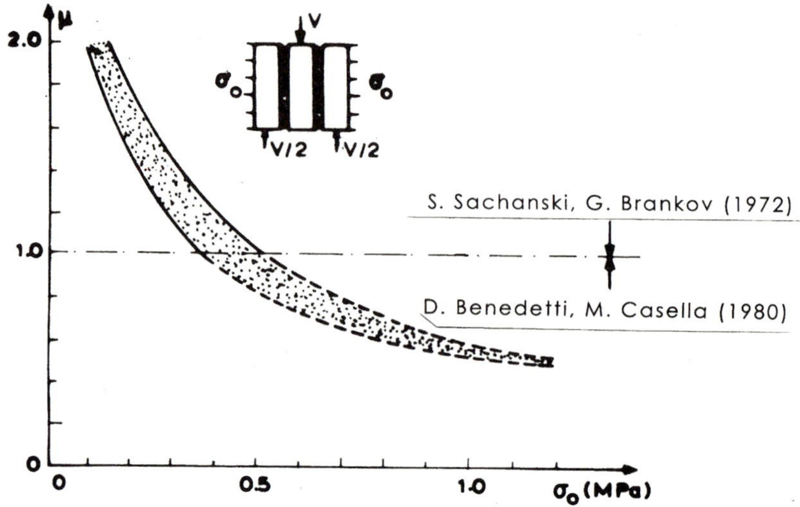

Figure 14.22 Friction coefficients for masonry as a function of the average normal stress.

er of the two. Therefore, in the case of concrete, its tensile stength f_{ctk} must be introduced in all calculations related to the load transfer through this interface. Of course this value must be divided by a γ_m at least equal to 1.5.

The local shear resistance generated along such an interface is a function of the local slip and the normal stress acting on the area under consideration. Figure 14.20 gives the constitutive law of the shear resistance as a function of the slip for σ equal to zero (adhesion).

(e) Clamping effect of steel across interfaces

The friction generated across a sheared interface transversly reinforced by well-anchored steel bars may be evaluated as follows (Figure 14.23) (Chung and Lui, 1978):

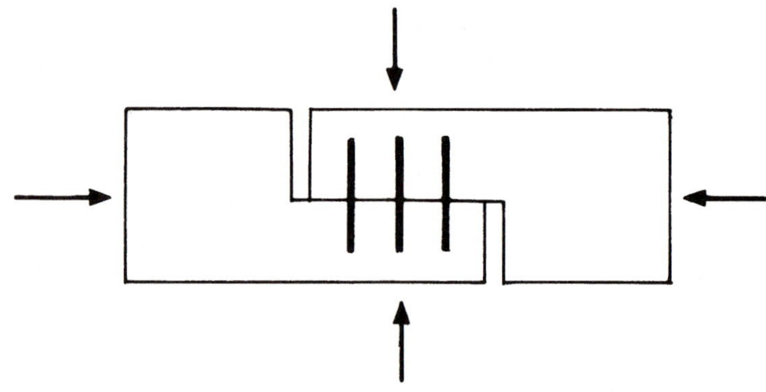

Figure 14.23 Clamping effect of steel across interfaces.

1. In the case of expected large relative displacement along the interface, the ultimate friction resistance may be estimated as

$$\tau_R = \mu\tau_{tot} \not> \tau_{u.m} \quad (14.3)$$

where μ denotes the friction coefficient available under normal stress (Figure 14.21) and

$$\sigma_{tot} = \rho f_y + \sigma_0$$

f_y is the yield strength of steel, σ_0 the external normal stress across the interface, ρ the effective steel ratio along the interface and $\tau_{u.m}$ the shear resistance of the material itself.
2. If large slips along the interface are not tolerated, the generated friction resistance is evaluated, taking into account the displacement compatibility on both faces of the interface.

(f) Dowel action

The design value of the maximum shear force which may be transferred by a bar crossing an interface may be calculated taking into account the strength and deformability of the dowel and the connected material as well as the distance of the dowel from the edges. According to Rasmussen (1963) for the plastic compressive stage

$$D_u \cong 1.3 d_b^2 \sqrt{f_c f_y} \quad (14.4)$$

while according to Vintzèleou and Tassios (1986) for the post-cracking stage

$$\frac{D}{D_u} \cong 0.7\sqrt[4]{s}\,(\text{mm}) \quad \text{for } d_b \cong 12\text{–}22 \text{ mm} \quad (14.5)$$

where D_u is the ultimate capacity of a dowel embedded in uncracked concrete, f_c the unconfined strength of concrete, f_y the yield strength of steel, d_b the dowel diameter, s the local slip at the interface (in mm) and D the dowel action for slip equal to s.

(g) Anchoring of new reinforcement

1. Anchorage lengths of steel bars in new concrete must follow the criteria of relevant codes such as EC2. In the case of bar anchorages in holes bored in old concrete where special grouts are used (e.g. high-strength concrete with shrinkage compensating admixtures, resin concretes) shorter anchorage lengths are needed. These are specified in the manuals of the material used and must be verified by pull-out tests performed by an authorized laboratory.
2. In most cases, the anchoring of additional steel bars is accomplished by welding them on to the existing bars directly or by means of additional welded spacers (Figure 14.17); such force transfers may be considered as rigid. In such cases it is necessary to verify that the bond ensured by the existing bar is sufficient to anchor the total force acting on both bars.

(h) Welding of steel elements

In designing steel-to-steel connections by means of welding, in addition to the checks of welding resistance, the following mechanical behaviour should be considered since the activation of force transfer depends on the concept of the connection:

- Direct welding of additional bars or steel profiles on existing ones ensures a complete generation of force transfer with almost zero slip.
- Intermediate deformable steel elements necessitate the introduction of proper models so that compatibility of deformations may be ensured (Tassios, 1983).

(i) Final remarks

From the preceding presentation the following conclusions can be drawn:

1. The constitutive laws of the transfer mechanisms need to be supported by additional experimental evidence covering several parameters related to the intervention techniques. This need becomes even greater if cyclic loading is to be included for the approach to energy dissipation or ductility issues.
2. It should be stressed that in designing the repair or strengthening of a structural element several force transfer mechanisms are generated, so that only an integrated model based on the finite element method (FEM) may take all of them into account, the interrelations among them and the level at which each of them is activated during loading, as happens with the analysis of original R/C elements or masonry walls (Ignatakis, Stavrakakis and Penelis, 1989, 1990).
3. Futhermore, even if such models based on the FEM were available, they would have to be verified through experimental evidence on repaired or strengthened structural subassemblages.
4. From the foregoing it is concluded that at present the formation of integrated analytical models cannot yet lead to dimensioning or safety verification methods for general use, suitable for practical applications. However, it is hoped that in the near future this procedure will lead to the derivation of reliable models.
5. For the time being the approach to the problem is based on a simplified estimation of resistances originating from practical rules which are verified by laboratory tests. Sometimes this approach is combined with oversimplified models of force transfer mechanisms as will be seen later. In the next subsection the basic concept of this semi-empirical method used in practice will be given in detail.

14.4.4 Simplified estimation of the resistance of structural elements

1. The basic concept in developing any repair or strengthening technique is to ensure that failure of the repaired structural element as a monolithic unit will precede any failure at the interfaces between old and new material. This is verified by tests and where failure at the interfaces occurs first, extra connecting means are provided on an empirical basis (e.g. closer spaced dowels, resin layer between old and new concrete). In order for this basic concept to be accomplished the specifications referring to each intervention technique should be rigorously followed during the execution of the work.
2. Having the above concept as a prerequisite, specimens of the repaired or strengthened structural elements are tested in the laboratory under monoton-

ic or cyclic loading to failure, and relevant displacement versus resistance diagrams are plotted. From these diagrams the basic values of ultimate strength $R_{u,rep}$, stiffness K_{rep} and energy dissipation $E_{u,rep}$ are determined.

3. The above values are calculated in parallel, based on the assumption that the structural element under consideration was constructed as a monolithic unit, including the initial element and its repair with the same cross-section and reinforcement. It is obvious that the respective values of $R_{u,monol}$, K_{monol} and $E_{u,monol}$ will be greater or at least equal to those of the repaired element. Therefore, 'model reduction factors' are introduced (EC8/Part 1.4/Draft), that is

$$\varphi_R = \frac{R_{u,\,rep}}{R_{u,\,monol}} \qquad \varphi_K = \frac{K_{rep}}{K_{monol}} \qquad \varphi_E = \frac{E_{u,\,rep}}{E_{u,monol}} \tag{14.6}$$

   The index 'monol' refers to a monolithic element consisting of the initial element and the repair. These factors allow the redimensioning and safety verification of the repaired element to be carried out as if it were a monolithic unit. In fact the results of the analysis which is based on monolithic considerations, are multiplied by the model reduction factors, in order to comply with those expected for the repaired or strengthened element. The whole procedure is accomplished with some additional simplified force transfer checks at the critical interfaces, as will be discussed later.

4. From the preceding presentation it is concluded that 'model reduction factors' have reliable values only for the special cases for which laboratory tests were performed. If the geometrical data of the original and the added sections are different, or the span or the height of the structural element change, there is no evidence which supports the notion that these values will still be valid. Therefore, it is clear that additional experimental and analytical research is urgently required to provide information about the seismic behaviour of structures repaired or strengthend by different techniques (Rodriguez and Park, 1991).

14.5 REPAIR AND STRENGTHENING OF STRUCTURAL ELEMENTS

14.5.1 General

Structural elements, depending on the desirable seismic resistance, the damage level and their type of joints, may be repaired or strengthened with resin injections, replacement of broken-off parts, glued-on plates, R/C jackets, or metal cages.

As mentioned in section 14.4.4, they key to the success of the repair or strengthening procedure is to attain a high degree of bonding between the old and the new concrete. This can be accomplished as follows:

- by roughening the surface of the old concrete
- by coating the surface with epoxy or other type of resin before concreting;
- by welding reinforcement bars;
- by using steel dowels.

The ductility of the repaired element is improved by proper confinement with closely spaced hoops, with steel jackets, with composite materials jackets, and so on.

It should be kept in mind that changes in the sectional area of the structural elements lead to a redistribution of stress due to resulting changes in the stiffnesses of the various structural elements.

Metal cages made of steel angles and straps are used exclusively for column repair. However, the repair of the joint between column and beam is not possible.

The gluing of metal or FRP plates on concrete is in general a technique easy to apply, whereby tension zones can be strengthened without altering the stiffness.

The last two methods require special means of fire protection, which is not the case with R/C jackets.

14.5.2 Columns

Damage to columns appears at different levels such as;

- fine cracks (horizontal or diagonal) without crushing of concrete or fracture of reinforcement.
- surface spalling of concrete without damage to the reinforcement;
- crushing of concrete, breaking of the ties and buckling of the reinforcement.

Depending on the degree of damage, different techniques may be applied, such as resin injections, removal and replacement or jacketing.

(a) Local interventions

Resin injections and resin mortars are applied only for the repair of columns with small cracks or peelings, without crushing of concrete or damage in the reinforcement. The degree of retrofit can be checked by comparing the force–displacement (H–δ) diagrams of the original column and the repaired one with epoxy resins (Figure 14.24) (Sariyiannis and Stylianidis, 1990; Sariyiannis, 1990). The results from such comparisons are very encouraging, with regard to the effectiveness of the repair.

Removal and replacement are applied in columns with high degree of damage, that is, crushing of concrete, breaking of ties and buckling of longitudinal reinforcement. Of course, before carrying out such work, a temporary support system is always provided to carry the column loads. Then, if concrete failure is only superficial, partial removal and repair are carried out (Figure 14.25); otherwise, if it is a total failure there is a complete removal of the material, placement of new logitudinal reinforcement with welding, placement of new closely spaced ties and concreting (Figure 14.26). It should be mentioned that in the first case good bonding between old and new concrete is absolutely necessary. In the second case, most of the time the construction of an R/C jacket follows the retrofit.

(b) R/C jackets

R/C jackets are applied in the case of serious damage or inadequate seismic resistance of the column. Depending on the existing local conditions, jackets are

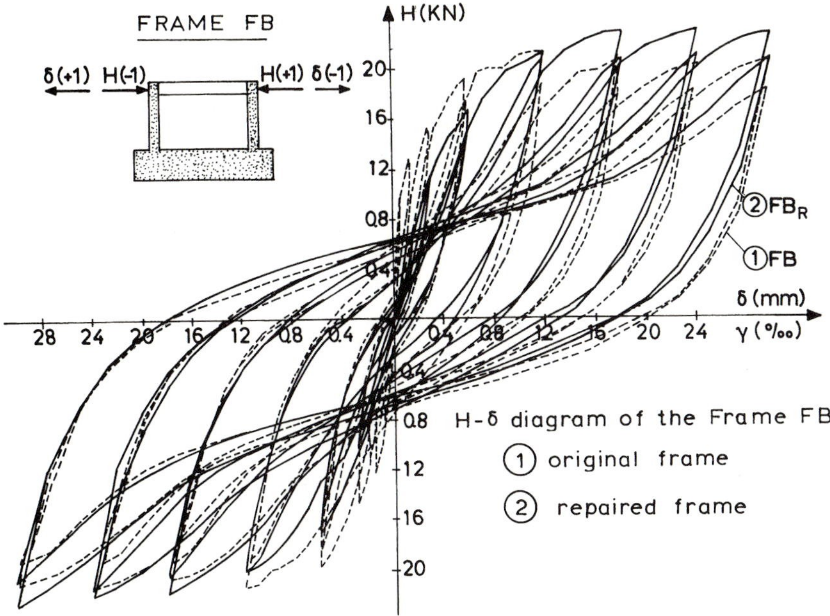

Figure 14.24 H–δ diagram of the original frame and then when repaired with epoxy resin injections.

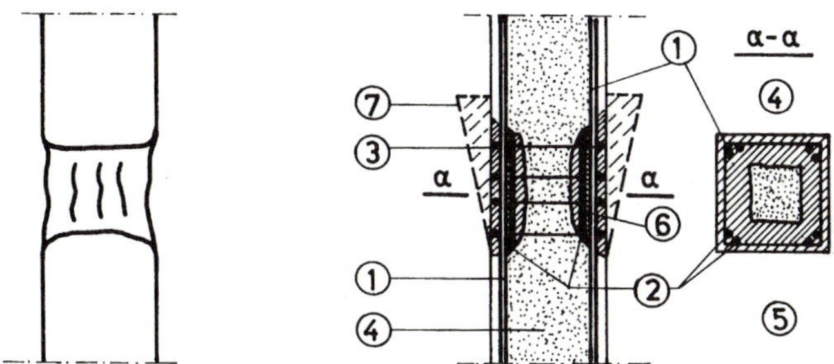

Figure 14.25 Column repair in the case of superficial damage: 1 = existing reinforcement; 2 = added new reinforcement; 3 = added new ties;4 = existing concrete; 5 = new concrete; 6 = welding; 7 = temporary cast form.

applied at the perimeter of the column, which is the ideal case, or sometimes on one or more sides (Figure 14.27). In cases where the jacket is limited to the storey height, an increase in the axial and shear strength of the column is achieved with no increase in flexural capacity at the joints. Therefore it is recommended that the jackets protrude through the ceiling and the floor slabs of the storey where column repair is necessary (Figure 14.28).

In the case of one-sided jackets, special care should be taken to connect the old with the new part of the section; this can be accomplished by welding closely spaced ties to the old reinforcement (Figure 14.29).

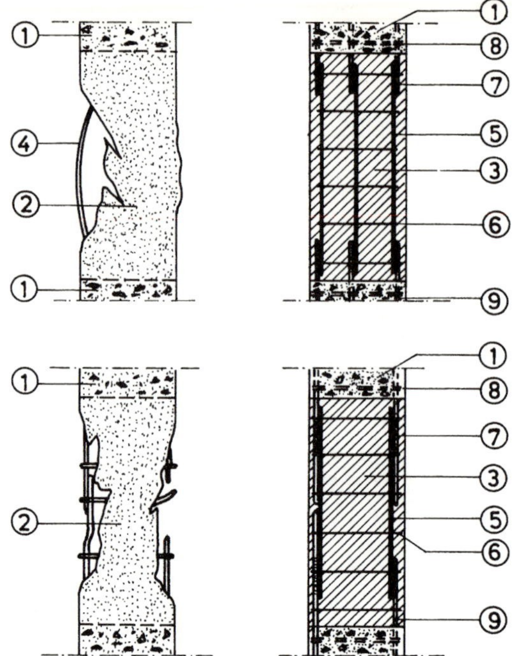

Figure 14.26 Repair of a seriously damaged column: 1 = existing undamaged concrete; 2 = existing damaged concrete; 3 = new concrete; 4 = buckled reinforcement; 5 = added new reinforcement; 6 = added new ties; 7 = welding; 8 = existing ties; 9 = existing reinforcement.

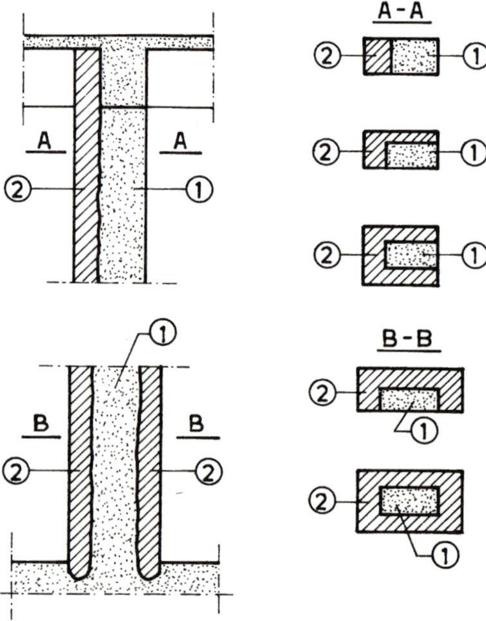

Figure 14.27 R/C column jacketing arrangement: 1 = existing column; 2 = jacketing concrete.

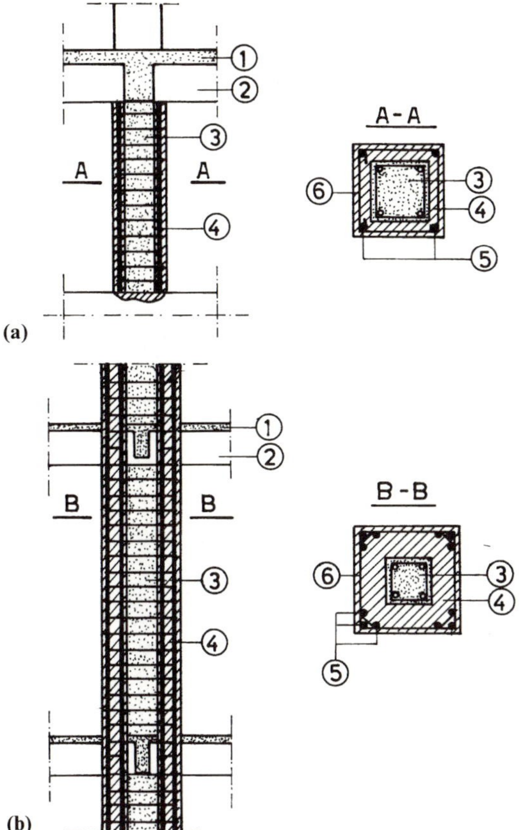

Figure 14.28 Column jackets: (a) jacket along the height of one storey; (b) jacket extended to the upper and lower storey, 1 = slab; 2 = beam; 3 = existing column; 4 = jacket; 5 = added longitudinal reinforcement; 6 = added ties.

In the usual case of full jackets, the composite action of the old and the new concrete is sometimes left solely to the natural bonding of the two materials, which can be strengthened with roughening of the old surface, and sometimes it is strengthened by welding some bent-up bars between the old and the new longitudinal reinforcement (Figure 14.30). This connection is necessary when the column has completly deteriorated or when its height is too large, in which case there is a danger of buckling of the longitudinal new reinforcement. However, laboratory tests have shown that, in general, the degree of composite action obtained is very satisfactory even without the strengthning of force transfer by welding the longitudinal reinforcement (Zografos, 1987).

(c) Steel profile cages

In general, this is a technique not widely used. The cage consists of four steel angles of minimum dimensions L 50.50.5, which are connected to each other with

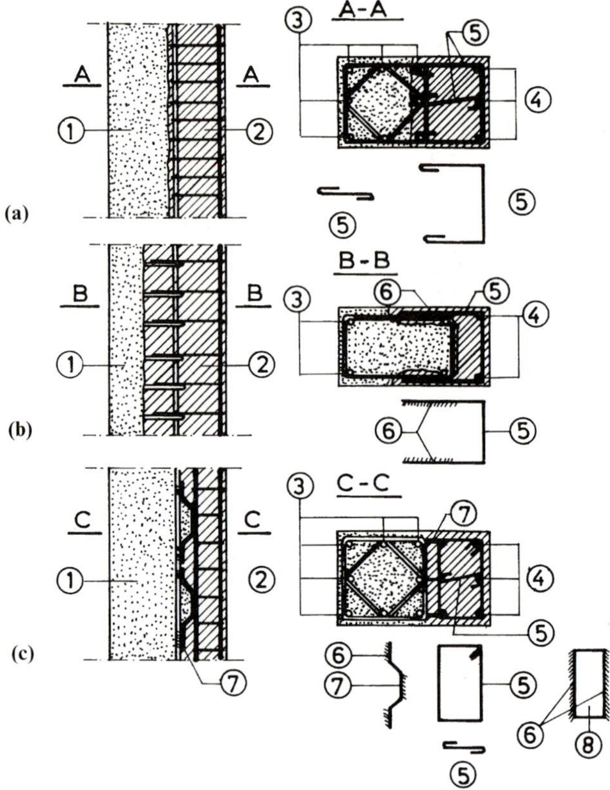

Figure 14.29 One-sided strengthening of a column: 1 = existing column; 2 = jacket; 3 = existing reinforcement; 4 = added longitudinal reinforcement; 5 = added ties; 6 = welding; 7 = bent bars; 8 = metal plate.

welded blades of minimum dimensions 25.4 mm (Figure 14.5). Prior to welding the angles are held tight on the column with the aid of transverse angles and pre-stressed ties (Figure 14.5). The voids between the angles and the concrete are filled with non-shrinking mortar (EMAKO, EMPECO, etc.) or resin grout and then the column is covered with gunite or cast-in-place concrete reinforced with welded wire fabric. It is obvious that with this arrangement, increase in the flexural capacity of the column at the joints with the top and bottom is impossible, due to the fact that the cage is not extended into the floors above and below.

(d) Steel or FRP encasement

Steel or FRP encasement is the complete covering of an existing column with thin steel or FRP sheets. This type of intervention offers the possibility of only a small increase in column size. Steel sheets (with 4–6 mm thickness) are welded together throughout their length and located at a distance from the existing column. The voids between the encasement and the column are filled with non–shrinking cement grout.

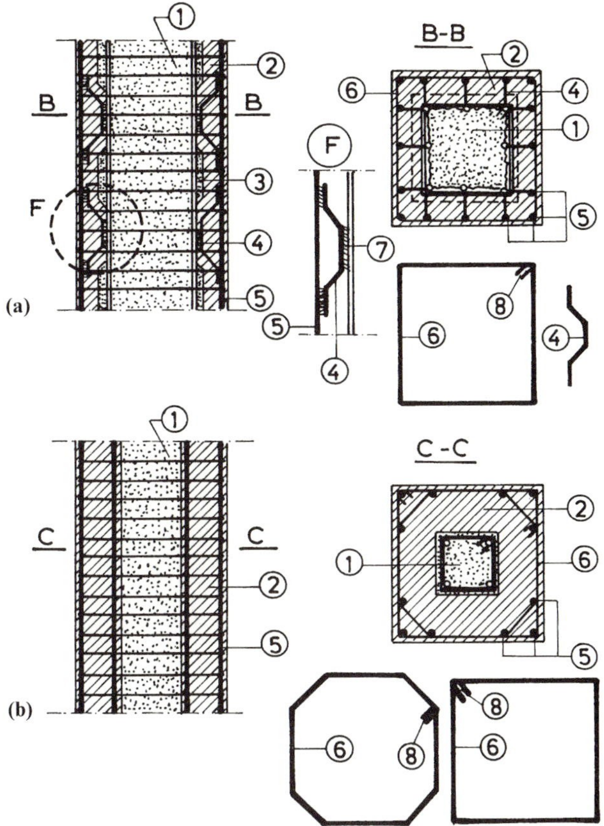

Figure 14.30 Connection of the old to the new reinforcement of the jacket: (a) protection of new bar against buckling with weldings; (b) protection of new bars against welding with octagonal ties. 1 = existing column; 2 = jacket; 3 = key; 4 = bent bars; 5 = added reinforcement; 6 = ties; 7 = welding; 8 = alternating corners.

The strengthening with FRP can be accomplished either with banding of the R/C column with continuous FRP straps glued on the concrete surface using epoxy resin, or with encasement using FRP sheets again glued on the concrete surface. In the case of banding, transverse prestress of the strap is possible, to increase confinement.

With the above interventions, ductility and axial force carrying ability can be considerably increased locally. However, the flexural strength of the frame structure cannot be improved because it is impossible to pass the encasement through the floors.

(e) Redimensioning and safety verifications

Recent experimental results (Sariyiannis and Stylianidis, 1990; Sariyiannis, 1990; French, Thorp and Tsai, 1990) regarding the dimensioning of repaired columns have shown the following:

1. In the case of repair with resin injections, the ratio of the strength of the repaired element to that of the original one is about 1. In general, the epoxy-repaired cracks do not reopen in tests; new cracks tend to develop adjacent to the repaired ones. The stiffness of the repaired column appears to exceed 85% of the original one, and the same happens with the energy dissipation capacity, that is

$$\frac{R_{d,rep}}{R_{d,orig}} \cong 1 \quad \frac{K_{rep}}{K_{orig}} \cong 0.85 \quad \frac{E_{rep}}{E_{orig}} \cong 0.85 \tag{14.7}$$

   The bond between reinforcement and concrete also appears to be restored, even for high inter–storey drifts exceeding 4%.
2. In the case of repair with reinforced jackets cast-in-place, the experimental results have shown (Zografos, 1987; Bett, Klingner and Jirsa, 1988; Bush, Talton and Jirsa,1990) that the lateral capacity of the strengthened column can be reliably predicted, assuming complete compatibility between the jacket and the original column. For jackets with gunite concrete, despite all the opposite estimation (NTU, 1978) the results fall slightly below those of conventional R/C jackets cast in forms. However, given the fact that the field conditions are not as ideal as those of a laboratory, the authors' opinion is that, on the one hand, the new concrete must have a strength 5 MPa greater than that of the original element, and on the other a model correction factor $\varphi \cong 0.80$ to the strength and the stiffness of the repaired element must be introduced:

$$\frac{R_{d,rep}}{R_{d,monol}} \cong 0.80 \quad \frac{K_{rep}}{K_{monol}} \cong 0.80 \tag{14.8}$$

   The index '*monol*' refers to a monolithic element consisting of the initial element and the jacket.
3. In the case of repair with metal cage of straps and angles (Arakawa, 1980; Tassios,1983) redimensioning may be done according to what was suggested in (2).
4. In the case of repair with glued steel or FRP sheets, the additional shear resistance V_{fc} of the column may be estimated by the following expressions:

   (a) For FRP (Triantaphillou, 1994; Balinger, Maeda and Hoshijima, 1993)

$$V_{fc} = 2tE_p\varepsilon_u h \cot\theta \tag{14.9}$$

   where t is the thickness of the sheet, E_p is the modulus of elasticity of the material, ε_u the deformation of the FRP (conservative values of ε may be considered as 0.005 for carbon FRP and 0.01 for glass or aramide FRP), h is the dimension of the column cross-section parallel to V_{fc} and θ is the angle between the column axis and diagonal cracks. It may be considered that $\theta \cong 30°$ (Priestley and Seible, 1991).

   (b) For steel

$$V_{fc} = 2tf_y h \cot\theta \tag{14.10}$$

   where f_y is the yield strength of the steel sheet. The coefficient 2 has been introduced to take into account that the plates are glued on both sides over the shear crack. In both cases a load transfer verification control is necessary on the glued interface between concrete and sheet (section 14.5.3).

14.5.3 Beams

As in the case of columns, depending on the degree of damage in the beams, several techniques are applied, such as resin injections, glued metal or FRP sheets, removal and replacement of concrete and R/C jackets.

(a) Local interventions

Resin injections are applied only for the repair of beams with light cracks without crushing of concrete.

Removal and replacement are applied to beams with a high degree of damage such as crushing of concrete or rupture of reinforcement, loss of bonding, spalling due to dowel action. Putting in temporary supports always precedes repair work of this type. The procedure which is then followed is similar to that described for column repair. However, at this point it has to be stressed that difficulties may arise regarding the compaction of concrete if it is not possible for casting to be carried out from the upper side of the beam with special openings in the slab.

(b) R/C jackets

Reinforced concrete jackets can be applied by adding new concrete to three or four sides of the beam. In the same technique one should also include the strengthening of the tension or compression zone of a beam through concrete overlays. In order to accomplish force transfer between old and new concrete, roughening of the surface of the old concrete is required, as well as welding of connecting bars to the existing and new reinforcement bars.

Reinforced overlays on the lower face of the beam (Figure 14.31) can only increase the flexural capacity of the beam. Existing reinforcement is connected to the new by welding.

Jacketing on all four sides of the beam is the most effective solution. The thickness of the concrete which is added to the upper face is such that it can be accommodated within the floor thickness (50–70 mm). The placement of the ties is achieved through holes which are opened in the slab at closely spaced distances, which are also used for pouring the concrete. The longitudinal reinforcement bars of the jacket are welded to those of the old concrete (Figure 14.32).

Jackets on three sides of the beam are used to increase flexural and shear capacity of the beam for vertical loading, but not for seismic actions, given that strengthening of the load-bearing capacity of the section near the supports is impossible. The key to the success of such an intervention is the appropriate anchorage of the stirrups at the top of the sides of the jacket (Figure 14.33). Due to the fact that using forms and pouring the concrete from the top is not possible, the only feasible solution is gunite concrete.

(c) Glued metal or FRP sheets

The technique for gluing metal or FRP sheets on to concrete was described in detail in a previous section. These sheets are glued either on the lower face of the beam under repair, for strengthening of the tension zone, or on the vertical sides of the beam near the supports, for shear strengthening. This procedure should be

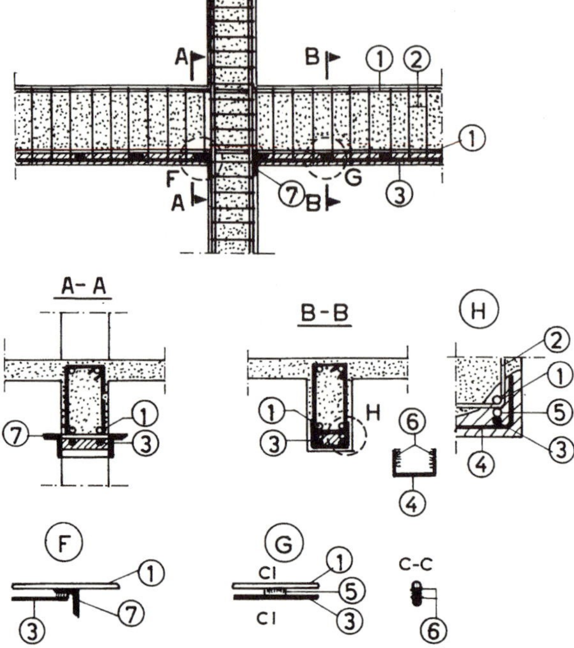

Figure 14.31 Strenthening of a beam on the lower face: 1 = existing reinforcement; 2 = existing stirrups; 3 = added longitudinal reinforcement; 4 = added stirrups; 5 = welded connecting bar; 6 = welding; 7 = collar of angle profiles.

preceded by crack repair with epoxy resin. The glued plates must be protected by welded wire mesh and cement plaster or shotcrete.

(d) Redimensioning and safety verification

(i) Resin injections Extensive laboratory tests (Popov and Bertero, 1975; French, Thorp and Tsai, 1990; Economou, Karayiannis and Sideris, 1994) have shown that if there is no concrete degradation, epoxy resin injections are very effective. The repaired beam is capable of resisting several loading cycles, the initial strength is completely restored, while stiffness and energy dissipation appear to be somewhat lower than those of the original beam.Consequently 'the model correction factor', φ, may be considered as equal to 1 in this case, that is

$$\frac{R_{\text{d,rep}}}{R_{\text{d,orig}}} \cong 1.0 \quad \frac{K_{\text{rep}}}{K_{\text{orig}}} \cong 1.0 \quad \frac{E_{\text{rep}}}{E_{\text{orig}}} \cong 1.0 \tag{14.11}$$

(ii) R/C overlays or jacketing Extensive experimental results have shown (Saiidi, Vrontinos and Douglas, 1990; Abdel-Halim and Schorn, 1989; Tassios, 1983; Vassiliou, 1975) that concrete overlays or jacketing are an effective technique for repair or strengthening. The additional layers and the parent concrete remain bonded throughout loading until failure, provided that construction spec-

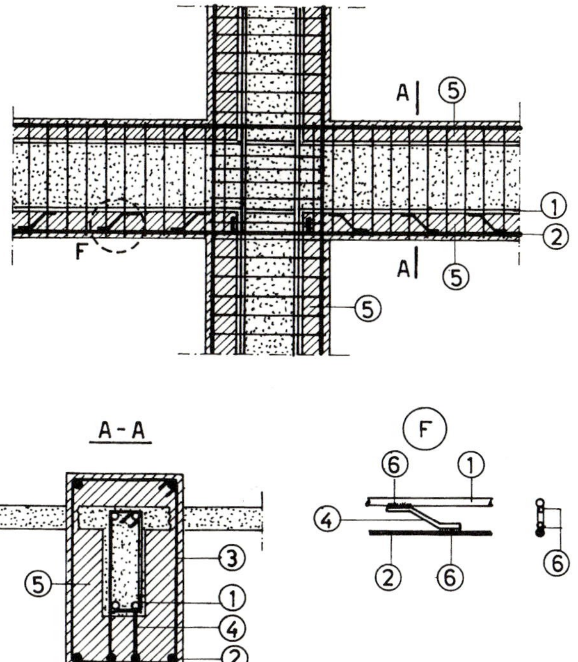

Figure 14.32 Jacket on four sides of a beam: 1 = existing reinforcement; 2 = added longitudinal reinforcement; 3 = added stirrups; 4 = welded connecting bar; 5 = concrete jacket; 6 = welding.

ifications given in the previous paragraphs are met. The reduction in strength of the repaired beam varies between 8 and 15% of the strength of the monolithic beam (initial+jacket). The reduction in stiffness of the repaired beam is somewhat higher (10–20%) with respect to the stiffness of the monolithic beam. Consequently 'the model correction factor', φ, may be considered as follows:

$$\frac{R_{\mathrm{d,rep}}}{R_{\mathrm{d,monol}}} \cong 0.85 \quad \frac{K_{\mathrm{rep}}}{K_{\mathrm{monol}}} \cong 0.75 \tag{14.12}$$

In addition to the general strength and stiffness verifications described previously, specific verifications for the force transfer mechanisms along the several interfaces between existing and additional material should be performed. In the case where adhesion between old and new concrete is proved to be inadequate, the transfer mechanism should be ensured with extra connectors on the interface. Two such cases can be identified:

1. *Interface of connection in the tension zone* (Figure 14.34a). The shear stresses developing on the interface between old and new concrete are given, according to the theory of strength of materials applied to reinforced concrete, by the approximate relationship (Tassios, 1984)

$$\tau_{02} = V_{\mathrm{d}} \Big/ \left[b_{\mathrm{w}} z \left(1 + \frac{A_{\mathrm{s1}}}{A_{\mathrm{s2}}} \, \frac{d_1 - x}{d_2 - x} \, \frac{z_1}{z_2} \right) \right] \tag{14.13}$$

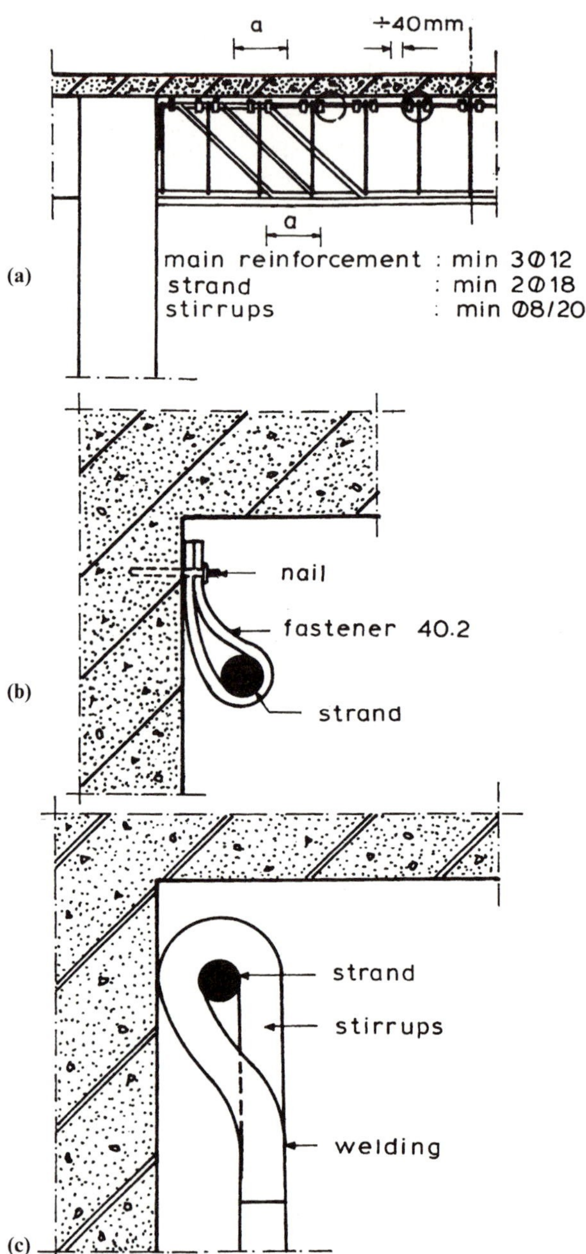

Figure 14.33 Jacket on three sides of a beam: (a) general reinforcement pattern (b) detail of fixing of the strand; (c) detail of anchoring the ties on the strand.

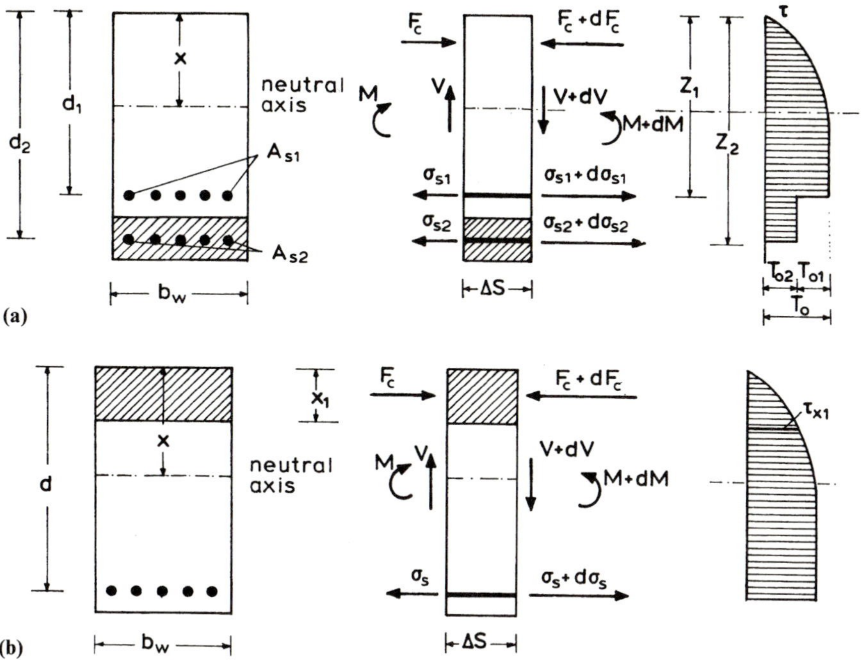

Figure 14.34 Shear stress between old and new concrete: (a) intervention in the tension zone; (b) intervention in the compression zone.

where

$$z = \frac{A_{s1}z_1 + A_{s2}z_2}{A_{s1} + A_{s2}}$$

Bearing in mind that special care is taken in securing the adhesion of the new to the old concrete through resin coats of higher strength than that of concrete, the value resulting from the above relationship at the interface must be compared with the basic concrete shear strength (see section 14.4.3 (d)). Therefore, if τ_{02} is greater than

$$\tau_{Rdl} = \tau_{Rd} k\,(1.2 + 40\rho_l) + 0.15\sigma_{cp} \qquad (14.14)$$

(see EC2 section 4.3.2.3) where τ_{Rd} is the basic design shear strength (Table 4.8, EC2)

$$\tau_{Rd} = 0.25\,\frac{f_{ctk}}{1.5}$$

$k = 1$ of $1.6 - d \nless 1$ (d in metres), ρ is the percentage of the longitudinal reinforcement in the concrete section

$$\rho_l = \frac{A_{sl}}{b_{wd}} \ngtr 1$$

and σ_{cp} is the axial stress due to axial loading or prestress (compression positive)

$$\sigma_{cp} = \frac{N_{sd}}{A_c}$$

that is

$$\tau_{02} \geqslant \tau_{Rd1} \tag{14.15}$$

then the total shear flow ($T = \tau_{02}b$) must be carried by welding of the new reinforcement to the old one.

Therefore, for a distance a between welding, welding thickness t and number n of new bars, the welding length l_{weld} must be equal to (Figure 14.17)

$$l_{weld} = \frac{\tau_{02} b_w a}{tn \times 0.8 f_{yd}} \tag{14.16}$$

where $f_{yd} = f_{yk}/1.15$ is the yield stress of the welding steel divided by the safety factor γ_s of the material (design strength). It is understood that a model correction factor equal to 0.8 has been introduced in formula (14.16).

2. *Interface of connection in the compression zone* (Figure 14.34(b)). The shear stresses developing at the interface between the old and the new concrete are again given according to the classic theory of strength of materials by the approximate relationship

$$\tau_{x1} = \frac{V_d}{b_w z} \frac{x_1}{x}\left(2 - \frac{x_1}{x}\right) \tag{14.17}$$

If the resulting value of τ_{x1} is larger than τ_{Rd1} as defined above, the total shear flow ($T = \tau_{x1}b$) must be carried by shear connectors (Figure 14.35). The ultimate shear carried by the two legs of such a connector is equal to (Tassios, 1984)

$$D_u \cong 2d^2\sqrt{f_{cd} f_{yd}} \tag{14.18}$$

where d is the diameter of the connector, f_{cd} the design strength of concrete and f_{yd} the design strength of the connector's steel. In equation (14.18) a model correction factor $\varphi = 1.30$ has been introduced (section 14.4.3(f)).

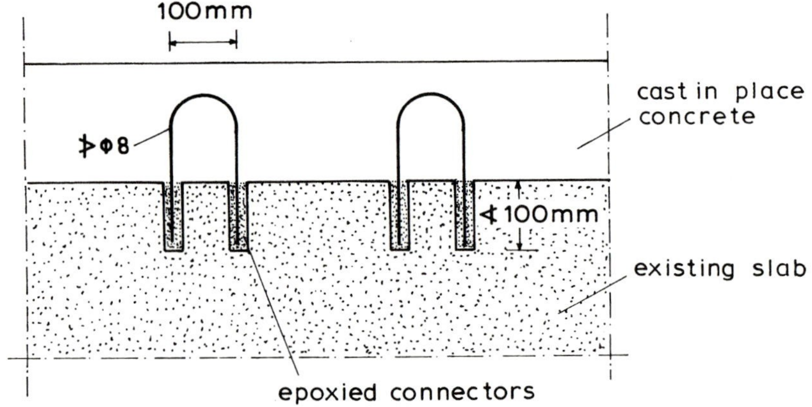

Figure 14.35 Shear connectors between old and new concrete in the compression zone.

(iii) Glued metal sheets The required section bt of a sheet in a flexural area results from the relationship

$$\Delta M_d \leqslant (bt) z f_{yd}$$

and hence

$$(bt)_{req} \geqslant \frac{\Delta M_d}{z f_{yd}} \tag{14.19}$$

(model correction factor equal to 1) where ΔM_d is the additional moment (strengthening) beyond the ultimate M_{du} carried by the original section (ΔM_d should not be greater than $0.5M_{du}$ for construction reasons), z the lever arm of the internal forces and f_{yd} the design strength of the sheet.

The required anchorage length of the sheet is given by the relationship

$$l_a = \varphi\left(\frac{f_{yd}}{r\tau_u} t\right) \tag{14.20}$$

where φ is the model correction factor ($\varphi \cong 1.3$), f_{yd} the design strength of the sheet, t the thickness of the sheet, τ_u the maximum local adhesion strength between concrete and the steel sheet for sheet thickness $t < 1$ mm, $\tau_u \cong 2f_{ctd}$, while for $t = 3$ mm, $\tau_u \cong f_{ctd}$ (Figure 14.20), f_{ctd} the tensile design strength of concrete and r the correction factor to take into account the non-uniform distribution of τ_u over the glued area, due to the different slippage from point to point, from the crack to the end of the sheet (Figure 14.20). Recommended value: $r = 0.40$ (Tassios, 1983).

The required thickness t of the sheets which are glued on both sides of a beam over shear cracks to carry additional shear forces may be given by the relationship (Tassios, 1984)

$$\Delta V_d \leqslant 2tzf_{yd} \cot \theta$$

$$t_{req} \geqslant \frac{1}{2} \frac{\Delta V_d}{2zf_{yd} \cot \theta} \tag{14.21}$$

For the meaning of the symbols included in the above relationships see section 14.5.2(e), item (4).

The safety verification of the force transfer through the glued interface may be carried out with the aid of the following expressions (Figure 14.36) (Tassios, 1983):

$$\frac{\Delta V_d}{2} \leqslant F_s = \left(\frac{1+\xi}{2}\right) d l_0 r \tau_u \tag{14.22}$$

taking into account that

$$l_0 = (1-\xi) d \cot \theta$$

equation (14.22) takes the form

$$\frac{\Delta V_d}{2} = \frac{1-\xi^2}{2} d^2 r \tau_u \cot \theta \tag{14.23}$$

For $\xi \cong 0.30$ and $\theta \cong 30°$ equation (14.23) takes the form

$$\Delta V_d \leqslant 0.7 \tau_u d^2 \tag{14.24}$$

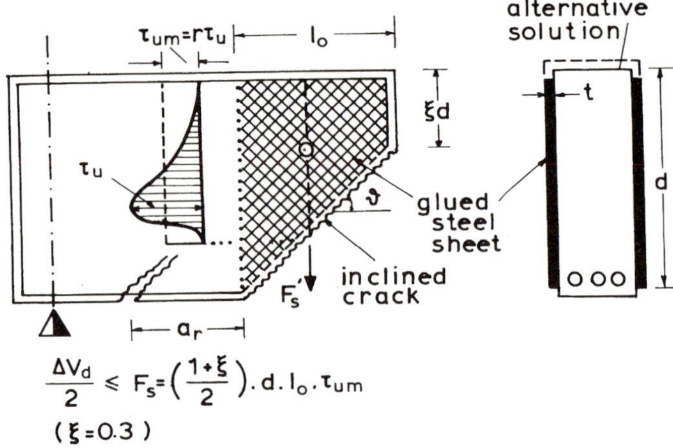

Figure 14.36 Shear force transfer through epoxy resin glued steel sheet.

Glued FRP sheets Equations (14.19)–(14.21) and (14.24) may be easily transformed to cover the case where steel is replaced by FRP in sheeets, by replacing f_{yd} of the steel with $\varepsilon_u E_p$ of the FRP (section 14.5.2(e), item (4)) as follows:

The required section of the sheet in a flexural area is given by the relationship

$$(bt)_{req} = \frac{\Delta M_d}{z\varepsilon_u E_p} \tag{14.25}$$

The required anchorage length of the sheet is given by the relationship

$$l_a = \varphi\left(\frac{\varepsilon_u E_p}{r\tau_u}\ t\right) \tag{14.26}$$

The required thickness t of the plates which are glued on both sides of a beam over shear cracks to carry additional shear forces may be given by the relationship

$$t_{req} \geqslant \frac{1}{2}\ \frac{\Delta V_d}{z\varepsilon_u E_p \cot\theta} \tag{14.27}$$

Safety verification of the force transfer through the glued interface may be done by using the following expression:

$$\Delta V_d \leqslant 0.7\tau_u d^2 \tag{14.28}$$

14.5.4 Beam–column joints

Depending on the degree of damage the following techniques are applied for the repair of beam to column joints:

- resin injection
- X-shaped prestressed collars

- glued steel plates
- R/C jackets.

(a) Local repairs

Resin injections are applied in the case of fine and moderate cracks, without degradation of concrete or buckling of the reinforcement bars. However, restoration of bond between steel and concrete with the aid of epoxy resin is questionable since controversial results appear in the international literature (Popov and Bertero, 1975; French, Thorp and Tsai, 1990). Therefore the joint should be strengthened at the same time with one of the techniques which will be presented next, especially in the case of frame structural systems without R/C walls.

(b) X-shaped prestressed collars

After the cracks are filled with resin injections, or after the decomposed concrete is removed and the voids are filled with epoxy or non-shrinking mortar, the joint is strengthened with external ties (collars) which are prestressed with tensioner couplers (Figure 14.37). Then the joint is covered with welded wire fabric and a jacket of gunite concrete. When four beams are framing into the joint the application of this technique is not feasible because the X-shaped collars cannot go through the joint (NTU, 1978).

(c) R/C jackets

The construction of R/C jackets to a damaged joint is the safest method for strengthening. This is generally a difficult technique given the fact that a jacket must usually be constructed for every structural element framing into the joint. It is obvious that roughening of the surfaces is required as well as punching of the slabs in order for the ties to go through, and gluing of the damaged joint area with resins must precede the construction of the R/C jackets (Figure 14.38).

(d) Glued metal plates

Glued metal plates can only be applied to plane joints, as the case is with X-shaped collars. This is a technique which provides strengthening to the joint without altering its dimensions. Local repair precedes the gluing of the plates and then the plates are tied with prestressed bolts (Figure 14.39). The thickness of the plates in this case must be at least 4.0 mm, which does not create any problems with the gluing process since the plates are kept tight to the concrete surface with the aid of prestressed bolts.

(e) Redimensioning and safety verification

The redimensioning of the joint is carried out under the assumption that complete compatibility has been achieved between the original element and the added material as happens with the columns. The internal force distribution is given in Figure 14.40. However, given the fact that the field conditions are not as ideal as those in a laboratory, the authors' opinion is that a model correction factor, φ,

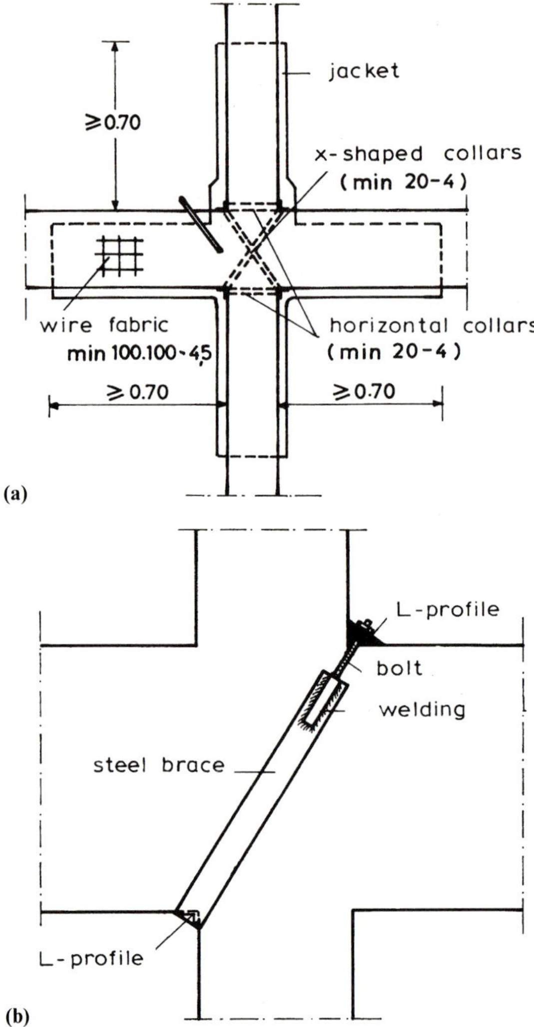

Figure 14.37 Strengthening of a joint with prestressed collars: (a) general arrangement of the strengthening; (b) detail of the prestressed collar.

should be introduced, equal to

$$\frac{R_{\mathrm{d,rep}}}{R_{\mathrm{d,monol}}} \cong 0.80 \qquad \frac{\mathrm{K}_{\mathrm{rep}}}{\mathrm{K}_{\mathrm{monol}}} \cong 0.80 \tag{14.29}$$

14.5.5 R/C walls

It is well known that R/C walls, due to their high stiffness and shear strength, are the most effective seismic-resistant elements of a structure. Therefore, the repair

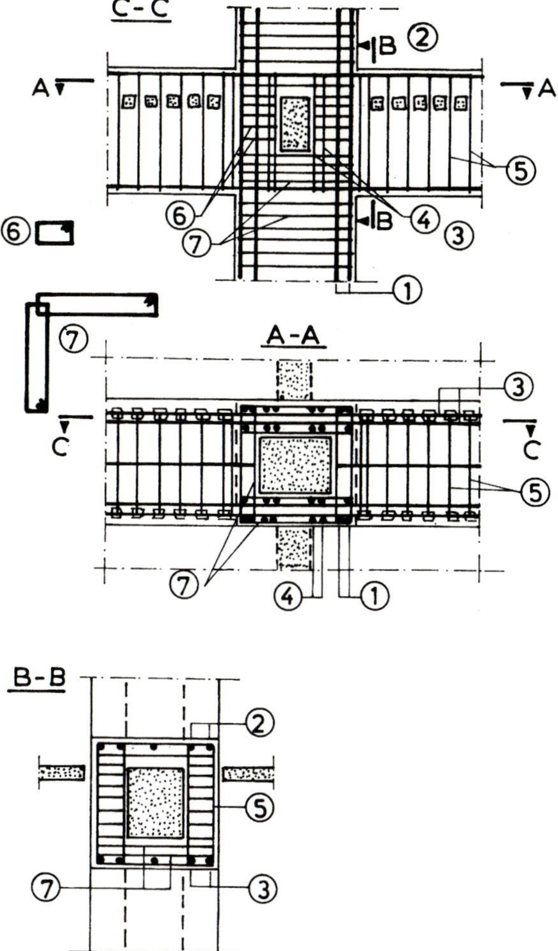

Figure 14.38 Strengthening of a joint with a jacket: 1 = column reinforcement; 2 = beam top reinforcement; 3 = beam bottom reinforcement; 4 = joint vertical stirrups; 5 = beam stirrups; 6 = column ties; 7 = column ties in joint.

and strengthening of a damaged R/C wall can drastically improve the seismic resistance of a structure.

(a) Local repairs

If a properly reinforced wall exhibits cracks with small width, without bond deterioration or concrete crushing, it can be repaired with epoxy resins. Laboratory tests have shown that such an intervention fully restores the strength of the wall, but not its stiffness and energy dissipation capacity due to the fact that resin cannot penetrate into the capillary cracks which accompany cracks with larger open-

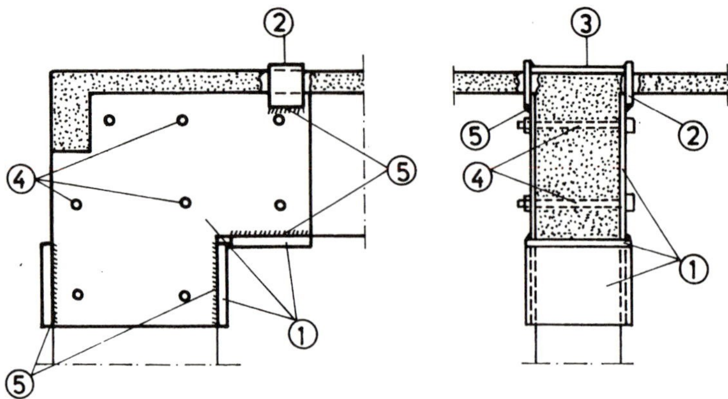

Figure 14.39 Glued metal plates on a joint:1: 1 = steel plate; 2 = steel plate; 3 = steel strap; 4 = prestressed bolts; 5 = welding.

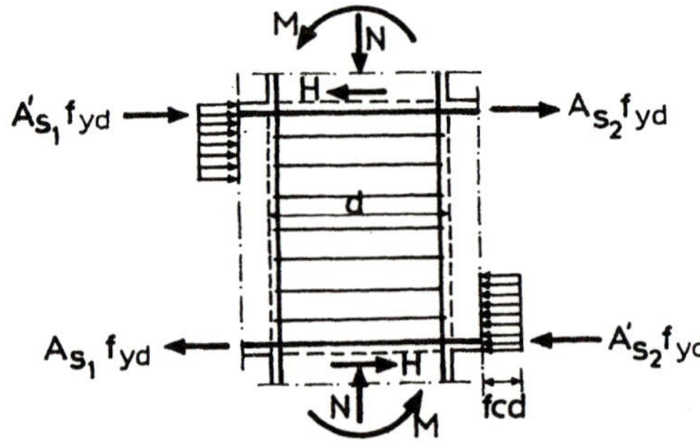

Figure 14.40 Schematic representation of the internal forces in a joint.

ings (Tassios, 1983; Lefas, Tsoukis and Kotsovos, 1990; Lefas and Kotsovos, 1990).

It should be mentioned here that most of the walls in older buildings have inadequate reinforcement due to code requirements applied in earlier years. Thus, a simple repair with resin injections is very often not enough. It needs to be combined with R/C jackets to strengthen the wall.

(b) R/C jackets

R/C jackets can have one of the forms shown in Figure 14.41. In the case of a jacket on both sides of the wall, the connection of the two layers with through thickness ties is necessary (at least 3ϕ 14/m^2).

At the points where the wall passes from one storey to the other it is necessary to punch holes in the slab and place diagonal reinforcement through them (Figure 14.42).

During the construction of the R/C jackets the following rules apply:

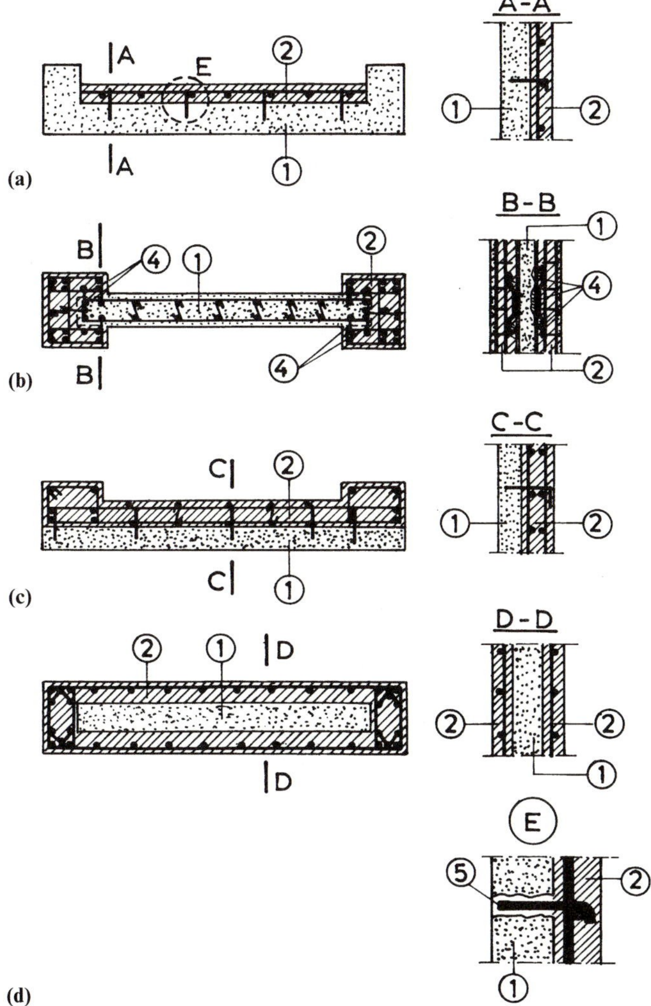

Figure 14.41 Strengthening of a wall with a jacket: 1 = existing wall; 2 = added wall; 3 = added columns; 4 = welding; 5 = epoxied bar.

- The strength of the new concrete must be at least 5 MPa greater than that of the old concrete.
- The minimum thickness of the jacket should be 50 mm on each side.
- The minimum horizontal and vertical reinforcement should be 0.25% of the section of the jacket.
- The minimum reinforcement with which the ends of the wall are strengthened should be 0.25% of the section of the jacket.
- The diameter of the ties at the wall ends should not be less than 8 mm with a maximum spacing not exceeding 150 mm.

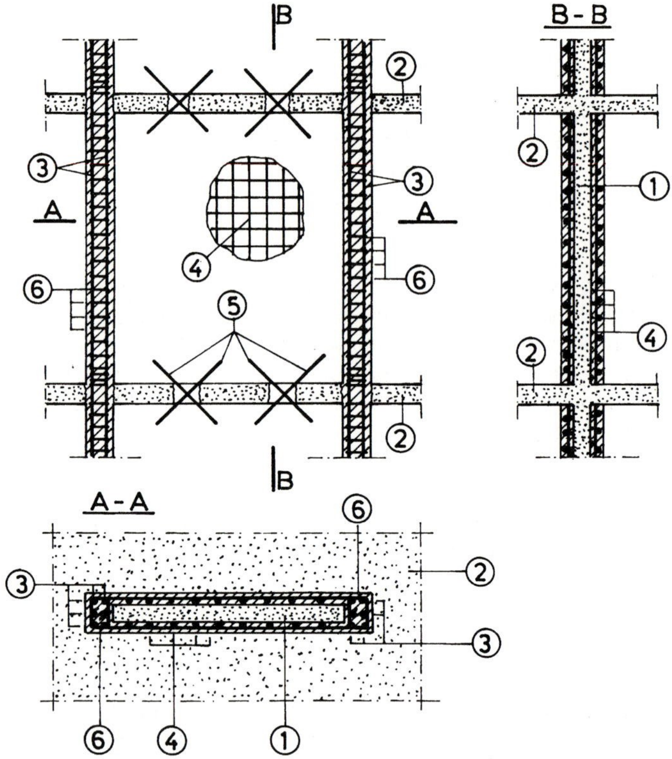

Figure 14.42 General arrangement for the strengthening of a wall: 1 = existing wall; 2 = existing slab; 3 = added longitudinal reinforcement; 4 = added wire fabric; 5 = diagonal connecting bars; 6 = added ties.

- The jacket must be anchored to the old concrete with dowels spaced at no more than 600 mm in both directions (NTU, 1978; AUT, 1978).

(c) Redimensioning and safety verification

1. In this case of repair with resin injections the ratio of the strength of the repaired element to the strength of the original element may be taken equal to 1, as discussed earlier, while the ratio of the stiffness and energy dissipation capacity may be taken equal to 0.85, that is

$$\frac{R_{\mathrm{d,rep}}}{R_{\mathrm{d,orig}}} \cong 1 \qquad \frac{K_{\mathrm{rep}}}{K_{\mathrm{orig}}} \cong 0.85 \qquad \frac{E_{\mathrm{rep}}}{E_{\mathrm{orig}}} \cong 0.85 \tag{14.30}$$

2. In the case of repair with jackets, provided that the damaged wall was repaired earlier either with resins or resin mortars or non-shrinking cement mortars, the behaviour of the repaired element does not differ from that of the monolithic one (original+jacket), as far as both strength and stiffness are concerned. However, as in the case of columns, walls are dimensioned based on the rela-

tionships

$$\frac{R_{d,rep}}{R_{d,monol}} \cong 0.80 \qquad \frac{K_{rep}}{K_{monol}} \cong 0.80 \tag{14.31}$$

The required number of dowels between the original wall and the jacket can be estimated by the relationship (Tassios, 1984)

$$n_d = \frac{[(V_d - V_{R\,orig}) - l_w\, h_w\, \tau_{adh}]}{D_u} \tag{14.32}$$

where V_d is the shear strength of the repaired wall (wall+jacket) (MN), $V_{R\,orig}$ the shear strength of the original wall after it is repaired, estimated to be 0.80 of the strength of the original undamaged wall (MN), l_ω, h_ω the dimensions of the wall under repair (m), τ_{adh} the average adhesion design strength of the new to the old concrete estimated to be equal to τ_{Rd1} (section 14.5.3(d), item 2) and D_u the dowl strength equal to

$$D_u = d^2\sqrt{f_{cd}\, f_{yd}}\ (\mathrm{MN}) \tag{14.33}$$

where d is the diameter of the dowel (m), f_{yd} the design strength of the dowel (MPa), and f_{cd} the design strength of concrete (MPa) (section 14.5.3(d), item (2)).

14.5.6 R/C slabs

It has been discussed in Chapter 11 that slab damage mainly appears in the form of cracks in the middle of large spans, near discontinuities such as corners of large openings, at the connections of stairs to the slabs, etc. Depending on the extent and the type of damage, a different degree of intervention can be applied.

(a) Local repair

If a properly reinforced slab exhibits cracks of small width without crushing of the concrete or bond deterioration, it can be repaired with epoxy resins.

In the case of local failure accompanied by crushing or degradation of concrete there can be a local repair for the full thickness of the slab (Figure 14.43). However, the need for such a repair is typically accompanied by the need to increase the slab thickness or to add new reinforcement.

(b) Increase of the thickness or the reinforcement of a slab

Where the computational check indicates that the slab resistance is insufficient, the slab can be strengthened either by increasing its thickness from the upper side with cast-in-place concrete, or by increasing its thickness and placing additional reinforcement on its lower side with gunite concrete (Figure 14.44). The force transfer between the old and the new concrete is the key to the success of the intervention. This can be accomplished by some other means besides roughening of the old surface or resin coatings on the interface, such as anchors, dowels etc. (Figure 14.45).

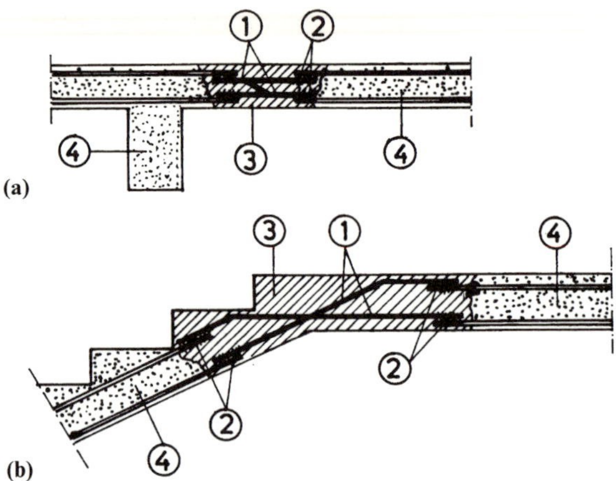

Figure 14.43 Local repair through the thickness of a slab: (a) repair in the span; (b) repair on the connection of a stair to the slab. 1 = added reinforcement; 2 = welding, 3 = added concrete; 4 = existing slab.

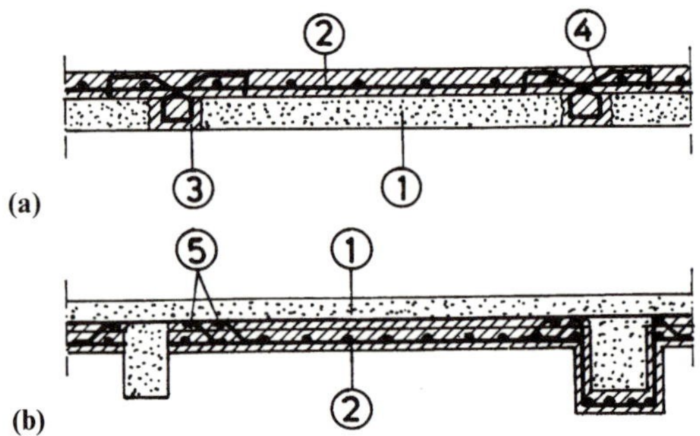

Figure 14.44 Increase of the thickness of the slab – addition of new reinforcement: (a) increase of the thickness on the upper face; (b) increase of the thickness on the lower face with the addition of new reinforcement. 1 = existing slab; 2 = added reinforcement; 3 = dowel; 4 = anchoring bent bars; 5 = welded connecting bars.

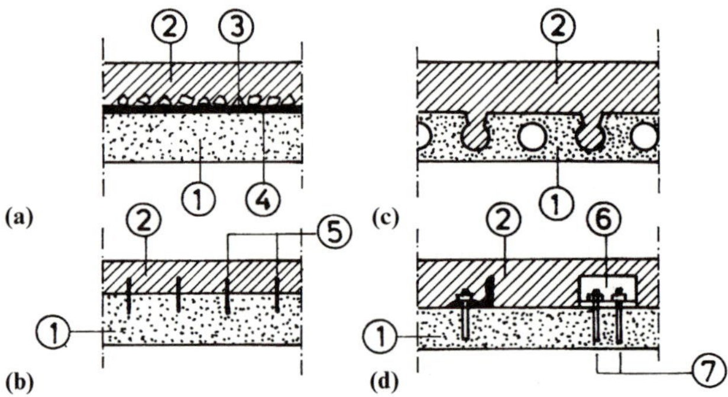

Figure 14.45 Details of connection of a new layer to the old concrete in a slab: 1 = existing slab; 2 = new slab; 3 = sand corner; 4 = epoxy glue; 5 = epoxied bolts; 6 = angle profile; 7 = anchor bolts or shoot nails.

(c) Redimensioning and safety verifications

The dimensioning of slabs which have been strengthened with additional reinforcement and increase of the thickness is carried out based on the assumption of a monolithic section (original+additional layer). The results are multiplied by the model correction factor, φ, which is taken as equal to 1.0 if the thickness of the new layer h is less than $h_0/3$ and $\varphi=0.65$ if h is equal or larger than $h_0/3$ (Tassios, 1983, 1984), that is

$$\frac{M_{rep}}{M_{monol}}=1.0 \quad \text{for } h<\frac{h_0}{3} \tag{14.34}$$

$$\frac{M_{rep}}{M_{monol}}=0.65 \quad \text{for } h\geqslant\frac{h_0}{3} \tag{14.35}$$

The proposed values for the stiffness ratio are

$$\frac{K_{rep}}{K_{monol}}=0.90 \quad \text{for } h<\frac{h_0}{3} \tag{14.36}$$

$$\frac{K_{rep}}{K_{monol}}=0.40 \quad \text{for } h\geqslant\frac{h_0}{3} \tag{14.37}$$

However, in addition to the general safety verifications there should be specific considerations for the force transfer mechanism through adhesion between the old and the new concrete, similar to those which were explained for the dimensioning of beams.

14.5.7 Foundations

The methods of repair or strengthening of foundations fall beyond the scope of this book, given the fact that they are related to interventions which belong to the field of foundation engineering. Indeed, when damage related to the foundations occurs, it is not unusual that the need arises for construction of retaining walls with anchorages to resist landslides, for construction of piles, for strengthening the soil with cement groutings, etc. Therefore, here only the technique of connecting the column jacket to the footing will be dealt with, as well as possible strengthening of the footing itself.

(a) Connection of column jacket to footing

Given the fact that the critical area of a column from the flexure point of view is at its top and bottom, the column jacket must continue beyond the point where the column frames into the footing, so that reinforcement bars will have the required anchorage length. This can be accomplished either with the arrangement of Figure 14.46 or with that of Figure 14.47.

(b) Strengthening of footings

Increase in the area of footing is decided either because of inadequate bearing surface due to poor original estimation of the soil-bearing capacity, or because larger axial forces are transferred to the foundation due to the addition of new

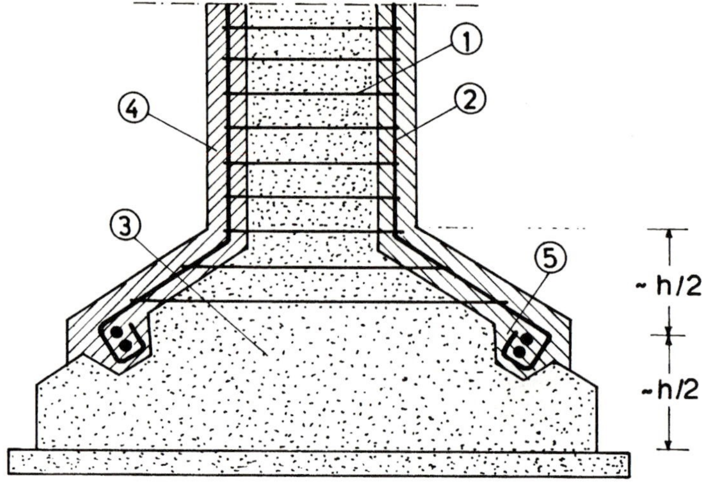

Figure 14.46 The end of a column jacket to the footing: 1 = new ties Φ12/100 mm; 2 = longitudinal reinforcing bars; 3 = existing concrete; 4 = added concrete; 5 = dowel in old concrete.

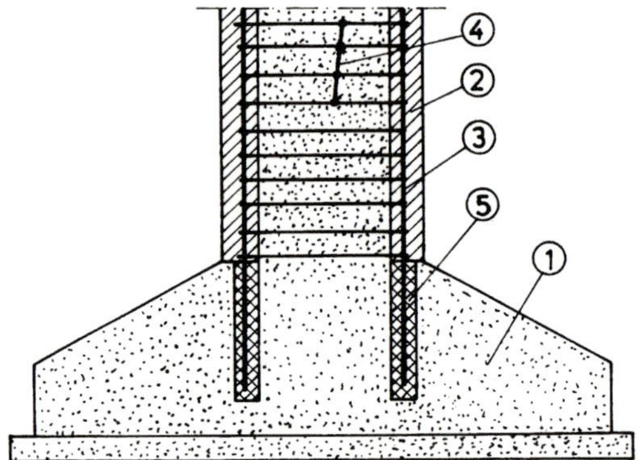

Figure 14.47 Anchorage of the column jacket reinforcement to the footing: 1 = old concrete; 2 = jacket; 3 = long. Reinforcement; 4 = new ties; 5 = epoxied connections.

structural elements. In these cases the increase in the area of footing is carried out according to the arrangements shown in Figure 14.48 and 14.49.

The first arrangement, which is simpler than the second, is applied when the strenghtening of the footing is extended in the form of a jacket to the column. In this case, the inclined forces for the transmission of the soil pressure to the column jacket (Figure 14.48) are carried by rectangular closed reinforcements rings, which are formed either with large overlaps or welding.

The second arrangement is much more difficult because excavation under the existing footing is required. It is applied in cases where strengthening of the footing is not extended in the form of a jacket to the column. In this case, a tempo-

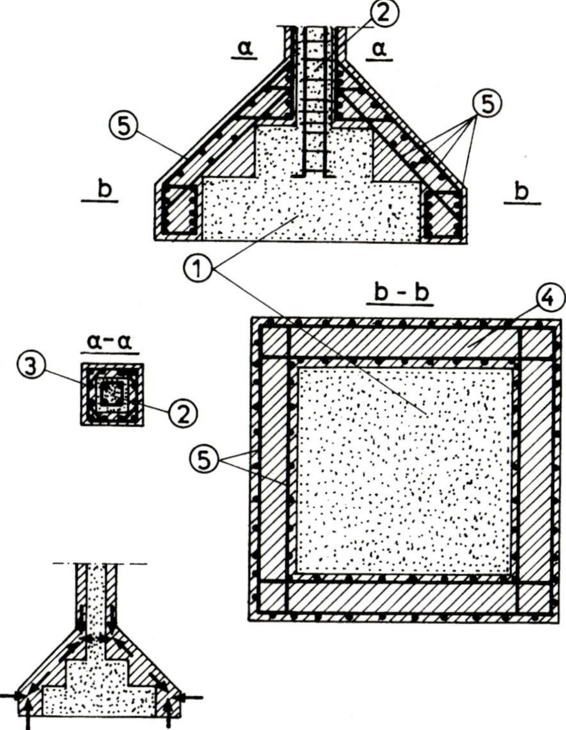

Figure 14.48 Strengthening of footing – column: 1 = existing foundation; 2 = existing column; 3 = reinforced jacket; 4 = added concrete; 5 = added reinforcement.

rary support is typically required and special attention should be paid to avoiding settlement due to underdigging.

14.5.8 Infill masonry walls

In previous chapters there was a systematic reference to the significance of the infill system to the seismic behaviour of the structures and it has been explained how important is their repair (Bertero and Brokken, 1983).

(a) Light damage

Cracks which do not go through the thickness of the wall but appear only on the plaster have already been characterized by the term 'light damage' (section 11.1.7). To repair this kind of damage a band of plaster of a width equal to 100–150 mm on each side of the crack is removed and it is replaced by new plaster after the wall is moistened with water. Very often, a band of light wire mesh is used as a reinforcement underneath the new plaster.

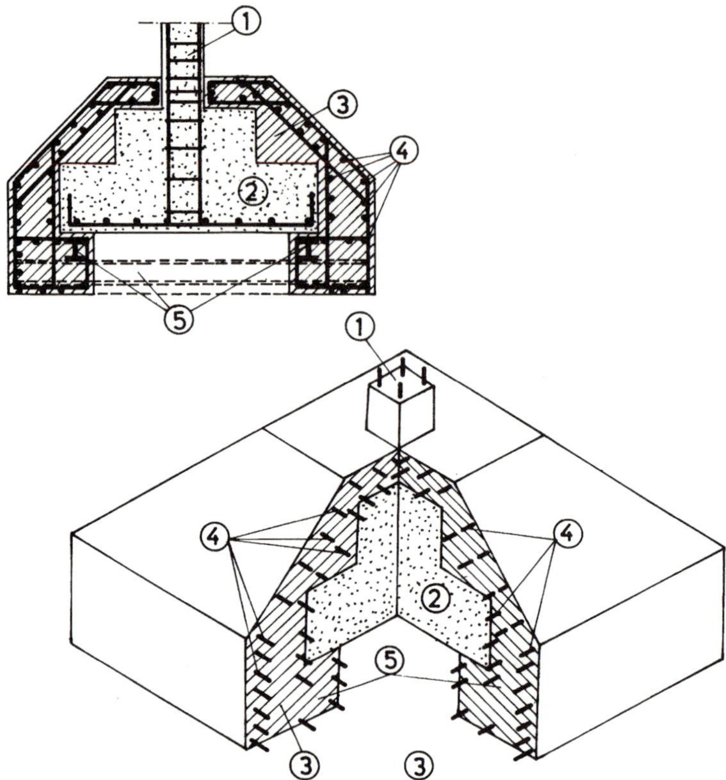

Figure 14.49 Strengthening of a footing without strengthening of the column: 1 = existing column; 2 = existing foundation; 3 = added concrete; 4 = added reinforcement; 5 = steel profile.

(b) Serious damage

This term refers to open (full thickness) cracks in the infill wall, independently of the crack width. In this case obviously, the strength, the stiffness, as well as the ability of the infill to dissipate energy has been reduced, and therefore an intervention more elaborate than the previous one is required. Therefore, if the crack width is only a few millimetres, after the plaster is removed in a band of 100–150 mm on each side, the crack is widened on the surface of the wall, it is washed using a water jet and filled with cement mortar of high cement content, pushing the mortar as deep as possible inside the crack with a thin trowel and smoothing the surface. Then a wire mesh is nailed on the area where the plaster has been removed and new plaster is applied (Figure 14.50).

If the cracks are wider, two solutions are possible: either the wall is removed and reconstructed, or the plaster of the whole surface of the wall is removed and the procedure of the previous paragraph is followed. The wire mesh in this case is placed on the whole surface of the wall and a plaster consisting of cement mortar of 20 mm thickness or a thin layer (about 30–40 mm) of gunite concrete is constructed. It is understood that interventions of this type lead to strengths and stiffnesses of

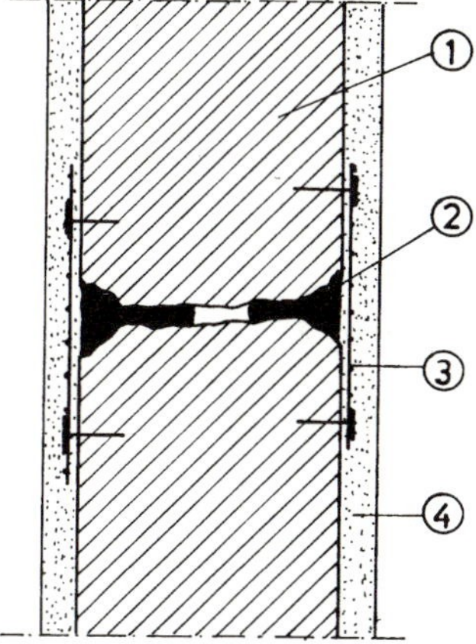

Figrue 14.50 Repair of a through-thickness crack in an infill wall: 1 = existing masonry walls; 2 = sealing of the crack with cement mortar; 3 = wire mesh; 4 = plaster.

the masonry wall higher than the original ones (Figure 14.51). Therefore, there should be a check of the relative strength and stiffness of the adjacent columns in order to avoid shear failure in the columns due to a new earthquake (Sariyiannis, 1990) in case the repaired masonry is not extended into the next span.

14.6 ADDITION OF NEW STRUCTURAL ELEMENT

The seismic resistance of a structure is drastically improved with the addition of new structural elements of great stiffness, able to carry large horizontal forces.

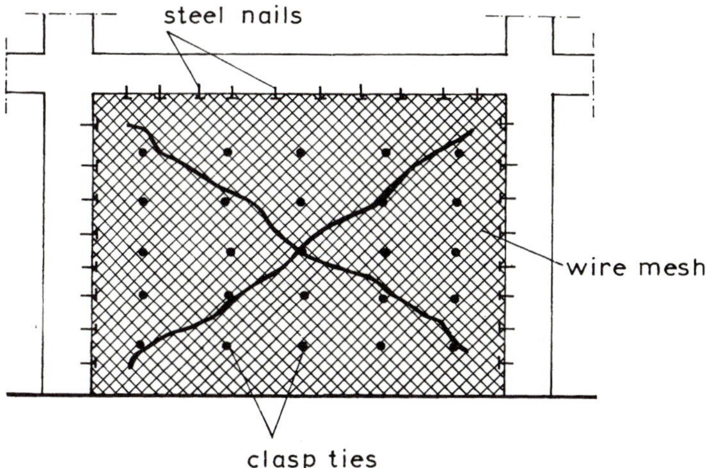

Figure 14.51 Repair of a seriously damaged infill masonry wall.

The new structural elements could be (Bertero and Brokken, 1983; Bush, Wyllie and Jirsa, 1991; Rodriguez and Park, 1991; Sugano, 1981):

- R/C walls inside the frames that are formed by beams and columns (Figure 14.52);
- additional R/C walls outside the frames (Figure 14.52(d));
- new frames;
- truss systems (made of metal or R/C) in the R/C frame (Figure 14.53).

The choice of type, number and size of the new elements depends on the characteristics of each structure. The most common type is the addition of R/C walls. Since with interventions of this type the stiffness elements of the stucture as well as its dynamic characteristics are altered, they must be performed with special care, and re-evaluation of the whole analysis and design of the building is necessary. This re-evaluation must also be extended to the foundations, since the addition of new stiffness elements (e.g. multi-storey walls) leads to a concentration of large shear forces and moments at the base of these elements which requires the appropriate strengthening of their foundation with widened footings or additional connecting beams or new foundation beams. Finally, it has to be stressed that very often the addition of new elements is carried out not only to increase the stiffness or the strength of the structure but mainly to alleviate some of the existing stiffness eccentricities which led to damage (Figure 14.54).

In the case of addition of new elements special care should be taken to ensure their force transfer with the existing elements. Especially in the case of addition of new R/C walls, the connection with the frame is made with dowels anchored with epoxy mortar or with reinforcement bars welded to the existing ones (Figure 14.55) (Bertero and Brokken, 1983; Bush, Wyllie and Jirsa, 1991).

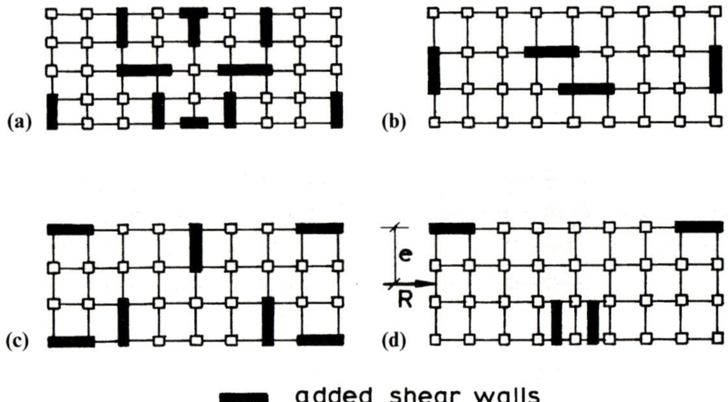

Figure 14.52 Addition of new R/C walls inside a frame or skeleton structure: (a), (b), (c) favourable layout (symmetric); (d) unfavourable layout (eccentric walls).

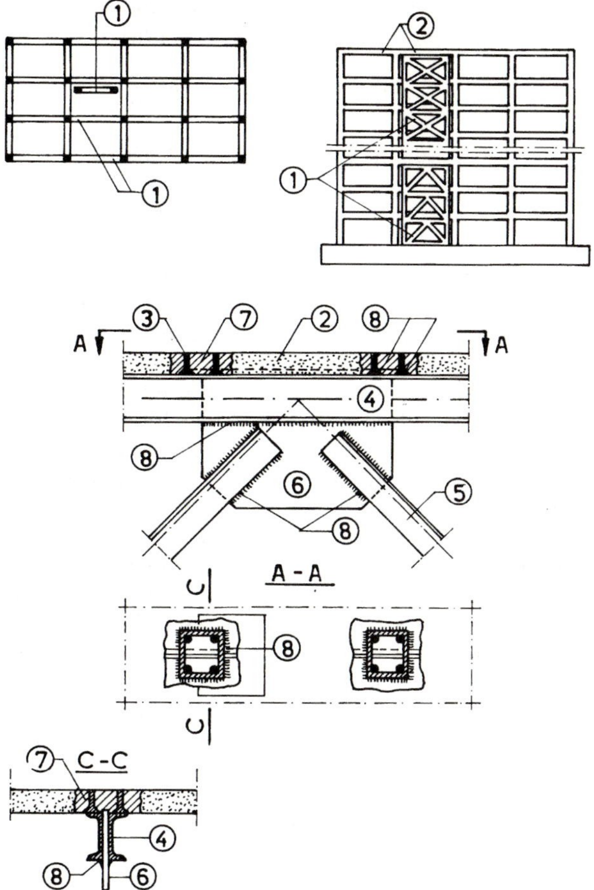

Figure 14.53 Addition of truss systems inside R/C frames; 1 = added steel truss; 2 = existing structure; 3 = steel dowel; 4 = horizontal steel rod; 5 = diagonal steel rod; 6 = steel joint plate; 7 = added concrete; 8 = welding.

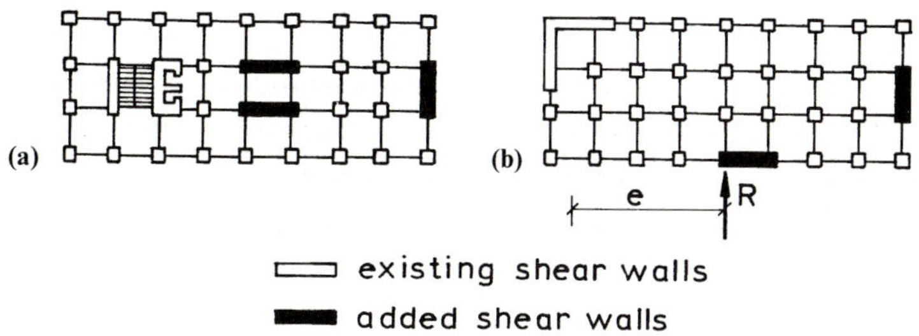

Figure 14.54 Improvement of stiffness eccentricities with the addition of new R/C walls.

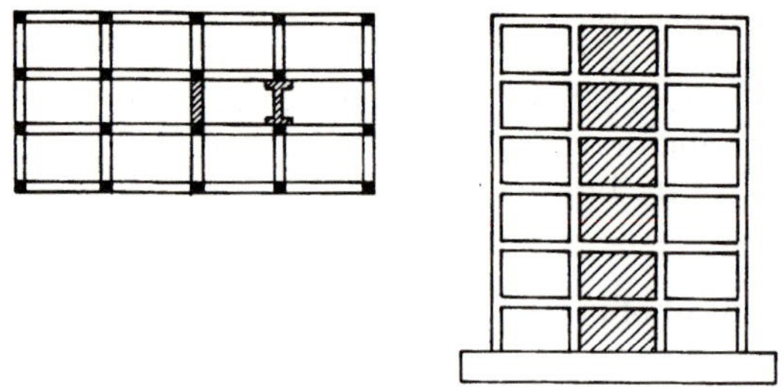

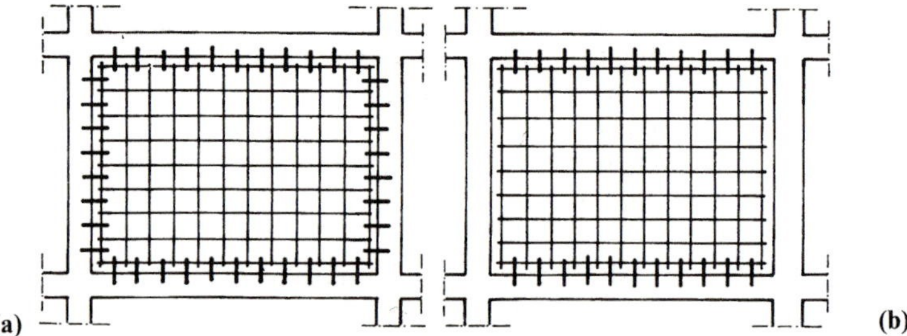

Figure 14.55 Addition of new R/C walls inside a frame: (a) connection along the four sides; (b) connection only with the beams.

14.7 QUALITY ASSURANCE OF INTERVENTIONS

14.7.1 General

For a successful structural intervention, additional measures are needed in order to ensure quality of design and construction (EC8/Part 1.4; UNIDO/UNDP, 1983). Quality assurance of the design includes a thorough review by the designer as well as by an independent reviewer in order to make sure that the design criteria and solutions are the proper ones and that the computational work and structural details have been properly prepared. Quality assurance of construction includes the inspection and testing of materials and procedures of construction and the assurance that the design has been properly implemented during the construction. While quality control is important for all constructions affecting the safety of the occupants, it is particularly important for seismic repair or strengthening due to the fact that these activities require a high degree of engineering judgement and careful attention to detail.

14.7.2 Quality control of design

Seismic repair and strengthening projects require an appropriate scheme of counter-checking of design documents. When the design has been completed and the designer responsible for the project has thoroughly reviewed the work, an additional check should be performed by an independent reviewer. This may be a governmental or private agency, responsible for verifying the criteria and checking the calculations and drawings to make sure that they conform with the criteria and regulations of the building codes.

14.7.3 Quality control of construction

Construction inspection is carried out by an individual agency or firm, similar to conventional construction. However, an experienced engineer, with an extensive knowledge of repair materials and techniques, should be appointed as construction inspector of repair and strengthening projects. The design engineer should continue to be involved with the inspection process and provide answers to questions arising during the implementation of the design details in the construction.This is extremely important for such projects, as many unexpected situations will be encountered during the construction, related mainly to hidden damage discovered after the finishes have been removed.

The quality of materials is verified by sampling and testing as in a conventional project. The differences involve only the verification of existing conditions and the testing of special materials such as resins, non-shrinking mortars, shotcrete, etc.

The design documents should include a detailed description of the work schedule related to the repair and strengthening, as well as detailed specifications for the materials and construction techniques.

14.8 FINAL REMARKS

From the preceding presentation the following final remarks can be made:

1. During the repair process of damage caused by an earthquake, due to demolition works on the structure, a strong temporary supporting system is required to avoid collapse.
2. During the repair process additional materials and techniques are used which are very rarely applicable to new structures. Therefore a detailed study of their characteristics is required, as well as very careful supervision during their application.
3. The form and exent of repairs cannot be completely foreseen during the design phase. The engineer is often compelled to improvise in order to adjust the materials and techniques to the needs of the existing special conditions.
4. Those interventions which drastically alter the original dynamic characteristics of the building must be applied with extreme care.
5. The redimensioning and safety verification of the repaired elements is achieved by more or less approximate procedures, firstly because no reliable analytical models based on laboratory tests have yet been developed for the variety of cases met in a damaged structure, and secondly because there is a high degree of uncertainty with regard to the achieved degree of composite action of the old element and the new material.
6. Finally, the repair cost of an element is much higher than the cost of its original construction, due to the fact that, on the one hand repair involves com-

plicated works such as demolition, supports, welding, injections and so on, and on the other it inhibits the unobstructed use of mechanical equipment. Therefore, the cost estimate of such an operation is generally difficult.

14.9 REFERENCES

Abdel-Halim, M.A.H. and Schorn, H. (1989) Strength evaluation of shotcrete-repaired beams. *ACI Struct. Journal*, **86**(3), 272–6.

Arakawa, T. (1980) Effect of welded bond plates on aseismic characteristics of R/C columns. *Proceed. 7th World Conf. on Earthquake Engng*, Istanbul, **7**, pp.233–40.

AUT (Aristotle University of Thessaloniki) (1978) *Earthquake Damage Repair in Buildings*, Thessaloniki (in Greek).

AUT(1979) *A Seminar for Earthquake Damage Repair in Buildings*, Thessaloniki (in Greek).

Balinger,C., Maeda, T. and Hoshijima, T.(1993) Strengthening of reinforced concrete chimneys, columns and beams with carbon fiber reinforced plastics, in *Fibre-Reinforced-Plastic. Reinforcement for Concrete Structures*, ACI, SP–138.

Benedetti, D. and Casella, M.L. (1980) Shear strength of masonry piers. *Proceed. of 7th International Conference on Earthquake Engineering*, Istanbul.

Bertero, V.and Brokken, S.(1983)Infills in seismic resistant buildings, *Journal of Struct. Engng, ASCE*, **109**(6), 1337–61.

Bett, B.J., Klingner,R.E. and Jirsa, J.O. (1988) Lateral load response of strengthened and repaired concrete columns. *ACI Struct. Journal*, **85**(5), 499–508.

Bush,T.D., Talton,C.R.and Jirsa, J.O.(1990) Behaviour of a structure strengthened using reinforced concrete piers. *ACI Struct. Journal*, **87**(5), 557–63.

Bush, T.D., Wyllie, L.A. and Jirsa, J.O. (1991) Observations on two seismic strengthening schemes for concrete frames. *Earthq. Spectra*, EERI Journal, **7**(4), 511–27.

CEB (1991) *Behaviour and Analysis of Reinforced Concrete Structures under Alternate Actions Including Inelastic Response*, Vol. 1. *CEB, Bul. d' Inf. 210.*

CEN Techn. Comm. 250/SC8 (1993) *Eurocode 8: Earthquake Resistant Design of Structures, Part 1. 4(Draft): Repair and Strengthening*, CEN, Berlin.

Chung, H.W. and Lui, L.M. (1978) Epoxy-repaired concrete joints under dynamic loads. *Journal of ACI*, **75**(7), 313–16.

DIN 1052 (1981) *Holzbauwerke – Berechnung und Ausführung*, Ausgabe 1969, Beton Kalender 1981, Part I, Wilhelm Ernst & Sohn, Berlin.

Economou, C., Karayiannis, C. and Sideris, K. (1994) Epoxy repaired beams under cyclic loading. *Proceed. of 11th Hellenic Conf. on Concrete*, Techn. Chamber of Greece, Corfu, **3**, pp. 26–37 (in Greek).

French, C.W., Thorp, G.A. and Tsai. W.J. (1990) Epoxy repair techniques for moderate earthquake damage. *ACI Struct. Journal*, **87**(4), 416–24.

GMPW (Greek Ministry of Public Works) (1978) *Recommendations for the Repair of Buildings with Earthquake Damage*, Thessaloniki (in Greek).

Gylltoft, K. (1984) Fracture mechanics model for fatigue in concrete. *Materials and Structures*, RILEM, **17**(97), 55–8.

Hanson. N.W. (1960) Precast–prestressed concrete bridges. 2. Horizontal shear connections, *JPCA, Research and Development Laboratories*, **2**(2).

Ignatakis, C., Stavrakakis, E. and Penelis, G.(1989) Parametric analysis of R/C columns under axial and shear loading using the finite element method. *ACI Struct. Journal*, **86**(4), 413–18.

Ignatakis, C., Stavrakakis, E. and Penelis, G.(1990) Analytical model for masonry using the finite element method. *Software for Engineering Work Stations, Int. Journal* **6**(2), 90–6.

Ladner, M. and Weber, C. (1981) *Geklebte Bewehrung im Stahlbetonbau*, EMPA, Dübendorf.

Lefas, I.D. and Kotsovos, M.D. (1990) Strength and deformation characteristics of reinforced concrete walls under load reversals. *ACI Struct. Journal*, **87**(6), 716–26.

Lefas, I., Tsoukis, D.and Kotsovos, M. (1990) Behaviour of repaired R/C walls to monotonic and cyclic loading. *Proceed of the 9th Greek Conf. on Concrete*, TCG, Kalamata, **2**, pp. 231–8 (in Greek).

Leonhardt, F.(1962) *Spannbeton für die Praxis*, Wilhelm Ernst & Sohn, Berlin.

NTU (National Technical University) (1978) *Recommendations on the Repair of Earthquake Damaged Buildings*, Athens (in Greek).

Penelis, G.G *et al.* (1984) Case study: the Rotunda, Thessaloniki, in *Repair and Strengthening of Historical Monuments and Buildings in Urban Nuclei*, **6**, UNIDO/UNDP, Vienna, pp.165–88.

Plevris, N. and Triantafillou, T.C. (1994) Time–dependent behaviour of R/C members strengthened with FPR laminates. *Journal of Struct. Engng*, ASCE, **120**(3) 1016–42.

Popov, E.P. and Bertero, V.V. (1975) Repaired R/C members under cyclic loading. *Earthq. Engng and Struct. Dynamics*, **4**(2), 129–44.

Priestley, M.J.N. and Seible F. (1991) Design of retrofit measure for concrete bridges, in *Seismic Assessment and Retrofit of Bridges*, Struct. Syst. Res. Rep., No. SSRP 91/03, Univ. of California, San Diego.

Rasmussen, B.H. (1963) *Betonindstobte Tvaer Belastede Boltes og Dornes Baereevne*, Bygningstatiske Meddelser. Copenhagen.

Rodriguez, M. and Park, R. (1991) Repair and strengthening of reinforced concrete buildings for seismic resistance. *Earthquake Spectra*, Journal of EERI, **7**(3), 439–59.

Saadatmanesh, H. and Ehsani, M. (1991) R/C beams strengthened with GFRP plates. I: Experimental study, II: Analysis and parametric study, *Journal of Struct, Engng*, ASCE, **117**(11), 3417-55.

Sachanski S. and Brankov, G. (1972) Investigation for determining the size of seismic forces on the evidence of damaged buildings. *Proceed. of the 4th European Symposium on Earthq. Engng*, London, pp. 187–95.

Saiidi, M., Vrontinos, S. and Douglas, B. (1990) Model for the response of reinforced concrete beams strengthened by concrete overlays. *ACI Struct. Journal*, **87**(6), 687–95.

Sariyiannis, D. (1990) *Seismic Behaviour of R/C Frames Filled with Masonry after Repair*, Reports of the R/C Lab, AUT, Thessaloniki (In Greek).

Sariyiannis, D. and Stylianidis, K. (1990) Experimental investigation of R/C one–storey frames repaired with resins under cyclic shear. *Proceed. of the 9th Hellenic Conf. on Concrete, TCG, Kalamata*, **2**, pp. 223–30 (in Greek).

Sugano, S. (1981) Seismic strengthening of existing concrete buildings in Japan. *Bull. New Zealand Nat. Soc. for Earthq. Engng*, **14**(4), 209–22.

Tassios, T.P. (1983) Physical and mathematical models for redesign of damaged structures. *Introductory Report, IABSE Symposium*, Venice, pp. 29–77.

Tassios, T.P. (1984) Repairs after an earthquake. *Proceed. of Conf. Earthq. and Struct. EPPO, Athens*, **1**, pp. 595–636 (in Greek).

Tassios, T.P. and Vintzèleou, E.N. (1987) Concrete–to–concrete friction. *Journal of Struct, Engng*, ASCE, **113**(4), 832–49.

Triantafillou, T.C. (1994) R/C structures strengthened with FRP sheets. *Proceed. of 11th Hellenic Conf. on Concrete*, Techn. Chamber of Greece, Corfu, **3**, pp. 69–82.

UNDP (1977) *Repair of Buildings Damaged by Earthquakes*, New York.

UNIDO/UNDP (1983) *Repair and Strengthening of Reinforced Concrete, Stone and Brick–Masonry Buildings*, **5**, Vienna.

UNIDO/UNDP (1984) *Repair and Strengthening of Historical Monuments and Buildings in Urban Nuclei*, **6**, Vienna.

Vassiliou, G. (1975) Behaviour of repaired R/C elements, Ph.D. Thesis, NTU, Athens (in Greek.).

Vintzèleou, E.N. and Tassios, T.P. (1986) Mathematical models for dowel action under monotonic and cyclic conditions. *Magazine of Concrete Research*, **38** (134), 13–22.

Zavliaris, K (1994) Repairs in R/C structures. General report. *Proceed. of 11th Hellenic Conf. on Concrete*, Techn. Chamber of Greece, Corfu (in Greek).

Zografos, P. (1987) Repair of columns with R/C jackets. Reports of the R/C Lab, AUT, Thessaloniki (internal report).

Index

Acceleration
  attenuation of 13–14
  peak ground 14, 64, 394
Accelerogram 8
Accelerographs 3
Addition of new structural elements 557–60
Adhesion 206, 524
Amplification factor γ_{Rd}, *see* Overstrength factor γ_{Rd}
Analysis of structural systems, *see* Modelling; Multimodal response spectrum analysis; Simplified modal response spectrum analysis
Anchorages
  beam bars at exterior joints 335–6
  interior joints 333–5
  of new reinforcement (in repairs) 527
  see also Beams; Columns; Walls
Assessment of seismic performance
  using inelastic analysis 390–2
  through inspection 389
  through testing 389

Base isolation *see* Seismic isolation
Base shear 27, 80–1
Bauschinger effect
  beams 238
  reinforcing steel 201
Beam-column joints
  beam stubs 339–40
  bond in the joint core 333
  damage patterns 445–6
  design criteria 318, 320, 327–8
  design example 341–5
    anchorage requirements 344–5
    hoop requirements 343–4
    shear design 342
  detailing 332–3
  diagonal compression 322–3, 325
  diagonal cracking 323–4
  diagonally reinforced 338–41
  eccentric 259
  effect of floor slabs 328
  effect of transverse beams 328
  elastic 337–8
  exterior joints
    anchorages 335–6
    damages 446
    shear forces 322
    special anchorages 339–41
  failure modes 318–20
  limitations on bar diameters 333–7
  limitations on shear stresses 330
  mechanical anchorages 340
  performance criteria 317–18
  reinforcement
    for horizontal shear 331
    for vertical shear 331
  shear
    design 329–33
    cyclic 324–7
    forces 320–2
    transfer mechanisms 322–8
  strut mechanism 322–8
  truss mechanism 324

Beam-column joints (*contd.*)
  types of 319
Beams
  anchorage of bars 259–60
  critical regions 237, 253
  cyclic loading 237–50
  example of ductility estimation 253–4
  damage patterns 440–5
  design example 296–304
  design for flexure 251–5
  design for shear 255–8
  diagonally reinforced 247–8
  eccentricity at joints 259
  flanges 255
  flexure-dominated 237–9
  lapped splices 260–1
  limits of beam width 258
  modelling 248–50
  monotonic loading 224–37
  plastic hinges 230
  reinforcement
    longitudinal 251–2
    transverse 257–8
  relocated plastic hinges 337–8
  shear-dominated 241–7
  slenderness criteria 258–9
  stiffness, effective 406
  supporting cut-off members 261
Behaviour factor 68, 70, 71
Bending moment envelope
  modified (tension shift) 299, 301
    see also Structural walls
Bond
  code provisions 217–19
  constitutive equations 205–6
  cyclic loading 211–16
  effect of hooks 217
  modelling 209–11, 215–16, 218–19
  monotonic loading 206–11
  in joints 333
Bond-slip, *see* Bond
Bracing
  with buttresses 509–12
  with diagonal X-braces 513
  with interior anchoring 513
  with tension rods and rings 513–14
Bridges 223, 236
Buckling of longitudinal bars 239–41, 272

Capacity design
  of beams 122–3
  of structural walls 128–30
  of columns 123–8
  of connecting beams of footings 130
  general criteria (EC8) 121
  philosophy 121
Centre of mass 76, 83
Centre of stiffness 61, 76, 83
Clamping effect of steel at interfaces 526–7
Columns
  anchorage of bars 290–1
  biaxial loading 279–80
  bidiagonal reinforcement 276–8
  brittle 456–7
  confinement requirements 274–6
  critical regions 285–7
  cyclic loading 269–80
  damage patterns 434–6
  design example 304–13
  design for flexure and axial load 280–3
  design for shear and ductility 285–90
  effect of axial compression 270–2
  effect of axial tension 272–4
  examples of ductility estimation 267–8, 283–4
  monotonic loading 263–9
  plastic hinges in 123–4, 263
  short 276–9, 460
  slenderness limits 290
  splicing of bars 291–2
  stiffness, effective 406
  transverse reinforcement 287–90
  uncertainties regarding design 261–3
Combination of gravity and seismic actions
  EC8 relationships 72–3, 109–11
  in modal analysis 102–9
Compression against precracked interfaces 524
Concrete, cast-in-place (for repairs) 515
Concrete, high-strength 177, 516
Concrete material properties, *see* Plain concrete; Confined concrete
Configuration of the structural system
  fundamental requirements 56–7
  recommendations 58–64
    see also Irregularities
Confined concrete
  code provisions 194–7
  cyclic loading 190
  example 186–90
  with hoops 180–90
  modelling 183–6, 193–4
  monotonic loading 180–3
  parameters affecting confinement 179–80
  with spirals 190–4

Confinement 150, 177–9
see also Beams; Columns; Joints; Walls
Continental trench system 3
Continuum damage models 173
Curvature
ductility 37, 224–9, 233, 265, 348
maximum (ultimate) 224–5, 7, 266–7
yield 224, 226–7
Cyclic loading 149
see also Beams; Columns; Diaphragms; Joints; Walls

Damage index 215, 391–2
Damage to buildings
assessment (post-earthquake) 471–3
effect of number of storeys 464
factors affecting the degree of 455–68
spatial distribution 452–4
see also Beams; Columns; Joins; Slabs; Walls
Damping
hysteretic 29–31
viscous 28–9
Demands, evaluation of 390–1
Design philosophy 52–4
see also Capacity design
Detailing, *see* Beams; Columns; Joints; Walls
Diaphragm action 90
Diaphragms
configuration 59–60, 383
connections to walls 381, 384
cyclic loading 384
resistance 384–5
see also Slabs
Displacement (deflection)
components 241, 243, 350–2
estimation of 229
see also Ductility
Dowel action 244, 247, 364, 365, 527
Dual systems
analysis 88–92, 96–98
design examples 137–47, 372–80
design actions 139–42
load combinations 142
structural analysis 142–7
wall design 372–80
seismic performance
design earthquake 409–15, 421–3
serviceability earthquake 415
survival earthquake 415–8, 424

Ductility class
comparisons of classes 72, 418–24
definition 69
existing structures 484
k_D factor 70
curvature 37, 224–229, 233, 265, 348
definition of 34, 149
displacement 34, 235–6, 348–9
empirical estimation 235
factor, 34, 37
foundation flexibility influence 236–7
member, *see* Beams; Columns walls
ratio, *see* factor
relationships between various factors 233, 235–237
rotational 37, 231–3
of a structure 492–4

Earthquakes
Alkyonides 393, 452
Bucharest 25, 452
El Centro 25, 393–4
intensity 8, 10–11
Kalamata 38, 393, 452, 457
magnitude 5
Mexico City 25, 433, 452, 456
Montenegro 452
origin of 3
Volvi (Thessaloniki) 25, 393, 433, 452
Effective duration (of a motion) 394
Elastic rebound theory 3
Emergency inspection after earthquakes
action plan 476–81
goals 469–71
inspection forms 474–5
levels 476
organizational scheme 473–4
personnel 476–7
Endochronic theory of inelasticity 173
Energy dissipation 32–37, 223, 224, 247, 318, 428–9
Epicentre 7
Epoxy resins, *see* Resins
Equivalent static analysis, *see* Simplified modal response spectrum analysis

Failure criteria (for analysis)
global 398–9
local 396–7
Flange width, effective, *see* Slabs
Flexible ground storey 425–9, 458–60
Flexural deformations 229–31
Flexural overstrength 123, 263

Flexural resistance
  beams 251
  columns 280–2
  walls 348, 361
Flexural strength, *see* Flexural resistance
Floor slab, *see* Slab
Focal depth 7
Foundations
  configuration 61, 63–4
  capacity design of connecting beams 130
  effect of foundation type on damage 464–5
  repair and strengthening 553–5
  specific measures 136
Frames (bare or pure)
  analysis, *see* modelling
  design example
    beam longitudinal reinforcement 296–301
    beam transverse reinforcement 299–301
    beam detailing 302–4
    column detailing 303, 313
    column logitudinal reinforcement 304–10
    column transverse reinforcement 310–13
    design actions 293–4
    structural analysis 294–5
  modelling 86, 93–6, 395–6, 406
  P–Δ effects 111–12
  plastic collapse mechanisms 122–4
  seismic performance
    design earthquake 399–404, 419–21
    serviceability earthquake 404–5
    survival earthquake 405–6, 423–4
  vulnerability of 463–4
Friction 335, 365, 371, 524–6
FRP (fibre-reinforced plastic) 521–3
  encasement 534–6

Gluing
  FRP sheets on concrete 521, 537–8, 544
  metal sheets on concrete 520, 537–8, 543–4
Grouts 520
Gunite, *see* Shotcrete

Hoops 177, 257–58
    see also Beams; Columns; Confined concrete; Joints; Walls
Hypocentre 7

Importance factor 70–1
Infill panels, *see* Masonry infilled frames
Inspection, *see* Assessment of seismic performance; Emergency inspection after earthquakes
Intensity, *see* Earthquake
Inter-storey drift
  allowable 133–4, 491
  sensitivity coefficient 112, 283
Interventions, structural
  criteria 500
  objectives 487–8
  types 500–1
    see also Repair of seismic damage; Strengthening
Irregularities
  in plan 58–9, 457–8, 461
  in elevation 60–1, 461
Isoseismal contours 8

Jackets, R/C 530–3, 537–42, 545, 548–51
Jacks
  hydraulic 514–15
  mechanical 510, 514
Joint core, *see* Beam-column joints
Joint, *see* Beam-column joints

Lateral seismic forces, see Simplified modal response spectrum analysis
Limit states
  collapse 52, 54
  damage limitation 54
  serviceability 52, 54, 133–4
  survival, *see* collapse
  ultimate 52, 54, 131–3
Linear-elastic fracture models 172
Linear elements, *see* Beams; Columns
Lithospheric plates 3

Magnitude, see Earthquake
Masonry infilled frames
  damage to infill panels 447–51
  design action effects (EC8) 114–15
  effects on analysis 113–14
  effects on the structure 113, 115–16, 424–9
  modelling 115, 425
  repair
    light damage 555–6
    serious damage 556–7
Mass
  centre of, 76, 83
  contributing to intertia forces 78

Medvedev, Sponheur, Karnik scale 8, 11
Mechanics of the non-continuum 523
Micromechanics models 173
Microzonation 8
Modal superposition, *see* Multimodal response spectrum analysis
Modelling
  dual systems 88–92, 96–8
  general remarks 116–18
  finite element 163–8, 171–6, 395, 528
  frames 86, 93–6, 395–6, 406
  member-by-member 395
  walls 87
Model reduction factors 529, 536, 538–9, 553
Modified Mercalli intensity scale 8–10
Moment redistribution 122
Multi-degree-of-freedom (MDOF) systems
  elastic 38–44
  inelastic 44–9
Multimodal response spectrum analysis
  proposed procedure 100–1
  range of application 98–9

Natural period, *see* Period of vibration
Nonlinear-elastic models 172
Non-structural elements 133
Normal modes 17–18
Normalization of input motions 24–7, 390, 394

Orthotropic behaviour 163, 174
Overlays, R/C, *see* Jackets
Overstrength factor γ_{Rd}
  for beams 123
  for columns 125–8
  for walls 130
Open ground storey, *see* Flexible ground storey

P-Δ effects, *see* Second-order effects
Period of vibration
  estimation (for design purposes) 82
  Rayleigh's method 82
  of vibration 18
Pilotis, *see* Flexible ground storey
Plain concrete
  biaxial loading 159–68
  code provisions 177
  cyclic loading 155–8, 165–8, 175–6
  failure modes 153, 159–60, 168–9
  modelling 153–8, 163–5, 171–5
  monotonic loading 151–5, 160–3, 170–1
  triaxial loading 168–76
Plasticity-based models 172–3
Planar elements, *see* Diaphragms; Joints; Walls
Plastic hinges
  in beams, *see* Beams
  choice of locations 123–5
  in columns, *see* Columns
  length 230, 233–5
  relocation of, *see* Beams
  in walls, *see* Structural walls
Polymer concrete 518
Pounding (collision) of adjacent buildings 136, 466–8
Power spectrum 72
Pre-earthquake organizing procedures 478–9
Prestressed collars 545
Pseudo-dynamic testing 389

q-factors, *see* Behaviour factors
Quality control 223
  of interventions 560–1

Reanalysis 497
Redesign, *see* Repair of seismic damage; Strengthening
Redimensioning and safety verification
  load transfer mechanisms 524–8
  revised γ_m factors 523
  simplified estimation of resistances 528–9
Regularity (structural)
  in plan 75–77
  in elevation 77
  see also Irregularity
Reinforcement, *see* Beams; Columns; Diaphragms; Joints; Walls
Remaining life of a structure 492
Repair of seismic damage
  beam-column joints 544–6
  beams 537–44
  columns 530–6
  criteria 488–9, 499–501
  definition 486
  design steps 498–9
  examples 501–3
  foundations 553–5
  masonry infills 499, 555–7
  objectives 487–8
  slabs 551–3
  UNIDO/UNDP procedure 489–95
  walls 546–51

Resins 518–19, 525
Resin-concretes 519
Resin injections 530, 537–8, 545, 547–8, 551
Response spectra
  design (EC8) 68–70
  elastic 21–27, 64, 66–8
  inelastic 37–8
Richter scale 5

Scaling of input motions, *see* Normalization of input motions
Second-order effects 111–2, 283
Seismic hazard 9, 12–14
Seismic isolation 55–6
Seismic performance of structures
  dual systems 406–18
  frames 392–406
  influence of ductility class 418–24
  influence of masonry infills 424–9
Seismic resistance (of existing structures)
  available 484–5
  loss of 485
  required 484
  residual 485
Seismic zones 64–5
Seismicity 12–13
Seismographs 3
Shaking tables 389
Shear
  deformations 229, 241–3, 350–2
  reinforcement, *see* Beams; Columns; Diaphragms; Joints; Walls
  reversal 247
  sliding
    beams 243–8
    columns 271, 276
    coupling beams 382
    walls 355–6, 364–6
  strength, *see* shear resistance
  stress, nominal 238, 246
  resistance, *see* Beams; Columns; Diaphragms; Joints; Walls
  transfer mechanisms 243–5
Shoring, *see* Temporary supports
Shotcrete 516–17
Shrinkage compensating admixtures 516
Simplified modal response spectrum analysis
  horizontal forces 80–1
  range of application 79–80
Single-degree-of-freedom (SDOF) systems
  equations of motion 19–21
  inelastic response 27–37
Slabs
  damage patterns 446–7
  diaphragm action 90
  effective flange width 255
  flat plate systems 461–2
  slab level in adjacent structures 466
Soft storey, *see* Flexible ground storey
Soil conditions 65–66
Spalling 178, 238, 244, 268, 271–2, 287, 319–20, 335
Specific measures (EC8 provision) 134–7
Spectral modal analysis 40–3
    see also Multimodal response spectrum analysis
Spectrum intensity 24–7, 394
Splices, *see* Beams; Columns; Walls
Stability, *see* Second order effects
Steel,
  reinforcing
  code provisions 204–5
  cyclic loading 201–4
  monotonic loading 199–201
  seismic performance requirements 197–9
Steel profiles
  cages 533–4
  for separation of infills 135
  for shoring 509
Stickers (for damage classification) 473–4
Stiffness
  centre, *see* Centre of stiffness
  degradation 454–5
  effective 406, 408
Storey drift, *see* Inter-storey drift
Storey shear 94
Strain hardening 199–200, 239, 261, 320–1
Strain rate effects
  concrete 155
  steel 203–4
Strain, ulimate
  concrete 186, 224–6, 265–67
  steel 197, 200, 225–6
Strength
  index 486
  of a structure 490–1
    see also Beams; Columns; Joints; Walls
Strengthening (of existing structures)
  beam-column joints 544–6
  beams 537–44
  columns 530–6
  criteria 488–9, 499–501
  definition 486

design steps
analysis and verification 496–7
cost estimate 498
drawings 498
redesign 497–8, 507
structural assessment 495–6
examples 501–3
foundations 553–5
of masonry infills 499
objectives 487–8
slabs 551–3
UNIDO/UNDP procedure 489–95
walls 546–51
Structural configuration, *see* Configuration of the structural system
Structural system
covered by codes 57–8
dual 87–9
flat-plate, *see* Slabs
frames 86
pseudospatial 85–98
walls 87
see also Frames; Dual systems; Strucural walls
Structural walls
advantages 345–6
anchorage of reinforcement 371
analysis, *see* Modelling
aspect ratio 349, 362
bending moment pattern 128, 147, 372
boundary elements 346, 368–70, 372–3
buckling, out-of-plane 347, 351–2
cantilever walls 87
capacity design of 128–30
choice of locations 58–9
confinement
boundary elements 370
criteria 368
critical regions 368–9
construction joints 371–2
cores 59, 86
coupled walls 358–61
coupling beams 358–61
crack patterns 353
critical regions 128–9
cross-section analysis 361
cross-section shapes 347, 350–1
curvature ductility 348
cyclic loading 349–56
damage patterns 436–9
degree of coupling 360
design example
boundary elements 372–3
confinement 377–8
detailing 378–80
shear design and reinforcement 374–7
design for flexure and axial loading 361
design for shear and local ductility 361–70
design of walls with openings
coupled walls 381–3
perforated walls 380–1
dynamic shear magnification factor 129–30
effective stiffness 408
failure modes 353–5
flexural strength 347–348
geometrical constraints 370–1
interacting
with each other 87
with frames 87–9
see also Dual systems
monotonic loading 346–9
openings 356–61
out-of-plane buckling, *see* buckling
plastic hinge length 349
reinforcement, *see* design; shear
rocking of foundation 353
role of 345
shear
determination of shear force 129–30, 374–5
diagonal compression 353–5, 363–4
diagonal reinforcement 356, 367
diagonal tension 353, 361–3
failure models 353–4
reinforcement 367
resistance 362–6
sliding 353, 355–6, 364–7
strength, *see* resistance
stress 349–51
web crushing, *see* diagonal compression compression
slender walls 349–54
splicing of reinforcement 371
squat walls 354–6
thickness 370–1
Strut and tie models 380
Supplies, evaluation of 391–2
see also Beams; Columns; Walls

Temporary supports
general principles 507–8
of lateral loads 509–14
of vertical loads 508–9
wedging techniques 514–15

Tension shift, *see* Bending moment envelope
Tension stiffening 229
Ties, *see* Hoops
Timber elements 509
Time-history analysis 40, 43–4, 72
  inelastic 44–9, 390–2
Torsion in buildings
  accidental 82, 99–100
  asymmetric stiffness in plan 457–8
  eccentricity 82–4, 100
Torsion of edge of beams 329

Unconfined concrete, *see* Plain concrete

UNIDO/UNDP manuals 476, 489, 507
Usability classification, *see* Emergency inspection after earthquakes

Vertical (component of) seismic action 64

Wall-frames, *see* Dual systems
Walls, *see* Structural walls
Welding
  of new reinforcement 521, 527–8
  in splices 260, 292, 371
  welded mesh fabric 371
  welded steel plates 340–1